CAMBRIDGE LIBRARY COLLECTION

Books of enduring scholarly value

Botany and Horticulture

Until the nineteenth century, the investigation of natural phenomena, plants and animals was considered either the preserve of elite scholars or a pastime for the leisured upper classes. As increasing academic rigour and systematisation was brought to the study of 'natural history', its subdisciplines were adopted into university curricula, and learned societies (such as the Royal Horticultural Society, founded in 1804) were established to support research in these areas. A related development was strong enthusiasm for exotic garden plants, which resulted in plant collecting expeditions to every corner of the globe, sometimes with tragic consequences. This series includes accounts of some of those expeditions, detailed reference works on the flora of different regions, and practical advice for amateur and professional gardeners.

The Suburban Gardener, and Villa Companion

The Scottish landscape gardener and prolific horticultural writer John Claudius Loudon (1783–1843) published this guide to suburban living in 1838. The book is intended to provide instruction on choosing a house or 'villa' (or the site on which to build one); on the furnishing of the house; and on the laying out, planting and general management of the garden and grounds. Loudon had also planned a section on horticulture, but was forced to postpone this to a separate volume, which was never written. Like most of Loudon's books (several of which have been reissued in this series), the work is detailed and didactic: for example, the precise construction of chimneys is discussed, with reference to the various existing styles and the pros and cons, aesthetic and functional, of each. Offering insights into the practical and social aspirations of the emerging middle classes, the book also contains numerous engravings.

Cambridge University Press has long been a pioneer in the reissuing of out-of-print titles from its own backlist, producing digital reprints of books that are still sought after by scholars and students but could not be reprinted economically using traditional technology. The Cambridge Library Collection extends this activity to a wider range of books which are still of importance to researchers and professionals, either for the source material they contain, or as landmarks in the history of their academic discipline.

Drawing from the world-renowned collections in the Cambridge University Library and other partner libraries, and guided by the advice of experts in each subject area, Cambridge University Press is using state-of-the-art scanning machines in its own Printing House to capture the content of each book selected for inclusion. The files are processed to give a consistently clear, crisp image, and the books finished to the high quality standard for which the Press is recognised around the world. The latest print-on-demand technology ensures that the books will remain available indefinitely, and that orders for single or multiple copies can quickly be supplied.

The Cambridge Library Collection brings back to life books of enduring scholarly value (including out-of-copyright works originally issued by other publishers) across a wide range of disciplines in the humanities and social sciences and in science and technology.

The Suburban Gardener, and Villa Companion

John Claudius Loudon

CAMBRIDGE
UNIVERSITY PRESS

University Printing House, Cambridge, CB2 8BS, United Kingdom

Cambridge University Press is part of the University of Cambridge.
It furthers the University's mission by disseminating knowledge in the pursuit of
education, learning and research at the highest international levels of excellence.

www.cambridge.org
Information on this title: www.cambridge.org/9781108074629

This edition first published 1838
This digitally printed version 2014

ISBN 978-1-108-07462-9 Paperback

THE

SUBURBAN GARDENER,

AND

VILLA COMPANION:

COMPRISING

THE CHOICE OF A SUBURBAN OR VILLA RESIDENCE,
OR OF A SITUATION ON WHICH TO FORM ONE;
THE ARRANGEMENT AND FURNISHING OF THE HOUSE;
AND THE LAYING OUT, PLANTING, AND GENERAL MANAGEMENT
OF
THE GARDEN AND GROUNDS;

THE WHOLE ADAPTED FOR GROUNDS FROM
One Perch to Fifty Acres and upwards in extent;
AND INTENDED FOR
THE INSTRUCTION OF THOSE, WHO KNOW LITTLE OF GARDENING AND RURAL AFFAIRS,
AND MORE PARTICULARLY FOR THE USE OF LADIES.

BY J. C. LOUDON, F.L.S. H.S., &c.
AUTHOR OF THE ENCYCLOPÆDIAS OF GARDENING, OF AGRICULTURE, AND OF COTTAGE, FARM, AND VILLA ARCHITECTURE AND FURNITURE.

Illustrated by numerous Engravings.

LONDON:
PRINTED FOR THE AUTHOR;
AND SOLD BY
LONGMAN, ORME, BROWN, GREEN, AND LONGMANS; AND
W. BLACK, EDINBURGH.
1838.

London:
Printed by A. Spottiswoode,
New-Street-Square.

PREFACE.

The object of the *Suburban Gardener* is sufficiently explained in the titlepage. The work, when commenced, was intended to include the *Suburban Horticulturist* in the same volume; but, this having been found impracticable, the author is under the necessity of producing another volume under the last-named title. This circumstance would occasion him much regret, were he not aware that such an arrangement will be found exceedingly convenient for the public; as the subjects treated of in the two volumes are quite distinct, and many persons who might require information on the one, might not require it on the other.

The second volume, which will be entitled the *Suburban Horticulturist*, will be devoted to gardening as an art of culture; and in it the whole science and practice of the propagation, cultivation, and improvement of plants will be rendered familiar to the uninitiated reader. It will treat in succession of the nature of plants, and the influence on them of climate, soils, manures, water, &c.; of the implements of garden culture, of horticultural operations, and of the laying out, planting, and culture of the kitchen-garden, fruit-garden, flower-garden, shrubbery, reserve garden, and the forcing department; and also of room, window, and house-top gardening; with selected lists of fruits, culinary vegetables, flowers, ornamental trees, and shrubs; and a monthly garden calendar of work to be done, and produce which may be obtained.

J. C. L.

Bayswater, July 15. 1838.

CONTENTS.

LIST OF THE ENGRAVINGS, IN THE ORDER IN WHICH THEY OCCUR.

LIST

OF

THE GROUND PLANS OF SUBURBAN RESIDENCES,

INCLUDING THOSE WHICH HAVE DETAILED PLANS OF THE HOUSE AND OFFICES.

THE SUBURBAN GARDENER, AND VILLA COMPANION.

INTRODUCTION.

THE enjoyments to be derived from a suburban residence depend principally on a knowledge of the resources which a garden, however small, is capable of affording. The benefits experienced by breathing air unconfined by close streets of houses, and uncontaminated by the smoke of chimneys; the cheerful aspect of vegetation; the singing of birds in their season; and the enlivening effect of finding ourselves unpent-up by buildings, and in comparatively unlimited space; are felt by most people: but it requires some little knowledge of the progress or decline of vegetation throughout the year, and of rural nature generally, to be able to derive much enjoyment from the recurrence of the seasons, and their influence on plants; and much more knowledge of vegetation, botany, natural history, and the art of gardening, to derive the greatest amount of advantages which a country house and garden are calculated to afford.

For these reasons, it is our intention to endeavour to give our readers a more intimate knowledge of the subjects treated of, than has hitherto been attempted in works of this kind. Instead of a mere didactic calendar, or a dictionary of plants to be cultivated, directing what works are to be done in each month, or what operations are to be performed in the culture of particular plants or trees, we shall endeavour to convey such a notion of the structure and nature of a plant, as to show the necessity and advantages of cultivating it; and such an idea of the nature of soils, as to show the improvements which may be made in them, by mechanical operations, and the application of manures. We shall treat of the various operations of gardening, whether performed on plants, or on the soil; and for the performance of every operation, from the most mechanical to the most scientific, we shall assign a reason. Let not the reader be alarmed, however, lest we should go into minute scientific details, only capable of being understood by the botanist or chemist, or by the professional gardener: so far from contemplating this, our intention is to confine ourselves to the simplest and most important features; to treat every subject in such a manner as to be understood by those who have little knowledge of either gardening or country affairs; and, more especially, to adapt it to the instruction and amusement of ladies. For the sake of those who wish to pursue any particular topic further than we have done, we shall always refer to what are considered the best works on each subject of which we shall treat.

Before proceeding to the garden itself, we shall endeavour to excite some interest in its favour, and to show the solid advantages which may be derived from a suburban residence; though on both these topics we shall bestow very few words.

There is a great deal of enjoyment to be derived from performing the different operations of gardening, independently altogether of the health resulting from this kind of exercise. To labour for the sake of arriving at a result, and to be successful in attaining it, are, as cause and effect, attended by a certain degree of satisfaction to the mind, however simple or rude the labour may be, and however unimportant the result obtained. To be convinced of this, we have only to imagine ourselves employed in any labour from which no result ensues, but that of fatiguing the body, or wearying the mind: the turning of a wheel, for example, that is connected with no machinery, or, if connected, effects no useful purpose; the carrying of a weight from one point to another and back again; or the taking of a walk without any object in view, but the negative one of preserving health. Thus, it is not only a condition of our nature, that, in order to secure health and cheerfulness, we must labour; but we must also labour in such a way as to produce something useful or agreeable. Now, of the different kinds of useful things produced by labour, those things, surely, which are living beings, and which grow and undergo changes before our eyes, must be more productive of enjoyment than such as are mere brute matter; the kind of labour, and other circumstances, being the same. Hence, a man who plants a hedge, or sows a grass-plot in his garden, lays a more certain foundation for enjoyment, than he who builds a wall or lays down a gravel walk; and, hence, the enjoyment of a citizen whose recreation, at his suburban residence, consists in working in his garden must be higher in the scale, than that of him who amuses himself, in the plot round his house, with shooting at a mark or playing at bowls.

To dig, to hoe, and to rake, are not operations requiring much skill; and the amateur gardener will, perhaps, chiefly value them for their use in preparing for crops, or in encouraging the growth of crops already coming forward: but the operations of pruning and training trees, when well performed, are not only interesting to the operator at the time, but the plants so pruned or trained afford him pleasure every time he sees them afterwards throughout the season, till the period returns when they must be again pruned and trained. The operation of striking plants from cuttings is performed in a variety of ways, according to the nature of the plants; and may truly be called one of intense interest, both in its performance, and in the expectation of its results. By the great majority of amateur gardeners, cuttings are made and planted at random; and their failure or success is, in consequence, a matter of chance: but a very little scientific light thrown on the subject leads to rules for operating, which will turn chance into certainty in almost every case that can occur to ordinary practitioners; and, consequently, will greatly enhance the pleasure of performing the operation, from the consciousness that the labour bestowed will not be thrown away. We need not here refer to the operations of grafting, layering, or sowing seeds; nor need we mention innumerable other operations which require to be performed in the course of the year, even in the very smallest garden; but we must be allowed to notice the watering of plants, which all persons can enjoy from the earliest infancy upwards. What pleasure have not children in applying their little green watering-pans to plants in pots, or pouring water in at the roots of favourite flowers in borders? And what can be more rational than the satisfaction which the grown up amateur, or master of the house, enjoys, when he returns from the city to his garden in the summer evenings, and applies the syringe to his wall trees, with refreshing enjoyment to himself and the plants, and to the delight of his children, who may be watching his operations? What can be more refreshing than, in a warm summer's evening, to hear, while sitting in a cool parlour, with the windows open, or in a summer-house, the showering of water by the syringe upon the leaves of the vines or fig trees trained under the adjoining veranda, or upon the orange trees and camellias, or other exotic shrubs, planted in the conservatory connected with it? What more delightful than to see the master or the mistress of a small garden or pleasure-ground, with all the boys and girls, the maids, and, in short, all the strength of the house, carrying pots and pails of water to different parts of the garden; and to see the refreshment produced to the soil and plants by the application of the watering-pan and the syringe?

Even the search after insects is a great enjoyment in a garden; and, in fact, opens up an entirely new field of exertion and interest to those who have not before made minute observations in this department of nature. Fifty years ago, the subject of destroying insects was scarcely considered as belonging to gardening; and their eggs, which now every young gardener recognises, in winter glued in rings to the branches of his fruit trees, or in spring deposited on the back of his gooseberry leaves, passed unheeded through their different stages of developement; and the ravages the larvæ committed on crops were considered as inevitable blights, produced by the atmosphere. In the present day, so much of the beauty and the value of the products of all gardens is known to depend on subduing insects, that a knowledge of the subject is considered essential to every gardener: but it is more especially necessary that the possessor of a suburban garden should know how to keep insects in subjection; both because he is generally his own gardener, and because insects are more abundant in such gardens than in those situated farther in the country, sometimes from the comparatively weak and crowded state of the plants, and, in other instances, from the absence of those natural enemies of insects, the small birds.

One of the greatest of all the sources of enjoyment resulting from the possession of a garden is, the endless variety which it produces, either by the perpetual progress of vegetation which is going forward in it to maturity, dormancy, or decay, or by the almost innumerable kinds of plants which may be raised in even the smallest garden. Even the same trees, grown in the same garden, are undergoing perpetual changes throughout the year; and trees change, also, in every succeeding year, relatively to that which is past; because they become larger and larger as they advance in age, and acquire more of their characteristic and mature forms. The number of plants, and especially of trees, which can be cultivated in a suburban garden at one time is necessarily circumscribed; but, if a suburban amateur chose to limit the period during which he cultivated each tree or plant to the time of its flowering with him for the first time, he might, in the course of a few years, more or less in number according to the size of his garden, have had growing in it all the plants in cultivation in the open air in Britain, with the exception of a few of the larger of the forest trees; and even these he might also have flowered, by making use of plants raised from cuttings or layers, or of miniature trees, made by ringing and rooting the branches of old trees in the Chinese manner. Independently, however, of the variety and change resulting from the plants cultivated, every month throughout the year has its particular operations and its products: nay, it would not be too much to say, that during six months of the year a change takes place, and is perceptible, in the plants of a garden, every day; and every day has, in consequence, its operations and its products. Even in winter, there is still something to do in every garden, however small may be its extent: the walks require to be kept in order, and some plants must be protected by litter or matting;

and, if there should be no trees to prune, no ground to dig, no manure to collect or to barrow out, no dung to turn and prepare for hotbeds, there is, at all events, the preparation of names or numbers for plants; the cutting and painting of rods to tie them to; the sorting of seeds; the making of baskets; and the search after information on the subjects of plants and their culture, in books.

But imagine that to the suburban garden there is added a small green-house, or a flued pit! What a source of amusement and interest does not either of these garden structures hold out to the amateur gardener, during the winter and spring! Exactly in proportion as, in autumn, the out-door operations become fewer, the in-door operations of the green-house or pit become more numerous; and, in midwinter, the citizen amateur, if he is detained in his shop or his counting-house till after sunset, will be under the necessity of shifting, cleaning, and watering his plants, and otherwise operating with them (as some of our friends are obliged to do), by candle-light. A green-house, from the quantity of glass that it requires, is, for some suburban residences, too costly to erect: but much of the produce of the green-house may be procured, at half the expense, by the use of a pit, which requires no other glass than the sashes which form its roof. The amusement and the products which such a pit, in the hands of an ingenious amateur, is calculated to afford, are almost without end. Small salading may be produced in it throughout the whole winter. Chicory roots (though this may be accomplished in a common cellar) may be made to throw out their blanched leaves, which form the most delightful of all winter salads, at least to our taste; tart rhubarb or sea-kale may be forced in pots; as may parsley, mint, and other herbs. Bulbs may be forced; and a bloom of China roses may be kept up throughout the winter. But, perhaps, the most important use to which such a pit can be applied, in a small suburban garden, is to preserve throughout the winter, and to bring forward in spring, pelargoniums, fuchsias, salvias, calceolarias, verbenas, and other fine exotic flowers, and also half-hardy and tender annuals, for turning out into the flower-garden, or into the miscellaneous border, in the beginning of summer. We are, however, going too much into detail; we shall, therefore, only further allude to the enjoyment which can be had in every suburban house, and even town house, without the aid of a green-house or a pit, by keeping plants through the winter in a garret (always the warmest, and, when furnished with windows, much the lightest, part of a house), immediately under a skylight, or other window; or by keeping them near a window in the ordinary rooms.

These are a few of the absolute enjoyments to be derived from a suburban house and garden; and we shall next notice another, which flows from the same source, but which may be called relative or incidental. The opportunity which a garden affords to its possessor of acquiring a scientific and practical knowledge of plants is a source of great interest, not only in his own garden, but wherever else plants may come in his way; whether in a wild state, in gardens, exposed for sale in markets, or delineated and described in books. Another source of incidental enjoyment is that which will arise from the acquirement of some knowledge of gardening, and of rural architecture as an art of design and taste. As it forms a part of our plan to assist our readers in the choice of a suburban house and garden, or in the building, laying out, and planting of these, when he prefers, or is obliged, to undertake the forming of them himself; so we trust we shall, in some degree, render him a man of taste, and, consequently, a critic, both in landscape-gardening and rural architecture. How great a source of enjoyment this is, and how great an interest it enables its possessor to take in suburban scenery, in landscape, and in architecture, generally; or, in short, wherever he sees a house or a tree; those only can know who have gone through the necessary preparation. We anticipate, in particular, that it will afford a great source of enjoyment to ladies; and that it will tend to the improvement, as far as relates to matters of taste, not only of the gardens under their care, but of the architecture of their houses, and of the style and taste of the furniture and finishing of their interiors.

There is scarcely such a thing to be found as a lady who is not fond of flowers; but it is not saying too much, to affirm that there are very few ladies indeed who are competent to lay out a flower-garden; though the skill required to do so is within the capacity of every woman who can cut out, and put together, the different parts of female dress: and, supposing a female to have grown up without the slightest knowledge of the art of cutting out a gown, or of tracing out a flower-garden, it would certainly be much easier for her to acquire the latter art than the former. The result, in both cases, might be obtained almost without instruction, provided the party desiring to form the dress, or the flower-garden, had a clear idea of what was wanted. But, while every female understands this in regard to dress, and, consequently, can succeed in making gowns when she is obliged to make the attempt, very few have any distinct idea of what a flower-garden ought to be; and, hence, we seldom or never see them produce a satisfactory design for one, without the aid of a professional man. We venture to assert that there is not a mantuamaker or milliner, who understands her business, that might not, in a few hours, be taught to design flower-gardens with as much skill and taste as a professional landscape-gardener; and so as to produce incomparably better results than are now generally to be seen in the flower-gardens of the great majority of British country residences.

If we can succeed in rendering every lady her own landscape-gardener, which we are confident we can do, we shall have great hopes of effecting a general reform in the gardening taste, not only of this country, but of every other for which this work is calculated: and we intend it for circulation in the temperate climates of both hemispheres.

We have shown, in the *Encyclopædia of Cottage Architecture*, in what way ladies may, and how desirable it is that they should, acquire a knowledge of and taste for the art of design in building and furnishing. It is not our intention to go much into architectural detail in this work; but, as we shall have to treat of the choice, or of the building, of a house, and of the style of furniture and finishing requisite to harmonise with the architecture, and be consistent with itself, we shall constantly have in view the instruction of the female part of our readers in these important particulars. Almost every lady has a taste for good furniture, rich carpets, bed hangings, and window drapery; and the consequence is, that, unless ladies have some principle to guide them in their choice, they are apt, when their means admit, to bring the most inconsistent things together in the same house, and even in the same room. The great object, both in building and furnishing a house, is to know what to aim at; and the next, to know what is practicable, and what is impracticable, within the limits to which we are circumscribed by locality, income, or other circumstances. We shall endeavour to show what is consistent, and what is inconsistent, in the various tastes and schemes of building and furnishing; and, if we can do this, we shall have no difficulty in convincing our readers that, to produce a consistent whole, however humble may be the means, and simple the result, is a much greater beauty, and argues a superior degree of mind, than inconsistency, however beautiful or exquisite may be some of the individual parts. We think we can succeed in teaching ladies the beauty of consistency in architecture, furniture, and finishing, as compared with inconsistency, by a comparative view of the furniture or dress of the house, with the dress of the person.

Next in influence on society, in every country, to the female sex, is the class of teachers; including under this class the two orders, ministers of religion and schoolmasters. The instruction of these orders in the science and practice of gardening, and their improvement in both gardening and architecture as arts of design and taste, shall be amongst our principal aims in the composition of this work; as well in the hope of adding to their own resources for comforts and enjoyments, as of enabling them to infuse a taste for these comforts and enjoyments into the minds of the rising generation. We can hardly conceive any rural pursuits more adapted for a clergyman than natural history and gardening: and what can better afford a relaxation to the schoolmaster, from the arduous and sedentary duties of his profession, than the cultivation of a field of useful vegetables, and of a garden of curious and ornamental plants, not only for his own amusement, but for the instruction of his pupils? A garden and a field are, in our opinion, as well merited by the schoolmaster as a glebe is by the clergyman; and we trust they will, in a short time, be considered as no less indispensable in Britain, (in the establishment of a national system of education) than they already are in most parts of Germany, and in many parts of North America.

We might enlarge, here, on the great advantages which would result from bringing up children with a taste for garden pursuits and natural history; and the vast influence which this is calculated to have on their future happiness, and on the welfare of society, by enabling them, instead of passing their leisure hours in a manner degrading to human nature, to interest themselves in recreations both agreeable and useful: but the field is too wide to be entered on within our limits, and we must therefore leave the subject to be worked out by the imagination of our readers.

Much of the enjoyment of a suburban residence depends on knowing what to expect from it; what, in short, is consistent, and what is inconsistent, with its limits and its local situation. We have shown, in the *Encyclopædia of Cottage, Farm, and Villa Architecture* (p. 8.), that all, in the way of house accommodation, that is essential to the enjoyment of life, may be obtained in a cottage of three or four rooms, as well as in a palace; and we shall prove, in this work, that a suburban residence, with a very small portion of land attached, will contain all that is essential to happiness, in the garden, park, and demesne of the most extensive country residence. Let us briefly make the comparison. The objects of the possessors of both are the same: health, which is the result of temperance and exercise; enjoyment, which is the possession of something which we can call our own, and on which we can set our heart and affections; and the respect of society, which is the result of their favourable opinion of our sentiments and moral conduct. No man in this world, however high may be his rank, great his wealth, powerful his genius, or extensive his acquirements, can ever attain more than health, enjoyment, and respect. The lord of an extensive demesne seeks after health by hunting, shooting, or other field sports, or by superintending the general management and cultivation of his estate; the lady seeks recreation in her pleasure-ground, or in airings in her carriage: and both find their enjoyment in their children, and in their house and garden, and other surrounding objects. Now, the master of a suburban villa finds health in the change it affords from his occupation as a citizen; or, if he has retired from business, in the personal cultivation of his garden. He also finds enjoyment, not only in his family, friends, and books, but in his garden, and in the other rural objects which he can call his own, and which he can alter at pleasure, at a trifling expense, and often with his own hands. It is this which gives the charm of creation, and makes a thing essentially one's own. Every one must have felt the infinitely greater pleasure which is enjoyed from the contemplation of what we have planned and executed ourselves, to what can be experienced by seeing the finest works belonging to, and planned by, another. Our own work is endeared to us by the difficulties we have met with and conquered at every step: every step has, indeed, its history, and recalls a train of interesting recollections connected with it. The master of a suburban residence, however small may be his demesne, may thus procure health and enjoyment at the same time, with more certainty than the possessor of a larger property; because his grounds lie more in his hands, and he can superintend every change himself. His wife's exercise consists in the personal management of her household affairs; and her enjoyments are the same as those of her husband. The respect, or the good opinion, of his friends and neighbours, or of the inhabitants of the parish in which he resides, are, to the citizen retired to the suburbs, of as great value as the respect of the inhabitants of a district, or of the whole nation, is to the wealthy landholder or the senator. The difference of the happiness of the parties will therefore depend almost entirely on the difference in the degrees of their ambition; for in every other respect they are equal. All the necessaries of life may be obtained in as great perfection by the occupier of a suburban residence in the neighbourhood of London, who possesses 200*l.* or 300*l.* a year, as by the greatest nobleman in England, and at a mere fraction of the expense. Nay, the markets of London will supply the citizen, at all times, with the larger and more important articles of vegetable and animal food, of a better quality than can be raised, with certainty, in the garden or farm of any private individual in the country, where it may reasonably be calculated on, that every year, from unforeseen casualties, some crop will be found to fail, and others to be of inferior quality. Whatever advantages are enjoyed by the possessor of a residence in the suburbs of London, may be enjoyed to nearly the same extent in the suburbs of any large town in Europe and America, where the communication between such town and the sea is by good roads, and where the interchange of goods is free from heavy and vexatious government duties. Hence it follows (though it would occupy too much time to explain the process in detail), that, ultimately, the possessors of suburban residences in every country, and in every part of that country, will have nearly equal advantages; and that such residences may be considered as the ultimatum, in point of comfort and enjoyment, of the great mass of society; not only at present, but even after society has advanced to a much higher degree of civilisation, and to a comparative equalisation of knowledge, wealth, and taste.

One immense advantage of a suburban residence over one isolated in the country consists in its proximity to neighbours, and the facilities it affords of participating in those sources of instruction and enjoyment which can only be obtained in towns: for example, public libraries and museums, theatrical representations, musical concerts, public and private assemblies, exhibitions of works of art, &c. The suburban resident, by his locality, has an opportunity of witnessing these spectacles with as much ease as the distant wealthy proprietor has, by means of his establishment of horses and carriages. The small proprietor and the farmer in the interior of the country, on the other hand, are comparatively shut out from participating in the enjoyments which constant intercourse with society procures; as well as from receiving that refinement of mien and of manners which it produces. Hence it happens, in the progress of civilisation, and the changes that society is continually undergoing, that, at a certain stage, mankind arrive at a similar point to that from which they set out, only in a more improved and refined form; and as, in a rude state, they congregated together in encampments or in villages, for mutual protection and security, so, in a refined state, they congregate together in towns and suburbs, for business and refined enjoyment.

We have now said enough, we trust, to lead to the due appreciation of a suburban residence, as compared with a house in the streets or squares of a town, on the one hand; or with one in the interior of the country, at a distance from a town, on the other. To sum up the whole in a few words, towns, by the concentration which they afford, are calculated essentially for business and facility of enjoyment; and the interior of the country, by its wide expanse, for the display of hospitality, wealth, and magnificence, by the extensive landed proprietor; or for a life of labour and health, but without social intercourse, by the cultivator of the soil. The suburbs of towns are alone calculated to afford a maximum of comfort and enjoyment at a minimum of expense.

From the view which we have here taken of the subject, some idea may be formed of the great importance which we attach to the forming, improving, and managing of suburban residences. The subject appears to us of immense importance; involving, as it does, so much of the comfort and happiness of by far the largest and most important class of society in every civilised country. We have long seen that the poor, by cooperation and self-cultivation, may insure to themselves all that is worth having of the enjoyments of the wealthier classes; and it has been our study, for many years past, to find out in what way all the improvements in architecture, gardening, husbandry, and domestic economy, may be brought to bear upon the residences of the working and middle classes of society. The result, as far as respects architecture and the comforts and elegances of a house, will be found in our *Encyclopædia of Cottage, Farm, and Villa Architecture and Furniture;* and, as far as respects gardening, agriculture, and domestic economy, but more especially gardening, in the pages which we are now about to submit to the reader.

We shall arrange this work in two parts, each forming a volume complete in itself. The first volume will be entitled *The Suburban Gardener and Villa Companion*, and will treat on all that relates to design and general management.

The second part, which will be entitled *The Suburban Horticulturist*, will embrace the whole subject of the cultivation of the suburban garden, and will comprise the science of cultivating the kitchen-garden, the fruit-garden, the flower-garden, the shrubbery, the nursery, the reserve garden, the forcing department, &c., with selected lists of fruits, culinary vegetables, flowers, shrubs, &c., and a monthly garden calendar.

CHAPTER I.

ON THE CHOICE OF A SITUATION FOR A SUBURBAN OR COUNTRY HOUSE AND GROUNDS.

THOUGH it does not fall to the lot of every one to build the house he lives in, or to lay out the garden or grounds he occupies or enjoys, yet there are few possessors of suburban residences who have not, at one time or other, had it in their power to make a choice of the house in which they would reside. This choice can only be made judiciously, by a consideration of all the different circumstances, general and local, with which the residence is connected, as well as of all the different points in the arrangement and building of the house, and in the laying out and planting of the garden or grounds, which are most conducive to convenience, use, and beauty. How much the locality, the aspect, and the surrounding circumstances affect the culture of a garden, of even a single perch in extent, can hardly be conceived by those who have not paid considerable attention to the subject. Few persons are even aware of the difference in point of comfortable enjoyment and health, and also economy in warming and ventilating, which there is between choosing a house in a street lying in the direction of north and south, and one in the direction of east and west. A house in a street lying east and west must necessarily have one front in shade the greater part of the year; and consequently the walls of that front, in the autumnal and spring months, will be cold, damp, and gloomy; and, as the direct rays of the sun will not enter by the windows to warm the apartments in winter, larger fires will be required at that season. On the other hand, a house situated in a street lying in the direction of south and north, will have the sun on both fronts every day in the year in which he appears; the outer surface of the walls will be dry at all seasons; both outside and inside of the house will be cheerful every day in the year; and every room, even without fires, will, in consequence of the sun's rays entering it, be comparatively warm and well ventilated.

In every point of view, therefore, we think the subject of this chapter may be treated with advantage, as introductory to all the others; instead of being, as it might at first sight be supposed, only calculated to be useful to such as have not yet taken up housekeeping in the country, and intend to build or plant there. We shall devote one chapter to situation, another to the choice or building of the house, and a third to the laying out and planting of the grounds.

Sect. I. *Considerations of a permanent Nature affecting the Choice of a Suburban Residence, or of a Situation for building one.*

The various matters which require to be taken into consideration in the choice of a situation for a suburban residence, or by a person already resident in the country, when he desires to change his abode from one place to another, may be included under three sections: those of a general or permanent nature, such as elevation, soil, air, &c.; those of a local and temporary nature, such as streets, roads, churches, schools, markets, &c.; and those of a personal nature, such as the individual's business, pursuits, or taste, and his health, income, connexions, &c.

The healthiness of the situation intended for a dwelling is obviously the first and most important consideration. Situations differ in regard to healthiness, not only with reference to the constitution of human beings generally, but they differ as to their suitableness for persons affected with, or constitutionally liable to, particular diseases. The subject embraces elevation, character of surface, exposure, aspect, soil, subsoil, climate and character of the weather, water, and a general capacity for improvement.

Elevation. In the neighbourhood of towns, where the air is always more or less charged with smoke, an elevated site will always be found to have the clearest atmosphere; and, for persons in good health, it is generally allowed, that the atmosphere should not only be free from impurities, but free from more than the average quantity of water held in suspension in the given climate. In valleys and low situations, there is always a larger proportion of water in the atmosphere than in situations which are high, and have a dry soil and subsoil; and for these reasons, an elevated situation, for strong healthy persons, provided all other things be agreeable, should be selected. On the contrary, where the constitution of the intending occupant or his family has a tendency to consumption, a low situation, with a somewhat moist atmosphere, is preferable; or, in the case of asthma, an intermediate position, sheltered, and with a comparatively warm climate.

In choosing a situation with reference to the height of the surface, its absolute elevation is sometimes of less consequence than the facilities which it affords for the free circulation of air. A situation may be hilly, and yet so covered with high trees and houses, that the air, except in winter, when the trees are not in leaf, becomes stagnated and charged with watery exhalations from the ground, which is kept damp by the shade, and by the transpiration from the leaves of the trees. Dry ground, on a lower level, but open and exposed to the influence of the sun and the winds, is obviously healthier than a situation of this description, when covered with houses and closely planted gardens; while a single residence, on a larger scale (which would render it unnecessary to crowd the surface of the ground with trees and buildings), though in the same situation, would be quite healthy.

A situation may also be elevated, and yet not healthy, from its soil being naturally damp; and, on this account, if the walls of the house have not been built in such a manner as to prevent the damp from ascending through them, it may be inferior in point of healthiness to a lower situation where the soil is dry.

A low situation near the sea, or close on its shore, is almost always healthy; whether on a coast, like that of the west of England and Scotland, exposed to moist winds, or like that on the east of both countries, which is chiefly exposed to dry winds. The reason seems to be, that the air, in both cases, does more good by its bracing properties, than injury by the excess of moisture in the one case, or the deficiency of moisture in the other. On the other hand, elevated situations on surfaces covered with peat bog, or with marshy, wet, springy soil, are generally unhealthy. The banks of a sluggish river are unwholesome, while those of a river with a rapid current are the contrary. The beauty of the view or prospect, or the circumstance of there being or not being a prospect, is a subject of consideration so obviously depending on elevation, that it only requires to be mentioned. Scarcely any object that can be created within the boundary wall of a small spot can compensate for the want of a distant prospect, to a general admirer of landscape; and, on the contrary, no distant prospect, however beautiful it may be, can compensate to a lover of plants, for the want of a good garden: the *beau idéal* is to unite both; but, as this can rarely be done, it is for the proprietor to seek after the nearest approximation to it that he can find, or to choose which of these desiderata he considers preferable.

Character of Surface. An irregular surface is always attended by an irregular climate. This is occasioned by the different influence of the sun on surfaces of different degrees of slope; and by the different degrees of interruption which hills or irregularities of different shapes and sizes give to the wind. There is scarcely such a thing as a perfect calm during bright sunshine in a hilly irregular country; because, while on one side of the hill the sun generates an intense heat, on the opposite side of it the soil and air remain cold. For persons of narrow chests and weak lungs, a hilly situation, therefore, is far from being desirable; for, independently of the irregularity of the climate, the fatigue of walking up and down hills or slopes, is greater to such persons than it is to others. The most even and regular climates are to be found over the most even and regular surfaces. Hence, other circumstances being favourable, a level plain, at a distance from mountains, generally affords the mildest and best climate for invalids. This is also the kind of situation in which a kitchen-garden is least subject to be injured by winds; and, if the subsoil admits of being thoroughly drained, and is not so near higher grounds, or a large surface of water, as to be liable to be inundated by night dews from them, it is preferable to every other.

A situation near the sea, near a lake, or near a broad river, has a climate resembling, in some respects, that of an irregular surface; for, the effect of the sun on the dry soil being very different from its effect on the water, a current of wind is created, in consequence of the difference of temperature. The air, in such situations, is milder in winter, in consequence of the heat given out by the water; and cooler in summer, in consequence of less heat being given out by the water than by the dry ground, and of the breezes which arise from the tendency to an equilibrium in volumes of air of different temperatures. Natural or accidental hollows, when they have an outlet for drainage, and for the escape of the heavy air which accumulates in them, afford sheltered, and sometimes very picturesque, situations for building in. Old stone quarries, gravel pits, and chalk pits are of this description; and, when they are open to the south or south-east, with, perhaps, a prospect in front, the most delightful little places may be made of them that can well be imagined.

Exposure. All elevated situations are more exposed to the prevailing winds of a country than plains; but, as the highest winds of any particular locality generally blow from one direction only, situations that are equal in point of elevation may be quite different in point of exposure. In the neighbourhood of London, where the most violent winds are from the south-west, a house placed on that side of a hill will be more exposed to the winds than on any other side. The exposure, also, is very much affected, in irregular hilly situations, by the influence of other hills, in directing currents of wind out of their natural course; so that, while the south-west side of an isolated hill may be the most exposed side, the south or the north side of a hill, situated among other hills, may be equally exposed, or more so, than the south-west side. Though a situation exposed to the prevailing winds of the district can seldom be considered the most desirable for a dwelling-house, and never for a garden, unless it be one where none but the hardiest plants are to be grown, yet there are exceptions, arising from local circumstances: for example, in the immediate neighbourhood of smoky towns, such as Manchester or Birmingham, any situation on the side of the town most exposed to the prevailing winds will be the best, as being that most free from smoke.

Aspect. By the aspect of ground, is generally understood the direction of its surface relatively to the different points of the compass. A south-eastern aspect, or that in which the surface of the ground has an inclination to the south-east, is, in Britain, generally considered the best, because it is the warmest. The winds from the south-east are never very violent: they are not so dry as those from the east, and they occur less frequently than winds from any other quarter. On a hill side, especially if the hill be high, the aspect least subject to high winds is decidedly to be preferred; but, in a flat country, a slight inclination of the surface, in any direction whatever, is a matter of no great consequence. An aspect to the south is, in some cases, less favourable than one to the north: for example, when the objects which are seen from the windows of the house are near; as, in that case, the worst side of the trees is seen, and that side will be wholly in shade, during the most agreeable part of the day, in winter and spring. An aspect to the north, on the other hand, shows the best side of the trees; the light side of all objects, during winter and spring; and the greatest variety of light and shade during summer, and, in short, thoughout the year. In the choice of a situation for a house and grounds, the views from the house should not alone be taken into consideration. Recreation in the open air is one of the great advantages of a country residence; and an aspect sloping to the south will admit of the grounds being walked in during winter, when, with one sloping to the north, they will be covered with snow, frost, or moisture. Land sloping to the south may also be much sooner walked on after rain, at all seasons; and it is a great source of enjoyment to persons living in the country, and fond of seeing the progress of vegetation, to examine their gardens and grounds immediately after a shower, while the trees and plants are in the very crisis of excitement, produced by a liberal supply of water after a long drought. Every one knows that a garden sloping to the south or south-east not only produces earlier crops, whether of culinary vegetables, fruits, or flowers, than any other, but that it may also be walked in through the greater part of the winter. Grass fields open to the south produce earlier pasture, and the corn grown on arable land so situated ripens sooner.

Soil. Though the soil, in small spots, may be very much changed by art; and though, in suburban gardens of only two or three perches, the soil may be entirely artificial, and, consequently be made, in a great measure, what the proprietor wishes it; yet it is always desirable, where it can be done, to choose a soil which is good by nature. If we were asked what was the single quality in a soil, the predominance of which would entitle it to be called good, we should say dryness. With reference to a small dwelling-house, and also to a small garden, this quality is more especially desirable, since it will almost always be found easier to alter the texture of a dry soil, so as to render it retentive of moisture, than to drain and alter the texture of a wet soil so thoroughly as to give it a character of dryness. A dry soil may either have sand or gravel as its prevailing quality: the latter affords the best foundation for a house, and the former the best subsoil for a garden. A soil, however, may be naturally either sandy or gravelly, and yet not be dry, from being placed on a retentive subsoil, or from the subsoil being connected with the watery subsoil of higher grounds. In the case of a retentive subsoil, the surface soil, though sandy in its original nature, from being long saturated with surface water, and from that water containing vegetable matter, will probably have become black and peaty in appearance; and, unless the water can be thoroughly got rid of by draining, such soils are decidedly unfavourable both for building and gardening. As chalky soil is almost always dry, it is favourable for building on; but, unless it has a considerable depth of soil over it, the expense of forming suitable garden ground is greater than on most other soils. A chalky subsoil in a valley is generally covered by a depth of loam on the surface, which loam forms one of the very best soils for growing every description of vegetables in the highest degree of perfection. Chalky soils, though dry, are invariably colder than most others, from the whiteness of the chalk not absorbing readily the sun's rays, and from the slowness with which a dense body like chalk is penetrated with the rain of summer, which is one of nature's chief modes of warming subsoils. A strong clayey soil is not to be desired; because it will cost a great deal to render it fit for garden purposes; and because it forms a dangerous foundation for small houses built on its surface, on account of its liability to shrink during the great heats of summer, and to expand when remoistened by rains, thus throwing the walls of the house out of their perpendicular. In various parts of Middlesex, to the north of London, brick cottages may be seen built on strong clays, with walls leaning to one side, or bulging out, or with deep cracks in them produced by this cause. However, where the foundation of the house is sunk to the depth of 3 ft. or 4 ft.; or where the surface of the clay is covered with a thick coat of gravel, small stones, or sand; or where it is paved, or laid with flag-stones, to the breadth of 3 ft. or 4 ft. all round the house; this disadvantage of a clayey soil will not be experienced; because the soil immediately connected with the foundation is thus, in a great measure, protected from atmospherical changes. A strong clayey soil can be walked on with pleasure fewer days of the year than any other, on account of its retentiveness of moisture; and it is the most expensive to cultivate, from its being alternately too hard and too wet. It is, therefore, the very worst kind of soil for houses having an acre or two of ground attached; and it must obviously be the very worst that can occur, either for kitchen-gardens or flower-gardens. There is a description of clayey soil which is almost always soft and moist, because it abounds in springs; and, though no soil whatever, in its natural state, can be worse for building on, or for gardening, than this moist clay; yet, if it admits of being thoroughly drained, it may be rendered better for either purpose than the tough strong clay above mentioned. The reason is, that this springy clay, from the very circumstance of its being springy, or, in other words, so porous as to admit of water rising up through it, is less retentive of moisture, and more freely and easily worked when drained, than the other. In some cases, however, the springs which rise from the subsoil cannot be effectually cut off, so as to render the surface perfectly dry; by which is meant, perfectly free from all moisture, but that which falls on it in the form of rain or snow. Loamy soils (which, for our present purpose, it will be sufficient to consider as intermediate, either between sand and clay, or between peat or soft black earth, and chalk) are not unfavourable for building on, and are highly favourable for every description of gardening. A sandy loam, with rock or sand, or even gravel, for a subsoil, may be considered as the most favourable of all for gardening purposes, and, indeed, as the best soil for a small country residence.

Subsoil. The substratum on which the surface soil rests, is, both for building and gardening, of more importance than the soil itself. No surface soil whatever, on a wet or springy subsoil, or even on a soil not springy, but clayey and retentive, can be rendered fit for either garden or field cultivation without considerable expense. On the other hand, any soil on a subsoil not absolutely retentive, and not too open, may, at a moderate cost, be highly improved. Rock, of almost every kind, forms a favourable subsoil, because it is generally neither too retentive nor too porous: it is proverbially the best foundation for building on; and, if the surface soil is of sufficient depth, it is fit for every purpose of cultivation.

There are some descriptions of decomposing sandstone rock, impregnated with oxide of iron, which form unfavourable subsoils for the culture of fruit trees, and produce canker in them; but such subsoils can be freed from the bad effects of the iron to a sufficient depth for the culture of herbaceous vegetables, by frequent dressings with quicklime. Dry gravel forms a good subsoil; but, unless care be taken not to found the walls of a house too deep, and to have a perfect exterior drainage, the surface water will, in rainy seasons, find its way into the floor of the sunk story, if there be one. A gravelly subsoil is not the best for a garden; because, in the summer season, it absorbs too rapidly the water of the surface soil. On marshy or peaty soils or subsoils, no house whatever should be built; but such surfaces and subsoils may be rendered peculiarly suitable for certain descriptions of garden culture, by judicious draining, or by a command of water combined with the power of laying the whole dry to a certain depth at pleasure. In the case of irregular rocky surfaces such marshy or peaty spots are not unfrequent; and, when the house can be placed on a rocky knoll, and the garden formed in an adjoining hollow, the concurring circumstances may be considered favourable.

The Climate, and the Character of the Weather, of the given locality are those circumstances of a permanent nature which, next to elevation and soil, ought most to influence the choice of a suburban residence. In warm climates, an open and airy situation is preferable to a close and sheltered one; and in hot climates, such as in Australia, situations with such an inclination of surface as is favourable for shade, will be sought rather than those with surfaces so inclined as to meet at a more direct angle, the sun's rays. In cold climates, sheltered situations will be preferred, for the sake of warmth; and, in moist climates, elevated dry situations, open to the south, and exposed to the breeze, with a dry subsoil, and good natural surface drainage, afford the only opportunities of counteracting or diminishing the natural inconveniences always attendant on continued rains and fogs. In almost every situation, the wind prevails, throughout the year, more in one direction than in all the others; and, hence, it is obvious that such winds must blow the smoke of the town or city, throughout the greater portion of the year, in the opposite direction. Thus, in the climate of Britain, in a great majority of cases, all other circumstances being equal, the best situation for a suburban residence is to the south-west of a town or city, and the worst to the north-east. Next to the south-west in preference is the west; and next to the north-east in disadvantages is the east. It will be borne in mind, however, that the natural currents of the wind, in any given locality, are always more or less modified by inequalities of the surface; and these inequalities, when they are so great as to become hills and mountains, will materially affect the theory here laid down. In such cases, the observation of what actually takes place in nature, will form the best guide. A large river, in a tolerably straight direction, more especially if acted on by the tides of the sea, has always considerable influence upon the direction of the wind, from the difference which exists between the temperature of the surface of the water, and that of the adjoining land and houses. In the heat of summer, the river has a tendency to create a current of air by its coolness, and in winter by its warmth; and this action is greatly increased when the motion of the river is rapid, or when it rises and falls with the tide. When the direction of a river, therefore, which passes through a town coincides with the direction of the prevailing winds, the banks of that river below the town must be much less fit for suburban residences than those above it. In the climate of Britain, it may be laid down as a general rule, that, where rivers run through a town in the direction of east and west, the banks above the town will always be preferable for a suburban residence to those below it; and, on the other hand, that where a river runs south and north, the banks both above and below the town will be found nearly equal in point of eligibility for building and garden purposes.

Water. This element, so essential both for the house and garden, is not likely to be lost sight of; because, from the earliest times, it has been an understood requisite both for a dwelling and a garden. Hence, as cities and towns formerly were placed by a river, or on some inlet of the sea, so small dwellings were placed by brooks or springs. With the present improved means of procuring water by boring, or collecting it on the surface and preserving it in tanks, few difficulties respecting the procuring of water can occur. In the neighbourhood of all large towns water may generally be procured from public works, erected and put in operation for the supply of the town; in which case there can be no difficulty, provided the party about to build or hire a house can afford the expense. It must, however, be taken into consideration, with reference to this expense, that, where there is a garden however small it may be, much more water is required than where there is only a house; and it also ought to be considered, that water fit for garden purposes may generally be got on the spot; while water fit for culinary purposes and washing linen, may require to be brought from a distance, or may not be attainable without great expense.

The Capacity of a Situation for Improvement, or the reverse, ought to be taken into consideration. A situation may be improved in various ways: it may be raised, in point of commercial value, by an increasing neighbourhood, or by ameliorations made in the soil; while at the same time, to an occupier fond of quiet and retirement, an increasing neighbourhood may be felt as a deterioration. In situations of very limited extent, there is always a risk of being overlooked, or of being prevented from looking out, by the operations of adjoining proprietors; as, for example, in the case of a house set down among fields not belonging to its occupier, there is the liability of these fields being built on, and the view spoiled, &c.: but these and other considerations belong more to the class of circumstances which will form the subject of the succeeding section.

Sect. II. *Considerations of a temporary Nature affecting the Choice of a Suburban Residence, or of a Situation for forming one.*

The considerations to which we have directed the attention of the reader, in the preceding section, can, in very few situations, at least in old countries, be acted on independently of those which are of a temporary nature. In many parts of America, and of Australia, a choice may be made with reference to natural circumstances alone. The newly arrived emigrant, who has obtained his allotment in the wilderness, may fix the site of his house on a high or low level, on a knoll, by a lake, by a river, or by a spring, as the case may be, without caring anything as to who will ultimately be his neighbours; without much reference to future roads or canals; and without regard to churches, schools, or markets, which do not yet exist: but the individual who desires to choose a suburban residence, or a situation for forming one in Europe, and especially in Britain, must take many circumstances into consideration, which are entirely of an artificial nature; and, in the neighbourhood of such a metropolis as London, or of such large and populous towns as Liverpool, Manchester, Birmingham, Sheffield, Bristol, &c., the artificial circumstances may truly be said to be of more importance than the natural ones. These circumstances, we think, may be included under the following heads; viz.: extent, shape or outline, position of the house relatively to the garden, artificial aspect, artificial exposure, distance from the place of business of the occupier, direction of roads, height of road relatively to the floor of the house, vicinity of churches, public buildings, and markets, vicinity of nuisances, state of repair, fashionable locality, and kind of neighbourhood.

Extent. The accommodation of the house, if one already exists, and the extent of the ground connected with it, or which may be obtained adjoining it, are among the first considerations; and here it must be taken for granted that the intended occupant or proposed builder knows what extent and accommodation will answer the ends which he has in view. The difficulty of attaining this knowledge, in the case of a suburban or other residence of the smallest size, is not great; because the occupant of a similar house, in the interior of a town, without a garden, may generally procure one in the suburbs with a garden, at little additional expense, and sometimes, indeed, for less. As the garden, in such a case, will be kept in order principally by the occupant himself, the outlay incurred on account of it will be trifling; and, if the individual has a family, in all probability less than the amount of his apothecaries' bills, when he resided in the interior of the town. The case, however, is different with the citizen, who thinks of hiring or forming a country residence, with a garden and pleasure-grounds occupying an acre or two of land; even though the house attached were no larger than the one he had occupied in town. Supposing a correct calculation made as to the fitting out of such an establishment, it is not so easy to foresee the different expenses of keeping it up; especially if the occupant be ambitious of keeping it in a very superior style. It would be tedious, and out of place, in this part of our work, to enter into details; but we may safely recommend every one who sets up such a country establishment for the first time, to make ample allowance for unforeseen expenses. Very few country houses on a large scale, in the neighbourhood of any large town, that have been built for any length of time, have remained long in the occupation of the same family; partly from the changes in circumstances to which all commercial men are liable; and partly because such residences are retired to at an advanced period of life; and, on the death of the head of the family, are necessarily sold in order that the proceeds may be divided among his children, or other heirs; but principally from the actual expenses of such residences generally far exceeding the previous calculations of their occupiers. In consequence of the frequency of these changes, however they may have been produced, most large suburban residences that are to let will be found in a state of indifferent repair; and the same may be said of their gardens and grounds. Hence, when a citizen hires or purchases such a residence, on the strength of his own knowledge, he is very frequently imposed upon, or deceives himself; and is unavoidably led into repairs and renovations, of which, on first taking possession of the residence, he had no idea. The cause of this evil is to be found in the rashness and imprudence of citizens in judging for themselves in such cases, instead of employing a professional man, a builder, a surveyor, or even an experienced carpenter, to examine the house; and a nurseryman, or other experienced gardener, to look at the garden and grounds, before coming to a determination respecting them.

The Shape or outline of the plan of the ground is the next consideration; and, in the case of residences of the smallest size, this is of great importance. A narrow slip of ground, as it must be enclosed, must always include an atmosphere more or less stagnated, and unfit for the purposes of breathing, or walking in, gardening, and even of drying clothes. A form approaching to a square or a circle will generally be found to admit, in the greatest degree, the influence of the sun and wind, and to be the most advantageous for domestic purposes, as well as for gardening. On a larger scale, the shape of the ground is comparatively of less consequence in point of health; though, with regard to variety within the enclosure, it is of the greatest moment. In this case, a long narrow slip may have advantages, instead of disadvantages, by increasing the extent of the home views from the house; by lengthening the straight walks of the garden; and even, in some cases, by affording an opportunity of forming an avenue of trees as an approach road. When the piece of ground is long, and the outline circuitous, then the opportunity of varying the interior by planting, and by extending the walks, is much increased; and when to length and circuitousness is added irregularity in breadth, every thing, as far as beauty is concerned, is obtained that can be desired in point of form. Where economy is the great object with reference to shape, then that form which requires the smallest quantity of boundary fencing will be the best; and this is obviously the square or the paralellogram. Theoretically, the circle or the polygon would be more perfect; but in practice the application of these figures would be found inconvenient.

The Position of the House relatively to the shape or outline of the plan of the garden in which it stands, is a matter of importance in residences of the smallest size. The reason is, the house may be so placed as to throw a shadow over a great part of the garden every day in the year; or it may be so placed as to tl row no shadow on it at any time, or only a partial shadow. The great disadvantage which every small garden has, as compared with a large garden, is the shadow thrown on it by the boundary walls, or by the house or adjoining houses; and, if an intending occupant will bear this in mind while looking out for a house, it will aid him in obtaining what ought to be considered a great desideratum in a suburban residence; viz. a garden but moderately shaded by walls and houses. In the case of a road in the direction of east and west, with houses and gardens on the north side of it, if the houses in these gardens are placed next or near to the road, it is evident that they will shade great part of the garden behind them every day in the year; and, on the other hand, if the houses are placed at the farther extremity of the garden, that they will shade no part of it any day in the year. In this latter case, the garden, even in London (along the New Road, for example), might be rendered a splendid scene, both from the street or road, and from the house of the occupant; while in the former, it never could (as is actually the case with the gardens on the north side of the New Road) be made productive of either much beauty or much use. In the case of the houses and gardens along the south side of a road running from east to west, if the houses are placed close to the road, the whole of the garden will be exposed to the sun every day in the year. We do not state these facts with a view of leading to the conclusion, that all houses on the south side of an east and west road should be built close it, and all those on the north side as far from it as the garden will permit; but, simply, to impress on the minds of our readers, who have a suburban residence yet to choose, the great importance of taking the subject of shade and sunshine into consideration when a garden is one of their main objects; and to enable those who already have suburban residences to determine whether their gardens are suitable for the culture of the finer plants, or only for more ordinary productions. (See, on this subject, the third chapter of the present Book, *On laying out and planting*, &c.)

The artificial Aspect, in the case of very small plots of ground, is even of more importance than the shape; but when the extent of ground amounts to an acre or upwards, the aspect can scarcely be said to be artificial, as it does not depend upon the boundary fences, or other artificial objects. If we imagine a narrow slip (say about 20 ft. in breadth, which is the general width of the gardens of the smallest, or fourth-rate, houses in the neighbourhood of London), placed in the direction of east and west, and that the fences are 10 ft. high, it is evident that the greater part of that garden will be in the shade every day in the year; and the whole of it will be under shade at least two months every winter. On the contrary, if a plot of ground of the same width, and with fences of the same height, be placed in the direction of north and south, the sun will shine on every part of it during the warmest portion of every day in the year. In the latter garden, in the climate of London, peaches and grapes might be ripened; while in the former nothing would thrive but ivy, and a few of the commoner shrubs and herbaceous plants. Wherever, therefore, the garden is an object in a suburban residence of the smallest size, and the piece of ground attached to the house is longer than it is broad, a house should be chosen, which has the garden belonging to it, running north and south, or as near to that direction as is practicable. Where narrow slips of garden ground are of necessity placed due east and west, the fences between them ought to be as low as can be permitted, if opaque, or to be composed of open iron railings. For the sake of seclusion, these railings may be covered, during the summer season when the occupants are most in their gardens, with deciduous creepers, such as scarlet runners, nasturtiums, sweet peas, or that splendid plant, the everlasting pea; but, during winter, they should be left naked, to admit the sun to dry and warm the soil. Where the form of a piece of ground on which a suburban residence of the smallest size stands, is a square, or approaches to that form, it should, if practicable, be chosen in a street, or other situation, where an imaginary diagonal line through the square, would be in the direction of north and south; for the obvious reason, that the shadows produced by the walls, would be diminished in their effect, by being produced during only one half of the period of sunshine during each day. In the case of a square garden so placed, the sun would shine on both sides of the four walls, during some part of every day in the year, excepting for a week or two before and after Christmas. Sometimes the occupiers of houses in towns have small gardens lying at a distance from their residences; and many acres of ground are laid out in such gardens in the neighbourhood of Birmingham, Wolverhampton, Lancaster, Paisley, and other manufacturing towns. These gardens are surrounded by thorn hedges, trained to the height of 7 ft. or 8 ft.; and, in order to avoid, as much as possible, the shade produced by such hedges, the same principles of selection should be adopted, as have been recommended for choosing the aspect of a suburban residence of the smallest size.

As the smallest houses and gardens, which form streets in the suburbs of large towns, are generally placed in a direction at right angles to the street, a person choosing such a residence should, if possible, avoid taking one in a street that runs north and south; because in that case the gardens of the houses that abut on it will be in the direction of east and west, and, consequently, shady and damp the greater part of the year. The very best direction for a street, bordered on each side by rows of houses of the smallest description joined together, and with small gardens before and behind them, is south-east and north-west; because, in that case, the sun during the greater part of the year, will shine, during some part of the day, on both sides of the houses on both sides of the street; in the morning on the north-east side, and in the afternoon on the south-west side. The aspect, in the case of larger houses, is of less consequence; as, from the greater width of the gardens, the space shaded by the walls bears a much smaller proportion to the entire area: but still, as it is desirable that the sun should shine on every side of a house, or of a line of houses, every day in the year, the sides of the house or street, should always, if possible, front the south-west, south-east, north-west, or north-east; and never the direct south, north, east, or west. This is a principle which ought always to be attended to in forming plans for new towns or villages, and in laying out new streets in the suburbs of old towns. It is a great object to get the sun to shine, if only for an hour or two, on every side of a house, most days in the year; for it not only adds greatly to the cheerfulness of the rooms, but dries the damp from the outer surface of the walls, and from the window-frames and roof. In the case of larger plots, amounting in extent to an acre or upwards, we have shown, in the preceding section, that the preferable natural aspect is to the south-east or south, so that the surface of the ground may be soonest freed from frost, and soonest rendered dry after snow or rain. In small suburban residences, which have been planted with the larger class of forest trees, the aspect of the ground may be naturally good, while its advantages are in a great measure neutralised by the shade of the trees: but the remedy for this evil is obvious; and consists simply in the removal of the trees.

The artificial Exposure, in the case of houses and gardens on a small scale, is no less affected by local circumstances than the natural exposure, in houses on a large scale, is by such as are of a permanent nature. We have shown that an irregular surface has always more or less an irregular climate, and that the side of a hill naturally sheltered may be exposed to a current of wind, reflected or turned out of its natural direction by an adjoining hill. On the very same principles, a house in a sheltered or favourably exposed situation may be injured by a current of wind directed against it by adjoining houses, or by trees; and this, as every one knows, is one grand cause of smoky chimneys. An attentive observer of the chimney tops of the houses in the suburbs of London will find that those which have most of these unsightly contrivances for preventing smoking, are generally small houses near large ones, or near large trees. In the case of regular streets, where the houses are all of nearly the same size, or of houses that are completely isolated, a common chimney pot will be found sufficient.

The Distance from the nearest town, or from the place of business of the occupant, is an object which requires only to be mentioned, for every one to be aware of its importance. Every individual must determine this for himself. In the case of many persons engaged in business as clerks or assistants in such towns as London and Birmingham, a suburban residence can hardly at present be indulged in; unless the head of the family foregoes the satisfaction of taking his principal meal at home. It is probable, however, that, when the advantage of living in a purer air, and enjoying greater opportunities for the recreation of children, are more powerfully felt by such persons, the manufactories and workshops in which they are engaged, will be placed in the suburbs, or at various distances from the town, round which they may live; while the goods manufactured may be sent to a warehouse or office in the town, at which warehouse or office they may be sold, and orders taken in. The payment of the office-keeper would be such as to enable him also to have his suburban residence, and to go to and return from it by some public conveyance. In all civilised countries, arrangements of this sort will be greatly facilitated by the railroads now becoming so general in Europe and America. In London, in the case of the printing business, for example, there would be no difficulty in having the work done twenty or thirty miles from town, provided the place were near a railroad, and there was an office in London centrically situated for receiving and transmitting orders and goods. This would enable some hundreds of families to live in cottages with small gardens, instead of being cooped up in close streets, to the great injury of their health.

The Direction of the Road leading to a suburban residence is not beneath notice. We have shown that, on a small scale, where the houses and gardens are placed in rows parallel to the road, the direction of the road is of the utmost consequence with reference to the enjoyment of sunshine. This will also hold good, to a certain extent, in the case of detached houses placed alongside a road, and with one of their sides parallel to it; for which reason, in the case of detached houses, we think that this parallelism ought not to be rendered imperious on the part of those who have the general inspection of streets and buildings, or what, in England, are called "district surveyors." But the direction of the road is of some moment, even in the case of villas of two or three acres, with reference to the dust produced on that road in summer, and the mud in winter. Where the road is in the direction of the prevailing winds of the country, the dust, in summer, will be blown in the direction of that road, and will consequently prove a much greater annoyance to the traveller, than where the road is in a direction contrary to that of the wind. About London, the prevailing winds are from the west, or from the east; and hence the roads to Edgeware, Hampstead, &c., on the north, and to Epsom, Croydon, &c., on the south, are much less disagreeable to travel on than those to Uxbridge and Brentford, or in the Essex direction. The dust from the roads running north and south is blown off them into the fields; whereas in the roads running east and west, whether the wind is in the east or in the west, the dust is blown along them. It must not be forgotten, however, that those roads which have the dust blown from them are the least eligible for having small houses and gardens placed close to them, as the dust from the road is blown upon such houses and gardens. The road which is most covered with dust in summer will be most covered with mud in winter, unless the dust be removed when it is first turned into mud. The wetness of the surface of a road during winter, all other circumstances being the same, will be great in proportion as it is in the direction of the wind, and is shaded from the sun. Hence, about London, those roads which are in the direction of east and west are decidedly the wettest during winter; because they are in the direction of the prevailing winds, which blow the moisture along them as the dust was blown in summer, instead of blowing it off them; and because they are all more or less shaded by fences, houses, and trees on the south side. Hence, also, the houses on the sides of roads which run east and west, ought to be placed farther from them than is necessary in the case of roads running north and south. A road which runs east and west affords shade to the pedestrian at least till between two and three o'clock of every day throughout the year; so that a citizen whose residence is along such a road may always walk to town in the shade; while, on the other hand, if, during the six months of summer, he returns from town to his residence between three and six o'clock, he will have the sun full in his face. A road in the direction of north and south affords shade to the pedestrian early in the morning, and in the evening; but, on going to town during the day along the road from his house in the north suburbs, the traveller will have the sun full in his face; while on going to town along a road from the south suburbs, it will be on his back. These remarks apply strictly to roads which are straight, and which have no other objects along them to produce shade but fences of 8 ft. or 10 ft. high; but, where the roads are winding, and bordered by trees or houses, or both, the influence of the sun is in a great measure neutralised; and, as far as respects the enjoyment of shade, roads in different directions will be nearly equal. The same remarks will also apply with respect to dust being blown along the road, or off the road; since, if we imagine the road in a serpentine or zigzag direction, the wind, from whatever quarter it may come, will blow the dust alternately off the road, and along the road. Nothing, however, can be more certain than this, that, in the climate of Britain, all roads and walks in the direction of north and south, are much sooner dried after being wet, and much more comfortable to travel on in winter, than roads in the direction of east and west. This may be observed even in the streets of towns in the winter season; and also in the gravel walks of walled gardens, and in the avenues through woods.

The Height or Level of the Road relatively to the Height or Level of the Ground Floor of the House, may seem almost too insignificant a subject to be noticed as a separate desideratum. In practice, however, it is of very considerable importance. No house that stands on a lower level than the adjoining road ever yet looked well; unless, indeed, the extent of the scenery was so great, and the house at such a distance, and so decidedly below the level of the road, as not to tempt the comparison. Every one feels that there is always something mean in looking down upon an object, or in descending to it; while, on the contrary, there is always something dignified and commanding in ascending to one. The truth or falsity of these positions any one may prove, by observing the front gardens of houses, even of the smallest size, in the neighbourhood of London, Edinburgh, or any large town. Where the front garden ascends towards the house, there is an appearance of dignity, dryness, and comfort; but, where it descends, we receive from it the impression of meanness, dampness, and unwholesomeness. Even in the choice of a street, or of a public road, on which to choose or build a residence, it is desirable, if it can be done without sacrificing more important objects, to fix on one that ascends towards the residence, rather than descends. In all cases of this kind the absolute elevation is, in a great measure, out of the question; the effect is produced by the elevation or depression of the spot on which the house stands, relatively to a near object; and it can, in all suburban residences of moderate extent, be completely effected by art, whatever may be the natural character of the surface. Let it be observed, however, that this evil can only be avoided before the house is built, and afterwards is past remedy, for, when once the house is set down on a piece of ground on a lower level than the adjoining road, and the ground floor is on a level with the surface, or even if it is only a little above it, it is beyond the art of man to give a character of dignity to the house. All that he can do is to raise the ground round it and between it and the house or road, in effecting which a portion of the house is necessarily buried; and, while a part of its dignity is lost by diminishing its height, the expression of meanness, and the actual unwholesomeness, are aggravated by having to descend to the parlour, or ground floor, by steps. Many villa houses and large mansions in England, have been spoiled, both in regard to effect and use, as well as rendered unwholesome, by earthing them up, in order to give them the appearance of being situated on a level, or on a gentle elevation. It is better, in general, to submit to the first evil, rather than to incur another which is greater. What we would recommend, in the case of small suburban houses built on a lower level than the adjoining street, would be to shut out the lower part of the house from the road, by a wall or open fence, so that the relative levels of the house and road might not be observed; then to scoop out the ground between the house, and the door opening to the road; and to arrange the walk so that the entrance-door to the house could not be seen till the person walking on it had arrived at the lowest part of the intervening walk, from which he might ascend to the house. A great deal more might be said on this subject; but to one class of readers we need only add, bear in mind that elevation is dignity, and depression meanness; and to another, who require principles to be reduced to rules, avoid a house to which you have to descend, either through the front garden or grounds, or by steps outside or within the house.

Places of public Resort, such as a church, a school, a literary institution, a library, a museum, a market, shops, a theatre, a public garden or park, a common or piece of waste ground, and private nurseries or commercial flower or fruit gardens, in which the public are allowed to walk, are important circumstances in the choice of a suburban residence; and more especially, in the case of such as are comparatively on a small scale. Much of the comfort of the occupants of a suburban residence of the smaller size must depend upon their ready access to these sources of instruction, provision, entertainment, and recreation. To live in the neighbourhood of a botanic garden, especially where, as in France and Germany, such gardens are open every day in the year to the whole of the public, is a very great source of enjoyment to all who are fond of flowers and plants; or rather, we should say, to all mankind, as there is scarcely a person to be found in any country, who does not attach the idea of enjoyment to walking in a garden. Next to a botanic garden, a large nursery-garden affords the greatest source of botanical and gardening enjoyment; afterwards commercial flower-gardens and public orchards are to be preferred; but commercial gardens, devoted solely to the culture of culinary vegetables are to be avoided, from the smells of decaying vegetables, which generally arise from them, and which are aggravated by the daily watering given to the plants. It is partly owing to the taste which the public have shown for botanic gardens adjoining town and suburban residences, that the idea of ornamenting the area of public squares with verdant scenery has been adopted; but, though such squares are very agreeable, and may contain many fine plants, trees, fountains, statues, and other objects, yet they want the peculiar animation of the cultivated botanic garden or the nursery; which cheerfulness arises from the labours continually going forward there. A public square laid out as a garden, in England, is never so varied and interesting, considered as a scene of public recreation, as a similar scene is on the Continent; because, in England, such squares are only accessible to the occupiers of the surrounding houses; whereas on the Continent they are open to all persons whatever. Even in new countries, and in remote situations, where gardens, squares, and other desirable public objects, do not now exist, the probability of their being instituted at a future period ought not to be lost sight of. The immediate neighbourhood of a churchyard, or of a cemetery, cannot be considered as very desirable; because familiarity with the interment of the dead must either deaden the feelings appropriate to such occasions, or it must cause a preponderance of melancholy injurious to healthy enjoyment. The immediate neighbourhood of a manufactory, to those having no interest in it, is generally considered a nuisance; but, if the branch of manufacture carried on be not of itself of an offensive or injurious kind, such as a gas-work, a soap-work, a tannery, &c., and if it be properly conducted, and the moral conduct of its workpeople properly enforced, we think the objections may be overcome. The exterior appearance of manufacturing buildings is now undergoing great changes in almost every part of the country; and, instead of huge masses, presenting a mean appearance, from the absence of architectural design, or deforming the landscape by their unsightly chimneys, they are becoming magnificent masses, in which the chimney, in the form of a lofty imposing column, is a feature seen at a great distance on every side, and serves as a general ornament to the country. A railroad is a new feature in the suburbs of large towns; and, though a residence close along the line may not be agreeable to many persons, yet, at a moderate distance, it is quite the reverse; and, indeed, from situations looking down upon the line of road, we should think that it would form one of the finest artificial features of the moving kind, that could be introduced into landscape, next to a canal or navigable river. We mention railroads, because we have been much struck with the effect of the carriages passing along the line of the Manchester railway, as seen from the beautiful villas, particularly one erected from a design by Mr.

Barry, situated on the high bank which overlooks the valley through which that railway is conducted.

In the neighbourhood of a large town, the most favourable situations for suburban residences are generally those on the borders of commons, or open pieces of ground under pasture, and free to all the public. Not only do such open places greatly contribute to the purity of the air, but they afford the opportunity of amusement and exercise to children; of walking to grown up people; of some knowledge of plants, insects, and scenery, to all; and of agreeable rural views from the windows of the dwellings. In these respects, some of the small suburban residences at Hampstead and Highgate, and at Wimbledon and Barnes, near London, are very favourably situated.

The Vicinity of Nuisances of every kind is to be avoided. Nuisances may be considered as either absolute or relative. Tanneries, soap-works, gas-works, candle manufactories, manure heaps, stagnant water, brick fields, lime-kilns, &c., and even the immediate vicinity of a market garden, exclusively devoted to the culture of culinary vegetables, we consider as positive nuisances. But there are some persons who object to the immediate vicinity of a school, on account of the noise made by the boys while at play, or by the girls while practising music; others to a church, on account of the ringing of the bells; others to an inn or public house; others to a large pond, from the danger of children falling into it, and so on; all which we consider as relative nuisances, because the annoyance they occasion depends upon the peculiarities and state of health of the individual.

The State of Repair; the validity of the tenure, its duration; the local outgoings, fixed, probable, or uncertain; the local history; and the character of the neighbourhood in point of healthiness, cheapness, and morals; and even that of the former occupiers of the house; are all subjects that require more or less to be enquired into by the person who would either choose a country residence already prepared to his hands, or fix on a situation in which to form one. With respect to the house, no prudent man would take one on lease, or build one, without employing a surveyor or an architect, however much he might know or suppose he knew of the subject himself; and it is equally advisable to consult a master gardener before taking or laying out a garden, if it is of any extent. Other information, implied in this paragraph, may possibly be obtained from a lawyer, the local medical man, and the clergyman, the schoolmaster, or the parish authorities; and the opinion of persons in the neighbouring parish ought, if possible, not to be lost sight of.

Fashionable Locality. In the neighbourhood of most towns there are localities which, as compared with others, are resorted to in preference by the fashionable world. Thus, in London, the most fashionable locality is familiarly called the West End; because it happens accidentally that the Court, and persons of the highest rank of the present day, reside in that direction. It will readily be understood that ground to build on, or houses already built, are higher in price in fashionable localities than in localities less fashionable; and hence, when a person wishes a good house cheap, he should not choose a fashionable locality.

The Kind of Neighbourhood is a consideration familiar to every one, and need not be enlarged on. Perhaps the best general principle to be followed in selecting a suburban residence, or a situation to build one, is to choose a neighbourhood where the houses and inhabitants are all, or chiefly, of the same description and class as the house we intend to inhabit, and as ourselves. To go into a neighbourhood where the prevailing character of the houses indicated a smaller income than that of the individual choosing a residence, would be to incur a risk of being marked out as the rich man of the locality; while to choose one where the majority of the houses was of a higher class, would be to incur the risk of either becoming unsocial on the one hand, or of being led into unsuitable expenses, by associating with persons of greater fortune, on the other. In the neighbourhood of large towns, many persons live in the midst of people with no one of whom, perhaps, they associate; but this arises from the inequality of education and morals which at present prevails, and is by no means a desirable state of things. Not only is social intercourse essential to the happiness of grown-up persons; but, as children must mix with the world sooner or later, it is better that they should do this with their neighbours, and under the eye of their parents, than that it should be deferred till they leave the paternal roof. To choose a neighbourhood where we can be neighbourly must, therefore, be far preferable to choosing one where we are likely to be isolated, either from an appearance of greater pretensions, or from actually having inferior means to those who are around us.

Sect. III. *Considerations of a personal Nature affecting the Choice of a Suburban Residence, or of a Situation on which to form one.*

Though, in practice, permanent, local, and personal considerations must be combined in their application; yet, in treating of a subject, there is always an advantage in submitting every kind of action, or of thought, to analysis. Personal considerations may be classed as those which respect fortune or income, objects, connexions, and constitutional peculiarities.

The Fortune of the individual, or the income which he derives from the exercise of his trade or profession, is by far the most important circumstance of a personal nature which requires to be considered in the choice of a suburban dwelling and garden. Perhaps the best advice, in this case, is the very common, but neglected one, of calculating our income rather under than over the truth; or, rather, what is much more manly and consistent, to take our income, as nearly as it can be ascertained, at what it really is, and to set apart a certain portion of it for incidental expenses. When a person has a fixed income, and can adopt a fixed expenditure, this rule of conduct is of comparatively easy application; but, when a man is engaged in business, with a necessarily uncertain income, and with an expenditure consisting partly of money laid out in his business, and partly of money consumed in household expenses, this rule can only be applied by limiting the expenses to a certain sum, and fixing that sum considerably under his estimated income. After all, such advice as to expenditure is of very little use: it is scarcely ever taken or acted on by men in business, or having much intercourse with society; and the actual practice of every individual in this, as in a hundred other cases for which there are regular rules of action, and calculations prescribed in books, is that of trial and correction. When we have spent too much in one year, we try to spend less the next. When a citizen, who has a suburban residence, contemplates having a dwelling-house in town, as well as in the country, neither can be so complete as in the case of his having only one dwelling, unless on the supposition of a greatly increased expenditure. The most economical mode of having both a town and country residence is, to live a part of the year entirely at one house, and the other part of the year entirely at the other; as one set of servants will then be sufficient.

The Object which a person has in view in desiring a country residence will necessarily influence his choice. Health and recreation are the most universal objects; but joined to these, or independent of them, is the love of distinction; of retirement; of seclusion; of horses, and dogs, which a country residence affords an excuse for maintaining; of astronomy, botany, gardening, and entomology; or of some other study which can be better carried on in the country, or in the suburbs of a town, than in the town itself.

The Connexions of an individual may influence his taste. A man in business is connected commercially with his employers or customers; by blood, with his relations; and by love and duty, with his immediate family. Under these circumstance, it will be for him to consider, whether proximity to either or all these connexions be desirable, or otherwise. Formerly, persons in the same way of business were averse from living near each other in towns, and equally so to having their suburban residences in the same quarter; and, though this feeling is much diminished at the present time, and sometimes even the reverse is the case, yet it is worth mentioning in a work like the present, when the object is to call up in the mind of the reader every idea in any way connected with the subject.

The constitutional Peculiarities of the individual, or of his family, must necessarily have some influence in directing his choice of a residence. Persons subject to pulmonary complaints will naturally choose a dry warm situation, with an equable climate, as far, at least, as their circumstances will enable them to do so. Other constitutional tendencies may direct individuals to situations where the air is humid; and others, again, to the sea side, where it is generally dry and bracing. A person may be constitutionally fond of seclusion, or the contrary; or he may be lame, or, from some bodily infirmity, compelled to seclude himself. In the case of any bodily infirmity, which precludes motion, it is a great source of enjoyment to occupy a residence commanding an extensive prospect; and having in the foreground, but not too near, a much frequented public road or railroad, or a canal, a river, or the sea shore.

CHAP. II.

ON THE CHOICE OR BUILDING OF A DWELLING-HOUSE, WITH REFERENCE TO ITS CONSTRUCTION, ITS EXTERNAL APPEARANCE, AND ITS INTERIOR ARRANGEMENTS.

In the preceding chapter, we have discussed the subject of the choice of the situation for a house, and a plot of ground of more or less extent; and in the present one we mean to confine ourselves to considerations which affect the house alone, independently of the grounds. We shall arrange our observations on this subject in four sections; viz. the classification of houses, the choice of a house, the arrangement of the domestic offices, and the arrangement of the living-rooms. Under each section the same observations, principles, and rules, which we shall apply to the choice of a house already built, will serve as guides in the building of one.

SECT. I. *On the Classification of Houses.*

Houses, in England, may be classed in four divisions; viz. those of the smallest size, those of the largest size, and two intermediate sizes. In London, the smallest are called fourth-rate houses, and the others third, second, and first-rate houses, according to their gradatión from one extreme to the other. On each of these we shall offer a few observations, commencing with the fourth-rate house, as being the lowest in the scale.

The fourth-rate House. The smallest of these are built principally for the occupation of mechanics, in the suburbs of London, in inferior situations. These houses consist of two rooms on a floor: they have generally from 12 ft. to 14 ft. of frontage, and are from 12 ft. to 14 ft. deep, having an entrance on the ground-floor in front into one of the lower rooms, and steps outside at the back leading into the upper floor. They have sometimes little gardens behind, and nearly always a few square yards of ground laid out as a garden in front.

The larger of the fourth-rate houses present a much better appearance, and have gardens of some size. These dwellings are mostly occupied by the families of respectable artisans, and others who are engaged in manufactories and counting-houses during the day; and are about 15 ft. wide in front, by 23 ft. deep; with a basement story, cellars, and washhouse; a staircase inside the house, a parlour floor of two small rooms, a drawingroom floor above it, generally consisting of two rooms divided by folding doors, and two bed-rooms over. The back room on the second story of a house of this description, in London or its suburbs, is obliged, by the Building Act, to be curbed (that is, contracted, by being carried up into the roof), which spoils the room; and the gutters are frequently so narrow at the bottom of the curb, that they convey the water into, rather than off, the house.

The third-rate House. The third-rate house is from about 17 ft. to 18 ft. wide in front, and from 28 ft. to 29 ft. deep. Houses of this class generally contain the same number of rooms as the largest size fourth-rate house, with an attic story over in addition. This story is sometimes partly in the roof; but, more generally, the walls are carried up to allow the rooms to be finished with square ceilings. At the back of the parlour floor there is frequently built a small room, used as a dressing-room or store-room. These houses have generally two windows in the width of their front; and are, perhaps, the most numerous of any class of suburban dwellings. They have frequently a detached washhouse; and, in the suburbs, have sometimes a gig-house and small stable at the back. The gardens to these houses vary exceedingly in size, but are never less in width than that of the front of the house; and they are frequently three or four times that width in depth.

The second-rate House. The second-rate houses are of a better and larger description, and possess conveniences that cause them to be occupied by wealthy tradesmen, professional men, and gentlemen of good fortune. This kind of house is usually 20 ft. or 30 ft. wide in front, by 30 ft. or 40 ft. deep, with additional rooms at the back. It can, and does, in many instances, contain all the apartments required by a family keeping their carriage, footman, housekeeper, &c.; and has attached to it, or in some mews or livery stables in the immediate neighbourhood, a coach-house and stable. These houses are usually built with two windows in the width of the front; but many of them have three windows in this width. The rooms are higher and better finished than in the third and fourth classes.

The first-rate Houses are those containing more than 900 superficial feet on the ground floor; and these include the residences of the nobility and gentry, and the wealthiest professional men and merchants. Houses of this class may be said to be unrestricted as to size, either in height or in width.

This technical classification of houses has been made by the British legislature, chiefly with a view to facilitate their assessment for taxes; and to regulate the thickness of party or division walls, with a view to prevent the spreading of fires. When a party intends to build, either in London or in its suburbs, he first gives notice to the district surveyor, to whom he describes the kind of house he intends to build; and this surveyor states to him certain conditions required by special laws, with which he must comply. Before the legislature interfered with the building of town and suburban houses, very little attention was paid to having the division or party walls of sufficient thickness; in consequence of which, when a fire broke out in any house, it frequently consumed the adjoining houses, and sometimes the entire street. Even now, the legal regulations on this subject appear to us far from being adequate to prevent the spreading of fires. In the first place, the builders by contract, and on speculation, frequently contrive to elude the vigilance of the district surveyor; and, in the second place, the party walls contain so many flues, and the spaces between these flues, and between the flues and the rooms, being seldom more than the width of a brick, are so slight as to be easily penetrated by fire. Hence, a fire taking place in the chimney of the house on either side of a party wall is in danger of setting fire to both houses. It is therefore much to be desired, not only that in all party walls the separation between the flues, and between these and the outer surface of the wall, should be not less than 9 in. in thickness, but that the brickwork of all party walls whatever should be laid in cement. In many cases of party walls which have been built with bad mortar, the mortar is found, on taking down any part of the wall, to have become little better than dust; in consequence of which the smoke readily penetrates from the flues into the rooms of both houses in these parts of the walls, and generally about the skirtings of the floors, which have not been plastered; and, when a chimney is on fire, the flames readily follow the smoke through these crevices. There is yet another cause which contributes to the defectiveness of party walls in London and its suburbs. It is, that during those paroxysms of rage for building speculations which occasionally take place, there is such a demand for bricks, that they are laid into the walls hot from the kilns; the consequence of which is, that they immediately absorb the whole of the moisture from the mortar, and leave it no better than a compressed powder, which gives no strength to the wall from its cementitious property, as mortar ought to do, and which is readily penetrated by smoke from the flues, or by moisture from the exterior. We could point out whole streets in which all the walls of the houses have been built with bricks in this hot state; and which, however well they may appear exteriorly, will crumble to pieces with the first fire that breaks out in them. On the Continent, and more especially in the south of France and in Italy, all bricks are steeped in water for some hours before they are used; and in Rome, every brick, immediately before it is laid, is dipped into a pail of water by the bricklayer. The same practice also occasionally takes place in Birmingham, and some other parts of England.

SECT. II. *On the Choice of a House.*

THERE are few persons, whatever may be their rank in society, who have not occasion, at some period or other of their lives, to make choice of a house. Perhaps we should not be far wrong, were we to say that this duty has to be performed by most men several times. How much of health, comfort, economy in living, and respectability of appearance depends on the choice made, few people, we believe, are aware; and few people have any idea of the seemingly trifling, and, we may almost say, invisible, circumstances which constitute the principal comforts of a dwelling-house.

The choice of a house will, in some respects, depend on the size and character of the house required, the purpose for which it is to be used, and the station in life of the party intending to occupy it. There are some things, however, common to all houses, whether large or small, which should be especially attended to, whether in a building intended solely for business, or for a private residence.

Soil. The first points to be considered are, the nature and quality of the soil, on which the house is to be erected, and whether it is effectually drained, or is capable of being drained, so as to be kept perfectly dry; for no advantages in other respects can compensate for a damp situation, both as regards health and property. A house built in a damp situation, even though the greatest care has been taken in making an artificial foundation of concrete (a composition of coarse gravel, sharp sand, and quicklime, which, when mixed well together, and moistened with water, is thrown into a foundation trench, and there forms one homogeneous mass), is still unwholesome; and, should the materials of the foundation be of inferior quality, such as of place (that is, of soft half-burnt) bricks, rubble, freestone, or other soft stone, and soft pine timber (also a common case), it will speedily decay, and be a constant and unavoidable source of expense. A gravelly soil is the best to build on, provided care be taken to keep out the springs, by drains below the level of the bottom of the walls; or hard sand, if gravel cannot be found: but soft sand or clay is to be avoided, if possible.

The Construction of the House is a matter of serious importance to any person about to take a lease; as, by doing this, he will probably render himself liable to reinstate dilapidations, many of which may be in an incipient state when he takes possession. It is therefore quite advisable, and, indeed, is imperative on every person who is unacquainted with the nature of building, to employ a respectable architect, surveyor, or builder, to examine the strength and durability of the house he is about to engage, in order to ascertain whether it is likely to remain strong and firm for a number of years. The intended tenant should also try to discover the nature of the soil; by which he will at the same time ascertain that of the air which he will have to breathe. In low damp situations, it is well known that the air is at all times charged with a greater degree of moisture than is the case in dry open situations. A moist air suits very few constitutions, even in our humid climate; and seldom fails to bring on rheumatism, more especially in those who cannot afford to live well, and take abundance of exercise.

Ventilation. Another important matter to be attended to, is the thorough ventilation of houses; for, should the air become stagnant from want of a free ventilation, particularly in houses that have a story underground, it is highly injurious to the persons living, and particularly sleeping, in them. There should, therefore, be windows both at the back and front; and, when possible, at the sides also. From rooms in the basement story, and from cellars that have neither fireplaces nor windows, there should be air flues carried up to the open air. Care should likewise be taken that the floor in the basement story is raised above the soil, and that air is freely permitted to circulate between the soil and the floor, whether that floor is of wood or stone. Where this has been properly attended to, rooms on the basement story may be used as sleeping-rooms; but, where it has been neglected, they are by no means fit or proper for any human being to sleep in.

The Roof is a part of the house which should be carefully examined; for, if it be badly constructed (too common a case with the houses built on speculation, both in the suburbs of London and in the country), with narrow gutters, and with gutters difficult of access, the wet will generally be found to have penetrated to the upper rooms after any fall of snow or rain. Many of the first-rate houses built in the neighbourhood of London are now covered with lead, which is by far the best covering. The next best is slate, if of good quality, and finished with wide lead gutters; with lead flashings (that is, strips of lead covering the joints) to them, and to those parts of the walls which are carried up higher than the slating. Zinc-covered roofs seldom keep out the wet many years; and, when tiles are used, they require a steep roof, and great attention to be paid to the gutters, otherwise the wet will find its way through between the tiles.

The Walls and Foundations. In the choice of a house, having ascertained that the site on which it is built is healthy, the drainage good, the roof properly constructed, and free of access, not merely for the purpose of keeping out the wet, but as a safeguard and means of escape in case of fire; the next portion of the building which requires to be examined is the substance of the walls, and the materials of which they are composed. The soft half-burnt bricks before mentioned, which are called place bricks by the builders, ought never to be employed in the walls of any building which it is desirable to keep dry. Whenever these bricks are found in the foundation of the party walls of a house, it should be rejected; and, if they are seen in the outside of any of the external walls, it may be looked on as a certain indication that every beating rain which falls will penetrate into them. Very porous sandstone is liable to the same objection. Such walls suck in water like a sponge, and give it out to all the interior fittings-up and finishings. Sound, hard, well-burnt bricks (called, about London, stocks) are the strongest, most durable, and best calculated to resist the weather, and to keep the inside of a house dry; provided the mortar used with them is composed of fresh-burnt stone-lime and sharp road-grit or sand, and is well mixed. The stock bricks absorb but little moisture, and that little is soon evaporated; whereas the place or soft bricks absorb a large quantity of moisture; and, allowing that to pass through them into the middle of the wall, are a long time wet; because the centre of a wall retains the moisture long after the surface is dry. It is particularly desirable, as we have before stated, for the walls of houses built on clay, or on any moist soil, to have the entire foundation, or, at all events, a few courses of the brickwork above the ground, laid in Roman cement. When a house, whether in a town or its suburbs, is detached on one, or on all sides from other houses, the preference should be given to one which has its walls built with bricks of the same quality and colour on every side. This is for the sake of appearance. It is very common, in the neighbourhood of brick-built towns, to find suburban houses quite detached, having the exterior surface of the walls of one kind of brick (that considered the best) in front, and another kind on the two sides, and the back. This is offensive; not only from the idea which it conveys of meanness and parsimony, but because it gives the effect of incompleteness, or of something being wanted to make it constitute a whole. In short, it appears to be a component part of some street of a town; which, to make it perfect, requires to be taken back, and fitted into the place from which it has been cut out. The trifling saving in the article of bricks which is made by building houses in this manner is not for a moment to be put in competition with the bad effect which such houses must necessarily produce as long as they continue to exist; unless, indeed, the walls on every side are coated over with cement, so as to give them a new character.

To illustrate the importance of attending to the foundation walls, let us suppose a new house, either large or small, most substantially built, and, in every apparent circumstance, eligible either for purchase or occupation; and that the intended occupier or purchaser has completed his bargain without examining the subsoil, and the manner in which the foundation walls are built. On the supposition that the subsoil is dry, all will be very well, and the house will turn out what it appears to be. But supposing, on the other hand, that the subsoil should be a clay, or a stratum of moist gravel, or of moist soil of any kind, and that the foundation walls should have been built with spongy bricks, or porous stones, and bad mortar, and not with good hard bricks, or stones, and Roman cement; the consequence of this will be, that the kitchen and other apartments on the ground floor will appear dry and comfortable for a year, or perhaps longer; but after this, from the bottoms of the walls acting like sponges in absorbing moisture from the soil, the damp will rise up through them more and more every year, till, at last, it will reach 6 ft. or 8 ft. above the exterior surface of the ground. We could refer to a house, in all other respects most substantially and judiciously built, and surrounded by dry areas as deep as the footings of the walls, but on a clayey soil, and without cement having been used in the foundations, through which the damp has, in the course of eight years, risen as high as the parlour floor; and the family occupying the house are now quite surprised at finding their furniture becoming mouldy there, after having been for years without experiencing anything of the kind. We know another house, built on a clayey soil, with very porous bricks, and with the outer walls only 14 in. in thickness; the inner part of the walls being built with what are called place bricks. In this house the moisture of the subsoil was sucked up in such quantities as to throw off the papering in the bed-rooms. As a proof that the moisture was sucked up from the foundations, we may mention that the walls were coated with Roman cement when the house was built, and that, consequently, the wet could not possibly penetrate through them from their exterior surface. Mr. Hawkins, the engineer, found a house, of which he had taken a lease, so damp in the foundations, that he considered it necessary to take a course of bricks out of all the walls immediately above the exterior surface of the ground, and to supply their place with a course or layer of bricks bedded in Roman cement, to prevent the moisture from rising higher. In many houses of the fourth class, the cesspools of the privies are made adjoining the foundation walls; in consequence of which, the moisture and smell are absorbed, and diffused all over the house. This is sometimes the case even in houses of a superior description; and, as these cesspools generally adjoin the foundation of the staircase, the annoyance they occasion in every room of the house, is only to be counteracted by extraordinary ventilation, or by introducing in the walls a layer of brick and cement in Mr. Hawkins's manner. These cases will, we think, prove our position (p. 37.), that some of the causes which ought to influence us in the choice of a house are almost invisible; and will show, even independently of the fear of the dry rot, the importance of examining the foundations of a house under its lowest floors, previously to entering on its occupation.

The Timber used in any building should be from trees of slow growth, such as the pine or fir of cold countries (Norway or Sweden for example), or oak. If for work under or near the ground, the oak should be of English growth; but the American oak may be used with safety above ground. Oak is the only timber fit for beams and sleepers (that is, joists laid on the tops of dwarf walls) next the ground, unless the soil be particularly dry, and the floor well ventilated; in which case fir may be used.

The Strength of the Joists and other Timbers of which the several floors are composed, is another subject of importance to every one about to take the lease of a house. If these are weak, they will necessarily shake if the tenant allows his friends to dance on them; and, though the walls may not absolutely give way, yet the ceilings and cornices of many modern houses have been known to fall down from this cause; while in the lease of many London houses there is a clause prohibiting dancing on the first floor. The floors of what, in England, are called the first and second class of buildings (or, in other words, of the largest and next largest mansions that are usually built), are usually pugged (that is, filled in, between the floor of one room and the ceiling of that below, with mortar, &c.), to destroy sound, and as a security against fire. When this is not done, it is an unpardonable omission on the part of the builder, as the expense is small, and the benefit great. All the partitions of a house should, if possible, be brick walls. At all events, no timber partitions ought to be admitted in the basement or other lowest story of any house; nor in any of the upper stories, except where, from the arrangement of the rooms, the partitions on the upper floors cannot be placed perpendicularly over the lower portions: even in this case the timber partitions ought to be trussed up so as to rest their weight upon the side walls. All timber partitions should be filled in with brick nogging (that is, a framework of timber, with the interstices filled in with brick and mortar); the width of the wall or partition being commonly that of the width of the brick; though in small houses the partitions are sometimes not wider than brick on edge. If this were universally done, and the party and other partition walls were plastered, so as to prevent all draughts of air, it would tend more to check the progress of fire than any other mode of construction. If a fire were made on the floor of such a room, it would probably burn itself out without communicating with the timber partition; or, at all events, so little would be the tendency of the fire to spread (for want of a current of air), that a very moderate application of water would put it out. But, where the floors are pugged with mortar, care must be taken that the timbers are well seasoned and dried, and not taken, as is customary even in some of our largest buildings, wet out of the Thames (where they have been immersed with a view of depriving them of their sap, but which deprivation cannot take place unless they are afterwards dried slowly by evaporation in the open air), sawed, fixed, and closed up in the building, in a few weeks, while they are yet reeking with wet, and in such a state as to exude moisture at their extremities after the weight of the superincumbent walls is put upon them. The dry rot, and premature decay, are the frequent consequences of this careless and ignorant mode of building.

Light. Having carefully examined the nature of the soil, the drainage of the premises, the strength and durability of the walls and timbers, and the ability of the roof to keep out wind and water, the next subject to be attended to is the manner in which the house is lighted; as the cheerfulness and healthiness of a house depend much on the admission of abundance of light, and on its proper distribution through the different apartments. As a general principle, it must be remembered, that light is favourable both to cleanliness and cheerfulness: but, in the distribution of light through a house, attention must be paid to the uses for which the several rooms are designed. A common sitting-room is always more cheerful when light is admitted at both ends; or, when this cannot be obtained, when the windows are placed in its longest side; or one window at the side and one at the end. If, however, the room is required for the purpose of a study or painting-room, a cross light will be found very injurious to the eyes; and the room should have only one window, or, at least, only windows on one side. A veranda has a great effect in softening the light, and taking off the glare. In this, however, as in most other cases, much must depend on the occupier. The happiness of some persons depends much more than is generally supposed on the cheerfulness or dulness of the room which they occupy during the day, and on its fittings-up and furnishing; and it is next to impossible for such persons to have buoyant spirits and agreeable thoughts in a dark half-lighted room. In all cases, a room should never be chosen to pass the day in that is lighted by a borrowed light, or a sky-light; or that has a dead wall nearly opposite.

The Windows, in order to light a room cheerfully, should be brought down as low as the nature of the occupation of the room will allow, and becarried up as high nearly as the cornice. None of the sashes should be fixed; and particularly all the upper sashes should be made to open, as the confined air is more speedily and effectually expelled by opening the upper, than the lower sashes; and, therefore, in sleeping rooms, the upper sashes should be opened every fine day throughout the year. The air, in a room where the windows are opened, and not the door, is changed by the external air entering by the window, and mixing with, and expelling a portion of, the air of the room. This admixture will be slow or rapid, according to the difference of temperature between the air of the room and the external air. If there were no difference whatever, no change whatever would take place: but there is always a difference, more or less; and, as this difference is greatest near the ceiling of a room, in consequence of heated air always ascending, there is a more ready escape of the heated air, and a greater admission of the external air to supply its place, when the upper sashes are opened, than when the lower ones are. Supposing the lower sashes only to be opened, and that there were some degrees of difference between the temperature of the internal and the external air, the latter would still enter, expel a portion of the former, and mix with the remainder: but the process would not go on with nearly so much rapidity as when the upper sashes were opened. Unfortunately, in houses occupied by the poor, the upper sashes are seldom hung; and the close unwholesome air is therefore never effectually dispersed, even when the windows are opened.

Artists are continually inveighing against having more than one large window in a room, and against bringing the windows down to the floor; but this proceeds from their viewing the subject solely with reference to their own art. It is certain that a picture or a statue, and, for the same reason, a living human being, will be seen to most advantage when the light is admitted from only one opening, and when that one is rather above, than on a level with the figure; but for the common purposes of living-rooms such lights are altogether inefficient. Where there are a number of persons in a room, for general purposes, the light must either be admitted from the centre of the ceiling (which, in houses of several stories high, is impracticable), or from at least two sides of the room; otherwise a part of the company in it must always be in the shade. In small rooms, lighted by one window, every one must have experienced that, while the party sitting with his face to that window has light enough, the party sitting with his back to it is deficent in light; but this will never be found to be the case in rooms lighted from two sides. The dictum of artists, therefore, on this subject, however much it may be suited for a painting-room, a study, or a picture gallery, is altogether unfit to be applied indiscriminately to all the rooms of a dwelling-house.

Shutters. All windows should be provided with shutters inside, where comfort is an object. They afford the only effectual means of completely excluding the light, which is sometimes desirable in cases of illness; and they keep the rooms warmer in winter and cooler in summer, to say nothing of affording security from thieves. In the best rooms, they should fold back into boxings; that is, cases formed in the jambs or sides of the windows, to receive them, during the day; but, where the rooms are small, the shutters may be made to slide up and down, in order that they may occupy less space, and not prevent there being curtains to the windows.

Fitting of Doors and Windows. Whenever it is observed that the soffit of a window or doorway is sunk so as to impede the door or shutter from closing easily, it should be examined; and it will generally be found that the defect arises from the weakness of the lintel, and that the arch in brickwork, which should be turned over every opening where it is possible, has been omitted. This must be remedied without delay, or it will get worse.

The Floors of the inferior kinds of houses are mostly laid with white deals, 9 in. wide, and often not more than three quarters of an inch thick. For the best rooms, they should be of yellow deal, and never less than 1 in. thick: at least, they should be invariably so in the basement story; and they should never be of soft spruce fir deals. These deals may easily be detected, by observing if there are any small round holes in the boards of the floor where the hard knots have fallen through; for, from the soft spongy nature of this timber, it shrinks so much when thoroughly dry, that the knots fall out. From the shrinking of this wood, the joists of floors laid with it often open a quarter of an inch wide, so as to require the floor to be taken up and re-laid, or slips or fillets of wood to be fitted into each joint; and, if the evil is not remedied, the water, every time the floor is scoured, finds its way through these open spaces, on to the ceiling of the room below, which it not only discolours, and even destroys, but, in time, occasions the decay of the laths, as well as the plastering. In good houses, batten floors (that is, floors laid with narrow boards 1¼ in. thick) are used in all the best rooms: inch deal floors are put in the upper servants' rooms; and 1¼ in. yellow deal floors in such rooms in the basement story as are boarded. Floors of the best kind are doweled; a mode of flooring which will be hereafter explained.

Staircase. The internal appearance and comfort of a house depend much on the kind of stairs; and there are two classes of persons (the very young and the very old) who are materially inconvenienced by an ill-constructed staircase. The best staircases are those without winders (those steps where the stair makes a turn without a landing, and which steps, of course, are broader at one end than at the other); where the height is divided into two or three flights of steps, by half paces or spaces, or quarter paces. (A pace, or space, in a flight of stairs is a level board or square step or platform, occuring between flights of steps, but not forming the landing-place or platform to a floor.) In a good staircase, the height of each step should not exceed from 6 in. to 7 in.; and the width of the tread, or upper surface of the step, should be from 11 in. to 12 in. Of the two evils (viz. a high rise and narrow tread, without winders, or an easy rise and broad tread, with winders), choose the staircase with the winders, as the easiest and safest; but never choose a staircase, the steps of which are continued straight up a whole story, without landing-places or winders. Stone staircases are very desirable: they are, when properly constructed and executed, much superior to wood, as they are noiseless, and afford a great security in case of fire. Knowing that the public always fully estimate their value, some builders have erected stone staircases in the houses they have built; but, to avoid expense, or from ignorance, they have executed them so improperly, that the lives of those who pass up and down them are in great danger. This remark refers to stone staircases let into 4-in. timber partitions, which are brick-nogged (wooden framework filled in with bricks), and have pieces of wood nailed above and below the steps, to keep them from falling. The half-pace landings are sometimes composed of two stones carried across the opening of a window, the window being part below and part above the landing. A stone staircase should never be pinned into a wall of less than 9 in. thick; but, when the steps are of a considerable length (that is, above 3 ft. 6 in. out of the wall), the wall should be 14 in. thick. Another piece of negligence is not uncommon: the party-walls, in such houses as have stone staircases (which are generally either first or second-rate houses), being lofty, and containing a great number of flues, and being only two bricks thick, it follows that a great part of the wall, instead of being two bricks thick, is only two half bricks thick, between which the flues pass; so that, when the mason cuts into the wall to fix the end of the stone step, the flue is broken into, and part of the step passes into the flue, where it is left for the soot to lodge on; and, if carelessly pinned in, the smoke, should the skirting be of wood, is thereby allowed to escape into the house. This evil is often occasioned by the carpenter cutting into a flue when making a hole to fix the bearers for his landings; which having wedged, he, without directing the bricklayer to fill up the spaces he has left round the timber with brick and mortar, fixes the skirting; and the evil is often not found out until the chimney takes fire. Where there is any doubt or suspicion of this being the case, it would be advisable to make fires in the flues, and put some damp hay or straw on these fires, so as to cause a large quantity of smoke to ascend; and then to open the windows; by which means the evil, if it exists, will most probably be detected, by the smoke appearing through the crevices.

Partitions. It frequently happens, even where the principal walls, roof, and floors of a house are sound and strong, that the middle partitions are infirm and weak. These, therefore, should be carefully examined; for it is only too common to find, either from the bad materials used in the walls in the basement story, or from neglecting the foundation, or, more properly, from the bad construction of timber partitions, that they sink. The consequences are, that the floors become out of a level; the doors do not fit their openings; the gutter-plates that rest on these partitions sink also, and cause the water to lie in a pool in the hollow part, and the plasterings and cornices crack. This, although a serious evil, is not always insurmountable, as it frequently occurs from the hasty manner in which buildings, particularly those built on speculation, are erected; the partititions being plastered before the building has settled to its proper bearing; and, when arising from this cause only, it may be remedied. One of the first questions asked by many gentlemen who are about to have a house built, or to have alterations made, is, "How soon is it in your power to get the works completed?"

and, generally speaking, the shortest possible time is allowed. This is a great error: a sufficient time to dry and season the materials (particularly in the bricklayer's and carpenter's work) should be allowed, before the joiner's work or plasterer's work is begun, in order that the defects arising from the shrinking of the timbers, and any unequal settling of the brickwork, may be remedied before the finishings are fixed. Should any settlement take place afterwards, the finishings, being attached to the brickwork and rough timbers, must necessarily partake of any unequal settlement or change of position that may occur in them.

Brick or Stone Partitions. Next to stone staircases, or cast-iron staircases with stone treads, brick or stone partitions are desirable in a house. Wooden partitions are necessarily hollow; and, as they are always in the interior of the house, and as the staircase is commonly included between two such partitions, they may almost be considered as flues, for the more rapid communication of fire from the lower part of the house to the upper part. We have known a case in which fire communicated to a wooden partition in the sunk story of a house has, in a few minutes, set fire to the roof; the reason of its rapid progress being, that, before the flame burst forth in the lower part of the partition, a portion of the wood composing it had become charred; and, in consequence of this process, all the upper part of the vacuity in the partition was filled with hydogren gas. This gas, the moment it was ignited, spread the flames as rapidly as the fire-damp in a coal pit; and, had there not fortunately been, in the case alluded to, very ample means of putting out the fire at the time, the house must have been burned down in the course of a few hours. Hollow partitions of this sort should never be suffered to exist in a house by any person who is alive to danger from fire. The vacuities should either be entirely filled up with clay or mortar, or, as a correspondent in the *Architectural Magazine* (vol. i. p. 40.) has practised, the skirting-boards of the rooms adjoining the partition should be removed, and the space behind them filled in solid with a layer of tiles, and Roman cement. This mode, by cutting off the communication between the partitions of different stories, will effectually prevent any fire which may break out from spreading by means of the partition, either upwards or downwards. If all partitions were of either brick or stone, or rendered solid by being filled up with mortar or clay; and if all floors were rendered solid by the same means, and all staircases were built of stone, or of cast iron, with stone steps and treads; the danger from fires, whenever they might break out, would be very greatly diminished.

Closets, when properly fitted up, and of a sufficient depth to be useful (that is, when the shelves are at least 12 in. wide), are a very great convenience; but, when the shelves are only

8 in. or 9 in. wide, the closets generally become what are familiarly called sluts' holes, and the receptacle of all the rubbish of the house. When they are put up independently of the plastering (that is, unconnected with the plaster, or with the walls not plastered at all), they should be lined all round with deal, and made air and dust-tight. If it can be avoided, they should never be placed against an external wall, especially one facing the north, unless the wall is battened, on account of the damp. (To batten is to fasten narrow strips of wood against a wall, on which plaster laths are nailed, or canvass stretched, so as to preserve a vacuity between the plaster, or the paper, and the wall.) External walls, indeed, should always be battened in good rooms, as there is but little dependence on freedom from damp, when the external walls are plastered on the brickwork, and the precaution of battening is neglected.

Fastenings. Attention should be directed to the fastenings, which are often of a cheap and inferior description, and thus are a source of continual annoyance, from the locks getting out of order, and the bars, bolts, &c., not acting properly.

Fireplaces. While examining the fireplaces of a house intended to be bought or hired, it is necessary to observe if the hearth-slabs are very narrow: if they are, the probability is, that the chimney-pieces are of an inferior description. Marble chimney-pieces should be carefully looked to, as it frequently happens that they are contracted for at some incredibly small sum compared with their appearance, and put together with old marble, the stains in which often become visible again when they have been some time exposed to the heat of the fire. The sides or profiles, the slips, and the soffits or under sides of the shelves, are often not more than one fourth of an inch thick, and sometimes they are even less than that. All these things it is of importance for the tenant to examine; as, should the chimney-pieces fail, and become dilapidated, he will be called upon by his landlord to restore them to a sound state and condition, at an expense, most likely, greater than their first cost.

Smoky Chimneys are an insufferable nuisance, and they are, in some situations, very difficult to cure. They arise from a great variety of causes: one is, building the opening to receive the grates too shallow, in some instances only 9 in. deep; while experience proves that it should never be less than 14 in., and is better if 18 in. These dimensions will allow the introduction of the circular wire fire-guard, which turns round in front of the fire, when it is required, and slides in out of the way behind the grate when it is not in use.

Innumerable modes of building chimneys so as to prevent them from smoking, and of curing such as do smoke, have been proposed by architects, builders, and persons of various descriptions. All the numerous causes which produce smoky chimneys may be included in two classes: the first, those which prevent the ascent of the smoke; and the second, those which drive it down after it has ascended to the chimney top. One of the commonest causes why smoke does not ascend freely is, the small diameter of the chimney in proportion to its height; in consequence of which, the column of smoke becomes cooled before it reaches the chimney top, and, by its gravity, impedes the ascent of what is below. Another, and a very common cause, is, the accuracy with which the windows and doors of the room are fitted; by which means there is not an adequate supply of air admitted into the room, to support the draught up the chimney. In second-rate houses, to which class most suburban villas may be considered as belonging, this will most commonly be found the cause of smoky chimneys; and there is no mode of counteracting it, but by admitting more air to the rooms, either by openings in the upper parts of the windows, or by what is now a very general practice, making a concealed opening under the upper part of the architrave of the door of each room. To counteract the effect of exterior currents of wind, in driving smoke down the chimney, is not so easy a matter as to supply more air to a room; but the latter operation seldom fails to have a considerable influence in preventing the smoke from being driven down. Caps or pots of some sort are generally resorted to; but they are all unsightly, unless when they form part of the architecture of the chimney top. This might always be the case; but, unfortunately for the exterior beauty of dwelling-houses, the defect which renders such caps necessary is seldom found out till the house is occupied; and, when this is the case, the occupier being unwilling to recommence the annoyance of building, submits to the unsightly appendages usually put up to cure smoky chimneys, rather than incur the trouble of any farther alterations or additions; especially when these are of such a nature as to interrupt the use of some of his principal rooms. Chimney-pots in themselves, when of geometrical forms, and when placed so as to be regular and symmetrical, cannot be considered deformities; and, when they appear to have answered their intended purpose, they are felt as satisfactory. But the deformity of chimney-pots appears, when, by their shapes, by the manner in which they are placed, or by the ornaments bestowed on them, all geometry, regularity, and symmetry, in the shape and arrangement of the chimneys, and all architectural character in the sky outline of the building to which they belong, are destroyed. It appears to us that it would be a desirable object, if architects and builders, when giving plans and undertaking contracts for dwelling-houses, should bind themselves specifically, to form all the chimneys in such a manner as to prevent their smoking; which, in our opinion, they might always do by proper architectural contrivances. Where more flues than one are carried up in a chimney top, and, consequently, more than one chimney-pot are required, all the pots ought to be of the same exterior form, in order to preserve the expression of regularity, which is one of the fundamental beauties of architecture. The principle of symmetry ought to be preserved, by having all the chimney-pots and tops alike. This principle requires particularly to be attended to in all houses where either regularity, or symmetry, is the prevailing feature; but, where irregularity, or the picturesque, is the leading expression, as in most cottages and cottage villas, several kinds of chimney-pots may be used on the chimney tops of the same house, provided that only one kind is used on each chimney.

Fig. 1. shows the general appearance of a right-angled chimney-pot, which has been found effective in preventing the smoke from being blown down a chimney. It has been in use at Poole Park, near Ruthin, Denbighshire, for several years; and we are informed (May 31. 1837) by Lord Bagot's agent there that it has answered, and continues to answer, in every instance in which it has been applied. It is formed of cast iron; and the appearance of it as seen from below, is shown in *fig.* 1. It is cast in eight pieces, which are put together as shown in *fig.* 2. *a*; by which it appears, that the upper part of the pot is 1 ft. 4 in. in height; and the neck, which is inserted into the flue, is 11 in. on the side. The length of the neck may be increased at pleasure, so as to produce the appearance shown in *fig.* 1. A vertical section of this chimney-pot is shown at *fig.* 2. *b.*: *c* and *d* are elevations of

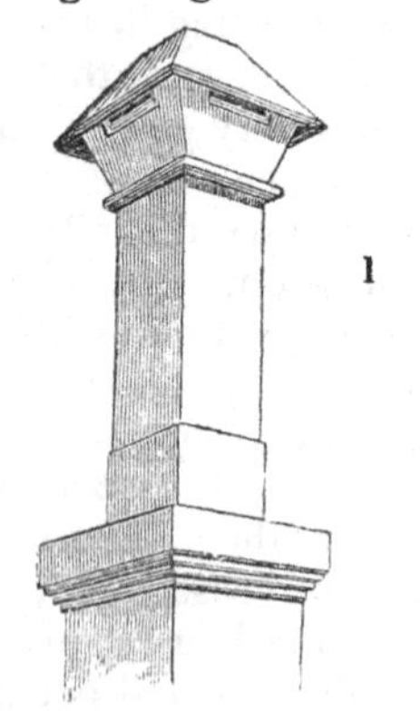
1

2

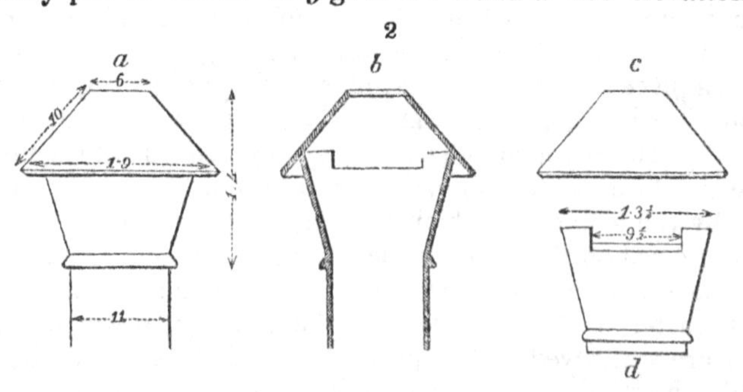

the two pieces, four of each of which, when screwed together form the pot. We have sent a model of this pot, which was kindly forwarded to us by Mr. Turnor, Lord Bagot's agent at Poole Park, to Messrs. Cottam and Hallen, iron-founders and manufacturers, Winsley Street, Oxford Street, London; from whom pots of this kind may in future be purchased.

Fig. 3. shows the form of a cast-iron chimney-pot, in use in the neighbourhood of Barnsley, in Yorkshire; where it has been tried for a number of years, and is considered to be an effective cure for a smoky chimney, where the smoke is blown downwards by wind. *Fig.* 4. shows the general appearance of a chimney top furnished with pots of this kind. The usual form of this pot is that of a truncated pyramid, as in *fig.* 3.; but it might easily be made to terminate in a Gothic pinnacle, or in any other ornament which was considered suitable to the style of the building to which it was to be applied. It might even be made circular, and used in the case of cylindrical chimney shafts built in the form of Grecian columns. Where the form is that of *fig.* 3., the four sides are cast separately, and bolted together; and, where the pot is to stand alone, or with a short distance between it and other pots of the same kind, each side is fitted up with a hinged door, as shown in *figs.* 3. and 4.; but, where the pots are to stand close together, these hinged doors are only placed on the two exposed sides. The top, in either case, is closed with an iron cap. Each hinged door is connected with the opposite one by a rod of iron, about 2 in. longer than the diameter of the pot; so that, when the weather is calm, the lower part of each of the four doors projects from the chimney-pot about 1 in.; while, on the other hand, when the wind blows, it closes the door on the side against which it strikes, and opens the door on the opposite side to the extent of 2 in. It is evident that the same kind of doors might be introduced near the termination of each flue, in an architectural chimney top without pots.

3

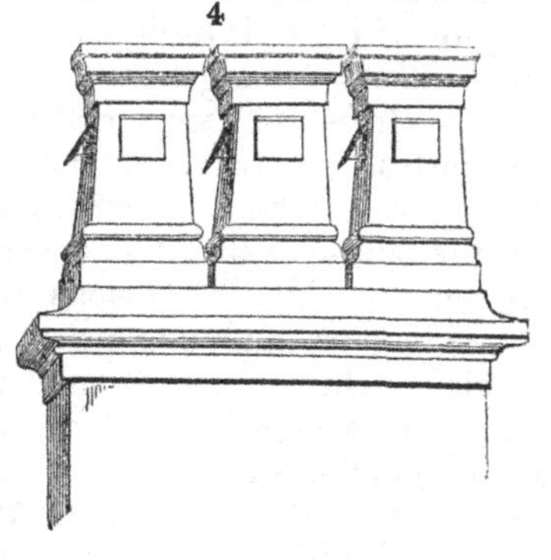
4

5

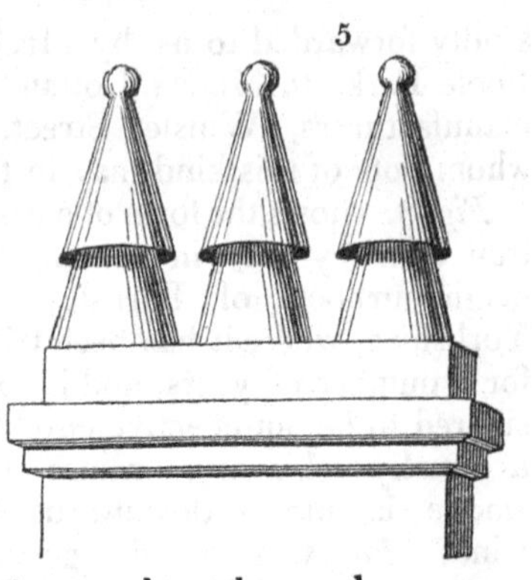

Fig. 5. shows the general appearance of a chimney top where the pots are formed of double cones of metal of any kind; the object being to prevent the smoke from being blown down the chimney. The lower portion of each chimney-pot is a truncated cone, in which is fixed a vertical spindle, as shown in *fig.* 6., on the upper extremity of which is suspended, by a caster joint, another cone, which moves in every direction when acted on by the wind. In calm weather, the smoke passes out under the upper cone, all round it; but, when the wind blows in any particular direction, the upper cone is forced to one side, and the smoke escapes by that opposite. This cone we have seen in use, in and around London, in various houses, since 1814; and, in general, with the most perfect success. Care should be taken, however, in the case both of this pot, and in that represented by *fig.* 3., to oil the joints occasionally. A good rule is, to have them oiled every time the chimney is swept. If made of copper, and varnished, or of cast iron, and painted over with oil when nearly red-hot, and afterwards varnished with a mixture of oil and Indian rubber, such cones will last for a generation.

6

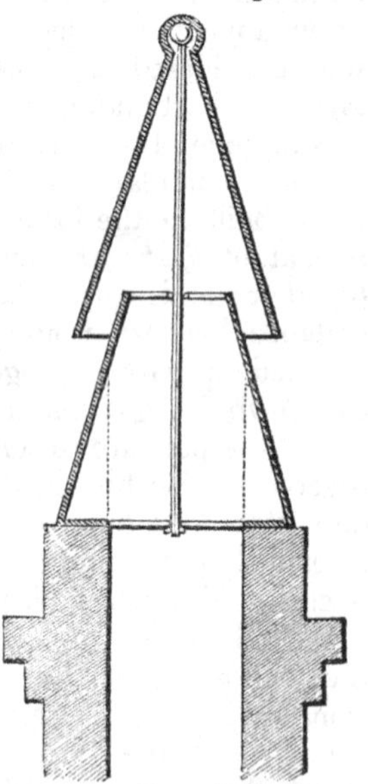

It is almost unnecessary to observe, that, instead of a truncated cone, a truncated pyramid may be employed; and that the cap, in such a case, must also be a pyramid. A less obvious improvement, however, and one which renders this kind of chimney-pot completely architectural, is to employ a truncated pyramid for the cap, finishing it with architectural mouldings, as in *figs.* 7. and 8. We have never seen nor heard of any cap of this sort having been tried; but we see no reason why it should not be as effective as the conical cap (*fig.* 6.). A cap with a pyramidal top we have seen as a termination to the large chimney shaft of a brewery, where it succeeds perfectly. If all the caps of this form were placed so close together on the chimney tops as almost to touch each other; but, at the same time, each kept distinct and properly suspended, the appearance might be made as architectural as if the chimney tops were finished in stone.

Chimney-pots, in houses where irregularity is the character of the architecture, may be introduced of more than one shape; but, at the same time, if one shape only could be employed, it would, in general, be found preferable. The reason is, the irregularity produced by the chimney shafts being differently placed, and by some of them containing more flues and pots than others, is already sufficiently great; and nothing can be more dangerous in a building, than, by too great a multiplicity of forms, to cause irregularity to degenerate into confusion. No irregularity in a building or other object can be satisfactory to the mind, unless it can be reduced by it, in a short time, to some principle of order and unity of design: for example, the irregularity in the chimney tops of a Gothic villa or cottage is accounted for by there being more flues in some chimneys than in others; and this, again, is accounted for by some of the chimney tops containing only the flues from one apartment in the roof; others, flues from two apartments, and so on; and the circumstance of the house containing so many stacks of chimneys is accounted for by the scattered disposition of the apartments, which is supposed to be produced in order to enjoy particular prospects, in some cases; and, in others, to be the additions made by different builders at different times.

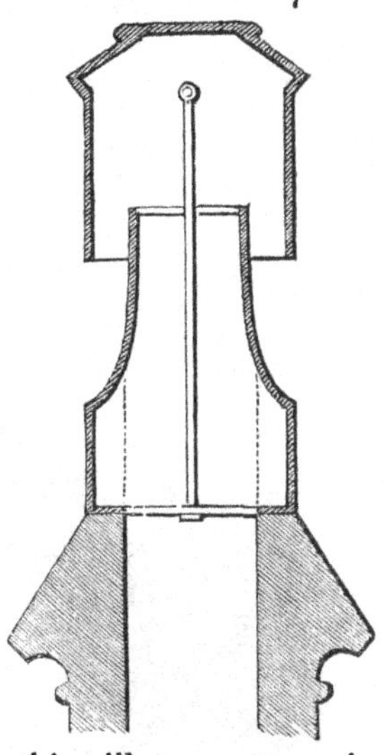

7

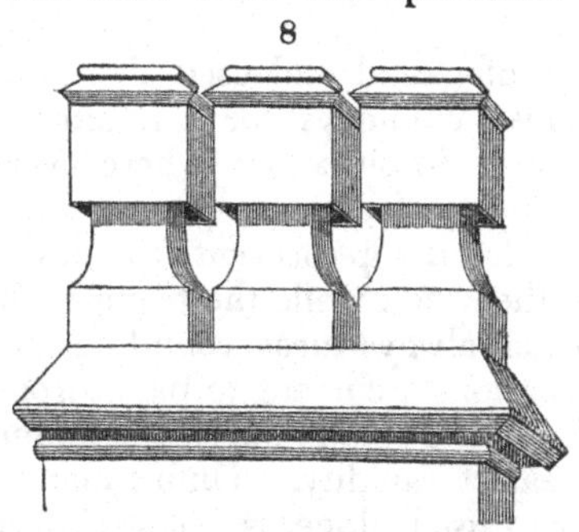

8

Fig. 9. is the general outline of a cottage villa roof, to show the manner in which several kinds of chimney-pots may be used. We repeat, however, that it is not often desirable to have more than one kind; we merely wish to state that, in houses where irregularity is a prominent feature, it may be extended to chimney-pots as well as to chimneys.

Figs. 10. and 11. show the nautical cowl, which is one of the most effective chimney-pots ever invented, for curing a smoky chimney caused by the wind blowing the smoke down the flue. Mr. Dawson, an ironmonger in Welbeck Street, London, has

9

manufactured and used this description of cowl for the cure of smoky chimneys for a number of years, and has found it to answer in situations where every other description of chimney-pot or cowl has failed. Its closed top prevents the downward action of the wind; while the trumpet-shaped mouth which always turns round to meet the wind, occasions a current to pass through it, which draws the smoke out of the chimney with the greatest rapidity. Unfortunately, this cowl, when used alone, is not architectural. The only mode by which it could be rendered admissible in regular or symmetrical buildings, would be by enclosing it in an architectural case, such as *fig.* 12., *fig.* 13. or *fig.* 14., formed either of brickwork, stone, slate, or cast iron. In *fig.* 14., it is surrounded by small columns, which disguise it from view; and in *fig.* 13., it is enclosed in open brickwork. These two forms, it is obvious, may either be made circular or square in the plan; or, where several nautical

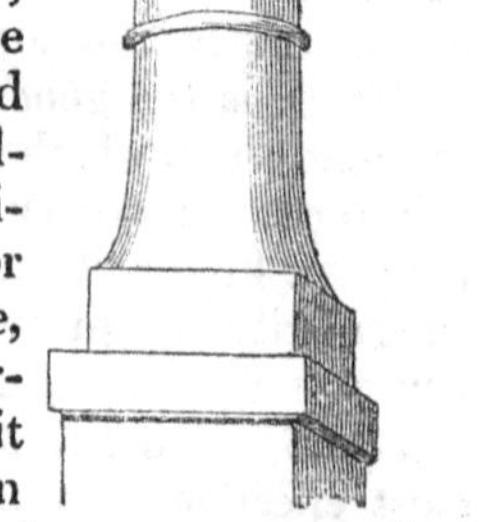

10

cowls are to be introduced, they may be made parallelograms, as in *figs.* 15. and 16.

11

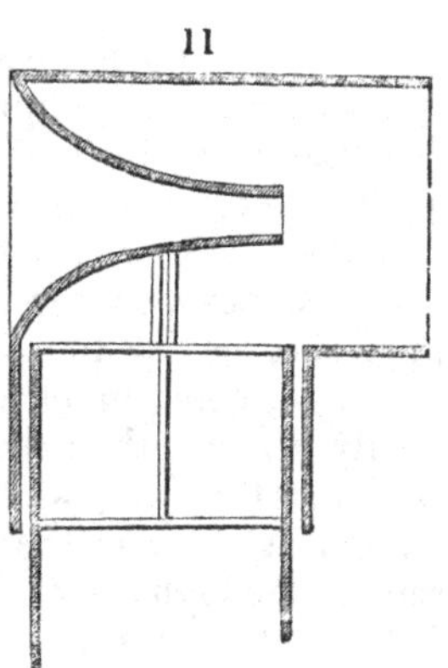

All these contrivances are only calculated for smoky chimneys where the cause is the action of the wind on the chimney top; and we think we may with confidence assert that a case can hardly occur in which one or other of these caps will not be effective. The other cause of smoky chimneys, the want of what is called a draught in the flue, is, in many cases, not so easy to be either discovered, or to be remedied. Often it proceeds from the flues being of too small dimensions; from their being placed in an outside wall, with only a thin stratum of masonry between them and the external atmosphere; or from some other defect in their original construction: and such evils hardly admit of a complete remedy, otherwise than by taking down and rebuilding the chimney. In by far the greater number of cases, however, in well-built houses, when the chimneys smoke, and the cause is not to be found in the wind blowing the smoke down the chimney, it will be found to proceed from the want of a sufficient supply of air to the fire, arising from the tightness or accuracy with which the doors and windows are fitted to their frames. This evil, as we have already observed (p. 49.), is best remedied by having concealed horizontal openings over the door of each room, communicating with the staircase, as shown in the section *fig.* 17. The air thus admitted, being somewhat raised in temperature by the heat radiated from the walls of the staircase and the stairs, and being admitted to mix with the air of the room several feet above the heads of the persons occupying it, can never be felt cold, and will never produce a draught in it; while, by mixing with the air of the room, it will greatly increase its salubrity, as well as contribute to increasing the draught in the chimney. This mode of admitting air from the staircase, by openings on purpose over the room doors, instead of allowing air to find its way into the room, to supply the draught of the chimney, through such chinks as there may happen to be in the

12

13

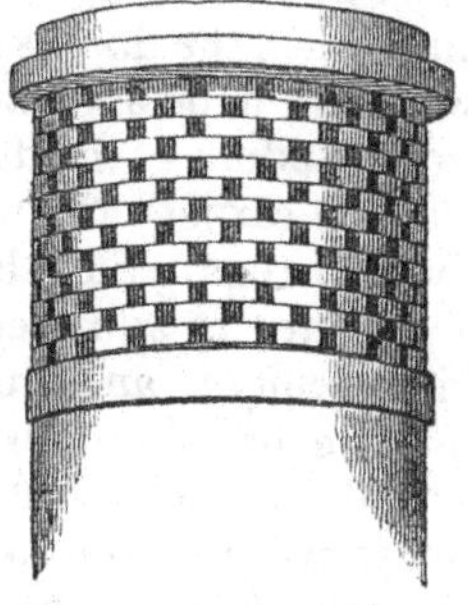

windows or doors of the apartment, is one of the greatest modern improvements that has been introduced in the finishing of houses. When the exterior air is allowed to come in by the windows, it commonly enters by the chinks between the lower sash and the upper one, or by the sides of the lower sash; and, being much colder, and consequently heavier, than the air of the room, it sinks to the floor, where it forms a cool stratum, as deep as the lower bars of the fireplace are high above the hearth. Almost all the cool air in the room above the level of the lower part of the grate is drawn into the fire, to supply the draught of the chimney; leaving the upper part of the air of the room in which the occupants breathe, to the chance of being freshened by such air as may enter by the upper part of the sides of the windows and doors, and by the occasional opening of both. Nothing can be more widely different in effect, than the admission of air to a warm room by openings at the tops of the doors and windows; and by openings at the bottoms. In the former case, the air admitted mixes with the air of the room, and greatly improves its wholesomeness to those who breathe it; in the latter case, it merely cools the lower extremities of the occupants, and increases the draught of the fire. For example, in ordinary houses, in winter evenings, when the window shutters are closed, and the curtains drawn, no air can find its way into the room but through the doorway; and, as the largest chink when the door is shut is between the door and the floor (in order to admit of the door sliding over the carpet), the whole supply of air required by the draught of the fire is admitted through this vacuity; and, unless the staircase has been heated, the natural consequence is that almost every person in the room has cold feet. The ladies, to mitigate this evil, generally use footstools, which keep their feet rather warmer, by removing them into the upper part of the

14

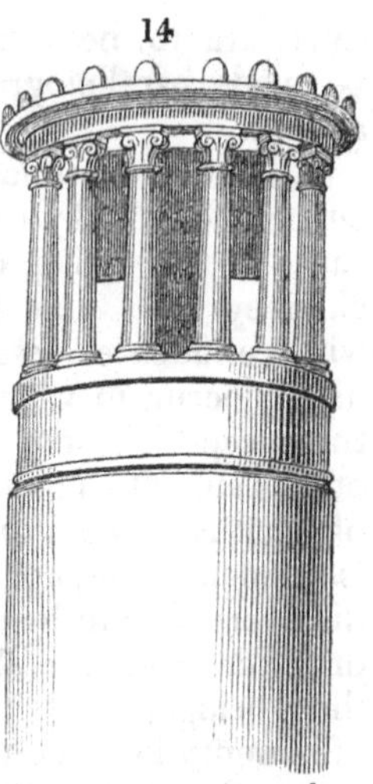

15

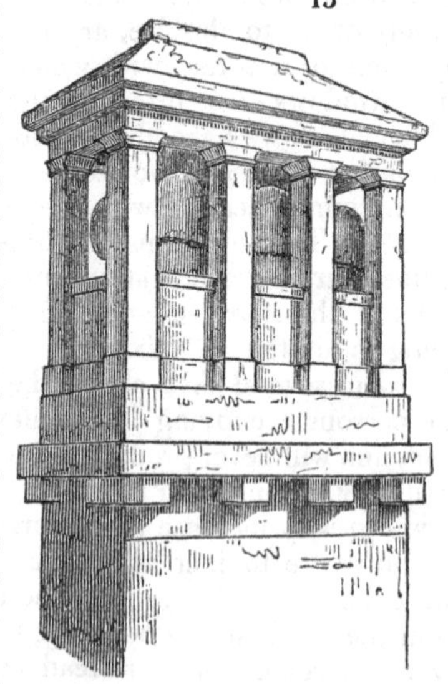

16

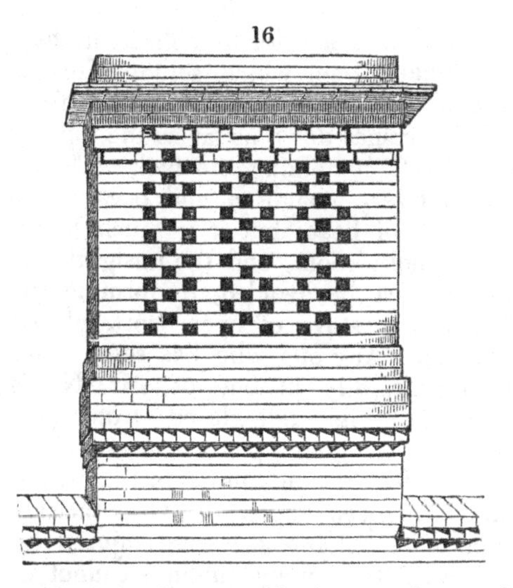

cold stratum of air, which is warmer than the lower part; or they raise them above it altogether, by reclining on a sofa; while little children playing on the floor are almost entirely immersed in it. If the air, however, were admitted only by the top of the doorway, instead of the bottom, even if the staircase were not heated, it would mix with, and be diffused through the other air, before it reached the floor, producing a volume of air every where of the same temperature; and the cold stratum over the floor could not be formed. The air of the room would thus be rendered much fitter for breathing, and would be equally adapted as before for supplying the draught to the fire. To prevent the air from entering at the bottom of the doorway, the door should be made to fit as tight as is consistent with its moving over the carpet; and, to do this in the most effective manner, Redman's, or some other description of rising hinges, ought to be used, instead of the common kind.

17

When a smoky chimney is neither produced by wind blowing down the smoke; nor by the want of air to supply the draught of the fire; nor by its being placed in an external wall, and subject to have the air in the flue chilled, either by cold, damp, or the admission of external air through the masonry; the evil may sometimes be remedied by placing pots, in the form of long cylindrical tubes, on the tops of the chimneys. Some of the largest mansions in England, both in town and country, are disfigured by pots of this kind. These are among the most unsightly of the appendages to the roof of a house, and we can in no case recommend them, unless when they are built into the masonry, either by thrusting them down a chimney top already formed, or by placing them on it, and carrying up masonry round them. In either case, the section across such a chimney-pot will be as represented in *fig.* 18. Finally, after all the above modes, and others of a similar kind, for curing a smoky chimney

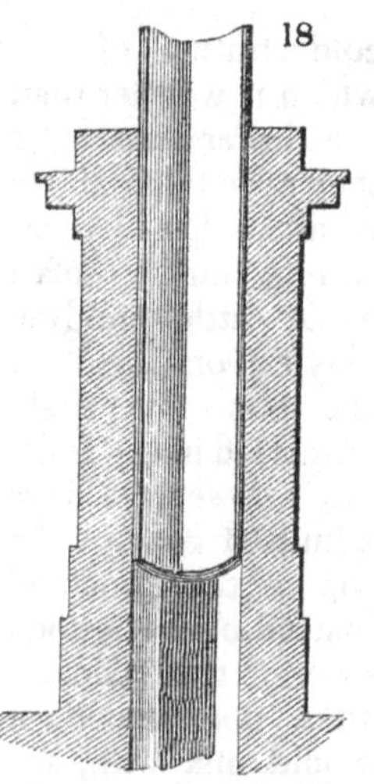

have been tried, and found ineffectual, there still remains one, which seldom fails of success. This is, to use a stove, the front of which comes down as far as the top bar of the grate containing the fuel. This is no doubt very unsightly, and occasions a great loss of heat; but it is much better than a smoky room, and the unsightliness may, in part, be avoided, by forming the front of the stove of open work glazed with talc. Any one who has seen a fire burning brightly in one of Mr. Nott's stoves, with its talc-glazed front (see *Encyc. of Cott. Arch.*, p. 1032.), will allow that the effect is much more brilliant than could have been expected; while the talc is not in the slightest degree injured by the heat. If a smoky chimney cannot be cured even by this means, the last and absolutely certain mode is to close the entire front with a cast-iron frame glazed with talc, and to supply air to the fire by a flue communicating with the external atmosphere; but, in this case, provision must be made for the ventilation of the room.

We have been thus diffuse on the subject of smoky chimneys, not only on account of its importance, but because it is intimately connected with those of warming and ventilating; subjects by no means so well understood as they ought to be.

Warming and ventilating the entire House. Various plans have been adopted, within the last twenty years, for adding to the comfort of houses, both in town and country, by the introduction of some general system of warming and ventilating them, so as to supersede, in a great measure, if thought desirable, the use of open fireplaces. The objects in view are, generally, a more equal diffusion of heat, and a saving of fuel; and both have, in many instances, been attained, in connexion with the renewal of the air, so as to render the houses so heated as wholesome to live in as those heated by open fireplaces; though, in by far the greater number of cases, the heating, and the economy of fuel, have been effected without due regard to ventilation, and the result has, consequently, proved injurious to the health of the occupants of the house. Up to 1814 and 1815, almost the only mode for warming the temperature of the air of halls and staircases was by the placing of German or Swedish stoves in them; the former being built of brick, and generally covered with Dutch tiles; and the latter being wholly of cast iron. The first grand improvement upon this mode was made by Mr. Strutt of Belper, near Derby, about 1807; who, soon afterwards, introduced this improved system of heating into his manufactories, the Derby Infirmary, and his own dwelling-house in Derby. It was afterwards made public by Mr. Sylvester, in 1820, in his work entitled the *Philosophy of Domestic Economy*. Mr. Strutt's apparatus was placed in the lowest part of the house or building to be heated, and consisted of an immense iron pot (called a cockle, from the circuitous passages formed round it for the passage of the air to be heated), placed in an inverted position, with the fire under it, and a constant stream of fresh air from the atmosphere passing over it. This air, being heated to 80° or 90°, was carried by air-flues into all the principal rooms of the house; while, the fireplaces being left open, a portion of the air in the rooms escaped by the chimneys, equal to that thrown in by the heating apparatus. It would occupy too much space in this work, to enter much into detail on this subject; otherwise, it would be easy to show that Mr. Strutt's principle, that of introducing fresh air along with fresh supplies of heat, is the only principle which combines a due regard to the health of the occupants of the house, with the other objects in view. Mr. Strutt's mode of heating (or, as it is generally called, Sylvester's mode) has been introduced into numerous private houses, with different degrees of success: but, partly from the management of the apparatus not being thoroughly understood; partly from the difficulty of introducing it, so as to be thoroughly effective, in houses already built; and partly from the difficulty of getting servants to attend regularly to apparatus of this, or of any other description; but chiefly to another mode of heating (by hot water) having come into fashion; it is now very generally given up. Another reason why it has been given up is, the liability of this plan to overheat the air, and deprive it of its moisture; more especially in cases where the persons managing the stove do not understand that, when an increased temperature is wanted in the rooms heated, it is not to be produced by raising the temperature of the air introduced (which is always understood to be about 80°, or a few degrees more or less), but by introducing the air in greater quantities, which is done by producing a more rapid current. Now, this additional quantity and increased current of heated air cannot be introduced, unless an increased quantity is abstracted from the rooms to be heated; and, as in most houses no sufficient provision is made for this purpose, or, if made, is neglected to be used, hence, all that can be done is to increase the temperature of the air forced in, by increasing the fire in the cockle. This is an objection to which, we fear, Mr. Strutt's mode of heating will always be liable; and, hence, the superiority of a plan, subsequently introduced, which may be described as heating the cockle, or inverted pot, with steam or hot water, instead of heating it by the direct contact of the fire. By this means, the temperature of the air heated can never be raised to an injurious height, and the air is never rendered insalubrious, by being deprived of its moisture. (See, on this subject, three important papers in the *Architectural Magazine*, vol. ii. p. 407., and vol. iv. p. 161. and p. 313.)

There is yet another objection to the working of Strutt's method, which, however, does not at all lessen the merit of the principle. In some instances, from the imperfect construction of the apparatus and the flues connected with it, houses have been set on fire; and in others, as in the Custom-House, for example, the apparatus has continued to be used after it was so far worn out as to admit, along with the heated air, smoke and ashes; or, as in the case of heating the same building, from the apparatus not being sufficiently powerful, the cockle is heated to such a degree as to deprive the air passing over it of its moisture, and even to decompose the organised matters suspended in it. Hence, in the case of the Custom-House, a most unwholesome atmosphere was produced, and various diseases were the consequence. The manner in which houses have been set on fire by hot-air stoves is easily explained. In consequence of the imperfection of some part of the apparatus, or of the flues connected with it; or of the over-anxiety of the parties attending the stove, to produce a sudden increase of temperature in the apartments heated; the fire is increased to such an extent as to render the iron pot red-hot. In this case, being acted on by the intensity of the fire on the one side, and the cool air on the other, the iron is very liable to crack; and when a crack takes place, however small it may be at first, it soon widens sufficiently to admit the flame, which thus passes into the air-flue, instead of into its own smoke-flue; and, of course, is discharged into the apartment to be heated, setting fire in its course to whatever it approaches that is combustible. It is perfectly true, that, under a proper system of management, no such accident could happen with Strutt's apparatus, nor with various others that have been introduced in imitation of it; but it is equally true, that such accidents do happen very frequently, and that they never could occur, if steam or hot water were employed to heat the cockle, instead of its being brought into immediate contact with the burning fuel.

We repeat, that, notwithstanding all these results, Mr. Strutt's principle, that of communicating heat by means of a fresh supply of air, is still the only true one: and, as a proof of this, we may mention that the systems of heating adopted by Dr. Desaguliers, Sir Humphry Davy, and Dr. Reid, in the House of Commons; and that recommended by Dr. Ure for adoption there, in the Custom-House, and in other places; proceed on the same principle; their several inventors only recommending different descriptions of apparatus for carrying that principle into effect. A main feature in the most recent of these improved apparatuses consists in the application of the heat of the fuel to steam or hot water in the first instance, over the tubes or plates containing which, as already observed, the air to be heated is made to pass. As neither steam nor water, under ordinary circumstances, can rise above the boiling point, there is no danger, where this mode is adopted, of burning the air, as the expression is; but, on the other hand, as the air entering the apartments to be heated, being at a lower temperature, will be more sluggish, greater care is requisite in providing a system of air-flues, with regulating valves, by which a portion of the air in the room equal to the portion entering may continually escape. To do this effectually, without opening the windows, it is necessary either to have a fan for blowing in the heated air, as Dr. Desaguliers and Dr. Ure recommend; or flues for the escape of the air of the room, carried up in close contact with a flue connected with a fireplace in constant use, in order to create a draught in it. When this mode of heating is contemplated at the time the house is building, such air-flues may be carried up with facility around, and in close contact with, the flue from the fire which heats the apparatus; by which means a draught will always be created in the air-flues when the apparatus is at work, in proportion to the heat applied to it; but, when a house already built, finished, fitted up, and furnished, without a view to any mode of heating, is to be warmed and ventilated by this, or by any other mode requiring air-flues, the flues, and other arrangements made, are almost always more or less imperfect, and the end seldom or never completely attained.

In the choice of a house, therefore, we would never attach much importance to the circumstance of a heating apparatus being either wanting or existing in it; at least, in the present state of our practice in this department of domestic architecture. The time, doubtless, will come, when a heating apparatus will be considered as an essential part of all houses of a certain size; and for smaller houses in streets, along a railroad, or at no great distance from one another, warm air may be supplied, as gas and water are now, by public establishments; but these improvements are too far distant for any practical use to be made of them in this work. All that we can safely recommend in the way of general heating is, a stove, which may be of cast iron, at the bottom of the central staircase, or in a central room on the cellar floor, to be used in severe weather. Such a stove will communicate a moderate temperature to the air of the staircase; which air will be renewed from time to time, in consequence of the frequent opening of doors by servants in the basement story, and of the main entrance door in the ground floor by the family and their visiters. Such a mode of general heating requires no other attention than that of supplying the fire in the stove with fuel; and this is as much as can be expected, in the present state of things, in respect to servants. Most of the large first-rate houses recently built in Belgrave Square, London, are heated in the manner above recommended, by immense cast-iron stoves placed in the basement floor.

With respect to the different modes of heating apartments in dwelling-houses by the introduction of steam or hot-water pipes, we cannot recommend any one of them; because, as they warm without ventilating, we do not believe that they are wholesome. Steam-pipes, however accurately they may be fitted, always permit the escape of water, even though it should be in the form of invisible vapour; and this must consequently deteriorate the air of the apartment, except in the case of persons subject to particular diseases, and for whom a moist air is desirable. These pipes, also, when the steam first enters them, make a disagreeable noise by their expansion; and the same thing takes place when the steam is withdrawn, by their contraction. Hot-water pipes, of every kind, are still more liable to communicate moisture to the atmosphere of rooms than steam-pipes; while, by neither of these modes of heating is a provision made for ventilation, or, in other words, for changing the air of the room. It is true, that, in all rooms having an open fireplace, with a fire burning in it, and having the doors and windows not more tightly fitted than usual, there is no occasion to introduce a change of air by any systematic arrangement; because that change will go on in the same manner as if there were no pipes for heating in the room; but, in general, where these pipes are introduced, one principal object is the saving of fuel; and, in proportion as this is effected, the ventilation is diminished, for want of a sufficient draught up the chimney. We repeat, therefore, that all we can recommend, under the ordinary instances of a villa residence in Britain and similar climates, is, to heat the air of the staircase by a common stove, and to trust the heating and ventilation of the rooms to open fireplaces.

In climates having the winters more severe than that of Britain; such, for example, as the greater part of Germany and of North America, the consumption of fuel, if heating were to be effected solely by open fireplaces, would be much too great for general use, even in first-rate houses; as, indeed, has been found to be the case, where open fireplaces have been tried, both in Hamburg and New York. All that can be recommended for similar climates is, in the case of cottages and houses of only two stories, to introduce a system of smoke flues under the ground floor; as explained in detail in the *Encyc. of Cott. Arch.*, book i. chap. i.; or in larger houses, or houses of several stories, to introduce the brick-built German stoves; the plans, sections, and various details for building which will be found in p. 1029, 1030. of the work quoted. It is true, that, by such a system of heating, the air of the room can never be so frequently changed, as either by open fireplaces, or the mode of heating by the introduction of hot air; and the consequence is, that, in all countries where such stoves are in use, the human countenance has a much paler hue than in Britain; and the faces of the women (who, of course, stay most in the house,) are proportionately paler than those of the men. These, however, are evils inseparable from the nature of such climates, at least, in our present state of knowledge, and in its application to useful domestic purposes.

Water. The next subject (and it is one of great importance to the comfort of a house) is water. The facility of obtaining water in the upper part of houses in London and its vicinity, and the moderate charges made by the water companies for high service (that is, for raising the water to the top of a house, in contradistinction to low service, which implies not raising the water more than 10 ft. above the level of the ground), may induce those who can afford it to place a capacious cistern in the roof, if possible; or, at all events, on the attic or second floor, of all houses containing six or eight rooms, and upwards. The expense of a cistern and pipes to convey water to the upper part of a house (supposing the house unfinished, or undergoing a thorough repair) is so small, and the advantages of having an abundant supply of water in the upper part of a house so great, that, where the difficulties to be overcome are not of a very serious nature, every one who consults his comforts, the saving of his servants' time, and the means a cistern affords, in cases of fire, of saturating the floors with water, and thereby preventing the fire from spreading downwards, will seriously consider this subject before he decides on allowing the opportunity of obtaining these advantages to pass by him. Besides saving the time of carrying water from the basement story up three or four flights of stairs, a cistern in the upper floor affords the means of having a bath at but little expense; of providing servants with a sink on the attic or two-pair floor; of having a water-closet in the upper part of the house (a great convenience, particularly where there are children, or two families in a house); and of supplying any or all of the best bed-rooms or dressing-rooms with water, by means of a fixed basin, with a supply-pipe and cock; which may also have a plug and waste-pipe, carried either into the trap of the water-closet, or into the rain-water pipe. This is a comfort that can scarcely be estimated by any one by whom the luxury of it has not been enjoyed.

Water Cisterns. In the choice of a house, therefore, the capaciousness, construction, and situation of the cisterns for the supply of water are matters to be carefully looked to, otherwise it may be necessary to incur a considerable expense in providing a sufficient quantity of water, should the house about to be taken not require to be repaired, and a convenient opportunity not be thus afforded of fixing a cistern in the upper story. In the present day, every privy should be a water-closet: let the construction of it be ever so common, still water should be supplied, and that liberally. Where the expense of a proper closet is inconvenient, or the occupancy of the house of an humble kind, an iron hopper, with a bent soil-pipe, which shall act so as to prevent the ascent of smell, set in brickwork, and supplied with water by a pipe and cock, will answer the purpose. Connected with the water-closet is the drain by which the water is carried off from it, and generally from the sinks and other parts of the house. From want of proper care in the construction of these drains, many houses, otherwise very comfortable and healthy, are rendered the reverse, by offensive and noisome smells. This arises either from there not being proper traps, or from the covering or sides of the drains being insecure, and permitting the impure air to escape. Enquiry should be made as to the construction of the drains, before a soil-pipe is permitted to enter one. Where soil is permitted to enter a drain, the drain should be rendered (that is, smoothly plastered over) with cement, half round the bottom inside, and all over the top outside, and covered with earth, of which clay is the best kind. In large houses, the cistern intended to supply the whole of the basement story with water may be placed over the scullery; or, in smaller houses, in the yard. The cistern, whether inside or outside of the building, should always be covered over, and have a flap in the cover, to give access to the pipes. The best kind of cistern is formed of quartering, framed together, with the bottom and sides boarded, and lined with lead. The lead at the bottom should, in ordinary cases, be from 7 lb. to 8lb. weight to the foot, and that at the sides from 5 lb. to 6 lb. to the foot; but, where the water has a tendency to corrode the lead (for instance, if it is at all impregnated with any kind of salt), the lead should be proportionably thicker. The pipes to supply the cistern, as well as the pipes from the cistern to serve the different sinks, &c., should be of the kind called by the lead-merchants extra-strong pipe; particularly those pipes that are laid under the ground, and those that are outside the building, and are exposed to the frost in winter. All pipes outside the building should be covered over, to protect them from injury by blows or frost. The cocks should also be of the very best manufacture, or they will be constant sources of annoyance and expense.

A Tank for the Reception of Rain Water is very desirable in situations where water springs are not to be found, and which are beyond the reach of water companies. The tank may be built in any convenient form, and arched or domed over at top, with a man-hole in the crown of the arch, or in the summit of the dome, having a stone lid to fit into a rebate in the masonry, in order to admit a man to descend by a ladder to clean out the tank. A pump must also be inserted to draw up the water. The deeper such a reservoir is sunk in the ground, so much the cooler will the water be kept. A very convenient form for a small tank is that of a circular well, which may be covered with a flat stone, over which there may be 2 ft. or 3 ft. in depth of earth, so as to keep the water cool. The lower extremity of the pump ought to be kept 2 ft. or 3 ft. above the bottom of the well, to allow room for the deposition of sediment. The rain water, whether collected from the roof of the house or from other buildings, or, as is done in some dry hilly parts of the country, from the surface of the ground after showers, should be filtered through a stratum of sand and charcoal before it enters the tank; and this filtering medium should be so placed as to be taken out, and cleaned, or renewed, at pleasure. In whatever form the tank is built, all the masonry or brickwork connected with it should be laid in cement, and plastered with that material on the side next the water.

A Pump-Well should, if possible, be established on every premises in the country. There may be no permanent spring in most cases; and, in many, the water collected from the surface springs may be unfit for culinary purposes and for washing; but, in the summer season, it will be useful for the garden, and at all times it will be a valuable resource in the case of the breaking out of fire.

An Artesian Well, when it is practicable at a moderate expense, is a most valuable addition to a suburban residence, which is out of the reach of water companies, and is not situated near a natural supply of surface water. Before any attempt is made to procure water by boring, however, some knowledge should be acquired of the locality; for, unless there is either a considerable substratum of porous materials, with the interstices filled with water, and supplied from the surface immediately over it; or a thin porous substratum, connected with a supply of water at a distance, the attempt to procure water will not be attended with success. Even when water is procured by boring, it will not rise to the surface, unless the stratum in which it is found is connected with a supply of water in a stratum lying on a higher level. In this case, it will rise above the surface, in proportion as the height and the supply are great, and the orifice of the perforation small. When the water will rise a few feet above the surface, it may be made to ascend into a vase, and flow over it into an open basin; provision being made by a cock to receive water direct from the ascending pipe for culinary purposes, more especially in summer when it is cool. When the water does not rise within several feet of the surface, it is necessary to insert a pump in the perforation; and if this pump be of the lifting kind, and not dependent on the pressure of the atmosphere, it may be made to raise water from the depth of a 100 ft. and upwards. In cases of pumps of this sort, and, indeed, in most cases where a pump is connected with an Artesian well, it is desirable to have a tank adjoining it; into which the water can be pumped at times when the labour can be most conveniently performed; or, when from the season of the year, rains, or the melting of snow, the supply is most abundant. It deserves to be remarked, however, that in districts most favourable for Artesian wells, and which are populous, in the Vale of London, for example, so many are generally made, that the supply to each is found to be insufficient; or, at all events, not to be adequate to the expense incurred. In all suburbs, it will probably be found more economical and convenient for the inhabitants to join together, and bring a supply of water in pipes from some natural source on the surface; than for each, or even every two or three families to join together, to form an Artesian well.

Balconies and Skylights. A balcony, independently of its ornamental effect, whether with or without flowers, is valuable as a place to go out on in case of fire. There ought, therefore, to be a large balcony for every story of the house above the basement story, into which the staircase window ought to open; so that the inmates of each floor of the house, if alarmed by fire in the night-time, might have immediate access to the balcony, where they could remain till assistance was procured. A skylight is not only useful for the purpose of repairing the roof and the chimney tops, and having access to the latter in case of the chimney being on fire, but it is often valuable as a fire escape. In examining the windows of a skylight, or, rather, the opening in the roof, the manner in which it opens and shuts ought to be particularly observed, with the view of ascertaining whether it is likely to be effective in excluding the rain and snow. The best construction is where the light or door is perpendicular, and opens out to the roof like a common door; but, as this can only be done in the case of large roofs, the more ordinary form is to have the opening in the slope of the roof, and in the same plane as the slating. In this case, the best mode of forming a cover, light, or door, to the skylight, is to make it with sides which fit on to raised ledges, in the same manner as the cover of a bandbox fits on the box. The ledges attached to the roof, on which the light or cover fits, should be rendered waterproof on their lower side, by lead flashings projecting over the slates. In whatever way the door or lid is formed, there should always be more or less of glass in it, in order that it may be readily found in the night-time.

Sect. III. *On the domestic Offices of a House.*

Though the observations in this section refer more immediately to houses of the first and second rate, than to those of a smaller character, or to cottages, yet they will be found not without their use when applied to the latter. They will show what it is desirable to obtain, when there are the means of obtaining it; and, by displaying all the good and evil points in the offices of a large and complete house to be desired or avoided on a large scale, they will impress them more forcibly on the mind, and contribute to their application in houses on a smaller scale. As a villa will be designed and executed in a more perfect manner by the architect who has rendered himself familiar with the arrangements of a mansion; so will a cottage, or a fourth-rate suburban house, be better planned by an architect or builder who is familiar with the requisites of the villa. In like manner, also, the intending occupant of a cottage or fourth-rate suburban house, who understands what constitutes a commodious, comfortable, and elegant villa, will be better able to make choice of any inferior description of dwelling.

All that is essential to a villa should be found in a cottage; and, though a cottage has not a housekeeper's room, a butler's pantry, a larder, a wine-cellar, a coal-cellar, &c., each separate, and of some size; yet in every comfortable cottage there is a closet, where the mistress of the house keeps her dry stores, such as linen, &c., and which supplies the place of a housekeeper's room; and a pantry, where she keeps her cooked provisions, and a safe, in which she keeps her meat; which last two serve instead of larders. She has a place, also, for liquors and ale, and another for fuel, &c. In short, there is nothing belonging to the mansion which has not its prototype in the cottage; and, as the former is only an amplification, or more complete developement, of the latter, it forms an excellent *beau idéal* for the cottage-builder, or the chooser of a cottage, to keep continually in his mind's eye.

The Kitchen is a most important apartment in every house; and, being chiefly used for the preparation of food, it should be furnished with every thing necessary to enable the cook to perform her duties. It should be as lofty and as well lighted as circumstances will allow: this is a most important point, and should be attended to in all houses. In small houses it is especially necessary; as low kitchens are generally dark, and light is essential both to cleanliness and comfort. In many third and fourth-rate houses in London, the kitchens are so low and dark, that the servants employed to wash up the tea-things, glasses, and other articles (generally, in such houses, washed in the kitchen), cannot see when they are clean. When cooking is going on, the whole kitchen is filled with steam; the tin covers, and other utensils which ought to look bright, become dim; and every thing has an air of dirt and untidiness. In houses of these classes, therefore, the kitchen ought always to be made higher than it is at present; and, instead of the lowest, it ought to be one of the loftiest apartments in the house. In most parts of England, the kitchen is in the basement story, or on the ground floor: but in the city of London, where the counting-house or shop is on the ground floor, the kitchen is often on the second or the third floor; which is attended with the great advantage of preventing any kitchen smells from escaping into the place of business. In Genoa, and many other towns in Italy, the kitchen is in the upper story of the house; the floor, in these cases, being generally laid with tiles. This practice is convenient for the escape of smells, but very inconvenient in almost every other respect. It would scarcely answer for being introduced into Britain, unless we could be induced to adopt the French mode of cooking, and use only charcoal as fuel. It might, however, do where cooking was performed by gas; which it might be, with great economy and cleanliness, wherever there are public gas establishments. Sometimes the kitchen is on the same floor with the dining-room, but detached from the house, and under a flat roof covered with lead; or what is termed, in London, a lead flat. When this is the case, the flat should be pugged (stuffed between the roof and the ceiling with some non-conducting substance) and ventilated, so as to keep out the intense heat, which would otherwise penetrate the lead. In houses of a superior description, the kitchen should not only be lofty and light; but also well ventilated, or supplied with air-flues for carrying off the steam and the effluvia from the food. When this precaution is neglected, even in lofty kitchens, the steam hangs like a cloud below the ceiling; and the smell of the food, when cooking, is often found very unpleasant. Sometimes the escape of these effluvia to the living-rooms of the house is prevented by pugging; but this only confines them to the kitchen, while the air-flues carry them away into the open air. In short, the air-flue ought to be considered essential to every kitchen, whether that of a large house or a small one. The introduction of these flues in building a house would add nothing to the cost; because the extra-labour is compensated for by the saving of material; and hence, in brickwork, unless under extraordinary circumstances, flues are never charged for.

Where practicable, the kitchen should look to the north or the north-east, as should all those domestic offices which require to be kept cool. When the kitchen is under ground, as is frequently the case in England, particularly in large towns, this is not of so much consequence; but it should always be as near as possible to the dining-room. In large houses, it is desirable to have a private communication between the kitchen and the dining-room, by a passage (and staircase, if the kitchen, &c., are under ground) leading into an anteroom, or waiting-room for the servants, adjoining the dining-room. This passage should be thoroughly ventilated, and have an air-flue, so that any escape of effluvia from the kitchen, &c., may be dispersed in summer, or carried off in winter, before it reaches the anteroom. The anteroom should be furnished with broad shelves fixed to the wall, with drawers beneath, to hold such articles as may be wanted by the servants waiting in the dining-room; and in some first-rate houses, there is a hot table, warmed by pipes of hot water, steam, or smoke-flues, to set the dishes on, when brought from the kitchen, before they are carried into the dining-room.

A kitchen should always be dry: the walls must, therefore, be preserved from damp, if the kitchen is under ground, by building them in cement, at least as high as the level of the floor; and, where ground outside is against any of the walls, it may be necessary to build an inner wall, one brick thick, and hollow, against the outer wall; or to build a dry area, carried down below the level of the floor of the kitchen, which area should be ventilated and drained. The floor should be of rubbed Yorkshire stone, laid on brick walls, at least two courses of bricks high above the ground. If the foundation is damp, concrete a foot or two thick, composed of clean gravel and fresh-burned stone-lime, should be thrown in first, to form a foundation on which the walls should be built. To keep the floor dry, air should be introduced, and made to circulate freely, under it; for which purpose, air gratings should be fixed in all the outward walls, and openings for the air to pass through left in all the walls on which the paving is laid.

In very large kitchens, a portion of the floor opposite the fireplace, about the centre of the kitchen, and from 6 ft. to 10 ft. square, according to circumstances, should be of wood; that is, of oak joists and sleepers laid on brick walls, with a deal or oak floor over them. On this wooden floor the table should be placed; and by this means the cook may generally be able to avoid standing on the stone floor. The skirting round the kitchen should be made of cement (not wood) wherever there are stone or brick floors; as wooden skirting is liable to rot in a short time, in consequence of the moisture which lodges about it from the frequent washing of the floor.

The ceilings of all kitchens, however small, should be lathed on the under side of the joists, and plastered; for, if this is not done, and only the space between the joists plastered, the effluvia arising from the cooking collect and remain between the joists, in spite of all the ventilation that can be given: the spaces between the joists also afford shelter for flies, spiders, &c. As the ceilings of kitchens should be scraped, cleaned, and whitened (or coloured) every year, the expense of plastering is soon repaid by the diminution of the surface; the sides of the joists making a surface of one and a half times more than the whole ceiling.

The kitchen doors should always be made to open towards the fireplace; otherwise the opening and shutting of them will be likely to cause the chimney to smoke by disturbing the current of the air.

The fireplace should be capacious in proportion to the quantity of cooking required; and it should be from 4 ft. to 8 or 9 feet wide, and never less than two bricks and a half, or 1 ft. 10½ in. deep. Large fireplaces should be from 2 ft. 3 in. to 2 ft. 7½ in. deep; the range can then have a proper boiler at the back, supplied with water by a pipe from the main cistern, and regulated by a small feeding cistern, so that the boiler will be always full of water. Where this plan is adopted, there will be a supply of hot, if not boiling, water, at night, as well as all day. In large kitchens, there should be a hot plate and stoves for made dishes and preserves on one side of the fireplace, and a boiler on the other. These should be covered over, at the height of 6 or 7 feet, by a projection or canopy from the wall, open in front, and communicating with a flue for carrying off the steam and effluvia from the meat. There should be one or two large closets in the kitchen, in a very dry situation, from 3 ft. to 4 ft. wide, and at least 18 in deep, for holding spices and other things that may be wanted by the cook.

The principal articles of furniture required in a kitchen are, a table, as large as the size of the apartment will allow, and made very strong, and a dresser or dressers. The tops of these dressers should be 2 in. thick, and the drawers about 2 ft. wide and 7 in. deep. The space under the drawers is sometimes enclosed with doors, and sometimes open, having a pot-board the whole width and length of the dresser, and raised 3 or 4 inches from the floor. There should be good locks on one or two of the drawers, and two iron or japanned handles fixed on each. A mill for coffee, one for pepper, and another for the finer spices, may be fixed to the ends of the dressers. Ranges of rails, furnished with hooks, should be fixed to the wall, for the dish covers; and round or jack-towel rollers should be attached to the backs of the doors.

Ash-Grate and Pit. There should be an ash-grate and pit made under the fireplace, that the ashes may drop through the ash-grate into the pit, and leave the cinders over it; which may thus be taken up and thrown on the fire without making any dust. As this appendage to a kitchen fireplace is both cleanly and economical, and not so generally known in some parts of the kingdom as we think it ought to be, more especially in cottages, we give a representation of it in *fig.* 19., which is a section across the ash-grate and fireplace. In this section, *a* is the kitchen fireplace; *b*, the ash-pit; *c*, the grating over it; and *d*, the boiler behind the kitchen fire. The bars of this grating should always be placed in a direction at right angles with the back of the grate or hearth, in order that the shovel may pass easily over them, when the cook takes up the cinders to throw them on the fire. The grating is neither hinged nor fixed, but fits into an iron frame set in the brickwork; or, where there is a hearthstone, into a rebate cut in the stone to receive it, from which it is lifted up occasionally, to take out the ashes. In large establishments, a smoke-jack is not only the best for roasting meat, but is always ready. Where there is no smoke-jack, a bottle-jack is generally employed as a substitute; and this, when not in use, should be hung on a hook purposely fixed in the wall, with a small hook near it for the key. A bottle-jack should always be used with an upright tin stand, or case, behind it, called in some parts of England, a hastener; which should be movable, and stand on feet, with a dripping-pan fixed in the bottom. This case is open to the fire in front, but is closed at the back and sides. It should be of block tin, and should be kept very bright inside, that the rays of heat may be reflected back on the meat. When the bottle-jack is not used with a tin stand, or, at any rate, is not fixed to it, there should be a small notched brass crane fixed to the mantel-piece for it to hang upon. In all kitchens where much cooking is required, if a smoke-jack is used, or even a bottle-jack, without a hastener, or upright tin case, a detached screen extending the whole width of the opening of the fireplace is desirable, in order to render the kitchen cool and comfortable in summer; and, in both summer and winter, to accelerate the roasting of meat. This screen is made of wood lined with tin, and is furnished with shelves, on which

19

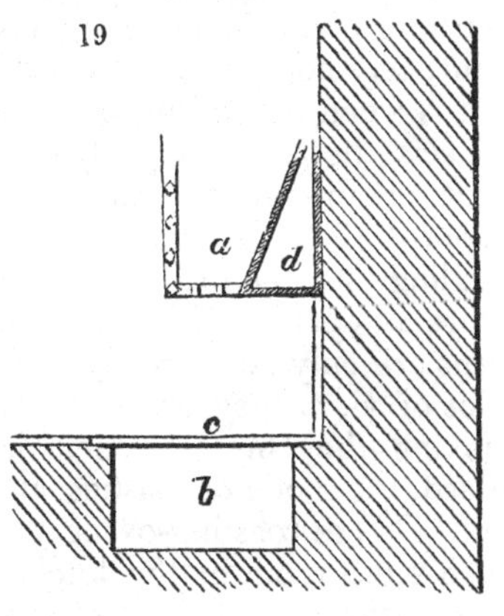

plates, &c., may be placed to be warmed, and cooked dishes to be kept hot.

Roasting-Oven. One of the modern improvements in the fitting-up and furnishing of kitchens is, the substitution of ovens of cast iron, placed by the side of the kitchen fire, for roasting-jacks, and their appendages of spits, hasteners, &c. These ovens, as improved by Count Rumford, are so contrived, by means of openings below, and air-flues from the top, as to have a current of air continually passing through them; and, in consequence, the meat is oxidised in roasting, in the same manner as it is when roasted before an open fire; and while, in all other respects, the meat is equally good, the first cost of the oven is not greater than that of a smoke-jack and its appurtenances, and the cost of the fuel consumed in roasting is incredibly less. Notwithstanding this, the prejudice in favour of the old mode of roasting before an open fire is so great among practical cooks, that these ovens are only occasionally introduced; though, in almost every point of view, they are so very desirable, that we cannot help thinking they must, sooner or later, become general; unless, indeed, the mode of roasting and cooking by gas should take the lead. Were roasting-ovens to become general, there would be much more room in the kitchen; much less occasion, or rather no occasion, for large open fires; and there never could be much smell, as that would be chiefly carried off by the air-flue of the oven; while any other smells floating in the open kitchen would be carried off by the air-flue close under the ceiling which has been already recommended. The whole business of cookery, by requiring less heat in the atmosphere of the kitchen, and by being comparatively unaccompanied by strong smells, would thus become much more agreeable, and better adapted for the mistress of a house and her daughters. The principal practical objection to an oven for roasting meat in a small establishment is, the constant attention it requires from the cook, to keep the fire in a state to give out constantly an equal degree of heat. Where there is only one servant, the fire, in so small a fireplace as that required for heating an oven, will be in danger of going out, or becoming too low, if it be left while the cook is attending to any of her other duties; or if, to avoid this, she makes a fire large enough to last the time she will be away, the meat will probably be burned by the great heat given out at first. Besides this, the smoke-flue to the fireplace of the oven, which generally opens into the main flue a few feet above the oven, frequently requires clearing out; and, though this can be done with ease by the cook, yet it is often neglected; and as, in that case, the fire does not draw well, the roasting or baking is spoiled. The smaller the oven, the more liable it is to this kind of mismanagement.

Cooking by hot Water. A remedy for these evils has been carried into execution by Mr. Weeks, who heats his oven by hot water, kept hot by the open fireplace in the centre of the kitchen-range; but whether this mode of heating an oven is ever likely to become general, is more than can be at present predicted. There can be no doubt whatever of the superiority of Mr. Weeks's mode of heating to that of placing a fire immediately under the oven; but the question is, whether the additional expense of first erection, and the possibility of the apparatus leaking after it has been some time in use, might not prevent its general introduction. Be that as it may, the superiority of this mode of heating an oven, hot plates, and other apparatus, for different purposes of cookery, is undoubted. A description of Mr. Weeks's apparatus will be found in the *Architectural Magazine*, vol. ii. p. 64.; and it may be safely recommended to all those who have much cooking to perform, as saving labour and fuel, and promoting cleanliness and the general comfort of the kitchen. Mr. Weeks has ranges combining his different improvements, of every description, from one adapted to a cottage of the smallest size, to one suitable for a mansion, or an inn, or other public establishment upon a large scale.

Cooking by Gas. The mode of cooking by gas, of which an account will be found in the *Encyclopædia of Cottage Architecture*, p. 725., might be adopted in any house, either in the kitchen or in any room between it and the garret. It is the most economical of all modes of cooking, and so entirely free from dust, dirt, smell, excessive heat, and other annoyances, that it might be performed by any lady in full dress, without soiling either her hands or her clothes. Ultimately, we should think, this mode of cooking will come into very general use in towns and their suburbs, and in all situations where gas is employed for lighting; but, in a work of this kind, and while there still exists a prejudice against even roasting meat in ovens, it is unnecessary to do more than to direct attention to the subject of roasting it by gas.

The Coal-Cellar should be placed as near to the kitchen as possible, and should be sufficiently large to hold coals for nine months' consumption. There should be a ready access to it from without, for the men to shoot the coals into it. In large houses, besides the common coal-cellar, a cellar for coke or for charcoal, or separate cellars for both, are desirable; and also a small and very dry cellar for wood to light the fire. In country places, where wood is abundant, there is commonly a large wood-house in the kitchen court, and a stack of wood made at some distance from it. The lighter part of the refuse of the stack, and of the wood-house, is collected and placed by itself, for the purpose of lighting the fires; and very little regard is paid by the housekeeper, or mistress of the establishment, to the quantity of wood used in this way. In suburban residences, however, the expense of wood for lighting the fires amounts to a considerable sum in the course of the year; and hence the advantage of some place for locking up the stock of this material, in order that it may be given out in small quantities, as wanted for use.

The Scullery should be as close to the kitchen fireplace as possible. It should be paved with Yorkshire stone, or with brickwork, and need not be more than 9 or 10 feet high, which will afford room for a cistern over it. In every scullery there should be a stone sink, as large as the space will allow, 7 in. to 8 in. thick, leaving room for a plate-rack at one end. Under the plate-rack, or by the side of it, should be a slanting drip-board, to convey the water that runs from the plates and dishes into the sink. A waste-pipe should be fixed in the bottom of the sink, and taken into the drain, with a bell stink-trap and grating over it; or the waste-pipe may be taken into a trap in the drain, and the grating soldered into the sink. Under the sink, or in some other convenient part of the floor, should be fixed a large air-tight stink-trap, to carry off the water when the floor is cleaned. There should be no sink stone (a stone pierced with holes to allow the water to run off, but without any cover), as it is by such stones that noisome smells most frequently escape from the drains. Hot water should be supplied to the sink by pipes taken from the cistern over the scullery, into the boiler at the back of the kitchen fireplace, and there coiled round several times; then conveyed to the sink, and afterwards to the washing-troughs in the washhouse, where they are situated below the level of the cistern; cold water should be also conveyed to the sink, which may be done in pipes direct from the cistern.

A Dust-hole, or place for containing the dust, and other refuse formed in carrying on the business of the house, is an essential part of every establishment, from the cottage to the palace. In general, small houses require more room for dust and refuse than large ones; because, in the case of the latter the dust is regularly carried away at stated intervals; whereas in the cottage it frequently remains till a convenient period, which sometimes occurs but seldom. The dust-hole, whether large or small, ought if possible, to have a northern exposure, and always to be furnished with a door, which ought to fit so tightly as to exclude smells. Attached to every dwelling there ought, properly, to be two dust or refuse holes; one for vegetable and animal matters, dust, ashes, &c., which are convertible into manure; and another (which may always be of much smaller size) for broken earthenware, glass, stones, &c., which are of no use, except for the bottoms of roads or walks, or for grinding into powder, to be used for forming Roman cement or anticorrosion paint. Bones, also, should if possible, be kept apart, in every house; because, when broken and fermented, they make a most valuable manure, as we shall show in another part of this work. As few materials thrown into the dust-hole produce more offensive and dangerous smells than recent bones, from the decomposing animal matter remaining on their surface; wherever it is determined to keep them apart, it will be well to throw some sifted ashes from the fireplaces of the house in along with them; because the ashes, by absorbing the decomposing matter, prevent it from giving out an offensive smell. In whatever way the dust, ashes, bones, and vegetable refuse of a house are kept, as little moisture as possible ought to be admitted with them, as this promotes putrefaction. In the case of cottages in the country, the dust-hole and the dunghill are most frequently combined; and, as the water, which in suburban town houses, is usually poured down the sink, is thrown into this pit, a very excellent manure is produced. In order that this manure pit may be as little injurious to health as possible, it should be at some yards' distance from the cottage; and, in warm weather, it should be covered with boards, or even with a straw hurdle, to prevent evaporation, and the diffusion of offensive smells.

The Larder, or safe, for keeping the meat and other provisions in, both before and after they are cooked, should be large, and, indeed, sufficiently capacious to contain all the provisions. It should be effectually protected from the sun's rays, and yet have a complete circulation of air all round it, if possible. It should be so placed as not to be near the dust-bin or beer-casks, or any other place from which dust, or putrid or other bad smells, can arise; it should be enclosed all round with fly-wire panels in wood framing, and should be raised above the paving, in order to admit air at bottom, and to keep it dry. The roof should be also of fly-wire, and it should be protected from rain, by a boarded roof over it. The larder should be lofty, and should have strong iron bearers, from one side to the other, with hooks to slide on them, so that the meat may hang above the head; and there should be shelves all round, to put dishes on. A separate safe should be provided for vegetables and fruits, and another for game, if there is likely to be any quantity of it. For large larders, there should be an inner door, with a space between the doors, to enable the cook to shut the one before she opens the other. For small families, one or two iron safes, hanging from some beam in the ceiling of the outer kitchen or cellar; or hung in an outhouse, or even in a shady place in the open air, contrived so as to be drawn up out of the way by lines and pulleys when not in use, with one or more shelves in each, and having fly-wire all round, are far better than safes placed against brick walls, as is now generally done.

The Beer-Cellar, and the Wine-Cellar. These cellars should not have any communication with each other. Both should be arched and dry, and neither should ever be placed, if it can possibly be avoided, under the yard; if, however, such a situation be inevitable, the cellars should have solid spandrils, and be covered with two or three courses of plain tiles laid in cement. The wine-cellar should be fitted up with bins, and, when large enough, divided, so as to form an inner and an outer cellar. Perhaps the best method of dividing the bins is by walls half a brick thick, carried up to the top, with horizontal York stone shelves every 3 ft., to divide the bins in their height. As some wines require a warmer temperature than others, the pipes that are to supply the several sinks with hot water may be carried through one of the wine cellars, or through a portion of the cellar which may be separated by a brick wall; and thus a warm cellar may be obtained at very small expense.

The wine-cellars of small houses, in the suburbs of London, are generally of the very worst description that can be imagined, with reference to the true use of a wine-cellar; which is that of keeping the wine in it of a uniform temperature throughout the year. For this reason, the outside wall of the house should never form one of the sides of the wine-cellar; because, even though sunk in the ground, it is liable to have its temperature changed by the warm rays of the sun in summer, and by the frost in winter, acting on the surface soil. Neither should the floor of the wine-cellar ever be on the same level with the floor of the kitchen, or of any room having a fire in it; because, from the tendency of temperature to an equilibrium, the atmosphere of the cellar will be constantly liable to be raised to the same degree of heat as that of the kitchen. Whenever a cellar is so circumstanced, it ought to be protected by double or treble doors with 2 or 3 feet of space between them, in the same manner as an ice-house. The best situation for a wine-cellar, in any house, is in the interior, so that there may be at least one set of cellars between it and the outside of the building. If the soil and drainage admit, it ought to be sunk so as to be on a lower level than the kitchen floor; but, even in that case, it ought to be protected with double doors, one being at the top of the staircase. When it must unavoidably be placed adjoining the outside of the building, and on the same level with any rooms having fires in them, recourse must be had to double exterior walls, and to double or treble doors; and, where these are impracticable, or where, from different circumstances, it is not considered desirable to incur the expense, the bins of wine may be kept at an average temperature within the cellar, with mats, straw, or thick masses of any other analogous material of the non-conducting kind; the outward ends of the bins being battened and lathed, and plastered; or lined with boards kept at 2 or 3 inches' distance from the wall.

In large or first-rate houses, *a Butler's Pantry, a Footman's Room, a Servants' Hall, and a Housekeeper's Room,* are indispensable. These rooms should have wooden floors in the centre, with a course of stone paving 2 ft. wide all round against the walls. The floors should be laid hollow in brickwork, as described for the kitchen, and the skirting should be of cement. An air-flue carried up from the hollow space under the floor, close to the chimney, would be very desirable to keep up a constant current of air under the floors. The sinks in these rooms (of which there should be two in each, fitted up in a window recess) should be lined with lead. They should be each about 1 ft. 6 in. or 2 ft. long, and 1 ft. or 1 ft. 3 in. wide, side by side, and about 1 ft. deep; each should be covered with a flap, and have the space beneath it enclosed with doors, to form closets, with a shelf in each. One of the sinks should have a brass grating, about 3½ in. diameter, soldered into the bottom, with a wooden drainer (a contrivance for retaining decanters on their sides, or in an inverted position, in order that the water may drain from them) for decanters, &c.; and the other a washer and plug with a chain. A waste-pipe should be soldered into each sink; but none of these pipes must be taken into a drain without the intervention of some trap, otherwise the foul smells from the drains will be a source of annoyance. Waste-pipes of this kind are sometimes allowed to empty themselves into the open area, over a sink-stone with a trap in it; but it is much better that they should be carried direct into the drain. The Housekeeper's Room should have a series of closets 2 ft. deep; some with shelves for linen; others for pickles and preserves; and, again, others with drawers for stores of various kinds. It is also customary to have in the housekeeper's room, or in a room adjoining, a stove or hot hearth for making preserves, &c.; and a marble slab or solid smooth table for making pastry. The Footman's Room, and Butler's Pantry, should have, in addition to the sinks, a large dresser fitted up with drawers and closets underneath; with a wide shelf continued all along the room, above the level of the top of the door. Adjoining the latter room, should be a fire-proof plate-closet, ventilated, and kept sufficiently warm to prevent it from being damp. The pipes that convey the hot water to the several sinks may be made to pass round this plate-closet, as well as through one of the wine-cellars; as, when damp gains admittance, it causes double labour to the footman, by tarnishing the plate. The plate-closet should be fitted up with shelves, lined with thick drugget.

The Passages should be well lighted and ventilated; otherwise the smells arising from the drains, the kitchen, &c., will be carried up into the body of the house, the air above being more rarefied than that below. The passages should be paved with stone, tiles, or bricks, with the skirtings of cement; and the walls plastered with stone-lime and sand, and lined out into blocks in imitation of stone, and coloured. Plastering of this kind is much more durable than plastering composed of chalk-lime and the sweepings of roads, such as is now generally used about London, and is very little more expense. The bell-board should be fixed in the passage between the kitchen and the servants' hall: each bell should have a pendulum attached to the spring, to make known which has rung, for some minutes after the ringing has ceased; and the names of the rooms should be written on the bell-board below the bells.

The Wash-house and Laundry, in small suburban houses, do not form distinct apartments; but the washing is carried on in the back kitchen, and the ironing in the principal one. In suburban houses of a somewhat superior description, the greater part of the washing to be done is given out to regular laundresses; and the remainder is carried on as in the smaller houses, in the kitchen and back kitchen. Where there is a large family however, it is very desirable to do the washing at home; and, in this case, it is necessary to have places set apart for that purpose, either connected with the house or detached from it. Where there is both a wash-house and a laundry, they should, if practicable, be separated from the main body of the house, and placed in the kitchen court; or in a court or yard formed on purpose, at no great distance from the house, and with a drying-ground adjoining. Where there is abundance of room in the ground floor of a square house, the wash-house and laundry may be placed on that floor; taking care to have air-flues from the ceilings of both wash-house and ironing-room, to carry off the soapy steam which rises from the boiling of the linen, and the moisture which is evaporated from the clothes in the processes of drying and ironing. If these steam or air flues are placed close to the flue from the copper or boiler of the wash-house, and to that of the fireplace or stove for heating the stoves in the laundry, a powerful draught will always be created in them, at the time when they are most wanted; and thus the whole business of the wash-house and laundry may be carried on in the ground floor of a house, without proving any greater nuisance to the persons inhabiting the rooms above, than the culinary operations of the kitchen and back kitchen. The wash-house generally consists of a room with the floor paved, and sloping towards one corner, where there should be a sink connected with a drain by means of a proper trap. The skirting should be of Roman cement; and the walls and ceiling plastered with that material, which will prevent them from being acted on by the moisture, as they are when common plaster is used. A large wash-house, when it can only be lighted on one side, should be longer than it is broad; and it should have the windows on its long side, for the sake of light. Where the wash-house is in the kitchen court, and connected with other buildings, it may be lighted on two or even three sides; and where it is quite detached, on all sides, with a boiler in the centre, and the drying and ironing room over entered by an outside or inside stair. Immediately under the windows, there should be a row of fixed washing-troughs, each having a supply of cold water by means of a pipe and cock communicating with a cistern, and of hot water by means of a pipe and cock from the boiler, and each supplied with a waste-pipe and plug to let off the dirty water. The boiler may be placed against the side of the room opposite to that of the washing-troughs, in order that the washers may be as little oppressed as possible by the heat of the fire. It should be supplied by a pipe from the cistern; and have a main pipe communicating with all the washing-troughs, with cocks over each trough. A recent improvement in wash-house boilers consists in setting them so as to burn the steam which rises from the boiling water; and this, while it keeps the air clear and cool, greatly increases the heat of the fire. The mode by which this is effected is by keeping the boiler closely covered with a lid; immediately under which, from the edge of the boiler, a flue is conducted downwards to the ash-pit, to which the steam is drawn by the draught of the fire, and, ascending through the red-hot fuel, is decomposed by it and consumed. The practice of burning waste steam is in use in various manufactories; but it appears to have been first applied to boilers used for washing and brewing by Mr. Reynolds, late steward to the Marquess of Westminster, at Eaton Hall. That the burning of the steam actually produces additional heat, has been proved in Jameson's *Journal*, vol. xxiii. for 1837, p. 173.; and *Gard. Mag.*, vol. xiii.

The Laundry and Drying-Room, in small establishments, are commonly united in one; but it is much more convenient, as well as more wholesome, for the laundry-maids to keep the drying-room apart. The operation of drying in a room is commonly performed by means of iron stoves; the upper parts of the windows and the lower parts of the doors having openings in them, with circular fly-ventilators. Steam and hot water are also frequently employed in drying-rooms; but a good stove with iron pipes is perhaps the simplest and best mode, and the one least liable to be misunderstood or put out of order by servants. The great object in drying-rooms is to create a draught; which shall at once be rapid, and carry off the moisture by the chimney tops, so as not to contaminate the air which enters by the windows of the living-rooms. For this purpose, there ought to be an air-flue carried up in close contact with the flue of the stove; and, in large laundries, there ought to be two air-flues, one on each side of the smoke-flue. The openings to these air-flues ought to be near the ceiling, and there should be one or more openings in the wall near the ground, or in the floor, communicating with an air-drain to supply fresh air.

In Germany, and generally in the north of Europe, the operation of drying is carried on in the garrets; which are large and lofty, the roof being high and steep, and with openings under the eaves, and without the tiles being laid in mortar. In the season of fires, the warm air of the house naturally ascends, and rising to the garret it passes through the interstices between the tiles, and carries off along with it the moisture from the clothes on the lines, which are suspended from one side of the roof to the other. In the summer season, when there are few or no fires, an intense heat is generated by the sun shining on the tiles; and this produces a draught through the interstices between them, which is supplied by the openings occasioned by the projection of the eaves. These openings, in the high and steep German roofs, appear to have been made chiefly with a view to the preservation of the timbers: but they form admirable drying-rooms; and not only in towns, but sometimes even in the country, many large families never have their clothes dried in the open air at any period of the year. In establishments of this kind, washing may be performed once or twice a week; the advantage of which is, that much less labour is required, the articles not being kept long in a dirty state; whereas, in France and in Scotland, where, in most families (especially in the former country), washing is only performed twice or thrice a year, the clothes require a great deal more labour to render them clean; and not only the processes of drying in the open air, and bleaching on the ground, but also that of bucking, become necessary, to remove the dirt and stains, which are, as it were, ingrained by time, and by that species of fermentation which takes place in consequence of the dirty clothes lying in heaps.

One of the great advantages, to a small family, of residing a short way out of town, is, that their clothes when washed may be dried in the open air; for this reason, no suburban residence ought to be without a drying-ground, in as open and airy a situation as circumstances admit; and it should be concealed from the house, and from all the principal walks through the garden or grounds. At the same time, with a drying-room properly ventilated by means of air-flues, and supplied with air from air-drains, clothes may be dried as effectually in the house as in the open air; with much less labour to servants, but with the additional expense of fuel.

The Brewhouse, in small establishments, is identified with the wash-house; but, in larger houses, it forms a separate building in the kitchen court. As, however, it is more a building for a farm or a country establishment, we shall confine ourselves to a very few remarks. The great object in the arrangement of the brewhouse is, to contrive the boiler, the coolers, and the vats for fermenting the liquor, so that the water, the wort, and the beer may be conveyed from the pump or cistern, to the casks in the beer-cellar, through pipes, and without the trouble of either lifting or carrying the vessels containing it. For this reason, the boiler and furnace should be placed in the upper part of the building, so that all the different processes may be performed in descending; the last being that of filling the casks in the cellar. For this purpose, the beer-cellar may be very conveniently placed under or near the brewhouse; but this is by no means necessary, as with leathern pipes having screw joints, &c., the beer may be conveyed to any distance. Some designs for the arrangement of very complete breweries of this description, will be found in the *Encyclopædia of Cottage Architecture.*

The outdoor Offices to a suburban house are chiefly the stable, coach-house, and harness-room. Where there are more than these, such as a cow-house, pigsty, poultry-house, &c., we consider them as belonging to the suburban farm; and they will be treated on in a future chapter. The smallest description of outdoor offices to a suburban house comprises a stable, gig-house, hayloft or room, and harness-room. The stable should have the floors perfectly dry; it should be well lighted and ventilated, and the windows should be furnished with shutters for use during winter. The stable windows, according to some, should be so high that a man cannot see through them; the object of which is to prevent careless or bad grooms from knowing when their master is coming to look after them. The floor of the stable ought to be perfectly flat; and in the centre of every stall there should be a small grating over a drain communicating with a general drain conducted lengthwise through the stable behind the stalls, so that every part of the floor may be at all times dry. There should be an opening in the lower part of the stable door to admit air; and ventilating tubes should be placed in the upper part of the walls, or carried through the roof, to allow the heated air to escape, but provided with sliding shutters to regulate its exit. The hay-room is generally placed over the stable; but a much better situation for it is on the ground floor in a separate house, a loft over the gig or coach-house, or even a spare stall in the stable, enclosed with a door, so as to exclude smells. At all events, when the hayloft is placed over the stable, the floor over the stalls should be made air-tight by being ceiled below; and no opening ought to be left for putting the hay into the rack, or which can by any means admit the air of the stable to ascend into the hayloft. When this is the case, the hay is contaminated by the effluvia proceeding from the stable litter, and, consequently, cannot be relished by the horse. When the loft is placed over the stable, there ought to be no communication between it and the stable whatever, except through a vertical tube, about 2 ft. 6 in. wide, and projecting about 1 ft. from the wall, behind the horses; by means of which, the hay can be let down as wanted, to a box with a lid, and thence taken out, and put into the rack. The entrance to the loft ought to be by a ladder outside, or by stairs from the gig-house. The best of all modes, however, of supplying a stable with hay, is to bring it direct from the rick, and never to cut more at any time than can be used during the next twenty-four hours. This, however, is impracticable in the case of small suburban residences without farms, the possessors of which, who have horses, must be content to purchase their hay in loads or half-loads at a time. The gig-house, and also the harness-room, ought to be well lighted and ventilated; and the harness-room ought never to be without a stove or an open fireplace for use in damp weather. In general, the harness-room and the gig-house should adjoin each other, and communicate by a door; and the fireplace or stove should be placed in the wall between them. The outside walls of all harness-rooms ought to be battened, and plastered inside, for nothing so soon injures harness as damp. We say little here respecting stables and coach-houses on a large scale; because those who have to choose or build such will doubtless have recourse to professional men, or works which treat more at length on the subject (such, for example, as our *Encyclopædia of Cottage Architecture*), than can be expected in a volume like the present.

For many of the remarks in this section, and in the preceding and following ones, we are indebted to some valuable papers on the choice of a house, by I. J. Kent, Esq., published in the *Architectural Magazine*, vols. i. and ii.; and nearly all the illustrative sketches in those sections have been made expressly for this work, by E. B. Lamb, Esq., F.I.B.A.

SECT. IV. *Of the Dwelling-Rooms of a House, their Fittings-up, and Furniture.*

In our introductory paragraph to the preceding section, we have shown that, in respect to domestic offices, the mansion is but a more ample developement of the cottage, and that the former should always be kept in view as the *beau idéal* of the latter. The same doctrine may be laid down with respect to those cottages which contain two or more living-rooms, besides the kitchen and bed-rooms. Even cottages of the smallest class, and fourth-rate suburban houses, may adopt the conveniences recommended, as far as their means will permit; and with regard to those for which the occupant has no occasion, if the knowledge of them has no other effect, it will at least tend to cultivate his taste; for taste, in any one class of buildings, is not to be cultivated so much by confining our studies to that class, as by bringing our knowledge of what belongs to all the other classes to bear on the particular one that we wish to improve. This holds good, not only in interior arrangement and exterior elevation, but in fittings-up, finishing, and furniture. To the humblest individual, therefore, into whose hands these pages may fall, this section is recommended, as a study connected with the two preceding ones; and to the citizen in comfortable circumstances, who can afford to have a suburban dwelling, or small villa with two or more sitting-rooms, we recommend it, as being full of the most instructive remarks. We do this with the less hesitation, since we are indebted for the greater part of them to the same able and experienced architect and surveyor before mentioned, Mr. Kent.

The Dwelling-Rooms of a complete Country House are, the breakfast parlour, the dining-room, the drawingroom, the library, the music-room, the gentleman's room or business-room, and the lady's boudoir or private sitting-room; while, as appendages, may be mentioned, the billiard-room, museum, picture and statue galleries, orangery, conservatory, green-house, and aviary. The breakfast parlour, dining-room, library, gentleman's room, store-room, and sometimes the footman's room, are usually placed on the principal or entrance floor; and, in country houses designed in the cottage style, the drawingroom, music-room, best bed-room, and dressing-room are likewise often placed on the same floor. The billiard-room is almost invariably placed on the ground floor, unless it is in a detached building in the garden. In small villas, the living-rooms generally consist of a dining-room a drawingroom, a breakfast parlour, and sometimes also a library, all of which are on the ground or entrance floor; and in suburban residences, of a breakfast and dining parlour on the entrance floor, with two drawingrooms, divided by folding doors, above. The orangery, conservatory, or green-house, generally communicates with the drawingroom, or with the library or breakfast-room; and, at the same time, with an open veranda, arcade, loggia, or colonnade.

The Breakfast Parlour is the smallest of the sitting-rooms. It should have the windows to face the south or south-east, the sun not having, at the early part of the day, when this room is in use, acquired an unpleasant degree of power, and only producing that genial warmth which excites cheerfulness and good-humour. In large houses, it should look into the flower-garden or grounds, and on the lawn; and, in small suburban villas, it should, if possible, look into the garden: but, where this is impracticable, a few flowers seen from the window, or inside the window recess, tend to produce agreeable thoughts; and much good is done, when we begin the duties of the day with favourable impressions. The floor of this room should be carpeted all over, and, therefore, need not be of an expensive kind; it only requiring to be, as, indeed, all floors ought to be, perfectly level. The boards should be quite flat, and the joints close; for, when the joints of the boards are open, the dust will work its way through the carpet, and spoil it; and when the edges are higher than the middle of the boards, the carpet will be cut through at those places, and consequently will not last half the time. The colours of the carpet should be light and cheerful, and should include a portion of green. The skirting should be from 10 in. to 12 in. high, and with a bold moulding on it; fixed so as to project 1½ in. from the wall, in order to prevent the chair backs from spoiling the paper. The wall should be papered with a light cheerful-patterned paper, perhaps a trellis pattern, with roses, &c.; but, at all events, with light and cheerful colours. The ceiling should be tinted either grey or of a light green shade: the cornices, plain or but slightly enriched, should be of a little darker tint than the ceiling. The woodwork in this room should be either green tints, relieved with light or dark mouldings, as may be most agreeable to the parties; or grained in imitation of green or grey hairwood, and varnished. The windows should be brought down to within 6 in. of the floor, and be taken up as high as possible: they should be sufficiently large to light the room thoroughly, as no one who has not attended to this would believe the difference an abundance of light makes in the cheerfulness of a room. There are various ways of closing the windows of sitting-rooms; but the best and handsomest is by shutters, folding back into boxes made to receive them during the day. They should never be made the whole height of the window, up to the soffit, or upper part of the architrave, as they generally are; but should either have the upper panel fixed, omitting the flaps, or there should be a capping fixed on the soffit, so that the shutters may be put to without any risk of rubbing, and thereby spoiling, the paint on the soffit, or of entangling the fringe or drapery of the window-curtains. The shutters should be fastened back in the day-time by brass knob mortise latches: when this is done, it will prevent the shutters from coming partly open, as they frequently do when not secured in their boxes by latches. They should be fastened at night with iron locking-bars, and not, as is most common, by latch bars. The sash fastenings should be strong, and capable of drawing the sashes closely and firmly together. The doors should be as far from the fireplace as possible, but not close in the corner of the room; and they should be hung with rising hinges, in order that they may open clear of the carpet, and yet, when shut, be close to the floor. The locks should be mortised into the door. In this room, closets may be admitted with propriety; and they should never be less than from 1 ft. 2 in. to 1 ft. 6 in. deep. The doors and handles of the locks of these closets should correspond with the other doors of the room; and the space above the closets should be enclosed up to the ceiling, in order that the cornice round the ceiling may be continued unbroken by any projection. The chimneypiece should be low, not more than 2 ft. 9 in. high in the opening, and may be of light-coloured British marble, or veined marble, plain in its design.

Lady's Store-room. Where the mistress of the house is her own housekeeper, it will be found very convenient to have a store-room communicating with the breakfast-room, but so placed that it may be accessible without entering that room. It will likewise be found desirable to have it near the dining-room. This store-room should have a dresser the whole length of the room; and, indeed, when wide enough, along both sides. Under the dresser should be abundance of drawers, some with locks and keys; and the space under the drawers may be open, or partly enclosed with doors. Closets, sufficiently capacious to hold all the china, glass, and dessert service, and also the pickles and preserves, should be provided; and broad shelves should be fixed all round the room above the door. Brass hooks of different sizes should be screwed into the edges of the shelves, on which to hang cups, jugs, &c. One or two large filters should be fixed in or near the store-room, to supply it with filtered water; and a feeding-cistern should be placed on a level with the filter, so that the latter may be always full of water. The supply to the feeding-cistern must be regulated by a ball-cock. Great care should be taken to have the store-room perfectly dry, and inaccessible to frost. To make this room complete, there should be a double sink, lined with lead; one half with a plug and waste-pipe, the other with a brass grating and waste-pipe, and having a drainer in it, with holes for decanters and glasses. Both sinks should be supplied with hot and cold water; and there should be a small fireplace in the store-room, that a fire may be lighted when thought necessary, for the purpose of expelling any damp that may have gained admittance. Where there are these conveniences, all china, glass, &c., may be washed up and put away, without being taken into the kitchen, and thus much risk of breakage will be avoided.

For store-rooms and other places where it is thought desirable to cover the floor with oil-cloth, we would advise old

Brussels carpets to be sent, before they are quite worn out, to the oil-cloth manufacturer, to be painted any pattern which may be preferred; by which 1*s*. 6*d*. per yard will be saved, as the painting of the carpet will be so much less than the price of new oil-cloth; and it will not only be much warmer than oil-cloth painted on canvass, but also much more durable, if it is kept until it is thoroughly dry and seasoned before it is used. A small press for tablecloths and napkins is very useful in a store-room; and there should, if possible, be a veranda or shade before the windows, to keep off the heat of the sun; or, where this is not practicable, an exterior blind.

The Dining-room. This, in England, is a room of great importance, particularly with the rich; but even in the smallest houses, this room should be a comfortable one, as much of the family enjoyment depends on it. We shall first treat of the dining-room in a mansion, and then of that of the small villa or suburban residence.

The dining-room in a mansion may be on the least agreeable side of the house, as regards prospect; because it is seldom wanted, except when the company are engaged in eating and drinking. It should be quiet, and unconnected with any other room except the library, or with the store-room; and with the latter only by an opening at the back of the side-board, or in some other part where it may be concealed, and through which the dessert, &c., may be handed, thereby preventing the too frequent opening and shutting of the door. The accessories to the dining-rooms of the rich are, the ante or waiting-room for servants, and a retiring-room for gentlemen: though they should not either of them communicate immediately with the dining-room, but should be approached through a lobby. No dining-room should ever be less than 14 ft. wide; and, even then, it will not admit of furniture being placed against the sides of the room. A width of 14 ft. is requisite to enable the servants to pass round the table safely with the dishes. In houses of a superior class, this room should be 18 ft. or more in width, in proportion to its length; and it should be lofty and well ventilated. The floor of this room, as of all good rooms, should be pugged, to destroy sound. The sides of the floor are often left uncovered; the carpet not reaching farther than to within from 18 in. to 24 in. of the skirting: but, even in this case, it should be bordered, as Turkey carpets are, to give the idea of completeness. The floor boards should be of the best description; and, where the sides of the floor are to be left exposed, such parts are frequently executed in oak or wainscot, and sometimes framed in panels, or parqueted, as described in the *Encyclopædia of Cottage, Farm, and Villa Architecture*, p. 1013. When so formed, they should be kept polished by dry-rubbing. Sometimes, instead of a bordering of oak, the part of the floor exposed to view is painted of the colour of that wood; which, in ordinary houses, both in the case of rooms and staircases, answers quite as well. There should be a dado round the dining-room, from 2 ft. 3 in. to 2 ft. 9 in. high, to preserve the wall from being injured by the furniture, and to give importance to the room. The general character of this room should be more subdued and sober than that of the drawingroom, or, indeed, of any of the dwelling-rooms, except the library. The most suitable colour is wainscot for the doors, dado, and windows; and some other warm tint, to harmonise with it, for the walls above the dado: the choice of the latter must depend on the taste of the occupier; but few colours look better than a deep crimson paper in flock, either plain or figured, with gold or wainscot mouldings, where paintings or prints in frames are to be hung. It is better, however, to have the walls painted than papered, and the pictures hung by strong cords, from brass or painted iron rods, laid on hooks close under the cornice. The ceiling and cornice should be very lightly tinted, to agree with the colour of the walls; and the enrichment of the cornice should be relieved. The ceiling in this room may be paneled, and enriched with casts of ornaments and flowers in papier mâché or plaster of Paris, for lamps or chandeliers to hang from. The cornice should be bold and enriched, in proportion to the size and style in which the other parts of the room are finished. Where sufficient height can be obtained, a cove, either plain or ornamented, may be introduced with good effect; not only in the dining-room, but in the living-rooms on the principal floor. No one, who has not had some experience in the effect of coved ceilings, could imagine how much more the addition of even a foot to the height of a room adds to its dignity when the ceiling is coved, than when it is square. A recess should be constructed for the sideboard, if possible: at all events, some architectural preparation should be made to receive it. The chimneypiece should be of dove-coloured, black, or black and yellow, marble, and it should be bold and massive. If carved at all, the carving should be in masses, and in high relief. The character of the chimneypiece should be architectural, and in keeping with the style or order used in the room. Means may be provided for ringing the bell without rising from table, in case of wanting anything after the servants have left the room; and this may be effected by pressing the foot on a knob under the table, or by lines passing through pulleys, and under the carpet, to the chair occupied by the master of the house. There should not be any closets in the dining-room. If it be thought necessary to have other means of putting things away in the room beyond those afforded by the sideboards, pieces of mahogany furniture, French-polished, such as wine coolers, sarcophagi, chiffoniers, &c., might be introduced. The doors, if not painted like the window and dado, may be of mahogany, finely polished; in which case they should be wide and folding; and there should be double doors to every doorway. The sashes and beads of the windows should then be of mahogany also. The windows should be large, and brought down to within about 2 ft. 3 in. of the floor. The knobs to the shutter latches and to the locks should be of turned dark wood, or of cocoa nut inlaid with ebony. The whole of the furniture, as the dining-table, chairs, sideboard, wine-coolers, and chiffoniers, should be of mahogany, French-polished. The frames to the looking-glasses may be of oak inlaid with ebony, unless there are gilt picture-frames hung in the room; in which case, the frames to the looking-glasses should be gilt also, as likewise the carved cornice over the curtains. The curtains should be a scarlet geranium, or of a fawn colour, with broad lace and fringes.

These observations apply more to the dining-room of the man of rank, (which is strictly and exclusively a room for dining in, or *salle à manger*,) than to the dining-room of persons in the middle class of life, where the dining-room is often also the library, and, indeed, the only family room except the drawing-room. The following sketch will give some idea of a dining-room which is capable of fulfilling all the purposes that are usually desired in a room of this description: — The dining-room for a small villa or suburban residence should be a comfortable family room, as is shown in *fig.* 20., from 20 ft. to 24 ft. long, from 15 ft. to 18 ft. wide, and from 11 ft. to 13 ft. high, with the bay window (*a*) looking to the north, or the north-west. This window should be large, and reach down to within 6 in. of the floor, so that a view of the adjoining flower-garden, or of the distant prospect, if in the country, might be commanded by it. The fireplace (*b*) should be on one side of the room, and on the opposite side the doors (*c c*). The recess at *e* is for the sideboard; but, if the occupier is a great lover of music, it may be filled by a pianoforte, placed as shown at *g* or *h*; in which case, the small stained-glass windows, *m m*, may be introduced at each end of the recess, to throw light on the instrument when it is used by daylight, and which may yet be sufficiently subdued not to interfere with the general tone of colour in the room. In this case, the sideboard may be placed at *d*. With regard to the furniture; the carpet should be a good Axminster or Brussels, of a well-covered pattern, in circles or octagons, with rich brown colours, and flowers, &c., of glowing tints, looking warm and comfortable, and like home. The walls may be crimson, or of some lighter but still warm tint, covered with engravings or other pictures, according to the wealth and taste of the occupier.

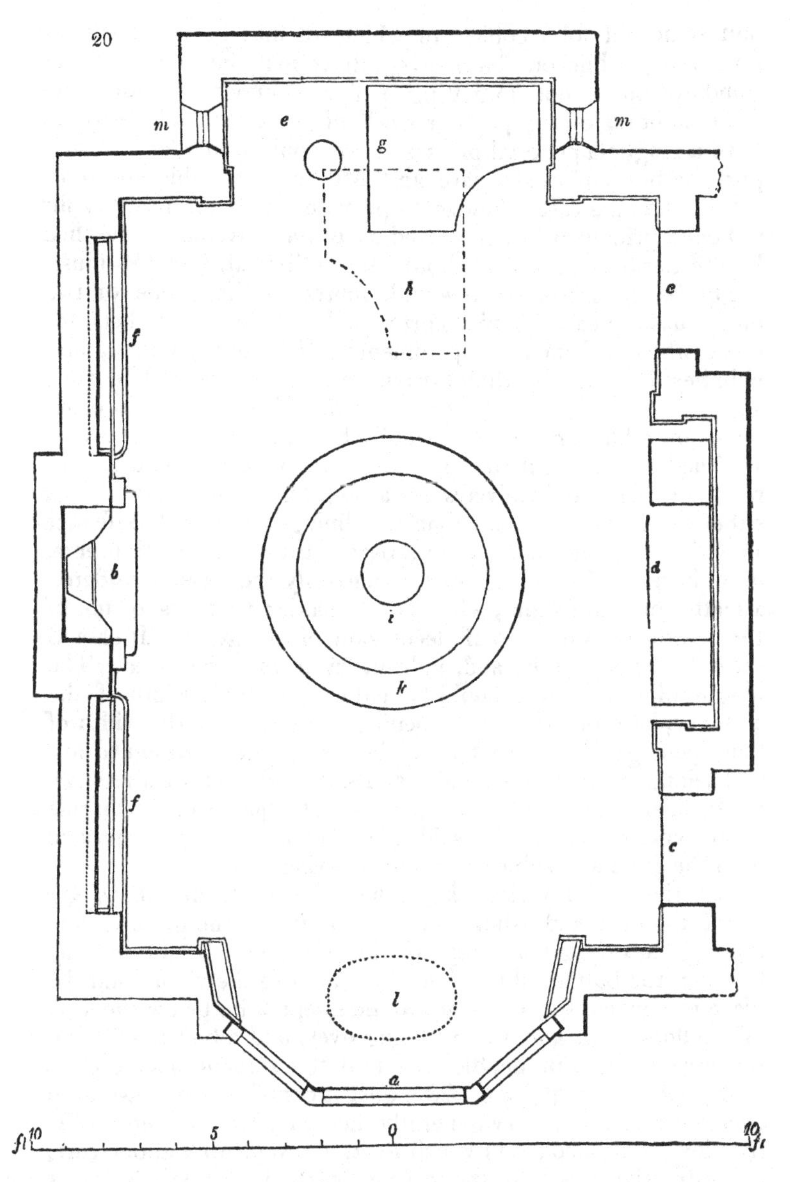

There may also be a few plaster brackets from the antique fixed against the walls, and painted in a warm tone of colour or bronzed, with some casts of the busts of favourite authors, and of two or three of the most celebrated antique vases, on them, and over the bookshelves. "On each side of the fireplace we would have handsome bookshelves (*f f*), the lower ones broad, to contain some old folio copies, of which we may be the fortunate possessor; while on the shelves above may be copies of our standard poets, not forgetting a few volumes of more modern authors, or any popular work of the day; and we would have a scagliola pedestal or two, if we could find room, to support the busts of Shakspeare and Milton, if possible, in stone. We would take care, likewise, to provide a secure repository for unbound printed music, so placed as to be conveniently reached by the performer; which should be so divided, that the music might be classified. We would, moreover, have one or two music desks near the pianoforte, made to fix on brackets, for those who perform accompaniments. The stove, fender, and chimneypiece, in the dining-room, should correspond in style and character with the apartment itself. Bronze or brass ornaments on a black ground, or wholly black, are most appropriate. Polished steel and cut steel should be reserved for the drawing-room, which should always be a step higher in every article, either of use or ornament, than the dining-room." The fender should, to a great extent, form a part of the stove: at all events, it ought not to be, as is now too generally the case, considered a matter of indifference, whether the fender partakes or not of the character and style of decoration of the grate. It should seem to belong to it, and, indeed, even to support it. The length of the fender should be either the whole width of the chimneypiece, or, where the opening is large, only the width of that opening: at all events, the fender should be designed for, and belong to, the chimneypiece and stove; and, above all things, never finish in an abrupt manner against the jambs of the marble chimneypieces, as may be said to be the case in nine tenths of both the town and suburban houses of England.

Contrivances of various kinds have been introduced, to get rid of the dust and ashes accumulating under the grate during the day: a very good receptacle for these may be formed, by keeping the bottom of the fender 2 in. above the floor, and the place into which the ashes are to be swept 2 in. below the level of the floor, with a movable grating over, and a box made to fit the receptacle, with handles, so that the servant may take it away, without any portion of the dust it contains being dispersed over the room, as is now generally the case every morning when the stove is cleaned; and which dust, if swept up without care, is destructive alike to the carpet, curtains, books, &c. A veranda, with the roof either glazed or open, may be placed against the outside walls of the room. To those who like a subdued light, glass is preferable; and under it, against the wall of the house, may be grown vines and fig trees; or, if abundance of light is desired, the veranda may be formed by narrow trellis standards, at convenient distances, with a trellis frieze, and movable ribs for an open roof. At the standards may be planted different kinds of clematis and vines, the evergreen and Chinese honeysuckles, the hardy bignonias, the hardy passion-flower, the wistarias, and climbing and sweet-scented monthly roses, &c. In winter, when the shade of the plants is not required, their shoots might be loosened from the roof, and tied in along the frieze of the veranda; and the ribs might be removed, to admit more light to the windows.

The furniture of the room may be of mahogany: consisting, besides what has been already named, of the dining-table and chairs, which should be solid and substantial, and of the finest wood, but plain and uncarved; the whole beautifully French-polished. The dining-table may be either circular, square, or a parallelogram. If the former be preferred, the table (*fig.* 20. *i*), in its ordinary form, should be capable of dining eight persons; and, by the addition of marginal rims (*k*), each 20 in. broad, of dining twenty persons. Of course, the size of these additional rims will depend upon the size of the room, and on the number of persons to be accommodated. If a square or long table be preferred, the smallest size should be 6 ft., and the largest 10 ft.; the additional size being gained by detached leaves. The centre of the table should be occupied with flowers, or some ornament not holding anything required for the dinner; and by this means the dishes will be brought within the reach of the company. Where much company is kept, it is almost impossible to accommodate them without long tables; which not only admit of more guests, but a much greater number of dishes, in a given space. In some dining-rooms, a tube, or speaking-pipe, is fixed out of sight, and made to communicate with either the kitchen, servants' hall, or waiting-room, according to the size of the house and establishment. The orifice of this pipe should be closed with a weighted slide, to be lifted up while the person using it is speaking or receiving an answer, and which will afterwards close of itself: for, if left open, the conversation in the dining-room will be distinctly heard in the kitchen, and the reverse. To prevent the possibility of this, such speaking-tubes ought always to be placed in a small recess in the wall, closed with a door about a foot square, having a spring hinge, and painted of the same colour as the wall. These tubes are principally useful for the servant in waiting to acquaint the cook with what is wanted. When they are used, it is customary for the servant in waiting to ring the bell, to excite the attention of the cook to listen at one end of the tube, while he whispers his instructions at the other.

The curtains of the dining-room should be of a scarlet or geranium-coloured moreen, with gold fringe and binding: they should be made to draw, and have a gilt cornice over them. In the bow of the window may be a flower-stand, or small oval table (*l*). The sideboard should be fitted up with every possible convenience (for, as we have before said, in a dining-room there should be no closets); and under it may be a sarcophagus for wine, having means within it for cooling the wine with ice in hot weather. The room may be lighted at night by lamps suspended from the ceiling, or a chandelier with candles and metallic wicks, having ground glass shades; with similar candles and shades for the instrument.

As some people may prefer an octagon-shaped room to a square one, *fig.* 21. shows the plan of one in that shape. In

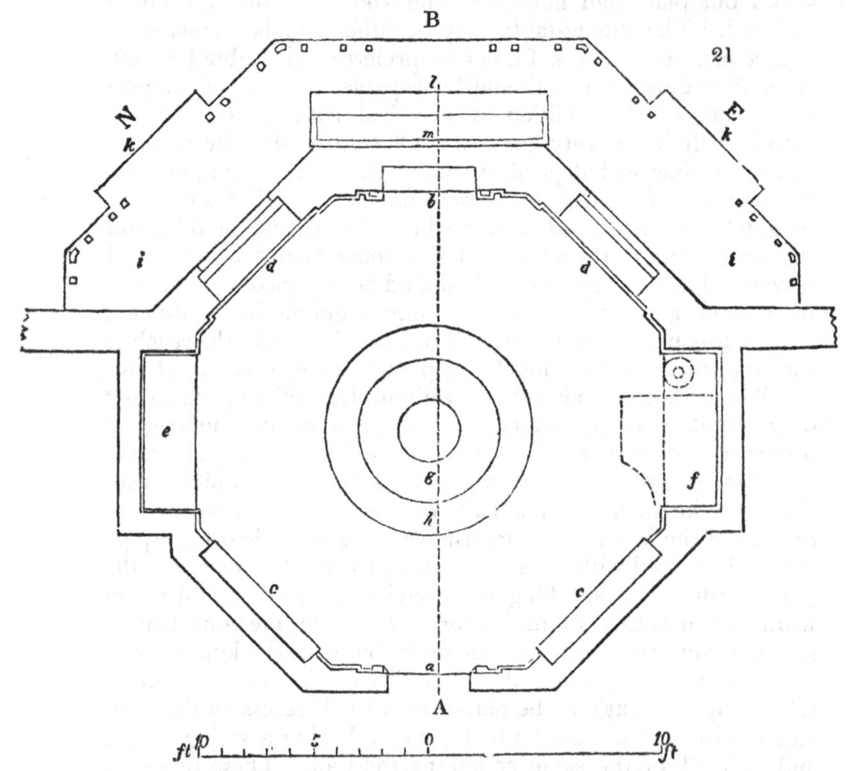

this figure, *a* is the doorway; *b*, the fireplace; *c c*, book-shelves; *d d*, windows; *e*, recess for the sideboard; *f*, recess for the pianoforte; *g*, circular dining-table, 6 ft. in diameter; and *h*, additional pieces, to increase its size to 9 ft. in diameter, when requisite. *Fig.* 22. is a section on the line A B, on the plan. The diameter of the room is 20 ft.; the height, up to the top of the cornice, is 12 ft.; and the height, up to the ceiling, 13 ft. 6 in. From the top of the cornice springs a cove, which

22

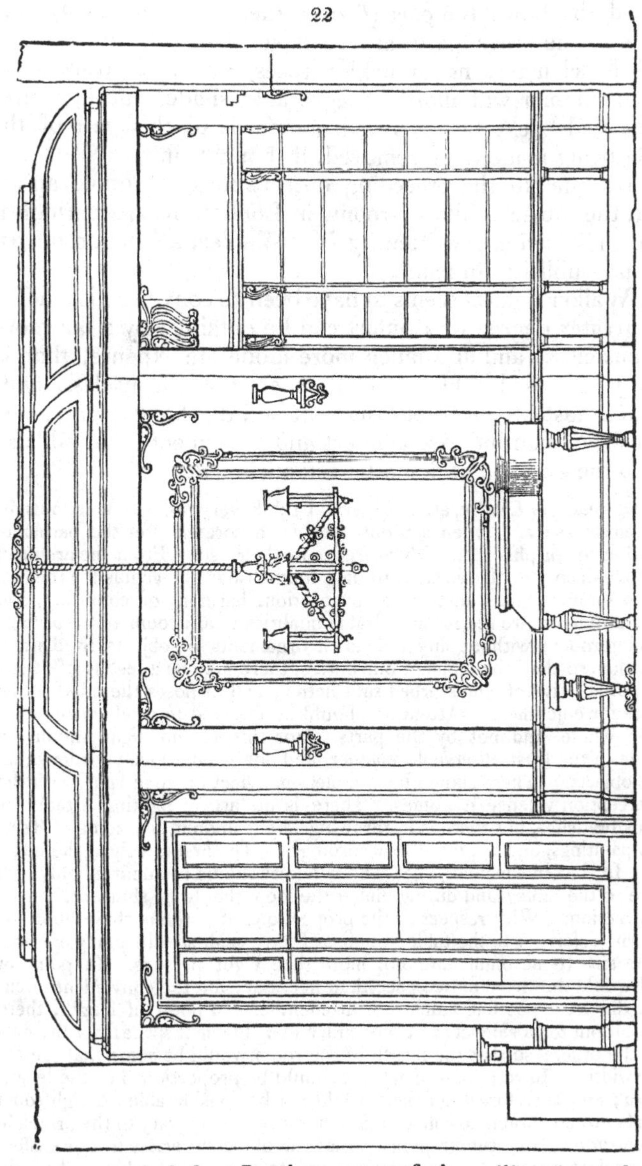

is shown as paneled. In the centre of the ceiling is a plaster or papier-mâché flower, 3 ft. in diameter, round which might be stenciled a Grecian honeysuckle ornament. From the centre of the flower is suspended a chandelier. The windows are folding casements, opening on a terrace of Portland stone (*fig.* 21. *i i*), with steps (*k k*) down to the gravel walk. A veranda is shown all round the bow: the part (*l*) over the garden chair (*m*) is to be roofed; the other parts are to have ribs only, and are intended for clematis, honeysuckles, roses, &c., to be trained on, which in summer will afford an agreeable shade, and in winter may be tied back, the ribs and standards to this part of the veranda being made to be removed, if thought fit.

As a sequel to the preceding observations, we quote an article on the subject of dining-rooms in London, from a periodical entitled the *Original*, written by Mr. Walker, a London magistrate, and published in 1836.

Mr. Walker's object seems to have been to convince his readers that a greater degree of comfort can be obtained by a far more numerous class, and at a much more moderate expense, than is generally supposed. He contends that it is the useless, and, very often, tasteless, love of ornament and display (which rather detract from than increase comfort and enjoyment) that adds so much to the expense of domestic establishments.

"*Dining-rooms, in London*, are in general, I think, very tasteless and uninspiring in themselves; and, when set out, they are decorated after the barbarian style, rather for display than with reference to their use. From the architect to the table decorator there seems to be a total absence of genius for the real objects to be aimed at. Justness of proportion, harmony of colouring, and disposition of light, are the most desirable qualities in any room, but especially in a dining-room; without any individual ornaments or objects to distract the attention, so that, the moment one enters, there may be a feeling of fitness which is productive of undisturbed satisfaction, and disposes the mind to the best state for enjoyment. Attention should be directed to produce an effect from the whole, and not by the parts. For this reason, light should be thrown in in the least observable manner, and not ostentatiously, from ornamented objects. There should be the pleasing effect of good light, with the least perception whence it comes. There is no art in lighting a table by cumbrous branches, but there is in throwing a light upon it, like some of Rembrandt's paintings, and the effect is accordingly. The first is vulgar, the latter refined. In the same manner, light from windows should be admitted only with reference to the table; and during dinner the view should be shut out, to prevent distraction. With respect to the proportions of a room, they should be studied with reference to the table, which, as I have said, should, in my opinion, be of the size to accommodate not more than eight persons. In point of width, I would not have more space than necessary for the convenient circulation of the least possible number of attendants. In point of length, there should be room for a sideboard at one end, and a sufficient space from the fireplace at the other; so that the length of the room would be somewhat greater than the width. In respect to height, it should be proportioned to the length and width; and, therefore, the height would not be considerable. A high room is certainly not favourable to conversation, because it is contrary to the principle of concentration; and the prejudice in favour of height arises from its effect considered with respect to large parties, and to overloaded tables. I would have the door in the side, at the end near the sideboard; and the windows on the side opposite. As to colouring, the same rule ought to be observed as in every thing else; that is, to study general effect. To suit all seasons best, I think the walls ought to be of rather a sober colour, with drapery of a warm appearance for cold weather, and the contrary for hot. Perhaps it may be thought by many that all these particulars are very immaterial, and that the consideration of them is very trifling; but my opinion is, that, in all our actions, whether with reference to business or pleasure, it is a main point, in the first place, to produce a suitable disposition: and as dining is an occurrence of every day of our lives, or nearly so; and as our health and spirits depend, in a great measure, upon our vivid enjoyment of this our chief meal; it seems to me a more worthy object of study, than those unreal occupations about which so many busy themselves in vain. But I am forgetting an important matter in the dining-room: I mean, the due regulation of the temperature, upon which comfort so much depends, and from the want of attention to which there is annually so much suffering, both from heat and cold. In hot weather, the difficulty is the greatest, and is best to be overcome by attention to ventilation and blinds. In winter, there is little difficulty, with due care, and no stinginess; which latter is apt to appear, both in having the fire only lighted just before dinner, and in not keeping it up properly to the end of the party: and I do here protest against the practice I have often witnessed, of letting the fire actually go out, in cold weather, before the guests; there is nothing more cheerless, or of more inhospitable appearance. On the other hand, a bright blazing fire has a very inspiring effect on entering the dining-room, and is an object worthy of special attention to those who wish their parties to succeed. Moreover, in such a room as I have described, the opening, after dinner, on a dreary day, to admit a cheerful fire, would be a very inspiring moment, with an agreeable party, brought into perfect unison by a well-imagined, well-executed repast: a scene to kindle, equally, attachment to one's friends, and love of one's country. The cultivation of the fireside is one of the greatest import, public and private.

"Having said, I think, every thing I have to say as to the arrangement of the dining-room, I will here dedicate a word or two to its necessary appendage, the kitchen, which I would have literally an appendage; and not, as at present, a distant and unconnected establishment. As I said before, I am now supposing the whole object to be the perfection of dinner parties, without reference to any other consideration; and, therefore, I put aside custom, fashion, and prejudice, as enemies to the true theory and practice; and I boldly advance my own opinions. I must beg the reader to bear in mind that I am speaking with reference to small parties, and that I am not an advocate for dinners which, as nearly as can be calculated, are just enough, and no more. I speak not of the bustle of preparation for twelve, sixteen, or twenty people, with about four times as much as they can possibly consume, and with a combination of overpowering heat and disagreeable scents. I have in view a quiet little kitchen, without noise, or annoying heat, or odour, save some simple savoury one, provocative of the appetite, and incapable of offending the most fastidious. Such an establishment would I have immediately adjoining my dining-room, and communicating with it by an entrance close to the sideboard, closed, during the process of dinner, by a curtain only, so that the dishes could be brought in without noise, or current of air, or constant opening or shutting of a door." (*The Original*, No. 17.)

The Drawingroom. The size of the drawingrooms in London houses necessarily depends upon the rate or class of the house; the front drawingroom in a second-rate house being about 18 ft. long by 15 ft. wide; and the back drawingroom about 15 ft. long by 11 ft. wide; both being from 11 ft. to 12 ft. high, and communicating with each other by folding doors; while in third and fourth-rate houses the drawingrooms are smaller according to the size of the house. In houses of the first rate, which houses may be built of any size and height desired, the drawingrooms are not unfrequently from 30 ft. to 40 ft. long, from 20 ft. to 25 ft. wide, and from 13 ft. to 18 ft. high. It should always be part of the arrangement of a drawingroom, particularly in houses belonging to persons of distinction, that it be lofty; for the principal use of this apartment being to receive large assemblies of persons at night, when lighted up, as it commonly is, by lamps suspended from the ceiling, the pure air becomes quickly exhausted; and the visitors, especially those in a delicate state of health, suffer much by breathing so noxious an atmosphere. It is, therefore, of great importance, where these rooms are intended for the purpose of receiving large parties, that they should be effectually ventilated; and this may be done at a comparatively small expense, when the house is being built, by having large tubes or tin pipes fixed between the joists, and communicating with the open air; and having openings between the leaves of the flowers on the ceiling, through which the heated air may escape into these tubes, and thus be carried off; also by having the same kind of air-pipes carried under the floor, to give admission to fresh air from without, the quantity of which may be regulated by gratings and valves, concealed in the skirting, or in some other more convenient part of the room. These contrivances are introduced into the more recently built first-rate London houses, and are used on the occasion of routes or very large assemblies, but closed at all other times. On ordinary occasions, in such houses, the ventilation produced by admitting air from the staircase over the tops of the doorways, as already described, p. 55., is found quite sufficient to preserve the air of the apartment in a healthful state.

The style of decoration employed in the drawingroom should be of the most cheerful kind, elegant, and even splendid, but chaste, and not gorgeous: the only room that will admit of a more brilliant style of finishing than the drawingroom is the ball-room, or, perhaps, the lady's boudoir. The floor of the drawingroom should be of the very best description; and, if of deal, it should be of the kind termed battens (narrow boards, 1½ in. thick), listed (sorted and matched in regard to width and colour), and free from sap (consisting entirely of heart wood). The boards are made free from sap by sawing off the edges, by which process the batten is reduced from its usual width of 7 in., to 4 in., or 5 in., leaving only the heart of the wood. This floor, when of a very superior kind, is doweled, so that the surface may be perfectly smooth, and without a nail-hole. A doweled floor is formed by nailing the first board next the wall, on one side, through the board to the joists, in the common manner, the nails being afterwards concealed by the skirting; and on the other, by nails driven in a slanting direction through half the thickness of the board to the joist below. Dowels, or pins, either of iron or wood, are previously let into the board thus nailed down, for the purpose of uniting it with the next board; into holes made in the thickness of which they are forced, to keep it steady on one side, while it also is only nailed on the other. The whole of the floor is laid in this manner, the only boards nailed on both sides being the first and the last. When the room is finished in the style of Elizabeth or of Louis XIV., the floor should be of oak or wainscot, with the sides bordered (that is, the boards composing them should be laid in a different direction to those in the middle of the floor); and parqueted (laid with small pieces of wood, fitted into each other in regular figures), if the expense be not an object. As the furniture in these rooms is generally of an expensive kind, and liable to injury by dust, the skirting, or plinth, should be tongued (that is, fitted into a groove) in the floor: when these precautions are taken, and the timbers under the floor boards are pugged, no dust can get through to spoil the carpet and other furniture. If the rooms are large and lofty, there should be a dado, or cornice, carried round them. A dado, in architecture, is the plain space of a pedestal between the plinth, or base, and the cornice; but, among carpenters and builders, the term dado is generally applied to the whole assemblage of these architectural members; and the cornice, which is purposely formed about the height of the backs of the chairs, is called the chair-rail. The original use of this dado was to keep the chairs from injuring the plaster-work or papering of the wall; and hence it was formed entirely of boards, or wainscoting; but, since timber boards have become much more expensive than plaster, an economical substitute has been formed, by merely making the chair-rail and the base of wood; and leaving the dado, or panel, of plaster, to be painted or papered. A still more economical substitute consists in forming a base, or plinth, projecting so far into the room, that, when the hind legs of the chair touch it, the back of the chair is at the distance of at least an inch from the wall. To lessen the appearance of projection in this plinth, or base, another, called the surbase, and which is smaller in all its parts, is sometimes placed over it; the two plinths being united by mouldings. In small rooms, the moulded plinth, skirting, or wash-board, as it is called, is commonly made from 1 ft. to 1 ft. 3 in. high, and projecting about 1 in. into the room. Where a double plinth is used, the lower part of it must not project farther than the single one usually does; unless the projection of the architraves round the doors and windows is so far increased as to admit of the lower plinth finishing against it, and not projecting beyond it. When there is a dado, it should be of the height a pedestal would be if an order of architecture were employed; so that, if the whole order were 14 ft. high, the pedestal, and consequently the dado, should be about 2 ft. 9 in.; if only 12 ft. 6 in. high, the dado would be 2 ft. 6 in., or about one fifth the height of the room from the floor to the top of the cornice. This is the case in the drawingrooms of the new palace at Pimlico, round which there is a peristyle of scagliola columns, of a grand and imposing character. Where columns are introduced, the only orders that can be employed with propriety in a drawingroom are the Corinthian and composite. The character of the enrichments on the ceiling and cornice should be light and pleasing; the compartments, when the ceiling is paneled, should not be so deeply recessed as in the dining-room or library; and the ornaments should be of a gayer and more fanciful description: but good taste alone can decide as to the extent to which these enrichments may be carried. A cove, springing from the top of the cornice, when there is sufficient height, will add much to the dignity of the room; this may likewise be enriched by paneling, with carved mouldings and pateræ or festoons, or by paintings of pleasing subjects, or wreaths of flowers. The principal objection to most of the enrichments generally used in drawingrooms is, their want of sufficient relief: so much is this frequently the case, that, where the enrichments are not thrown out by difference of colour or gilding, their effect, when seen on a lofty ceiling, is almost lost. The enrichments that form or surround the several compartments should be well marked and in bold relief. If it be thought necessary to ornament the panels, this may be done by the artist's pencil. When gilding is employed, it should be used sparingly; for, if overdone, that which would otherwise have been elegant will become gaudy and vulgar. If the enrichment to be gilded is large, the prominent parts only should be etched (edged) with gold; but of the plain mouldings, the smaller ones, or those only that will have a good effect, and will mark more distinctly the form of the general mouldings, should be selected for gilding. When the ceiling of the drawingroom is plain, the colour should harmonise with the prevailing tint of the room, and the cornice be relieved by other tints of the same colour, deeper and brighter than that of the ceiling. When the room is low, an acroter, as it is called by builders, or margin, with light and shade lines, is often run round the ceiling, to make the room appear higher than it really is. The walls are occasionally painted with some warm tint in distemper, or in turpentine colour, of one shade, with margins, and light and shade lines, to form the sides of the rooms into panels; frequently having ornaments of flowers, foliage, &c., painted in relief, in the angles of each panel, and gold mouldings under the cornice and on the dado, and up the four internal angles of the room; with a smaller moulding round the doors, windows, and chimneypieces. A small gold moulding is likewise sometimes put to the internal edge of the margin round the panels. But a more general style is, to cover the walls with a rich-patterned satin paper; the more expensive kinds being highly relieved with gold, or some inferior metal. This kind of paper is, in a few instances, in good taste, but more frequently in bad. A flock paper, with or without a pattern, relieved with gold, may be adopted; or a plain flock may be laid on the walls, of any shade of colour the employer may approve, giving them the appearance of being covered with a beautiful cloth without a seam: this is often rich and pleasing in its effect; but the deep crimson colour, so frequently seen on the walls of drawingrooms in London, too nearly resembles the colours appropriated to the dining-room. It is adopted, because it is considered becoming to the complexion of the ladies, who are to assemble in it; but it destroys the effect of the furniture, and the light, gay, and cheerful character of the room, and leaves all the recesses, and other parts of it not strongly and powerfully lighted up, dull and gloomy: light, warm, sunny tints are much more suitable; or even green, which, though very unbecoming to most complexions, is better for a drawingroom than crimson. When the walls are hung with flock papers, gold mouldings are almost invariably used; but with satin papers, a patterned border in flock, designed to suit the paper, is generally substituted. In drawingrooms of a more stately kind, the walls may be hung with figured silk, or with rich damasked or striped satin; but, when so hung, the walls should be first battened (that is, have projecting pieces of wood fixed to the wall at given distances), to which laths may be nailed, and then plastered and covered with canvass; or which may be lined with deal boarding, on which canvass should be strained, and then papered, so as to prevent as much as possible all risk of stains from damp or dust, which would very soon destroy the beauty of the satin.

There are many ways of painting these rooms: some are painted of the prevailing tint of the paper, making the styles to the doors, windows, &c., a dark tint, the panels lighter, and the mouldings either darker still, or of a French white, or a very pale tint of the same colour as the styles; other rooms are painted in imitation of some choice wood, as satin wood, wainscot, hairwood, maple, or rosewood. Of all these, the most appropriate is the satin or the maple wood, either of which, when highly varnished, has a rich and pleasing effect. The rage for painting the mouldings in a different kind of wood from the general graining, such as cross-banded, or black, or gold, was very general some years since; but this mode of painting mouldings is now (1837) discontinued. The most chaste and appropriate style in which a drawingroom can be painted is, however, of a French white, with gold mouldings; and, where the mouldings are carved, with the carved parts slightly etched with gold. The doors to this room, where the expense is no impediment, should be of mahogany or maple wood, veneered with choice veneers, and highly French-polished; they should be hung with strong rising brass hinges, so that they may easily be taken off, and put on again, without taking the screws out of the hinges. In smaller houses, they may be painted to correspond with the other parts of the room, or in imitation of some kind of fancy wood. The locks should be mortised into the doors. Where brass handles and furniture are not considered handsome enough, cut glass knobs, or turned or carved hard wood, or ivory or ebony knobs, for the locks and shutter latches, may be used. The sashes, when required to be superior to painted deal, should be of a light-coloured mahogany; the bottom rail having a groove in it to receive a metal bar, which is to be let into the oak sill of the sash-frame. The beads should be tongued into the sash-frame all round, and the sash fastenings should be effective in drawing the sashes firmly together. The boxings, or cases into which the shutters fold back, should be splayed (made slanting, with the widest part towards the room), to give a freer admission of light. They should not reach quite up to the soffit (the upper part of the architrave, or frame in which the window is set, or generally the under part of the arch, cover, or lintel, of any opening), and they should be kept securely in their boxings during the daytime by shutter latches. When the windows open on to a lead flat (the flat roof, if covered with lead, of any projecting room, or other addition to the ground floor of a house, which is not carried up any higher,) or balcony, the sashes or casements are usually hung with hinges, folding to act as doors; but great care is required, when windows are so hung, to prevent the water from finding its way in. The sashes, in all rooms of a superior kind, but more especially those of the drawingroom, should be glazed with polished plate glass. The fireplace should be ample in its size, and sufficient for its purpose; viz. that of expelling all damp, and effectually warming the air in the room. Where the room is 30 ft. or more in length, there should be two fireplaces, either on the same side of the room, or one at each end, as circumstances may direct: the opening being rather high in proportion, than low, so as to cause greater heat to be thrown into the room. The chimneypiece in a modern drawingroom should be of statuary marble, the purer in its whiteness, and the freer from veins or stains, the better: this, in a moderate way, is usually moulded, with, perhaps, a carved tablet in the centre of the frieze, and carved caps or blocks on the pilasters; but in more stately rooms the design should be in accordance with the rest of the apartment. The carvings should consist of fanciful and graceful groups of flowers, foliage, or fruits, or graceful mythological figures; or, in mansions, it may embody some interesting and agreeable episode connected with the history of the family, in which figures may be introduced in bold relief. Whatever it may be, the subject should be such as may appear likely to call forth in the beholder agreeable and pleasing sensations. The stove and fender for the drawingroom should be of polished steel with the projecting parts richly cut; and the stove should be set in marble. The design must depend upon the style of decoration in which the room is fitted up; for to put a stove composed of Gothic arches, pillars, and ornaments in a room otherwise Grecian or Italian in its finishing and furniture, would be as repugnant to good taste as to see a Doric or Corinthian portico in front of a fine old Gothic church or chapel.

The furniture for the drawingroom being generally chosen by the lady, a tolerably fair opinion of her taste and good sense may be formed from the manner in which it is displayed and arranged. The carpet may be considered as one of the most important articles of furniture, and much care is necessary in its selection. If the colours be gaudy and overpowering, they will injure the effect of the other furniture; and, if too cold, they will give an air of cheerlessness and discomfort. For use and economy, the pattern should be rather diffusive than set and formal; the principal shade being interspersed with gayer and richer toned colours, and such as will not easily be injured by dust. The carpets in most general use are of the kind called Brussels; but the Axminster carpet, thick and without a seam, is the most splendid article of the sort made in England. In some of the houses of the rich are carpets manufactured in Persia, splendid and gorgeous beyond description. Here again the design for the pattern should be guided by the architecture of the room: the same carpet that would harmonise well with the character of a room in the Elizabethan style, or of the time of Louis XIV., would hardly suit with the modern light style of fitting up a drawingroom. The carpet, when not fitted to the room, should have a border, and should not leave more than from 1 ft. 6 in. to 2 ft. of the floor uncovered. The hearth rug should be of a pattern to agree with the carpet, but bolder; it should be thick, warm, and ample in its dimensions, for if too small it will give a mean character to the fireplace; the fringe round the rug, if there be any, should be thick and massive. The card tables, and the occasional sofa tables, as well as the frames of the sofas of the present day, are usually of rose wood, plain and not inlaid; but, if of a superior description, they are richly carved. The seats to the chairs, sofas, &c., are usually covered with the same kind of material as the curtains are made of; and stuffs, or moreens, of lighter colours than those of the dining-room, either plain or damasked, are frequently used, particularly for Elizabethan or Louis XIV. rooms; but for a modern room, printed chintzes, lined with silk or glazed cotton, are thought by some to have a lighter and more tasteful effect. Whatever may be the material of the curtains, they are generally accompanied by white muslin under curtains. In larger houses, plain or figured silks, perhaps with a worked border of flowers, or damasked or striped satins, may be used, with rich binding and deep massively fringed drapery and gold cornices. For chairs, some may be fauteuils, with the wood-work painted of a cream or blush colour, some of the plain mouldings gilded, and the carved parts richly relieved with gold; the seat being covered with embossed or damasked silk or satin, like the curtains, and fringed, or with white velvet stamped to imitate paintings of flowers in their natural colours; and others, of elegant forms, to suit the style of decoration in the room, with similarly covered seats or cushions. Besides these, in modern drawingrooms, light chairs are frequently introduced, like those in bed-rooms, but made of costly fancy woods, with their seats formed of willow in different patterns, as shown in the *Encyclopædia of Cottage Architecture* (p. 1061. to p. 1063.), There should also be sofas, and other cushioned seats and footstools, placed in convenient parts of the room. The pianoforte is now to be found, in one shape or other, in almost every drawingroom, from that of the humble tradesman, to that of the palace; and, in form and kind, it must depend upon the taste of the proprietor; but, where an upright piano is chosen, the fluted silk in its upper part should correspond, or harmonise in colour with the curtains of the room. This remark applies also to the chiffoniers, screens, and all other articles of furniture in which silk is introduced. The card table, either plain or carved, should be in keeping with the other furniture; and if the chairs and sofa are carved, the card table should be carved also. The top, when open, should exhibit a smooth cloth, its colour harmonising with the other colours in the room; or the table may be formed entirely of rosewood, or some other fancy wood, instead of its inside being covered with cloth.

On the walls of the drawingroom are hung the choicest paintings of the owner; and a French clock, a beautiful vase or two, with cut-glass lustres, and a few other choice ornaments, may be placed on the chimney-shelf. On marble slabs, with gilded frames or brackets, may be placed the sculptured busts of celebrated persons, or of some one dear to us. Silvered mirrors of polished plate glass, in gilded frames, cannot be too profusely employed in a drawingroom; and, where it can be contrived, they should be placed exactly opposite one another, by which means the reflection of every thing that comes within their focus is endlessly reproduced; and when the cut-glass chandeliers are lighted at night, and ladies in elegant dresses are moving within the scope of their operation, the scene becomes fairy-like and brilliant beyond description. Frames for looking-glasses are sometimes covered with velvet and gold mouldings; and these frames have a beautiful effect while fresh; but they soon soil, and, on the whole, are not equal to frames gilded all over. In the *Encyc. of Cottage, Farm, and Villa Architecture*, numerous designs will be found of every description of furniture, hangings, pianofortes, &c., suitable for drawingrooms; and several views are there given of the interior of drawingrooms, completely fitted up and furnished in different styles of architecture.

Some of the preceding remarks apply chiefly to houses of a large size; but by far the greatest part of them are applicable to every house which contains a drawingroom. They may, of course, be modified according to the taste, habits, and resources of the individual; but two of the general rules laid down should never be deviated from. The first, and most important, is, that the drawingroom should always be more elegantly fitted up than any other living-room of the house, particularly the dining-room, the fittings up and furniture of which should always be plain and massive; and that it should contain furniture of a lighter, and yet superior, description. The other rule is, that the style of the fittings up and furniture should, if possible, bear some relation to that of the house. As a further illustration of this last remark, we may give the following description, by Mr. Kent, of a drawing-room in a country house belonging to a London professional gentleman. "The style was that of Elizabeth, or, perhaps, of rather a later period. It was approached by two doors, both on the same side of the room, the one leading from an entrance hall or passage, the other from the best staircase. Between the doors, centrally situated, was the fireplace; ample in its dimensions and massive in its character. It was of native marble, but plain, with carved oak above, reaching up to the ceiling; on the opposite side were three windows; the centre one was an oriel window, deeply recessed, and approached by an arched opening; the plan of the window was that of five sides of an irregular-sided octagon, the centre one wide and divided by mullions; the sides next the centre were filled in with oak transom frames, and all had oak casements; the remaining two sides were wainscoted with oak, and were recessed five or six inches, with well-filled bookshelves from the floor to the ceiling. Round under the window was a continued oak seat; and in the centre of the recess was a small carved oak table. The diameter of the whole window might be about 10 ft.; the opening into it about 6 ft. The soffit, or ceiling of the architrave, was paneled in oak. The size of the drawingroom was about 45 ft. long, 22 ft. wide, and 13 ft. high. These proportions rendered the room too low, which was its great defect. The other windows were canted bows (projecting windows, curved in the middle, and having the sides straight), with seats round them, and transom frames (that is, with a strong perpendicular bar up the middle, and one crossing it at about a third of its length from the top,) and casements. These windows looked direct on the lawn and flower-garden, the roses in which were growing up to the windows; at one end of the room was a large transomed window, wide and down to the floor; its plan was half a decagon, with sashes all round; and the whole of the sashes were glazed very tastefully in lead. The window at the west end led into a conservatory of about the same dimensions as the drawingroom, having a broad walk along the centre, terminated by folding sash-doors leading on to the lawn. The room was wainscoted from the floor to the cornice with old oak framing; the cornice was likewise of oak, very richly and boldly carved, and the whole was of a fine tone of colour, and highly varnished; the ceiling was plain, except that two oak girders formed it into three compartments, and it was tinted with a warm tone of colour. The furniture, consisting of tables, chairs, pianoforte, sofa, screen, &c., was all of oak, carved and polished; the seats of the chairs were covered with tapestry; and the curtains were a crimson moreen, with carved oak cornices." (*Kent*, in *Arch. Mag.*, ii. p. 358.)

The Library. This is a room which was at one time to be found only in public buildings and institutions, monasteries, palaces, universities, and other places devoted to learning, and in the houses of the wealthy and noble; but, in the present day, no villa, or suburban residence, having more than two sitting-rooms, can be considered complete without a library. Where there are only three living-rooms, the library and breakfast-room may be combined in one; and this, in fact, is frequently the case in suburban houses.

The situation of the library in a large house should be sufficiently retired, so as not to be interfered with, either by the family or by visitors. It should be farther removed from the entrance than any of the sitting-rooms; it should be lofty, and every possible care should be taken to preserve it from damp. For this purpose, whether in use or not, in moist and foggy weather, it ought to be aired, either by a fire in the grate, or by warm air introduced from a heating apparatus; for, if once the damp affects the leaves or the binding of the books, they will very quickly mould, and become wormeaten. The floor of this room, in particular, should be pugged to destroy sound, and all the walls battened. Where the books are very numerous, and the library is intended for a person of wealth, it should be carried up the height of two stories; that is, from 25 ft. to 30 ft. high, with a gallery all round, or on the sides only; the staircases to which should be so contrived as not to destroy the uniformity of the room, and they would be better, if possible, placed outside of the room, and communicating with it by doorways.

The floor of the library should be of oak, or Dutch wainscot, and the sides, and, perhaps, a centre pathway, paneled or parqueted: but, where economy is to be studied, a deal floor, painted in imitation of oak, and varnished, may be substituted. The woodwork and furniture in this room should be of oak polished, or of deal painted in imitation of oak, and varnished. The walls, where not covered by book-cases, may be of a warm stone colour or crimson. The same tone of colour should prevail in the ceiling, so that the whole may harmonise, and impress on the beholder a feeling of quiet and repose. The styles the most appropriate and effective of all for a library are, the Elizabethan and the Gothic. In no situation does the deeply recessed bay, or oriel, window tell with better effect than in a library. The linings of the deeply recessed doors, and of the archways, should be paneled and parqueted; the stained glass windows admitting a quiet and reposing light, however rich, or even gorgeous, their colours may be. The ribs on the ceiling should form it into recessed panels, which may be designed to any degree of richness; with the larger and smaller bosses formed of foliage, fruit, and flowers, all subdued and made to harmonise, as before observed, with the prevailing tone of colour in the room. The doors should be hung with hinges that work in sockets, to permit them to be opened without noise. This method will allow of their being lifted off, and put on again, without removing the hinges. The locks should be of the very best construction. Those of Chubb, and also those of Mordan, are excellent. The windows, if the library be of large dimensions, should be placed on both sides of the room; and it should likewise receive light from lanterns above the roof; otherwise, if deep recesses, leading up to each window, should be formed for the reception of the books, so much shadow would be thrown across the room, that the centre of it would be gloomy; and, although a library should be quiet and sober in its style and effect, it should on no account be rendered gloomy and cheerless. The books in every library, be it ever so small, should be classified, so that they may be referred to readily; and a catalogue of them should be made, when the collection is large, so as to afford not only a list of the books, with the names given them by their authors, but directions for finding every work the library contains on any given subject. A list of authors, alphabetically arranged, should be added, with the names of the books written by each, and indicated by the number in the catalogue. Maps, in cases, on spring rollers, placed at a convenient height, so that they may be readily examined, or merely mounted in cases and hung up, are almost essential to every library; and if on spring rollers, they may, where there is little room, be hung up in front of the books. Portraits of the authors most prized might be hung against the walls to cover any vacant space; and vases, busts, and statues on pedestals, might be introduced in very large libraries.

For members of either of the Houses of Parliament, it would be very useful to have the bookcases or shelves made with two sets of shelves, one behind the other; the front shelves having a boarded back, and a small ledge on them in front, to prevent the books falling off, and being hung with hinges, or on centres; and the back shelves arranged to receive the statutes, acts of parliament, and other voluminous works not usually bound handsomely enough to be exposed on the bookshelves. A very agreeable and convenient shape for a library in a mansion is the octagon, fitted up all round with bookshelves, the lower part dado high, projecting so as to give depth for large folio books, with marble slabs: four of the sides might be enclosed with glazed doors, for the more valuable books, and the other four sides fitted up with bookshelves only, without being enclosed. The door and fireplace might be, as it were, cut out of the shelves; or shelves might be fixed on the door to contain books, and appear like the other sides of the room. If the room were to be warmed by hot air, the fireplace would be unnecessary; and the room might be lighted by an octagon lantern light; but if it be preferred that the room should be warmed by an exposed fire in a grate, the bookcases may be continued above the chimneypiece, so that the cornice over them may be uninterrupted. Over these bookcases may be placed the busts of celebrated authors, vases, models, &c.; and on the table in the centre of the room might be arranged models of celebrated buildings, and some fine classical groups of statuary and other works of art. Under the table it might be well to lay a carpet; but the floor, generally, should be uncovered, and either of oak, or painted in imitation of oak, and varnished.

In small villas, the library is frequently an addition to the house on the ground floor, without any room over it; in which case it is much better lighted by a skylight than by windows; because the walls can then be entirely covered with bookshelves; and the light coming from above is much better for the eyes. Such a library is much more suitable for a place of study, than one with windows on the sides of the room; because the objects seen from these are apt to distract the attention of the student. Hence it is, that, when the library and the breakfast-room are combined, the room cannot be considered as a place of study; but merely as one for light, or occasional, reading or writing.

Music Room. It is seldom that a room is set apart entirely for music, except in very large houses; but, where this is the case, the fittings up and furniture of the room should, as far as possible, be composed of sonorous materials. The form of the room may be octagon or circular, or simply a cube; because in that case the sound will be nearly equal in every part. The most perfect construction would be, to have the walls, and even the ceiling, lined with wood; but, if not lined, the walls should be battened, and then lathed and plastered as smoothly as possible, for painting. The ceiling should be coved, or formed into a dome. We have known an instance of a domed ceiling in a music room being lined with copper, on account of the sonorousness of that metal. The floor should be of oak, rubbed bright; or, if this be objected to, it may be covered with oil-cloth, which may be painted with a handsome Turkey carpet pattern. If the room can be heated without an open fireplace, it will be better, as much of the sound is otherwise carried up the chimney. The walls may be painted of a fawn colour, in panels, with wreaths enclosing groups of musical instruments; and the ceiling may be painted in compartments to correspond. The carpet (if there be one) should be as thin as possible, and may be of a fawn colour, with a rich pattern in crimson, and a handsome border, to disguise the thinness of the material. The curtains may be of crimson silk, without draperies, supported on large rods of brass, with handsome knobs. The chairs and sofas should be without cushions, with the frame-work of ebony, and the seats of cane, gilt, as cushions are found to deaden sound. There should be one or two long rosewood tables, placed against the walls; and music-stands, tripods for lights, and musical instruments, according to the taste and habits of the occupants.

A Museum may, in the case of the particular taste of the occupier, be considered desirable in a villa. If it is for a herbarium, or for specimens of woods, or seeds, or other parts of plants, or for shells, minerals, or fossils, it may be adjoining the library, or on any part of the same floor that may be found most convenient: but, if it is for stuffed birds, or other zoological preparations, it should either be in a room distinct from the other apartments; or separated from them by double doors, with means provided for the thorough ventilation of the space between them, in order to prevent the disagreeable smell, which there always is in such museums, from being diffused through the house. For the same reason, the museum may be placed on an upper story, as smells being always of a volatile nature, and generally lighter than the atmospheric air, seldom or never descend. Where the museum forms an important feature, it is best placed in a wing, separated from the main body of the house by a corridor or a veranda; because in that case there can be no danger from smells; and, while it is separated by the covered passage, it may be approached at pleasure during every kind of weather. Every museum should have a fireplace or a stove; and, if it can be lighted from the roof, as well as from one or more of the sides, it will be more favourable for ventilation, as well as for giving room along the sides for cases and shelves. Whatever may be the articles kept in a museum, they ought to be placed on shelves, enclosed in presses with glazed doors; and the glass of such doors should always be plate glass, in order to admit of a clearer view of the objects through them.

A Green-house, Orangery, or Conservatory, ought, if possible, to be attached to every suburban residence. The custom of rearing plants in pots, and keeping them in the windows of dwelling-houses, is of great antiquity; though it is only in modern times, and chiefly since the days of Louis XIV., that a house for plants has become a conspicuous feature in the elevation of a mansion. The most ancient description of plant-house is what is called an orangery; in which, formerly, orange trees, planted in large boxes or tubs, were kept during the winter, and set out of doors during the summer, season. Such houses almost always fronted the south: the back wall was of masonry, the roof covered with slates, tiles, or lead, and the front contained a range of large glass windows. Beneath the floor there was sometimes a flue for heating; and at other times this purpose was effected by means of German stoves. As the object of the orangery was merely to keep the trees from the frost, and they were not expected to grow while in the house, this description of building suited them perfectly; and it might still be very properly added to a mansion, provided no other plants were placed in it than orange trees, and a few other evergreen trees or shrubs, and succulent plants; such as myrtles, olives, cactuses, agaves, aloes, &c. Such houses, however, are totally unfit for plants which grow or flower in the winter season; such as camellias, heaths, acacias, and all those Cape and Australian trees and shrubs which, by their flowers and newly produced foliage, constitute the great charm of British conservatories during the winter months. To render an orangery fit for keeping such plants, it is necessary that the roof should be entirely of glazed frames, to admit perpendicular light, without which no plant in a growing state can thrive; and when this is the case according to the common usage of gardeners, the building is no longer called an orangery, but a conservatory, a word which appears to have been first applied to plant-houses by Evelyn, in his *Calendarium*. A further alteration or improvement in such houses consists in forming beds of earth in the floor, and planting the trees and shrubs in them, instead of keeping them in tubs and boxes. This, indeed, is the description of the modern conservatory, which is almost the only kind of plant-house now attached to first-rate mansions. The term green-house is now generally confined by gardeners to houses having glass roofs, which are kept at the same temperature as the orangery or conservatory, but where the plants are grown in pots, which are usually small, and elevated upon stages, so as to bring them at once near the light and near the eye of the spectator. The characteristic of a conservatory is, that it grows a few plants to a large size, and so as to produce scenery of a magnificent exotic aspect ; while that of a green-house is, that it produces a great many different kinds of plants, of small size, which may be considered as merely living botanical specimens of exotics. The green-house is, consequently, much better adapted for the smallest description of suburban residences than the conservatory; and a modification of the green-house, which may be called a plant cabinet, or cabinet green-house, in which a few choice plants are kept, and always taken away and renewed as they begin to fade, is, perhaps, still more appropriate. What are called plant-stoves, tropical plant-houses, or hot-houses, in the proper sense of the word, are unfit for being attached to dwelling-houses, from the great heat and moisture required to render their atmosphere fit for the plants of hot climates.

The enjoyments afforded by a green-house, however small, to the female part of the family are very considerable; and, where there are children, these enjoyments may be mingled with useful instruction, by teaching them in it the names and nature of plants, and their culture and management at seasons, or during weather, when it cannot be done out of doors. A green-house also affords exercise, in shifting, potting, tying up, pruning, &c., in cold and wet weather, and at periods of the year when nothing can be done in the open garden. At the same time that we recommend a green-house, it is proper to state that, where the mistress of the house has not a taste for plants, and is not in the habit of working among them herself; and where this taste does not exist in any part of the family, and no gardener is kept; a green-house is in danger of becoming a nuisance, rather than an ornamental appendage. In such a case, where it is determined to produce the effect of a green-house, for the sake of fashion, or the reputation of being fond of plants, or some similar motive, the best mode is to engage with the nearest nurseryman or florist to keep the green-house furnished with plants, at so much per annum or per month. By this means it will always look well; but, as none of the beauty which it presents will be the result of the care and attention of any part of the family, of course the enjoyment derived from it cannot be anything like so great as where the contrary is the case.

In the smallest description of suburban houses, the common substitute for a green-house is the window-sill; and the greatest extent to which this kind of gardening can be carried is by having the sill made to project 2 ft. or 3 ft. from the wall of the house, and enclosing it with an outside bow window. Into the space between the two windows, the warm air of the room may be admitted at pleasure; and, if the panes of both windows are large, and kept at all times perfectly clean, the view into this plant cabinet from the interior of the room will be agreeable, and create an allusion to the green-house of the villa, or the conservatory of the mansion. For houses that have a garret, a sort of green-house may be established there, by forming glass windows in the roof. An enthusiast amateur might, indeed, have the roof of his house entirely of glass, and train vines or creepers under it; which might be planted in the ground, and their stems brought up against the outside wall, and covered with a wooden case. In such roofs, the panes of glass should not be more than 2½ in. or 3 in. wide, or plate glass should be used, in order to prevent breakage from hail. The most fitting suburban residences for having green-houses are such as are either quite detached, or in pairs; which last-mentioned houses, in the neighbourhood of London, are called double detached houses. In cases of this kind, whether the houses are of the fourth rate or the first rate, they may always have a plant-house of some kind attached to them. One of the most ordinary modes of connecting a green-house with a small house is by placing it against the gable end; it being understood that this gable end fronts the south, the south-east, or the south-west: though, even if it fronts the direct west or direct east, such a green-house will answer for many kinds of green-house plants; and for all kinds whatever, with an extra-allowance of fire heat during winter. For double cottages or houses, where the gable ends front the south-east and north-west, a green-house may be placed against each; but where the one gable fronts the south and the other the north, then the one green-house should be placed on the east side of the house, and the other on the west side. In single houses, the green-house may be placed in a great variety of ways, and may be of many forms, as will be seen hereafter, according to the arrangement of the ground plan, and the style of elevation, of the house. In whatever manner a green-house, or plant-house of any description, is attached to a house, means ought always to be provided for warming, ventilating, and watering the plant-house, altogether independently of the dwelling-house; for few things are more disagreeable and unwholesome to human beings, as well as injurious to furniture, and the walls of the room, than the close damp effluvia from the earth, water, and plants of a conservatory. For this reason, the plants grown in conservatories immediately attached to drawingrooms should be such as are natives of very dry climates (for example, the Cape of Good Hope, Australia, &c.), and, consequently, require very little water; and the gardener should contrive to give his waterings either late in the evenings, or very early in the mornings, when there is no chance of the conservatory being in use by the family. Previously to the hour when it is expected the family will walk in the conservatory, it ought to be thoroughly ventilated, so as to carry off the damp; and the surface of the ground ought never to be kept very moist, in order to produce as little evaporation from it as possible.

In large houses, few objects connected with them produce a more splendid effect, or contribute more to luxurious enjoyment during winter, than a large well kept conservatory. It should be of considerable length and breadth; at least twice the length of the drawingroom, and broad rather than narrow; with a glazed roof, and with both sides glazed, if the length is in the direction of south and north; but with a wall on the north side, if it is in the direction of east and west. It should be separated from the drawingroom by glass doors or windows, opening down to the floor: the piers between these ought to be as slender as is consistent with strength, and the style of architecture of the house; while the glass should be in very large plates, and, if possible, one plate only to each window or door, so as to give the idea of the drawingroom and conservatory forming but one room. A drawingroom, in the form of a pentagon or an octagon, projecting from the body of the dwelling-house, might be enclosed, on every side but that on which it is entered from the house, by a conservatory; and this, if 100 ft. deep, and from 30 ft. to 40 ft. in height, with a broad walk radiating from the window in the centre of each of the outer sides of the drawingroom, would render the allusion to a summer-house in an Eastern garden complete. Some of the paths might be covered with gravel, and others with turf; the gravel being kept perfectly smooth and firm by frequent rolling, so that no part of it might stick to the shoes, and be carried into the drawingroom. In order that the gravel paths may be kept perfectly dry, the flues and hot-water pipes by which the house is heated should be carried under them; and, to contribute to the same end, when the plants are watered with the syringe, the walks should be covered with canvass. While this attention to dryness would contribute to the enjoyment derived from the conservatory in the winter season, a certain kind of attention to moisture might be made to contribute equally to enjoyment during the spring and summer months. What we propose is to be effected by producing artificial rain, in the manner invented, successfully carried into execution, and practised for the last ten years, by Messrs. Loddiges, in their magnificent palm-house at Hackney. The mode by which artificial rain is produced is both simple and unexpensive; requiring merely that lead pipes should be conducted horizontally close under the glass roof, pierced with small holes, and supplied from a cistern at a higher level. The holes may be pierced in the pipe with a common sewing needle; and the distance of the pipes from one another may be from 8 ft. to 12 ft., according to the height of the roof above the heads of the plants. The water can be turned on and off such pipes at pleasure, by stop-cocks properly placed. All the details for executing a work of this kind will be found given in a future page (along with various designs for green-houses and conservatories, adapted to suburban gardens); but, indeed, they are so simple and obvious, that any plumber may carry the idea into execution from merely having it suggested to him. In warm summer evenings, when the weather is dry and sultry without, to sit in a drawingroom, and see and hear a shower falling in the conservatory, cannot fail to impart a sensation of refreshing coolness, as delightful to the spectator as it is invigorating to the foliage and roots of the plants. In the case of all conservatories which are immediately connected with the drawingroom, a superior gardener ought to be kept, with abundance of assistance; otherwise the plants cannot be maintained in that high degree of order and keeping, which is essential to the effect of this appendage, and without which, as we have already observed, it becomes a nuisance, great in proportion to its pretensions.

For this reason, where a first-rate gardener, and abundance of hands, are not kept, we would recommend the conservatory to be separated from the drawingroom by a lobby; in which there should be openings, so arranged as to secure a constant ventilation: or it may be connected with the house by means of a corridor, or other covered way; in which case, however, it is seldom completely seen from the drawingroom; and, consequently, a great luxury of the conservatory, that of looking down its main walk from the drawingroom window, and feeling as if we were sitting in a bower in an Eastern garden, is lost.

From these remarks respecting a green-house or conservatory, the reader will be able to appreciate such as he may see appended to a house which he intends to purchase or hire; and he will also be able to form some idea as to how a conservatory or green-house is to be placed, when he is building a house. In suburban houses which have been for some years built and occupied, the conservatory, or green-house, will generally be found to be neglected and out of repair; with the plants drawn up for want of air and light, and eaten up with insects for want of attention; or, perhaps, only the more common and hardy kinds will be found to occupy the house. The flues, also, or whatever mode of heating has been adopted, will be found out of order, and unfit to act; and all the gutters and water-pipes stopped up; and, perhaps, great part of the glass in the roof broken. This evil, to which a great number of plant-houses with glass roofs are subject, might be materially mitigated by two features in their construction, to which we would beg the reader to give his most particular attention.

First, the slope of the roof ought never to be less than 45°; and, in order that this slope may not be carried so high as to be too conspicuous over the parapets, the roof of every wide house ought to be in several spans, with wide gutters between. This is beautifully exemplified in the magnificent conservatories at the Grange, of which a section will be found in the first volume of the *Gardener's Magazine;* and also in that at Woollaton Hall, near Nottingham; and in several other conservatories put up in different places by Messrs. Clark and Jones of Lionel Street, Birmingham; and Messrs. Cottam and Hallen of Winsley Street, London.

The second feature in the construction of every description of plant-house roof, to which we would direct attention, is, the advantage of having the glass in very small panes. When the panes of crown glass are 8 or 9 inches wide, they cannot be made air-tight without a lap of from a quarter to half an inch; and, whether this lap is puttied or not, the water, in the course of a year or two, is apt to lodge in it, and in winter to freeze, and, by its expansion, to break the glass; while, in summer, such panes are liable to be broken by very heavy hail showers. On the other hand, when the panes are not wider than from 3 in. to 5 in., they may be made air-tight by a lap of from one sixteenth to one eighth of an inch, without the aid of putty; and, consequently, the laps can never get filled with dirt, so as to obstruct the light; and can never retain water so as to freeze, and crack the panes. Another great advantage in having panes of this small size is their superiority of strength; in consequence of which, no hail that ever falls in this country, or, we believe, in any other, will break them. Roofs glazed in this way doubtless require a greater number of sash-bars, and hence a good deal of light is excluded; but, if the bars are made of wrought iron, the proportion of light excluded will not be much greater than in the case of glass roofs with the broad black laps usually found when the panes are 8 in. wide. Admitting, however, that more light is excluded by the use of small panes than by the use of large ones, still the advantage of the former, in point of strength for resisting hail, and in the impossibility of their being broken by frost, is so great, that we should never for a moment hesitate in giving them the preference. There is yet another advantage attendant on the use of small panes; which is, that their first cost is not half that of the large ones. The use of small panes has now become general among British nurserymen; and it is one of the greatest improvements that have been made in glazing roofs, since they were first introduced. In Germany, there being no duty on glass, the panes in hot-houses used formerly to be made from 1 ft. to 18 in. wide, and, unless covered, they were broken by every hail storm. The consequence was, either that no person could have a green-house, who could not afford to keep a gardener to be in readiness to cover the roof on the first appearance of a storm; or that, if he did venture on a green-house, the roof must be opaque, and only the front and ends glazed. Of late years, however, small panes have been introduced throughout the Continent; and the glass roofs in which they have been used at Vienna, Berlin, and even at Copenhagen, are found to resist hailstones as effectually as they do in England. These small panes are now also becoming common in the green-houses of France, Belgium and Holland.

On the supposition that Kyanised wood is as durable as it is said to be, we would recommend all the wood used in plant-houses, after having been accurately fitted together, to be Kyanised; in which case it need not be painted, except for ornament, or to give uniformity of colour. It is a sufficient recommendation of this practice to add that it is adopted by the Messrs. Loddiges, at Hackney. (See *Gard. Mag.*, vol. xiii. p. 281.)

SECT. V. *On the Choice of a House, with Reference to its external Appearance and architectural Style.*

MANY persons, who have not had much experience in the choice of a house, are captivated by the exterior; and are more influenced by its picturesque effect, than by any property in the dwelling connected with habitableness. One person is an admirer of the Gothic; without considering that, unless the number of windows in such a house is greater than in a building in the Roman or Italian style, the rooms will be ill lighted, from the thickness of the mullions, and the smallness of the panes; and probably, also, not well ventilated, from the defective manner in which Gothic windows generally open. Some prefer a cottage with latticed windows, and surrounded by creepers; not considering that the rooms in such buildings are very frequently low, ill lighted, and ill ventilated; the floors subject to the dry rot, and the walls to damp: for, notwithstanding the beauty of flowering creepers, there is not a single kind of creeping plant, which will grow against a house, that does not make the walls damp, with the single exception of the ivy. Some prefer a house with a veranda all round it; and, no doubt, such an appendage will keep the house warmer in winter, and cooler in summer, and will afford a dry walk during rain; but those who take a lease of a house with a veranda, which they are, of course, bound to keep in repair, should first well consider if it admits sufficient light to the rooms on the ground floor; and, secondly, whether it is well put together, and made of durable materials. Some houses which are surrounded by arcades, and which are exceedingly handsome and architectural externally, are yet more gloomy within than their possessors would, perhaps, like to acknowledge.

A thatched cottage is an object of admiration with many persons who have not had much experience of country life; and, accordingly, we find several in the neighbourhood of London. Such cottages have, perhaps, the gable end covered with ivy, the chimney tops entwined with Virginian creepers, and the windows overshadowed by roses and jasmines. The ivy forms an excellent harbour for sparrows and other small birds, which build there in quantities in spring and early in summer, and roost there during winter. In June, as soon as the young birds are fledged, all the cats in the neighbourhood are attracted by them, and take up their abode on the roof of the house every night for several weeks; the noise and other annoyances occasioned by which we need only allude to. We say nothing of the damp produced by the deciduous creepers and the roses, as we have already mentioned that: but we must here notice another evil, which is not so obvious, though quite as serious, and this is, the numerous insects generated in the decaying thatch; and more especially that loathsome creature, the earwig, which, in autumn, whenever the windows are open, comes into the house in quantities, and finds its way into every closet, chink, piece of furniture, and even books and papers. All cottages of this kind harbour snails and slugs in the ivy, and spiders under the eaves of the thatched roof; and wherever there are spiders, there are also abundance of flies. As there is always a garden attached to such cottages, it is almost certain, if on a clayey soil, to abound in snails, slugs, worms, and, if the situation is low, perhaps newts. Some of these, from the doors, or, at all events the back-door, being generally kept open, are quite sure to find their way not only into the kitchen, but even into the pantry and cellars. Slugs, when very small, will enter a house through a crevice in the window, or a crack in the door; find their way to the moist floor of the pantry or the cellar, and remain there for weeks, till they are of such a size that they cannot retreat. There are few persons indeed who do not experience a feeling of disgust at seeing the slimy traces of a slug in any part of their house, not to speak of finding them on dishes in which food is kept, or even on bread; or at discovering an earwig in their bed, or on their linen. The kitchen, in low damp cottages of every kind, almost always swarms with beetles and cockroaches, and the pantry with flies; while, from the closeness and want of ventilation in the rooms, it is almost impossible to keep fleas, &c. from the beds. If a large dog be kept in or near the house, as it frequently is, or if a stable or cow-house be near, the fleas from the dog, the horses, or the cows, which are larger than the common kind, will overspread the carpets, and find their way to the sofas and beds. Having lived in cottages of this kind ourselves in the neighbourhood of London, we have not stated a single annoyance that we have not ourselves experienced; and we have purposely omitted some. Two of these, offensive smells and rats, are the infallible results of the want of proper water-closets and drainage; but these evils, great as they may seem to be, are much easier to remedy than the others already mentioned, which are, in a great measure, inseparable from the kind of house. Two others, the danger of setting fire to a thatched roof, and its liability to be injured by high winds, are sufficiently obvious; but it would hardly occur to any one, who had not lived in a house of this description in the neighbourhood of London, that a thatched roof is, of all roofs the most expensive, both when first formed, and afterwards to keep in repair. A plumber or a slater, to repair a lead or a slate roof, may be found every where in the suburbs of large towns; but a professional thatcher must be sent for from the interior of the country. For example, the nearest cottage thatchers to London are in the Hundreds of Essex on the east, and in Buckinghamshire on the west.

We have dwelt more particularly on the evils incident to a thatched cottage, because in it all cottage annoyances exist in an extreme degree; but the truth is, that all cottages which have not their ground floors so much elevated above the surrounding surface as to be perfectly dry, and their rooms lofty and well lighted and ventilated, are subject to the same evils, though not quite to so great an extent.

Notwithstanding all that we have stated, we do not recommend our readers never to take a thatched, or other fancy, or ornamental, cottage; we only wish to point out the inconveniences and extra-expense to which their doing so will render them liable. We think we may safely assert that the same family that would want two servants under ordinary circumstances, would require three in a cottage of the kind we have been describing.

What, then, it may be asked, is the kind of suburban house least liable to these inconveniences? To this we answer, one that stands high, dry, and free; that is compact in its general form; that has the diagonal line of its general plan south and north, so as to obtain the sun on every window, on some part of every day that it shines, throughout the year; or, in other words, that has no front or side pointing directly either east, west, north, or south; that has the rooms, and especially the kitchen story, lofty, and well lighted and ventilated; that has a secure roof of slate, lead, or flat tiles; few creepers on the walls; and that is not choked up with trees and bushes. These conditions being complied with, the architectural style of the building may be left to the taste of the occupant.

As the cubic form is known to enclose more space with the same quantity of walling and roof than any other, so it is an established rule, that a house square in the plan is preferable in all that regards comfort, habitableness, and economy of heating, keeping clean, and in repair, to one which is irregular in its plan. The next best form to a square is that of a parallelogram; and the worst form that can be adopted is that of a long, narrow, irregular building. A square house is more compact within, and, from its form, it is warmer in winter and cooler in summer than any other; it is more easily heated; it has less space occupied by passages, and is, consequently, more easily cleaned; and, externally, it exposes less surface to the atmosphere, and is, consequently, more easily kept in repair, than any other. When economy is the main object, therefore, a square house ought to be chosen; and, that it may combine architectural beauty with economy, both in first cost and in future management and repairs, one should be chosen in which the same description of brick or stone, the same style of workmanship, the same magnitude, kind, and disposition of windows, the same facings to them, the same kind of cornice, and, in short, the same architecture, is adopted on all the four sides. Above all things, as a matter of taste, a house ought to be avoided which has any one of its sides decidedly inferior to the rest, in respect either to architectural design or execution. We should say, also, avoid, in point of habitableness and comfort, every house, the diagonal line of the general plan of which is not south and north; were it not that this maxim would condemn all those houses which have been built along, and parallel to, streets or roads which run directly east and west, or north and south. Unfortunately, the custom of placing suburban houses that are near streets or roads, with one of their sides parallel to that street or road, and without any reference whatever to its direction, is almost universal, even where there is a distance of 100 yards or more between the road and the house, though it is productive of two serious evils, which admit of no remedy. The one is, that the opposite side, or front, of the house to that which faces the road is considered as the back, and is, therefore, generally designed and finished in an inferior style: and the other is, that no attention can be paid to placing the diagonal line of the plan of the house due south and north; and that, whether this is the case or not, depends on the direction of the road, and not on the will of the builder. The latter is much the greater evil; for so numerous are the advantages of this disposition of the plan, in point of solar light, warmth, ventilation, and cheerfulness, and even dryness and healthy vegetation in the garden or adjoining grounds (see p. 23.), that, in our opinion, it ought to be made the governing principle in the placing of every detached house, whatever may be the direction of the street to which the house may be said to belong. In the suburbs of towns, according to the present parallel mode of building, the only houses that are properly placed, relatively to the sun, are those along streets which run from north-east to south-west; and from south-east to north-west. Near London, where houses are built of brick, and where different kinds of brick are employed in the same house, it would be difficult to find a suburban house of moderate size, in which an inferior kind of brick was not used on the back front, and even on the sides; but the bad taste of this mode of building only requires, we are convinced, to be pointed out to the occupiers and builders of suburban houses, to cause it to be avoided. It has, no doubt, originated in the practice of building street houses, in which the best bricks are always used on the side next the street; and which practice the town builder employed to build in the country has not been able to separate from his mind. We may observe here, incidentally, that, where houses are built in continuous lines, the sides of one house forming the sides of those adjoining it, the sun can only shine on the two exposed sides; and, that it may shine on these exposed sides, it is necessary that the line of the houses should be diagonal to a square formed by the cardinal points, as recommended above for single houses. Hence, no streets in towns ought either to be due east and west, or due north and south; and, as this arrangement would not hinder all the streets from being at right angles as at present, the greatest advantages in point of light and ventilation would result from it, without any disadvantage whatever. It is surprising to us that this disposition of the streets has not been attended to, as a principle, in the laying out of new towns or villages

We know it will be said, by some architects, that a square house affords less architectural beauty than any other form, from the sameness of the general shape; but this is partly a mistake, and is chiefly believed to be true by such as consider variety to be one of the main beauties of architecture. Now, our opinion is that variety, however prominent a beauty it may be in landscape, is only a subordinate one in architecture; and that the grand characteristic beauties of that art are magnitude and symmetry. We would not exclude variety; on the contrary, we would produce it to as great an extent in the details as was consistent with symmetry; and, of any two buildings equal in magnitude and symmetry, we should say that the one which possessed the greater amount of variety was the most beautiful.

Two great beauties in architecture, and without which an

edifice can hardly merit to be called architectural, are the appearances of solidity and of magnitude. Now, no form whatever gives the appearance of solidity to so great an extent, in proportion to the quantity of matter employed, as the square or the circular form. Both these forms are equal in point of solidity; but the square (the quantity of matter being the same) has greatly the advantage in point of magnitude, by presenting, especially when two sides are seen at once, a much greater surface to the eye. A building in the form of a parallelogram, if seen only in front, may have the appearance of magnitude; but, when one end is seen, and that is found to be narrow, the impression of the want of solidity is immediately felt in a high degree, and the impression of magnitude is proportionately diminished. Hence, a building in the form of a parallelogram, in whichever manner it may be viewed, is never so satisfactory as one in the form of a square, or cube. Now, that a square, or cubic, building not only possesses, by the nature of its form, the essential and fundamental architectural beauties of solidity, magnitude, regularity, and symmetry, but also may be made to display the accessary beauties of variety, harmony, character, and style, might be proved by various examples; but it may be sufficient here to refer to the Gothic and Elizabethan mansions of former times; the noblest of which, and those which make the greatest figure in the history of our domestic architecture, were almost always built either in squares or quadrangles, or in the form of three sides of a quadrangle.

In choosing a house that is irregular in the plan and elevation, there are some advantages, but many disadvantages. The advantages are, that, in particular situations, particular distant prospects may be better displayed; and that in all situations, even on a flat surface, a greater variety of home views, that is, views within the grounds, may be created. This, it is obvious, is to be effected by placing the rooms in such a way that the principal windows in them will look on the view at a right angle. The disadvantages of an irregular house are, that it is always colder in winter, and warmer in summer, than a square house, from the exterior surface of almost every room being exposed to the weather on two or more sides; whereas, in a square house, only the corner rooms are exposed on more than one side. The chimneys in an irregular house do not draw so well, because the greater part of them are in the outside walls. On account of the great surface both of walls and roof, and of the greater number of gutters in the latter, it is clear that irregular houses must cost more at first, and require more to keep them in repair, than square, or parallelogram, houses; the quantity of decoration on both being equal.

In choosing a house, with regard to style, the Roman, or, as it is commonly called, the Grecian, is obviously preferable to the

Gothic, from its greater compactness, and from its having comparatively few ornaments. The Roman may, indeed, be called the regular, and the Gothic the irregular, manner of building. The Roman also deserves the preference on another account; viz. that all the interior finishings and furniture, necessary to correspond with that style, are of a less expensive description, and are more easily kept clean and in repair, than similar articles in the Gothic style. In the present day (1837), from Gothic furniture not being in general use, it is always less easily procured, and dearer, than Grecian or Roman furniture; and the same may be said of furniture in other less common styles or manners; such as the Louis XIV. manner, &c. It is but just, however, to add, that there are some modifications of the Gothic style of building, which admit of being employed in cubical masses, and with almost as little ornament as is required in Roman or Italian edifices; and the kinds of Gothic furniture adapted to such houses are also comparatively simple.

Sect. VI. *On the Choice of a House with Reference to the Taste displayed in its interior Fittings-up, Finishing, and Furnishing.*

The interior Finishing of a house affords, in general, a very good criterion of the skill and taste of the architect by whom it has been designed, and under whose inspection it has been carried into execution. It has long been too common for the employers of architects to consult them only on the general arrangement and external elevation of a house; neglecting, in a great measure, the finishing of the interior in such a manner as to fit every room for the furniture peculiar to it, and for no other furniture. As we have enlarged on this subject in our *Encyclopædia of Cottage Architecture*, we shall not here repeat anything that we have there stated, but proceed at once to point out some of the most glaring errors in the interior finishing of moderate-sized houses, in almost every part of Britain.

Breaks in the Walls. On entering the sitting-rooms of almost every house in London, or its suburbs, of from 40*l.* to 80*l.*, or even 100*l.*, a year, and where the rooms are from 12 ft. to 16 ft. square, or from 20 ft. to 24 ft. long, by from 10 ft. to 12 ft. wide, the end containing the fireplace presents almost invariably the appearance shown in *fig.* 23. Here we have the projection, containing the fireplace of the room and the flue of the chimney from the kitchen or other apartment below, not quite in the middle of the side of the room; which is the effect of carelessness on the part of the architect, when designing the direction of the courses of the flues of the different chimneys, and which direction, in the case of many buildings erected by builders, without any architect being employed, is very frequently left

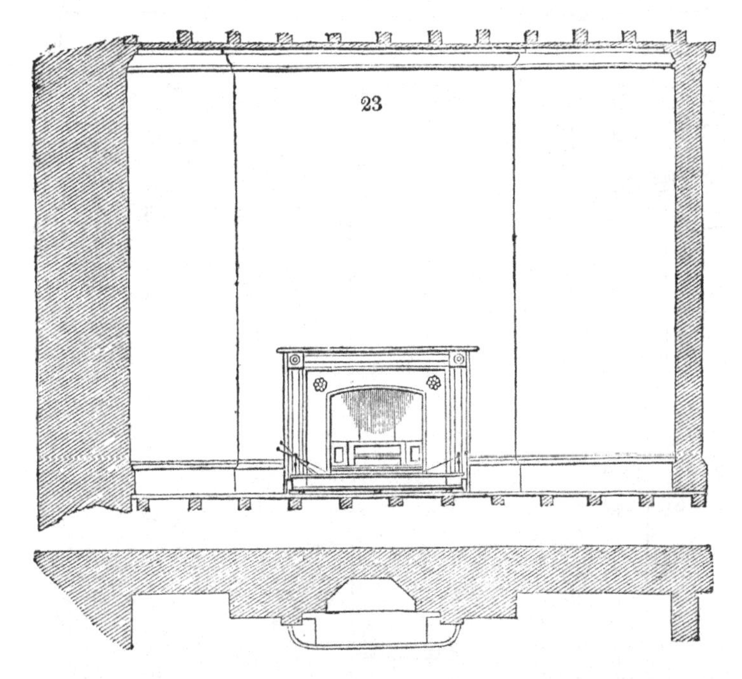

to the operative bricklayers. The recesses on each side of this projection, if no piece of furniture is to be placed in them, may, in other respects, pass without censure; but, if they are to contain bookcases or tables, there will be an obvious want of fitness to an architectural eye, not only in the bookcase or table required to fill one side being of a different size from that required to fill the other side; but on account of the bookcases or tables fitting against the chimney pier on one side, and not against any pier on the other.

Recesses and Breaks. Fig. 24. shows another very common mode of finishing rooms in the kind of houses described. Here, we have the exterior wall (*a*) with a window (*h*) having its jambs battened out beyond the thickness of the wall, as shown at *b*, so as to produce a sufficient projection for containing the boxing for the shutters; while, to save the expense of battening on the remaining part of the wall, the recess *c* is formed. Between this outside wall and the projection containing the chimney, there is another recess at *d*. Now, though this is not very architectural, because the small pier at *e* is not symmetrical with that at *f*, yet, if no marked piece of furniture, such as a table, bookcase, musical instrument, or sideboard, is to be placed in either of the recesses, they may be allowed to pass without censure: but imagine any of the pieces of furniture mentioned introduced in

24

either of these recesses, and the want of fitness of the piece of furniture so introduced, to the situation in which it is placed, will be in a moment obvious. To adapt both recesses for con-

25

taining furniture, the small pier at *e* should be enlarged, as shown by the intersection of the dotted lines at *g*; in which case bookshelves might be introduced in the one recess, and a pedestal sideboard in the other, as shown in the plan and perspective elevation forming *fig.* 25. In this figure, it will be observed that the angles of the piers are taken off, as shown in the plan at *h h h*, which not only has the effect of separating the bookcase from the sideboard more definitely, but of producing a sort of architectural character in the supports to the lintels over the recesses.

The Chimneypieces, in such houses, have generally somewhat of the appearance of *fig.* 26., which, it must be allowed, when considered as a mere assemblage of lines and forms, is very neat; but which, when considered architecturally, is an absurdity. The two sides, or jambs, of a door, window, or chimneypiece, are

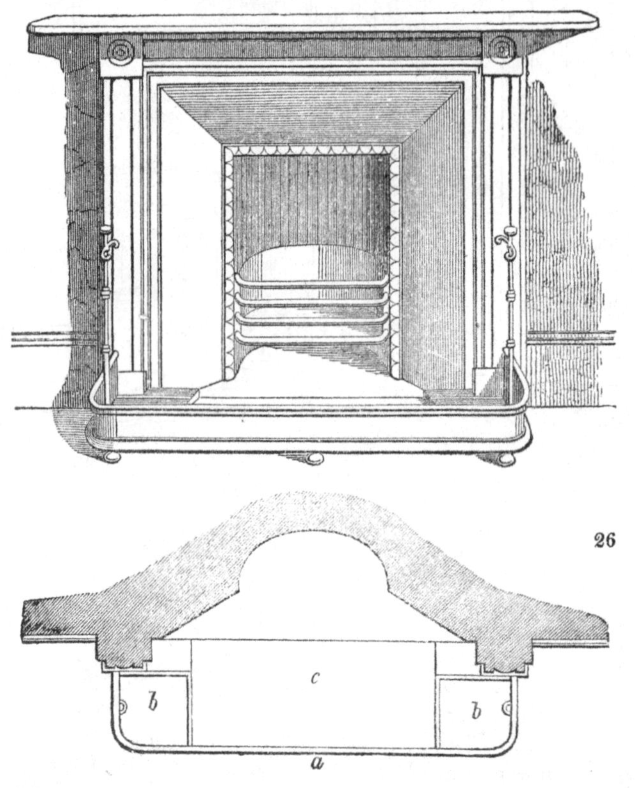

placed there, according to the architectural system, for supporting the lintel, or horizontal piece that connects the two; and, whether they really do support the lintel, or not, they ought at least to have the appearance of doing so; but in such chimneypieces as that shown in *fig.* 26., instead of supporting the lintel, they only support a small square tablet on each side, while the lintel is left suspended in mid-air between them. The obvious remedy for this evil is to dispense with the square tablet, and to introduce a plain pilaster and architrave, as in *fig.* 27. Of two chimneypieces, the one in the taste of *fig.* 26., and composed of the most

27

valuable marble, and the other in the form of *fig*. 27., and composed of common stone, the latter is by far the higher in point of architectural rank; because, in all the plastic arts, form is higher in the scale than material: the one argues mind and taste, and the other merely wealth and power.

The Architraves of the Doors and Windows, both internal and external, and also the frames of mirrors, and other pieces of furniture, are frequently finished in the same manner as the marble chimneypiece in *fig*. 26.; viz. with the jambs supporting only square tablets; and this, in exterior architecture more particularly, is very offensive to a mind imbued with architectural knowledge and feeling. In some streets in London, elevations may be found, not only of street doors and shop fronts, but where even the pilasters, which form the main feature of the elevation of the entire house, support nothing more than square tablets.

The Manner in which Fenders are fitted to Fireplaces is particularly unsatisfactory in almost all houses; even those which, in other respects, have considerable pretensions to good taste.

Modern fenders are badly fitted, by their iron rims abutting abruptly on the plinth, or lower part, of the marble chimney jambs; and by the raised platform within, with its ledge for containing the lower part of the fire-irons, not fitting against anything. This kind of fender may be seen by turning back to *fig.* 26. in p. 124.; in which *a* is the outer rim of the fender; *b*, the space on which the lower extremities of the fire-irons rest; and *c*, the sloping iron plate in front of the fire. The awkward manner in which the raised ledge for enclosing the fire-irons protrudes itself, both in the plan and in the elevation, is obvious; and also the want of fitness in the upper bar of the rim of the fender abutting against the outer edge of the marble jamb, instead of fitting into, or against, some point or place adapted for it, and which would be incomplete without it. It will also be observed, in the same figure, that the height of the plinth of the marble jamb differs from the height of the rim of the fender, and also from the height of the plinth and wash-board of the room; whereas, to be architectural, it ought to have fitted to the top line of one or other of these members. In common fireplaces, there is a very simple mode of overcoming this architectural deformity: it is by projecting the two blocks seen at *d d* in the plan of *fig.* 27. in p. 125 (which may be of marble or stone, or of iron cast hollow); attaching them to the hearth by two wooden pins in the under side of the blocks, which should drop into two holes in the hearth; raising the supports to the fire-irons on these blocks; and placing a straight fender between them; as shown in the elevation of *fig.* 27. A very common fault in the construction of fenders, and in the position and height of the standards for holding the fire-irons, occasions the handles of the latter to strike or rub against the marble jambs; and hence we often see chimneypieces of black marble, with a portion of the jambs scarred with white. The remedy for the evil is to have the standards higher, or at a greater distance from the jamb, as in *fig.* 28.; or to adopt the plan of placing the fire-irons in an oblique position, as in the fire-place shown in *fig.* 23. p. 121.

Bellpulls, in the principal rooms, are generally intended to be rendered ornamental, but are too often only tawdry. So great a deviation from the spirit or system of architecture, as that exhibited by a long cord or a riband hanging down from the ceiling, ought never to be resorted to without a powerful necessity; and this necessity need never occur in bell-hanging. The wire should always be concealed, and only a knob or handle, by which it is to be pulled, exhibited; and this ought only to be done in an architectural manner. In some houses where this manner of fitting up bells is adopted, the knob is of brass, and is made no larger than one would be on which it was intended to hang a hearth-brush, or some similar article; but this kind of knob is unarchitectural. A rosette, as shown in *fig.* 28., or some equally architectural ornament, in the centre of which the knob is placed, is more according to art. The knob, or, indeed, handles to doors or drawers in the best rooms of a house, ought never to be of metal, but of ivory, or of ebony, or some other kind of ornamental wood.

Finishing. A good finish is an expression in very common use among mechanics and artisans. In order to be able to test the fitness of furniture to the parts of the room in which it is placed, the fitness of the finishings of the room to their situation, and the fitness of the various details, both of buildings and furniture, to one another, it is necessary to have a clear idea of the meaning and application of the word finish as applied to arts and manufactures. This word is as common among artisans and mechanics, as the word effect is among artists and dramatic authors. To resolve the expression, a good finish, into an abstract principle, we should say that it consists in definiteness; that is, that every part of the details of an object should be clear and obvious, and such as to leave no room for wishing any of them to be more in one way than in another; more in one place than in another place; higher or lower, broader or narrower; or, in short, in any way different from what they are. It is extremely difficult, however, for a person to comprehend clearly what is meant by a good finish in any art, without having a knowledge of the elementary details of that art. Thus, a joiner or a cabinet-maker understands perfectly well what is considered a good finish in joinery or in furniture, in which the details belong to classical architecture; because he has drawn on paper, and worked at the bench, all the different details, or members, as they are called, which belong to that style of art. But a person, neither a mechanic nor an artisan, and one who has not drawn on paper the different details of that style of architecture, will not be able to comprehend what constitutes a good finish in it, without some elementary instruction. In like manner, the mechanic or artisan whose studies have been confined to classic art will not be able to understand, with equal clearness, what is a good finish in the Gothic style, from never having gone through the necessary details. Further, such mechanics and artisans may know what constitutes a good finish in the details of Grecian, and Gothic architecture, without being able to detect the want of a good finish in the adaptation of furniture to the breaks and recesses of a room, or of a fender to a chimneypiece. The reason is, they have been too much occupied with the details, to have had any occasion to generalise on the subject of their extensive application. Perfectly to understand what is a good finish, therefore, requires not only a knowledge of the details of the art employed in the kind of production, but of the general principles of composition. Thus, a critic, to form a sound judgment on the finishing, fitting up, and furniture of a room, ought to unite the theoretic knowledge of the joiner and cabinet-maker with the science of an architect, the taste of an artist, and the power of a philosopher of generalising, not only on them, but on the purposes to which art is applied.

On Furniture and Furnishing, we add only a few general remarks, supplementary to those which we have already given, when treating of each room separately. The principle of order requires that the finest rooms should have the best furniture; and the principle of fitness, that the furniture should be adapted, in point of strength, plainness, or ornament, to the room in which it is to be used. In general, the furniture of the entrance porch, or hall, ought to be lower in the scale of design than that of any of the living-rooms, or even bed-rooms; but, on account of the uses to which it is applied, it ought to be more massive. Next in the order of art ought to be the furniture of the business-room, then that of the breakfast-room, next the library, then the dining-room, and then the drawingroom, with its adjuncts the music-room, and ladies' boudoir. The characteristics of the furniture of the principal of these rooms we have already noticed; but it may not be amiss to add, that, for the desks and presses of the business-room (where there is one), the bookcases of the breakfast-room or the library, the side-board of the dining-room, and the principal couches of the drawingroom, there ought to be appropriate recesses in the walls. In bed-rooms, also, there ought to be small recesses, indicating the proper situation of the bed, and of the commodes, dressing-tables, wardrobes, &c. The style of all bed-room furniture, where the room is only used as a sleeping-room, ought to be simple rather than gorgeous; because, where display forms no part of the enjoyment to be derived from a room, nothing can be more superfluous than finery in its decorations. On the Continent, where the bed-room is used for an apartment in which the lady of the house receives company, the case is quite different, and the furniture may be as ornamental, or nearly so, as that of the drawingroom.

Such of our readers, therefore, as wish to acquire a knowledge of what is right and wrong in the finishing of the interior of a house, and its fittings-up and furniture, ought to study the details of the style of architecture in which it is built; and such as wish to understand when the breaks and recesses of a room exhibit architectural fitness, can only acquire this knowledge by studying the general principles of architectural composition. Grown up persons cannot be expected to undertake either of these studies; but it is highly probable that, at no distant period, the principles of architecture, and the study of the composition of lines and forms, will be generally taught in public schools, and this will be the commencement of a new era in the taste displayed in building and furnishing.

SECT. VII. *On keeping a House in Repair.*

EVERY one who occupies a house, either on lease, or which is his own property, must necessarily be aware that the annual expenses of repairs are very considerable; so much so, indeed, that, in London and its suburbs, it will invariably be found much cheaper to live in unfurnished lodgings, at a much higher rent than would be paid for the same number of rooms in a house. The case is different in Edinburgh and Paris, and in most of the cities of the Continent; because there the houses have common staircases, and each floor is let out on lease to one or two separate families. The expenses of repairs are in a great measure unavoidable; but some of them may be diminished by attending rigidly to the following rule, which is founded on the principle of prevention, rather than cure.

Agree with each of the different tradesmen that are employed to do repairs or make additions to a house, such as the carpenter and joiner, the bricklayer and slater, the painter, plumber, and glazier, the smith and bell-hanger, and the locksmith and ironmonger, to look over every part of the house quarterly or half-yearly, on a certain fixed day; and either to repair or report on what is necessary to be done the day following.

In small houses, the carpenter alone will be sufficient to point out what is wanting, and the other tradesmen need only be called in if wanted: but, in very large establishments, all the tradesmen mentioned, and even some others, may be necessary. The carpenter, in small houses, will perform the part of the locksmith, in oiling all the locks and hinges; and he will also examine the roof, the boarded floors, the fittings up, such as the window blinds, curtains, &c., and in general all that, strictly speaking, belongs to the joiner and cabinet-maker. The plumber and glazier will attend to cracked or broken glass, and to every thing connected with lead gutters, painting, traps for smells, &c.; so that, in small houses, these tradesmen, with the assistance of the bricklayer and smith, to repair the roof, locks, and bells, will be found sufficient.

Inside Painting and Papering. A considerable part of the annual expenses of houses in London and its environs is increased by the frequency with which they are painted and papered, or coloured. These operations, when frequently repeated, are not only expensive, but they destroy all the sharpness of architectural ornament, whilst they are quite unnecessary as a preservative against decay. On examining some of the old houses in London that have wooden staircases, with richly carved rails and balusters, which have been painted every three or six years for two or three generations, the beauty of their workmanship will be found to be entirely destroyed; while, if the accumulated paint, perhaps a quarter of an inch thick, were to be removed, they would be found to be of very great beauty as works of art. The ornaments on the ceilings of such houses are equally obliterated by repeated coatings of whitewash. The excuse offered for this frequent painting and whitewashing is, the dirt which attaches to the walls and ceilings of London houses, in consequence of the soot continually floating in the atmosphere; but a much better mode, both for the beauty of the rooms, and the pocket of the occupier, and even for cleanliness, would be to wash all the work painted in oil with flannels dipped in warm water, and carefully to wipe with a dry cloth, and afterwards to rub over with a piece of crumb of bread, the paper, if not adapted for washing, and the ceilings. To lessen the trouble of reaching the upper part of lofty rooms, the cloth for wiping the paper may be put on the extremity of a brush attached to a long handle, and a contrivance might easily be devised for substituting a piece of bread, or a large piece of India-rubber, for the brush. In general, it will be found the best economy to have the ceilings of the living-rooms painted in oil, and to have the walls either painted, or covered with what is called washing-paper, the colours of which, being in oil, admit freely of being washed with flannel, or sponge, and water.

To guard against Danger from Fire, the chimneys of all the rooms in which fires are kept ought to be swept at stated periods; the outlet to the roof ought to be easily accessible, and the means of opening it ought to be familiar to every inhabitant of the house; and there ought to be a stepladder constantly kept on the roof, for reaching the tops of the chimneys in case of one taking fire. On the ground floor of the house, immediately within the outer door, one of Reid's syringes, or some other equally efficacious, ought to be kept, and every male person in the house instructed how to use it on the first breaking out of a fire. Precautions of this kind are useful, as leading to habits of carefulness and forethought, which, after all, are the surest means of avoiding accidents by fire, or of any other description whatever. More of the misfortunes which befall us in our passage through life are the results of our own imprudence, than even the most candid of us are willing to allow.

CHAP. III.

ON LAYING OUT AND PLANTING THE GARDENS AND GROUNDS OF SUBURBAN AND OTHER COUNTRY RESIDENCES, OF FROM ONE PERCH TO TWO OR THREE ACRES.

HOWEVER paradoxical it may seem, it is nevertheless true, that the difficultiesof building, planting, and gardening, on a small scale, so as completely to attain the objects in view, are greater than on a large one. This will be found acknowledged by the late eminent landscape-gardener, Mr. Repton, in various parts of his works; and it is also well known to every architect, and to every gentleman's gardener who has been accustomed to lay out grounds. There is scarcely an architect who does not find it much easier to satisfy himself in devising a design for a mansion, than one for a cottage residence; or a landscape-gardener, who would not have more confidence of success in laying out and planting a park of a thousand acres, than a ground plot of half an acre. The difficulty, in the case of small places, arises from the deep consideration required to produce the greatest possible result from very limited means. In building or planting on a large scale, the means are generally ample; and, if not unlimited, they are, at least, frequently indefinite: the results obtained are, therefore, generally considerable, and such as to afford ample enjoyment to the possessor. Should they fail of this, however, his wealth and his resources will enable him to alter, amend, and improve, till he has succeeded to his wishes; or, should he ultimately not succeed, will prevent him from being ruined by the attempt. The builder of a small house, on the contrary, whether it is for his own occupation, or for that of another, undertakes a task of great moral responsibility; since the result may either be inadequate to the means employed, and thus time and money which can ill be spared may be thrown away; or the expense may be greater than was desired, or could be afforded by the party, and might thus blight his fortune, and, consequently, mar his prospects of happiness. In short, when an artist undertakes to arrange a country residence on the smallest scale, he may be considered as intrusted with the whole of what his employer can afford to expend in that manner; while, on a large scale, he is only intrusted with a part. Hence the serious consequence of failure in the former case, as compared with the latter; and, hence, the deep consideration required in designing, and the great anxiety felt in executing, a cottage residence, as compared with a mansion or a palace. Those who have had most experience in matters of this kind will be the most deeply impressed with the truth of what is here stated.

Before entering into the details of laying out and planting gardens, we shall briefly submit to our readers a few remarks on the principles by which we mean to be guided. In every work which man undertakes, the first thing to be considered is the end in view; and the next, the means of attaining that end. The ends proposed to be attained by a country residence are, comfort and enjoyment; and the means which are to be employed for the attainment of these ends are, the construction or the improvement of a house, and the arrangement, planting, and management of a certain portion of ground surrounding it, and belonging to it. What relates to the house we consider as already sufficiently treated on in the preceding chapter; and we shall here confine ourselves entirely to the grounds.

From the Grounds of a Country Residence, whether small or large, two kinds of products are desired by the possessor; viz., those which are necessary for comfortable occupation, and those which contribute to elegant enjoyment. In the earlier stages of civilised society, the comforts and the elegances of life were necessarily few; but, as civilisation and refinement proceeded, both must naturally have increased. The first house was, doubtless, a mere hovel; and the first garden a piece of dug ground, surrounded by a rude enclosure; and as mankind advanced in knowledge and taste, the hovel and the enclosure were gradually improved, till they became what they are at present. It is the privilege of every existing generation to take advantage of all that is considered excellent in the practices of those which have preceded it. The first architects could have only one manner of architecture to study, and the first gardeners only one mode of laying out their gardens; but the architects of the present day can have recourse to all the different styles of design, and manners of building, which have been practised by different nations, from the earliest ages of human records to the present time. The modern architect may build a house in the classic or in the Gothic style; or he may adopt the historical and geographical variations of these styles, as exhibited in the Hindoo, Elizabethan, Italian, English, and other manners of building. In like manner, the landscape-gardener, who would lay out grounds at the present day, may adopt either the oldest, or geometrical, style, in which the forms and lines of the house are reflected in the garden in front of it, and which, as it has been recently shown, was practised by the ancient Egyptians more than 3000 years ago; or he may adopt the modern, or irregular, style, in which the forms of nature are brought into immediate contrast with the forms of art: and he may, farther, combine the two styles in such a manner as to join regularity and irregularity in one design. In a word, both in architecture and in gardening, the artist of the present day has it in his power to adopt the style or manner of any former age, or of any other country, and adapt it to the wants of the present age, in the country in which he lives.

Whatever style or manner is adopted by the architect and the landscape-gardener, there are certain principles common to all the arts of design and taste, by which both artists must be guided; and certain others, or rather, perhaps, certain rules, deduced from fundamental principles, which are peculiar to each art. Whatever, either in a building or a garden, cannot be justified on fundamental principles must undoubtedly be wrong; and whatever cannot be referred to preestablished rules must necessarily be new, and may either be right or wrong, according to its consistency or inconsistency with fundamental principles. Hence it is that all the productions of the fine arts, no less than of the mechanical arts, may be subjected to reason; and, consequently, that when any part is produced, either of a building or of a garden, for which no sufficient reason can be given, that part must be either a superfluity or a deformity.

Though it cannot be expected that every one who contemplates hiring or building a country residence should know as much of the fundamental principles of architecture and gardening as those who make these arts their study; yet the great leading principles of both are easily comprehended by every reader, and may be most usefully taken into consideration and applied by whoever intends either to build or plant a new residence, or to make choice of a residence already formed. For this reason, we shall devote the first section of this chapter to the developement of the great leading principles of architecture and landscape-gardening, and to some of the more obvious rules which may be deduced from them; while the remaining sections of the chapter will treat of laying out and planting gardens and grounds adapted to the four different rates of suburban houses.

SECT. I. *Of the fundamental Principles of Architecture and Landscape-Gardening.*

THE various arts which are necessary, or have been invented, to supply the wants of civilised life and refined society, may be divided into the useful, such as baking, brewing, &c.; the agreeable, such as music, painting, &c.; and the mixed, such as architecture, gardening, &c. The main object of the merely useful arts is to supply our physical wants; that of the agreeable arts is to supply the wants of the mind; and, consequently, that of the mixed arts is to supply jointly the wants both of the body and of the mind. It is not denied, that the humblest physical art may, to a certain extent, be also made an art of taste; but it is sufficient for our purpose here, that architecture and gardening are recognised as mixed arts; having for their object not only the supplying of certain physical wants, but also the gratifying of the imagination and the taste. It is evident, therefore, that there must be two classes of principles and rules applicable to each art; the first having reference to mere use, and the second to use and beauty combined. We shall consider each separately.

SUBSECT. 1. *Of the Principles common to Architecture and Gardening as useful Arts, and the Rules deduced from them.*

THE *Principle of the Adjustment or Fitness of the Means employed to the End to be obtained* will comprehend all the subordinate principles and rules applicable both in architecture and gardening as useful arts.

In Architecture, the dimensions, the arrangement, the situation, and all the various details of the interior, must be regulated by the uses to which the building is to be applied. In like manner, in its construction, the foundation must be adjusted to the intended superstructure; the thickness and the materials of the walls, to the height to which they are to be carried, and to the weight which they are to sustain; and the quality of the materials throughout, to the intended durability of the edifice. The reader may easily trace this principle through the whole of a building, and test by it, not only the size of the building, and of all its different parts; but its situation, foundation, walls, roof, floors, and even the fittings-up and furniture of the interior. Apartments may be unfitting for use, from being of too small dimensions in respect to length and breadth, or in respect to height only: doors may be too narrow or too low for easy ingress and egress: windows too small for affording light and for ventilation; or they may be large enough for the former, and yet badly constructed for the latter purpose: the walls may be off the perpendicular, or they may be formed of heterogeneous materials, such as clay and stone; or of combustible materials, such as wood; or of dissolvable, disorganisable materials, such as salt-stone, vegetable earth, &c.; or the weaker materials may have been so placed as to be made to support the stronger; and so on.

The Rules which, in Architecture, are derived from the Principle of Fitness are numerous, and they are more or less absolute. As some of the more obvious of them, we may state the following:— Every wall should be broadest at its base; openings, such as windows, should be made above one another; solid parts of a wall above solid parts, rather than above openings; wide openings should be arched; lintels should be of one piece of wood or stone; all the stones forming a wall should have horizontal surfaces on their under and upper sides; the largest stones of a wall should be nearest the foundation; all openings should be on the same horizontal plane; and, with a few exceptions, all angles should be right angles, &c.

The Fitness of a Building for the End in View ought not only to be real, but apparent. This, indeed, is partly a matter of taste, though it may be more conveniently treated of here as a matter of utility. A dwelling-house, for example, ought not to be so constructed as to be mistaken for a barn, or a chapel, or a manufactory; nor a chapel, nor any public building, so designed as to be mistaken for a private house. Supports should not only be, but they should appear to be, adequate to the weight to be supported. Hence, when cast-iron pillars are introduced in a brick or stone building, they ought either to be cased, or cast hollow, so as to appear of the usual dimensions of brick or stone pillars: or, if the small dimensions required in an iron column are preferable, in order to admit more light, as in shop fronts, the metallic character of the material ought to be rendered obvious by painting or bronzing, or by some such means; the material of which they are formed ought always to appear obvious at first sight. Stone lintels over openings, when they are not formed of one piece, ought to have the joints of the pieces in such a direction towards a centre as to give them the appearance of being the joints of an arch, as in *fig.* 29., even though the real source of strength should be an unseen iron plate or beam let into the soffit. The principle

29

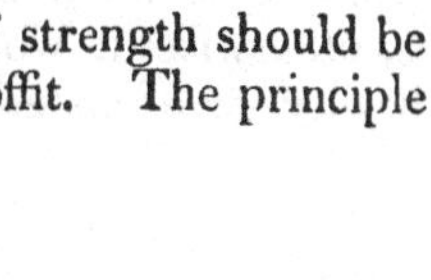

of fitness, applied to the roof, requires either that it should be always shown above the parapet walls, where there are any, or, where there is a flat roof, that there should be a far-projecting cornice, either forming the termination of the wall, or the basement of the parapet, to indicate its place. Much more might be said on the subject of fitness as applied to architecture and building; and the reader will find it treated at length in our *Encylopædia of Cottage, Farm, and Villa Architecture*, book iv. ch. i.

In Gardening, the principle of fitness, or the adjustment of the means to the end, may be applied to the situation of the garden; to the fitness of its soil for the articles to be cultivated; to the fitness of the forms of the compartments for carrying on the processes of cultivation; to the fitness of the culture for the particular plants cultivated; and so on.

The Rules which, in Gardening, are derived from this Principle of Fitness are, that, in the latitude of Britain, and in the climate of the neighbourhood of London, the best situation for a kitchen garden or fruit-garden is on a level plain, open on all sides, and at a distance from hills; that in hilly districts, in the same latitude, the best aspect is on a declivity to the south-east; that the best soil for general purposes is a sandy loam; that the best form of compartments is a square or parallelogram; and that the best form of culture is in rows; and so forth.

Subsect. 2. *Of the Principles common to Architecture and Gardening as Fine Arts, and the Rules deduced from them.*

In order to render this subject as plain as possible, it may be advisable to commence by endeavouring to point out what a fine art is. The synonymes to this term are, elegant art, art of imagination, art of imitation, art of taste, art of design, art of beauty; and each of these terms is applied to architecture, landscape-gardening, painting, sculpture, and music. One quality, common to all the terms, is the word art; and another is understood, viz. that the end of that art is to please. A fine art, then, may be said to be a creation, or composition, intended, through the eye or the ear, to please the mind. To please is to gratify the various faculties or modes of the mind; such as to satisfy the reason or the judgment; to awaken the sentiments of approbation, of love, of reverence, or of admiration; to surprise, delight, astonish, or wholly to absorb the attention and the imagination. The two essential qualities are, to create and to please: the work produced must be a creation of the artist, and must be acknowledged as such, otherwise it would be no work of art; and it must excite pleasing emotions, otherwise it would be no fine art, no art of imagination, of beauty, or of taste. If this chain of reasoning is correct, and the conclusion is fairly derived from the premises, the fundamental principles of the fine arts would appear to be two; artificialness in appearance, and mental excitability in effect. We shall endeavour to develope these principles, to trace their ramifications into others, and to lay down the principal rules which may be derived from them.

Artificialness of Appearance. Any creation, to be recognised as a work of art, must be such as can never be mistaken for a work of nature. It is true that art may create a work which shall be mistaken for nature; but in such a case the object created could afford no pleasure as a work of art, because it would be without the first condition, viz. artificialness; or, in other words, that kind of aspect by which art is at once recognised. Much is very properly said about the imitation of nature, because no work whatever could produce an effect on the human mind, but such as was in accordance with those works from which the human mind receives all its impressions, and to which impressions alone it is accommodated. A work of art, therefore, that is not composed in imitation of, or, in other words, upon the same principles of composition as, the works of nature, can no more give pleasure to the human mind, as at present constituted, than an article not fit to be taken into the stomach as food can give nourishment to the human body. The great object of all human exertion, after satisfying those wants which are essential to our existence, is to procure the approbation or applause of ourselves or others. To imitate nature in such a way as that the object produced should be mistaken for nature, could never excite much approbation for the artist, because its very perfection, by deceiving the spectator into a belief of its reality, would prevent it from being considered as a work of art. On the contrary, when an object is imitated in a totally different material from that in which it appears in nature, and the imitation is successful, the applause of the spectator is great in proportion to the degree of skill displayed.

Thus, suppose a person to succeed in making artificial flowers of coloured paper, so like real flowers as to deceive the spectator; and another person to carve an imitation of such flowers in wood or stone: what would be the merit which the common sense of mankind would assign to each of these two persons? The artificial flower maker would be considered in the light of a manufacturer, or mechanical producer, or repeater, of an object; while the carver in wood or stone would be considered as possessing a superior degree of mind, from his having produced the resemblance of a flower in a material so unlike the texture of flowers as wood or stone: he would, in short, be considered an artist.

Artificialness of appearance, or the principle of the recognition of art, is thus, we think, proved to be founded in human nature: it is recognisable in every description of human improvement; and it is no less essential in the case of the fine arts, than in those of common life, or of any of the mechanical arts and manufactures. We have considered it necessary to insist on this principle here, in order that our readers may go along with us when we come to make the application of it to the modern style of landscape-gardening. This style is said to be an imitation of nature; and, in consequence of this expression, many persons have argued in favour of imitating nature so closely as to produce scenes which might be mistaken for natural ones; or, in other words, for those which result from causes operating independently of man. If we are right in our principle, however, such fac-simile imitations of nature, even of the most beautiful nature that can be selected, constitute but a very inferior style of art; and the landscape-gardener who should produce a piece of water surrounded by grass and trees, with its margin fringed by bushes and water plants, and varied by gravel and stones, in such a natural-looking manner, that it might be selected for copying from by a landscape-painter, and mistaken by him for a piece of natural scenery, has exactly the same pretensions to the character of an artist, as a manufacturer of artificial flowers or wax figures, who should produce a flower of tinted paper, or a dressed figure of a man or boy, so complete a fac-simile of nature, that a botanical painter, or a cursory observer, might be desirous of making a drawing from the one, and of speaking to the other, believing both to be alive.

This may seem a very bold conclusion, and, doubtless, it is at variance with many of our preconceived opinions and acknowledged prejudices. Let us, however, try the application of the principle from which it is derived, to the modern, or natural, style of landscape-gardening. A landscape which might be mistaken for one created by nature, we have stated to be low in the scale of art on that account. What is to be done, then, in the case of laying out the grounds of a proprietor who is desirous of displaying only the highest style of art? Must we have recourse to the ancient geometrical style, and have all our trees in straight rows, all our lawns in regular levels or regular slopes, and our water bounded by mathematical lines? Certainly, we must, if no other mode can be found of conveying a high character of art. But a mode may be found, by which a scene in imitation of nature shall be so like nature, as to be copied by a landscape-painter as such; and yet so different from the natural scenery of that part of the country, as to be as instantly recognised for a work of art as a geometrical garden; and never for a moment to be mistaken for the result of fortuitous circumstances. How is this to be accomplished? Simply, let foreign trees and shrubs, or such as are totally different from the trees in the given locality, be planted, instead of indigenous trees; let the same be done as to the water plants; the same as to the stones and gravel; the same as to the slopes of the turf; the same as to the outline of the water; and, as far as practicable, the same even as to the grasses composing the lawn. It is evident, that a landscape so produced might have all the beauties proper to landscapes of this kind, and might be selected by a landscape-painter to copy from, and yet never for one moment be mistaken for a work of nature. But, it may be said, what greater merit can there be in the artist, in planting one kind of tree or shrub more than another; and in choosing plants not common in the locality, rather than those which are common there? To this we answer, that the merit is not, certainly, very great; but neither does landscape-gardening rank very high in the scale of the fine arts. Who, for example, would ever think of putting it in competition with painting or sculpture? However much the scenes it produces may be loved and admired, it is certain that their beauty depends more on causes which are, in a great measure, beyond the control of man, than on those which are controllable by him: on the beauty of the growth of trees, for instance; on the effect of the seasons; on the weather; and on the accompaniments of buildings, animated objects, &c.

The Recognition of Art, in Architecture, is so obvious a beauty, and one so inherently essential to it, that it is not in the power of the artist to avoid it. He may, however, counteract it by such deviations from correct art, and from artist-like shapes, as may injure the expression, and recall, if the expression may be allowed, natural forms in a state of nature. Thus, an architectural column of wood or stone can never be mistaken for a natural object; but, if the artist were, in a stone column, to make a coloured imitation of the trunk of a tree, or, in a wooden one, to leave the bark on, the impression of artificialness, and of artist-like composition, would be materially injured.

The Rules which, in Architecture, are derived from the Principle of the Recognition of Art are few, because the principle is never for a moment lost sight of in a building. Some of them, such as the squaring of stones, may be considered as included under the rules derived from the principle of fitness: but the squaring of the stones, so far as is required by this principle, has nothing more for its object than that they should fit together properly, so as to insure the strength and durability of the wall; whereas, the principle of the recognition of art goes farther, and requires the faces of the stones to be hewn smooth, or, if kept rough, to be hewn in a regular manner. Common rustic-work and rubble walls belong to the lowest degree of art, as does building walls with land stones of different sizes, with their natural rounded forms, as gathered from the earth's surface. Ashlar and rubbed work belong to a higher degree of art.

The Rules which, in Landscape-Gardening, may be derived from the Principle of the Recognition of Art are numerous. With respect to ground, it must either be reduced to levels, or slopes of regular curvatures, as in the ancient style; or, in the modern style, to polished curvatures and undulations, which shall be, either from their beauty of form, or from their clothing of herbage, distinguishable at first sight from the natural surface of the ground by which the work of art, that is, the lawn, park, or pleasure-ground, is surrounded. Wood, if the common trees of the locality are employed, must be either planted in lines, or massed in geometrical figures; or, if foreign trees and shrubs only are used, they may be planted in irregular masses or groups, and as single trees. If indigenous trees and shrubs are at any time introduced in the modern style, of landscape-gardening, the greatest care must be taken not to crowd, or even group, them together in such a manner as that a stranger might conclude they had grown up there naturally. They must be placed so as to stand distinct from other trees and shrubs, and so as to take forms more perfectly developed than what the same species are found to have in a natural or accidental state in the surrounding country. For example, in a country abounding with the common English oak, no artist, who understands his art, would employ that tree in his artificial plantations, unless at their boundaries, so as to harmonise them with the natural woods of the country; or unless in an avenue, or in some other way in which they could at once be recognised as having been planted. But, supposing that one indigenous oak existed in the midst of his artificial plantation, which he was obliged to retain; or that he were absolutely required to plant one; or that he were desirous of having one to complete a collection; how is that tree, supposed to be common in the neighbourhood, to be treated, so as to subject it to the principle of the recognition of art, and yet so as not to violate its natural form, by clipping it in the ancient manner? To answer this question, it is necessary to consider the state of the common oaks in the neighbourhood: these are, in all probability, either crowded in oak woods, or pruned or otherwise mutilated in hedgerows. Give the oak, in the artificial plantation, therefore, ample room on every side; preserve it from cattle, and let its branches stretch out all round, and hang down upon the ground; and you have at once a tree of art, and of great natural beauty. Water, bounded by the formal lines of the ancient style is easily recognisable as artificial: but how is this end to be attained in a style which professes to be an imitation of nature? There are two kinds of water, in imitations of natural scenery.

One is, where there already exists a brook, or a lake, or a river, which is to be appropriated, and rendered a work of art; and the other is, where there is no visible water naturally, but where excavations are to be made, and to be filled with water, which is to assume the character of a lake or river. In the first case, the brook, lake, or river, is readily appropriated as a work of art, by planting exotic, woody, and herbaceous plants along the margins, in a natural-looking manner; carefully removing all that are indigenous, with the exception, perhaps, of such as are not conspicuous, as the usual grasses which compose turf. Thus, the cut-leaved alder might be substituted for the common species; the weeping poplar, or the weeping willow (if not too common in the neighbourhood), and the paper birch, for the common willow and birch; and that truly elegant exotic aquatic tree, the deciduous cypress, might be made the prevailing species.

The Application of the Principle of the Recognition of Art to rocky Scenery may, at first sight, appear to present some difficulties. By rocky scenery is here meant scenes of nature in which rocks are predominant features. By what means, for example, are the terrace walk and the perpendicular rocks on the banks of the river Wye, at Piercefield in Monmouthshire, to be rendered a work of art? By substituting another kind of rock for the indigenous one? No; for not only is the scale too large to render this practicable, but, if it were accomplished, the very largeness of the scale would make it be still considered as the work of nature; unless, indeed, rocks, which every one knew did not exist in the country at all, were substituted for the natural ones; and even in this case, though the character of art would be maintained, yet the associations connected with solid natural rocks would be so effectually obliterated, that, on the whole, the interest of the scene would be destroyed. Neither in landscape-gardening, nor in any other art, ought every thing to be made to give way to one view of a subject. The most complete mode of appropriating the rocky bank at Piercefield, as a work of art, would be to remove all the indigenous vegetation above and below it, and to supply its place by foreign vegetation of a similar character. Instead of the common oak, which is the prevailing tree at present, substitute the evergreen and the Turkey oaks; and, instead of the common brambles, sloes, and thorns, which protrude from the rocks, substitute the thorns of Greece and America, and the laurustinus, the arbutus, and the cistus of Italy. Let it be observed, however, that we by no means say that it would be desirable to effect these changes in such a scene as that at Piercefield; we merely undertake to show how the principle of the recognition of art may be applied to rocky scenery, as well as to grounds, woods, and water. The recognition of art, in such a scene as that at Piercefield, may be effected, to a very considerable extent, by the artificial manner of conducting a walk through it, and by the manner in which this walk is constructed. Such a walk should be carried along on a level, or on a regular slope or slopes; and the contrast of such a line, and the rough and ever varying natural surface, would alone indicate the employment of art. It must never be forgotten that there are exceptions to all rules, and that a natural scene, preserved and recognised as such, may, in many situations, be more appropriate than a scene of art, to which it will often form a very desirable contrast.

In the Case of Buildings in the artificial landscapes created by landscape-gardening, art is always recognised in the building itself, that being indisputably an artificial object; but the principle of the recognition of art is not always perceived in the placing of the building: on the contrary, nothing is more common, in arranging the grounds round an edifice, than a violation of this principle. Wherever a building, whether a house, an obelisk, a column, or a statue, is set down among trees and plants, and appears to rise up among them as if it were itself a tree or a plant, there will be found a want of the artificial principle. This want is produced by the vegetation being placed too near the artificial object, and by the trees and ground not having been treated according to art. An artificial object ought, surely, not to appear to grow out of the ground, like a natural one: it ought to grow out according to art, — which, as will be hereafter shown, indicates that it ought to rise from an artificial basement; and that the ordinary vegetation of the spot ought to be kept at some distance from it. May not creepers be planted against it? Yes; because the planting and training of these indicate design and intention, and enable the spectator to recognise art. We do not say that it is always advisable to plant creepers against a building; because there are a variety of circumstances to be taken into view before any one point can be determined: we merely say that there is nothing in the principle of the recognition of art at variance with the use of creepers, where their use is not at variance with any other principle.

Art is easily recognised in all Walks and Roads; but not always artist-like art. The uniformity of the breadth, and the evenness of the surface, of a walk may secure it the character of art, while this character may be counteracted by the footpath-like junction of one walk with another as in *fig.* 30., while the artist-like junction is shown in *fig.* 31. The same remark will apply to the forms of flower-beds on gravel or turf: they are always easily recognised as belonging to art, but not always to high art; that is, the shapes of the beds are not always artist-

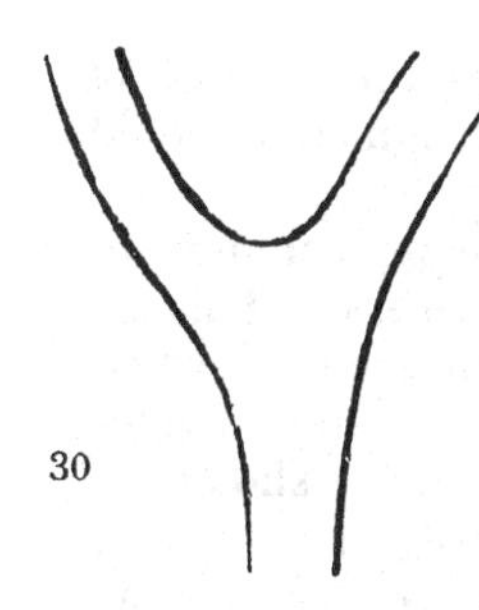

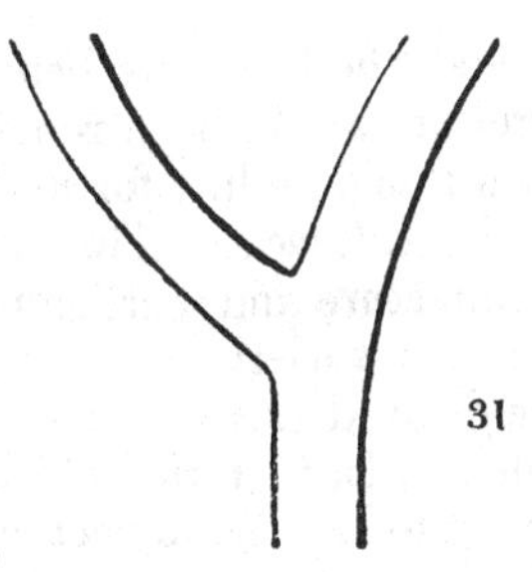

like. In *fig.* 33. the forms of the beds resemble those of common cordate leaves, thrown down in a natural manner, some in one direction and some in another, as if they had dropped off from a dried specimen in a herbarium. In *fig.* 32. the same leaves are disposed of, as a whole, in an artist-like manner. In *fig.* 34. the shapes, considered separately, are artist-like; but they are thrown down without the slightest regard to symmetry. In *fig.* 35. they are disposed of symmetrically, that is, according to art. Even a straight line, in gardening and in architecture, may be laid out or formed in an unartist-like manner: for example, a line of box, or a brick edging, to a walk, or to a bed or border, which, instead of being perfectly straight, is bent to one side, will be much more offensive to the eye of an artist, than a line perfectly straight in the direction of all its parts, but some parts of which are wanting. It is not that either line could have been formed by nature, but that the

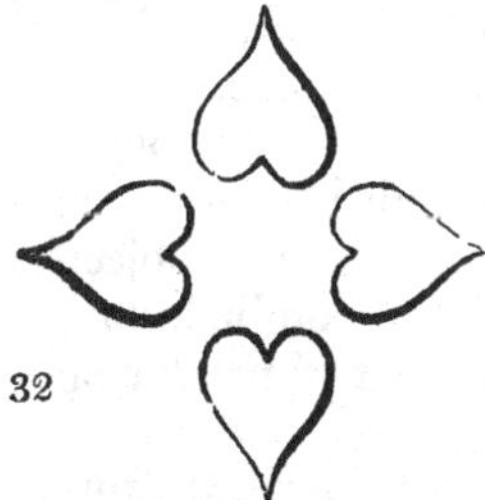

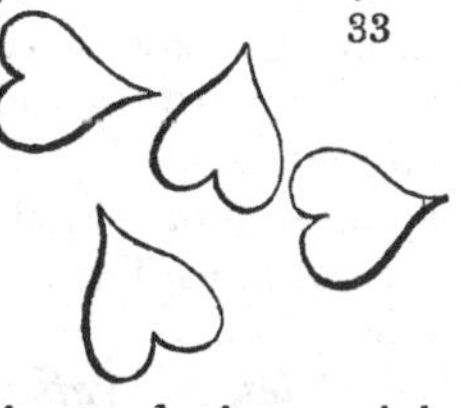

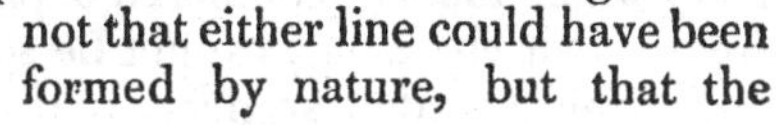

evidence of art is more decided in the one case than in the other. The imagination easily supplies the parts which are wanting; but it will not so easily set that part of the line straight which is bent to one side. If, indeed, the line were bent equally to both sides, the absence of rigid art would be less offensive, because the imagination would form a middle line for itself.

Agreeableness. The next fundamental principle, common to architecture and gardening, is that of agreeableness. This includes a variety of subordinate principles, from each of which are derived various rules. The first essential requisite to agreeableness is, that the object possessing that property should be understood. An object may overwhelm, astonish, or confound, by some unknown quality, or by the gross exaggeration of some quality with which we are familiar; but it never can please or be agreeable, unless it can be comprehended by the mind, through the eye or the ear. Now, the eye can only see, and the ear only hear, one thing at a time. The object seen, or the sound heard, may be composed of a great many minor objects or sounds; but they must all be united or blended together, in such a manner as to be seen or heard at one time, as one object, or as one sound. Hence, the first principle in all combinations, whether of lines, forms, colours, or sounds, is that of producing a whole.

The Rules, or rather, subordinate Principles, derived from the principle of a whole are very numerous, both in architecture and landscape-gardening. In architecture, a building is generally considered as forming a whole of itself, without reference to the scenery with which it may be surrounded; but, in landscape-gardening, a building is only considered as forming a whole in combination with the scenery by which it is surrounded. Hence, as every whole must be composed of parts, a building in a town, to aspire to that character, cannot be so simple as it may be in the country, amidst verdant scenery. In the town, it ought, with a view to its effect as a whole, to be broken into parts, one of which should prevail in effect over the others, which ought to be subordinate to it, while they cooperated with it in forming a whole. Thus, two pavilions joined together, without a centre or main body, could not form a whole; but, with the main body larger than either pavilion, the whole produced would be acknowledged as such by every eye accustomed to look at objects otherwise than in detail. In the country, the plainest form of a house, a mere cube of masonry, may form a whole, if judiciously surrounded by trees. These trees must, if planted near the house, be either considerably lower than the house is high, or, if the trees are of the same height as the house, there must not be more than one or two of them, or there must be so many as to render the trees the main feature of the whole, and the house only a subordinate feature. Wherever the house is surrounded, or even embraced, on three sides, with a mass of trees of the same height as itself, the view fails to produce the effect of a whole: no one object in the picture has the ascendency; and, if it were not for other counteracting associations, such as that of the wealth and dignity of the proprietor, and the comfort and splendour which are known to exist in and about such dwellings, the bare impression, as a landscape, would be disagreeable. On the other hand, when a house is surrounded, or embraced on three sides, by a mass of wood, either a good deal lower than itself, or a good deal higher, a whole is produced, in which the character of architectural dignity prevails in the former case, and of sylvan dignity in the latter. A square house, in the country, in an open plain or pasture, unsurrounded by trees, or by other buildings, can never form a whole; because it has no object of any kind to group with it.

A house may form a whole by itself, without the addition of trees, and so may trees, without the addition of any other objects; but as, in that case, the house must be rendered independent of exterior objects by being broken into parts, so must the wood. In the one case, as in the other, one part must take the lead from one point of view; and all the other parts must obviously belong to it, and yet be subordinate. In the case of a park sprinkled over with trees, if these have been judiciously disposed, they will form a whole with almost every change of the position of the spectator; that is, those near the eye will group together, and form the principal mass; while those which are more distant will form subordinate masses, and unite in supporting the first. For this purpose, the trees in the park must not be uniformly scattered over the surface, but planted in such a manner as to exhibit connexion and grouping, even in the ground plan. In *fig.* 36., the trees are too far apart, and at too uniform distances from one

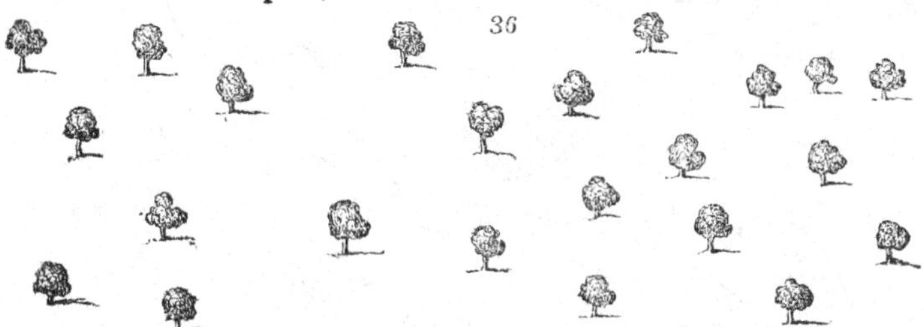

another, to group, or fall into expressive wholes; but in *fig.* 37. they will group agreeably with every change of the spectator.

The expression, "a group of objects," merely implies that these objects form a whole. Nearly the same remarks will apply to a lawn varied by flower-beds, or by beds of low shrubs. The beds, if distributed uniformly over the lawn, will never group so as to satisfy the eye of a spectator who is either walking in it, or on a gravel walk round it. The defect will be rendered obvious by comparing *fig.* 38. with *fig.* 39. The shapes of the former are

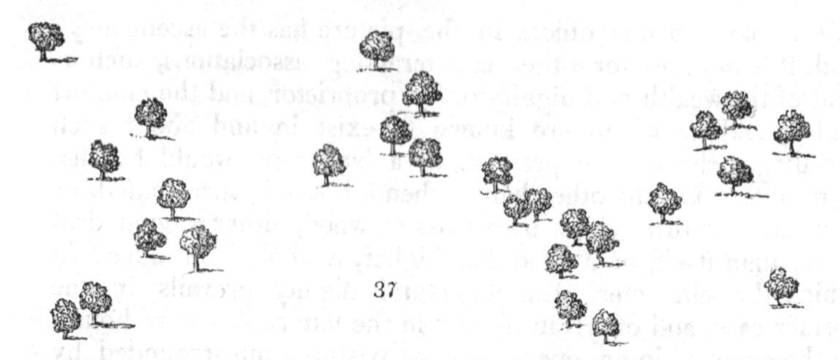

37

unartist-like, as well as too uniformly distributed over the surface; those of the latter are artist-like, and group or unite both with the turns of the walk, and with their reciprocal shapes.

Trees in a park may form a whole relatively to one another, and yet not relatively to the surface of the ground: for example, they may be placed on the levels only, and not on the hills; in which case, the hills will not group with the trees; and, when the height of these hills approaches nearly to that of the trees, the effect, both of the hills and trees, will be, in a great measure, counteracted. On the other hand, by planting trees on the heights

38

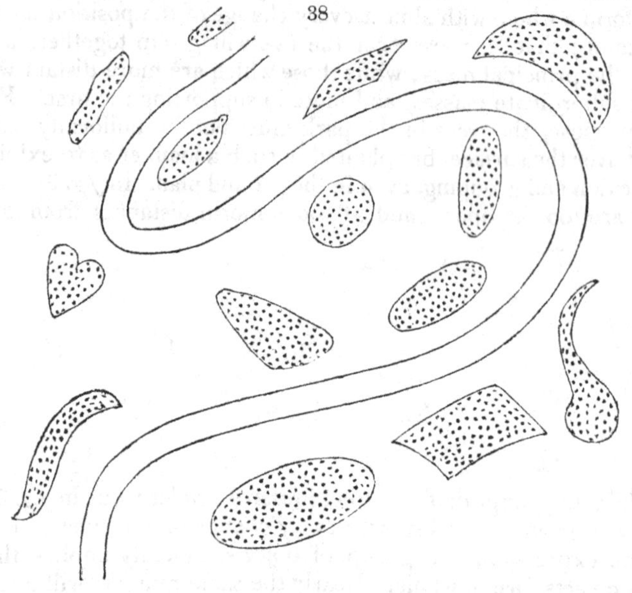

as well as on the plains, the views would present groups as effective as if the whole park had been a plain; and, if the hills were chiefly planted, their effect would be much more striking than anything that a plain could possibly produce. Even the magnitude which trees are calculated ultimately to attain, relatively to the extent of the surface on which they are to be

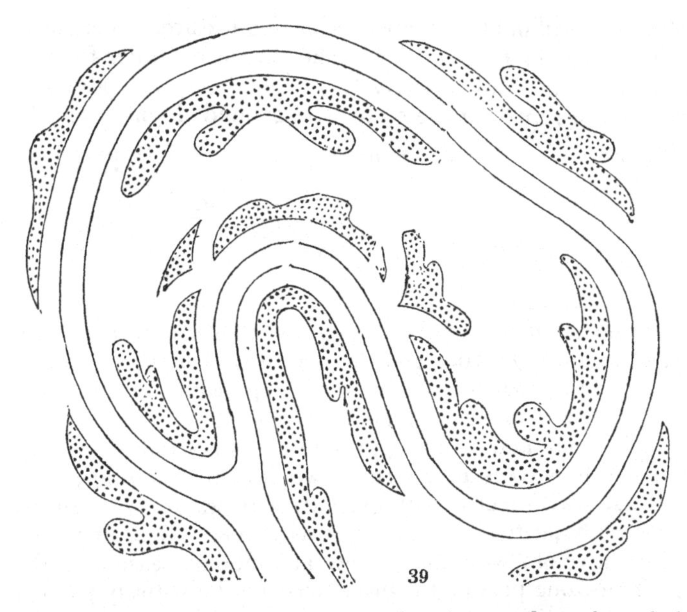

planted, should be taken into consideration, no less than their magnitude relatively to that of the buildings which are near them. Thus, a small park would be injured in effect if planted with the highest and most bulky trees, because they would not form a whole with any object in it; and, though they might group together, and form a whole among themselves, yet that whole would be utterly disproportionate to every thing else in the park. On the same principle, the apparent magnitude of water, relatively to the size of the park in which it is placed, may be diminished or increased according to the size of the trees planted near it. Perhaps one of the practices most adverse to the formation of a whole in planting trees is, to plant one part with very large trees, and another part, seen in the same view, and at the same distance from the eye, with small ones. Hence, groups of aged trees among groups of shrubs do not unite so as to form a whole, without the introduction of trees of an intermediate size. In planting trees, even the kind of tree requires to be noticed, with reference to the production of a whole. An equal number of spiry-topped trees with round-headed ones in a group will not form a whole, from the incongruity of their forms; while a number of round-headed trees of the same bulk, and equidistant from the eye, will not form a whole, from the sameness of their forms and magnitude. Even in sloping and smoothing the surface of ground, the principle of a whole must constantly be kept in view; for, if all the curves and the slopes are of the same curvature and inclination, and of the same magnitude, they will not group;

because there will not be a central or leading feature. There must be a prevailing slope; one which takes the lead, either from its magnitude, or its position relatively to the others. Suppose *figs.* 40. and 41. to represent the sections of ground sloping from the

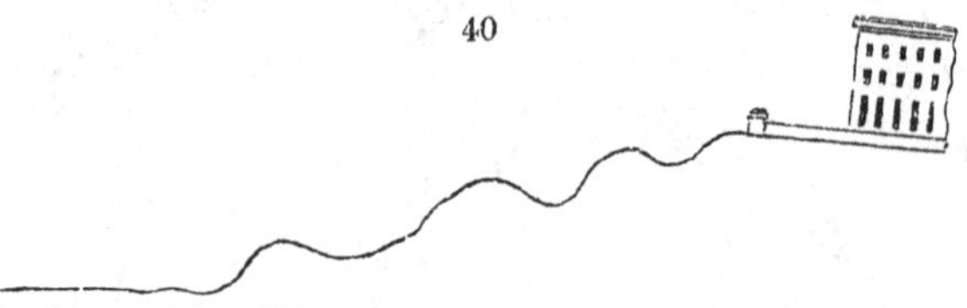

front of a house, it will not be denied that there is more of effect in *fig.* 41. than in *fig.* 40.; and the reason is, that there is a feature in *fig.* 41., produced by the large slope which occupies the place of the two smaller undulations in *fig.* 40.

In laying out and planting grounds, it is not only necessary to consider how trees may form a whole with buildings, with themselves, with shrubs, with ground, with water, with rocks, and even with fleeting objects, such as animals; but how they may form a whole with the objects at different seasons of the year. Thus, one part of the place must not be entirely planted with evergreens, and a corresponding part, which is seen at the same time, planted with deciduous trees. In looking down from the windows of a house, whether on an extensive park, or on a

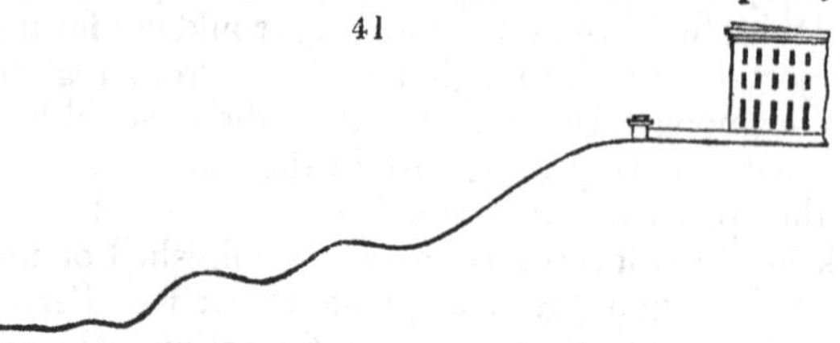

lawn of a few acres, it would be unsatisfactory, during winter, to see the principal masses of plantation, on the one hand, all, or even chiefly, evergreens; and, on the other, all, or chiefly, deciduous trees. It would also be unsatisfactory to see evergreens equally mixed together throughout the view, instead of being so distributed, and yet so connected, as, at a distance, to unite in forming one grand whole.

Regularity and Symmetry. In the modern, or irregular, style of landscape-gardening, as well as in the irregular style of architecture, which, whether under the name of Gothic or Italian, is the style of country houses now most prevalent, the production of a whole requires a much greater knowledge of art than in the ancient style, either of landscape-gardening, or of Greek or Roman domestic architecture: in both of these styles an attempt was seldom made to produce a whole, except by means of regularity and symmetry. It is almost unnecessary to state, that, in the ancient style, whether in ground, in wood, in water, or in buildings of every description, and in roads, regularity or symmetry were the governing principles. The place, as a whole, was generally symmetrical, one half reflecting the other; and the details were always regular. In an age when the beauties of irregularity, and the variety produced by wild scenery, prevailed throughout the country, those of regularity and symmetry would be found to be characteristic of art and civilisation; and they were preferred by our ancestors, with a taste as just and correct relatively to them, and to the circumstances in which they were placed, as our widely differing taste is to us, and our circumstances.

Though symmetry may appear to be a beauty exclusively employed in architecture, and in the ancient style of laying out grounds, yet this only applies to symmetry when it is joined with regularity. In every irregular whole, that is satisfactory to the eye, there will always be found a certain balance or proportion, which one side of the centre of the picture bears to the other, and which balance is nothing more than symmetry. It will be recollected, that the essential principle of symmetry is the union of two parts as a whole, which do not form wholes separately; in opposition to uniformity, where, the parts being regular, each taken separately forms a whole. Now, in every pleasing landscape it will be found, that, if it were bisected perpendicularly by an imaginary line, something like an equal body of scenery would be found on each side. The same may be said with reference to any irregular building which is pleasing as a picture, and also to any irregular flower-garden, or the planting of an irregular park. A pleasure-ground, which, viewed from the drawingroom windows, appeared to have all the shrubs on one side, and only flowers and lawn on the other, would not be so satisfactory as one where they were more equally balanced. Neither would the views from the house, over a lawn the surface of which formed a hill on one side and a hollow on the other, be agreeable. Hence, a view across a slope, as in *fig.* 42., is never so satisfactory as one either up or down the declivity; but a view across two slopes intersecting each other, as in *fig.* 43., is satisfactory; because, in this last case, the one balances the other. Single objects, that are not regular, such as a tree, are never satisfactory, unless they are symmetrical; that is, unless the quantity of branches on one side appears to balance the quantity on the other. Thus, those trees which, being the most irregular by nature, are symmetrical at the same time, are more pleasing than those

42

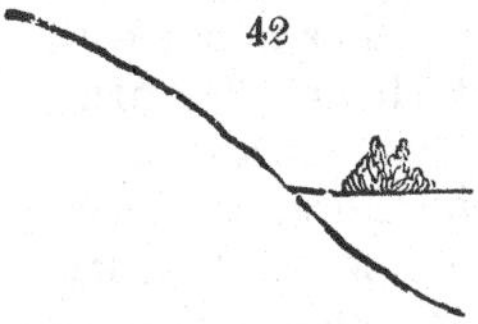

43

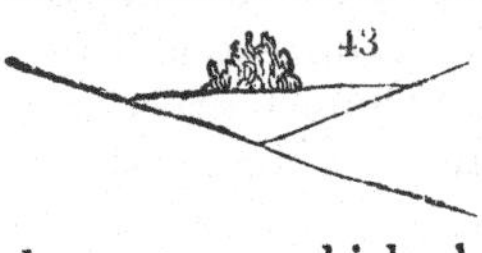

which are comparatively regular and symmetrical; because they show a greater amount of variety, combined with symmetry. In this point of view, an oak, an elm, and a sweet chestnut are more pleasing trees, and higher in the scale of beauty, than a silver fir, a spruce fir, or a larch. The same observation will apply to shrubs, and even to herbaceous plants. One of the most interesting results of symmetry, as applied to trees, is, where the trunk is thrown, by nature or accident, into a position where it requires extraordinary forms of growth in its branches to adjust them to the natural symmetry belonging to its species: for example, when a tree, planted on the banks of a river or lake, has its trunk inclined over the water, and its head in an erect position, and balanced by branches on each side, in the usual manner.

Variety is the next beauty common to architecture and to landscape-gardening; and it must be acknowledged to be a great addition to symmetry, or to any whole in which there is a complete unity of expression, whether that whole be irregular or symmetrical. In architecture, variety is produced in symmetrical and regular buildings, by a difference in the details; and, in irregular buildings, by a difference even in the smaller parts of the composition, as well as in the details. In regular and symmetrical buildings, variety is chiefly confined to what may be considered the ornamental parts of the edifice; such as architraves, mouldings, cornices, &c. The component parts which are essential to the symmetry, or the regularity, of the whole must remain the same; whereas, in irregular buildings, not only may there be a variety in the mouldings and ornaments, but in the component parts; such as projections, recesses, towers, &c. Hence, irregular buildings admit of a degree of harmony and intricacy, and what, in common language, is called ornamental effect, of which regular and symmetrical buildings are not susceptible. Hence, too, as far as effect is concerned, irregular architecture is in a higher style of art than that which is regular; but, on the other hand, with a view to durability, convenience, and use, which are the principal objects of architecture, the regular and symmetrical styles have greatly the advantage. Such styles, therefore, are more properly denominated architectural than the other; and hence, however well calculated the irregular style, whether Italian or Gothic, may be for cottages and villas, which are always expected to be more or less temporary in duration, because more or less subjected to the casualties attending the occupants, and to the influence of fashion or change of taste, it cannot be recommended for public buildings, unless the situation, or the uses, of such buildings should require it.

In the ancient Style of Landscape-Gardening variety was very seldom attempted, except in flower-gardens; because, as we have already observed, it was a beauty so common in the surrounding scenery, as not to be in demand. The reverse is the case in the modern style of laying out grounds; for, the country being now under regular cultivation, in right-lined enclosures, variety is, in a great measure, banished from general scenery; and, hence, it is sought for as a rarity in artificial scenery, in the same manner as uniformity was in ancient times.

In Planting, variety is produced in two ways: by varying the dispositions and distances of trees of the same kind relatively to one another, which may be said to produce simple variety; by varying the disposition and the distances of trees of the same kind with shrubs principally of one kind, which may be said to produce variety joined to intricacy; and by the use of trees and shrubs of many different kinds, which may be said to produce harmony. The addition of intricacy to variety, by the use of shrubs, is equally easy, since it is nothing more than a repetition of the mode of using trees without shrubs; but, where harmony is to be introduced by the use of trees and shrubs of many kinds, some knowledge of botany and of the art of plant culture is required, as well as a knowledge of art. The reason is, that this variety of trees and shrubs is known to comparatively few; and, many of them having been but a few years in the country, it is only in consequence of botanical knowledge, and a knowledge of the art of cultivation, that an estimate can be formed of what, in point of bulk and character, they will afterwards arrive at. Hence, an artist may have a knowledge of architecture, and of the principles of picturesque beauty, in the general disposition of landscape scenery, without being at all adequate to produce the higher beauties of which landscape-gardening is susceptible; and hence, also, it will almost always be more easy to form a perfect landscape-gardener out of a gardener and botanist, than out of a landscape-painter or an architect. The cause is, the science of botany, and that practical application of it which constitutes a knowledge of plants and their culture, consisting of numerous details, and requiring considerable time, are best attained in youth.

In the Disposition of a great Number of Sorts of Trees and Shrubs in a Plantation, some principle of order must be adopted, to produce either variety or harmony. If the sorts are mixed together indiscriminately, the result is left to chance; if they are mixed together as equally as possible, then the result must necessarily be monotony, by the same number of kinds appearing in every part of the plantation; and, if each kind is kept in a group or mass by itself, there will be a risk of want of connexion, and, consequently, of unity in the general result. In what way are all these difficulties to be overcome? Chiefly, we should say, by keeping each sort by itself, and placing all those sorts nearest each other which are most alike;

avoiding all formality in the outlines of the spaces allotted to each sort; and allowing these spaces to indent or ramify into one another. For this purpose, a knowledge of the natural system of botany is of the greatest use to the landscape-gardener; since it teaches him that all those trees and shrubs that belong to the same natural family, order, or tribe, admit of being brought together in the same group. The chief difficulty, therefore, will be in joining and connecting the groups together; and, here, a general rule may be given. This is, that, when the groups have a near general resemblance to each other, such as those of *Quércus* with *U'lmus*, or *Cratæ gus* with *Pỳrus*, then the union may be comparatively abrupt; that is, with a slight intermixture of the trees of both groups at the points where they join. But, when the groups are very dissimilar, such as when the pine and fir tribe adjoin deciduous trees, the union must be very gradual, by means of numerous indentations and ramifications of the one group into the other. For example, suppose it were desirable to join a large mass of different species of oak (*fig.* 44. *a*) with a large mass of different species of pine (*b*); then, adjoining some of the evergreen oaks

in the one group (*a*), place one or two pines, which grow in large and compact forms; and against some of the half-evergreen oaks, such as the Fulham or old Lucombe oak, place one or two pines of comparatively slender growth. In like manner, in the opposite group, evergreen and half-evergreen oaks may be planted among the pines, as at *c*, so as to form the extreme points of the oak group in that direction; and, as the one group approaches the other, not only evergreens and half-evergreens, but deciduous species, in equal numbers, may be introduced. In all this, care must be taken to avoid a regular progression, and every thing like formality. By such means, a degree of union is at first indicated, and afterwards gradually increased, till, at a point equidistant between both groups, the number of plants belonging to each group will be equal. In this way, trees of the most opposite kinds may be introduced in the same plantation, even adjoining each other; but, where the plantation is of any extent, and many kinds are to be introduced, the introduction of opposite kinds adjoining each other can never be required. It can never, for example, in such a plantation, be requisite to unite the weeping willow with the Scotch pine; nor the poplars with the larch or the weeping birch; nor very large-growing trees, such as the elms, with trees of small size, such as the pyrus or the cratægus; nor trees with pinnated leaves, such as the ash and the acacia, with such as have narrow, grass-like, or needle-like leaves, such as the deciduous cypress, and pines and firs. We repeat, that, for assembling trees and shrubs in a harmonious manner with reference to general appearance, excellent hints are afforded by the natural system of botany; and that the artist who has a general knowledge of this system, and understands, at the same time, the principles of composition as applied to lines and forms, which we have been endeavouring to lay down, can never be at a loss how to proceed.

Variety in the Views obtained from the House, and from the different Walks conducted through the Grounds, is one of the grand desiderata in every place laid out in the modern style, whether its extent may be large or small. With respect to the views from the house, the first thing to be attended to is, the disposition of the rooms, so that their windows may look in different directions. Unless this has been studied by the architect, it will be impossible, even in the finest situation, to produce much variety in the views. Suppose a house placed on a slope, commanding an extensive prospect; if all the rooms looked towards that prospect, all of them would have good views, but these views would not be varied; whereas if, from one side of the house, the windows of one room (say the drawingroom) looked out on a level flower-garden; and if, on another side, those of the dining-room looked up the slope; while, on a third side, those of the library, or breakfast-room, commanded the distant prospect; there would be three distinct characters of view. Now, in very small places (say of a quarter of an acre, or even less, in extent), this varied disposition of the rooms, or, rather, of the manner of lighting them, ought never to be lost sight of; because, altogether independently of distance, or of any object beyond the boundary fence, the views may be rendered of different characters by the different kinds of trees and shrubs planted, by their different disposition, by a difference of form in the ground, and by a difference in the architectural ornaments, or by the absence of architectural ornaments altogether. Even a difference in the form and size of the window, or the absence or presence of a balcony or veranda, will altogether alter the character of the scenery. Wherever, therefore, a house stands isolated, and has a clear space of a few yards on each side of it, it may always have at least four different characters of view, independently of the effect produced by balconies, verandas, or other changes in the windows or foreground.

Hence, also, in limited plots of ground, whatever is their shape, greater variety of view will be produced by placing the house nearer one end, or nearer one side, than in the centre. In the latter case, it is impossible to get depth of view from any side, and thus a great source of beauty is lost. A deep view includes a greater number of objects, and, consequently, admits of a greater variety of effect of light and shade; it increases our ideas of extent, and, by concealing more from the eye than can be done in a confined view, it gives a greater exercise to the imagination. Add to this, that, in a small place, depth of view is not expected; and, consequently, when it does occur, its effect is the more striking, by the surprise it occasions, as well as by its contrast with the other views, which must necessarily be very limited. In *fig.* 45., *a* is the house, placed at one side of a plot; *b*, the drawingroom, having a view the whole length of the garden: *c* is the dining-room, having a very confined view, and,

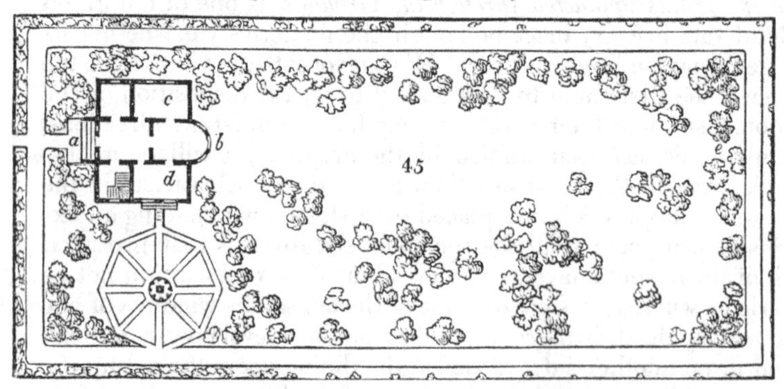

in short, looking across some bushes, to a screen of evergreens (say hollies or evergreen oaks); *d* is the breakfast-room, or common sitting-room of the family, looking on a flower-garden, to which there is a descent from a balcony by three steps. The other small room may be used as a business, waiting, or gentleman's room; and the situation of the staircase is indicated. The central hall is large for the size of the house, and may, in summer, be used occasionally as a music-room, or as a play-place, or dancing-room, for children. All the offices are on the basement story, and the first and second floors are bed-rooms. If the garden were larger, or even of its present size, if circumstances were favourable, a small piece of water, supplied from a dripping rock, at *e*, would have a good effect; and there might be a statue on a pedestal, surrounded with tazza vases of flowers, in the centre of the flower-garden; or, if water were abundant, a fountain might be substituted for the statue. The rest of the garden, with the exception of the surrounding border between the walk and the boundary wall, is entirely of turf, varied by choice ornamental trees and shrubs, including some fruit trees and fruit shrubs.

The standard roses, and the fruit shrubs, such as gooseberries, currants, raspberries, vacciniums, &c., of which there cannot be more than two or three plants of each kind, stand in small circles, kept dug and manured, in order that they may produce their flowers and fruit of good size; but the mulberry, the quince, the medlar, and the few apples, pears, plums, &c., for which space can be afforded, may stand on the grass. Against the walls are planted one or two peaches, nectarines, and apricots; and against the house, a fig tree and a vine. The remainder of the walls and of the house may be varied by roses and flowering creepers; except the more shady parts of the surrounding wall, which may be covered with the common, the giant, and the variegated ivy. The surrounding border between the walk and the boundary wall is wholly devoted to bulbs, in spring and the beginning of summer; with a row of Russian violets inside the box, for producing fragrance in winter; and patches of mignonette at regular distances, to scent the air during summer. Among the groups of trees, and close by their roots, common cowslips, snowdrops, wild violets, and wood anemones may be planted, to come up among the grass; and, being only planted in a few places, and these near the roots of the trees, they may be easily avoided by the mower. In such a garden as this, small though it be, a very great variety of trees and shrubs might be grown; and the flower-garden is sufficiently large to produce a very good display of the finer kinds of hardy flowers.

Where walks have been judiciously laid out, and the grounds on each side of them planted with the proper degree of skill, the views from them will continually vary, even in grounds of limited extent, and without any distant prospect. This result, however, is not to be obtained otherwise than by a very carefully studied plan for disposing of the walks and the trees: it can neither be attained, even in a tolerable degree, by a gardener who knows nothing of the principles of composition, nor by an architect or landscape-gardener who is not well acquainted with the forms, sizes, modes of growth, times of flowering, &c., of trees and shrubs. In proceeding to lay out a small place with the view of making the most of it in this manner, the first thing is, to ascertain what extent of walk may be produced without ever showing any portion of the walk to the pedestrian, except that which is immediately before him. This is evidently more difficult to accomplish in a small place, with a level surface, than in one where it is sloping or undulating; but, by creating artificial undulations, even on a very small scale, much may be accomplished in a very little space; and the effect of such undulations can always be determined beforehand. Thus, in *fig.* 46., a spectator at *a*, 63 ft. from a walk at *b*, would see the gravel of that walk, unless the view of it were intercepted by a small flower-bed,

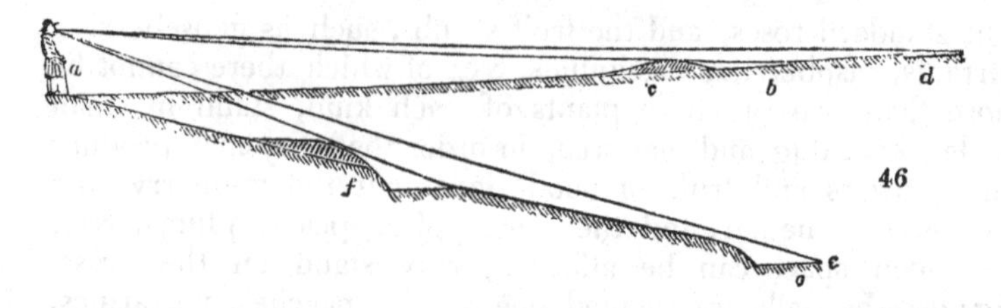

or gentle rise in the turf, 18 in. high, such as *c*; whereas, by removing the walk to *d*, the mere distance would conceal it. This is supposing the general surface of the ground to be a perfect level; but, if it were to slope from the house in the degree indicated by the line *a e*, then the walk would be concealed by a raised bed at *f*, at little more than half the distance from the eye; and by extreme distance at *g*.

Another means of concealing walks is by chains or continuous groups of low shrubs; but these, when not carefully introduced, are, in very small places, apt to injure the view, by interrupting the continuity of the surface of lawn, and by producing too many parts for so small a picture. These groups of shrubs may, however, be resorted to for purposes of concealment, at a great distance from the eye, at the sides of a view, or at its farther extremity.

In some cases, walks may be concealed, or prevented from obtruding themselves on the eye, by forming the more conspicuous parts of them of blocks of earth-coloured stone, which will not have so glaring an effect as bright-coloured gravel. Walks paved with blocks of wood, or of grey bricks, have also a subdued tone of colour, and are not offensive to the eye; but, of course, they can only be adopted on a small scale, or in particular parts of a design, on account of the expense.

The walks about a place may be increased and varied by occasionally crossing each other, either in grotto-like tunnels under ground, or by disguised bridges above the surface. This is a source not only of great variety in the character of the views, but a means of procuring views of a very striking description. In forming short tunnels in pleasure-grounds, the greatest care must be taken to have them straight in direction, so that, on entering at one end, the light may be seen at that opposite; and never to construct them in a circuitous direction, because, in that case, they are necessarily dark, in consequence of the light not entering the tunnel directly. This kind of circuitous tunnel is also objectionable, because it might raise in some minds an idea of the tricks of grotto-work, which were thought so much of in the ornamental gardening of the seventeenth century, but which are now justly considered as puerile conceits. The kind of tunnel that we allude to as desirable in small gardens is, in point of use, nothing more than a bridge, carried over a walk instead of over water; and with the architecture so disguised by vegetation as not to be obtrusive, or to interrupt the idea of the continuation of the walk. It is almost needless to observe that perfect dryness, both in the walk and in the masonry, is essential to bridges or tunnels of this kind; and this can readily be obtained by covering the arch, after it is built, with a layer of clay. The effect of even one tunnel of this kind in a limited space, in adding to the length of walk, and in giving variety to the scenery, may be seen in the public terrace garden at Gravesend. Open and covered seats are also sources of variety in the views of artificial scenery, both as foregrounds to look from, and as objects to look at. In small places, however, they require to be introduced with the greatest caution, and never as conspicuous features; because more than one, for the sake of being able to read or work in the open air in fine weather, can seldom be requisite. In large places, on the other hand, seats in different situations are not only useful as affording resting-places, but as indicating particular points of view, which might otherwise pass unnoticed by a stranger.

Architectural Ornaments, such as Vases, Statues, &c., water in different forms, pieces of rockwork, and other objects of the like kind, independently altogether of trees, shrubs, and plants, and of the view of objects beyond the boundary fence, form sources for varying the views from the walks of a small place. Statues, vases, and other architectural ornaments, ought to be very sparingly introduced at a distance from the house, in gardens in any style; but more especially in such as are laid out in the irregular or modern manner. Mixed up with groups of flowers and shrubs, they divide the attention between the beauties of art and the beauties of nature; and, as the mind can only attend to one sensation, and experience one emotion of pleasure, at a time, it becomes distracted among so many. The true situation for statues is on an architectural terrace, or in an architectural flower-garden adjoining the house, the conservatory, or some other architectural structure, where architecture and sculpture are the main features, and flowers and vegetation are altogether subordinate.

When the artist has brought together in his imagination all the different artificial sources of variety that his subject is capable of receiving, his next business is to arrange these according to some kind of order. He will not think of having all his shady walks at one place, and all his open walks at another; neither will he distribute shade and sunshine equally over the whole scene. He will contrive to have at least one shady walk for midsummer, and one perfectly open to the sun, for midwinter; and he will dispose of the shady parts of the walks so as generally to form separations between striking scenes, that the one may be forgotten before the other is entered on; and that, by means of shade near the eye, the objects at a distance on which light is thrown may appear the farther off. In studying the succession of views, the artist will operate on the principle of contrast so far as to make them different; and on the principle of union, so as not to render them inconsistent with each other. The principle of contrast will direct that the same description of ornaments should not be distributed over the whole place; and that neither flowers nor trees and shrubs should prevail every where, any more than that the same kind of tree and shrub should. The same turf and the same description of gravel may prevail throughout a whole place, as fundamental principles of union, as the same sky and clouds appear in the atmosphere; but every thing else should be varied, by passing from one character of scene into another, till we have made the circuit of the entire garden; by which is meant not merely walking round it, which may suffice in small places, where the means are limited; but walking round it twice, the second time being in a contrary direction to the first. This is to be effected by the mode before mentioned, of making one walk cross over another, so that the spectator shall not be aware of being carried twice over the same ground. In places of the very smallest kind, even of a few perches in extent, there are always two characters of view, one looking towards the house, and one looking from it. Hence, in a larger place, there must be a series of views of the one character, and another series of the other; and, consequently, to see all the beauties of such a place, a stranger would be required to walk twice over the same ground; that is, after he had been once over it, to turn round and retrace his steps; but, by the system of bridging and tunneling, the eye of the spectator is carried twice over the same ground without his knowing it, and without his passing twice over the same walks; and thus he can hardly fail of giving the place credit for a greater extent than it really possesses.

Variety may be added to small places laid out in straight lines, by breaking these lines; and this may be done in many different ways. The straight lines of the coping of a wall may be broken by clusters of ivy; by allowing some of the trees or flowering climbers, planted against the wall, to grow above it in some places; or by allowing plants to grow out of crevices in it. The lines of walls may be broken, so as to display much variety and intricacy, by allowing the branches of adjoining shrubs to extend over them so far as to interrupt continuity in some places, and in others to produce masses of shade and concealment. Even flower-beds and compartments enclosed by box edgings, which, when newly formed, are perfectly regular or symmetrical, may have the additional beauty of variety conferred on them in a similar manner. All objects distant from the eye, in a formal garden, may be varied by placing trees in the foreground; and the more slender these trees are, the greater the number that may be planted. Variety and intricacy may be introduced, even on flat surfaces, by the addition of ivy, and allowing it to trail along the ground, or by planting evergreen herbaceous plants; care being taken that the masses so produced shall always be connected with something else with which they may form a whole. Rockwork, whether natural or artificial, derives a great part of its beauty from the variety produced by a partial clothing of vegetation; and even houses and naked walls may be rendered interesting by the shade, intricacy, and variety produced by trees or other plants trained against them; altogether independently of the botanical interest which these plants will create. Water depends for its variety on the vegetation placed along its margin, as much as a naked wall depends on the trees or plants trained against it. Water is also varied by its general outline, and by islands, and projections and recesses; all of which may themselves, in their turn, be varied by vegetation. Trees and shrubs, which confer so much variety on other objects, may even have their own variety increased by pruning. Heavy compact forms, and orbiculate lumpish shapes, may be broken and lightened by the judicious removal of branches, so as to admit light and air; and to show, in part, the trunk, and the ramifications of the tree.

Harmony, in landscape-gardening and architecture, may, for our present purpose, be considered as a more intense degree of variety; a variety of a more complex kind, in which a greater number of different forms, lines, and colours are introduced, than in common variety. In Grecian architecture, where almost all the lines are either perpendicular or horizontal, there is very little variety; but in Roman architecture, where the arch is introduced, there is a much greater degree of it; and, when sculptures and statues are added, there is so much of this quality, that it may be said to constitute harmony. In artificial landscape, this effect is produced in various ways: for example, where the scene consists only of wood and ground, the trees may be of such a great number of kinds as to present something more than a merely varied appearance; and, if the result were agreeable, we should call it harmony: while, when disagreeable, if produced by an excess of kinds, it would be discordant; or, if from the want of kinds, or their equal mixture in every part, monotonous. Harmony may also be produced in landscape-gardening by the introduction of architectural and sculptural objects; and also by water, rocks, and stones, and by greatly varying the surface of the ground.

Style and Character. A house, or the scenery of a country

residence, may be pleasing from its regularity, its symmetry, its variety, and the harmonious effect of the whole, and yet have nothing in it to distinguish it from other agreeable places of the same extent; that is, it may be without exhibiting any particular style of design or expression of character. Thus, a house may have an agreeable general form, and windows so ample as to indicate large and lofty apartments within; it may be placed on an architectural basement; and it may be terminated by architectural chimney tops; and yet neither be decidedly Roman, decidedly Italian, nor decidedly Gothic nor Elizabethan. A house may also have a particular character given to it by some feature more or less striking; such as a veranda, a far projecting portico, a terminating cupola, &c.; or it may be enriched so as to be expressive of some degree of character, by sculptures, statues, or vases, &c.; or it may assume the character of a cottage, of a castle, or of an ecclesiastical building. It is not difficult to give character to landscape scenery, when there exists naturally considerable irregularity of surface; but on level surfaces, where character is to be given chiefly by trees and buildings, more skill is required. In such a case, the first consideration is, to produce something that shall powerfully contrast with adjoining residences, similarly circumstanced with reference to natural features and trees. If, for example, the trees in the adjoining residences are chiefly deciduous, those in the residence which is to be rendered expressive of character may be chiefly evergreens; and among these evergreens some striking kinds ought to prevail; such as cedars, silver firs, &c. But a still more immediate expression of character may be produced by employing the geometric style of laying out roads and planting the trees; or by adopting some style of laying out, different from that adopted in the adjoining residences.

The prevailing Style of laying out Grounds in England is what is emphatically called English gardening; to which epithet a vague general idea is attached, of grounds and plantations formed in flowing lines, in imitation of nature; as contradistinguished from ground formed into regular slopes and levels, or plantations in straight lines, or included in plots, bounded by lines always decidedly artificial. What is called the English, or natural, style of gardening, however, may be divided into three kinds: the picturesque, the gardenesque, and the rustic. By picturesque gardening is to be understood the production, in country residences, of that kind of scenery which, from its strongly marked features, is considered as particularly suitable for being represented by painting; while by the gardenesque style is to be understood the production of that kind of scenery which is best calculated to display the individual beauty of trees, shrubs, and plants in a state of culture; the smoothness and greenness of lawns; and the smooth surfaces, curved directions, dryness, and firmness of gravel walks: in short, the gardenesque style is calculated for displaying the art of the gardener; while the picturesque style has a constant reference to what would look well in a picture; and the rustic style to what is commonly found accompanying the rudest description of labourers' cottages in the country. The object of this last-mentioned style, or rather manner, is also to produce such fac-simile imitations of common nature, as to deceive the spectator into an idea that they are real or fortuitous. It thus appears that there are four distinct styles in which grounds may be laid out, without any reference to the natural character of the surface, the kinds of trees planted, or the architectural or other objects introduced. We shall briefly notice the leading features of each of these styles.

The Geometrical Style of Landscape-Gardening, or, as it is frequently called, the ancient, or architectural, style, is evidently taken from the architectural lines of houses and fortifications; and it was accordingly, when it was the sole style in use throughout Europe, considered to be as much the province of the architect as the house itself. The antiquity of this style is evidently as great as that of house-building, with the progress of which it must have kept pace in every country. As the first beauties attempted in house-building would be those of regularity in the walls and roof, and in placing the doors and windows; so, in gardening, the fences, the surface of the ground, and the roads or walks, would be made regular. The next step in the progress of both arts would be rendering the house, and also the garden, symmetrical; one half of the elevation of the former, and of the ground plan of the latter, reflecting the other. In general, the house was approached directly in front, by a straight avenue, bordered on each side by one or more rows of trees; and whatever plantations or enclosures were made on one side of this avenue, were repeated on the other side. Immediately round the house, and commonly enclosed by a half-sunk wall (the excavation or ditch for which frequently formed a canal, or moat, filled with water), was the garden, in which were cultivated not only flowers and fruits, but, for a long period, even the culinary vegetables. Numerous improvements and variations were made in this mode of arranging grounds in the architectural style; but the characteristic of all these variations, and that which at once distinguished them from surrounding nature, was, the prevalence of geometrical lines as the boundaries of forms, and of regularity in the distance of objects from one another. The lines might be straight or curved, or combinations of straight and curved lines: but they never deviated into lines which might have been supposed to be the work of chance. Trees, whether in hedgerows, avenues, or masses, were always planted at regular distances; and never, in any case, so as to leave any room for mistaking them for trees which had belonged to some natural wood or forest. The reason of all this is sufficiently obvious. The whole country abounding in natural and irregular forms, there would have been no credit for either wealth, skill, or taste, obtained by imitating these; while, by employing artificial forms which are nowhere to be found in nature, a mark of distinction was at once obtained, which conveyed the idea of wealth, skill, and taste; and which thus answered every purpose of the possessors.

The Irregular, Natural, or English, Style. As the lands devoted to agriculture in England were, sooner than in any other country in Europe, generally enclosed with hedges and hedgerow trees, so the face of the country in England, sooner than in any other part of Europe, produced an appearance which bore a closer resemblance to country seats laid out in the geometrical style; and, for this reason, an attempt to imitate the irregularity of nature, in laying out pleasure-grounds, was made in England, with some trifling exceptions, sooner than in any other part of the world; and hence the style became generally known as English gardening. The English, or natural, style of gardening was first called landscape-gardening by Shenstone; it was also called natural gardening by Bosc, Chinese gardening by Sir W. Chambers, and picturesque gardening by Gabriel Thouin; though none of these authors, unless we except Chambers, attempted to give a correct definition of what they meant by the terms they used. The words landscape-gardening are evidently applicable to the geometrical style, as well as to the natural style; because landscapes are produced by both: but these landscapes differ in being, in the former cae, a geometrical disposition of nature, and, in the latter, an artistical imitation. By an artistical imitation of natural scenery, or, in other words, by natural scenery imitated according to art, is to be understood a representation of nature, but in other materials than those which nature employs. Thus, an artistical imitation of a landscape is made by the landscape-painter, by using colours upon a flat surface; and a similar imitation of man and other animals, either by colours on a flat surface, or by solid forms of wood, marble, or stone, carved in imitation of the animals themselves. If, instead of marble, wood, or stone, of their respective natural colours, being employed as the material in which to imitate an animal, an attempt were made by the aid of colours and of various other materials to make so correct a representation of it as that it might be mistaken for the animal itself, this would be a fac-simile imitation of nature, and would not be entitled to rank as a work of imaginative art. It might have great merit as a piece of mechanical skill and patient labour; but it could never be considered on a par with those productions in which the thing to be represented was imitated in a manner, or by a material, totally different from that which was natural to it. Now, in the imitation of natural landscape in garden scenery, it will very properly be asked, what it is that makes that kind of imitation artistical, since a garden landscape consists of trees, shrubs, and ground, no less than a natural landscape. Here we must confess that the difference between a natural landscape and a garden landscape being much less than the difference between such a landscape and a painted representation of it, it follows, that the art of landscape-gardening must rank much lower in the scale of imitative art than landscape-painting. It has, however, the same advantage over landscape-painting which architecture has over sculpture; viz. that of combining the useful with the agreeable.

Artistical Imitations of Natural Scenery. Having shown the rank which landscape-gardening holds among other arts, we shall now point out in what manner imitations of natural landscapes in gardening may be rendered artistical. We cannot substitute anything for trees, shrubs, and ground, which shall render our imitation as different from the scene imitated as is the imitation of the landscape-painter. We must plant trees and shrubs on ground, and these must live and grow as they do in nature: but we may plant them differently from what is done by nature in the given locality and country; and this may be accomplished in three different ways. First, we may employ the same sorts of shrubs and trees which are common in the surrounding country; but, in this case, we must arrange them differently. Thus, if the whole country be covered with wood, treated as coppice-wood, the artistical scenery may consist wholly of groups of full-grown trees, surrounded by, and interspersed with, smooth turf; or, on the other hand, if the surrounding woody scenery be composed wholly of masses of full-grown timber trees, such as thick close-growing pine groves, we may employ the same kind of trees sparingly, on an extensive breadth of smooth turf, as before. There are various other modes in which a difference might be made between the object produced and the object imitated; but these will suffice to give an idea of this first or lowest degree of artistical imitation, which may be called the imitation of indigenous landscape.

The second kind of imitation consists in employing trees and shrubs of kinds totally different from those of the surrounding country; and which, whatever may be the mode of their arrangement, will, at first sight, distinguish the landscapes of which they form a part from those of the given locality or country. Thus, in Britain, the trees and shrubs of North America, the Continent of Europe, and the temperate parts of Asia, afford ample resources to the landscape-gardener. Now, the trees employed in this imitation of nature may be disposed in two ways, which we have already designated as the picturesque and the gardenesque; the first being the imitation of nature in a wild state, such as the painter delights to copy; and the second the imitation of nature, subjected to a certain degree of cultivation or improvement, suitable to the wants and wishes of man.

Picturesque Imitation. To design and execute a scene in either of these styles of imitative art, the artist would require to have the eye of a landscape-painter; to a certain extent, the science of an architect and of a botanist; and the knowledge of a horticulturist. Every part of nature, whether rude or refined, may be imitated according to art. For example, an old gravel pit, which had become covered with bushes and indigenous trees, and contained a hovel or rude cottage in the bottom, with a natural path worn in the grass by the occupants, would be improved according to imitative art, if foreign trees, shrubs, and plants, even to the grasses, were introduced instead of indigenous ones; and a Swiss cottage, or an architectural cottage of any kind that would not be recognised as the common cottage of the country, substituted for the hovel. To complete the character of art, the walk should be formed and gravelled, at least, to such an extent as to prevent its being mistaken for a natural path. Rocky scenery, aquatic scenery, dale or dingle scenery, forest scenery, copse scenery, and open glade scenery, may all be imitated on the same principle; viz. that of substituting foreign for indigenous vegetation, and laying out artificial walks. This is sufficient to constitute a picturesque imitation of natural scenery.

Gardenesque Imitation. Where the gardenesque style of imitating nature is to be employed, the trees, shrubs, and herbaceous plants must be separated; and, instead of being grouped together as in forest scenery (where two trees, or a tree and a shrub, often appear to spring from the same root, and this root is accompanied by large rampant herbs), every gardenesque group must consist of trees which do not touch each other, and which only become groups by being as near together as is practicable without touching, and by being apart from larger masses, or from single trees or rows of trees. It is not meant by this, that in the gardenesque style the trees composing a group should all be equally distant from one another; for in that case they would not form a whole, which the word group always implies. On the contrary, though all the trees in a gardenesque group ought to be so far separated from each other as not to touch, yet the degrees of separation may be as different as the designer chooses, provided the idea of a group is not lost sight of.

In *fig.* 47. the trees are arranged in the gardenesque manner; and in *fig.* 48., in the picturesque style. The same character is also communicated to the walks; that in the gardenesque style

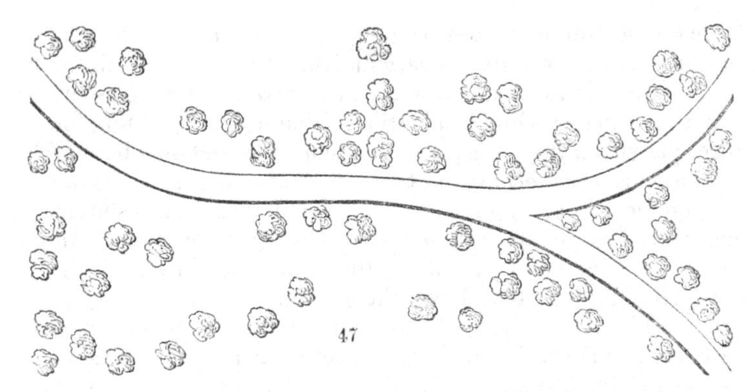

47

having the margins definite and smooth, while the picturesque walk has the edge indefinite and rough. Utility requires that the gravel, in both styles of walk, should be smooth, firm, and dry; for it must always be borne in mind, that, as landscape-gardening is a useful as well as an agreeable art, no beauty must ever be allowed to interfere with the former quality.

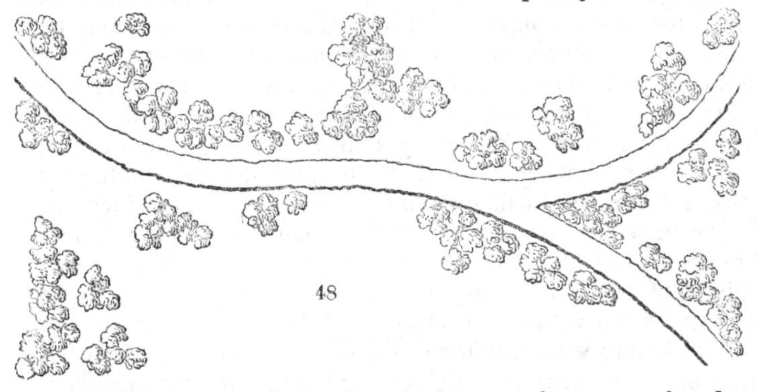

48

In laying out grounds, or in criticising such as are already formed by eminent artists, it is necessary always to bear in mind the difference between the gardenesque and the picturesque; that is, between a plantation made merely for picturesque effect, and another made for gardenesque effect. Gardenesque effect in plantations is far too little attended to for the beauty of the trees and shrubs, whether individually or collectively; and picturesque effect is not generally understood by gardeners: so that the scenery of suburban residences is often neutralised in character by the ignorance of professional landscape-gardeners of the gardenesque, and of professional horticulturists and nurserymen of the picturesque. To make the most of any place however small, all the styles of art ought to be familiar to the artist; because there are few places in which, though one style prevails, some traits of other styles may not be advantageously introduced.

In planting, thinning, and pruning, in order to produce gardenesque effect, the beauty of every individual tree and shrub, as a single object, is to be taken into consideration, as well as the beauty of the mass: while in planting, thinning, and pruning for picturesque effect, the beauty of individual trees and shrubs is of little consequence; because no tree or shrub, in a picturesque plantation or scene, should stand isolated, and each should be considered as merely forming part of a group or mass. In a picturesque imitation of nature, the trees and shrubs, when planted, should be scattered over the ground in the most irregular manner; both in their disposition with reference to their immediate effect as plants, and with reference to their future effect as trees and shrubs. In some places trees should prevail, in others shrubs; in some parts the plantation should be thick, in others it should be thin; two or three trees, or a tree and shrub, ought often to be planted in one hole, and this more especially on lawns. Where, on the contrary, trees and shrubs are to be scattered in the gardenesque manner, every one should stand singly; as in the geometrical manner they should stand in regular lines, or in some regular figure. In the gardenesque, there may be single trees and single shrubs; but there can be no such thing as a single tree in the picturesque. Every tree, in the picturesque style of laying out grounds, must always be grouped with something else, if it should be merely a shrub, a twiner, or a tuft of grass or other plants at its root. In the gardenesque, the beauty of the tree consists in its own individual perfections, which are fully developed in consequence of the isolated manner in which it has been grown; in the picturesque, the beauty of a tree or shrub, as of every other object in the landscape, consists in its fitness to group with other objects. Now, the fitness of one object to group with another evidently does not consist in the perfection of the form of that object, but rather in that imperfection which requires another object to render it complete.

In Rustic, Indigenous, or Fac-simile Imitations of Natural Scenery, the object, as we have already observed, is to deceive the spectator, and make him believe that the scene produced is of a fortuitous origin; or produced by the humble exertion of a country labourer. Such scenes differ from those of the geometric style, and also from those of artistical imitation, in this, that the same person who contrives them must also execute them. They can have no merit in design, and only mechanical merit in the execution. They scarcely require the aid of either a professional landscape-gardener, or a professional horticulturist; but, at the same time, they could not be executed by every common labourer. The imitation of such scenes must be made by a sort of self-taught artist, or a regularly instructed artist who will condescend to accept of this kind of employment.

Those suburban residences in which it might be desirable to

produce a fac-simile imitation of fortuitous scenery ought to be situated in or near the heart of a large town, in order that the scene created may contrast the more advantageously with every thing around it. In many spots in the neighbourhood of London, and other towns which are built of brick, and where gravel is found for forming the roads, there are often clay pits or gravel pits on the ground which is to be let for building on; as in other situations there are old chalk pits or stone quarries. Suppose a pit of either kind to be in some part of a piece of ground of an acre or two in extent, which is to be laid out as a suburban residence; and that it were thought advisable, as an episode to the general scenery of the place, that a fac-simile imitation of nature should be created in this pit. As a first example, we shall suppose that the pit is a clay pit, and not fit for a human habitation at the bottom. In this case, let the bottom of the pit be covered with turf, smooth in some places, and in others mixed with nettles, thistles, and other weeds, and varied by thorns, briars, brambles, elder bushes, and other trees and shrubs that generally spring up on waste ground. In one or two parts of the bottom of the pit let there be pools of water, with rushes and other aquatic plants, and some alders and willows of the commonest kind for shade. These and other details being executed in the bottom of the pit, surround it on the outside by a thick plantation of one or two kinds of trees and shrubs, such as are generally found in copse-wood; and let there be a winding straggling path through this copse-wood, of such a length as to obliterate for the moment the impression of the artificial scenery of the other parts of the pleasure-grounds on the mind of the spectator while he is pursuing the winding slightly marked path among the bushes to the bottom of the pit. If the plantation were surrounded by a hedge or other fence, and the entrance to the path were through a gap in this fence, the deception would be the more complete.

The second example we shall suppose to be a dry gravel pit, and that in the bottom of it a dwelling-place might be formed for a workman and his wife, with a hovel to serve as a cow-shed, in which cows might be kept for the family, and in which also an ass might be kept for the use of the gardener, in rolling his walks, carting manure and weeds, and for other purposes. Instead of a crooked footpath entering through a gap in a hedge, as in the first example, a rough winding road might be formed, by which it might be supposed that the gravel had been carted out of the pit, but which, owing to the lapse of time, had become principally covered with grass; and this might be entered through an old rickety gate; while in the bottom of the pit there might be the cottage dwelling, and the hovels, which, though comfortable within, ought to appear in a half-ruined state without; and a hayrick rudely fenced round, with a small stack of fagots for fuel, &c. The reader can easily supply the rest.

Both these examples would be fac-simile imitations, which might easily be mistaken for nature itself, or what we call rustic scenery; and, though they might, and doubtless would, afford pleasure in themselves, and as contrasted with the scenery around them, yet that pleasure could in no respect be considered as resulting from them as works of art, unless we were told that they were artificial creations.

Comparative Merits of the different Styles. We have been thus diffuse on the different styles of laying out grounds, in order to show the distinctive character of each; and that each has its peculiar uses and beauties. Since the introduction of the modern or irregular style of laying out grounds, it has been customary to consider that style only as exclusively beautiful; and the geometrical style as unnatural and altogether in bad taste. In consequence of this opinion pervading, we may say, all ranks, and being found in all books that mention gardening as an art of taste, we see constant attempts made to introduce the irregular manner of planting, and serpentine lines, in places where they are altogether unsuitable. For example, it is not uncommon, in the suburbs of London, to find a garden, or a public square, with the boundary fence in a straight line, and parallel to it a serpentine gravel walk. If the reader has understood the remarks in this and the preceding subsection, he will feel and comprehend the utter want of harmony which exists between the straight line and the serpentine line, independently of the impossibility of walking with comfort, when, at every five or six steps, the walk makes a turn; and where the turns, for some distance before the eye, may be all seen at once. For such scenes, and for all small gardens, bordered by straight lines, the ancient or geometrical style is unquestionably the best; and as a proof that this is the case, it may be mentioned, that all small gardens are laid out in this style involuntarily, by those who have no preconceived ideas on the subject, doubtless from an innate feeling that it is the most suitable to the boundary fence. Another class of persons, who know just enough of gardening to be aware that there are two styles, and who have been accustomed to hear the ancient style decried by all the authors who have written or even touched on gardening since the time of Horace Walpole, look at the straight walks of their small walled gardens as deformities; and regret that they cannot, for want of room, indulge in that style which alone they have been taught to esteem as beautiful. Such ideas are entirely the result of prejudice in favour of opinions expressed by men considered as authorities, and who, at the time they wrote, were so much captivated by the novelty of the irregular style, that they could not allow themselves to do justice to any other. Hence, they condemned the geometric or regular manner, without stopping to enquire into its merits; or, in their admiration of the modern style, being able to separate what was merely the effect of novelty, from what was due to its permanent characteristics. If what we have said on this subject has been understood by the reader, he will consider the ancient style as equal in point of merit to the modern one, and possessing quite as much claim to be called natural, and in good taste, as it does. In fact, the geometric style may be said to be the more natural of the two; since the first indications which children and rude nations give of buildings, sculpture, gardening, or any other ornamental or mixed art, are always marked by regularity, or some disposition of the materials which shows the object produced to be the work of man, and not the result of nature.

As a general rule, it may safely be laid down, that whenever a piece of ground to be laid out as a garden is small, and bounded by straight lines, the geometrical style is that which ought to be employed; that when the ground to be laid out as a garden is large, it may be laid out in any style, or partly in the regular, and partly in the irregular styles; and that where the surface of the ground is varied, the irregular style is most suitable; while the geometrical style should be preferred when the surface is even or flat.

With respect to those modifications of the irregular style which we have described as the picturesque, gardenesque, and rustic or rural, the first, as it requires least labour in the management, is best adapted for grounds of considerable extent; the second is more suitable for those persons who are botanists, rather than general admirers of scenery, because it is best calculated for displaying the individual beauty of trees and plants, and the high order and keeping of lawns, walks, &c.; and the third for persons of a romantic or sentimental turn of mind, who delight in surrounding themselves with scenery associated with a station in life strongly opposed to that in which they are really placed; or to attract attention by producing a striking contrast to refined and artistical scenery, whether in the irregular or geometric styles.

SUBSECT. 3. *Of those Principles which are peculiar to Architecture and Landscape-Gardening respectively.*

HAVING treated on those principles which are common to architecture and landscape-gardening, it seems reasonable that something should be said on such as are peculiar to each. These, it must be at once obvious, are very different in the two arts; since in the one, viz. architecture, the artist has to operate with inanimate matter, in consequence of which his productions are most perfect the moment after they are finished; while, in the other, he has to operate with materials which live and grow, and the effect of which depends on time and future management.

Permanence may be said to be the principle which is peculiarly characteristic of architecture, and which leads to all the various subordinate principles and rules taught by the art of building; such as stability of position, durability of materials, strength of construction, &c.

Progress may be said to be the peculiar principle which is characteristic of landscape-gardening, and which leads to all the subordinate principles and rules which regulate gardening as an art of culture. Among these are, providing for growth, by suiting the trees to soil, situation, and climate; providing for their future increase in size, by thinning and pruning; providing for all the various points of culture and management; and providing for removal and decay, by the introduction or substitution of young plants.

In comparing any house with any garden, it will be found that the one is nearly always the same; and that the other, not only changes every year, but changes with every season. Every natural change which takes effect on an architectural structure after it is completed, may be considered as a step in its progress to decay; while every natural change in a garden, after it is completed, and up to a certain point, is a step in its progress to maturity. It is the business of the architect, therefore, to bring his work as near to perfection as he can himself, and to construct it so that it may remain as long as possible without change; while it is that of the landscape-gardener to form his plan and execute it in such a manner, as that it may admit of all those future alterations, such as the removal of trees by thinning, altering their shapes or sizes, pruning, &c., which the progress of trees and shrubs generally, and the growth and habit of particular kinds, will require.

Our object in mentioning these two principles is, to show that inasmuch as a knowledge of the art of building is essential to the architect, so is that of gardening, as an art of culture, to the landscape-gardener.

SECT. II. *Ground Plans for laying out and planting Suburban Gardens, with Directions for their Management, and Estimates of their first Cost and yearly Expense.*

SUBURBAN gardens may be classed in the same manner as houses, viz. into first, second, third, and fourth rates.

First-rate Gardens we shall consider as including all those which have a lawn and pleasure-ground, and also a park or farm. In residences of this class, the house stands at some distance from the entrance gate; and the extent of the grounds may be from ten acres upwards.

Second-rate Gardens we shall consider as those in which the house stands at some distance from the entrance gate, but to which there is no park or farm attached. These residences have a pleasure-ground, a walled kitchen-garden, and stable offices; and their extent may be from two acres upwards.

Third-rate Gardens may be such as have the house at some distance from the entrance gate, but in which the lawn, pleasure-ground, and kitchen-garden are combined; and they may be an acre or more in extent.

Fourth-rate Gardens are those in which the house forms part of a street or row; and their extent may be from one perch to an acre.

The first three of these classes may be considered as villa gardens; and the gardens of street houses and cottage gardens are included in the fourth. It may be observed here, that the size or rate of the garden is not necessarily a criterion for the size and rate of the house; for, in the immediate vicinity of a town, it is very common to find a first-rate house forming one of a row of houses placed along a road, with a fourth-rate garden; and, farther in the country, it is equally common to find a cottage removed to some distance from the road, and standing in a third, or even second, rate garden. We shall commence with the lowest class, or fourth-rate gardens.

SUBSECT. 1. *On laying out, planting, and managing Fourth-rate Suburban Gardens.*

IN treating of this class of gardens, we shall first take into consideration those circumstances which are common to the class; next, those which depend on particular circumstances of situation and extent relatively to the house; and, thirdly, those which depend on the taste and wishes of the occupier.

The general considerations are those which relate to permanent out-door arrangements and aspect.

The relative considerations include the gardens of houses in a connected row or street; the gardens of double detached houses, that is of houses built in pairs, and forming part of a row; and the gardens of houses which are detached on every side, but which still form part of a row, or line of houses.

The personal considerations are, laying out and planting with a view to, 1. Economy in the first cost; 2. Economy in the after-management; 3. Profit; 4. Exercise and recreation for the occupier and his family; 5. Growing fruits and culinary vegetables, with some flowers; 6. Growing only ornamental trees, shrubs, and flowers; 7. Growing tropical fruits and flowers, and forcing; 8. Florist's flowers; and, 9. Forming a botanical collection.

Division i. *General Considerations.*

§ 1. *Permanent out-door Arrangements.* — The house being built, we shall suppose the whole plot of ground, including the front garden, the space on which the house stands, and the back garden, to vary from 50 ft. to 300 ft. in length, and from 20 ft. to 100 ft. in breadth; for the same modes of planting and laying out are applicable to all the plots between these two extremes, and even to larger and smaller ones attached to houses, forming part of a street or row. The soil of such a plot, whether large or small, we may suppose to be a loam more or less light or heavy; that soil being more common than any other in the neighbourhood of towns, because they are generally built in valleys or plains, where the soil is most commonly alluvial or clayey, though sometimes it is sandy or gravelly. The surface of plots of ground within the given limits can scarcely be otherwise than tolerably even; for, though it may slope in some directions, and have some holes and protuberances, from accidental circumstances, it can scarcely have any natural undulations. The extent and the kind of ground being given, the first things to be considered are, the supply of water or of gas to the house, if by pipes that come through the garden; the pump or well, if water is not laid on: the outbuildings, if any; the under-drainage of the ground; the surface drainage; the number and direction of the walks; the improvement of the texture of the soil; and the arrangement for posts for drying clothes. The rate of the house is not an important object of consideration, as it is chiefly a guide in so far as it respects the kind and the magnitude of the outbuildings.

Pipes through the Garden, for Water or Gas, &c. As most houses in the neighbourhood of large towns are supplied with water from public sources, by means of small pipes laid from a main pipe in the street to the cistern or cisterns in the house, one of the first things to be attended to, in contriving the arrangement of the front garden, is, to fix on the place where the service-pipe, as it is called, shall be laid down. The object should be, to secure the pipe from frost throughout the whole of its length, and to admit of its being laid bare, when necessary, for repairs, with as little derangement to the garden, and especially to the walks, as possible. The service water-pipe commonly enters the front garden under the sill of the street entrance or gate, and is conducted along one side of the walk, or through that part of the lawn, or front garden, on which there are few trees or shrubs, to the cisterns in the house or in the area. The depth of the trench, in the bottom of which the pipe is laid, ought not to be less than 3 ft., in order to secure it from frost; and if, in any part, it passes through ground which is likely to be occasionally trenched, there ought to be a line of bricks or tiles placed immediately over the pipe, to protect it from the spade or mattock. Second and third-rate suburban houses are frequently supplied with gas from the mains, which are carried along the streets for the public lamps, for a lamp outside the front door; and, probably, gas will ere long be required, even in fourth-rate houses, for the purposes of cookery. The service-pipes for conveying gas need not be laid deeper in the soil than 1 ft., as they are not liable to be injured by frost; but, like the water-pipes, they ought to be protected, by bricks or tiles, from the risk of the spade, when the gravel of the walks is turned or flower-beds are dug; and, like them, they should be laid down in such a direction as that, when repairs are wanted, they may occasion as little derangement of the surface of the ground as possible. In general, the gas-pipe should not be laid directly over the water-pipe; because, in that case, the latter could not be laid bare and examined without disturbing the former. If, however, the water-pipe is laid so deep as to be out of the reach of frost, or of being disturbed by the spade, it is not likely to require repair for many years, or even a lifetime; except at its junction with the main pipe in the street, and with the cistern in the house or in the area; and, in such a case, the gas-pipe may be laid over it. That part of the water-pipe which rises out of the ground to supply the cistern, if not carefully protected by a thick casing of charcoal, ashes, or sand, enclosed in brickwork, will be liable to be frozen and burst every winter; more especially where the cistern is placed in the area. The connexion, indeed, of the service water-pipe with the cisterns of a house is one of the most difficult points which the builder has to manage. It is very commonly got over by directing the servants of the occupier, when the winter approaches, to wrap straw round the pipes where they join the cistern; or, perhaps, a cock is so placed as to enable the servants to empty the service-pipe, or that portion of it next the cistern, every time that the cistern is filled; but, as these precautionary measures are commonly delayed till after a severe frost has occurred, the pipes are generally burst once or twice every winter, and the plumber requires to be called in accordingly. There are, indeed, very few second and third-rate suburban houses, the occupiers of which have not a plumber's bill to pay every year. The greatest care, therefore, is requisite, to have the service water-pipe rendered perfectly secure from frost, from the point where it leaves the main in the street, to the orifice of the cock which delivers the water to the cistern; and builders and occupiers should insist on their plumbers making sure of this result.

Out-buildings, Pump or Well, &c. In gardens to third and fourth-rate houses a privy is commonly required outside the house, and a dust hole is essential. Both these ought to be placed close to the house behind, and near the back door, in order that they may be as conveniently reached as possible. At the bottoms of gardens, even of the smallest size, it will generally be found useful to have a shed for lumber, clothes-posts, old wood for fuel, &c.; and a pit, 3 ft. square and 2 ft. deep, lined with brick or stone, for such rubbish as will rot into manure. There should, also, wherever it is practicable, be a pump or well, because, even if it does not supply water fit for the kitchen, it will, at least, be useful for watering the garden, and in case of fire. If the occupier is much attached to gardening, and intends to cultivate culinary vegetables, a manure tank, for liquid manure (which might serve, in some cases, as a cesspool for the privy), would be found useful; and, if it were properly constructed of brick laid in cement, with a closely fitted covering stone, it would be attended with no offensive smell. When the family washing is performed at home, the soapsuds should always be emptied into this tank, as, after they have fermented for some time, they make excellent manure. It is convenient to have the shed, well, &c., at the lower part of the garden; or, in whatever part of it they may be, to have them hidden from the windows of the house, and from those of the adjoining houses, by evergreen trees and shrubs. In fourth-rate gardens to first or second-rate houses, there may be one or more bowers, or covered seats; and these, and also architectural ornaments and green-houses, or other plant-houses, if they are required, ought to be taken into consideration before determining on the position of the drains, and the direction of the walks.

Drainage, Walks, and Texture of the Soil, &c. The ground, where not naturally dry, ought to be perfectly under-drained, by drains down the middle of both the front and back gardens, communicating with the public sewer, or with some other main drain; or, if the back garden be large, and the soil very wet, there may be a drain along one side of each of the side walks: to these drains there ought to be communications, with gratings at regular distances, if the walks are above 100 ft. in length; but, if they are shorter, one grating, in the lowest point of the garden, will be sufficient. If the garden be 50 ft. or 60 ft. broad, the soil a very strong clay, and the surface flat, there ought not only to be drains to the walks, but drains to the part of the garden kept under kitchen crops or flowers, and also to those parts which are kept under grass. Nothing contributes more to the comfort afforded by a garden on a strong clayey soil, than thorough drainage; without which, it is impossible to walk out in it after a shower of rain, though this is generally one of the most delightful times in which a garden can be viewed by a person fond of gardening. In very small gardens, where it is not considered necessary to have under-drains, it will be sufficient to direct the water by the surface-drainage to one point, where it may sink into the soil. The proper execution of the surface-drainage is of the utmost importance to the enjoyment of every description of garden, whether large or small; more especially of all such as are on loamy or clayey soils. For this reason, even, in some cases, before determining on the under-drainage of the back garden, the surface-drainage, ought to be attended to by smoothing down all the inequalities of the ground, and by forming it so as to give it an inclination, or slope, to some one point in the garden, generally that which is the most distant from the house. To this point all the rain water that falls on the walks should run, and find an exit either by a porous subsoil, or by a grating communicating with an under-drain. In back gardens of great length, it may happen that the middle is the lowest part of the surface; and when this is the case, the surface water ought to be directed thither from both extremities; but in fourth-rate gardens it seldom happens that the surface may not be made to slope from the house to the extreme end of the back garden, even though it should be 200 ft. in length. In regard to front gardens, if the walk and the lawn have an inclination from the house to the street, that will generally be found sufficient. The surface-drainage, and also the under drainage, being determined on in the mind of the artist, we shall suppose that he has had the surface of the ground, both before and behind the house, sloped accordingly. The next point is to determine the number and direction of the walks; because it is alongside of them that the under-drains are commonly formed with the greatest advantage. In all small gardens, the object is to get as long a walk as possible; and the longest line within any given space must necessarily be that which goes round it. The direction and position of the walks in fourth-rate gardens is therefore easily settled. The next point is to determine where the drains are to be placed, and to have them dug and filled in, the manner of doing which will be treated of under the head of garden operations. The walks may be then hollowed out, in the manner which will also be hereafter described; and the ground which is to be planted or covered with turf may be trenched to the depth of 2 ft. or 3 ft.; and, in performing this operation, if the soil requires to be rendered lighter, or more porous, which is always the case with a strong clay, it ought to be mixed with lime rubbish, gravel, or sand, or with all of these, to such an extent as to render it readily pervious to water. It is more especially necessary to do this in those parts of the garden which are most in shade; particularly on the north side of the house, when it fronts the south. The ground being trenched, and the walks hollowed out, the whole should remain some weeks, and, if convenient, one whole winter, to settle, before the final smoothing of the surface, the graveling or paving of the walks, the edging of them, and the planting of the trees and shrubs. When so much time cannot be allowed for the ground to settle, it is better not to trench it under where the walks and their edgings are to be placed at all, but merely to dig it; because, when trenched ground does not settle evenly, which it very seldom does, the materials of which the walk is formed, and its edgings, require, after a few months, to be taken up and relaid.

Arrangements for Posts for Clothes-Lines. One considerable advantage of a suburban residence to most families, and particularly to such as have children, is, that they are enabled by it to wash at home, and have their clothes dried in the open air. In the country, clothes are generally dried on hedges or bushes; but in suburban gardens they are commonly hung on lines which are stretched from post to post along one side, or round the entire garden. In gardens to houses of the smallest size, these posts are commonly fixed in the ground, and the lines fastened to them on washing days, and taken down when not wanted; but, in all the better description of gardens, sockets, which have been previously fitted to the lower part of the posts, are fixed in the ground; and into these the posts are inserted on washing days, and the lines attached to them. When the clothes are dried, and the lines are also dry, and have been taken off and laid in a dry part of the wash-house or back kitchen, the posts are taken up out of the sockets, and put somewhere under cover. If there is no shed, they may be laid on brackets projecting from the boundary wall, with a coping board close over them to throw off the rain.

49

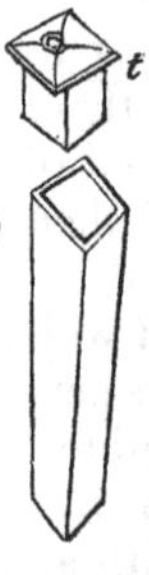

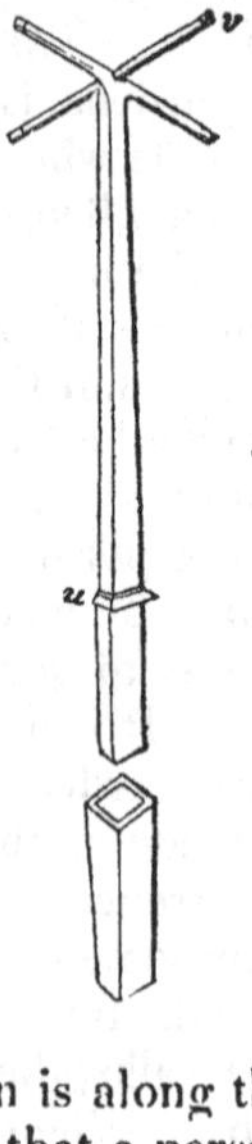

Fig. 49. shows the socket for the clothes-post: it is made of four pieces of board, forming a rectangular tube, rather narrower at the lower end; and *t* is a cap, or cover, also of wood, with a ring in the top for lifting it off, which is put on when the posts are not in use, to prevent the socket from being filled up with dirt and stones; the length of the tube forming the socket is generally about 18 in., and the width inside about 4 in. at top and 3 in. at bottom.

50

Fig. 50. represents a clothes-post: it has a shoulder at the lower end (at *u*), to prevent it from being wedged too firmly into the socket; and two pins (*v*) passed through the top, in opposite directions, for the purpose of fastening the lines. In some gardens the lines are fastened to trees, or stretched across the garden from hooks in the side walls; but the most convenient position is along the sides of the walks, over the margin of the turf, so that a person may stand on the walk, while hanging up and taking down the clothes.

§ 2. *Laying out and planting Fourth-rate Suburban Gardens, relatively to Aspect.* As gardens of this class are always of comparatively small size, they are more influenced by their aspect, and the position of the house relatively to them, than any others. The first point, therefore, to be taken into consideration, in contriving how they are to be laid out and planted, is, to ascertain how far the surface of the garden, and the walls or other fences by which it is bounded, are exposed to the rays of the sun throughout the year. On the degree of exposure to the sun all the future operations depend; for this must determine the kinds of trees and plants that will thrive in the garden; and on these trees and plants necessarily depends the kind of soil that ought to be procured for growing them in, or the improvement that ought to be bestowed on the soil already there; as well as the future culture of the whole. *Fig.* 51. is a diagram, representing the four leading directions of streets which occur in towns and their suburbs; and the bare inspection of this figure will show what positions are most favourable for the admission of the sun to the surface and boundary walls of the gardens, as well as to the different sides of the houses, throughout the year. In this figure, *a a* represent two rows of houses placed in a direction parallel to a street running north and south; in the gardens of which there will be no shadow at midday throughout the year, but those of the separation fences: if these are low, or if they are of open work, the shade will do very little injury; and, unless the situation is surrounded by a very smoky atmosphere, peaches and grapes may be ripened in the climate of London, on the wall which faces the south. Ivy or ornamental creepers may be grown on the wall having a northern exposure; and almost any kind of plant thought desirable may be cultivated in the interior of the garden; provided such trees or shrubs as may be planted there are not allowed to grow to such a height as to shade the peach wall.

At *b b* two double detached houses are shown; the fronts of which are also parallel to a north and south road, and the gardens of which have nearly the same advantages as those shown at *a*, except that a small portion of the garden of the house next the north will be in the shade during the greater part of the winter season; as shown by the triangular shadow thrown upon that garden in the diagram.

At *c* are two rows of houses similar to those at *a*; but, as their direction is south-west, the sun will shine on both sides of these houses, and both on their front and back gardens, every day on which he appears, throughout the year. The small proportion of shadow which these houses will throw upon their

51

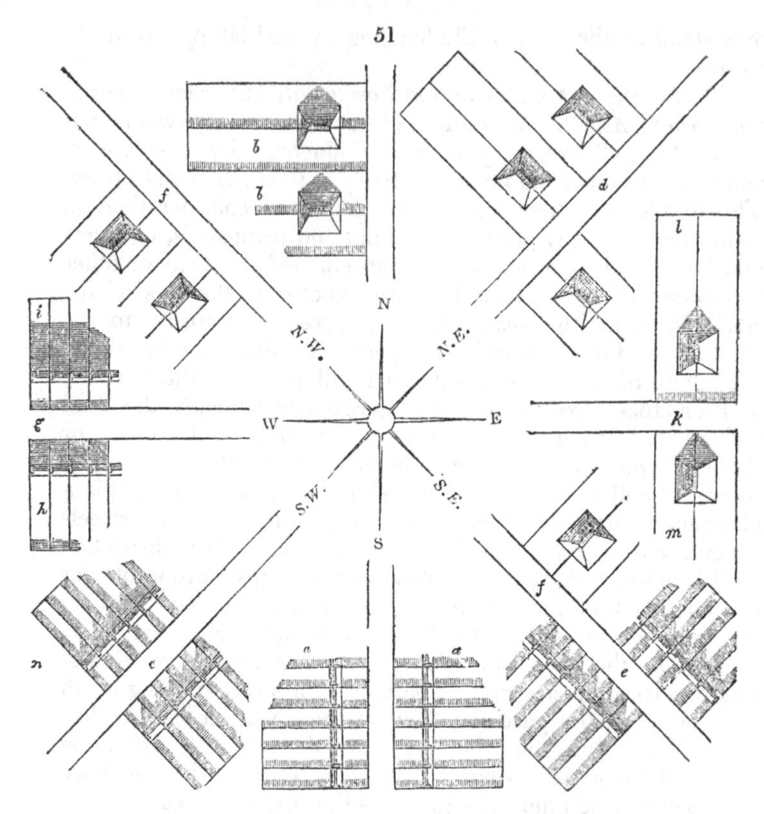

gardens may be considered as represented by the triangular masses of shade on the front gardens on one side of the road; and on the back gardens on the other. The actual shadow thrown by a row of connected houses on the ground would, of course, be a parallelogram, and not a series of triangles; but our object in giving this diagram is not so much to show the actual shadows that will be thrown at any particular time upon the ground, as to indicate the proportion of shadow which one position of the house and garden fences will throw on the ground relatively to what will be done by another position. In the double detached houses, *b*, *d*, *f*, &c., the proportion of shadow which we have shown in the diagram also indicates the part of the garden which will be most in shade throughout the year. The walls of those gardens which have a south-east exposure will ripen figs, apricots, and the finer plums; while those having a north-west exposure may be devoted to currants or morello cherries; or to roses, and the finer kinds of ornamental creepers.

At *d*, on both sides of a road in the same direction, are shown double detached houses, the gardens of which have similar advantages to those at *c*, but increased in consequence of the houses being detached, and their diagonals being south and north, which admits of the sun shining all round them on every day on which he appears.

At *f f*, on a road in the direction of south-east and north-west, are shown double detached houses, square in the plan; the diagonals of which, being directly south and north, their exterior walls and their gardens will have nearly the same advantages as those of the houses shown at *d*; but with this difference, that a wall having a south-west aspect is not considered quite so good for ripening the finer fruits, or growing the more delicate creepers, as one facing the south-east.

At *g* are shown two rows of houses on the north and south sides of a road running east and west; by which it will be seen that the front or street gardens of the houses on the south side of the road, will be entirely in the shade during the greater part of the winter; while the gardens behind these houses will be entirely in the sun throughout the year. As the side walls of these gardens (at *h*) can only enjoy the sun one half of every day, neither of them are fit for ripening the finer fruits; but both of them will grow plums and cherries, and both are admirably adapted for ornamental climbers. Against the wall at the further end of the garden, which will be in the shade almost every day in the year, ivy and some very hardy climbers may be planted, such as the Virginian creeper and the Ayrshire rose: and against the house at the opposite end, which will be fully in the sun every day in the year, may be planted a vine and a fig; the latter to cover the lower part of the wall, as high as the first floor windows, and the former to clothe the upper part as high as the roof. The walks of such gardens will be equally dry on both sides of the garden, because one boundary wall does not throw more shade on the walk running parallel to it, and near it, than the other; and, provided the cross walk at the south end is kept at such a distance from the end wall, or boundary fence, as not to be shaded by it, it will be as dry as the side walks. Hence, these gardens are more favourable for invalids to walk in during winter, and immediately after rains, than those can be where one of the side walks is chiefly in shade, as will be the case in the gardens of the houses at *a a*. The front gardens of the houses on the north side of this east and west road will enjoy the sun throughout the day during the whole year, and are, therefore, unless the road is very dusty, suitable for the culture of the finer flowers and shrubs. The gardens at *i*, behind these houses, are, however, the most unfortunate of all in point of aspect; for one half of them will be entirely under shade during the whole of the winter months, and also during a considerable portion of both the spring and autumn. Only the small portion of wall at the bottom of this garden, which forms the end of it, is fit for the finer fruits; the remainder should be covered with ornamental climbers: and the walks, during the greater part of the year, will unavoidably be cold and damp.

At *k* are shown two double detached houses, one on the south and the other on the north side of an east and west road, which partake more or less of the character of the houses shown at *g*: but the two gardens at *m*, being open and airy, will be very suitable for taking exercise in throughout the year; and a vine or a fig may be placed on the south side of the house. The gardens at *l* will not be much inferior to those at *m*, in consequence of the sun being admitted to the sides of the house as well as to its front. In this case, as in the gardens of the houses at *d*, *b*, and *f*, the great advantage of detached houses, in point of the enjoyment of the sun and air, is obvious.

Single detached houses, it will be sufficiently obvious, are subjected to the same laws respecting shade as detached houses which are in pairs.

Division ii. *Relative Considerations.*

§ 1. *The Gardens of Houses in a connected Row, or Street,* include a front and back garden; neither of which can ever be of greater width than the frontage of the house. The row of houses being, for greater convenience, generally placed near the road, the front gardens are seldom of much depth; but the depth of the back garden may be carried to any extent that the situation will allow. The side walls, or fences, are always parallel to each other; and they are generally low, so as to obstruct the free circulation of the air as little as possible, and to produce but little shade. The end of the back garden, opposite to the house, is necessarily of the same width as the house; and frequently, though not always, it has a door into a back road or lane. In the back gardens attached to second and third-rate houses, the coach-house and stables, or other out-door offices, are frequently placed at the bottom, or farther end, of the garden, and are entered by a gate opening into the lane; but by far the greater number of fourth-rate gardens belong to houses the occupiers of which do not keep horses or carriages. It must be obvious, that, in gardens of so regular a shape, whether large or small, there can be very little variety produced in laying them out; that the style adopted must be regular, as indicated by the shape and the boundaries of the house; and that the chief interest must depend on the trees and plants introduced, and their culture.

§ 2. *The Gardens of double detached Houses* are, like those of connected rows of houses, necessarily bounded by straight and parallel lines; and they also consist of some ground in front of the house, with a larger portion behind it; but they differ in having their front and back gardens connected by another portion of ground along one side of the house, as at *b*, *d*, &c., in *fig.* 51. p. 178. As this arrangement requires a greater width of frontage, fourth-rate suburban gardens, belonging to double detached houses, are generally larger in proportion to the size of the house, than those belonging to houses in a connected row; and their outline frequently approaches more to that of a square, than to that of a parallelogram. In other respects, this description of garden differs very little from the preceding one, except in making a greater display towards the street, from its greater width of frontage. In laying out and planting such a garden, the chief source of interest still consists in the kinds of trees and other plants employed, and their careful culture, rather than in their disposition for a picturesque and varied effect, though more of this can be given than in the preceding kind of fourth-rate suburban gardens.

§ 3. *The Gardens of Houses which are detached on every Side* differ from the preceding kind principally in having a still greater frontage; a portion of the garden extending on each side of the house, in addition to that space which is directly in front of it. The general outline of gardens of this kind, from the circumstance of the houses to which they belong still forming part of a street, is necessarily rectangular; but, if the depth be considerable, the back garden will be large. Gardens of this kind make a handsome appearance from the street; and, though they do not admit of much variety in the disposition of the walks, from the boundaries being still in straight and parallel lines, yet the walk from the street entrance to the front door of the house may be so conducted as to display an interesting variety in the disposition of the trees and shrubs.

Having briefly characterised these three kinds of fourth-rate suburban gardens, we shall now take each kind separately, and show the manner of laying out and planting them, so as to suit the personal considerations of different occupiers.

Division iii. *Personal Considerations.*

The situation, aspect, and extent of the ground to be laid out being given, the artist is thus rendered aware of what kind of garden it is capable of being made; but it is necessary, in addition to this, that he should know what kind of garden the intended possessor is in want of. He may require a garden which shall be kept at the least possible expense; or one which shall be devoted chiefly to culinary vegetables, fruits, or flowers. He may wish one which shall never require any labour or attention from him; or one which will afford him something to do throughout the year. He may wish a good deal of grass for his children to play on, or a good deal of gravel, to save the expense of mowing the grass. He may be a botanist, and wish to have a collection of plants; or a florist, and desire to grow florist's flowers; and, finally, he may wish to indulge in forcing either fruits or flowers, or to grow hot-house or green-house plants. These, and similar temporary circumstances, must all be taken into consideration in laying out and planting a garden, even of the smallest size, no less than the permanent ones of situation and aspect; and, of course, they apply, in a greater degree, to gardens of the other kinds above mentioned.

§ 1. *Gardens of Houses forming Part of a connected Street, or Row.* — 1. *Laying out and Planting, where the Object is the saving of Expense in the First Cost.* If nothing is to be attended to but the saving of money, there can be no more economical mode of laying out and planting a suburban garden than that frequently adopted by London builders; viz. to level and dig over the ground, to line out the walks, and cover their surface with a thin coating of gravel, and to leave the planting of the back garden to the taste of the person who may purchase or rent the house; while a few trees and shrubs, of the cheapest kinds that can be got, are planted in the plot of ground in front. A person, without any experience in gardening, intending to rent or purchase a house, would not be aware of the falseness of the appearance thus put on the ground; and he might commence the occupation of the premises, and probably continue there several years, without finding it out. If the soil were dry, he would not have much positive suffering, though, as the things he planted could never grow freely, he could not have much enjoyment. On the other hand, if the soil were a strong clay, the whole garden would be so damp, during the greater part of the year, as not only to preclude all enjoyment, either from walking round it, or cultivating plants in it, but, from the surface being at all times saturated with moisture, the dug ground soddened, and the gravel of the walks blackened, and covered with lichens, a kind of malaria would be produced, which might bring on fever and other diseases. All this the builder might have prevented by proper under-drainage: but this would have added considerably to his expense, and not told in the eyes of his customers; and, besides, it sometimes happens that suburban houses are built in situations where there are no public drains to convey away the water collected in the under-drains of private property. Hence, when a garden is examined by an intended purchaser or occupier, the first points to enquire into are, the nature of the soil, whether it admits of being thoroughly under-drained, and whether this has been successfully accomplished. This subject, however, is one which very few persons who rent or purchase suburban gardens, especially of the fourth-rate class, can be supposed to know anything about; and, therefore, we would recommend them always to consult a commercial gardener, or some gardening friend, such as we trust this book will prove to be to them, before concluding their agreement.

As we do not consider the mode above described, however cheap it may be, as worthy even of the name of laying out a garden, we shall now proceed to show the cheapest mode of doing this, in such a manner as shall be conducive to health and comfort, and also agreeable to the eye. We shall suppose the piece of ground to be a parallelogram 150 ft. in length, and 30 ft. in breadth; and to lie in the direction of north and south; though, if it were in any other direction, the laying out would be the same, and only the planting somewhat different. The front garden (*a*, in *fig.* 52.) is a square of 30 ft., and the house is 30 ft. by 20 ft., exclusive of a sunk area behind, which is 6 ft. wide, and which

52

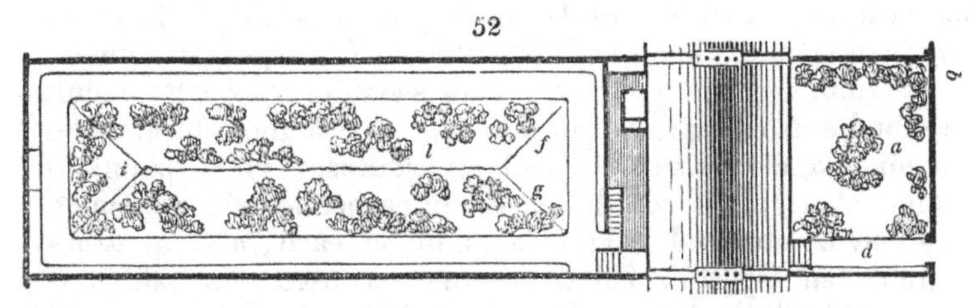

contains a privy and a dusthole at one end, and a landing to the back door of the house at the other, under which is the water cistern, and from which is suspended a cage safe. The service-pipe for water being laid down in the front garden, the walk in it is next to be formed, from the street entrance to the front door of the house. It is 4 ft. in width, and has a gradual inclination, from the steps leading up to the house door, to the street; and it may be laid with gravel or flagstone, whichever may be cheapest in the given locality. The surface of the plot in the front garden should have a direction given to its slope, so as to carry the water to one angle (say to *b*), which should be 6 or 8 inches lower than the opposite angle (*c*), and, consequently, some few inches lower than the margin of the walk (at *d*). This will provide for carrying the surface water from the house, and from the margin of the walk to the point *b*, where, if the subsoil is sand or gravel, it will sink into the ground; but, if it is clay, a drain must be formed from that point, and connected with the surface by a mass of loose stones or gravel. If this garden lies on the south side of an east and west road, it will be under shade the greater part of the winter, as will be seen by turning to the diagram, *fig.* 51. in p. 178. We would, therefore, not recommend, in this case, any dug beds to be in it, but simply one evergreen shrub to be placed in the centre of the plot, and the whole of the remainder to be sown down with grass. The shrub may be either a laurustinus, aucuba, box, variegated holly, or any similar kind that will grow in ordinary soil, and not cost more than 1*s.* or 1*s.* 6*d.* We would not recommend a deciduous shrub, because the effect of that in the winter season, when it was without its leaves, would be to increase the apparent coldness and dreariness of the situation. On the other hand, if the garden be situated on the north side of an east and west road, it will be exposed to the sun throughout the winter, and there may be a few shrubs of very low growth planted round the walls, and one or more deciduous or evergreen flowering shrubs in the centre, as indicated in *fig.* 52. The plot may then be sown down with grass, as in the other case. The number of shrubs required will not be more than a dozen; and, as the great object is economy in the first expense, these may be of kinds that will not cost more than 6*d.* or 1*s.* each. Against the walls, there may be a common honeysuckle, a Virginian creeper, a common ivy, a white and a purple clematis, an Ayrshire rose, a Boursault rose, and a Noisette rose; and, as a group in the centre, *Rìbes* sanguíneum, a Persian lilac, and an arbutus. These plants, if purchased large, will cost double the price mentioned; but, small, they may be obtained at an average of 6*d.* each; and, in so favourable an aspect, small plants may be used with safety, as there will be little danger of their not growing. Those planted against the side walls will, of course, be trained to them as they advance in growth; but those against the front wall, which we conclude to be about 2 ft. high, with a railing over it, may be allowed to take their natural shapes, as, indeed, may those on the wall facing the east; because the irregularity of their outline, and their projection over the turf, as shown in the figure, will add to the variety and intricacy of the scene, and be felt as a pleasing contrast to the formal and solid character of the surrounding walls. If the occupier should prefer flowers to shrubs for the centre of his front garden, then he may form a circular dug patch, or bed, 3 or 4 feet in diameter; and in this, where the garden faces the north, he may plant, in the centre, *S*axífraga crassifòlia (which has large thick leaves, and red flowers appearing in April), or any other strong-growing evergreen flowering plant; and, in the summer time, he may introduce round it different kinds of annuals, such as mignonette, candytuft, ten-week stock, Nemóphila insígnis, or any other dwarf neat-growing plant, with bright flowers. If the front garden is on the north side of the road, and faces the south, then the occupier may plant his central bed with any kind of flowers that he chooses; but, as the main object in planting this design is economy in the first cost, purchasing a few new annual plants every spring from an adjoining nurseryman, and sowing a little mignonette round the margin of the bed, will be found the cheapest mode, as well as that which requires the least cost in management.

The back garden may be formed with a walk round it, and a border 18 in. wide between the walk and the walls, except at the bottom, where it may be 2 ft. wide, to admit of the projection of a board to cover the clothes-posts, and of a pit in the ground, with a cover (as described in p. 189.), for containing garden refuse, as shown in *fig.* 53. *e.* The walks here need not be more than 3 ft. in width, and they may be laid with gravel, or whatever material is cheapest, that will make a solid smooth walk. In some situations, the scrapings of roads make excellent walks; and, though they have in general a dark earthy look, very inferior to that of gravel or flagstone, yet, as they cost little, they merit recommendation in this place. The mode of forming these, and every other kind of walk, will be described when treating of garden operations. The plot of ground surrounded by the walk should not only slope outwards from the house, but, if the soil is a loam or strong clay, in order that it may be kept dry on the surface, without incurring much expense of under-draining, it ought to slope on every side, from the walks to the middle of the plot, as indicated by the line *f g l*; the lowest point being at *i*, whence there may be an under-ground drain to a main drain, or the public sewer. If the house should be in a newly formed neighbourhood, and there should be no main drain or public sewer formed when the garden is laid out, the drain from the garden may be conducted to a well, which should have a pump; and, during and after rains, the water ought to be pumped out of this well to a surface gutter outside the garden, and thus conducted away. If this mode be considered objectionable, then the entire surface of the garden must be raised, so as to be higher than that of the ground exterior to it at the further end, and the water conducted along the surface of the garden till it reaches the outside. These are the only substitutes for public drains; and, as they are both expensive and inefficient, we would recommend every one to avoid occupying a house and garden where they are rendered necessary.

If the soil is a very strong clay, it will be advisable to keep the margin of the central plot of the back garden rather under the level of the walks than over it; in order that, after every shower, the water from the walks, and, indeed, every part of the garden, may immediately find its way over the surface to the middle, from which it will be carried off by the under-drain. Had the saving of expense in the first cost not been the main object, we should have recommended effecting this purpose by under-draining along the sides of the walks, instead of resorting to surface-draining to an under-drain conducted from one point: but these side under-drains, in a very strong clayey soil, if made so as to be effective, would cost as much, or more, money than the walks. The surface of the plot in the centre may be sprinkled over with low flowering deciduous and evergreen trees and shrubs, as indicated in the plan, and then sown down with grass seeds. The number of trees and shrubs required may be about forty-eight; and the most suitable way to procure these for a garden, where the main object is economy in the first cost, is to buy a mixture of this number of plants, leaving the kinds to the nurseryman, who, on the supposition that he is a respectable tradesman, will charge less for them than if an order for forty-eight specified sorts, of even the commonest trees and shrubs, were given to him; not only because he would have less trouble in selecting them, but because he could send most plants of those kinds of which he had most, and thus get rid of those articles with which his nursery might be overstocked. Where, however, the occupier would like to make a choice himself, we may recommend to him, as cheap and handsome-growing or flowering plants, the following: —

*A*mýgdalus commùnis, the common almond.
*B*étula álba, the common birch.
*C*ércis *S*iliquástrum, the Judas tree.
*C*órnus sibérica, the coral-wooded dogwood.
*C*ratæ`gus *O*xyacántha var., the double and scarlet-flowering hawthorns, and other species or varieties of *C*ratæ`gus.
*C*ýtisus alpìnus, the Scotch laburnum.
*D*áphne *Mezèreon*, the common mezereon.
*F*àgus sylvática purpùrea, the purple beech.
Halès*ia* tetráptera, the snow-drop tree.
*H*ibíscus syrìacus, the althæa frutex.
*H*ippóphae argéntea, the silver-leaved buckthorn.
*H*ypéricum Kalm*iànum*, Kalm's St. John's wort.
I'lex *A*quifòlium, the common holly.
*L*aúrus nóbilis, the sweet bay.
*L*igústrum vulgàre, the common privet.
*P*hiladélphus coronàrius, the common syringa.
*P*hillýrea angustifòlia, the narrow-leaved phillyrea.
*P*ỳrus baccàta, the Siberian crab.
*P*ýrus coronària, the garland-flowered crab.
*P*ýrus *M*àlus transpàrens, the transparent crab.
*P*ýrus spectábilis, the showy-flowered crab.
*Q*uércus *C*érris, the Turkey oak.
*Q*uércus *I*'lex, the common evergreen oak.
*R*hús élegans, the stag's-horn sumach.
Ribes aúreum, the yellow-flowered currant.
Robín*ia* viscòsa and híspida, the glutinous and rose-flowered acacias.
*S*àlix *a*mygdálina, the almond-leaved willow.
Sophòra japónica, the Japan sophora.
*S*órbus aucupària, the mountain ash.
*S*pártium *j*únceum, the Spanish broom.
*S*pártium multiflòrum, the white broom.
*S*piræ`a *h*ypericifòlia and *s*alicifolia, the hypericum-leaved and the willow-leaved spiræas.
Symphòria racemòsa, the snowberry tree.
*T*hùja occidentàlis, the arbor vitæ.
*V*ibúrnum *O*'pulus, the snowball tree.
U'lex europæ`a plèna, the double-blossomed furze.
Xylósteum tatáricum, the Tartarian honeysuckle.

If forty-eight sorts in mixture were ordered from a nurseryman, they would cost 9*s*. per dozen; but, if the above list were given to the same nurseryman, he could not afford them for less than 1*s*. each at an average, though the sorts named are found in all nurseries. The difference of cost, however, is so trifling, and so much of the beauty of the back garden would depend on the trees and shrubs planted in it being ornamental, that we think an exertion ought to be made to procure those we have enumerated, rather than plant a miscellaneous assemblage, in which there may be several of one kind, and that too, perhaps, not the most ornamental.

The side walls of the back garden are alike eligible for ornamental plants, or the commoner kinds of fruit trees, the one being exposed to the morning, and the other to the afternoon, sun every day throughout the year; but, as the soil is supposed not to have been trenched, under-drained, or otherwise properly prepared, it would not be advisable to plant fruit trees; and, therefore, we would recommend ornamental climbers and creepers, or planting the whole with ivy. If climbers are chosen, the sorts which may be procured at 1*s*. or 1*s*. 6*d*. a plant are: —

Ampelópsis quinquefòlia and cordàta, the five-leaved and heart-leaved Virginian creeper.
*A*trágene austrìaca, the Austrian virgin's bower.
Técoma (Bignòn*ia*) radìcans, the common trumpet flower.
*C*lématis flórida, Vitálba, Viórna, and Viticélla, different kinds of virgin's bower.
Wistàr*ia* frutéscens, the shrubby wistaria.
*J*asmìnum officinàle, common jasmine, of which there may be several plants.
Lonícer*a* flàva, japónica (flexuòsa), gràta, impléxa, pubéscens, and sempervìrens; all different kinds of honeysuckle.
*L*ýcium bárbarum and ruthénicum, the Duke of Argyll's tea tree.
*R*ùbus fruticòsus flòre plèno, and fòliis laciniàtis, the double-flowered and cut-leaved brambles; or *R.* nutkànus and *R.* spectábilis.
*R*òsa Bánks*iæ*, white and yellow; Boursaúlt*ii*, Grevíll*ei*, índica màjor, multiflòra repánda, Noisette, and a variety of other roses: which may be obtained at 1*s*. each.

As the roses would require some preparation of the soil, unless it were naturally dry and good, they might be planted at the farther end of the garden, in the border which is 2 ft. wide, and in the narrow border which is formed between the walk and the sunk area, both these borders being previously prepared with rich soil; or this last border may be planted with common and variegated ivy, which would cluster over the area wall, and have a lively verdant appearance at every period of the year. Common ivy may be obtained at 3*s*. per dozen, or 4*d*. a single plant; and the variegated ivy is 1*s*. 6*d*. a plant.

Management. All the management that this garden would require would be the mowing of the grass, which for the first two or three years, if the ground were in good heart, might require to be performed six or eight times during the growing season; but, if it were kept from growing by being frequently mown, the roots would soon become so weakened, that in three or four years they would not require mowing above three or four times in the course of the year. It is a great mistake to suppose that any thing is gained in the end, in the way of economy, by suffering the grass of lawns to grow long before mowing, in order to save the expense of once or twice mowing during the season; for, in proportion as the grass is allowed to grow long, in the same proportion are the roots strengthened, and enabled to send up still longer leaves and stems; whereas, if a lawn were kept short by frequent mowing for two or three years in succession, the plants of grass would at last become so weak that not one half the mowing usually required for slovenly-kept lawns would be necessary, and the turf would be much finer and neater in appearance. The trees, being planted with a view to picturesque effect, would require very little attention of any kind for a number of years, more especially if the soil were dry and not over-rich; circumstances very desirable in order to keep the trees and shrubs in a comparatively dwarf state. When they begin to grow too large, the best way to reduce them is to pinch out their young shoots when they first make their appearance, and before they have produced leaves; which will weaken the plant in the same way as grass is weakened by mowing before the leaves have attained maturity. The climbers (including the roses) round the walls will require to be pruned, trained, and nailed, unless ivy be planted, which will need no attention whatever. On the whole, the management of such a garden is so simple as not to require any further observations.

Expense. There is a considerable saving of expense in laying out this garden by not trenching and manuring the soil; and, if it were a strong clay, in not forming under-drains on both sides of the walks. Some expense will also be saved by using the cheapest materials for covering the walks. Supposing the garden situated in the neighbourhood of London, the first cost might stand thus:—

	£	s.	d.
Levelling the front and back garden before forming the walks -		10	0
Digging out the walks, and filling those of the back garden with road stuff, and of the front garden with gravel - - -	2	0	0
Cost of the road stuff and gravel - - - - -	2	0	0
Adjusting the levels of the grass-plots in both gardens, so as to produce effective surface-drainage, and preparing the surface for receiving the trees and shrubs - - - -	1	0	0
Cost of the trees and shrubs - - - - -	3	0	0
Planting - - - - - - - -		5	0
Purchase and sowing of the grass seeds - - - -		5	0
Forming the refuse pit, the sides built of 4-in. brickwork, and the cover a large slate - - - - - -		15	0
Fixing eight sockets for clothes-posts - - - - -		2	6
Fixing brackets and cover for protecting the clothes-posts when not in use - - - - - - - -		10	0
Cost of the clothes-posts and their sockets - - - -	1	0	0
In all	£11	7	6

The yearly expense of management, supposing the occupier not to be his own gardener, may be:—

	£	s.	d.
The lawns six times mowed, at 2*s*. 6*d*. each - - -		15	0
The shrubs against the walls four times looked over and dressed, at 2*s*. 6*d*. each - - - - - - - - - -		10	0
In all	£1	5	0

Remarks. This garden would be very suitable for an occupier who had no time to spare for its culture, and who did not wish for flowers. It would not suit a lady who was fond of gardening: but for one who was not, or had no time to attend to it, and who had several children, this garden would be very suitable, because it would afford the children abundance of room to play in without doing injury; added to this, the garden would have an umbrageous, varied, and rich appearance throughout the year; and, in the months of May and June it would be particularly ornamental, from most of the trees and shrubs being then in blossom.

2. *Economy in the after-Management.* The following is the mode of laying out, planting, and managing a fourth-rate suburban garden, the back garden of which is entered through the house; and which is required to be always neat, and yet to be kept at the least possible expense. *Fig.* 53. shows a plot of

53

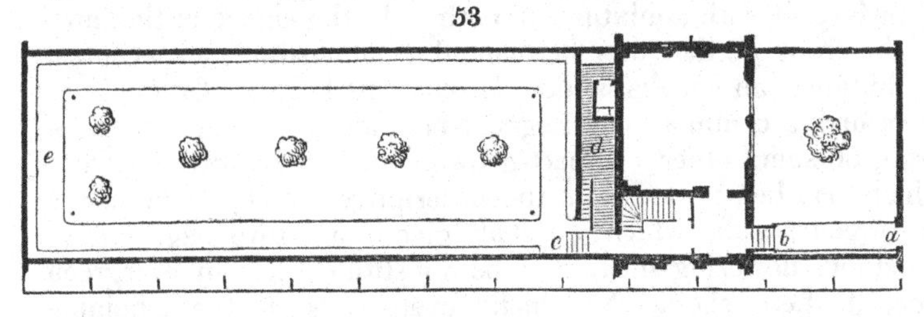

ground, in which the house, which is 30 ft. by 20 ft., has a front garden of nearly the same size, and a back garden about 90 ft. in length by 30 ft. in breadth. The kitchen of the house is half sunk under ground, and the ground floor is ascended to, from both gardens, by five steps. There are a privy and a dusthole in a sunk area (*d*) at the back of the house, on a level with the kitchen; and at the bottom of the garden there is a board projecting from the wall or fence, which will be covered with ivy, and under which, on two brackets, may be laid (by pushing them in endwise) the clothes-posts, when not in use. No manure tank or well is shown in this garden, because no part of it will be under the spade; but there may be a small pit sunk in the ground, with a cover fitted to it, at the bottom of the garden, at *e*, under the clothes-post bracket, for any decayed leaves or twigs, which may require to be removed in the intervals between the stated times that a man comes to mow the grass. From the entrance gate, or door (*a*), a walk 4 ft. broad is formed to the front door of the house (*b*): along one side of this walk, under the grass, the water service-pipe is laid; and on the other side, the gas-pipe, if any. *c* is the landing to the back door, under which are a safe or small larder, and also the water cistern. A small cistern over the privy, at the other end of the area, protected by a double roof, with the interstice stuffed with hay, would render this an excellent water-closet. A walk 3 ft. wide is conducted round the garden, the space between it and the wall forming a border 18 in. in width; except at the bottom, where it is 2 ft. wide, in order to make room for the refuse pit at *e*, over which is the clothes-post bracket. The plot of ground in front, and also the whole of that behind, with the exception of the walks, may be sown with grass seeds, or laid down with turf; and the walks laid with flagstones, or slates, or paved with bricks on edge, instead of being gravelled, in order to save trouble in keeping them in order. For the modes of performing all these operations, directions will be given in another part of the work. (See *General Index.*)

Planting. The boundary fences of the front garden may be planted with gold and silver-leaved ivy, intermixed with a plant or two of the common ivy; and the boundary fences of the back garden may be wholly planted with either the common or the giant ivy, or with a mixture of both. In the centre of the lawn in the front garden may be planted a laurustinus, an arbutus, a phillyrea, an aucuba, a double-blossomed furze, Cotoneáster U`va úrsi, a common or variegated box, an evergreen rhododendron, or some other compact-growing hardy evergreen shrub, which may be selected from the descriptive catalogue of hardy evergreen shrubs which we shall give in a future page; or, a deciduous flowering shrub may be substituted for an evergreen tree, if there should be chiefly evergreens in the adjoining gardens. Among the beautiful deciduous shrubs of moderate growth which require little or no pruning and management may be mentioned, the *C*ydònia japónica (either the pale or the deep red-flowered variety, or a plant of each put into one hole), the Persian lilac, and the *Rìbes* sanguíneum. Of all these plants the two most suitable are the laurustinus and the *C*ydònia japónica; because neither require any pruning, and both flower in the winter season. As these plants, however, from their beauty, cheapness, and easy culture, may possibly be common in the adjoining gardens, if expense should not be an object, one of the evergreen berberises or mahonias, such as *Bérberis* dealbàta or Mahòn*ia A*quifòlium, or Gárry*a* ellíptica (a most invaluable winter shrub, as yet very rare, but as hardy as the common holly), may be selected as the evergreen; or, if a deciduous shrub be preferred, *S*piræ`a ariæfòlia, or S. bélla, or some other species of that genus, or a yellow azalea, may be substituted. These comparatively rare evergreen and deciduous shrubs are as hardy as the others; and, like them, require no pruning whatever, further than cutting off dead wood or dead flowers. But if all the adjoining front gardens are planted with the more rare and beautiful foreign trees and shrubs, and the occupier should have the laudable desire of increasing the general variety in the street, he may step from the garden into the fields, and place in the centre of his grass plot, for an evergreen, the common spurge laurel, the butcher's broom, or the dwarf furze; and for a deciduous shrub (if he should prefer one), *Myrìca Gàle*, Genísta ánglica, or any dwarf British willow, may be made choice of. Should even these be already introduced, he may have recourse to the pine and fir tribe, and take one of the dwarf varieties of the common spruce, such as *A*'bies excélsa Clanbrasil*iàna*, or a dwarf pine, such as *P*ìnus sylvéstris pumílio, or *P.* s. *Mùghus.*

In the back garden, we would merely introduce a few standard low flowering trees, or fruit trees, placing them along the centre of the lawn, that they may not interfere with the walks, along the lawn side of which clothes-lines will probably occasionally be placed in the manner to be hereafter mentioned. The tree nearest the house should be a double-blossomed hawthorn, because it comes sooner into leaf than any other low tree, and the flowers being double, are not succeeded by fruit, so that the tree is in no year so exhausted, but that it can flower abundantly the year following; whereas a single-blossomed thorn, or tree of any kind, in which the flowers are succeeded by a large crop of fruit, seldom blossoms well two years in succession. Such trees, therefore, should never be chosen for points of view, where it is wished to have a fine show of blossoms every year; but rather trees which, like the above-mentioned variety of thorn, bear double blossoms. The next tree may either be a *P*ỳrus spectábilis, or transparent or Siberian crab; or some description of apple which has showy blossoms and bears abundantly, such as the Hawthornden. The third tree may be a perfumed cherry, standard all-saints cherry, double-blossomed cherry, an almond, or a Cotoneáster frígida, C. affìnis, or some similar tree, which may be selected from the catalogue of hardy trees and shrubs given hereafter. The next tree may be a mulberry, which thrives and bears abundantly in the very heart of London, and which should always be planted on grass; because, as the fruit drops the moment it is ripe, it can be picked up clean for use, which it cannot if it falls on dug ground or gravel. The two succeeding trees may be a laburnum and a scarlet thorn; or, if the occupier prefers fruit trees, they may be two pears, say a glout morceau, and a Marie Louise, or a beurré de Capiaumont; or they may be two plums, or cherries; or, if he prefers evergreen trees, they may be two variegated hollies. We recommend the variegated holly, because it is one of the most cheerful of evergreens, and is in no danger of growing out of bounds, so as to require pruning. Next to it, for the climate of London, the cedar of Goa may be planted; or, for colder climates, the common red cedar: but, as the cedar of Goa is somewhat tender, perhaps a preferable plant for a smoky situation would be the *Quércus I'lex*, of which the willow-leaved and the beech-leaved varieties may be selected. We have here shown only one line of trees down the centre of the lawn, because they will there have abundance of room; they will not require pruning for many years; and their leaves will drop on the grass, and not litter the walks. When low-growing trees are planted near walks, their branches hang over them; and, every year, those which inconvenience persons passing along the walks require to be cut off, or tied up, and this would occasion expense in keeping, which it is one of the desiderata in this mode of laying out and planting to avoid as much as possible. For this reason no tree or shrub is directed to be planted against the house; because that would be to incur the expense of training and pruning. The trees should be procured of 6 ft. or 8 ft. in height, so that their tops may be, when planted, out of the reach of injury from children; and the grass may either be sown, or turf may be procured and laid down. The latter produces the more immediate effect, though it is by much the more expensive, and, in the end, the turf is inferior, from its usually containing a mixture of unsuitable grasses and broad-leaved plants. The grass seeds will produce a close verdant surface in about three months, and, in a year, a much finer lawn than turf brought from common pasture, or meadow, or an old grass field. The flagstones or slates should be laid on brick piers, built on a solid foundation; so that their surface may be at all times level and even at the joints, for walking on. The surface of paved walks, like that of all others, as far as it is practicable, should be rather higher than the adjoining surfaces; otherwise, in heavy rains, they become receptacles for water, which, being often muddy, disfigures the stones, the pavement, or the gravel: but on this subject we shall enter at length, when treating of walks, in a future page. If the shrub planted in the centre of the front garden be one of the more rare kinds, the natural loam, which we have supposed to be the soil in both gardens, may require to be mixed with a little sand, peat earth, or vegetable mould, to lighten and enrich it. The kinds of grass seeds proper for sowing the lawn, and the mode of procedure, &c., will be given in a future page.

The Expense of carrying into Execution this manner of laying out and planting a suburban garden will depend chiefly on the price of flagstone or paving slate. In London, this, taking a quantity of the material together, would be from 4*d.* to 6*d.* a foot; so that every foot in length, of a walk 3 ft. broad, would cost from 1*s.* to 1*s.* 6*d.*, independently of the expense of building the piers by which the flagstones are to be supported. The whole expense of making these walks will be about three times that of walks laid with gravel; but the saving of expense in after-keeping will be so very great, that no one who can afford the first cost will ever consider that as a reason for not having them. The trees will cost 2*s.* 6*d.* each, and the grass seeds for both gardens 2*s.* 6*d.* The whole expense of laying out and planting, including the paving, may be between 20*l.* and 30*l.*

The Expense of keeping a Garden so laid out and planted, even if it were 200 ft. long and 50 ft. wide, would not cost, in the neighbourhood of London, 20*s.* a year; and yet it would at all times look neat. The ivy would grow up against the walls, and cling to them, without any expense of nailing or pruning, except when it infringed too much on the lawn in the front garden, or on the walk in the back garden, which it would not do for ten or twelve years, when it might require to be trimmed a little at the roots once a year. In consequence of there being no box or other planted edgings, there would be no expense of keeping them in order, and occasionally renewing them; and the margin of the lawn would only require to be prevented from spreading over the flagstones, by clipping the grass with a pair of shears every time the surface was mown. Neither the trees, nor the shrubs in the front garden, would need any pruning whatever for ten or twelve years, except the cutting out of such dead wood as might appear among the branches. We do not say that the fruit trees would not be improved by pruning, but merely that, as far as neatness of appearance is concerned, it would be quite unnecessary. The walks would be cleaned by every shower of rain, and would not even require sweeping, except when the edges of the grass were clipped. The sole expense would thus be that of mowing the lawns and clipping their edges, which might be done six or eight times in the course of the year, at 2*s.* or 2*s.* 6*d.* a time; or the occupier might cut the grass himself with Budding's mowing-machine, and the edges with a pair of small hedge shears.

Planting a Garden of this Kind in the Heart of a great City. The planting, respecting which we have gone into detail, is calculated for a situation not so completely enveloped in smoke as if it were in the middle of a city; but supposing it to be in the very centre of London, and economy of management and neatness of appearance were the great objects, then we should proceed exactly in the same manner, only using a kind of shrub, and kinds of trees, which experience proves will thrive even in the midst of smoke. Thus, for the plot in front, we should recommend *Aúcuba* japónica as an evergreen, because this remarkable plant, though a native of Japan, endures the smoke of London better than any indigenous evergreen shrub whatever; and, as a deciduous shrub, the common purple lilac, which is both hardy and beautiful, and comes early into leaf. The trees in the back garden might be the double-blossomed and scarlet thorns, both of which will grow and look well for at least eight or ten years; the laburnum, the almond, the mulberry (which thrives admirably in the most smoky places), and the weeping or allsaints cherry (which is one of the few flowering trees that prosper in the the gardens of Lambeth Palace, though enveloped in the smoke of numerous houses and manufactories. Ivy, whether common, giant, or variegated, will thrive in the very heart of London. Grass will not live, and look well, in smoky situations, for any length of time; but, if the *Pòa* ánnua be used, it will ripen its seeds and sow itself every year; and it has this advantage, that during winter it is greener than any other grass that will grow in a town. Should it fail in any part, and leave bare patches, seeds may be procured from the seed-shops, and, being sown at any season, will come up in a few days.

To lay out and plant a larger Garden of the same Kind, and for the attainment of the same objects, all the difference, would be that, instead of one shrub in the front garden, there should be several; and, instead of one row of trees in the back garden, there should be two, or perhaps three. The walks would still be laid with flagstones, or some description of pavement; the walls would be planted with ivy; and the house would still be left without a vine, a fig, or a rose, trained against it. *Fig.* 54.

54

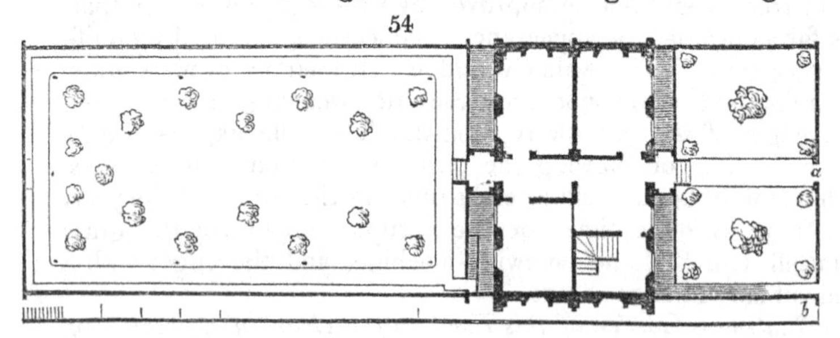

is the plan of a suburban street residence, the house and front and back gardens of which occupy a space 60 ft. in width, by 200 ft. in length. Here the entrance walk (*a*) passes through the centre of the front garden, on each side of which there is a grass plot, with a large shrub in the centre, and smaller ones at each angle. There is a servants' entrance at one side at *b*, and a sunk area, both before and behind; that in front being narrow, and serving merely to keep the walls dry, while that behind is broad, and contains a larder, bottle-rack, and similar conveniences required for a house of the second rate. The back garden is planted with four rows of low trees, two near each walk, in quincunx, leaving a broad space in the middle, about 100 ft. in length, well adapted for a party walking backwards and forwards on in the summer season, for a dance, or for placing a tent on, for sitting under, at the farther end. A garden of this kind might be laid out and planted for 30*l.* or 40*l.*; and kept perfectly neat for 30*s.* or 40*s.* a year. We shall now proceed to other desiderata, also depending on the taste or circumstances of the individual.

3. *Laying out and Planting, where the Object is to render the Garden profitable.* The mode of laying out and planting a fourth-rate suburban garden, entered through the house, where the object is to save expense to the occupier, by raising vegetables to be consumed in his family, differs considerably from those recommended where the object is economy in the first cost, or economy in the after-management. It may be necessary to premise, that, by profit, we do not here allude to the sale of articles (though the same directions will be found equally applicable where this is intended), but to the production of such vegetables as shall be most useful in the household economy of the occupier. We shall suppose the extent of the back garden to be the same as in *fig.* 53. or in *fig.* 54.; because the same directions are alike applicable to both, or to any other garden similarly circumstanced. We shall also suppose that the drainage, levelling, service pipes, &c., and also the walks, are completed, and the front garden sloped, as in the preceding figures; and the ground trenched, improved if necessary, and thoroughly manured. In addition, there ought to be a manure tank formed, and so connected with the privy and the sink of the back kitchen, as to receive the drainage from them. To this there ought either to be a fixed pump, or a movable cover to admit of readily dipping a bucket into the tank. If a pump be employed, it ought to be one of large bore, so as to bring up mud as well as water. In using this liquid manure, great care must be taken never to put it on the leaves of the plants, and either to follow it by watering with clear water, so as to prevent the surface of the soil from being disfigured; or, what is preferable, to use it chiefly during or immediately before rain. As the supply of liquid manure will be regular throughout the year, it ought to be regularly used; and at those seasons when it may not be proper to water annual herbaceous vegetables with it, on account of disfiguring or dirtying their leaves, it may be applied to perennials, such as tart rhubarb and sea-kale, and to the roots of fruit trees and fruit shrubs. The liquid manure from a house where the family consists of five or six persons,

and where they wash at home, if used as it is produced, so as to allow none of it to run off by the drain, will be quite sufficient for a garden 200 ft. in length and 60 ft. in breadth. Liquid manure, however, though powerful in a recent state, is always more efficacious after being a week or two fermented; but for this purpose two tanks are necessary, as will be hereafter described, when treating of the arrangements suitable for large gardens. All the laying out being completed, we next proceed to the planting.

The Front Garden we would devote chiefly to ornamental flowers or plants, which should be, at the same time, useful in cookery. The general surface we would keep in turf, forming round it a narrow dug border, and, in the centre, a bed in the form of a circle, square, diamond, or any other regular figure. In these borders, and in the central bed, we would plant no trees or shrubs, but only such ornamental herbaceous plants as could be rendered useful in the kitchen: for example, in the centre bed we would plant an eatable gourd, or vegetable marrow, the fruit of which is one of the most useful of summer vegetables, either boiled or fried, and serves either to mix with apples, or use alone, flavoured with lemon and sugar, for fruit pies. The best kind of gourd for using, when young, is the vegetable marrow. In the same bed we would plant a mammoth gourd, or American butter-squash pumpkin, the fruit of which should be allowed to ripen, for the purpose of being used for soups and pies, and also as a vegetable, when boiled, in the winter time. The fruit of both these species, when the plants are regularly watered every warm evening with diluted liquid manure, or even with simple water, will sometimes attain an enormous size, weighing from 100 lb. to 200 lb.; and, as it will keep the greater part of the winter, even though cut, it is a most valuable resource for soups, and is so used in some of the first families in England. When cut, a circular orifice, of about 4 in. in diameter, is made on one side, and the piece taken out is, after cutting off part of the flesh, preserved as a stopper to exclude the air. When a piece is wanted for soups or pies, the stopper is taken out, and a sufficient quantity scooped out of the inside with a knife or an iron spoon. This may be practised throughout the whole winter, and the fruit will still continue quite fresh. The mammoth gourd is much used in soups by the French, even of the humblest class, and in the public hospitals, as well as by the Italians and the Americans; and though in England it is as yet scarcely known, except at the tables of the nobility and gentry, it is well deserving of general cultivation. It gives a fine flavour and creamy richness to soup, and is very nutritious. The flowers of all the gourds and pumpkins are delicious fried in butter; and the points of the young shoots, boiled, are equal to spinach in tenderness and in flavour.

In the border next the house, in order to be trained against it, if facing the south, we would plant a vine or a fig; and, if it faced the east or the west, a beurrée de Capiaumont pear, which is a never-failing bearer, of excellent flavour, and a good keeper. If the front faced the north, we should not recommend any tree or shrub to be trained against the house, unless it were ivy, which keeps the house warm, excludes rain, and always looks well. As ornamental flowers in this border, and also in those of the two side walls, we would plant scarlet runners, which would give a length of 70 ft. of this vegetable, and would afford an ample supply for a family of six or seven persons during the whole summer. For the runners to twine on, pieces of pack-thread should be nailed to the walls, and to the house, at about 6 in. apart, and reaching from within 1 ft. of the ground, to from 4 ft. to 6 ft. above it, as may be suitable to the height of the side walls. The lower ends of the strings are fastened to a horizontal rod, supported by props at about 6 in. from the wall. The kidneybeans, being sown between the wall and the rod, will attach themselves, as soon as they come up, to the strings, and continue to twine round these till they have reached the top, flowering and fruiting as they advance. As the nails in the side wall next the path might be liable to catch the dress of ladies passing near them, instead of nails, a rod or a wire might be fixed there: and, indeed, the most elegant mode of training kidney-beans against a wall is, to have an iron wire of the fourth of an inch in diameter, fixed horizontally on studs let into the wall both at top and at bottom, and painted green, and to stretch the threads on these wires from the one to the other. Every April, when the kidneybeans are planted, a person taking a ball of twine, and beginning at one end, by passing it over and under the bars at 6 in. distance, would soon complete this string trellis, the lines of the strings not being perpendicular, but in a diagonal direction. By going over the space a second time, and crossing these lines, the effect would be improved, and the support for the twining stems of the runners increased. If thought necessary, a slight wire fence might be placed along the walk, to protect the grass plot from dogs, and, on this, also, scarlet runners might be grown; or an arched wire trellis, from the street entrance to the door of the house, might be formed over the walk, and gourds and scarlet runners might be trained on it. In the border on the fourth, or street, side of the front garden, we would sow nasturtiums, which would soon grow up against the dwarf wall and the railing, and also spread over the grass. Their flowers would make a fine appearance all summer, and, with the young leaves and the tender points of the shoots, might be used in salading; while the fruit, if gathered before it becomes too old, is well known to make an excellent substitute for capers. In the side borders containing the scarlet runners nothing else should be planted, unless it were a vine for the purpose of training its shoots along the tops of the walls; because these borders will require to be dug and manured every year, and the soil renewed every three or four years; and because no flowering shrub could thrive under the smothering influence of the foliage of the runners. The same may be observed of the border containing the nasturtiums. The only culture required for both the nasturtiums and the scarlet runners, in the summer time, in addition to the usual routine of watering, weeding, and keeping down insects, &c., would be, when any of the plants began to cease bearing, to cut them down, and water freely at their roots, in consequence of which treatment they will send up fresh shoots, and bear a second crop. In mild winters, the roots of the runners, if the soil is dry, sometimes survive, and spring up again the following year; and, in this case, fresh seeds are unnecessary, but the ground must either be manured during winter, or frequently watered with liquid manure whilst the plants are growing. The kidneybean is one of the most valuable of culinary vegetables, being always ready to gather during the whole summer, and requiring very little cookery. That kind known as the scarlet runner is by far the most profitable that can be planted; not only from its producing a greater quantity of fruit than any of the dwarf varieties, but because its pods are tender to the latest period of their growth, even when the seeds within are full grown; whereas the pods of the dwarf varieties become stringy, hard, and unfit to eat, even before the seeds are half-grown. The mature seeds of all the varieties, taken from the pods, and well boiled or stewed, form a farinaceous and most nutritious food; the quantity of gluten in them being nearly as great as it is in the best wheat. Before Miller's time, the scarlet runner was chiefly cultivated for the beauty of its flowers, and on account of their being produced by the plants during the whole summer. The plants were regularly sown every spring in the flower border, among the other ornamental annuals, and the flowers were eagerly sought after by ladies to put into their nosegays and garlands; but Miller having brought the pods into general use for the table, the scarlet runner has disappeared from the flower border, and has now almost ceased to be considered as an ornamental plant.

The bed on the grass plot adjoining the walk between the entrance gate and the front door may be planted or sown with such plants as are at once fragrant flowers and sweet herbs; such as lemon and common thyme, mint, marjoram, sage, winter savory, fennel, and tarragon. These would never require any manure, and need not be taken up and replanted oftener than once in two years.

If it should be desired to confine the culture of useful plants to the back garden, the front garden may be laid down in grass, and only dwarf fruit trees planted in it, as shown in *fig.* 55. The centre tree may be a filbert or a berberry. The filbert is most ornamental during winter, with its long male catkins moved by every wind; but the berberry has the advantage of being beautiful, not only when in flower in spring, but also in autumn when covered with its bright fruit; which is useful both for garnishing and making a delicious preserve. The bushes in the angles may be dwarf apples on paradise stocks, and the kinds may be the Hawthornden, if profit be the object, because no other apple tree bears so abundantly as a dwarf: but, for ornament, the Alexander, a very large apple; the red Siberian crab, nearly as small as a cherry, but an excellent bearer, and making an elegant sweetmeat when preserved in apple jelly; and the transparent crab, a most beautiful apple with a skin like transparent wax, also an abundant bearer, and good for preserving may be added. Other fruit trees are either not ornamental, such as the pear and medlar; or of too short duration, such as the cherry, which never looks well after the month of July, or the plum, which ripens its fruit in August. The quince is one of the most ornamental of fruit trees, and might be used for the centre bed, but unfortunately it is rather an uncertain bearer.

The Back Garden. If the back garden lies in the direction of east and west, then the wall having a southern exposure may be planted with peaches and nectarines, or with grapes; but the most profitable mode, if the family are fond of British wine, would be to plant it with grapes for wine-making. The wall facing the east or west might be planted with baking plums and cherries; and the wall with an aspect to the north with baking apples, including some of the earliest sorts, and some of the latest. If the back garden lies in the direction of north and south, then the only wall having a southern exposure will be one of the end walls, and on this grapes may be planted; while the two side walls may be covered with pears, apples, plums, and cherries, with a gooseberry or a currant between each, to be removed as soon as the fruit trees require more room. We recommend only kitchen fruits, as being the most profitable for pies, puddings, &c., to a family; but some table fruits may be introduced, if the occupier prefers them.

Selection of Fruit Trees. As the object in view in planting this garden is rather to produce the fruits most useful in a family, than to grow a great variety of sorts, the first point is to consider what kinds are best adapted for the kitchen. Of these apples are the most wholesome for children, and are not only useful for pies and puddings, but excellent roasted, or boiled down with honey, so as to make a kind of sweatmeat, resembling that called by the French resiné, to eat instead of butter with bread. Pears, when of the melting kinds, are both very agreeable and very wholesome, to eat raw; but they are seldom used in England for any purposes of cookery, except stewing; and this dish, when made palatable, is generally too rich for children. Of stone fruits, damsons are the most wholesome, and most easily preserved; and morello and Kentish are the best cherries for cooking. Gooseberries, raspberries, and red currants, with a few black ones, are also extremely useful for all purposes of cookery, particularly preserving. One of the most useful apples for the kitchen is the Hawthornden, which comes in early, bears abundantly, and falls (softens) well in boiling; but the fruit does not keep, and, in some soils, the tree dies off at an early age. The Keswick codlin is also a good and early kitchen apple. The Ribston pippin is an excellent fruit, either for the kitchen or dessert; but the tree is not quite so hardy or so good a bearer as either the Bedfordshire foundling (a very large apple), or the king of the pippins. The Brabant bellefleur and Wormsley pippin are, also, very excellent apples. The best keeping apples for winter use are the northern greening and the French crab, the latter of which will keep two years. For pears, the best for the table are the beurré de Capiaumont, the glout morceau, and Marie-Louise, all excellent bearers. The earliest of the fine-flavoured pears is the jargonelle; but it will not keep longer than a week or ten days. The best pear for keeping is the Easter beurré. The best cherries for the kitchen are the Kentish or Flemish, for the early crops, on account of their juiciness, and the smallness of their stones; and, in the autumn, the morello, for making cherry brandy and preserving. The wild, or black, cherry, which ripens between the kinds mentioned, is a very rich fruit for pies or puddings; but the stones are large in proportion to the pulp, and it continues in season but a very short time. Of plums, the green gage, the early and late Orleans, and the Shropshire or prune damson, are the best. The wine sour and mirabelle are also frequently used for preserving; and Gisborne's plum is an extraordinary bearer. Of grapes, the most productive, in the open air, are the esperione and the white muscadine; the best for wine-making is the black cluster. Of peaches, the best bearers are the grosse mignonne, the Royal George, and the Bellegarde; the last a most excellent sort. The best nectarine is the elruge, and the best apricot for the table is the Moorpark; though the Breda, which is a very abundant bearer, is most useful in the kitchen, as it not only makes a delicious preserve when ripe, but excellent tarts when green. The best bearing fig is the large blue or purple fig, which ripens well in the open air, and early; but the brown Brunswick is also much recommended. The best gooseberries for preserving green, and making green gooseberry wine, or British champagne, are the rumbullion and the white Dutch. The Warrington, the whitesmith, and the early rough red are great bearers, and afford excellent fruit for the table, and for preserving when ripe: the roaring lion is the best bearer of the large sorts. The common, or Dutch, red currant is the only one used for pies and puddings, and is the best for preserving. The black currant is used for making a kind of jam, called rob, which is thought good for sore throats; the black Naples produces the finest fruit. The best white currant is the white Dutch. The red Antwerp raspberry is the best, both for the table and preserving; and the white Antwerp has the finest flavour for the dessert. The only strawberry that is suitable for preserving is Keen's seedling, and it is also the best and most regular bearer. The red alpine is high-flavoured, and continues in bearing many months, but the fruit is small. The scarlets are the only kinds used for flavouring ices, and the variety called the Duke of Kent's scarlet is esteemed the best. The old pine is the finest-flavoured of all the strawberries, but it is a very uncertain bearer.

Walks and Borders. On the supposition that the walks are of flagstone, supported on brick piers, then the wall borders need not be more than 18 in. wide; as the space under the flagstones (the latter not touching the soil) will be almost as available for a border for the roots of the wall trees, as if it were fully exposed to the sun; while, in effect, as we have elsewhere observed, it adds all the space occupied by the walks to the superficial area of the garden. If the walks are of gravel, then the width of the borders between them and the wall on which peaches and grapes are to be grown ought not to be less than 5 ft.; and those for

55

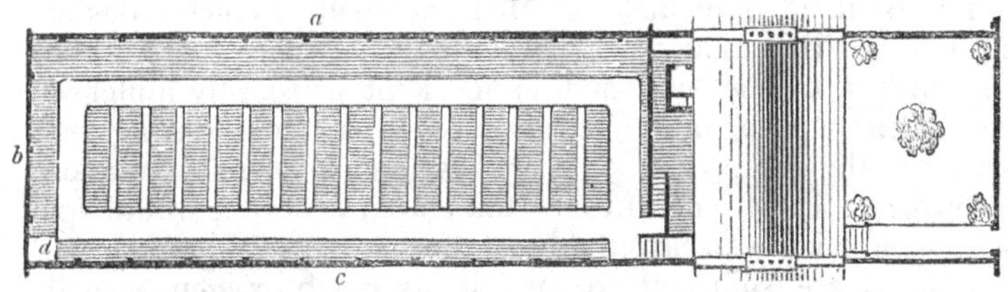

the other hardier fruits not less than 2 or 3 feet. In *fig.* 55., the direction of the side walls of the back garden is east and west: and, the walks being supposed to be of gravel, the border to the wall *a*, having a south aspect, is 5 ft. wide; that to the end wall *b*, having an east aspect, is 3 ft. wide; and that to the south wall *c*, having a north aspect, is 2 ft. wide; *d* is the rubbish pit. The interior of the garden is laid out in beds for the cultivation of kitchen crops. The preparation of the fruit borders will be fully treated of in another part of the work, under the head of Garden Operations.

The Mode in which the Central Plot of the Back Garden is to be planted and cropped will depend on its size. The object being to grow culinary vegetables in it, on no account would we recommend standard fruit trees to be planted there; because the shade of them is injurious to the flavour of the vegetables grown beneath them; and because the trenching and digging of the ground necessary to bring the vegetables to a large size, and succulent texture, is highly injurious to the fruit trees; by preventing them from ever being able to send up their roots to the surface, so as to enjoy the warmth of the sun, and the vivifying influence of the oxygen of the air. Hence it is that we so seldom see abundant crops on fruit trees growing as standards among culinary vegetables, and so frequently find the trees cankered. The want of crop arises from the roots getting down into the subsoil, which, if moist and rich, will occasion them to produce a great luxuriancy of wood, without blossom buds: and the canker is owing to the roots getting into a bad subsoil, or to the trees being too deeply planted in any description of soil. If the garden is sufficiently large to spare room for fruit trees in the middle bed, then the best mode of growing them is on espaliers; because, when so treated, being kept low, and within definite bounds, they produce no injurious shade; and, by planting them in a border by themselves, and only stirring the soil of that border with a pronged fork, and never farther than 2 or 3 inches deep, a sufficient number of the fibrous roots will establish themselves closely under the surface, to derive all the benefit that is necessary for them from the sun and air. It may be useful to observe here, that, though it is always an advantage to fruit trees to have a space around them undug to the same extent as that which is covered by their branches, yet that they will produce crops of fruit with less than that space; provided what space there is be fully exposed to the sun, and either kept constantly mulched with rotten manure, or stirred two or three times during the summer to the depth of 2 or 3 inches, and kept quite loose on the surface. Experience proves this; and, hence, it would appear that a few roots, favourably circumstanced in regard to the atmosphere, are sufficient to take in as much oxygen as will supply the whole plant, in the same manner as a few roots in contact with water will supply the whole plant with moisture.

In back gardens, where the centre plot is so large as to admit of espaliers, the east and west walls may be devoted to table fruits. In gardens too small for espaliers, there may still be room for growing a few gooseberries, either on espalier trellises, in which state they occupy least space; or as bushes, by which mode they occasion least trouble, and are more certain of producing a crop; because the upper and exterior parts of the bush protect the under and interior parts from spring frosts in the blossoming season. Gooseberries are so useful a fruit, that, however small the garden may be, we would plant a few of them in the centre plot; and, if there were room, to these we would add a few currants and raspberries. We would also plant a few beds of strawberries, if room could be found; but, in deciding as to these fruits, we would always bear in mind, that, where profit is the object, they ought never to be allowed to come in competition with the more useful culinary vegetables.

The Kinds of Culinary Vegetables which should be planted in a small garden, where profit is the object in view, must depend on its extent. If that is such as to admit of growing all the vegetables likely to be used by the family, except winter potatoes, then all that is necessary is, to determine the proportion of space that shall be allotted to the fixed or perennial crops: and that which shall be set aside for the movable or annual crops: but if, on the other hand, there is not space enough for growing all the vegetables required by the family, then the point is to determine which sorts ought to be cultivated in preference. On the supposition that there is space enough for growing all the vegetables required, and that the family use asparagus, sea-kale, tart rhubarb, and artichokes, then we would allot a sixth part for the smaller fruits, such as gooseberries, raspberries, strawberries, &c., leaving two thirds for the annual crops. Dividing these two thirds into 25 parts, we would dispose of them as follows:—

	Parts
The cabbage tribe, including Savoys, Brussels sprouts, cauliflowers, brocoli, borecole, &c.	6
Leguminous plants, including peas, beans, and kidneybeans	8
Roots, including turnips, carrots, parsneps, beet, and early potatoes (but not a main crop of potatoes for winter use)	6
Spinaceous plants, such as common spinach, white beet, and New Zealand spinach	1
Alliaceous plants, including onions, leeks, chives, garlick, shalots, &c.	1½
Acetarious plants, including lettuce, endive, chicory, celery, mustard and cress, radish, &c.	2
Pot and sweet herbs; thyme, sage, chamomile, &c.	¼
The gourd tribe	¼
	25

The above is given as an approximation to the usual space occupied by these crops in gentlemen's gardens; but, as most families have some particular vegetable for which they have a preference, the space for that kind may be enlarged, and that allotted to some other, for which they care little, diminished. We have thought it advisable to give in this place a slight general outline of the crops grown in kitchen-gardens, and the proportionate space allotted to each class of productions, in order to enable the occupier of a suburban garden to form some idea of what he would like to grow. If, for example, he would like to supply his family with peas, it would be evident to him, from the above view, that, if his garden is very small, he will scarcely be able to grow anything else; on the other hand, if he confines himself chiefly to acetarious plants and sweet herbs, however small his garden may be, there is little doubt of his being able to find room for all of those kinds of plants that his family can possibly require.

The System of the Succession of Crops in kitchen-gardens (which is as necessary to be understood as the extent to be allowed to each crop) will be discussed hereafter; but it may be useful to state here, that, as far as possible, plants of the same natural order, and especially of the same genus, should never follow each other in direct succession; nor should taprooted plants follow each other in succession, even though of different orders: for example, peas should never follow beans, or kidneybeans, or the contrary; and carrots should never follow beet. Notwithstanding the soundness of these principles, they require to be taken in connexion with another principle, no less important; viz. that of expediency. Thus, it frequently happens, from the lateness of the season, that a crop is longer than usual before it is ready to be removed from the ground; or, from the dryness and warmth of the summer, that a crop is removed sooner than usual. When the summer crop is not all removed at the proper time, the winter crop (such as spinach, for example, after a crop of cabbages) might be endangered, if sowing it were delayed; and, therefore, some ground which is empty, though according to the plan of succession laid down it should have been cropped with something else, must be used for the spinach. When, on the contrary, the crop is ripe, and the ground cleared before the expected time, as it is never desirable to see ground without a growing crop, one may be made to follow, for the sake of covering the surface, that did not enter into the regular course of succession. The grand secret of getting the greatest quantity of culinary vegetables from any given surface of ground is, to have one crop coming forward in the same bed, or compartment, to be ready to succeed that which is about to be removed: and, for this purpose, it is always desirable to grow the crops in drills; a mode which is also attended with several other advantages, as will be hereafter explained.

On the supposition that only a part of the vegetables consumed by the family can be grown in the back garden, the next point is to determine those which it is advisable to grow. This depends on various considerations; such as those kinds in which the occupier is most curious in his taste; those which the soil and aspect are best calculated for producing; those which are dearest in the given locality, &c: but the most general and influential consideration is, the keeping or non-keeping properties of culinary vegetables, after being gathered. All vegetables exposed for sale in markets or shops must necessarily have been gathered some time, and, consequently, deprived of their natural succulency, or even injured by fermentation; and this will be the case, more or less, according to the nature of the vegetable, and the mode in which it is brought to market by the grower, and kept by the seller. For example, lettuces, peas, and spinach are brought to market in cart-loads, in large baskets, or in sacks; and, if the distance is great, or the time between gathering and consumption is more than a day, or, at most, two days, fermentation will have taken place. Cabbages and turnips are also brought to market in cart-loads; but though, when in a young state with all their leaves about them, these vegetables ferment in the course of two or three days; yet, when they are full grown and firm, they may, when deprived of their outer leaves, be kept in heaps for a week or more without undergoing fermentation. The vegetables which are most injured by keeping are such as are eaten raw; and hence it is always desirable, if possible, for every man to grow his own salading, such as lettuce, endive, cress, mustard, radishes, young onions, &c., in order that they may be brought to table fresh, and newly gathered. These ought, therefore, to be grown, if possible, in even the smallest back garden of a suburban residence. The excellence of celery also depends much upon its being eaten crisp and fresh from the ground; and, therefore, if room can be found, this also ought to be grown in small gardens. Next to salading, it has been found from experience that peas are most injured by being kept after they are gathered, and, therefore, they are the second class of articles which should be grown in every suburban kitchen-garden if possible. Tender leaves, such as those of spinach and spring cabbages, are next liable to injury from the same causes; and, therefore, they ought also to be grown, if room can be found. A small bed can scarcely be more profitably employed than under winter spinach, which (the Flanders variety being chosen), in a rich loamy soil, not wet below, and in a sheltered situation, will continue growing almost all the winter, and afford occasional gatherings at that season, and an abundant supply for the table from April till August; care being taken to pick out the flower stems as soon as they begin to appear, and to water when the ground is dry.

Next to the vegetables which will not keep after being gathered, those which it is convenient to have always at hand, in case of an emergency, and which a cook may gather for herself, ought to be planted. Thus, parsley, mint, fennel, thyme, and chives ought to be in every suburban garden plot, however small, as supplying garnishings and seasoning, which are constantly wanted; but no one would think of growing horseradish in a fourth-rate suburban garden where a gardener was not kept; because, however frequently it might be used, it requires considerable labour to dig it up; and, as it keeps perfectly well, it may be procured from the greengrocer's at any time, and kept till wanted in a cellar, or in the garden, covered with soil. Where there is room, an asparagus bed, if properly formed at first by deep trenching, thorough drainage, and an abundant supply of manure, is a very convenient luxury; since the heads are easily gathered when wanted, require little aid from cookery, and are generally considered a delicacy. Asparagus beds, to those who prefer the shoots green, and, consequently, do not cut them till they are 2 or 3 inches above the ground, do not require to be earthed up every winter, and of course are less expensive in the management than those where the asparagus is eaten blanched, as this effect can only be produced by heaping up earth on the beds. A bed of sea-kale, though it will sooner afford a supply for the table than a bed of asparagus, yet requires more attention from the gardener, and fails sooner, and, therefore, is not so well adapted for a suburban garden where the saving of labour is an object. Tart rhubarb, if planted in soil deeply trenched, and well manured, occasions less trouble than any other perennial vegetable; and, if watered in dry weather, it will continue producing leaves the whole summer. In winter, the ground on which it stands ought to be well soaked with liquid manure. For pies and puddings, rhubarb stalks are preferred by many to either gooseberries or apples; and they are so easily gathered, so wholesome, and so well adapted for children, that no suburban kitchen-garden ought to be without its rhubarb bed.

Selection of Culinary Vegetables best adapted for small Suburban Gardens. Of the cabbages, Knight's early dwarf, introduced in the year 1835, and the early Battersea are the best kinds for the early crops; and the latter will be found suitable for the secondary summer, autumn, and even winter, supply. In most cases, these will be found sufficient, as late cabbages are seldom grown in suburban gardens; but, where they are thought desirable, the late sugar-loaf may be chosen. Red cabbages are useful for pickling and stewing in the Flemish manner. The kind usually cultivated is the Dutch red; but the dwarf red, which is not much known, is more delicate for the table. The dwarf Savoy is a very useful winter vegetable, as its flavour is improved by frost if not too severe, and it will supply the table with greens from November till spring. Brussels sprouts are a delicious vegetable, and should, if possible, be raised from seed ripened in Flanders, as the plant is apt to degenerate in this country. Scotch greens, or German kale, are very useful winter vegetables, as, when the heart is removed, the stalk continues for some months to send out side shoots or sprouts. The cauliflower is rather a difficult vegetable to grow well, as it requires an exceedingly rich soil, and careful culture. To have it early, it must also be protected during winter; and, as it is a vegetable that is brought to market in very great perfection, not being much injured by keeping a few days, it is better to purchase it when wanted, rather than to attempt to grow it, unless there is abundance of room. Purple Cape broccoli is not liable to the same objection, as, if sown in May or June, it will produce a good crop from August till December, unless destroyed by frost.

Of the leguminous vegetables, the best kinds of peas for a small garden are the blue Prussian, the dwarf marrowfat, and the white Prussian, or poor man's profit; or the white rouncival and Knight's tall marrowfat, where tall-growing varieties are wanted. We would not recommend any early pea to be grown, as these are generally very inferior in flavour to the others; but, where one is thought desirable, the early Warwick is said to be the best. For a late pea, the late blue dwarf Spanish has been strongly recommended. For beans, where it is thought desirable to grow them, the early mazagan, and long pod are the most fruitful; but the broad and green Windsors are generally preferred for the table. We have already mentioned that the scarlet runner is the most profitable kidneybean. For potatoes, the ash-leaved and Rufford kidneys are to be preferred for the first crop; but Shaw's early and the champion, though they are both very prolific, and are those generally grown for the London market, we can by no means recommend, either for their mealiness or their flavour. For turnips, the early white Dutch and the yellow Dutch may be chosen; but the Teltow, or French, is well worth cultivation, on account of its sweetness, and the richness it gives to soups. The Teltow is the smallest of all turnips, not having more leaves on it than a radish; and, when used, it should be only scraped, and not pared, as a great deal of the flavour is in the rind. For carrots, the early horn is best for an early, and the Altringham, or Altrincham, for a late, crop. For parsneps, the Guernsey grows to a large size, and the Siam is particularly well flavoured. The dwarf red beet is the kind best grown in gardens. The Flanders spinach is greatly to be preferred to the common kind, and will be sufficient for a small garden; but, where there is room, the white beet and the New Zealand spinach may be grown, to afford variety and a crop in summer, when annual spinach runs too rapidly to seed. For onions, the early silver-skinned and the Deptford may be recommended. The Portugal and the Spanish are mild, and grow to a great size, but do not keep well. Of leeks, chives, garlic, and shallots, only one kind is in general cultivation. Of the plants used in salading, the best are, the union cabbage, and the large Brighton Cos lettuce, the Batavian endive, the variegated chicory, the curled-leaved and broad-leaved Normandy cress, and the scarlet and white turnip radishes. The best celery is the Italian. The curled-leaved parsley is the handsomest for garnishing, and answers equally well with the common for all the purposes to which parsley is usually applied.

As the greater part of culinary vegetables are raised from seed, sown partly where they are to remain, and partly for transplanting, something requires to be said here on that subject. In our catalogue of culinary vegetables, we shall state under each kind the quantity of ground which a given quantity of seed will sow; and also the smallest quantities of such seeds which are generally ordered from the seed-shops; and we shall only here state that the larger seeds, such as peas, beans, and a few others, are purchased by the pint, and almost all the small seeds by the ounce. There are very few gardens to suburban houses in which half a pound or a pound of any seed is required; and of many articles, such as potherbs, parsley, celery, &c., half an ounce, or threepenceworth, is quite sufficient for a small garden. In general, it is always safe to order seeds in small quantities, because the chance is greater of getting them fresh.

Management. A back garden, 200 ft. in length, and 50 ft. in breadth, with a front garden forming a square of 50 ft., cropped in the manner we have described, would occupy one person from April to September, during a fourth part of his time, or at least two hours a day; and besides this time, when much watering was required to be done, or insects to be got rid of by collecting, he would need the assistance of the female part of the family. Where there is a steady man-servant, this is just the sort of garden that he could take care of, directions being given to him by his employer as to the quantity of particular crops, and the season of sowing or planting, on the supposition that he was not a reading gardener. If he were, and took an interest in having the garden in good order, and in raising large crops, we would recommend him to be as little interfered with as possible; for every man likes to have something on which he prides himself, and, to keep up that pride, it is necessary that it should be as much as possible his own work.

Expense. The first cost of laying out such a garden as that shown in fig. 53. p. 189., with a view to the culture of culinary vegetables and table fruits, supposing the walks to be paved, may be from 25*l.* to 30*l.* Tools may cost 2*l.*, and seeds, for the first year, 1*l.* The annual expense afterwards, supposing a commercial gardener employed, will be, on an average, a day and a half per week throughout the year; an arrangement being made that this time shall be bestowed on the garden, according as it is wanted, and not formally at stated intervals of once a week; because in some weeks two days' work will be necessary, while in others half a day will suffice. The expense will depend on the rate of wages paid per day, which, about London, being from 4*s.* to 4*s.* 6*d.* (the gardener finding his own tools), will amount to 15*l.* or 20*l.* a year. But the garden, if managed by a hired gardener, working in it only at stated intervals, cannot possibly be turned to so much account as if cultivated by the occupier, or his in-door servant; because, during the months of June, July, and August, there are operations which require to be attended to daily; such as the removal of insects, watering, &c.; and which, if deferred, as they must necessarily be if a hired gardener only be employed, will endanger the crops, and, at all events, deteriorate their value, both in quantity and quality. We would not, therefore, recommend any one to cultivate his garden with a view to profit, unless he could do so with his own hands, or with those of a servant whom he was obliged to keep at any rate; because, assuredly, for the 15*l.* or 20*l.* a year which he must pay a hired gardener he might purchase as much fruit and vegetables as he could grow in a garden of the extent we have mentioned.

Remarks. A suburban garden, to grow culinary vegetables well, ought to be in the outskirts of a town, and neither surrounded by buildings, where it will be injured by smoke, nor so near a public road as to be liable to be covered with dust. At all events, in such situations we would not advise lettuce, cabbages, or any article to be grown, the leaves of which form a close heart, and consequently enclose within them, as they advance in growth, whatever soot or dust may have fallen on the leaves; but there can be no objection to growing every description of roots, and to stalks and stems, such as asparagus, sea-kale, celery, tart rhubarb, &c., which can easily be washed clean before using. Even at a distance from smoke and buildings, we would not recommend any one to cultivate culinary vegetables of a great variety of kinds, unless they could attend to them themselves, or have them attended to by some part of the family. A garden, in which all kinds of culinary vegetables are grown, if left to the care of a person who can only work in it once or twice a week, can hardly be kept sufficiently neat to be an ornament to a house placed in the midst of it; and, if the gardener were to come much oftener, the expense would be far greater than the produce was worth.

In describing this garden, as in the preceding one, and most of those which follow, we have introduced remarks on culture, and other gardening topics, which might have been dispensed with if we had merely intended to describe a ground plan. Our object, however, is to infuse into our readers by degrees, as we go on, a scientific and practical knowledge of gardening as an art of culture, as well as one of design and taste; and we think we are more likely to do this by noticing incidentally subjects likely to arrest attention, than by presenting all that we have to say on any particular topic at once. Thus, for example, in directing in what manner fruit trees ought be planted, we have given the reasons for preferring espaliers in borders by themselves, to standards grown among culinary vegetables; and, in recommending certain varieties of both fruits and vegetables, we have said something of those of their qualities and uses in the kitchen, or in the dessert, which induced us to give them the preference over others. In this way, the reader will become gradually initiated into the most important facts connected with plants and their culture; so that, when he comes to enter on that part of this work which treats of the operations of gardening, and the reasons on which they are founded, he will enter on the study with more profit, as well as pleasure, than he could have done without the previous preparation which we are now giving him.

4. *Exercise and Recreation for the Occupier and his Family.* Where these are the objects, the taste or no taste for gardening of the occupier, and the number, age, and sex of his family, require to be taken into consideration, as well as the manner of laying out and planting the garden. We shall suppose the extent of the ground to be the same as in the preceding case, the surface drained and levelled, and the main walks laid out exactly in the same manner as in *fig.* 53. p. 189. In neither of the gardens would we have any turf, because more work, and consequently more exercise and recreation, would be created by having the ground under the spade. Neither would we have the walks of flagstone, but of gravel, because in the latter case they require box or other edgings, and those create labour in clipping them, and in keeping them in repair; whereas paved walks, if they are raised an inch above the surface, do not require any edgings, and occasion no labour in weeding, rolling, &c.

The Front Garden we may suppose laid out as in *fig.* 54., 56., or 57. In either case, the borders next the house and walls, *f*, *g*, *h*, in *fig.* 57., may be planted with shrubs for being trained against them; including a vine or a fruit tree, for training against the upper part of the house, if the aspect be the south; and a Wistàr*ia* sinénsis if it be towards the east or west. If any side of the house be directly north, the Virginian creeper, or common ivy may be planted against it. For training against the lower part of the house, if the aspect be to the south, there may be a myrtle next the door; then a camellia, and an aloysia (lemon-scented verbena), which plants would require protection with mats during winter.

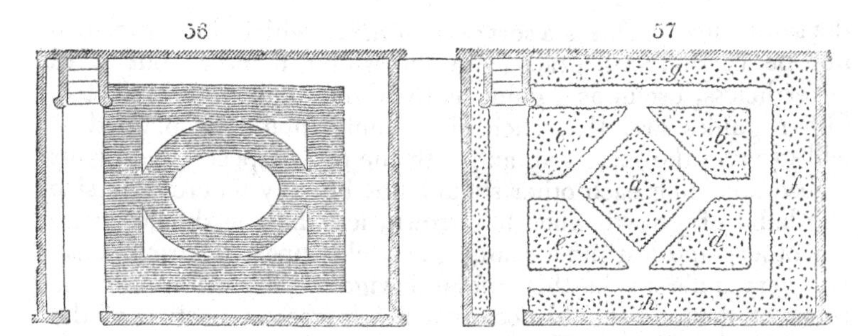

For training against the upper part, Passiflòra cærùlea and *Técoma* (Bignòn*ia*) capreolàta (which, as fast growers, would need a good deal of training), and *R*òsa rùga, sanguínea, Noisett*iàna*, and other China roses, might be planted, which, during summer, would not only require training, but constant attention to keep down the aphides, and to remove the decayed blossoms and their stalks. If the aspect were east or west, *Técoma* (Bignòn*ia*) radìcans, Lonícer*a* japónica (flexuòsa) and gràta, Wistàr*ia* frutéscens, camellias, myrtles, and climbing roses, might be employed. The wistaria, the tecoma, and the evergreen climbing roses would reach to the top of the house, and the others would cover the lower part. The myrtles and the camellias would require to be matted during the severest weather in winter, and the other plants would occasion a good deal of pruning and training during summer. On the supposition that the occupier of the house was a lady, we would not recommend any climber to be planted against the house that would grow higher than 4 ft. or 5 ft.; but such as myrtles, camellias, pelargoniums, fuchsias, &c., as these would not need the use of a ladder to train them. The side and front walls of the front garden may be covered with China roses, different varieties of honeysuckles, and *J*asmìnum officinàle. Against one of the side walls, if there be room, a double-flowering pomegranate may be planted in the centre; and against the others Magnòl*ia* conspícua or purpùrea. In the beds in the interior of this garden, we would plant nothing but flowers; or we might devote the centre bed *a* in *fig.* 57. to moss roses, or to some other kind or kinds of rose belonging to sections different from those which include the China roses. In the centre of this rose bed there might be a standard rose. At the end of this work we shall give not only catalogues, but selections of roses, as well as of herbaceous perennial, biennial, and annual flowers.

In choosing the flowers for the beds that are to be exclusively devoted to them, the objects ought to be to have an equal number of kinds of flowers in bloom, if possible, every month in the year, and to have an equal number of the different colours displayed at the same time. This is a sort of *beau idéal*, which, however, cannot be carried into practice in the winter months; but it is, nevertheless, useful as a guide as to what ought to be selected. Other guides are, that such plants only should be planted as will grow in the same soil, and with the same aspect; that plants placed adjoining each other should not be very different in size and habit of growth when full grown, lest the one should choke the other, and lest the assemblage should appear inharmonious; nor very different in their natural vigour of growth, lest the roots of the one should occupy a much greater portion of the soil than the others. To plant pæonias, everlasting peas, asters, *P*apàver orientàle, and such like plants in juxtaposition with daisies, hepaticas, and pinks, would be attended with the suffocation of the latter diminutive kinds; but the very tallest and the very lowest plants may be included in the same bed, provided it is very large, and there be a gentle gradation observed between the one and the other. If the bed is to be seen from every side, then the tallest-growing kinds can be placed in the centre, and the lowest in the circumference, and the intermediate sizes between them, so as to rise in gradation from the lowest at the margin, to the highest in the middle.. If the bed is a border, parallel to a walk, and to be seen only on one side, then the lowest plants should form a row next the walk, and the others should be placed behind them in quincunx, according to their sizes, the tallest being placed behind, and forming the last row. The distance at which one plant should be kept from another depends more on the habit of the plant with respect to lateral extension, than on the height to which it grows; thus a peony, which seldom grows above 18 in. high, but which spreads its large leaves over a much greater surface than is occupied by its roots, will require more room than a tall-growing phlox, which will reach the height of 4 ft., and not cover above a square foot of the surface of the border. Creeping plants also require more room than such as grow compact and bushy, or narrow and erect. The rule which we have just given, of not associating plants in the same bed that have different habits, directs that creepers, and climbers or twiners, should always be planted by themselves, or, at least, not introduced indiscriminately among other plants. At the same time, a plant with one of these habits, introduced occasionally into a miscellaneous border, has a good effect by contrast. Where creepers alone are employed, if each plant be of a different kind, they ought to be kept as distinct from each other, as bushy or tall plants ought to be; but where a bed is entirely filled with creepers of the same kind, then they may be allowed to cover the whole surface of the bed; because the object of the planter in this case must necessarily be to produce a mass of one kind of form and colour. In general, small plants ought to be chosen for small beds: beds of simple and regular shapes, such as squares, parallelograms, circles, triangles, &c., as in *figs.* 56. and 57., ought to have the plants at regular distances; while small, irregular, curvilinear beds, such as those exhibited in *fig.* 65. p. 227., ought to be entirely covered with low plants. As some of the leading rules for laying out and planting flower-beds, more especially in front gardens, we give the following: —

Leading Rules for laying out and planting Flower-Beds in the Front Gardens of Street Houses.

1. Where the space is small, and surrounded by trees and high walls, so as not to be open and airy, it is not desirable to form beds or borders round the margin of the plot, but rather to have only one bed in the centre, and the rest in grass.

2. Where the space is open and airy, either large or small beds may be formed; and it will generally be desirable to surround the whole plot with a narrow border. If the ground floor of the house is 2 ft. or 3 ft. above the level of the plot, then a figure, or collection of beds, may be laid out, which shall be looked down upon from the window as a whole; and, consequently, to aid this purpose, the beds ought to be planted with low-growing plants, and, in general, to have the surface covered by them; each bed, in this case, being of only one kind of plant.

3. Where the rooms on the ground floor are on a level with the surface of the front garden, or nearly so, large plants may be employed in the beds, provided the beds also are large; because, as in this case the beds cannot be looked down upon, and, consequently, their plan can never be taken in at one glance, they never can be seen as a whole from above. It is better, therefore, to use large plants, which, by growing of such a height as to form a whole, or group, when looked at laterally, will prevent the idea of a whole formed by the shapes and lines of the beds, when seen from above, having been intended by the planter.

4. In general, no figure or assemblage of beds of any degree of intricacy, and where the beauty is dependent on the shapes of the beds and their connexion together, should be formed where it cannot be looked down on so as to be seen all at once. In general, also, parterres, or assemblages of figures of this kind, should only be planted with very low plants, which will not obstruct any part of the outline of the figures; with the exception, however, of an occasional tall plant, such as a standard rose, to produce effect by contrast. It may be further observed, that, when plants are to remain permanently, such should be chosen as continue in flower for a long period (say two or three months), in preference to such as complete their time of flowering in a short period, say two or three weeks. On the other hand, when flower beds are furnished with plants in pots plunged in the soil, with a view to changing them, and replacing them by others as soon as they have done flowering, plants which remain a short time in flower should be chosen; because these have, in general, a greater number of blooms expanded at the same time, and, consequently, while they last, they have a more brilliant effect. This is particularly exemplified in the case of bulbous flowers, and in certain annuals, such as candytuft, ten-week stock, &c.

5. Where a symmetrical figure is employed, beds which answer to each other in form and position ought to be filled with plants, either of the same kind, or of the same general appearance, and which flower at the same time: for example, a bed of mixed hyacinths can only be properly opposed to another bed of mixed hyacinths; but the mixtures need not be the same in both beds.

A bed of the small dwarf blue lobelia, may be opposed to a bed of the blue anagallis, and so on.

6. Where it is desirable not to have more than one plant of a species, in a symmetrical figure consisting of various beds, the principle of symmetry may be preserved in planting, by placing each colour by itself. Thus, a bed of white flowers, consisting of ten plants of as many different species, may be opposed to another bed of ten other different species, also with white flowers.

7. The dug surface of beds formed on a grass plot ought either to be decidedly under the surface of the grass, or decidedly above its level, in order to increase the expression of art, and to take away from the commonplace idea of merely digging down a portion of the turf of a particular shape, and planting it with flowers.

8. In the case of flower-beds on turf or lawn, where a regular gardener is not kept, the outlines of the beds ought to be formed by concealed brickwork, masonry, tiles, or slates; otherwise, every time the grass is mown, and the edges of the bed trimmed with the spade, and, especially, every time the beds are dug, their outlines will be liable to be put out of shape. Where common bricks and plain tiles can be procured, the simplest mode of fixing the outlines of the beds or borders is by forming an underground outline, if the expression may be used, by tiles or bricks laid on their broad sides, at an angle of 45°, as shown in the section, *fig.* 58.; in which *a* is the brick, *b* the surface of the grass, and *c* the surface of the dug bed, decidedly under the level of the grass plot, agreeably to Rule 7.

58

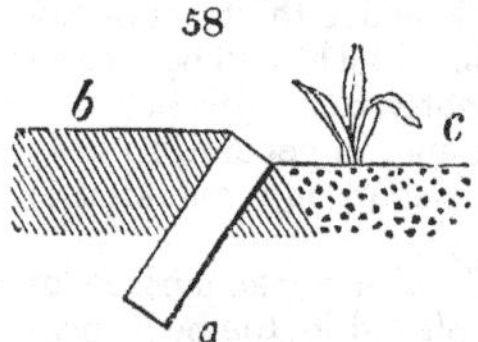

59

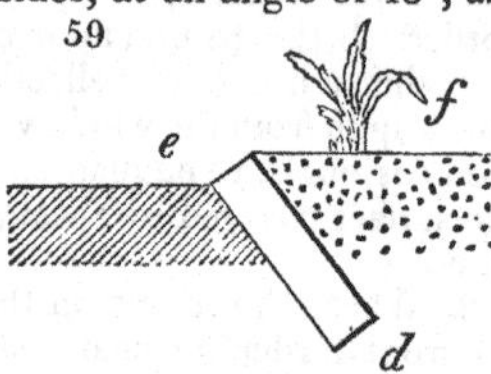

On the other hand, when the surface of the dug bed is to be above the level fo the turf, agreeably to the same rule, the brick should be placed as at *d* in *fig.* 59.; the turf kept to the level shown at *e*, and the surface of the bed to the level shown at *f*. When plain tiles are to be used instead of bricks, they may either be let into the ground perpendicularly, and their upper edge kept to the level of whichever surface is to be the highest (viz. that of the grass plot, or the bed); or they may be laid sloping in the same manner as the bricks, which will form a very delicate outline, well adapted for beds in small front gardens. By using semicircular tiles, and inserting them perpendicularly in the soil, very neat curvilinear outlines may be formed; but such tiles cannot be used in a sloping direction, like bricks or plain tiles. *Fig.* 60. shows the tiles let in perpendicularly, forming a bed 2 in. higher than the surrounding turf. *Fig.* 61. shows the tiles let in in the same manner, and the bed made 2 in. lower than the surrounding turf. *Fig.* 62. shows the tiles laid in, in a sloping manner, so as to form a bed the thickness of the tile only under the level of the turf; and *fig.* 63. shows the tiles so arranged as to form a bed only the thickness of the tile higher than the surrounding turf. *Fig.* 64. is a surface plan, showing the manner of forming the outlines of beds with semicircular tiles.

60

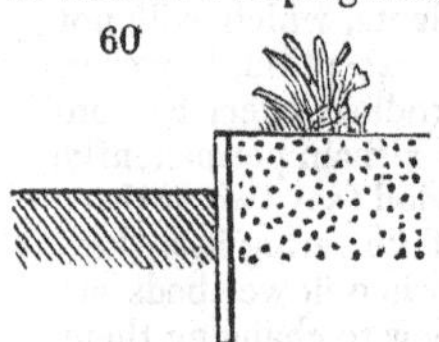

61

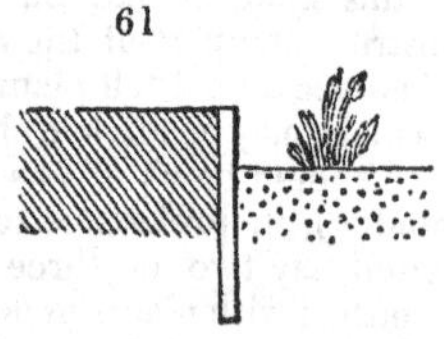

62

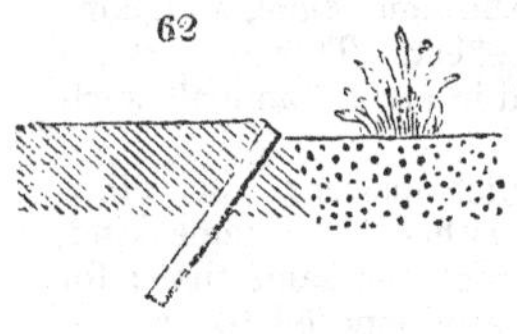

63

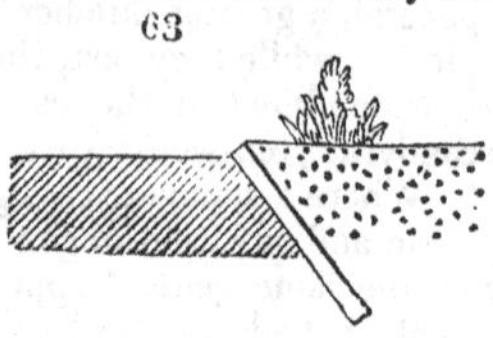

9. Where it is desired to express high art, every bed or border ought to be

surrounded, or, if the border be against a wall, bordered on one side, with either a raised or a depressed framework of turf; or of stone, or bricks, or pebbles, flints, &c. Where the beds are raised, the plants grown in them should be such as do not require excess of moisture, and which thrive best in a free air; such, for example, as pinks. Where the beds are lowered, an excellent opportunity is afforded of growing plants which require more than ordinary moisture, especially when they are in flower; such as ranunculuses, polyanthuses, heart's eases, &c.

64

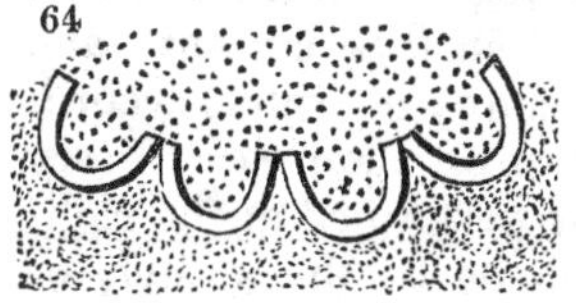

10. Borders of brick or stone, or other architectural materials, ought, in general, to be narrower than borders of turf, lest the force of contrast should be too great for the general effect.

11. When borders of box or other plants are employed to form margins to flower-beds on turf, they ought to be of several times the breadth which they are when employed to separate walks from gravel, in order to give them a distinctive character, and to produce sufficient force of effect to justify their use. Nothing looks worse than a narrow edging of box, surrounding a bed on turf; the narrow edge of box, appearing, in that situation, to be quite superfluous; and, its colour not contrasting with that of the grass, it has a dead dull appearance.

12. Where beds are surrounded by gravel walks, and edged with box, the latter ought always to be of such a breadth as to form a strongly marked line; and, though the sides of the edging may be clipped so as to give them a slope, and prevent their getting naked close to the surface of the ground, yet the top should always be cut quite flat and level.

13. Nothing looks worse in a flower-garden, than to have the box edgings narrow and high; except having the edges of turf margins so pared by the spade as to show the raw naked earth.

We might add considerably to the number of these rules, but we shall defer doing so till they are called for by some other design. It will be recollected that rules differ from principles, in being liable to exceptions; and, also, it ought to be further borne in mind, that in every art, as in every thing in life, cases will occur which are beyond the reach of rules. Such cases can only be treated by those who understand principles, while rules are merely guides for routine practice. Many persons left to act as servants, who even understand something of principles, yet, if they have not a rule deduced from these principles to go by, will, in the absence of their employers, be apt to neglect their duty. For example, every labourer, who has been employed in a garden, knows that bunches of grapes ought to be thinned after the fruit is set, in order to give the remaining berries room to swell; and that camellias require a great deal of water: but, unless rules are given to him as to how much, and when, the grapes should be thinned, and the camellias watered, the chances are that the first operation will be delayed too long, or performed imperfectly; and that the latter will depend rather upon the leisure of the labourer, than on the actual state of drought of the plants. On the other hand, by laying down a rule that the grapes shall be thinned when they are of the size of the heads of corking pins, and that at least one third shall be cut out; and that the camellias shall have a certain quantity of water, according to the size of the plant, say half a pot of water each, every other day, both operations are sure to be performed at the right time, and to the right extent; and it will be for the master, or for the regular gardener, where one is employed, to determine when exceptions and variations are necessary.

Keeping the Plants distinct. Where there is abundance of room, it is most convenient for culture, to keep perennials, biennials, annuals, and bulbs in beds or compartments by themselves; but, whether the space for flowers be limited, or of considerable extent, when the object is to have an equal display of flowers in spring and autumn, all these different kinds of plants must frequently be mixed together, to make up the colours required: for example, in February, March, and April, very little could be effected in a bed or border of mixed flowers, without the aid of crocuses, scillas, and hyacinths; and, late in autumn, there would be no great display without dahlias, lupines, China asters, marigolds, &c., all which are either tuberous-rooted plants or annuals. Whatever kinds of flowers may be admitted into a flower-bed or border, one principle of planting must never be lost sight of; that is, distinctness, or the keeping of every particular plant perfectly isolated, and, though near to, yet never allowing it to touch, the adjoining plants. This is merely the principle of the gardenesque applied to flowers; and it is so decidedly preferable in point of convenience for culture, to planting so close together as that the plants will soon join together and cover the surface of the soil, that we should never for a moment think of recommending what may be called the picturesque in flower-planting, either for a flower-garden or. for flowers in borders. We except, however, creepers and low plants, and perhaps plants generally, where the object is to produce one dense mass of any particular colour; because this object cannot be effected without allowing the plants to cover the whole bed. On the same principles of distinctness and suitableness for culture, no shrubs that are not either of very low growth, such as dwarf roses, or that have not very small heads, supported on stems 4 ft. or more in height, should ever be admitted into flower-beds or borders. Dwarf roses, for all practical purposes, may, in effect, be considered as herbaceous plants; since they flower best, when kept low by pruning, and when taken up and replanted in fresh soil, every year.

Planting the Design fig. 57. p. 211. As the main object in this design is to provide exercise and recreation for the occupier, by having an ample reserve garden, and bringing forward a stock of select flowering plants in it, he may contrive to have a continual bloom in the beds of his front garden, during the whole summer, by taking up every plant as it goes out of flower, with a ball of earth attached (for which there is an appropriate tool, which will be hereafter described), carrying it to the reserve ground, and bringing back a plant in flower, with a ball of earth attached, to supply its place. If the plants are grown in large pots, the labour of taking up and replanting will be considerably lessened, and the success rendered certain, though the exercise to the occupier will be less; and, indeed, there is something extremely interesting in removing plants in full flower by the transplanting machine, and watching the result. We have known persons, who have had no love for gardening or plants, so captivated by this operation, as to practise it for the sake of the excitement it produces; and such persons have ultimately become enthusiastic florists. When successional culture, whether by pots or the use of the transplanting instrument, is pursued, the beauty and gaiety which may be displayed in even a very small front garden is astonishing; and not less so the labour or exercise that is by this means created for the occupier. This labour, indeed, may be increased or diminished, according as the plants chosen are those which remain a long time, or only a short time, in flower. Some plants of great beauty, such as the scarlet verbena, continue blooming three or four months; and others, also of very great beauty, such as the cyclamens, produce all their flowers in the course of a fortnight or three weeks. The choice of flowers for front gardens is almost endless, and we therefore refer to our catalogue; but we shall here assist the reader by showing how selections may be made for four different designs for front gardens, and the same principle will apply to any flower-garden or flower-plot whatever.

Planting with fibrous-rooted Perennials of the lowest Growth. In *fig.* 57., the whole of the plot is dug, and the borders surrounding it and the beds are edged with thrift, or sea pink (Armèria vulgàris), and the walks gravelled. If paved, the flagstones should be laid hollow, and supported on piers, in order that the space beneath may serve as a drain for absorbing the water which falls on the beds, and also on the walks. The surface of the soil of the beds we may suppose to be 2 in. below the surface of the walk, and quite flat.

In the centre bed (*a*) there may be the 16 following plants, all under a foot in height, with a standard Noisette rose in the centre: —

*H*epática tríloba, red hepatica; February and March.

*S*axífraga oppositifòlia, purple saxifrage; February and March.

*P*hlóx subulàta, awl-leaved phlox; red, April.

Prímula vulgàris, fl. pl., the double white primrose; April.

Dràba aïzöìdes, whitlow grass; yellow, April.

Verónica alpìna, alpine speedwell; blue, May.

*A'*ster alpìnus, alpine starwort; purple, May.
Diánthus alpìnus, the alpine wild pink; red, June.
Coronílla ibérica, Spanish coronilla; yellow, June.
*Œ*nothèra *r*òsea, the rose-coloured evening primrose; July.
*A*lýssum montànum, mountain madwort; yellow, July.
*C*ardámine nítida álba, cuckoo flower; white, July.
Gypsóphila prostràta, creeping gypsophila; white, August.
*S*axífraga *a*ïzöìdes, the small mountain saxifrage; yellow, August.
*P*hyteùma orbiculàre, the round-headed rampion; blue, August.
Epilòbium alpìnum, the alpine willow herb; red, September and October.
Liàtris pùmila, the dwarf liatris; purple, September and October.

The beds *b*, *c*, *d*, and *e* will hold 7 plants each, if the plants are under 6 in. in height; and they may be as follows: —

In bed *b*, there may be: —

*B*éllis perénnis, the red daisy, February and March.
*R*anúnculus auricomus, the wood crowfoot; yellow, April.
*A*nemòne baldénsis, the Mt. Baldo anemone; white, May.
*A'*juga alpìna, alpine bugle; blue, June.
*A*nthýllis montàna, the mountain kidney vetch; purple, July.
Gaúra coccínea, the scarlet gaura; August.
*A*chillè*a* crética, the Cretan milfoil; white, September and October.

In bed *c*, there may be: —

*H*elléborus nìger, the Christmas rose; white, January to March.
*C*orydàlis longiflòra, the long-flowered fumitory; red, April.
Eròdium alpìnum, the alpine heron's bill; red, May.
Alètris aúrea, the golden-tipped aletris; yellow, June.
*A'*cynos alpìnus, alpine thyme; blue, July.
Arenària montàna, mountain sandwort; white, August.
Prunélla pennsylvánica, the Pennsylvanian self-heal; yellow, September and October.

In bed *d*, there may be: —

*M*andrágora vernàlis, spring mandrake; white, February and March.
Chrysosplènium alternifòlium, the alternate-leaved golden saxifrage; yellow, April.
*L*ýchnis alpìna, alpine lychnis; red, May.
Globulària cordifòlia, the heart-leaved blue daisy; June.
*E*rýngium alpìnum, the alpine eryngo; purple, July.
Silène marítima fl. pl., double white catchfly; white, August.
*S*tátice reticulàta, the netted sea lavender; purple, September and October.

In bed *e*, there may be: —

*V*iola trícolor, heartsease; variegated, February and March.
*I'*ris hùmilis, dwarf iris; purple, April.
Convallària majàlis, the lily of the valley; white, May.
*M*yosòtis palústris, forget me not; blue, June.
*T*eùcr*ium* pyrenàicum, Pyrenean germander; white, July.
*M*ímulus Lewís*ii*, Lewis's monkey flower; purple, August.
Solidàgo hùmilis, dwarf golden rod; yellow, September and October.

The border which surrounds the front garden will, in the wide part (*f*), contain two rows of plants, independently of the roses, or other flowering shrubs, that may be trained against the wall. The front row may be dwarfs, like those already enumerated; and the back row, plants growing from 6 in. to 1 ft. Those for the front row may be as follows: —

*T*hláspi alpéstre, alpine shepherd's purse; white, February and March.
*O*nósma stellulàtum, the starry onosma; yellow, April.
*G*enti*àna* brachyphýlla, short-leaved gentian; blue, April.
*A*spérula arcadiénsis, Arcadian woodruff; red, May.
*H*ieràcium aúreum, golden hawkweed; yellow, May.
*T*elèph*ium* Imperàt*i*, true orpine; white, June.
Tróllius asiáticus, Asiatic globe flower; dark orange, June.

Scutellària orientàlis, eastern helmet flower; yellow, July.
*C*entaurèa áspera, rough knapweed; purple, July.
Gratìola officinàlis, common hedge hyssop; white, tinged with blue, August.
*S*álvia Forskòhl*ii*, Forskohl's sage; blue, August.
*S*iderìtis *h*yssopifòlia, hyssop-leaved ironwort; yellow, September and October.

The plants in the back row, which grow higher, may be the 10 following, planted in quincunx with the others: —

Valeriàna trípteris, the three-leaved valerian; white, February and March.
*D*odecàtheon integrifòlium, entire-leaved American cowslip; April.
*L*àmium rugòsum, rough dead nettle, or Archangel; red, May.
Cheiránthus alpìnus, dwarf wallflower; yellow, May.
*G*erànium sanguíneum, red crow's bill; June.
*T*rifòlium comòsum, tufted trefoil; white, June.
*L*ýthrum diffùsum, spreading willow-herb; purple, July.
*L*òtus Dioscórid*is*, yellow bird's-foot trefoil; July.
*A*nchùsa rupéstris, rock bugloss; blue, August.
Stèv*ia* *h*yssopifòlia, hyssop-leaved stevia; red, September and October.

The side border *g* will contain 10 plants, which may be 1 ft. high, and of the following kinds: —

*I*sopýrum *t*halictròides, meadow rue-leaved isopyrum; white, February and March.
*T*ussilàgo díscolor, purple coltsfoot; April.
Díclytra formòsa, red fumitory; May.
Potentílla spléndens, yellow potentilla; May.
Aquilègia alpìna, alpine columbine; blue, June.
*P*olýgonum vivíparum, white knot grass; June.
Melíttis *M*elissophýllum, melissa-leaved bastard balm; red, July.
Fúnk*ia* ovàta, oval-leaved day lily; blue, July.
*N*épeta incàna, the hoary catmint; white, August.
Physotègia speciòsa, showy dragon's head; purple, September and October.

The side border *h* will also contain 10 plants, which may be as follows: —

A'sarum canadénse, Canadian asarabacca; dark brown, April.
Pulmonària officinàlis, red lungwort; May.
*P*olemònium mexicànum, Mexican Greek valerian; blue, June.
*C*áltha palústris flòre pleno, double-flowered marsh marigold; yellow, June.
Leonùrus supìnus, trailing mother-wort; white, July.
*A*strágalus tenuifòlius, slender-leaved goat's beard; purple, July.
Melíssa *N*épeta, white-flowered balm; August.
*L*upìnus perénnis, perennial lupine; blue, August.
*C*hrysócoma nudàta, golden locks; yellow, September and October.

The total number of plants that this front garden will contain is 86; and we have chosen them purposely of as many different genera, that we might show how much botanical interest a very small garden may afford, as well as floricultural enjoyment. No space is here allowed for bulbs; and we shall therefore next show how room may be found for them.

Planting with fibrous-rooted Perennials and Bulbs alternately. The advantage of introducing bulbs in flower-gardens is, that their flowers make a greater show than those of fibrous-rooted plants generally do in spring; and as, in small suburban residences, it seems more desirable that the gardens should look well in spring than in summer (because at the latter season many families go out of town for a few months), the use of bulbs appears very desirable. The manner of introducing them may either be in beds by themselves, to be succeeded by fibrous-rooted plants when they go out of flower; or intermixed with fibrous-rooted perennials, by using only half the number of the latter, and those of larger growth; and by placing the bulbs and the perennials alternately. As the bulbs come all into flower in March, April, or May, they will have faded before the perennials have come to their full growth; and hence, notwithstanding the increased size of the perennials, the bed will not appear crowded. In the selection of both perennials and bulbs, we shall give, as far as practicable, only one species of a genus, in order to produce as much botanical variety as possible within the given space.

The perennials for the bed *a* may be the 8 following kinds; none of which come into flower earlier than June and which are all about 1 ft., or from that to 1 ft. 6 in., in height: —

*B*etónica grandiflòra, large-flowered betony; red, June.
Campánula *a*riæfòlia, the beam-tree-leaved bell flower; white, June.
*Œ*nothèra undulàta, the waved-leaf evening primrose; yellow, July.
*D*elphínium élegans, perennial larkspur; blue, July.
Diánthus carthusianòrum, Carthusian pink; red, August.
Scutellària peregrìna, white helmet-flower; August.
*G*enti*àna* Saponària, the soapwort-leaved Gentian; blue, September and October.
*A'*ster diffùsus, white Michaelmas daisy; September and October.

The bulbs may be the 8 following: —

*S*cílla bifòlia, blue two-leaved squill; March.
*C*ròcus albiflòrus, white-flowered crocus; March.
*A*nemòne pavònia, the peacock's eye anemone; red, April.
*H*yacínthu*s* orientàlis, white hyacinth; April.
*N*arcíssus mìnor, small narcissus; yellow, April.
Fritillària tenélla, slender fritillary; purple, May.
Erythrònium Déns cànis, the dog-tooth violet; reddish lilac, May.
Muscària pállens, pale-blue musk hyacinth; May.

The bed *b* may be planted with the 3 following perennials: —

*O*nònis rotundifòlia, the round-leaved restharrow; red, June.
Silène chloræfòlia, the Armerian catchfly; white, July.
Commelìn*a* erécta, the upright commelina; blue, August.

The bulbs may be: —

Tùlipa suavèolens, the sweet-scented wild tulip; variegated, March.
Galánthus nivàlis, the common snowdrop; white, March.
*A'*llium amœ`num, the red-flowered allium; April.
*N*arcíssus tenuifòlius, the slender-leaved narcissus; yellow, May.

The perennials for the bed *c* may be: —

Valeriàna dioíca, the diœcious valerian; red, June.
Gypsóphila arenària, the sand gypsophila; white, July.
*T*agèt*es* lùcida, French marigold; yellow, shining, perennial, August.

The bulbs may be: —

*C*orydàlis albiflòra, the white-flowered fumitory, March.
*I'*ris tuberòsa, the tuberous-rooted iris; variegated, March.
Gàge*a* bracteolàris, the yellow star of Bethlehem; April.
*L*ílium cóncolor, the self-coloured lily; red, May.

The perennials for the bed *d* may be:—

Papàver orientàle, the scarlet poppy; June.
Campánula *persicifòlia*, the peach-leaved bell-flower; white, July.
Verónica cándida, the white-leaved speedwell; blue, August.

The bulbs may be:—

Leucòjum vérnum, the snow-flake; white, March.
Eránthis hyemàlis, the winter aconite; yellow, March.
O'xalis floribúnda, the abundant-flowering wood sorrel; red, April.
Scílla peruviàna, the Peruvian squill; purple, May.

The perennials for the bed *e* are:—

Lýchnis sibírica, Siberian lychnis; white, June.
Potentílla Russell*iàna*, Russell's potentilla; scarlet, July.
Baptísia tinctòria, the dyer's baptisia; yellow, August.

The bulbs are:—

A'llium Chamæmòly, the dwarf moly; white, March.
Muscària botryöìdes, the grape hyacinth; blue, April.
Tùlipa præ'cox, early tulip; red, April.
Narcíssus Bulbocòdium, the hoop-petticoat narcissus; yellow, May.

The perennials for the bed *f* may be the 6 following:—

Clinopòdium vulgàre, common wild basil; red, June.
Lýchnis neglécta, white lychnis; June.
Hypéricum púlchrum, yellow St. John's wort; July.
Campánula carpática, the Carpathian bell-flower; blue, July.
Phlóx suavèolens, the sweet-scented phlox; white, August.
Málva moschàta, musk-scented mallow; red, August.

The bulbs may be the 5 following:—

Scílla itálica, Italian squill; blue, March.
Fritillària lùtea, yellow fritillary; March.
Ornithógalum umbellàtum, the umbel-flowered star of Bethlehem; white, April.
Narcíssus *Jonquílla*, common jonquil; yellow, April.
Lílium aurántium, the orange lily, May.

The perennials for the bed *g* may be the 5 following:—

Anthýllis coccínea, the scarlet kidney vetch; June.
Diánthus cæ'sius, the grey pink; white, June.
Geránium ibéricum, Spanish crane's bill; blue, July.
Œnothèra serótina, late evening primrose; yellow, August.
Pentstèmon diffùsus, spreading pentstemon; purple, September and October.

The bulbs may be:—

Fritillària præ'cox, early fritillary; white, March.
A'llium incarnàtum, flesh-coloured moly; red, April.
Tùlipa sylvéstris, wood tulip; yellow, April.
Scílla campanulàta, bell-flowered squill; purple, May.
Narcíssus poéticus, the poet's narcissus; white, May.

The perennials for the bed *h* may be:—

Diánthus deltöìdes, the deltoid pink; red, June.
Verbèna sulphùrea, the yellow verbena; July.
Erígeron *bellidifòlius*, the daisy-leaved erigeron; purple, July.
A'ster *conyzöìdes*, the fleabane-like aster; white, August.
Scabiòsa austràlis, the southern scabious; blue, September and October.

The bulbs may be as follows:—

Bulbocòdium vérnum, spring bulbocodium; purple, March.
Leucòjum æstìvum, summer snowflake; white, April.
Gladìolus commùnis, common cornflag; red, May.
Fritillària imperiàlis, crown imperial; yellow, May.

By this mode of planting we have 36 species of perennials, and 38 bulbs; and the advantage that it has over the preceding mode is, that a much greater show will be made in the months of March, April, and May; because, as already mentioned, the flowers of bulbous-rooted plants are much larger in proportion to the foliage than those of fibrous-rooted plants. The flowers are also much more conspicuous; because, in general, they expand before the leaves have attained their full size. On the whole, however, the culture of bulbs in mixture with perennials is inconvenient; except when florist's bulbs only are employed, that come into flower and fade all about the same time, and the roots of which may be taken up annually in June or July, and replanted in November or December. This admits of taking up the perennials every year, or every other year, stirring, refreshing, or renewing the soil, and pruning or otherwise reducing the plants, and then replanting them. On the other hand, if what may be called botanical bulbs were introduced, as these are best allowed to remain in the soil for several years, the perennials cannot be so conveniently taken up, reduced, and replanted, when they get too large. Another reason against intermixing permanent bulbs with perennials is, that the moisture required during summer to keep the fibrous-rooted plants in vigorous growth, has a tendency to rot the bulbs, they being at that time in a dormant state, and, in their native habitats, comparatively dry; almost all bulbs being natives of countries which have alternate seasons of drought and moisture, and flowering only in the latter. The best florist's bulbs to intermix with perennials are the different varieties of the common hyacinth, the crocus, the tulip, and the narcissus.

Planting with showy Perennials, which are common and cheap. In the above selections, both of fibrous-rooted plants and bulbs, our principal object has been to display a number of species, and thus to combine a botanic garden with a flower-garden; but, if the object had been to make a showy display, without regard to species, then we should have had recourse to popular and well-known garden species or varieties, which keep in flower two or three months together; and, as these generally grow much larger than the kinds we have selected above, one third the number will be sufficient. The following list consists of showy species and varieties, quite hardy, that will grow with ordinary care, in any common soil, and may be procured in any good nursery, at prices varying from 3*d.* to 1*s.* each, when purchased by the single plant; or from 2*s.* 6*d.* to 10*s.* per dozen, as will be seen by the priced lists at the end of this work. Those who wish to know something more of any particular kind than what has been here stated, may refer to our descriptive catalogue, in which they will also find short directions for their culture; and to the priced lists for their prices.

February and March.

*V*ìola tricolor, different varieties of heartsease.

*V*ìola odoràta, the sweet-scented violet.

*B*éllis perénnis, the double red, double white, variegated, and hen and chickens, daisies.

*G*enti*ā*na acaúlis, the dwarf gentian; purple.

Hepática tríloba, the double red, double white, and double blue, hepaticas.

Prímula vulgàris, the double white, double red, and double lilac primroses.

Prímula elàtior, the double oxlip.

Prímula Auricula, different border varieties of auricula.

Prímula vulgàris Polyánthus, varieties of double and single polyanthus.

*A'*rabis *r*òsea and álbida, white and red arabis, or wall-cress.

*A*dòni*s* vernàlis, spring-flowering adonis; yellow.

*A*nemòne apennìna, alpine anemone; blue.

*I'*ris pùmila, dwarf iris; blue.

Omphalòdes vèrna, spring-flowering Venus's navelwort; blue.

*O'*robus vérnus, spring bitter vetch; purple.

*C*orydàlis bulbòsa, the bulbous-rooted fumitory; red.

April.

*P*hlóx subulàta and setàcea, awl-leaved and bristly-leaved phlox; red.

*C*orydàlis longiflòra, long-flowered fumitory; red.

Aquilègia canadénsis, Canadian columbine; red.

*D*odecàtheon Meád*ia*, Mead's Virginian cowslip; lilac.

*L*ýchnis Viscària, the bladder Lychnis; red.

*A'*rabis alpìna, alpine arabis, or wall-cress; white.

*A*nemòne nemoròsa, the wood anemone; white.

*P*hlóx nívea, the snowy phlox; white.

*D*ràba *a*ïzöìdes, whitlow grass; yellow.

*A*lýssum saxátile, rock madwort; yellow.

*P*hlóx divaricàta, spreading phlox; blue.

*A*nemòne praténsis, meadow anemone; blue; and *A. Pulsatílla*, pulsatilla anemone; purple.

*A*nemòne coronària, the common anemone; variegated.

May.

*P*æòn*ia* officinàlis, numerous varieties of the common pæony; crimson, red, and white.

*P*apàver bracteàta and orientàlis, scarlet poppies.

Campánula glomeràta, double and single white and blue bell-flower.

Diánthus Caryophýllus, different varieties of the common pink, clove, and carnation; white, red, and variegated.

*L*upìnus polyphýllus, purple and white lupine.

*I*bèris Tenore*ā*na, Tenore's candytuft; white.

*L*àmium álbum, white dead nettle.

*O'*robus angustifòlius, narrow-leaved bitter vetch; white.

*R*anúnculus *a*conitifòlius fl. pl., the double white batchelor's buttons, or fair maid of France.

Tradescánt*ia* virgínica, white and purple spiderwort.

*G*erànium praténse, purple and white-flowered crane's bill.

*A*lýssum créticum, Cretan madwort;

*M*ímulus moschàtus, yellow musk plant.

*M*ímulus lùteus rivulàris and guttàtus, varieties of the monkey flower; yellow and brown.

Galárd*ia* bícolor and aristàta, yellow and red galardias.

*A*sphódelus lùteus, yellow asphodel.

*R*anúnculus àcris fl. pl., the double yellow bachelor's buttons.

Linària alpìna, blue toadflax.

*P*hlóx procúmbens, trailing phlox; blue.

Aquilègia grandiflòra and vulgáris, purple and common columbine.

*L*áthyrus venòsus, blue and purple pea.

*L*upìnus perénnis, perennial lupine; blue.

Pulmonària virgínica, Virginian lungwort; blue and purple.

*N*épeta grandiflòra, large-flowered catmint; blue.
*A*nchùsa itálica, Italian bugloss; blue.
A'ster alpìnus, alpine starwort; purple.
*L*àmium *Orvàla*, purple dead nettle.
Pentstèmon atropurpùreus, campanulàtus, and diffùsus, dark purple, bell-flowered, and spreading pentstemons.
*A*nemòne horténsis, garden anemone; variegated.
*V*erbáscum cùpreum and ferrugíneum, the copper-coloured and rusty mullein; variegated.

June.

*D*iánthus alpìnus, deltöìdes, and átrorùbens, varieties of pinks; red.
Silène acaúlis, the stemless catchfly; red.
Valeriàna dioíca, common red valerian.
O'robus sylváticus, wood vetch; red.
*A*ntirrhìnum màjus, double, single, white, red, and variegated snapdragon.
*L*ýchnis chalcedónica, the scarlet lychnis.
*L*ýchnis diúrna, the rose campion; red.
*A*conìtum Napéllus, the common monk's-hood; purple.
*L*áthyrus grandiflòrus, the large-flowered pea; red.
Campánula rotundifòlia, purple and white bell-flower.
*Œ*nothèra speciòsa, the showy evening primrose.
*H*ésperis matronàlis fl. pl., double white rocket.
*D*ictámnus álbus, white fraxinella.
*L*ìnum perénne, perennial flax; white.
*P*olemònium cærùleum and álbum, the common purple and the white Greek valerian.
*A*sphódelus ramòsus, the branchy asphodel; white.
Epilòbium angustifòlium, white and red French willow-herb; red.
*C*áltha palústris fl. pl., double-flowered marsh marigold; yellow.
*Œ*nothèra macrocárpa, and other species of evening primroses; yellow.
*A*conìtum grandiflòrum, and other species and varieties of monk's-hood.
Chrysèi*s* (Eschschóltz*ia*) califórnica, Californian eschscholtzia; yellow.
*A*sphódelus lùteus, yellow asphodel.
Tróllius europæ'us, common globe flower; yellow.
Pentstèmon confértus, crowded pentstemon; yellow.
*G*enti*àna* lùtea, yellow gentian.
A'juga pyramidàlis, the pyramidal bugle; blue.
*D*elphínium élegans, and various garden species and varieties of larkspur; blue and purple.
I'ris germánica, the German iris; blue and white.
Pentstèmon speciòsus, showy pentstemon; blue.
*L*àmium maculàtum, spotted dead nettle; purple and reddish lilac.
*L*ýchnis diúrna, rose campion; red and white.
*V*erbèna Lambért*i*, Lambert's verbena; purple.
*V*erbáscum phœníceum, dark purple mullein.
*D*ictámnus Fraxinélla, purple fraxinella.
*A*nchùsa angustifòlia, narrow-leaved bugloss; blue and purple.
*G*erànium lancastriénse, the Lancaster crow's bill; variegated.

July.

*Œ*nothèra *r*òsea, red evening primrose.
*P*hlóx stolonífera, glabérrima, pyramidàlis, and various other phloxes; red, lilac, and purple.
Verónica incarnàta; the flesh-coloured speedwell.
Saponària officinàlis, and fl. pl., single and double soapwort; red and white.
Monárd*a* dídyma, scarlet monarda.
Chelòne barbàta, the bearded chelone; red.
*L*áthyrus tuberòsus, the tuberous-rooted vetch; red.
Campánula *p*ersicifòlia, double and single white peach-leaved bell-flower.
*G*enti*àna* Saponària, soapwort-leaved gentian; white
*P*hlóx suavèolens, the sweet-scented phlox; white.
*A*lýssum montànum, mountain madwort; yellow.
Galárd*ia* aristàta, bristly galardia; yellow.
Coreópsis grandiflòra, large-flowered coreopsis; yellow.
*H*ypéricum elàtum, St. John's wort;

Dracocéphalum grandiflòrum, large-flowered dragon's head; blue.
Campánula carpática, Carpathian bell-flower; blue.
Véronica azùrea, marítima, and others, different kinds of speedwell; blue.
*A'*ster alpìnus and *a*mellöìdes; purple asters.
*S*tátice reticulàta, purple sea lavender.
Pentstèmon atropurpùreus, and various purple-flowered species of pentstemon.
Lobèl*ia* speciòsa, showy lobelia; purple.
*L*ýthrum Salicària, willow herb; purple.
Monárd*a* fistulòsa, purple monarda.

August.

Verónica cárnea, flesh-coloured speedwell.
Pentstèmon angustifòlius, narrow-leaved pentstemon; red.
*P*hlóx pyramidàlis, and other red phloxes.
Gypsóphila prostràta, trailing gypsophila; white.
*A'*ster álbus, white aster.
Coreópsis trípteris, aúrea, and verticillàta, and different kinds of yellow coreopsis.
*G*enti*àna* *a*sclepià*dea*, asclepias-like gentian; blue.
Commelìn*a* erécta, upright commelina; blue.
*A'*ster spectábilis and Nòvi Bélgii, New York asters; blue.
*V*erbèna venòsa, veiny verbena; blue.
*A'*ster alpìnus, purple asters.
Stenáctis speciòsa, showy stenactis; purple.
*L*ýthrum virgàtum, twiggywillow herb; purple.
*A'*ster cóncolor, self-coloured aster; purple.
Campánula versícolor, variegated bell-flower.
Caléndula stellàta, starry marigold; yellow.
*P*olýgonum orientàle, persicaria; red.

September and October.

Epilòbium alpìnum, alpine French willow herb; red.
*G*enti*àna* incarnàta, flesh-coloured gentian; red.
*A'*ster vimíneus, twiggy aster; red.
*P*hlóx triflòra and Wheeleri*àna*, three-flowered and Wheeler's phlox: red.
*A*ster hùmilis, diffùsus, and others, dwarf-spreading and other asters; white.
*A*chillè*a* crética, Cretan milfoil; white.
Boltòn*ia* *a*steröìdes, aster-like boltonia; white.
Coronílla mínima, the least coronilla; yellow.
Solidàgo hùmilis, dwarf golden rod; yellow.
Œnothèra serótina, late evening primrose; yellow.
*G*enti*àna* Catesb*æ'i* and others, Catesby's and other gentians; blue.
Scabiòsa austràlis, the southern scabious; blue.
*A'*ster corymbòsus, spectábilis, and others, different kinds of asters; blue.
Verónica elàtior, the taller speedwell; blue.
*S*tátice reticulàta, the netted sea lavender; purple.
*G*enti*àna* intermèdia, intermediate gentian; purple.
*P*hlóx Carolìna and suffruticòsa, Carolina and suffruticose, or half-shrubby, phloxes; purple.
*A'*ster Nòvæ A'ngliæ and others, New England asters; purple.
Verónica altíssima, the highest speedwell; purple.

Planting with Biennials and Annuals. The flowers of a front garden may consist solely of biennials, which require to be sown every two years; or of annuals, which must be sown every year; or there may be a mixture of these two kinds. The advantage of biennials is, that they are generally plants of great bulk (such as the hollyhock, which is one of the most splendid of biennial plants), and, consequently, few are required for a considerable space; and the disadvantages are, that the first year of their growth no blossoms are produced; unless, indeed, they are transplanted from a reserve ground, in which case they never flower so strongly as when allowed to remain where they are sown. Their large size renders them, in general, unfit for small front gardens, unless we except an occasional hollyhock, Canterbury bell, French honeysuckle, marvel of Peru, &c.; or unless the occupier should prefer temporary bulk and show, to permanent and more minute beauty and variety. The advantage of annuals is, that they occasion little trouble in cultivating the ground; because, as the greater part cease flowering at the commencement of frost, they may be then removed; and the ground, being dug, will require nothing further to be done to it till February, when a fresh supply of annuals may be sown; or till March, when, if annual plants are purchased from a commercial gardener, or brought from a reserve garden, they may be planted. The disadvantages of using annuals are, that the ground is naked during winter, and that there are among them no early spring flowers; with the exception of those Californian kinds, which will be hereafter enumerated, such as Calliópsis, Collíns*ia*, Gíli*a*, Clárk*ia*, and others, which, if sown in September, will stand our ordinary winters, and flower in May. Other disadvantages of annuals for a small garden are, that many of the common kinds are rampant, straggling, weedy-looking plants, especially when sown in patches, and not thinned out; and that there are but few of them, such as the mignonette, which continue in flower all the summer. Some, indeed, as the China aster, require the whole summer for their growth, and only come into flower in the autumn. Annuals, therefore, are not well adapted for making a perpetual display in a very limited spot of ground; though many of them, especially the dwarf and trailing kinds, such as Eschschóltz*ia*, Clárk*ia*, Gíli*a*, Nemóphila, *A*nagállis, &c., make a very splendid show during the time that they are in flower. This, however, is scarcely a sufficient compensation for the various flowers produced in March and April by such a selection of fibrous-rooted perennials and bulbs as that shown in the lists from p. 220. to p. 225. We shall not, therefore, say much on the subject of annuals, till we come to treat of villa, or first and second-rate, gardens; to which, from their greater extent of space, annuals and biennials are much better adapted than they are to fourth-rate gardens.

Planting a Front Garden with Roses and herbaceous Plants. In the front garden (*fig.* 57. p. 211,), the beds are on turf; and they may either be planted with dwarf perennials, twelve in the centre bed (*i*), and five in each of the other beds, selected from the list p. 217., so as to have an equal number of species in bloom every floral month in the year; or they may be planted with perennials and bulbs; or, what is preferable, with bulbs to be succeeded by annuals or perennials, brought forward in pots, and turned out into the soil immediately after the bulbs are removed. A

65

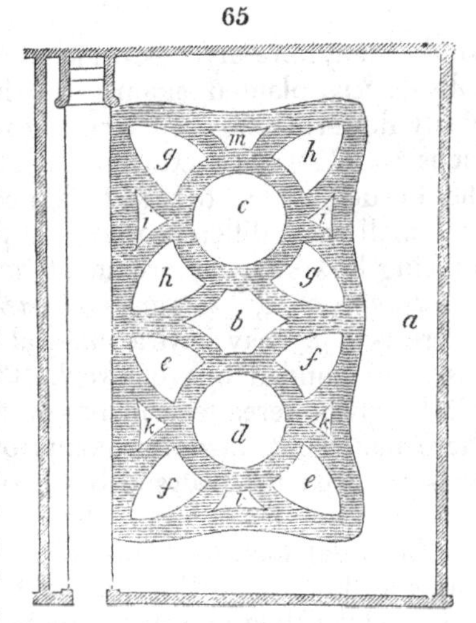

very good display of flowers might be made by planting the centre bed with China roses, and the four surrounding ones with either other sorts of roses, or mixed herbaceous plants; but this, though it might be as gratifying to a stranger who only saw the garden occasionally, as any other mode, yet, as the beds of roses would present the same appearance for nearly six months, the effect would be rather monotonous to the occupant, who looked down on it every morning from his parlour window. Variety, therefore, is a grand desideratum in front gardens: besides, roses thrive much worse when exposed to smoke than herbaceous plants; nor would this mode of planting create so much exercise and recreation for the occupant (which, it will be recollected, is the principal object in view in this mode of laying out a front garden) as if herbaceous plants alone were used.

Planting a Front Garden so as to produce symmetrical Masses of Colour. Fig. 65. is calculated for a larger space than any other design for a front garden which we have yet given. It may be supposed to face the south, like some of the front gardens on the north side of the New Road, London; and to be 30 ft. broad, and 45 ft. long. It is surrounded by a broad border (*a*), separated from the lawn, or grass-plot, by a line, which is slightly waved, in order to harmonise it with the curvilinear forms of the beds; but, at the same time, not so much so as to be out of harmony with the straight wall and the straight walk. These beds we shall suppose to be planted in masses of one kind in a bed; the object being to make a great display of colour from the windows of the house, and from the walk from the street entrance to the front door, and yet to preserve in the colours the symmetry shown in the forms and disposition of the beds. The border (*a*) we would plant with rhododendrons, and other low evergreens, to form a dark background to the flowers on the lawn; intermixing the rhododendrons with laurustinuses, autumn and spring flowering mezereums, and Cydònia japónica, for late and early flowers; and planting the wall with ivy. The border next the street may have three variegated hollies planted in it, to break the view of the street from the house; or one holly, one scarlet-flowered arbutus, and one variegated tree box may be chosen for this purpose. The border beneath the parlour window may have myrtles, camellias, maurandyas, passion-flowers, sollyas, &c., planted against the house, and some dwarf half-hardy flowering shrubs, such as escallonias, *Ceanòthus azùreus*, fuchsias, *Céstrum noctúrnum*, a dwarf fan palm, &c., planted in the border, so as to create a necessity for winter protection. We shall show different modes of planting the beds; previously noticing how statuary ornaments may be introduced in them.

Introduction of statuary Ornaments in Front Gardens. The centre bed (*b*) may have a sun-dial in the middle, rising from the base, surrounded by rockwork (the manner of forming which will be given hereafter), planted with select creeping plants; and there may be a vase for plants on a pedestal in each of the beds *c* and *d*; the pedestal being surrounded by climbers, to be trained to it, but not higher than the base of the vase. Throughout the floral months, these vases might be filled with pots of plants in flower, and throughout the winter with evergreens. The beds *e e* may then be filled with red-flowering low plants, allowing an equal number for every floral month; *f f*, with white-flowering plants; *g g*, with blue-flowering plants; and *h h*, with yellow-flowering plants. The small beds may have crocuses round their margins, and in the centre mignonette. As this plant is never fragrant except in poor soil, these beds should consist almost entirely of lime rubbish, except round the margins where the crocuses were planted. Among the shrubs we would plant bulbs of as many low-growing kinds as room could be found for, but no fibrous-rooted herbaceous plants whatever. Bulbs among evergreens make a very fine appearance in spring; and, as their foliage dies off altogether in summer, they have not that littery disorderly appearance which herbaceous perennials in shrubberies usually have when they have done flowering.

Planting with Bulbs, to be succeeded by showy Annuals. Instead of this arrangement, the three central beds might be planted with three different kinds of showy perennials, or with three different kinds of bulbs, or with three different kinds of annuals. The bulbs for the centre bed might be crocuses, as coming first into flower; and for the two others hyacinths; for the beds *e e* and *f f*, tulips; and for *g g* and *h h*, polyanthus narcissus: the smaller beds might be solely devoted to crocuses. This mode of planting would produce a fine display from March till the middle of May; when the bulbs should be taken up, and the beds, having been dug, should be immediately planted with annuals, previously brought forward to nearly a flowering state in pots. In planting these annuals, not more than one kind should be introduced in each bed; and the kinds in beds opposite should be as like in colour, height, and general appearance as possible. Thus, if one of the beds, *e*, were planted with the ten-week stock, the other ought to be planted with purple candytuft, or any similar flower that came into bloom at the same time; and, if one of the beds, *h*, were planted with Clárk*ia* pulchélla, the other should be planted with Lobel's catchfly, &c. On the whole, however, a more striking effect is produced by planting opposite beds of exactly the same kinds; and those for *fig.* 65. may be the following, it being understood that the plants are to be so close together as to cover each bed entirely; and that, whenever any gap appears in a bed, the adjoining plants are to be pegged down to cover it. For *b*, *L*upìnus mutábilis, which would produce its fine blue, white, and yellow flowers all the summer; for *c* and *d*, German stocks in mixture, purple, red, and white; for *e e*, German larkspurs in mixture, purple, red, and white; for *f f*, new scarlet candytuft, scarlet; for *g g*, Calliópsis bícolor, yellow and brown; for *h h*, Collíns*ia* bícolor, blue and white; for *i i*, Clárk*ia* pulchélla, lilac; for *k k*, Eschschóltz*ia* califórnica, yellow and orange; for *l*, *M*álope grandiflòra, dark crimson; for *m*, Zínn*ia* élegans coccínea, scarlet. All these annuals are new, and eminently beautiful; and seeds of them may be procured in most of the principal seed shops.

Planting with Bulbs, to be succeeded by half-hardy Annuals and Green-house Plants. The third mode is adapted for gardens where there is not only a reserve ground, but pits or hot-beds in which to preserve some of the plants which we shall recommend during winter, and to bring forward others in spring. During winter, the beds are supposed to be filled with bulbs in the manner before mentioned; or, what would produce a truly brilliant effect, wholly with hyacinths, and the margins of the beds with a line of crocuses. The bed *b* might be planted with mixed hyacinths, one of a sort; the beds *c* and *d*, with white hyacinths; *h h* and *e e*, with red; and *f f* and *g g*, with blue. The smaller beds might be filled entirely with crocuses. In the course of the month of May, the bulbs should be removed, and the ground dug and planted as follows: — *b*, with variegated pelargoniums (geraniums); *c*, with Bath scarlet pelargoniums; *d*, with Frogmore scarlet pelargoniums; *e e*, with *V*erbèna *c*hamædrifòlia (*Melíndris*), scarlet; *f f*, with Lobèl*ia* lùtea, yellow; *g g*, with *Petùnia* nyctaginiflòra, white; *h h*, with *Petùnia* phœnícea, dark purple; *i i*, with Lobèl*ia* grácilis, blue; *k k*, with Nierembérg*ia* grácilis, white; *l*, with Nemóphila aurìta, blue; and *m*, with *S*enècio élegans fl. pl. rùber, red. All these plants are easily procured; and they may be preserved in pits or frames, with a very slight assistance, during cold and damp weather, from fire or dung heat. All of them ought to be planted so close together as to cover the beds by the middle of July; and any shoots that rise above 6 in. from the bed should be cut off or pegged down, and any naked space that may occur on its surface should be covered by pegging down shoots from the adjoining plants.

Planting a Front Garden with Florist's Flowers. Fig. 66. is a front garden of the same size as *fig.* 65. in p. 227., laid out with a view to the culture of florist's flowers. The side walls should be planted with ivy; and in front, next the street wall and railing, the ivy may be trained to wires, one wire being introduced between each rail. The plants trained against the house should be select climbing roses. The borders *a* and *b* should be devoted entirely to dahlias in summer, and bulbs in winter and spring; the bed *c*, to chrysanthemums; *d*, to pinks and carnations; *e* and *f*, to tulips; and *g* and *h*, to hyacinths; the other bulbs, such as ranunculuses, anemones, &c., being grown in the dahlia borders before the latter are planted; or, if the symmetrical effect is dispensed with, tulips, hyacinths, ranunculuses, and anemones, may be grown in the beds *e f g h*. These beds, after the bulbs are removed, may be planted with showy annuals, or other plants brought forward in the reserve ground. As the foliage of the carnations and pinks, which occupy the centre bed, is evergreen, it will look well throughout the year.

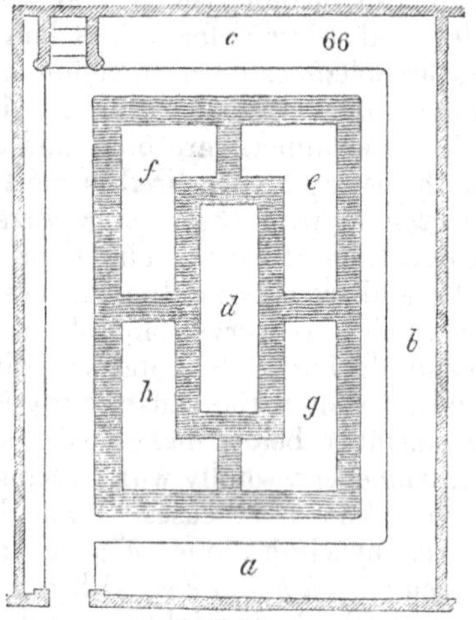

It will be evident, from the examples which we have given of the manner in which front gardens may be planted so as to make a fine display, and to give their occupiers ample exercise and recreation to keep them in order, that the variety of modes in which this may be done is almost infinite. We shall not, however, pursue the subject further here, having done enough, we trust, to show the reader how he may exercise his own taste and judgment, not only in a small front garden, but in one of any extent whatever.

The Back Garden, where the object is to give exercise and recreation to the occupier, may either be laid out and treated as a kitchen-garden, in the manner already described p. 199.; or it may be treated as one large flower-garden, in the same manner as the front garden of the same design; or it may be arranged as an exotic garden, with hot-houses, &c., or as a botanic garden, as will be described hereafter. Perhaps the best mode would be to plant and manage it as a kitchen and fruit-garden; because the practices requisite for its cultivation would thus form a greater contrast to those required for the front garden. If laid out as a culinary and fruit-garden (being surrounded in the usual manner with a walk, and a reserve ground being portioned off at the farther end, with a pit for forcing, and another for preserving plants for the front garden), the interior plot should have an espalier trellis; and this, in order that it may occasion the more labour, ought to be double, and have trees trained on both sides. A double espalier trellis may either be formed of Kyanised wood, or of wire; and, in either case, the width at the bottom should be 18 or 20 inches, with the sides tapering to a point at the top, like the section of the roof of a house, or like the letter A. If the trellis be made of iron, the standards may be of hoop iron, and the horizontal wires of ¼-inch iron rods, put together in the manner that we shall shortly explain.

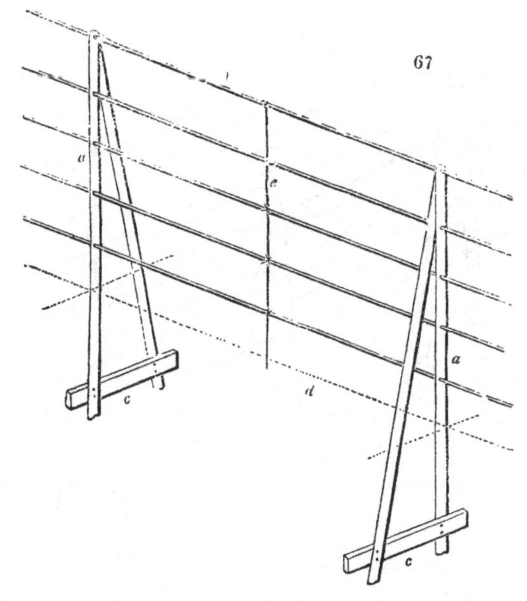

A single Trellis formed of Hoop Iron and Iron Wire. *Fig.* 67. is a perspective view of a single hoop and wire trellis;

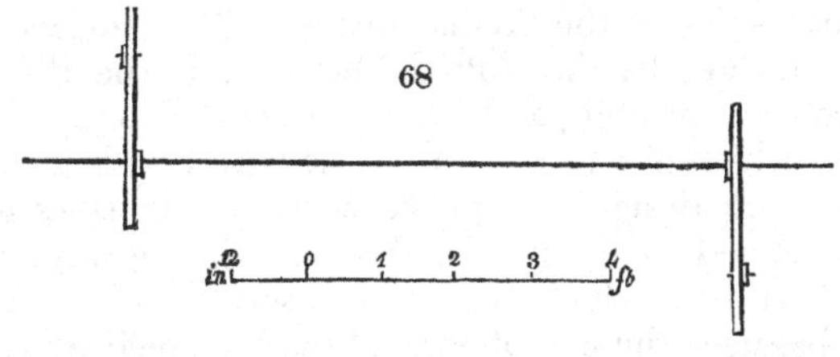

of which *fig.* 68. is the ground plan. The standards (*a a*) are of hoop iron, and are kept together at top by the rod *b*, which passes through them; and their lower extremities are nailed to pieces of wood (*c*), which, when the trellis is put up, are buried in the soil to the depth of 2 or 3 feet, as indicated by the dotted line *d*, so as to keep the espalier firmly in its place. In the middle of each portion (such as that indicated in the figure) of the espalier, and equidistant from the standards of hoop iron, is a standard of rod iron (of the same thickness as that used for the horizontal rails,) *e*; to which these rails are fastened with wire, merely for the purpose of keeping them steady, and at regular distances from each other.

A double Trellis, formed of Hoop Iron and Iron Wire. Fig. 69. has the standards (*a*) double as in *fig.* 67., with their lower extremities fastened to the pieces of wood, *c*, as in that design, but

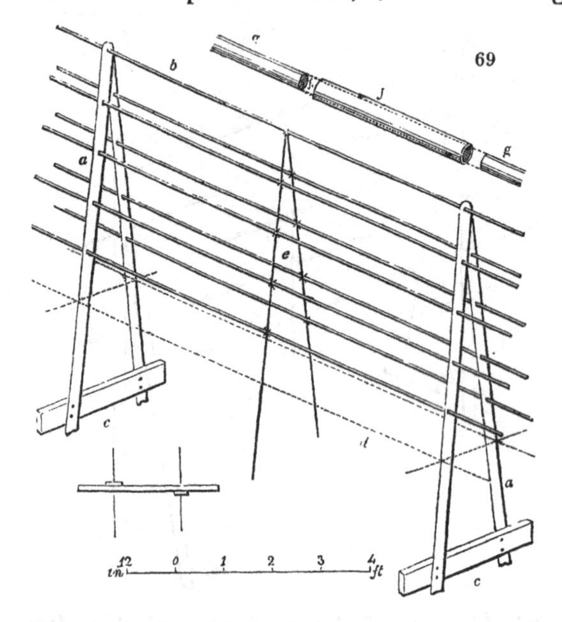

the wires, *b*, are put through both limbs of the standard; and there is a double wire at *e*, instead of a single one, to keep the rods on both sides of the trellis, steady. The surface of the ground is indicated by the dotted lines *d*; and the distance of the wires from the ground, and from one another, may be found by the scale, which is 4 ft. to 1 in. The wires, both in this and in the preceding design, are joined by inserting their ends in cylinders of tinned iron, like that shown at *f*; *g g*, representing the ends of two rods which are to be inserted in it. A notch should be filed near the end of each of the rods before insertion; and, after insertion, the tin cylinder should be struck smartly over the notches, with the wedge end of the hammer, so as to indent it into the wires, and keep them firmly in their places. They ought, also, before insertion, to be covered with gas tar, or white lead; and after the whole is completed, they ought to be coated over with gas tar, common paint, or, what is greatly preferable, anti-corrosive paint. We have had an espalier of this kind at Bayswater, covered with gooseberry bushes, since 1824. It has only been twice painted over with gas tar, and it is now, 1837, nearly as strong as when it was first put up. The total expence was about 1*s.* 6*d.* for 1 ft. in length. Espaliers of this kind, being broader at bottom than at top, admit of the rain falling on all the leaves from the top to the bottom; and, whether they are covered with shrubs or trees, for the sake of their flowers or for their fruit, they are alike handsome and advantageous. Different trees may be grown on each side of the espalier; or only one sort may be placed in the centre, half its shoots being trained on the one side, and the other half on the other side. The space enclosed by the trellis should be laid out in beds, in the direction of north and south, or in the direction of N. E. and S. W., or N. W. and S. E; and the crops grown in these beds, should be in rows, in the same direction as the beds. This will occasion the beds to be made sometimes lengthwise of the plot, instead of across it; in which case, it will be convenient to divide the plot by cross alleys into different lengths, to avoid the inconvenience of very long rows, and to facilitate crossing from one part of the plot to another. If this mode of disposing of the beds should not be liked, they may be placed diagonally across the plot; and thus a back garden, lying in the direction of north and south, may have its beds in the direction of N. E. and S. W.; instead of being lengthwise, that is, N. and S.; or directly across, that is, E. and W. If the crops are not to be grown in rows, the direction of the beds is of little consequence: but so great are the advantages of cultivating plants in rows, in every description of garden culture, that we can on no account recommend this mode of cultivation to be dispensed with in suburban kitchen-gardens. The advantage of having the rows in the direction of north and south consists in the sun shining on both sides of every row, and on the soil between the rows, every day on which he appears throughout the year.

If the back garden is to be laid out for the display of flowers, the wall facing the south should be flued; and it should be planted with half-hardy ligneous plants, which will require covering with mats every night throughout the winter and spring months. The plants on such a wall, during summer, would unite the great desiderata of the botanical florist — beauty and rarity. The central plot may either be of turf, with dug beds, as in *fig.* 70.; or there may be no turf, and the beds may be edged with box,

or brick; or they may be without edgings, the walks being all paved, as in *fig.* 71. If florists' flowers are to be cultivated the beds should be simple parallelograms, as in *fig.* 55. p. 201. If it were thought necessary, additional labour, beauty, and interest might be created, by placing boxes on the copings of the side walls, and the end wall; and in these, alpines might be grown in pots. The boxes might be of wood, of the same width as the coping; and a foot deep, so as to admit of plunging the potsof alpines in them in sand or gravel, so as to retain moisture, and act as a non-conductor to the heat produced by the sun shining on the sides of the box. Before this plan, however, could be carried into execution, it would be necessary to get the consent of the occupiers of the adjoining houses. But, supposing this obtained, and the garden 200 ft. long, and 50 ft. broad, there would be 450 ft. of coping; and, if the plants were grown in pots of 4 in., or 5 in. in diameter, there would be room for four in every foot, which would give 1800 species; nearly as great a number as was supposed to exist of every kind of plant in the world in the time of Miller; and, at all events, as many alpines as can be purchased, even now, in any British nursery.

Management. If there be a reserve ground, and two flued pits, with a flued wall for half-hardy plants, a suburban garden of this kind, 200 ft. long and 50 ft. broad, would afford occupation for a gardener throughout the year. In the winter, he would have the pits and the flued wall to look after, and to prepare tallies, compost, &c.; and his labours during spring and summer, if the object were to keep the garden in the very highest order, would be incessant. In winter, all the spare time that he had would be employed in looking at the new books on gardening and practical botany, in collecting soils and various other articles, and in going about to visit the neighbouring gardeners, and in arranging with them for making exchanges of plants, cuttings, and seeds in the ensuing spring. At the same time, so different are the ideas both of gardeners and their employers as to what constitutes high order and keeping, and as to what a place is capable of being made, that some gardeners, in a garden of the extent we mention, would not find enough to do, while others would want assistance; the difference depending on their knowledge of plants, and on their ideas as to what constitutes perfection in their art. At all events, we have described a mode of laying out and planting a garden, which will afford abundant employment and recreation to an amateur, whatever time he may have on his hands.

Expense. The cost of laying out such a garden would not be much greater than any other of the same size; and, if the occupier, or his gardener, were known as botanists, the annual expense of keeping it supplied with new plants would not be so great as might be imagined; because the cultivator of rare and beautiful species will generally be found to have some plants to spare, which another cultivator of like species is in want of. We may set the first cost of laying out and planting at 50*l.*, or, if the occupier has no botanical connexion, and is obliged to purchase all the plants, at 100*l.* As to the expense of managing, if a first-rate gardener is kept (which, however, is not supposed to be the case, as the avowed object of the garden is to afford exercise to the occupier), his wages will be at least 100*l.* a year; and 50*l.* more may be required for fuel, pots, soil, tallies, &c., exclusive of the annual purchase of seeds and plants, which, we suppose, would be very trifling. If the occupier manages his garden himself, with the assistance of a man servant, kept at any rate, and has the whole of his own time to devote to it, it need not cost him above 50*l.* or 60*l.* a year. Every thing, in this case, would depend on the botanical connexion of the occupier, and the manner in which the garden was kept up.

Remarks. This garden, may be regarded as the *beau idéal* of what a citizen, who retires from business rather early in life, and is fond of plants, might endeavour to realise. The principle of conducting it would be the same even if nothing were grown in it but indigenous or wild plants, which a British botanist might collect in their native habitats; or if only a small number of plants were grown, which might be purchased, either as seeds or in a growing state, from any nursery at a low price. Supposing there to be no pits, no flued wall, and no half-hardy plants, and that the back garden were laid out as a kitchen-garden, and only the front plot as a flower-garden, there would still be a good deal of exercise for the occupant; and the expense need be only a very few pounds a year; in short, little more than the repairing or renewing of the tools, and now and then purchasing a load of fresh soil.

5. *To lay out and plant a Fourth-rate Suburban Garden, where the Occupier intends to cultivate some Culinary Vegetables and Fruits, with a few Flowers.* The walks we should lay out as shown in *fig.* 72.; but in the back garden, the borders be-

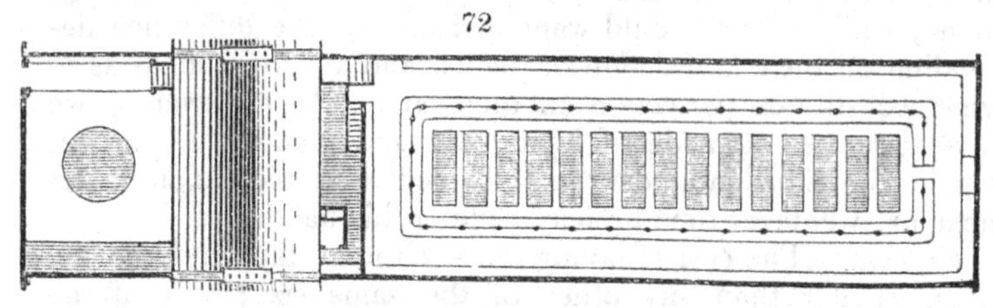
72

tween the walks and the fences (on the supposition that the latter were brick or stone walls, or fences of boards, or of some other material, on which fruit trees can be trained), we would make of double or treble the width, according to the height of the walls, their aspects, and other circumstances. If, indeed, the walks were formed of flag stones, supported from the ground on piers of masonry, we should not increase the width of the borders, because it is one of the advantages of flag stone walks, supported on piers, that the roots of adjoining shrubs and trees can spread freely beneath them, and thus, a garden, having such walks, may be said to be increased in size, to the extent which the walks occupy. This will sometimes, in effect, add a fourth, or a fifth, to the extent of the cultivatable part of a back garden of the kind we are now treating of. If the walls are 6 ft. or 7 ft. high, and the walks are laid with gravel, the breadth of the border where the aspect is to the south, and peaches, nectarines, figs, and grapes, can be grown, should not be less than 5 ft., as shown; but as every garden of this kind, that has a wall with a southern aspect, must necessarily have one on the opposite side of the garden, of the same length, facing the north; and as, on this wall, only the more ordinary fruits, such as cherries and baking apples and pears, can be grown, the border, on that side of the garden, need not be made wider than 2 ft. or 2½ ft. The end wall of such a garden will, necessarily, face either the east or the west; and as, on a wall of either of these aspects, may be grown apricots, and the finer plums and pears, the width of the border may be 3½ ft. or 4 ft. The depth of all south, east, and west borders of gardens of this kind, should not be more than from 18 in. to 2 ft.; because, the walls being low, the trees should not be encouraged to produce wood, but should be rather checked, so as to throw them into fruit. It is also more desirable, in a small garden, to have several small trees of different kinds of fruit, than one or two large ones, of only one or two kinds; because, in the former case, there may not only be a successional supply, by planting the earlier and later varieties of each kind of fruit, but, as some seasons are more suitable for one kind of fruit than another, there will be the greater chance of a crop every year. The bottom of the borders should be prepared in such a manner as to render them impervious to the roots of the trees; or, at least, not of materials so loose, rich, or moist, as to tempt them to enter far into it, because in proportion as they do so, in the same proportion will the branches run to wood, and not to fruit. The bottom of the border ought to slope outwards, and be continued as far as the drain, which should generally be formed under the walk, or on one side of it; in order that the rain water, after it percolates through the stratum containing the roots, may not be stagnated, but find its way readily to the drain. The soil for the border where the peaches and nectarines are to be placed, ought to be a fresh loam; that for the vines and figs, rich; and that for the apples and pears, more tenacious: but the soils proper for these trees will be hereafter treated of. The soil of the centre plot may be trenched to the depth of 3 ft., the stones and very bad soil removed; and if it be naturally clayey, sand, lime rubbish, or other materials, calculated to keep it open and pervious to moisture, added. On the other hand, if the soil be dry and sandy, either clay, or peat earth, whichever can be most readily procured, ought to be mixed with it, as both these materials have a tendency to retain moisture. Whether the soil be naturally clayey or sandy, it ought to have a good coating of manure intimately mixed with it from the surface to the bottom of the trench. If the soil of the garden be on a chalky subsoil, it will almost always be found that the surface part, fit for growing plants in, is very shallow; in which case, the subsoil ought to be dug out to the depth of at least 2 ft.; and good soil procured, mixed with manure, and put in its place. The ground, being trenched and levelled, may be left to settle for two or three months; during which time, a rail, or trellis, for training gooseberry and currant bushes on, may be prepared in the manner described in p. 231., and fixed round the bed at the distance of 2 ft. or 2½ ft. from the walk; leaving an opening as an entrance to the interior, at the lower part of the garden, as shown in *fig.* 72. A trellis of this kind is not essential to any garden; but in a small garden, where either gooseberries, or any other kind of fruit shrub is to be grown, it saves much room. If, however, the centre compartment of the back garden is less than 10 ft. or 12 ft. in width, we would not recommend a trellis; because it would confine the interior, and render it too close and shady for growing culinary vegetables. The height of the trellis should always be such that a person of ordinary stature can barely see over it, in order that he may have an idea that there is something within to be seen. This trellis may be either formed of wood or iron, and in either case, if a hoop and wire trellis, such as we have described p. 232., be not employed, the posts or standards, which are introduced at regular distances to support it, ought to have their lower ends let into stones, or fixed into some brickwork, which may consist only of four bricks, so far under the surface as to rest on the solid ground. Instead of being painted, the standards and the rods may be coated over with tar or gas liquor; or if paint is used, it ought to be of the kind called anticorrosive; but if Kyanizing wood succeeds in preserving it to the extent that it is said it will do, it will supersede both modes of protecting from the weather. Immediately within the trellis, and at about 18 in. distance from it, there should be an alley formed 18 in. wide, which need not be gravelled unless the soil be a tenacious clay; and on each side of it may be planted strawberries as edgings.

The front garden should be trenched, and the soil treated in the same manner as that directed for the back garden; but as it is chiefly for turf and flowers, it need not be manured, except in the flower bed in the centre. The form of this bed may be that of a circle or a square. Having now shown the mode of laying out the gardens, we shall proceed to planting them.

Fruit Trees. In no kitchen-garden whatever, whether large or small, would we introduce standard fruit trees among the kitchen crops; because, as we have already stated, p. 202., the digging and trenching necessary for the perfection of the latter, prevent the roots of the trees from spreading near the surface of the ground; and being forced to seek their nourishment in the subsoil, they run to wood from excessive moisture, become cankered, and do not produce either abundant crops or well-flavoured fruit. In a small garden of the kind we are treating of, standard trees would shade the ground so much, that, independently of other consequences, they would render it altogether unfit for the culture of culinary vegetables. We therefore propose to have no standards, but only trees trained against the walls, or boundary fences, and on the trellis-work. On the latter, we propose to train gooseberries, currants, and raspberries, and one or two apples and pears; and on the walls, peaches, nectarines, grapes, figs, apricots, cherries, and plums.

The Walls. We shall suppose that one of the side walls faces the south, and is 100 ft. long, and 7 ft. high. On this may be placed ten trees, to each of which there will be 70 superficial feet of walling; and a vine may be placed at each end, and one in the middle, from each of which a single shoot may be trained close under the coping of the wall, so as never to interfere with the other fruit trees. The peaches, in order to prolong the season of that fruit, may be the red nutmeg, which ripens in July, and is hardy, but bears small fruit; or the early Anne, which ripens in August, and bears very good fruit, but is rather tender; the grosse mignonne, and the bellegarde, both ripening in September; and the late admirable, ripening in October. The nectarines may be the elruge and the violet hâtive, both ripening in August; the Pitmaston orange (September); and the late yellow, which, though not common in this country, is of good flavour, and a very good bearer, ripening in October. The best dark fig is the brown Turkey; and the hardiest and most abundant bearer amongst the light coloured figs, is the small green; or the Brunswick, or the large blue or purple fig, both which produce very excellent fruit, and are very hardy and very prolific, may be substituted. The grapes may be the royal muscadine, the black sweetwater, and the esperione which is one of the greatest bearers in the open air. In favourable situations, the red Frontignan will also ripen against a wall with a south aspect in the climate of London, and it has a very excellent flavour, even when grown out of doors. The wall having an east or west aspect, may be planted with the large early apricot, which ripens about the middle of July, the Moorpark (beginning of August), and the Turkey, which ripens about the end of August, or later. For the wall with a northern exposure, we would recommend the May duke and morello cherries, the one the earliest, the other the latest sort; the bigarreau couleur de chair, and the black Tartarian, a large cherry of the very richest flavour, and which will hang on the tree, if covered with a net, till September. The best and most useful plums are, the green gage, Washington, Coe's golden drop, Orleans, and the white magnum bonum, or the diamond plum, the latter being a very large and handsome fruit.

The trellis will be above 200 ft. in length; and, as we suppose it to be only a single trellis, that is, one which is calculated for having the trees trained on only one side, it may be planted on the north side of the garden, where it is most exposed to the sun, with pears; at the two ends with apples; and on the south side with gooseberries, currants, and raspberries. Eight pears will be sufficient; and these may be the jargonelle, Marie Louise, beurré de Capiaumont, beurré Diel, glout morceau, Chaumontelle, beurré de Ranz, and Easter beurré; all excellent pears, quite hardy, and good bearers. The first ripens in August and the last may be gathered in November, and will keep in sand or in fern packed in jars set in a cool cellar till May or June, or indeed till pears come again. The other sorts are placed in the order of their ripening. The apples may be of four different kinds, viz. the golden and Ribston pippins, and the Hawthornden and Keswick codlin. The two latter are most abundant bearers, and the fruit is not only good to eat, but falls well in boiling. As the trees of these varieties grow with great luxuriance, they should be carefully attended to during spring and early summer, to pinch out a large proportion of the young shoots before they have matured their leaves, in order to prevent the plants from becoming too luxuriant. The gooseberries, and other small fruits, may be planted quite thick, and only two shoots trained in a vertical direction from each plant. They may be placed 2 ft. apart, which will require forty-five plants. The sorts we would recommend are:—*Red:* the red champagne, an early and excellent fruit; the early rough red; the ironmonger, an excellent fruit for tarts and bottling when green, and for the table when ripe; and Aston's Warrington, an abundant bearer, which retains its fruit till October if protected: to these may be added for those who like large fruit, the huntsman and the roaring lion, the former having won 141 prizes in the third and fourth years of its being exhibited, and the latter above 800. *White:* Whitesmith, early, prolific, and excellent; crystal, much esteemed for bottling; Taylor's bright Venus; and the white Dutch, which is preferred to every other kind for making British champagne. *Yellow:* Rockwood, a large early gooseberry, which gained 777 prizes in five years; and rumbullion, reckoned the best in the markets for preserving. *Green:* Massey's heart of oak, early and a great bearer; greenwood; and the Pitmaston green gage, remarkably rich and sweet. In all fourteen sorts, and the quantity planted of each may be proportioned to the taste of the occupier. The currants may be the red Dutch, Knight's early red, and Wilmot's large red; the white Dutch; the champagne, which is of a pale flesh colour, which, though it ripens very early, will hang on the trees without protection till October, and with a covering (such as will be hereafter described) till December; and the black Naples, which is remarkably fine and large. The raspberries may be the early prolific, the red and yellow Antwerp, and the double-bearing or late cane.

The walks in the back garden we propose to be edged with strawberries; and, if space could be spared, the alleys inside the espalier trellis might be edged on both sides with strawberries also. These edgings would in all be about 800 ft. in length, and they might be planted with the following kinds, the large sorts being placed next the broad outer walk, and the small sorts on each side the alleys. The strawberries for the margin of the outer walks may be the old scarlet and Grove End scarlet, which are the earliest; Keen's seedling, which is a great bearer, and, taking it altogether, the best strawberry in cultivation; the roseberry, which is a very great bearer; the old pine, which is the best-flavoured of all strawberries, though a shy bearer; and Wilmot's superb, which, though it has not much flavour, is remarkable for its size. For the margins of the alleys we would recommend, as having small leaves and bearing the fruit on high erect stalks, which are not likely to be injured by the feet of persons walking along the alley, the prolific or conical hautbois, the large flat hautbois, the red and white alpine, and the red and white wood. The last four kinds, if regularly supplied with water, will continue in bearing all the summer.

Culinary Crops. The space left for these, in a back garden of the size shown in *fig.* 72., and laid out in the manner we have described, with flagstone walks, will be but small; being merely a bed about 12 ft. wide by 75 ft. long. To make the most of this plot, it must be divided into smaller beds, say fifteen, each 4 ft. wide, with one foot alleys between, as shown in the figure; and that, at the lower end, as nearest the entrance through the trellis, may contain the perennial potherbs, such as chives, garlic, sage, marjoram, winter savory, thyme, fennel, tarragon, burnet, and mint, the last three being the most useful (tarragon for vinegar, and for flavouring salads and steaks; burnet, at all seasons, for salads; and mint for boiling with peas, and for sauce to lamb, &c.). Annual and biennial herbs, as parsley, chervil, and pot marjoram, may occupy an adjoining bed, double space being required for them; because, while one crop is in use, another must be sown to succeed it. For a very small family, not particular in their cookery, one bed, 12 ft. in length and 4 ft. in breadth, will be quite sufficient for the whole of these herbs, perennial, annual, and biennial. The remaining part of the plot allotted to kitchen crops may be occupied by those vegetables which it is always convenient to have at hand; and by such as never can be so well procured at market, or from the green-grocer, as direct from the garden. Among these, lettuce may be included as always convenient for making a salad, and we would, therefore, devote the beds 2 and 3 to a succession of this vegetable; bed 4 is for small and other salading, such as mustard and cress, radishes, lamb's lettuce, American cress, &c.; in beds 5 and 6 we would sow winter spinach, which, if the Flanders kind be chosen, will afford a dish now and then throughout the winter, and in spring twice or thrice a week, from April to August; beds 7 and 8 may be planted with scarlet runners, two rows lengthwise in a bed, and these will afford a small dish of kidneybeans every other day, from June till they are destroyed by frost; and beds 9 to 15 may be devoted to peas, that being a vegetable never to be obtained good, except direct from the garden; because, in a few hours after peas are gathered and put in baskets or sacks, they begin to heat and ferment, and when dressed they will be found to have become tough and vapid. The most suitable sorts of peas for a very small garden are the blue Prussian, Knight's marrow, and the dwarf marrow; because these have the seeds, or peas, large. The frame and the Charlton are early varieties, but both the seeds and the pods are small, and the peas are neither so succulent nor so sweet when cooked as the kinds before mentioned. Whatever kind of pea is sown ought to be staked as soon as the tendrils appear; because, in this way, the plants occupy less room. The kind of staking or supports proper for peas in a small garden is, a line of rods on each side of each line of peas, and at about 3 in. distant from them, with pieces of string stretched horizontally from rod to rod, so as to leave a clear space between the line of rods 6 in. wide, in which the peas will grow up, and support themselves by clasping the strings and rods. The rods may be from 3 ft. to 5 ft. apart, and from 3 ft. to 4 ft. in height; the first string may be 18 in. from the ground, and the others 9 in. apart. The rods, if Kyanised before being used, will last several years; or light iron rods, formed by bending quarter-inch iron rods in the form of a hair pin, as shown in *fig.* 73., and which will not cost above 2*s*. a dozen, may be used. There are wire hurdles sold for the same purpose, and also iron stakes, but both are much more expensive, though not more efficient, than the light iron rod we have mentioned.

73

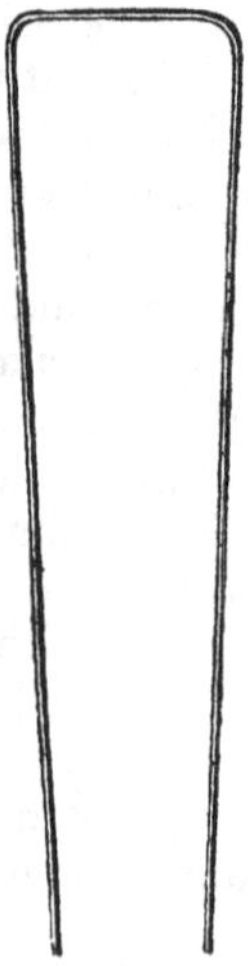

We have now planted and cropped the whole of the back garden, unless we except the wall borders, in which there is the row of strawberries that forms the edging to the walk on that side;

and the narrow space between the strawberry edgings to the other side of the main walks and the espalier trellises. In this last space, we would not plant or sow any thing, except for the first year, when the trees and plants were young, and then only a few onions, radishes, or lettuces. Against the south wall, between every two trees, we might introduce also, during the first and second years, a tomato, a vegetable marrow, and a nasturtium, and train them against the wall for their fruit; and against the others some scarlet runners, or crown or rouncival peas, in the same manner. The third year we would admit of nothing more in the wall borders than one flowering plant to be trained against the wall, in the centre of the space between every two trees. There being 25 spaces, this would give 25 kinds of plants, which would be the total amount of the flowering, or purely ornamental plants, that we would admit in the back garden. A few of them, instead of being herbaceous, might be China or Noisette roses; which, trained against the south wall, so as not to occupy a space of more than a foot in width, and not to interfere with the trees, would continue flowering the whole of the summer. Chrysanthemums, wallflowers, and stocks would form, when trained against the south wall, most beautiful autumn and early spring flowers: a tree carnation and a tree lupine might be placed against the end wall; and, against both the end and the south walls, pelargoniums, fuchsias, petunias, salvias, and other plants generally kept in green-houses, or considered only half-hardy, and which, being easily propagated, may consequently be procured at a cheap rate from an adjoining nurseryman, might be trained between the trees in a similar manner. In general, however, we would recommend hardy plants, perennial, biennial, or annual; but, in all cases, such kinds should be chosen as grow to some height, or spread to some length, in order that they may be trained against the wall, and not lie in a disorderly manner on the border, from which they would soon spread over the walk, and be liable to be trodden on by the passers by. Biennials, and the taller and more elegant annuals (descriptive catalogues of which will be found in their proper places), should be preferred, in most cases, to perennials; because they admit of changing the kinds every year or every other year, and thus every year may be made to produce something new. We would make it a rule, in the management of this garden, to change the sorts of the annuals every year, and the biennials every second year; with the exception of a few kinds, which we should continue to grow every summer, and one of these should be the mignonette. This annual, when trained to a wall, and protected from frost during winter, will last two or three years, and attain the height of 5 or 6 feet, perfuming the garden with its fragrance during nine months in every year. By thus frequently changing the greater number of the sorts, all our finer flowers may, in time, be grown in succession, even in a very small garden.

The Front Garden, it will be observed by the plan *fig.* 72. p. 236., consists of a circular bed for flowers, in the centre; and a border, also for flowers, on one side. If the situation be open and airy, four shrubs may be planted on the grass, one at each angle. These shrubs may be the *D*áphne póntica and *D*. neapolitàna, Mahòn*ia A*quifòlium, and *R*hododéndron azaleöìdes; or, if deciduous shrubs are preferred, a common and an autumn-flowering mezereon, *P*ỳrus spùria, and *C*ydònia japónica, may be chosen. Against the side wall which faces the south, a border 3 ft. wide is shown; and in this may be planted, for the purpose of being trained against the wall, several roses, red and white, a white jasmine, Lonícer*a* japónica, and Wistàr*ia* sinénsis, the last three of which may have some of their branches trained against the house. On the supposition that the front fence consists of a wall about 2 ft. high, surmounted by an iron railing, we would plant it with the different kinds of ivy, intermixed with the Virginian creeper, the everlasting pea (*L*áthyrus latifòlius), and the large-flowering pea (*L*. grandiflòrus). The bed in the centre we would devote to a miscellaneous assemblage of evergreen herbaceous plants, in order that it may look well during winter; bulbs to flower during spring; and plants of various kinds, including annuals and half-hardy flowers, with some pelargoniums, fuchsias, &c., to make a display during summer. The principal evergreen herbaceous flowering plants are, pinks, carnations, sweet Williams, statices, saxifrages, especially *S*. crassifòlia, &c. The bulbs should be chiefly hyacinths, which are at once some of the most beautiful and the most fragrant of flowers; but the bed should be bordered with crocuses, aconites, and snowdrops, and should contain, also, polyanthus narcissus, and other kinds of bulbs sold in the seed shops. Among the fixed herbaceous plants, some flowering in every month throughout the year ought to be chosen: for example, the Christmas rose for January; the hepatica and the primrose, for February; *A*'rabis álbida and *r*òsea, for March; Aquilègia grandiflòra, for April; *I*bèris Tenore*àna*, for May; lupines and cockspurs, for June and July; *Liàtris* corymbòsa, for August; pentstemons and phloxes, for September; and asters, and dwarf dahlias, for the remainder of the year, till they are destroyed by frost. This, or any other, selection may be varied or changed at pleasure, by consulting the catalogues given in a future page.

Estimate of Expense. Having now completed the laying out, planting, and cropping of both the back and front gardens, we shall next make an estimate of the first cost of the whole. The operations on the ground and the walks, including the espalier trellis, manure, &c., may be set down at from 30*l.* to 40*l.*; the fruit trees, two years trained, will cost, at an average, 2*s.* 6*d.* each, say 5*l.*; the gooseberries, currants, &c., 4*d.* each, say 20*s.*; the strawberries, 4*s.* per 100, 32*s.*; the potherbs and seeds of the culinary vegetables, about 15*s.*; the flowering plants for the back garden, about 2*l.*; and the shrubs, plants, and bulbs, for the front garden, 5*l.* In all, the first cost will be from 45*l.* to 55*l.*; the difference depending principally on the cost of the labour and materials expended in forming the walks, and in trenching and manuring the ground. This estimate is only to be considered as a rough approximation to what would be the actual cost; because so much depends upon the nature of the soil and subsoil, and the drainage and new soil required, the price of flagstones, of manure, &c., independently of the price of labour, that it is impossible to make an accurate estimate, unless a specific case were given. On turning to our priced lists, the cost of trees, plants, and seeds, in London, may be ascertained with something like certainty; but the exact price of every other particular must depend on the quantity and kind of work to be done, as well as on local circumstances.

The Mode of Management of a garden thus laid out and planted requires to be next noticed. We shall suppose the occupier to be desirous of managing it himself, and to be able to devote to it his spring, summer, and autumn evenings. We may premise that, in such a limited space as we have been treating of, and so closely planted, if the greatest and most unremitting attention be not paid to its management, it will soon become a mass of confusion; and, while some trees are in full vigour and overgrowing the others, some will be choked by shade, so as to produce no fruit worth eating; and others will be eaten up by insects. The great object that must be constantly kept in view, in the cultivation of such a garden, is, to keep every tree and plant within the limits assigned to it. This must not be done by merely cutting off superfluous branches in the usual manner, but by preventing them from being produced, by lowering the strength of the tree. This is to be done by disbudding, and by pinching off the young shoots at the extremities of the branches, and wherever else they would require to be cut off, if allowed to grow to maturity. Another important point to attend to in the management of the fruit trees and flowering shrubs is, to preserve their foliage, at all times, in a perfectly healthy state, and to keep it fully exposed to the sun and air. For this purpose, the strictest watch must be kept for the appearance of insects; and means taken to destroy them, if possible, in the egg state, or, at all events, as soon as they are hatched. During the whole summer, every tree and bush on the premises will require to be syringed with water in the evening of every day that it does not rain; and, when insects appear, instead of common water, lime water, soapsuds, or tobacco water must be used. The strawberries will require watering every evening, from the time they come into blossom till the fruit is set; and the alpine and wood kinds, as they bear fruit the greater part of summer, will require additional watering in proportion. The flowers in the wall borders, and the ground between the wall and the walk, and also the ground under the flagstones, will require frequent watering; and, indeed, the wall border and the espalier border ought to be mulched with half-rotten dung, to keep the ground constantly cool and moist. Neither should ever be dug, but merely be pointed on the surface with a three-pronged fork; except in so far as may be necessary for renewing the strawberry edgings. The runners from these strawberry edgings must be constantly taken off, as soon as they extend 6 in. from the plant over the walk; or 12 or 14 inches over the border, except when young plants are required to substitute for the old ones. The crops of culinary vegetables in the centre beds will require constant watering every evening during the whole summer, except when it has rained; more especially to render the peas productive and succulent, and the lettuces and other salading, crisp and tender. As soon as one crop of any article is ready to be removed, another should be sown or planted to succeed it, the ground being previously well dug, and manured if necessary. A constant look out must be kept for insects on the culinary vegetables, herbs, and flowers, as well as on the fruit trees; and, throughout the whole garden, snails, slugs, and worms must be destroyed as soon as they are perceived. The plants in the circular bed in the front garden should be taken up late in the autumn of every year, and the ground trenched; a fourth part of the soil removed, and a fourth part of fresh soil added, with a portion of thoroughly rotted manure, if necessary; after which the plants and bulbs should be replanted, and new ones substituted for such as may have become shabby, or are dead, or for the sake of change. This mode of reinvigorating the soil, if not performed every year, should be performed every second or third year, otherwise the stronger plants will overrun the weaker ones; and the bulbs, especially the hyacinths, which ought to constitute the principal beauty of the bed in spring, will degenerate, and cease to be ornamental. The roses against the wall in the front garden will require to be constantly watched, lest the aphides should make their appearance; and, the moment any are seen, watered with weak tobacco water, being washed immediately afterwards with pure water, to prevent the tobacco water from disfiguring the plants. A receipt for the preparation of this tobacco water, and more particular directions for using it, will be given hereafter. The roses will require to be taken up, the soil renewed, and all the old wood and roots cut out, every three or four years; that is, if it is wished that they should flower freely, and display themselves to the greatest advantage. Both the flowers and rose trees in the front garden will require constant watering; and the vine against the back front, and the wistaria, honeysuckle, and jasmine, against the street front, will require to be watered over the leaves with the syringe, as well as at the root; and to be summer-trained and pruned. Many other minor operations might be enumerated, but the above are enough to show that, where the most is intended to be made of even the smallest-sized fourth-rate suburban garden, if planted in this manner, there is work enough for one person, every evening, say from six till it grows dark, throughout the spring, summer, and autumn months; besides work that must be done previously to six o'clock, in the months when it is dark at that hour; such as digging and cropping the culinary part of the garden, &c.

Expense of Management. If the operations of culture and keeping are performed by the occupier himself, the annual amount of money required to be laid out will be very trifling. The first year, one of Reid's syringes, some watering-pots, a pruning-knife, hammer, nails and list, a spade, rake, hoe, trowel, and three-pronged fork, a garden line, baskets, and ladder for the vine, &c., against the house, amounting in all to about 10*l.* or 12*l.*, will require to be purchased; but the second and succeeding years there will only be wanted a few garden seeds, manure, tobacco, or other articles for destroying insects; nails, list, matting (for tying the espaliers, and for protecting any of the tender ornamental plants during winter, and the blossoms of the tender fruits in spring), and some occasional repairs and renewals; the cost of the whole of which cannot exceed 1*l.* or 2*l.* To employ a gardener to do all the work requisite in such a garden, during the summer months, would cost at least 1*s.* or 1*s.* 6*d.* per day; and thus, supposing him to work, at an average, three hours every day, except Sundays, from the 1st of March to the 1st of November, the cost will amount to about 10*l.* at 1*s.* a day, and 15*l.* at 1*s.* 6*d.* The remaining part of the year, viz. from the 1st of November to the 1st of March (about seventeen or eighteen weeks), need not cost, on an average, more than 2*s.* a week. The whole expense would thus be about 20*l.* a year. Where a man-servant was kept who was fond of gardening, he might, with occasional instructions during the first year from a professional gardener, perform all the work requisite.

The Produce of such a suburban garden would, if a hired gardener were employed, probably not be worth more than the expense, if so much: but the great satisfaction of seeing the things in all their different stages of growth, and of being able to procure fresh salading and herbs nearly all the year; and, at the proper seasons, spinach, kidneybeans, and peas, when wanted, quite fresh out of the garden; will more than compensate for the outlay required, to any person who can afford it. To those who cannot, or who do not wish to incur the necessary expense, we shall suggest a cheaper mode of planting and management.

A more economical Mode of laying out and planting a Back Garden of the size of that shown in *fig.* 72. In this case, we would advise the trellis and strawberry edgings to be omitted, and the 15 cross beds to be extended to the margins of the paved walks as in fig. 55. p. 201. One of these beds may be planted with gooseberries, another with currants, and a third with raspberries; and the other beds with kitchen crops and herbs, much in the same manner as indicated above, except that there will be only three beds for peas, instead of six. Against the house, there need not be any vine or other plant trained; and the circular bed in front may be planted solely with bulbs and annuals. This will lessen the first cost of the garden one fifth or one sixth, and the labour and expense of keeping it one half.

6. *To lay out a Fourth-rate Suburban Garden, where the main Object is a Display of ornamental Trees and Shrubs and Border Flowers.* We shall suppose the extent and form of the ground to be the same as in *fig.* 72. in p. 236.; and that the walks are flagged, and the ground thoroughly drained and prepared. The walls we would plant with flowering shrubs, instead of fruit trees; and the trellis we would plant solely with roses. The centre of the garden we would lay out in beds, in which the finer kinds of border flowers might be cultivated; or, we would form a border on both sides of the rose trellis, and lay down the centre of the garden in grass. In selecting such flowers as might be grown in the beds which we have supposed laid out within the space enclosed by the rose trellis, various objects may be kept in view, according to the taste of the occupier. He may have a favourite colour, or a favourite height; he may prefer climbing plants, or trailers, or bushy plants, or bulbs; or evergreen-leaved herbaceous plants, such as the pink, &c., to look well in winter. He may choose to make the greatest display in a particular month; or to cultivate plants which will continue in flower for two or three months at a time; or to grow only perennials or annuals, and so on. One of the most general objects of gardeners, in cases of this kind, is, to have an equal number of plants in flower during every month of the floral year, which consists of nine months, rejecting the three winter months. Of those in flower in each month, the next object is to have an equal number of each of the most prevalent colours; and more particularly of red, scarlet, orange, purple, blue,

violet, yellow, and white. Where this is the object in a small garden like that which we propose to plant, we would recommend a bed for each month; or, if the plants are to be arranged in borders, a row for the same period; or, rather, an imaginary row, so that there might be an equal quantity of plants in flower, at the same time, in every part of the border. Both in borders and in beds, it is desirable to place the lowest plants next the walk, and the tallest at the greatest distance from it, so as to produce a sloping surface of vegetation; in which mode it will be found that the most effectual display is made; the green foliage of the plants not yet come into flower, or that of the plants which are gone out of flower, contrasting advantageously with those in full bloom. The front garden may be surrounded by a border, and have a small circular, square, or diamond-shaped bed in the centre; or it may be laid out in many different ways, some of which are shown in *fig.* 74., care being

74

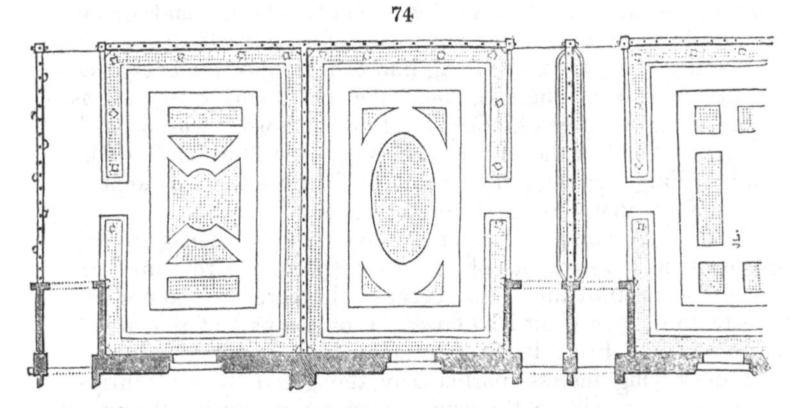

taken to employ artist-like shapes for the beds, and never to have less than 1 ft., or, what is still better, 2 ft., of turf between one bed and another; and between the beds next the walk and the wall. The wall of the front garden facing the south may be planted with the more showy evergreen and deciduous shrubs, the planter being guided in his choice of kinds by the desire of displaying different sorts from those prevalent in the adjoining gardens, in order to increase the general variety of the street. As there will be 230 ft. of walling in the back garden; and 30 ft. of walling in the front garden, if 4 ft. be allowed for each plant, 60 different sorts may be introduced; which will include all the finer climbers, twiners, and showy flowering shrubs in cultivation in British gardens; exclusive of the rhododendron and heath families, which, in general, are unsuitable for training against walls. From this general rule we except the stronger-growing rhododendrons, the tree hybrids of which may be placed against the wall with a south aspect; and *R.* pónticum and *R.* catawbiénse, with their varieties, against the wall with a northern exposure. Some of the more hardy azaleas might also be planted against this wall. The rose trellis being above 160 ft. in length, 80 sorts, allowing 2 ft. to each plant, might be displayed on it; in which might be included all the very best kinds; and in addition there might be standards placed along the trellis at regular distances, which, at 10 ft. apart, would give 16 sorts more. The 15 beds contain 50 square feet each; and, allowing two square feet to a plant, this would give 350 herbaceous or flowering plants; so that the collection in the back garden would stand thus: — 60 select climbing, and other deciduous and evergreen flowering, shrubs; 96 kinds of choice roses; and 350 kinds of choice flowers. Besides these, there might be a border of candytuft, or ten-week stock, slightly mixed with mignonette, on each side of the main walk. The choice of the shrubs and plants we shall leave the reader to make for himself from our descriptive lists; recommending him to introduce the evergreen shrubs here and there among the deciduous ones, so as to make the evergreen kinds prevail on the border with a north aspect, and to place the tenderer kinds on the wall which faces the south. Fragrance and the beauty of foliage should be kept in view, no less than the beauty of the flowers.

Mode of Management. This garden may be managed with greater ease and certainty of success than one where fruit trees are grown against the walls; because the care and labour requisite to bring a plant into flower is only a part of that which is necessary to bring its fruit to maturity. Pruning, watering, and destroying insects, particularly the green fly and thrips, from the roses, will be the principal operations during the summer months; and, in autumn and spring, the greater number of the herbaceous plants will require to be taken up and replanted. At these seasons, also, pruning and training will be required both for the roses and shrubs against the wall. During winter, there will be scarcely anything to do, unless to mat up any of the more tender kinds that may have been introduced; and, hence, this description of gardening is particularly adapted for being carried on by a lady, with her own hands. Instead of mats for covering any of the more tender trees against the wall, old newspapers, dipped in, or painted over with, hot oil, will be equally effective; the object being to reflect back the rays of heat radiated from the wall. The newspapers may be fastened on with small nails; or they may be tied to nails already in the wall, or to the shoots of the trees.

Estimate of Expense. This will depend chiefly on the rarity and value of the shrubs and plants chosen, which may vary from an average of 1*s*. to 3*s*. each. Taking them at the first sum, and estimating the total number at about 506, the amount will be 25*l*. 6*s*.; but the same number of plants might be chosen from our priced lists, of such prices as would bring the amount above 100*l*., or under 10*l*. In the latter case, however, there would only be the more common kinds; and both the roses and the herbaceous plants would be purchased at so much per score, or per hundred. On the whole, the expense of laying out and stocking a garden of this sort would be much the same as if it had been planted with fruit trees in the manner of *fig*. 72. p. 236.; but it has a great advantage over a fruit-garden, in admitting of being kept both with less labour and less skill. Supposing a commercial gardener employed to look after it, he might be required for two hours a day during June, July, and August; two hours, for three days in the week, during May and September; and two hours per week during the remainder of the year.

The Produce would consist entirely of flowers, and the enjoyment in looking after them. The roses would demand a lady's attention during the whole summer. Every day some roses will have faded, and will require to be cut off; and every evening, except when it rains, the syringe, or the barrow engine, should be brought into use. To keep down the insects will also require constant vigilance.

Another Design for laying out and planting a Fourth-rate Suburban Garden, where the Object is chiefly to make a picturesque Display of ornamental Trees and Shrubs, with a few Flowers. For this example, we shall take the back garden of a friend, which was laid out and planted from a design of ours in January, 1835. The situation is a piece of flat ground, on the border of a common, within two miles of St. Paul's, London: it contains, altogether, about a quarter of an acre; and the soil is a strong loam. The house forms part of a row; and, from a stable, chaise-house, and poultry-house being placed alongside of the dwelling, the frontage, and, consequently, the width of the back garden, is upwards of 60 ft. There being no common sewer, or general system of drainage, in the neighbourhood, the most defective part of this residence is the dampness of the surface of the ground in wet weather. The drains of the walks, however, are conducted to a well at the further extremity of the garden; whence the water, after heavy rains, may be pumped up to a gutter, which is conducted along the surface of the common. The situation, though flat, is not without some distant prospect, that is intended to be seen from the floor of the living-room (which is about 5 ft. higher than the surface of the garden), broken and varied by the trees which are to be planted. The shrubs will be, for the most part, below the eye; and the finer flowers are intended to be chiefly confined to the front garden;

in which, in order to form a greater contrast with the back garden, it is only intended to have climbers against the house and the side walls. The portion of the front garden before the coach-house and stable is separated from that before the house by a wall; thus giving a yard for the poultry, and room for cleaning the horse, chaise, &c. The only trees in the front garden are a row of thorns of six different kinds, placed close by this partition wall, in order to hide the view of the poultry and chaise-yard from the parlour window.

The house, the boundary fences, the walks, and the outdoor buildings, were designed and carried into execution before our advice was asked, and were to be considered as unalterable. Some asparagus beds, sea-kale beds, and gooseberry bushes were planted, and were also not to be disturbed. All that was left for us to do, therefore, was to arrange the beds on the plot of lawn, or turf, which formed the principal part of the area of the back garden; and to indicate the kinds of trees and shrubs to be planted in these beds and in their side borders. This we did on a working plan, of which *fig.* 75. occupying p. 254, 255. is a copy, accompanied by the following explanatory references; to which we shall subjoin a list of the trees and shrubs planted, with their prices:—

a The door of the house, which opens on a landing covered by a trellised porch; from which a flight of seven steps descends to the garden walk. *b*, Privy. *c*, Summer-house.

d, Pit for cucumbers, heated by dung from the stable; thrown into a vault, through a door at one end.

e, Raised cover to a well. *f*, Pump. *g*, Door to the stable.

h, Situation of the dining-room window, being the only window of a sitting-room which looks into the garden; and with reference to the view from which window all the trees are planted.

i, Border of shrubs and flowers; the fence on this side being wooden pales about 5 ft. high; the aspect south by east.

k, Border for rhubarb, sea-kale, chives, parsley, and other annual and perennial kitchen herbs: the fence here is a brick wall 5 ft. high. *l*, Asparagus beds.

m, Two rows of gooseberries, with strawberries between. The margin of the walks is, on one side, a continuation of the lawn, 1 ft. broad; and, on the other, box.

n to *o*, Part of the wall, on which currants may be trained.

p, Border of shrubs and flowers, with some trees, facing the north; the wall brick, and about 5 ft. high, covered with fruit trees of different kinds; but which, from the aspect, are of little use as such. Beyond the fences, on the right and left, are similar gardens; and at the extreme end there is an open common in grass.

q, Dark circles, indicating the fruit trees which are already planted, and are not to be removed.
r, Open circles, indicating the situation of trees to be planted.
s, Marks thus *, indicating the situation of evergreen shrubs to be planted. The dots, thus ·, indicate situations for herbaceous plants, annual or perennial.
t, Marks thus ×, indicating the situation of deciduous shrubs to be planted.
u to *v*, Twelve posts in the fence of pales, against each of which a China rose is to be planted, and trained on each side; and also allowed to overtop the wall, so as to break its formal outline.
w, Situation where a vase on a proper pedestal, a statue, or other architectural object, might be placed; taking care to connect it architecturally with the walk.
x, Situation where a small circular basin and fountain might be introduced. *y*, Situations where chairs may be placed.

The different tools required for the garden, including the wheelbarrow and roller, are kept in a division of the stable; the flower-pots, &c., under the summer-house; and the mould and compost heaps near the east end of the pit.

In the disposition of the trees, the object is, to preserve an irregular-sided vista along the centre of the lawn; to break the formality of the straight lines of the walks and fences on each side of it; to conceal the termination of the lawn, and hide the asparagus beds; and to vary and partially conceal the scenery of the neighbouring side gardens and of the country beyond.

The principle of guidance in the selection and disposition of the shrubs is, partly to cooperate with the above object; but principally to produce an agreeable variety of flowers and foliage throughout the whole space, and during every month in the year. For this purpose, certain evergreens (such as the laurustinus), and certain flowering shrubs (such as the China rose), are distributed throughout; the same variety of the species not being repeated, but different varieties. There are also shrubs for flowering at every season of the year: such as the chimonanthus and *Cydònia japónica* for autumn and winter; the mezereon for early spring; the common azalea and rhododendron for the beginning of summer; the clethra for August; and the arbutus and wych hazel for the latter part of the season. The whole of the trees and shrubs are of kinds which do not require peat earth, and may be purchased at moderate prices.

The trees are almost all of the low-growing and flowering kinds; under 20 ft. in height; and purchasable, on an average, for cash, at 1*s*. 2*d*. each. Their names are as follows. The prices were kindly put to them by a respectable London nurseryman.

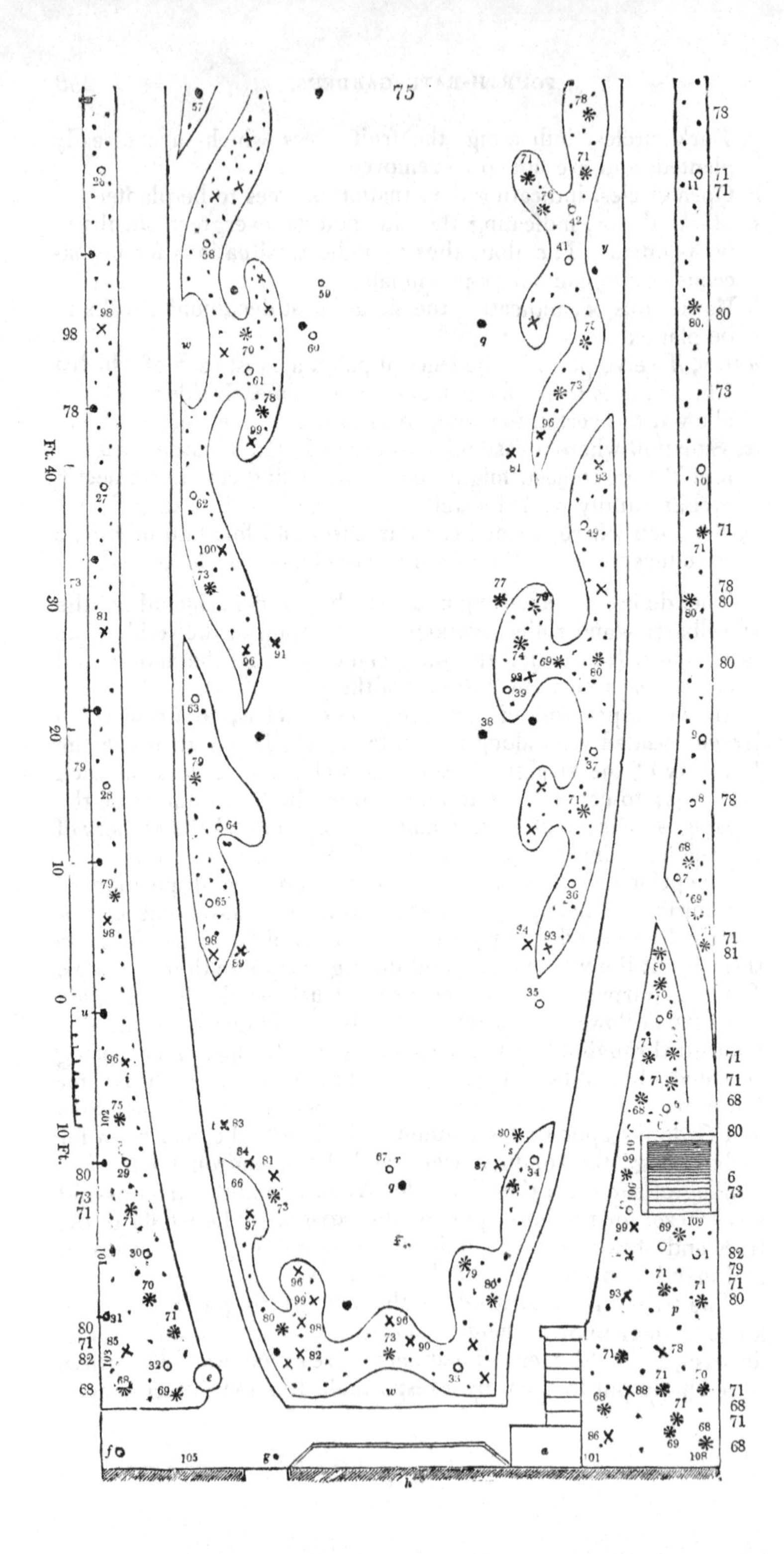

75
Ft. 40
30
20
10
0
10 Ft.

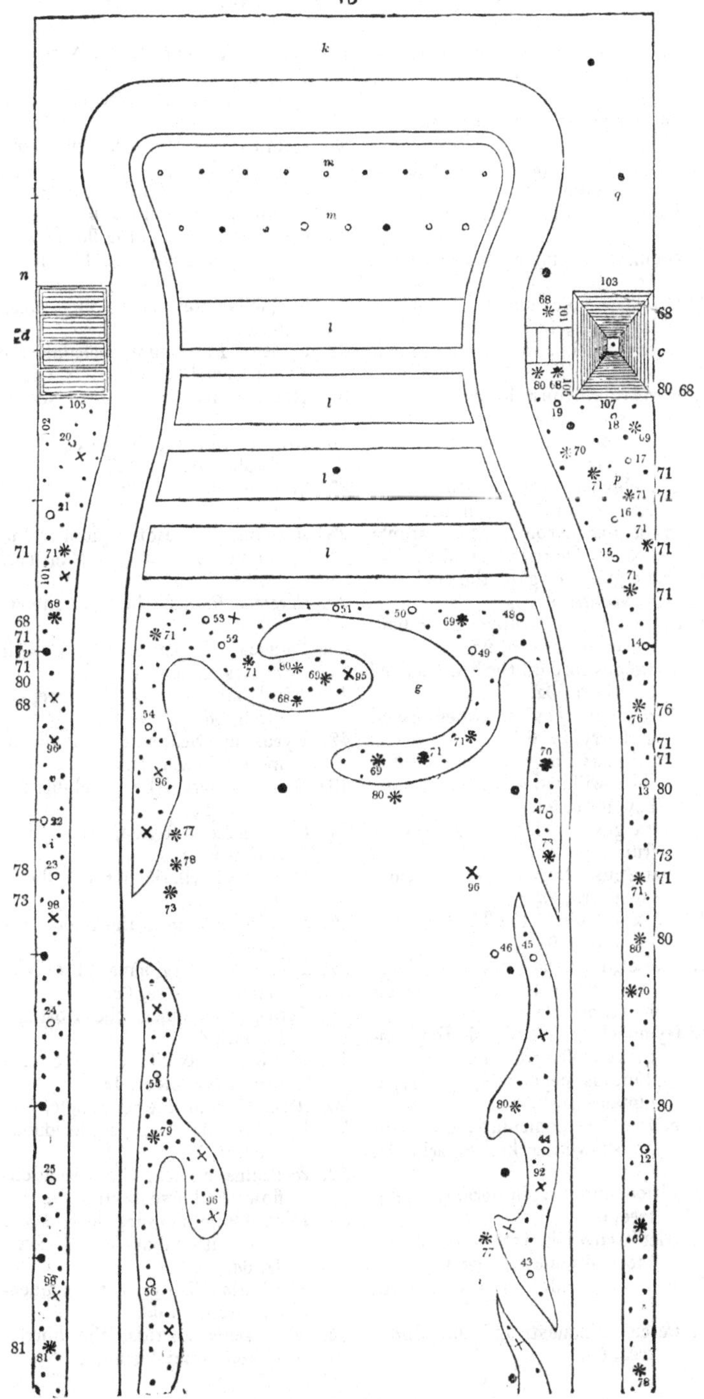

1. *P*ỳrus spectábilis, the showy-flowered Chinese crab tree, 1*s*.
2. Quércus *I*'lex, the evergreen oak, 1*s*. 6*d*.
3. *T*hùja occidentàlis, the American arbor vitæ, 9*d*.
4. *L*aúrus nóbilis, the sweet bay, 1*s*. 6*d*.
5. *J*uníperus virginiàna, the red cedar, 1*s*.
6. *C*ýtisus *L*abúrnum, the common laburnum, 1*s*.
7. *P*ỳrus aucupària, the mountain ash, 1*s*.
8. Pàv*ia* rùbra, the red-flowered small horsechestnut, 1*s*. 6*d*.
9. *P*ỳrus pinnatífida, the cut-leaved sorb, 1*s*. 6*d*.
10. *C*ratæ'gus odoratíssima, the sweetest-scented hawthorn, 9*d*.
11. *C*ratæ'gus Arònia, the aronia (yellow-fruited) hawthorn, 9*d*.
12. *C*ratæ'gus Crús-gálli, the cockspur hawthorn, 9*d*.
13. *C*ratæ'gus tanacetifòlia, the tansy-leaved hawthorn, 9*d*.
14. *C*ratæ'gus cordàta, the heart-leaved hawthorn, 9*d*.
15. *Bérberis* aristàta, the awned-leaved berberry, 2*s*. 6*d*.
16. *C*ratæ'gus Crús-gálli var. *s*alicifòlia, the willow-leaved cockspur hawthorn, 9*d*.
17. *C*ratæ'gus coccínea, the scarlet-fruited hawthorn, 9*d*.
18. *C*ratæ'gus *Azaròlus*, the azarole hawthorn, 9*d*.
19. *C*ratæ'gus nìgra, the black-fruited hawthorn, 9*d*.
20. *C*ratæ'gus *O*xyacántha var. flàva, the yellow-fruited common hawthorn, 1*s*.
21. Gymnócladus canadénsis, the Kentucky coffee tree, 1*s*.
22. Piptánthus nepalénsis, the Nepal piptanthus, 2*s*. 6*d*.
23. Kölreutèr*ia* paniculàta, the panicled-flowering kœlreuteria, 1*s*. 6*d*.
24. Liriodéndron Tulipífera, the tulip tree, 6*d*.
25. Gledítsch*ia* triacánthos, the three-thorned honey locust, 6*d*.
26. *Ailántus* glandulòsa, the ailanto, 6d.
27. Cércis *S*iliquástrum, the Judas tree, 6*d*.
28. *C*érasus virginiàna, the Virginian bird-cherry, 1*s*.
29. *C*ýtisus alpìnus, the Scotch laburnum, 1*s*.
30. Robín*ia* viscòsa, the glutinous locust, 1*s*.
31. *C*ratæ'gus *O*xyacántha, the scarlet-flowered hawthorn, 9*d*.
32. *C*ratæ'gus *O*xyacántha flòre plèno, the double-flowered hawthorn, 9*d*.
33. Magnòl*ia* conspícua, the Yulan magnolia, 3*s*. 6*d*.
34. *C*ýtisus *L*abúrnum incìsum, the cut-leaved laburnum, 1*s*. 6*d*.
35. Robín*ia* híspida, the rose acacia, 1*s*.
36. *P*tèlea trifoliàta, the three-leaved shrubby trefoil, 9*d*.
37. *C*érasus *Mahàleb*, the perfumed cherry, 1*s*.
38. *Amelánchier* Botryàpium, the snowy-flowered amelanchier, 1*s*.
39. *C*érasus *P*àdus, the bird-cherry, 66.
40. *C*érasus semperflòrens, the All Saints' cherry, 1*s*. 6*d*.
41. *B*étula álba péndula, the weeping birch, 3*d*.
42. *P*ỳrus americàna, the American mountain ash, 1*s*.
43. *C*érasus nìgra, the black-barked cherry, 1*s*.
44. Cotoneáster frígida, the frigid cotoneaster, 1*s*.
45. *P*ỳrus bollwylleriàna, the Bollwyller pear, 2*s*. 6*d*.
46. *Sophòra* japónica, the Japan sophora, 6*d*.
47. *D*iospỳros virginiàna, the Virginian lote tree, 6*d*.
48. *C*érasus lusitánica, the Portugal laurel, 6*d*.
49. *Negúndo f*raxinifòlium, the ash-leaved box elder, 1*s*.
50. *A*'cer rùbrum, the red maple, 1*s*.
51. Taxòdium dístichum, the deciduous cypress, 1*s*.
52. *Æ*'sculus cárnea, the carnation-flowered horsechestnut, 1*s*. 6.
53. *I*'lex *A*quifòlium var., the common holly, with smooth-edged leaves, 1*s*. 6*d*.
54. Salisbùr*ia a*diantifòlia, the maidenhair tree, 2*s*. 6*d*.
55. Gledítsch*ia* hórrida, the horrid-spined honey locust, 2*s*. 6d.

56. Aristotèl*ia* *Mácqui*, the Macqui tree, 2*s*. 6*d*.
57. *P*rùnus doméstica myrobálana, the myrobalan plum, 2*s*.
58. Halès*ia* tetráptera, the snowdrop tree, 1*s*.
59. *Catálpa syringæ*fòlia, the catalpa, 6*d*.
60. *E*uónymus latifòlius, the broad-leaved spindle tree, 1*s*. 6*d*.
61. *I* lex opàca, the opaque-leaved holly, 1*s*. 6*d*.
62. Virgíl*ia* lùtea, the yellow-flowered virgilia, 2*s*. 6*d*.
63. *Negúndo f*raxinifòlium críspum, the curled-leaved box elder, 1*s*.
64. Magnòl*ia* acuminàta, the pointed-leaved magnolia, 2*s*. 6*d*.
65. Liquidámbar styracíflua, the maple-leaved liquidambar, 1*s*.
66. *A*mýgdalus commùnis, the common almond, 1*s*.

The following is a List of Evergreen Shrubs: —

67. *C*upréssus sempervìrens, the evergreen cypress, 6*d*.
68. *A*'rbutus *U*'nedo, the common and scarlet arbutus, 1*s*., 8 plants.
69. *P*hillýrea angustifòlia, the narrow-leaved phillyrea, 1*s*. 6*d*.
70. *R*hámnus *A*latérnus, the common alaternus, 1*s*. 6*d*.
71. *I*'lex *A*quifòlium var., variegated hollies, 1*s*., 23 plants.
72. *C*érasus Laurocérasus var., variegated laurel, 6*d*.
73. *R*hododéndron pónticum and catawbiénse, hardy rhododendrons, 6*d*., 6 plants.
74. Escallòn*ia* rùbra, the red escallonia, 1*s*. 6*d*.
75. Cratæ'gus *P*yracántha, the evergreen hawthorn, 6*d*.
76. *Aúcuba* japónica, the common aucuba, 6*d*.
77. *J*uníperus suécica, the Swedish juniper, 1*s*.
78. *B*úxus sempervìrens var., variegated box of different sorts, 6*d*.
79. *C*istus sp., the rock rose of different sorts, 1*s*., 2 plants.
80. *V*ibúrnum *T*ìnus, the laurustinus of different sorts, 6*d*., 8 plants.

The following is a List of Deciduous Shrubs: —

81. *D*áphne *Mezèreum*, the common mezereon, white, red, and autumn-flowering, 1*s*. 6*d*., 3 plants.
82. Chimonánthus fràgrans, the fragrant chimonanthus, 2*s*. 6*d*., 2 plants.
83. *H*amamèlis virgínica, the wych hazel, 1*s*.
84. Calycánthus flórida, the Carolina allspice, 1*s*.
85. *Syrínga* vulgàris álba, the white lilac, 6*d*.
86. *Syrínga* vulgàris purpùrea, the purple lilac, 6*d*.
87. *Syrínga* pérsica, the Persian lilac, 6*d*.
88. *V*ibúrnum *O*'pulus *r*òsea, the Guelder rose, or snowball tree, 6*d*.
89. *C*ýtisus álbus, the white broom, 3*d*.
90. *S*piræ'a béllа, beautiful spiræa, 6*d*.
91. *C*olùtea cruénta, the bloody bladder senna, 6*d*.
92. Coronílla E'merus, the scorpion senna, 6*d*.
93. Azàlea póntica, the common yellow azalea, 1*s*.
94. Symphòria racemòsa, the snowberry, 6*d*.
95. *S*pártium *j*únceum, the Spanish broom, 3*d*.
96. *C*ydònia japónica, the Japan quince, 1*s*. 6*d*.
97. *C*lèthra *a*lnifòlia, the alder-leaved clethra, 6*d*.
98. *H*ibíscus syrìacus var., the althæa frutex, 6*d*.
99. *Rìbes* sanguíneum, the red-flowered currant, 9*d*.
100. *Rìbes* aúreum, yellow-flowered currant, 9*d*.

All the crosses marked in the plan, which are not numbered, are for different sorts of roses; and the number of these may be increased at pleasure, diminishing the number of herbaceous plants in proportion, according to the taste of the owner; 6*d*., 25 plants.

The following is a List of the Climbing Shrubs for covering the privy, summer-house, a part of the boundary wall and porch, and part of the walls of the house: —

101. *C*aprifòlium flexuòsum, the Japan honeysuckle, 1*s*. 6*d*., 4 plants.

102. Common twining honeysuckles of sorts, 6*d*., 6 plants.
103. *Cl*ématis and *A*trágene of sorts, 6*d*., 6 plants.
104. *L*ýcium bárbarum, the Duke of Argyll's tea tree, 6*d*.
105. *J*asmìnum officinàle, the common jasmine, 6*d*.
106. Ampelópsis *h*ederàcea, the five-leaved ivy, 6*d*.
107. *R*òsa multiflòra, and Grevíll*ei*, the many-flowered rose, and Greville's rose, 1*s*. 6*d*., 2 plants
108. Wistàr*ia* Consequà*na* (Glýcine sinénsis *B. R.*), Consequa's wistaria, 1*s*. 6*d*.
109. Giant ivy, and Ayrshire rose, 3*d*., 4 plants.

According to the above enumeration, there need not be a dozen duplicates in the garden; for, though there are some of the species repeated (such as the laurustinus, the *C*ydònia japónica, the Chinese rose, &c.), different varieties of each species may be chosen. The herbaceous plants may be selected on the same principle; so that, in this small garden of not quite a quarter of an acre, nearly 800 different kinds of ornamental plants may be exhibited.

The herbaceous plants, both for the front and back gardens, we left to be chosen by the lady of the house; and the reader may make choice of them from the lists given from p. 217. to p. 225. It is proper to observe, however, that, as the trees and shrubs in the back garden advance in growth, the room for flowering plants will be diminished. After three or four years, there will not be much space within the beds fit for bringing fibrous-rooted herbaceous plants to perfection; because, for this purpose, it is necessary that the plants should have unobstructed light, and free air, on every side. As the trees and shrubs advance, therefore, they must either be thinned out to make room for the fibrous-rooted herbaceous plants, or a smaller number of these must be grown. The same remark would apply to the roses planted in the beds; because, to flower well and look well, they require as much light and air as the others; and, like them, as has been before observed, they require to be taken up every second or third year, in autumn or spring, and parted, pruned, and replanted in fresh soil. Supposing the trees and shrubs, exclusive of the roses, not to be thinned out, or reduced by pruning, then, in five or six years, both roses and fibrous-rooted herbaceous plants would be choked. The best mode of proceeding, in that case, would be, to cease to dig the beds, and reduce or rake them to the same level as the turf, and sow any spots not covered with the branches of the shrubs with grass. This would look remarkably well, both in a picturesque and in a botanical point of view, for another five or six years, when it would become absolutely necessary to root up some of the larger trees, and to prune in, or cut over near the ground, some of the larger shrubs. This process of keeping the beds and groups in shape, by pruning and cutting down, might be carried on for an indefinite period, as may easily be believed by observing the great duration of hedges which are continually cut, and of coppice-wood. In order that the garden may always look well, an equal amount of pruning, thinning, and cutting over should, as nearly as possible, be performed every year; and regard should be had, in doing this, always to preserve the same proportion between trees and shrubs, and between plants both of kinds which are evergreens and those which are deciduous; unless, indeed, it is thought that an improvement might be made by altering these proportions.

Hitherto, we have supposed only fibrous-rooted herbaceous plants to be planted in the beds; but, if bulbs are mixed with these, the bulbs may be allowed to remain after the fibrous-rooted flowers have been removed: because they will thrive with much less light and air than the fibrous-rooted plants; and because they spring up and grow with great rapidity; flower early in the season, before the leaves of the deciduous trees have expanded so as to shade them; and, when they have done flowering, they fade speedily, and their foliage, when removed, leaves no trace of the plants behind, and, consequently, causes no unsightliness on the surface of the ground throughout the summer. Besides, there are certain kinds of bulbs, such as the scillas, some kinds of hyacinths, the snowdrop, and the narcissuses, which thrive better under a slight degree of shade than when fully exposed to the atmosphere. Bulbs, therefore, may be planted among groups of trees and shrubs, and in close shady places in suburban gardens, where fibrous-rooted flowering plants are inadmissible.

In the borders under the side fences, in this design, flowers of all kinds may be cultivated, during the entire existence of the garden; because, being unmixed with shrubs, except those which are trained against the fences, they would be freely exposed to the light and air, and might be taken up and replanted, and the soil renewed, at pleasure.

With respect to that part of the garden which is cropped with asparagus, sea-kale, strawberries, gooseberries, and other fruit shrubs, very little need be said. The asparagus and sea-kale will require to be liberally supplied with manure; the strawberries taken up and replanted every two or three years; and the gooseberries carefully pruned annually, so as to keep the bushes open for the admission of light and air; and, when they cease to bear abundantly, they should be taken up, the soil renewed, or refreshed with new soil, and young plants planted. The renewal should take place by degrees, say one row at a time; so that the garden may never be without full-grown bushes, and, consequently, every year have its crop of fruit. The herbs should be taken up and replanted every second or third year. The roses, and other plants trained against the fences and the house, will require to be regularly pruned, trained, and kept clear of insects, and, as they become unsightly, renewed. The box edgings, also, will require to be annually cut in June, and renewed every seven or ten years.

Estimate of Expense.

	£	s.	d.
The preparation of the ground, including the draining and the formation of the walks, in this garden; will cost about	10	0	0
The trees required amount to 67; which, at the above prices, average 1*s*. 2*d*. each (cash), and come to - - - -	3	18	1
Shrubs and roses, 108, - - - - - - -	4	2	6
Climbing shrubs, 26, - - - - - -	0	19	3
Herbaceous plants, annuals, and biennials, 546; and, supposing the greater number of them to be annuals, they may be purchased for - - - - - -	6	6	0
In all -	£25	5	10

The plants were p ocured, at the prices stated in the list, of Mr. Donald, nurseryman, Woking, Surrey.

Management. A garden laid out in this manner may be managed with great ease, and at very little expense. During the spring months, such training and pruning as are required may be effected; and, during summer, the chief operations will be mowing the grass, training the herbaceous plants and roses, and keeping the latter free from insects, by frequently syringing them with clear water, and daily hand-picking the aphides, caterpillars, snails, &c. The culinary department of the garden will require nothing more, in the summer season, than to be kept clear of weeds; and, in the winter, a dressing of manure. All this may be done by the man-servant necessarily kept for the horse, with the direction and assistance of the occupier himself, and the female part of his family; with the exception of mowing the grass, which, during the summer months, will require one day's work of a jobbing gardener every fortnight; and, during the spring and autumn months, once every three weeks. This is supposing that the fallen leaves in autumn are swept up every two or three days by the man-servant, or some one of the family; in performing which operation, in this garden, as in every other, a soft birch broom must be used, so as not to raise and sweep away the gravel from the walks. The mown grass, the prunings, the weeds, and the leaves, and such like refuse, may be thrown under the pit among the horse dung, so as to increase the quantity of fermentable matter, and thus to keep up a constant heat in the pit. The walks should be rolled at least every time the grass is mown; and, every three or four years, that part of the surface of the gravel which has become black should be removed, and the remainder stirred up, fresh gravel added, and the whole rolled with a heavy roller, till it has become as hard as a surface of pavement. The articles grown in the pit may be cucumbers and melons, the management of which will be given hereafter; or it may be used for raising and protecting pelargoniums (geraniums), or such other green-house or pot plants as the lady of the house may prefer. Gourds may be raised in the pit, and planted between the asparagus beds, so as to yield a supply, during the whole summer, of a very delicious vegetable. If there is any room against the side wall that faces the south, tomatoes, previously brought forward in the pit, may be planted against it, and trained so as to ripen their fruits, which are excellent, either plain boiled, or made into sauce. A row or two of scarlet runners may also be planted between the beds. The total expense of managing such a garden need not exceed from *3l.* to *5l.* a year.

Remarks. Such a garden is well calculated for a person of taste, who gets his chief supply of culinary vegetables from a market-gardener or a green-grocer. It will look well with very little care and keeping; more especially if a due attention be paid to give sufficient room to the arbutus, the laurustinus, the autumn-flowering mezereon, and other winter-flowering shrubs; and the Cydònia japónica, the common mezereon, and the *Ribes* sanguíneum, double-blossomed furze, and other spring-flowering shrubs. The dying off of the foliage of so many kinds of trees and shrubs in autumn, and their expanding foliage in spring, will produce a great variety of tints; exhibiting every morning something new, refreshing, and delightful to the lover of picturesque beauty, even if he should be no botanist. In this garden, as actually existing, all the trees and shrubs are named with zinc labels, suspended from their branches with metallic wire. The zinc is in pieces about 1 in. broad and 3 in. long; not painted, but written with a prepared ink; and, in addition to the scientific and English names, the native country of the plant is added. Such labels, the wire included, cost little more than one farthing each; and they may be easily procured from any of the London seedsmen; they add greatly to the interest of the garden, and have a tendency to give young persons a taste for plants.

7. *To lay out and plant a Fourth-rate Suburban Garden, where the Object is to have a Green-house, or to force Fruits and Flowers.* Wherever plant-houses, pits, or frames, are introduced into suburban gardens, unless these are to be kept in order, and supplied with plants, &c., by a commercial gardener residing in the immediate neighbourhood (by far the cheapest mode), it will be found necessary to have a small portion of the garden

separated from the rest as a reserve ground. In this space the proper soils and manures may be kept, a dung-bed for bringing forward plants, or a pit for the same purpose, and for forcing cucumbers, growing melons, &c. It is also necessary, wherever plants are grown under glass or kept in pots, that this reserve ground should have a back door, or communication with a public road, otherwise than through the house, for the supply of dung for hot-beds, soils, fuel, and other articles required. The possession of a cucumber or melon frame adds greatly to the interest of every suburban garden; and there need scarcely be one, however small, without it; for, on a small scale, where there is only one bed, if there is no back door, the dung may be carried through the house in baskets. On a larger scale, dung, except as manure, may be dispensed with; and the heating of cucumber-frames and pits effected by flues, or, what is greatly preferable, by pipes of hot water. Wherever there is a green-house, it ought, if possible, to be connected with one of the living-rooms of the house; and it might frequently be so arranged, that in a vault or cellar underneath this green-house, rhubarb, sea-kale, chicory, and other vegetables that are eaten in a blanched state, might be forced; or mushrooms grown throughout the year. When this is attempted, however, there ought to be no communication between the cellar and the green-house; nor any openings in the former that will admit the air from it into the living-rooms; such air being always overcharged with moisture, and having generally an earthy disagreeable smell. Some possessors of suburban gardens have a taste for forcing different kinds of fruit, more especially grapes and peaches; and some even might wish to grow pine-apples. All this may be effected in a suburban garden almost as well as farther in the country; because the operator has a greater command of the air enclosed by the glass case, than he has of the exterior smoky atmosphere; and, however paradoxical it may seem, it would be easier to grow a good crop of pine-apples under glass in a back garden in Cheapside, than it would be to produce a good crop of grass in the open air in the same garden, or to keep there a smooth closely covered turf. The principal consideration with respect to forcing-houses in small suburban gardens is, the difficulty of placing them so that they shall obtain the full influence of the sun, from its first appearance in the morning till sunset. In fourth, or even in third, rate gardens in large towns, it is next to impossible to accomplish this, from the proximity of houses that either prevent the morning sun from shining on the glass so soon as it otherwise would do; or, what is still more injurious, that intercept its rays between two and four o'clock in the afternoon, during which period, in towns, they have generally, from the comparative clearness of the atmosphere, the greatest power. Trees, also, in adjoining gardens, are often very injurious; and there is yet another drawback, which is the road dust, and small particles of soot, which, in dry windy weather, are floating in the atmosphere, and, settling on the glass roof, lessen the quantity of light that penetrates it. All these circumstances ought to be taken duly into consideration, before the occupier of a fourth-rate suburban garden ventures to erect forcing-houses for ripening fruits.

The Green-house. Whatever may be the aspect of the house, (as we have already observed, p. 108.), a green-house may be projected from it, unless it be due north; and even in that case there are many exceptions. We shall take the same extent of ground that has already come under consideration, and shall suppose a green-house projected from the living-room, as shown at *c*, in *fig.* 76.; in which *a* is the entrance-hall, and *b* the prin-

76

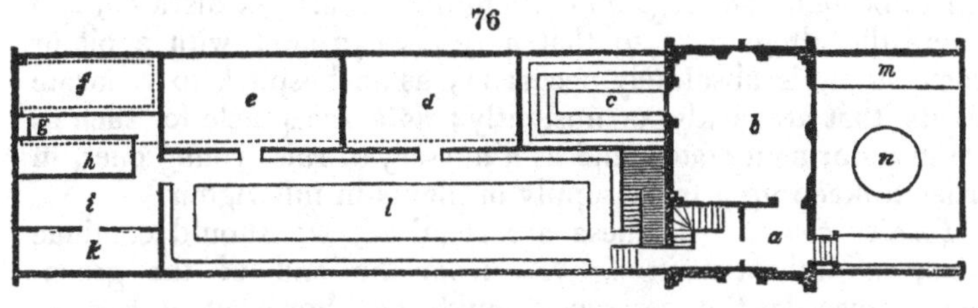

cipal living-room. Underneath the green-house there is a vault, in which is placed the furnace; and from a boiler over it hot-water pipes are conducted up into the green-house, where they are concealed behind the shelves, or stage, on which the pots stand. In the vault, various articles which require little or no light may be forced, as already mentioned. The communication between the green-house (*c*) and the living-room (*b*) is by a glass door, the view through which, from the room, is along the front of the stage, and, consequently, brings into perspective all the finest plants. Plans of green-houses will be given hereafter.

Pits, Frames, and the Reserve Ground. At the lower end of the garden, a space is shown, walled off, in which pines and melons may be grown in pits, either heated by dung or by hot water; and here, also, cucumbers, and roses and other flowers, and rhubarb and other culinary articles, may be forced at pleasure. One fireplace and boiler, centrally placed, will, with ease, heat all these pits and frames, even if they were of three or four times the extent we have supposed them to be. Plans for such pits will be given hereafter. In the ground plan, *fig.* 76., the boiler is supposed to be placed beneath the green-house (*c*); and pipes conducted from it, as indicated by the dotted line, will pass through and heat the vinery (*d*), the general forcing-house, or stove (*e*), the small pit (*h*), the propagating box (*g*), and the pine pit (*f*). There is a place for compost at *i*, and a potting-shed at *k*. The long bed (*l*) in front of the houses is supposed to be in turf, or devoted to exotic flowers and shrubs kept in the forcing-house and vinery in the winter season, and turned out during summer. The enclosure containing the pits will also serve for a reserve ground for bringing forward articles proper for decorating the green-house, and for containing soils, composts, pots, and various articles. The potting-shed is used for shifting and potting in, and for other operations required to be performed with house plants; and also for containing the pots, tools, &c. Without an appendage of this kind, no green-house can ever be made to look well for any length of time together; for some of the plants require to be removed as soon as they have left off flowering, on account of their unsightliness; others become too large and straggling; some get diseased, and others die altogether; so that a reserve ground, with a pit or frame in it, is absolutely necessary, as an hospital, to renovate plants that are sickly or unsightly; as a receptacle for such as are in a dormant state; and as a nursery to raise young ones, in order to keep up a fresh supply of plants in full vigour.

Forcing-houses. If these are required, we should continue them, as indicated in *fig.* 76., from the end of the green-house towards the reserve ground, as shown at *d* and *e*; and, if this were done, the same fire which heated the green-house, might, as already mentioned, heat the forcing-houses also. This is easily done by having a separate set of pipes for the green-house, the circulation of the water in which can be stopped whenever heat is not wanted there. The fire-place being in a vault beneath the green-house, there would be no danger of its ever communicating more heat through the green-house floor than what would be salutary for the plants. Indeed, by forcing culinary productions or growing mushrooms in the vault, and having the green-house over it, scarcely a particle of heat generated by the fuel would be lost. The forcing-houses, in a suburban garden of this kind, cannot, in general, be made higher than the party wall of the garden; because this would be to produce a greater shade on the adjoining garden than would be submitted to by its occupier; whose permission would be necessary even to raise the green-house, so that its floor might be on a level with that of the sitting-room.

A small suburban garden, like that shown in *fig.* 76., with the wall having a south aspect covered with glass, with two small forcing-houses, and a reserve garden with pits and frames at the end, would require a first-rate gardener, or a zealous and skilful amateur, to manage it to the greatest advantage; and, in our opinion, unless this is done, it is always better to dispense with glass in a garden altogether.

The open Area of this garden will be reduced by the plant-houses and the reserve ground to a very moderate space, which may be kept entirely under turf, or may be occupied as a bed of miscellaneous flowers, surrounded by a margin of turf 2 ft. wide.

The Front Garden may be treated in the same manner as in *fig.* 72.; or, as there will be abundance of green-house plants, the centre bed (*n*), and the border (*m*), the rest being in turf, may be devoted entirely to the more showy kinds of these during the summer months, and in spring entirely to bulbs. In order to make the most of the front garden in point of effect, the circular bed (*n*) may be sunk with a framing of turf round it, as indicated in *figs.* 77. and 78. In *fig.* 78., *a b* is the ground plan, or vertical

77

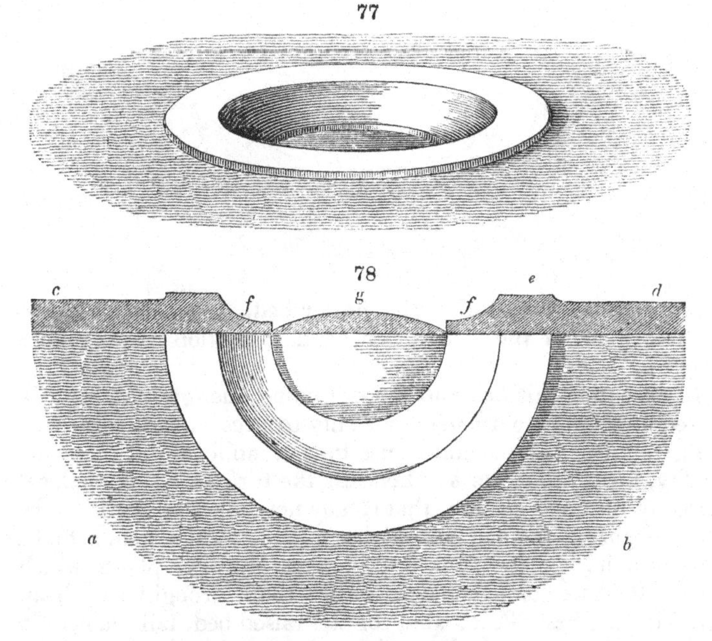

profile, of half the circular bed; *c d* is a section taken across the centre of the bed, showing the raised part of the turf frame at *e e*; the hollow moulding or side of this frame is shown at *f f*; and the dug bed in the centre at *g*. The perspective view is shown in *fig.* 77.

Instead of a sunk bed, a raised bed, also with a turf frame, may be adopted, as indicated in *figs.* 79. and 80. In *fig.* 80. *h i* show the ground plan, or vertical profile, of half the bed, and its frame, with part of the surrounding grass-plot. In this section, *m m* represent the lower part of the frame; *n n*, the

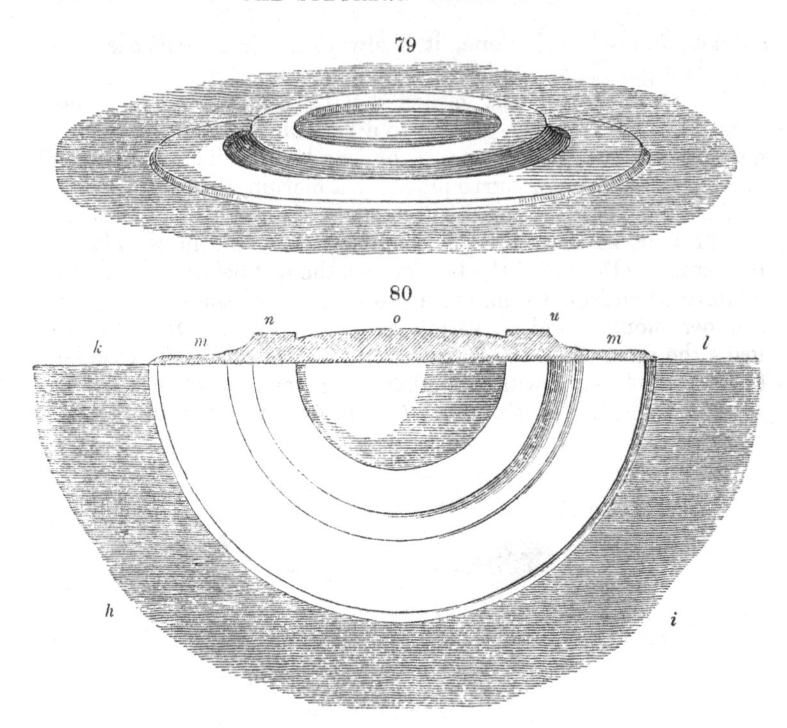

upper part, or rim; and *o*, the flower-bed. *Fig.* 79. is a perspective view of the entire bed, and a portion of the lawn round it.

In executing this bed and the preceding one, great nicety will be required in the gardener; not only to preserve the form with the greatest exactness, and to use turf of uniform thickness and quality, but to beat the soil beneath the turf, so as to render it perfectly firm and smooth, that it may not sink afterwards. The soil composing the bed should be suitable to the plants to be grown in it: for the sunk bed, low or creeping plants, which prefer moisture and coolness rather than drought and heat, will have the best effect; and, for the raised bed, tall plants. In some cases, the sunk bed may be a cistern of water for aquatic plants; in which, if thought desirable, and expense be not an object, there may be a fountain.

Another mode of forming raised or sunk beds is, by making the frame or border partly or entirely of brickwork, as indicated in *figs.* 81. and 82. *Fig.* 82. is a section of a portion of the frame of a raised bed of this kind; in which *p* is the surface of the lawn; *q*, a raised band of turf, to serve as preparation for, or base to, the brickwork; *r*, a circle of bricks laid flat, and half-sunk in the turf; *s*, a circle of bricks all laid to one slope, and

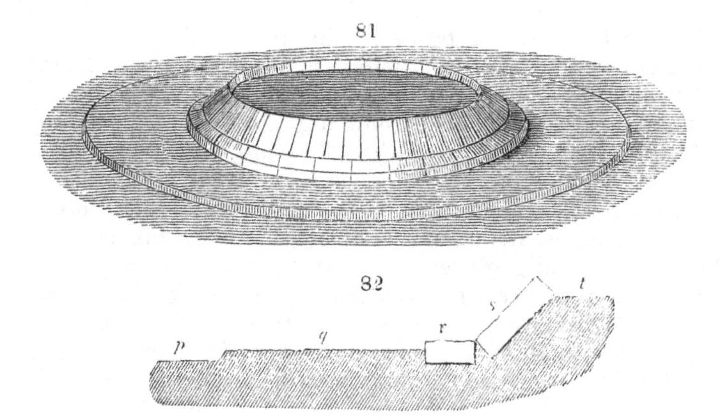

81

82

touching with their lower edges the circle of flat bricks; and *t*, the surface of the bed. The perspective view of this bed is shown in *fig.* 81. Such a bed ought to be planted in the centre with the most choice summer-flowering green-house plants; and, round the margin, with mignonette, *V*erbèna *c*hamædrifòlia (*Melíndris*), or some other fragrant, or brilliant-coloured creeper which will hang down over the bricks. A large fuchsia in the centre, surrounded by variegated pelargoniums, with a border of mignonette intermixed with blue anagallis, will have a good effect; as will a large Brugmáns*ia* suavèolens(*Datùra* arbòrea) in the centre, with the surface of the bed entirely covered with the *V*erbèna *c*hamædrifòlia, which would hang down with its brilliant scarlet flowers over the brick frame. The brugmansia (*fig.* 83.), when well grown, is a particularly suitable plant for this purpose.

83

It may be kept in the frame in the reserve ground during the winter; and, if turned out into rich soil, and kept well supplied with water during summer, it will produce a profusion of its fine trumpet-shaped, pure white, fragrant flowers, from June till the beginning of October; when the plant ought to be taken up, repotted, and returned to the pit, to prevent it from being injured by frost. In the garden of —— Durant, Esq., at Putney Hill, brugmansias are treated in this manner by the very intelligent and skilful gardener there, Mr. Spence; and grown and flowered to a degree of perfection which would hardly be credited by those who have not seen the plants. Some of them are nearly 6 ft. high, with woody stems 3 in. in diameter; and one, in the summer of 1837, bore upwards of 2000 flowers.

Estimate of Expense. The plant-houses, including the glass roof, front, and ends, with the water-pipes, and all the fittings up within, may be estimated at 3*s.* per superficial foot of the area on which the buildings stand. This will amount, on the houses shown in *fig.* 76., to from 150*l.* to 200*l.* The reserve garden, with its pits and frames, potting-shed, and other requisites, may be estimated at 50*l.*; and the paved walks, central bed, and plants for the wall facing the north and for the end wall (which we shall suppose to be the same fruit trees as those recommended for *fig.* 72. in p. 236.), together with the expense of the front garden, tools, &c., may amount in all to 25*l.*; thus making the total expense of this garden about 275*l.*, or say from 250*l.* to 300*l.*

Management. If the occupier intends to manage this garden himself, he must be at home every day throughout the year; with the exception, perhaps, of a month or two during summer, when his forced trees, having yielded their crops, are in a dormant state; and when he may commit the care of his reserve department, and of his green-house, to a commercial gardener in the neighbourhood. If he employs a gardener, it must be one who thoroughly understands his profession, and who is so decidedly steady and regular in his habits, that the utmost confidence may be placed in him. The neglect, for one night, during severe frost, of the fire which heats the hot-water apparatus, or the making of too large a fire in a mild night, would be sufficient to destroy a whole crop of either peaches or grapes. For this reason, not only a remarkably steady man must be employed as gardener, but one who lives either in the house, or near at hand; so that he may be able to attend the fires when necessary, at any hour of the night. Even during the day, there is great danger of overheating forcing-houses, either by sun heat alone, or by the joint effect of sun heat and fire heat; so that in the day time the occupier or his gardener will require to be as vigilant as during the night. Some very ingenious apparatuses have been invented for regulating the temperature of plant-houses, and of apartments generally, by a thermometer so constructed as to act on mechanism which opens and shuts the sashes, or valves made on purpose, as the temperature of the house rises or falls. Apparatuses of this kind are, however, unfortunately as yet not much in use, otherwise they might be considered as safety-valves for plant-houses, since they would effectually prevent extremes of heat. Some of the simplest of these apparatuses will be described hereafter.

Cost of Management. Even on the supposition that the occupier is his own gardener, this description of suburban garden would cost a considerable sum yearly. The fuel would amount to 25*l.* or 30*l.*; breakage of glass and flower-pots, keeping watering-pots, syringes, &c., in repair, say 10*l.* Fresh loam, peat soil, dung, and other articles for the reserve garden, 10*l.*; keeping of the open garden and the front garden, 2*l.*; in all 50*l.*, or from that to 60*l.* If a gardener were employed, a suitable one could not be obtained under 100*l.* a year; or 80*l.* a year with a house, coals, &c.; so that the total expense of this garden would not be less than 150*l.* a year; or, as every garden where there are much glass and constant fires is liable to many accidents, it would, perhaps, be safer to say from 150*l.* to 200*l.*

The Produce, independently of green-house plants and forced flowers, would be two or three dozen of peaches, and two or three dozen pounds of grapes, melons, strawberries, cucumbers, kidneybeans, tart rhubarb, mushrooms, &c., when these articles are rare, and bear a very high price. The open garden will produce some cherries, apricots, and plums, from the end wall enclosing the reserve ground, and from the side wall facing the north; besides flowers from the central bed; or, if the central bed were cropped with culinary plants, it might supply summer salading. Mustard and cress, during the winter season, would be raised in the forcing-houses; and figs, cherries, &c., might be brought forward in these houses in pots. In short, almost every table fruit and culinary production grown in first-rate gardens might be produced, to a limited extent, in this one; since, in the reserve ground, pine-apples might be grown in one of the pits; and even that new and most delicious fruit, *Mùsa* Cavendísh*ii*, the Duke of Devonshire's banana, which is easier cultivated than either the pine-apple or the melon. Notwithstanding the value of these products, they would not half pay the expense; and, therefore, the enjoyment and interest produced by seeing the operations carried on by the gardener, or by the occupier himself, must be considered as the main result.

8. *To lay out a Fourth-rate Suburban Garden, where the main Object is the Culture of Florist's Flowers, Prize Fruits, &c.* It may be advisable to notice here, that by border flowers are

meant such, whether perennials, biennials, annuals, or bulbs, as may be planted or sown in beds or borders, without any extraordinary preparation of the soil, with reference to an ordinary, or natural, standard of beauty; while by florist's flowers are to be understood, flowers chiefly of the perennial kind, which require extraordinary preparation of the soil, and to be cultivated in a particular manner, separately from any others, with reference to a highly artificial standard of beauty. Tulips, hyacinths, auriculas, carnations, &c., dahlias and hearteases, are included in this class; and for each of these flowers there is a set of laws, or conditions, to which they must conform in their shape and colour, to enable them to rank as florist's flowers. The most important of these laws, or canons, is, perfect symmetry of form, with reference to the entire flower; perfect distinctness of outline, in regard to the parts of which it is composed; and perfect clearness of colour, and definiteness of outline of the colours, where the flower consists of more than one. Prize fruits are, in like manner, such as are grown to an extraordinary size, symmetry of form, clearness of colour, freshness of bloom, and richness of flavour; and though, at present, prizes are given by the horticultural societies to every kind of vegetable production, according to its merits, relative or absolute, yet what are considered prize fruits, by the great mass of society who have suburban or cottage gardens, are chiefly gooseberries, cucumbers, and melons. The culture of florist's flowers is, next to the culture of plants in houses, that which requires the greatest and the most unremitting attention on the part of the cultivator. Wherever this culture is attempted in a suburban garden, there must be a reserve ground of considerable extent (unless, as is generally the case, the whole garden is treated as a floricultural workshop), in order to keep in it a stock of the different soils, manures, &c., which are necessary; and also for the purpose of raising seedlings, and bringing forward plants under glass. Florist's flowers may be cultivated to great perfection, even in the smallest suburban gardens; as a proof of which we may refer to the tulips, auriculas, pinks, heartseases, &c., growing in the small plots occupied by the Spitalfields weavers; and by the Paisley weavers, and those of the manufacturing towns of Lancashire. The culture of these flowers, and also of prize fruits, is conducted on the principle that an extraordinary amount of nourishment, of the richest kind, will produce extraordinary magnitude; and that great care in selecting, and in cross breeding, will produce superior beauty; both the magnitude and the beauty being far removed from ordinary nature. For example, the auricula, in its wild state on the Alps of Switzerland, seldom exceeds 3 in. in height, with flowers rarely measuring more than half an inch across, and frequently with some of the petals wrinkled, or of an undecided, or muddy colour; but in the garden of a Lancashire weaver, where the auricula is nourished with the richest essence of vegetable and animal manure, and protected from inclement weather, and from all extremes of temperature, it reaches the height of 1 ft., with leaves almost as broad and succulent as those of a small cabbage, and with flowers one of which is frequently nearly 2 in. across, flat and symmetrical in form, with colours of a clear and decided tone, and with their boundaries distinctly marked. The gooseberry, in like manner, in its wild state, in the valleys of Switzerland, is an insignificant bush, 2 or 3 feet high, producing a few fruit, seldom larger than a pea, and weighing not more than 1 dwt.; but, under the care of the same Lancashire weavers, by means of highly enriched soil, keeping the branches near the ground, thinning out all the fruit on a bush except two or three, and placing saucers of water beneath these, so as to keep them continually in a moist atmosphere, they have been made to grow as large as pigeons' eggs, and to weigh above 30 dwts. Cucumbers, under ordinary culture in frames, are generally from 6 in. to 8 in. long, and crooked; but the cucumber growers for prizes produce them from 20 in. to 27 in. in length, and quite straight; and, while the gentleman's gardener is not very particular in handling his fruit when gathering it, the prize grower contrives to cut it, and carry it to the exhibition, not only with its surface covered with a fine glaucous bloom, but with the withered flower still attached to its extreme end. Florist's flowers and prize fruits may, therefore, when compared with the same kinds of flowers and fruits in their natural state, be considered as monstrous productions, like prize cattle or crammed poultry. A florist's flower or a prize fruit, from the beginning of its growth, till it arrives at perfection, is in a state of training; it is therefore unfit to be seen, till it is ready to be taken to a public exhibition; where the grower, as he is called, receives the reward of his unwearied labours in the form of a prize cup or medal, or in the applause of the multitude, and the envy of his competitors. We have considered it necessary to state this much respecting florist's flowers and prize fruits, to show that it is a kind of gardening not at all adapted for the possessor of a fourth-rate suburban garden, who has a taste for neatness, or for a garden which is intended to look well, and be fit for walking in every day of the year; unless, indeed, there is a very large reserve ground. On the other hand, there is a class of persons who take an intense delight in this species of culture, and who have no taste for any other. Such persons, if they grow cabbages or onions, apply the same modes of culture and training to them; and they are not satisfied with cultivating any plants or crops, unless they can produce them of extraordinary size. We know engravers, gentlemen in public offices, and many tradesmen, who have suburban gardens not above 50 ft. in length by 30 ft. in breadth, some of whom spend hundreds a year in this kind of culture; and, in the summer season, work daily in their gardens before and after their hours of business; rising early inthe morning, and remaining in their gardens till it becomes dark. We have known an amateur florist (not overburthened with money) take the only sheets he had from his bed to form an awning for his tulips during the day, and his blanket to cover them during the

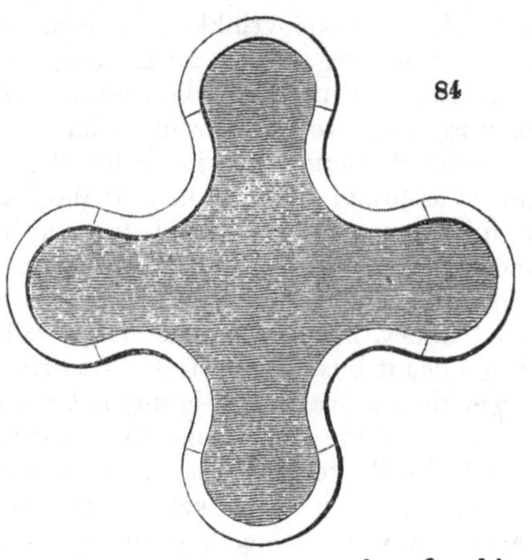
84

85
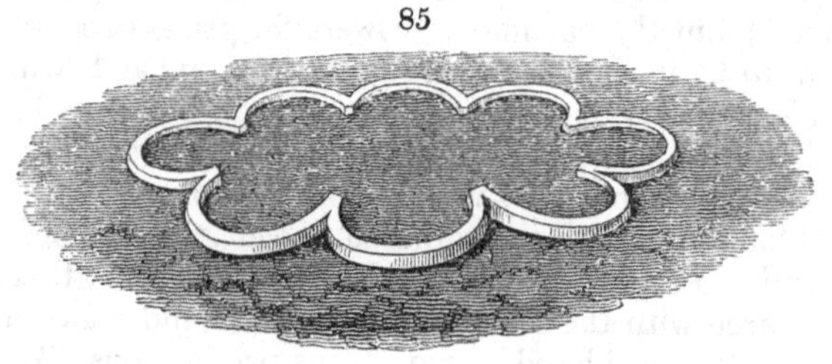

night. The taste of florists of this kind may be considered as springing more from a love of gambling than of flowers; but it may be the means, in consequence of the public exhibitions and competitions for prizes, of creating a love of flowers in others.

The Front Garden, where the main object is the culture of florist's flowers, may be considered as the show ground; not

86

indeed for the finer florist's flowers, which require to be covered, during sunshine, rain, and any kind of weather tending towards

any extreme, by awnings; but for the more ordinary florist's flowers, such as mixtures of hyacinths, tulips, anemones, ranunculuses, pinks, heartseases, &c. To enhance the beauty of these, they should not be grown on common flat beds, with grass or box edgings in the usual manner; but in beds either on grass or gravel, surrounded by brick frames, like those shown in *figs.* 81. and 82. in p. 267.; or surrounded by semicircular tiles in the manner shown in the ground plan *fig.* 84., or in that shown in the perspective view *fig.* 85.: or, where several distinct masses of florist's flowers are to be exhibited at once, a star-like bed, in a raised framework of turf, may be formed on the lawn or area of gravel, as shown in *fig.* 86. The centre of this bed may be filled with red hyacinths of different kinds; and in the four rays white and blue hyacinths may be arranged symmetrically. Whatever kinds of flowers are planted in such a design as that shown in *fig.* 86., they should (in conformity with Rule 5. of the rules already laid down in p. 213.) be all of the same size, and all come into flower at the same time.

If the possessor of a fourth-rate suburban garden, with such a bed in his front garden as *fig.* 86., could afford to grow his florist's flowers in pots, then his bed in front might exhibit the following succession throughout the year: —

November to February. Common primroses in the centre, and Russian violets in the rays. These are not florist's flowers; but the beds are supposed to be filled with them at this season, as being better than exhibiting the naked earth.

February. The centre bed may be filled with snowdrops, and the rays with winter aconites; or all the beds may be filled with red, blue, and white hepaticas; or the centre bed may be filled with red scillas, and the rays with white and blue scillas. The above are not florist's flowers, but they make the nearest approach to them to be found at this season.

March. Yellow crocuses in the centre, and white and purple crocuses in the rays.

April. Red hyacinths in the centre, and blue and white in the rays; and, after the hyacinths, auriculas.

Part of April and May. Scarlet flake tulips in the centre, and yellow and brown bybloemens in the rays.

June. Dark purple, white, and yellow ranunculuses; or heartseases of different sorts.

July. Pinks; succeeded by carnations.

August. Balsams; or purple, red, and white stocks.

September. Dahlias; or German asters.

October, and till the blossoms are destroyed by frost, Chinese chrysanthemums.

This mode of culture does not so much require expense as trouble; but no other mode whatever of managing a front garden can equal it in point of splendour of effect. Supposing forty pots of each kind of flower be required to fill the centre and the four rays of *fig.* 86; then at least fifty pots ought to be grown to provide for accidents; such as the death of some of the plants, others not growing to a proper height, or not being sufficiently advanced with their flowers, &c. In order to bring forward any pot in which the flowers are not so far forward as the rest, it may be watered with warm water, or set for a day or two in a frame; for it is essential to the beauty and completeness of this mode that the plants exhibited at any one time should be all of the same size, and all be equally in flower.

The Back Garden should, if the possessor wishes to have any part of it neat, have a reserve ground at one end, communicating with a public road, for composts, manures, &c.; and also for frames, and other contrivances, for bringing forward seedlings and offsets, protecting auriculas, &c. The remainder of the garden may be surrounded with a walk, as in the preceding plans, and the central compartment laid out in cross beds, as in *fig.* 55. in p. 201., for tulips, ranunculuses, carnations, &c. Tulips, carnations, and auriculas being, when in full bloom, liable to have their colours faded by the sun, are always exhibited under awnings, which are sometimes fixed, and sometimes temporary. When the latter is the case, there should be sockets fixed in the ground in the proper places, to receive the lower ends of the posts; and these should have caps, as in the case of the sockets for clothes-posts (see p. 176.), to prevent earth from falling into them. When the awning is fixed, the posts and framework remain all the year; and the tulip or hyacinth bed, being made along the centre of the space covered, comes first into flower; while the carnations, being grown in pots, are brought in when the tulip or hyacinth bulbs are taken up, and set on boards placed over the bed. The auriculas, in this case, as they come into flower before the tulips, require a separate awning; or, as a substitute for an awning, to be exhibited on a stage facing the north. Auriculas and carnations of the finer kinds, being grown in pots, and exhibited on stages with platforms, may be carried to these; while tulips, ranunculuses, hyacinths, &c., being grown in beds, the awning must be carried to them.

If the occupier of a back garden, of the dimensions of *fig.* 55. in p. 201., be desirous of cultivating florist's flowers, one third part at the farther end may be walled off, as a reserve garden; and a fixed framework for an awning erected along the middle of the remaining part of the ground, say 10 ft. wide, 8 ft. high at the sides, and 50 ft. in length. Along the centre of this space, a bed 6 ft. broad may be formed, with a walk 2 ft. wide on each side of it, and the space between the awning and the walk which surrounds the garden, on both sides, may be devoted to the culture of polyanthuses, hearteases, dahlias, and other florist's flowers. The bed under the awning may be divided into three parts of equal length; viz. one for hyacinths, one for tulips, and the other for ranunculuses; and, as these come into flower at different seasons, where economy is an object, there need not be more canvass than will suffice to cover one third of the bed at a time; it being understood that there are movable ends, of frames covered with canvass, and with doors in them, which can be shifted so that each kind of flower, when it is in bloom, may be completely enclosed. A very convenient stage for auriculas may be formed against either the east, west, or south wall of the reserve garden; the object of the structure being to protect the plants from rain and from the direct rays of the sun. The roof of such a stage is generally of green glass, or crown glass, with a canvass blind under it; and the doors, which should open in front, may either be of glass, or of slight frames filled in with canvass. Where economy is an object, a very excellent mode of displaying auriculas consists in placing them on a temporary shelf, on the north, east, or west face of a wall, at the height of about 4 ft. from the ground, so as to be under the eye, with a sloping board, fixed temporarily over the shelf to throw off the rain, and to protect the plants from perpendicular cold; or, in other words, to obstruct the radiation of heat.

Florist's flowers may be grown, on a moderate scale, by the use of hot-bed frames alone. These may be used for protecting from the cold, and the rain, tulips, hyacinths, polyanthus narcissus, and every other description of florist's bulbs; and also carnations, auriculas, ranunculuses, heartseases, and Chinese chrysanthemums. When the plants are to be shaded from the sun, it is only necessary to cover the glass with mats, or to remove the glass, and substitute mats. Should the tulips, carnations, or chrysanthemums grow too high, so as nearly to touch the glass, the frame may be raised by a substructure of 4 in. brickwork laid without mortar, or by boards, or merely by propping up the frame at the corners, and filling in the open space with soil. There is no disadvantage, that we are aware of, attending this mode of cultivating florist's flowers; except that those who come to admire them must be content to stand in the open air, instead of being under an awning; and that to examine closely the lower growing kinds, such as auriculas and heartseases, it will be necessary to stoop down a little; though even this might be prevented by placing the frames on platforms of earth, 3 ft. high, surrounded by brickwork. Besides the economy of this mode, in saving the cost of an awning, its frame, pullies, lines, &c., and auricula stage, the unsightly appearance of these objects when not in use is altogether avoided; and the glazed frames, when not wanted for the florist's flowers, may be applied to protecting alpines or other plants, or to raising tender annuals, cucumbers, melons, &c.

Expense and Management. If this description of culture is pursued with enthusiasm, even in a plot of ground 100 ft. in length, and 30 ft. in breadth, it will require a considerable outlay, independently of the expense of forming the walks, and the separation fence between the show garden and the reserve ground. The frames to be used as substitutes for an awning, and for bringing forward articles in the reserve garden, will cost at least from 20*l.* to 30*l.*; and, if framework, with an awning over it, and an auricula stage, be employed, the expense will be still greater, according to the length of the awning and the size of the stage. Besides these, there must be a cabinet for keeping bulbs, divided into some hundreds of compartments, which will cost 5*l.* or 6*l.*; different kinds of soils and manures, pots, frames, tools, &c., which will cost 10*l.* or 12*l.* every year; and a stock of roots and plants to begin with, which may amount to any sum the occupier chooses, from 5*l.* to 100*l.* The management will consist in unremitting attention, and in the performance of operations not generally familiar to ordinary gardeners; and which can not be delegated by one florist, except to another. In general, therefore, every cultivator of florist's flowers in a small garden, must be his own gardener. The produce, where the grower exhibits for prizes, consists in cups, medals, &c., and in the excitement produced by competition. Sometimes, also, when new kinds are raised by amateur florists, they are sold by them for considerable sums: a new tulip, a new heartsease, or a new dahlia will sometimes bring from 10*l.* to 50*l.*, or even a great deal more, from a commercial florist.

Remarks. We have dwelt longer on the subject of the cultivation of florist's flowers in small suburban gardens, than we otherwise should have done, because, although we have known some persons cripple their fortunes by this taste, or fancy, as it is called, we have known others who have been first attracted to gardening and botanical pursuits by having accidently had their attention directed to a florist's flower; and who, by devoting themselves to the culture of these flowers late in life, have not only derived from it a new and intense source of enjoyment, but have acquired in the end no small degree of botanical knowledge, and a taste for gardening in all its departments. The details of the culture and management of florist's flowers, and also of prize fruits, will be found under the proper heads.

9. *To lay out a Fourth-rate Suburban Garden, where the Object is a Botanical Collection.* A garden laid out and cultivated for this object has, like those which have preceded it, certain advantages and disadvantages. The advantages are, that, if only hardy plants are grown, it requires little skill in the management, and may always be kept neat at a trifling expense; and that, if the occupier be known as a botanist, he will be able to procure a considerable part of the stock, or plants to be grown, from his botanical friends. The disadvantages are, that this

kind of garden does not afford so much bodily exercise to the occupier as almost any of the other kinds we have mentioned; for, even in that (*fig*. 53. in p. 189.) which is laid down in grass, with only a few trees, the occupier may employ himself part of every day during summer with the mowing machine. In a purely botanic garden there is no digging, no mowing, and even very little watering, as it is not at all desirable that botanical plants should be rendered more luxuriant than they are in a wild state. Still, the botanical interest of such a garden is so great, that it must necessarily have considerable charms for all who have any knowledge of, and love for, plants.

The Front Garden may be laid out in many different ways. In *fig*. 87. are shown three front gardens, laid down in turf, with

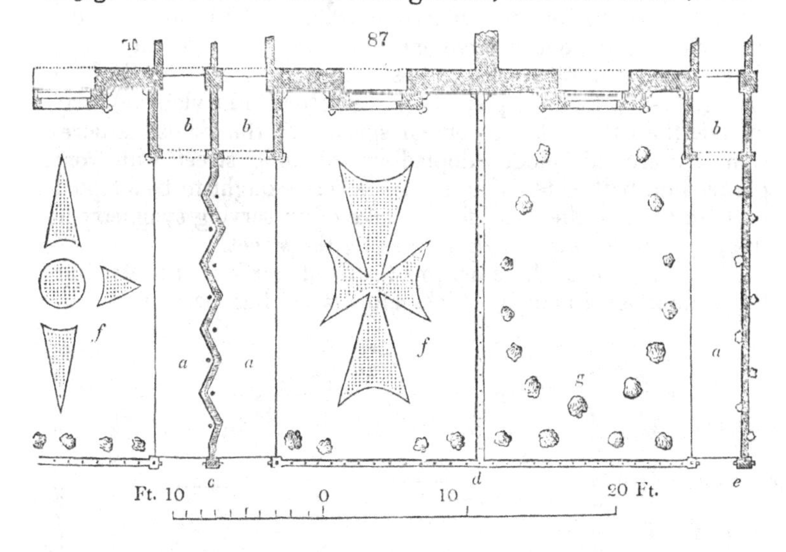

the walks (*a a a*) paved, and the houses entered by porches (*b b b*). The separation wall at *c* is zigzag in the plan, for the sake of admitting standard trees and hollies in the angles. The hollies are intended to fill up the angles, so as to form an evergreen hedge, in upright panels, separated by brick piers. The standard trees may be thorns of different kinds; or tall-growing flowing shrubs, trained to a single stem; such as privet, arbutus, phillyrea, purple and white lilac, syringa, snow-ball tree, &c. At *d*, the separation fence is a low wall with an iron railing, on which may be trained the Virginian creeper, and Ayrshire and other roses; and at *e*, it is a common wall, with standard trees, and ivy, planted against it; the ivy being for the purpose of clothing the brick-work with perpetual green, and the standards for throwing a shade upon the walk. The standards may be low trees, such as laburnums, *Pỳrus spectábilis*, *A'cer monspessulànum*, *Amelánchier* Botryàpium (the snowy mespilus), &c. The beds at *f f*, forming symmetrical figures, must be treated symmetrically, as laid down in Rule 5. in p. 213.; and the shrubs at *g* may be alternately evergreen and deciduous, all cut (not clipped) into regular, fastigiate, gardenesque forms. Every plant should be a different species or variety; and, if it is thought fit to change these every three or four years, in the course of a twenty-one years' lease, as there are twenty-one standard plants in this garden, all the more ornamental species may be passed in review, and rendered familiar to the occupant of the house and his family; thus rendering this garden a school of practical botany to the young people of the house, in accordance with the object of this ninth mode of laying out and planting a suburban garden. With respect to the trees along the front of these gardens next the road or street, they may form part of a series, which, if the street is long enough (such, for example, as the New Road, Marylebone), may include the whole of the popular species of the British arboretum. Were this mode adopted in any long street, with front gardens on both sides of it, the same series ought to be adopted on both sides of the road, for the sake of preserving symmetry to the public in walking or driving along the street.

Fig. 88. exhibits three designs for front gardens, adapted for the same kind of houses, which, after what has been said, will

88

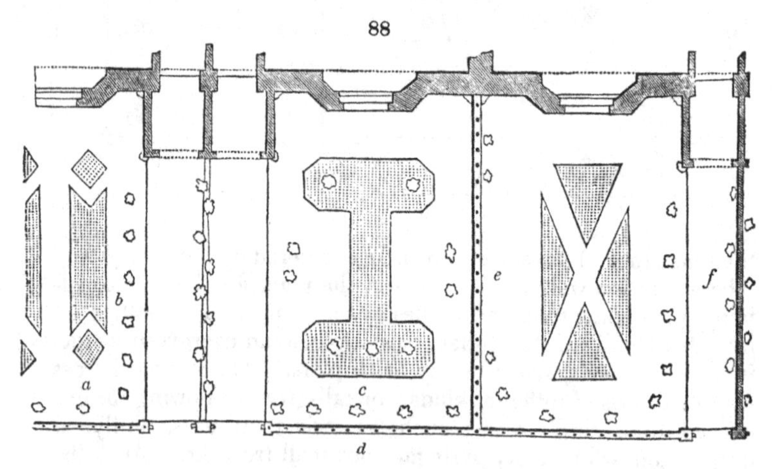

require little or no description. In the design *a*, the trees represented on the turf at *b* may be standard roses; and in the next design, those at *c* may be trailing plants, such as Cotoneáster U'va-úrsi, Spártium tríquetrum, *Méspilus floribúnda*, &c., grafted standard high. The five plants in the dug bed in this design may be standard rhododendrons. In the third design, the plants on the grass may be five different sorts of the genus *Pìnus*; viz., *P. Pináster*, *P.* sylvéstris, *P. Stròbus*, *P.* Larício, and *P. Tæ`da*. For variety's sake, the trees along the boundary fence *d* may be of the fastigiate-growing kinds, such as *Cratæ`gus Oxyacántha* fastigiàta, *Quércus* pedunculàta fastigiàta, *Cupréssus* sempervìrens, *Juníperus* virginiàna, &c. Those against the boundary fence *e*, as there is grass on both sides, and no walk to be impeded by them, may be drooping plants, such as the weeping sophora, the weeping laburnum, weeping horsechestnut, weeping thorn, &c.; or, should these exist in adjoining gardens, the spruce and balm of Gilead firs, and the larch, may be introduced, for the sake of keeping up the botanical interest. Those against the boundary wall *f* may be American oaks, which will have a splendid effect in spring and autumn; and which, with very little pruning, are easily kept within bounds, and rendered fit for any suburban garden, however small.

In laying out a Back Garden for a Botanical Collection, a shed at the bottom of the garden, and a small piece of reserve ground, will add greatly to the convenience of carrying on the operations; more especially if the rarer plants are grown, or such as require different kinds of soils. Indeed, in every description of garden where operations on the soil are to be carried forward, there can hardly be such a thing as perfect order and neatness, without some kind of reserve garden. In some cases, this may be so small as to be merely a square, 6 ft. on the side, enclosed by a privet hedge, or a boarded fence, covered by creepers; in others, it may be a portion of one end of the garden separated from the rest by a wall, and entered by a close door. A botanical collection may either be limited or general: if the former, it may be of hardy plants only, or of British plants, European plants, or the plants of any particular country; or it may consist only of ligneous plants, or only of herbaceous plants, or only of house plants; and of these last, it may be again limited to the plants requiring the temperature of the green-house, the stove, or dry stove; or to those of some particular country, which, in Britain, must be kept under glass. What is designated by botanists a general collection, does not necessarily embrace the plants of all counries; but, rather, is not limited to those of any country, and includes house, as well as hardy, plants.

Where a garden has its greatest length from east to west, and, consequently, a side wall with a south aspect, the occupier, if he thinks fit to erect glass structures along that wall, may cultivate in them a collection, not only of green-house plants, but of such as require the temperature of the stove. If his best aspect is to the east or west, or to some point between these quarters and the south, he may still cultivate green-house plants with success. It is one great advantage of botanic culture in small gardens, that the shady parts of them which face the north may be rendered as valuable to the botanist as those which face the south; and, consequently, are sunny. In these shady parts, houses might be erected for growing ferns or orchideous plants; or hardy ferns and alpine plants might be grown in that part of the garden, in pots, or in the free soil. Where ferns, alpines, or any other description of plant, is grown in the shade, whether under glass or in the open air, there must always be the free admission of perpendicular light; and, hence, it is never advisable to attempt to grow plants which delight in shady situations under the cover of the branches of trees, or under a wall, the top of which is overhung with ivy, or which has a very broad coping. The natural result of shady situations, in nature, is chiefly the interruption of the direct rays of the sun, and the comparative prevention of evaporation; and, if this is effectually done, in gardening, by a perpendicular wall in the direction of east and west, and by supplying water, in dry weather, for evaporation, and, in some cases, covering the plants with a glass frame to retain the moist vapour, the plants will be found to thrive better under culture than in their native habitats. If, then, the occupier of a suburban garden lying in the direction of east and west were to cover all his walls with glass, he might cultivate to the highest degree of perfection the tropical ferns and *Orchídeæ* on the wall having a north aspect; the plants of the Cape of Good Hope and of Australia, on the east wall; the *Cácti*, and other plants which do not require a moist atmosphere, against his house; and the other tropical plants, which require a hot and moist atmosphere, against the wall with a south aspect. All this might be done in a garden 100 ft. by 30 ft., leaving a space of 10 ft. by 80 ft. in the centre; in the middle of which there might be a walk, with a border against the glass on each side, for half-hardy and Cape bulbs, &c. So much glass, however, would involve considerable expense, both at first erection and in stocking, and annually, for fuel and repairs; and would, besides, require the constant attendance of a first-rate gardener, with one or more assistants.

A collection of hardy plants, ligneous and herbaceous, may be very conveniently grown in a back garden, laid out in the same manner as *fig.* 72. in p. 236.; covering the walls and trellises with trees and shrubs, and filling the beds and borders with herbaceous plants. None of the trees and shrubs could be expected to attain their natural size and shape; but the greater part of them would flower, and all of them would exhibit their foliage; which, alone, would enable any person of observation to recognise them again, when he met with them in natural or artificial plantations. These trees and shrubs, which, planted at 2 ft. apart, would give 200 species, might follow each other on the wall and trellis in some recognised botanical order; the most useful of which, at the present time, is undoubtedly that of the Jussieuan system, with its modern improvements. The same system should be followed in arranging the herbaceous plants in the beds; and, in order that these may contain the greater number of species, such as are of low and compact growth ought to be chosen in preference. In the front garden, the circular bed might consist of peat earth, and in it a collection of hardy heaths might be grown, some of which would be in flower every month in the year.

A botanical collection might be combined with picturesque effect, by having no trellis, and by having broad borders next to the walls; planting close to the latter the trees and shrubs, and allowing them to assume their natural shapes; and planting herbaceous plants in front, and also, in natural-like groups, in the centre bed. But not one half of the number of species could be grown in this way that there could in the other; and, while the stronger-growing trees and shrubs would have a constant tendency to choke the weaker, both trees and shrubs would injure the growth of the herbaceous plants in the same border.

After the kind of botanical collection to be cultivated is fixed on, the next thing is to make such alterations as may be necessary in the quality of the soil. If the more common hardy plants are chosen, and the soil is, as we have hitherto supposed it to be, a loam, rather light than stiff, no change will be necessary; but, if tender plants, requiring very sandy soil, light rich mould, or peat earth, form a part of the collection, then a stock of these soils must be procured, and masses of them placed in the proper situations for the plants requiring them. One of the great difficulties attending this practice in crowded gardens is, that the roots of the stronger-growing plants, adjoining masses of peat, or light rich soil, generally penetrate these soils, and grow in them with such vigour as to render them of little value to the delicate-rooted plants for which they were intended. For this reason, in all small gardens, intended to be crowded with plants, the first principle of arrangement, in certain cases, instead of botanical relation, ought, for economy's sake, to be those of size and duration; and the second, soil. A scientific arrangement, however, has such considerable advantages, with reference to the study of botany, and to the conveying of a general knowledge of plants in masses, that it is often desirable to attempt it, even under great disadvantages. When this is done, as different soils from that common to the garden are required for the more delicate and hair-rooted plants, it is necessary to enclose these soils in large pots, so that each plant may have the entire possession of the soil appropriated to it. These pots should be very deep, and should have a stratum of coarse gravel, or similar materials, at the bottom, to serve as drainage, and they may be plunged so deep as to cover their rims, and thus protect them from the weather.

A garden, containing a botanical collection, may be rendered ornamental by training the trees and shrubs against the walls, and having a border all round the garden, between the walls and the walks, of 3 ft. or 4 ft. in width, for dwarf herbaceous plants and bulbs; laying down the centre bed in turf, and forming a series of beds on it, as shown in *fig.* 70. in p. 234.; and planting them either with a mixed collection of perennial plants, or sowing them every year with annuals. This last mode will have the advantages of creating more work for the occupier, and giving him an oppertunity of exercising his taste, as well as showing his botanical knowledge; and it would also be more generaly admired than a plot of ground laid out entirely in parallel beds, such as *fig.* 55. in p. 201.

Management. It can seldom be desirable to have a small botanic garden managed solely by a hired gardener; because there are few such that would be competent to the task. We therefore conclude that the occupier will be his own gardener, more especially if he keeps his plants named or numbered. The operations necessary during the summer season will be, chiefly, tying up flower stems, reducing plants which are growing with too great luxuriance, by cutting out a part of their stems and leaves; cutting off decayed flowers and flower stems; pulling up weeds, destroying insects, stirring the surface of the soil, and, when necessary, watering. In early summer, disbudding the ligneous plants must be carefully attended to, in order to prevent them from becoming too luxuriant, and overtopping one another; and, throughout the season, the progress of insects must be carefully watched, and means taken to destroy them as soon as they appear. Decayed leaves and flowers must be removed; and, in order to preserve the foliage in a healthy state, watering with a syringe, two or three times a week, late in the evenings, will be advantageous. In autumn and spring, the ground about the plants may be slightly dug, and the stools, or stocks, of the more luxuriant kinds reduced; any plants that require particular kinds of soil may also be examined at the roots, and a fresh supply added if wanted, &c. At these seasons, the pruning and training of all the ligneous plants should be attended to, and the eggs of insects removed when they are observed on the trees and shrubs, or on the walls, or the trellis. In winter, the ground, where it is not under turf, may be slightly dug, overgrown stocks or stools reduced, names or numbers renovated, and sticks for tying up prepared.

When common jobbing gardeners are employed, for a certain number of hours or days in a week, to look after collections of this kind, they are apt to mistake weeds for plants, and the contrary; and even when gardeners who know plants are engaged from a nursery, or even botanic garden, to put in order a collection to which they have not been accustomed, from not knowing the condition of some of the plants; such as their being sickly, or lately transplanted, not having yet come through the soil, &c., they are apt to commit mistakes. We repeat, therefore, that the possessors of botanical collections, who cannot afford to keep a regular gardener, must necessarily manage their plants themselves; or, if they employ an assistant, they must be almost constantly present during his operations.

Expense. All the preliminary works, such as laying down pipes, underdraining, levelling, forming the walks, &c., we shall suppose to cost about the same as in Design 2. (p. 189.), or say 25*l.* As botanists, though they may exchange or receive presents of herbaceous plants, can seldom procure trees and shrubs from one another, these must be purchased from the nurseries; and 100 kinds, for common soil, will cost 10*l.* But the wall and the trellis may be made to contain two hundred kinds, which, supposing 50 of them to be of roses, will cost, at least, 25*l.* Other expenses of laying out and planting may amount to 5*l.*; and a similar sum may be allowed for the purchase of herbaceous plants, that cannot be procured from friends. The total may thus be set down at from 50*l.* to 60*l.*

The yearly expense of management, supposing a jobbing gardener only employed to mow the grass in the front garden, and allowing 3*l.* or 4*l.* a year for nails, shreds, tools, tallies, &c., need not exceed 5*l.*

Remarks. The kind of garden we have been describing may be considered as suitable for a person requiring recreation for the mind more than for the body. It does not offer exercise enough for a person who has been sitting all day in a public office, or counting-house; but for one who has been attending courts of law, or engaged in pursuits that require a good deal of walking during the day, it offers gentle bodily exercise, and considerable mental interest. To a person decidedly fond of botany, or of natural history generally, we should think this would be the best of all gardens. There are not many gardens of this kind in the environs of London contained within such a limited space as is indicated in our plans *figs.* 55. and 72.; but still there are a few in even smaller spaces. Of these last, those of the late eminent botanists, Mr. Haworth and Mr. Sweet, were the richest which we have seen. The late R. A. Salisbury had, in 1820, a very choice collection of alpines, in pots, in a small back yard in Queen Street, Edgware Road; and Miss Kent, about the same time, had a garden of British plants, in pots, on the roof of the house in which she lived in St. Paul's

Churchyard. Where plants are grown under glass, very choice collections may also be cultivated, in a very limited space. Mr. Ward of Wellclose Square, a distinguished botanist, grows a considerable hardy flora, in troughs, or boxes, on the tops of the walls which enclose his back yard, and on the roofs of the out-buildings in it; and in the different rooms of his house, including his own bedroom, he grows upwards of a hundred specimens of ferns, indigenous and exotic. We know a gentleman whose back garden is hardly 30 ft. square; nevertheless he has a dry stove in it, in which he cultivates a collection of succulent plants, one of the richest in species of any in the neighbourhood of London. This gentleman leaves his home every morning at eight o'clock, and does not return till six o'clock in the evening. No one attends to his plants during his absence, to give air to the house or take it away; but so well adapted are succulent plants to bear extremes of heat and cold, as well as of drought, that they never suffer from neglect. The house is heated from the back of the parlour fire. We have already alluded to this gentleman (p. 5.), as working in his green-house, during the winter season, by candle light. We notice these instances to show the immense resources which even the smallest plot of ground affords, where there is a taste for plants and their culture; and some degree of knowledge as to the best mode of gratifying that taste under very limited circumstances. In the case of our friend above mentioned, had he chosen any other description of green-house or stove plants but succulents, they would necessarily have required attendance during the day; and had Mr. Salisbury, or Miss Kent, chosen plants of more luxuriant growth than they did, they could not have been grown in small pots, and in such a limited space.

Miscellaneous Designs for Front Gardens. Before proceeding to treat of the renovation of old and neglected fourth-rate suburban gardens, we shall submit a few more designs for front gardens to street houses; and include among them designs for front gardens common to several houses. It may have been observed, that the fourth-rate back gardens, from their small size and long narrow shape, are susceptible of very little variety in the form or dispositon of the beds and walks; without cutting up the space in such a manner as to render it unfit for the culture of culinary crops, or fruit trees or espaliers. With regard to the front garden, however, the case is different; the object in it being to display the taste, fancy, or skill of the occupier, in the form and disposition of the beds; and the laying out and planting it being solely intended for the production of beauty, there is no artificial restraint whatever as to the design, except that imposed by the limits of the space. For this reason, we shall give a few more designs for front gardens, for the sake of furnishing ideas as to the form and disposition of the beds, and the occasional introduction of sculptural ornaments among them. We think it of importance, however, previously to remark, that the main source of interest in a fourth-rate suburban garden, or of any garden whatever which is of small extent, will ever consist much more in the kinds of trees, shrubs, and plants with which it is furnished, and the high degree in which these are cultivated, than in the form of the beds in which they are grown; and that even in this case, unless the owner knows something about the plants, and takes an interest in their culture, and in their progress from spring to autumn, the enjoyment which he derives from his garden will be very far short of what it is calculated to afford. We cannot, therefore, sufficiently impress on the mind of all who either possess, or intend to possess, a small garden, the necessity of making themselves more or less acquainted with the articles to be grown or cultivated in it; with the processes and products of vegetation; and with the operations, times, and seasons of culture. It is for this reason, and also to show that, in fourth-rate gardens attached to street houses, every thing may be grown, though on a smaller scale, and even brought to perfection, that is cultivated in the first-rate gardens attached to the mansions and palaces of the nobility, that we have gone so much into detail as to the kinds of plants which may be planted in fourth-rate gardens, their culture, and their products. We hope by these means to excite an interest in gardening pursuits in our reader; but, if this should not be the case, we would recommend him, if he should be ambitious of having a garden which shall make as good a display as any of those of his neighbours, without having personally any trouble about it, to keep a good gardener. If he be not at all anxious to make a display; but merely wishes to keep his garden neat at little expense, then we recommend him to adopt the mode described in § 1. p. 182., and exemplified in *fig.* 52.; or the mode recommended in § 2., and illustrated by *fig.* 53. in p. 189., and *fig.* 54. in p. 194.

Figs. 89. and 89.* are front gardens to street houses, designed by Mr. Rutger. They consist of one perch to each garden, and contain scarcely any turf; the beds and borders being edged with stone or brick kerbs, about 2 in. high; and the walks being paved, or of gravel. The beds are planted in the mixed style, with evergreen and deciduous shrubs and flowers.

Figs. 90. and 90.* are four designs for front gardens, of a perch and a half each, by the same artist; who observes respecting them, that he "should prefer the beds to be surrounded with grass, rather than with gravel, as in the preceding designs; and the shrubs introduced to be of low and compact growth."

89

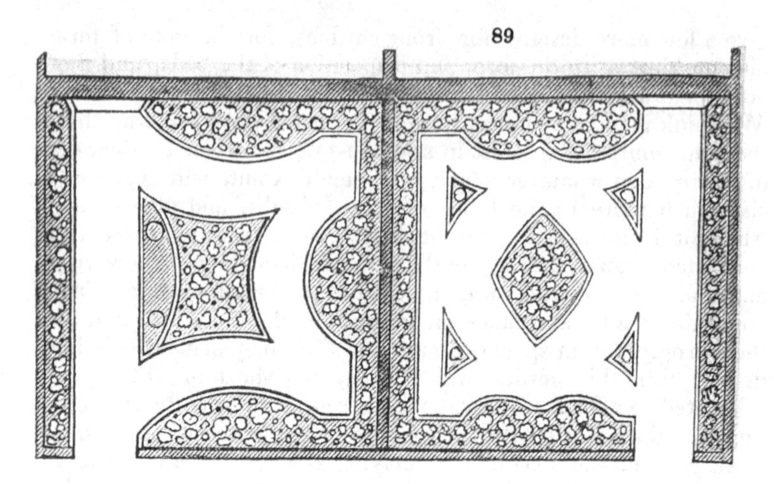

In these and the preceding designs by Mr. Rutger, the small circles indicate the situations for ornamental vases, and other sculptural and statuary objects, to be filled or decorated with

90

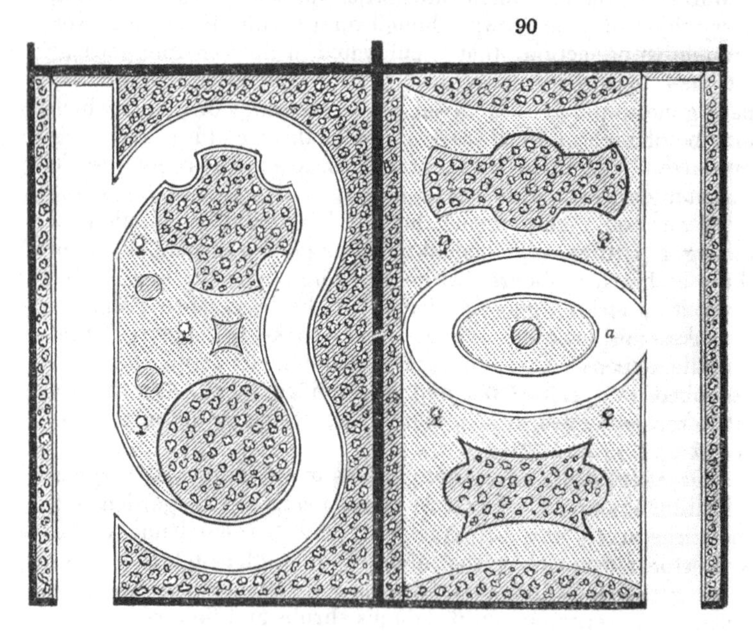

flowers, and of which there are now many elegant forms in artificial stone and earthen ware, which may be purchased in the metropolis at very moderate prices. In *figs.* 90. and 90.*, in the beds *a* and *b*, fountains may be introduced. With respect to the planting of these gardens, Mr. Rutger advises all persons, who

89*

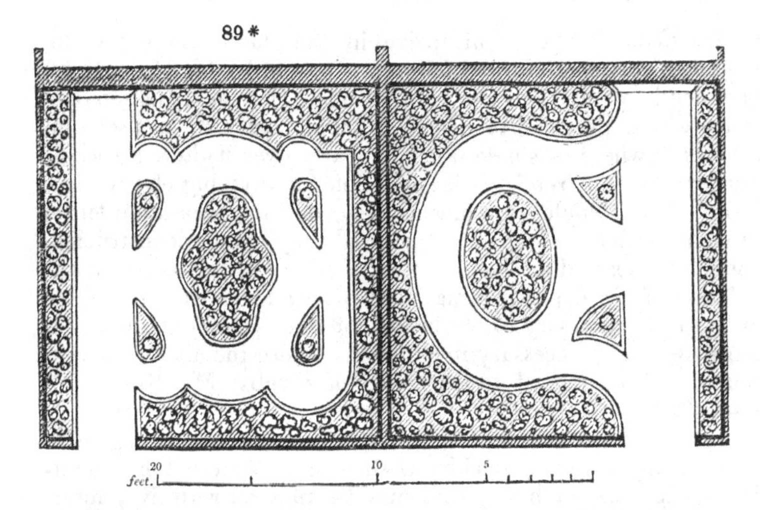

are about to lay out and plant front gardens in the vicinity of the metropolis, to inspect the nurseries which are nearest to

90*

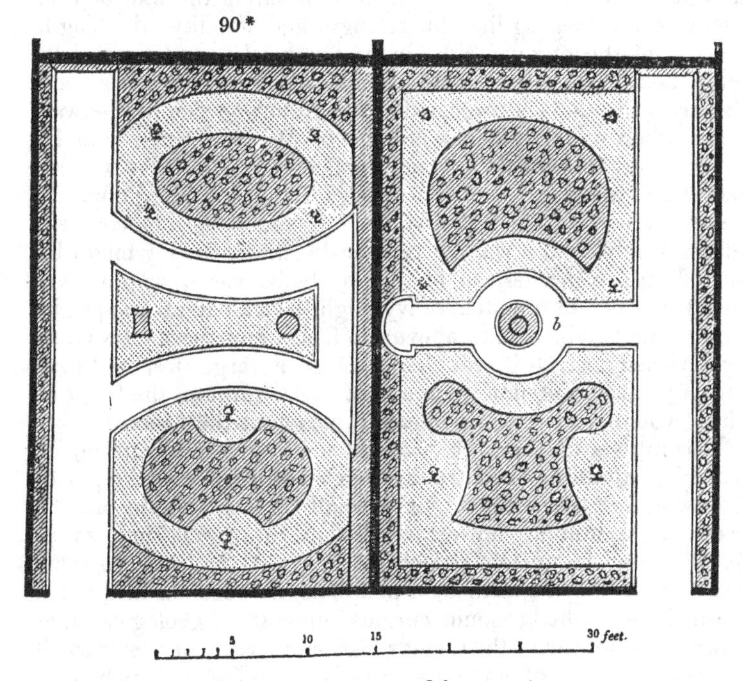

town; and also the front gardens of houses in or near town in different directions; and, having done this, to make such a selection of evergreen and deciduous trees and shrubs as they have ocular demonstration will thrive in the places where they intend to plant them. (*Gard Mag.*, xi. p. 503.) He strongly recommends plants of sizes proportionate to the situation and space which they are to occupy; adding, that he has seen instances "where a single tree has nearly overshadowed a whole garden, and thus rendered it impossible for anything else to grow in it." He "would recommend, also, that the separation fences of front gardens, of whatever material they may be constructed, should not exceed the height of 6 ft., in order that as free a circulation of air as possible may take place; for a free circulation of air is as necessary to the health of plants as light; and it is more especially necessary in situations where the air is not very pure." With regard to the choice of shrubs, Mr. Rutger advises "the introduction of as great a variety of evergreens as possible, that will thrive; but always interspersing among these some spring flowering deciduous shrubs." Where the separating fences are of brick, they may be planted with ivy, intermixed with the Virginian creeper, the Ayrshire rose, and, above all, with some of the climbing China roses. A very cheap and durable ivy hedge might be formed by building the wall of 4 in. brickwork, in a zigzag line, for strength and stability; the length of each of the angular sides being two bricks, and the bricks being laid in the pigeon-hole manner, to save materials, and also to allow the ivy to pass through from one side of the wall to the other. Only half the number of plants of ivy would be required for this mode of planting; and, if the plants were placed 3 ft. apart on each side, those on one side being opposite the spaces on the other, the whole wall would soon be covered. On the top of such a wall, at regular distances, the ivy might be trained on wire frames, so as to form balls, vases, or other artificial figures; or one stem of ivy might be trained up an upright rod, to the height of 6 ft. above the fence, and there allowed to form a head; which it would soon do of a large size, and produce a profusion of flowers and fruit every autumn, the latter of which would remain as ornaments throughout the winter.

Decorations of the architectural, sculptural, antiquarian, or geological kinds may be introduced in these gardens at the places represented in the designs by the small circles. There may be vases for containing flowers, either of the tazza kind, as in *fig.* 91., or of the upright or goblet form, as in *fig.* 92. In some cases, there may be statues on pedestals, or busts on therms; a sun-dial, or a shell; some curious mineral or geological specimen, as a portion of the Giant's Causeway, &c.; or even a small fountain. Pieces of antiquity are also admissible; such as a stone from some celebrated building now taken down; as, for example, from London Wall or London Bridge, or from the cellar in Bishop Atterbury's house at Hammersmith, where the

91

Protestants in 1556 (many years previous to the bishop's occupation of it) were confined. In the flower-garden at Cashiobury, there are china vases, specimens of various Derbyshire spar, large masses of plumpudding stone, groups of large shells, corals, corallines, madrepores, tufa, lava, petrifactions, ammonites, and different sorts of scoria. At one place there is a pedestal, surmounted by two large granite balls, weighing 700 cwt. each; and a plate on the pedestal informs us that they were shot from the Castle of Abydos in the Dardanelles, and fell on a ship of war which was under the command of a brother of the Earl of Essex, in the squadron of Admiral Duckworth, and killed or wounded fifteen men. Along the New Road, London, some of the front gardens are very richly and judiciously decorated with statuary; and the front garden of Mr. Sowerby's house, Mead Place, Westminster Road, in which is a rich museum of antiquities and geology, is most appropriately distinguished by a stone cross, and a column of basalt from the Giant's Causeway. Where no other artistical or curious mineral, or sculptural object, is introduced, an evergreen shrub, such as a privet, a juniper, or a yew, may be cut into the shape of some artistical object; as at Hendon, where, in the front garden of one of the houses in the village, there are two Louis XIV. vases in box. In short, there is scarcely any limit to the kind of objects that may be introduced into a front garden; provided three conditions are kept in view in introducing them: first, that they never interfere with the unity and propriety of the expansion, or effect, of the scene to which they belong; secondly, that they be not such as will be much injured by the action of the weather; and, thirdly, that due preparation be made for them by some description of architectural or gardenesque basement, plinth, or pedestal. The only objects in which plants are to be grown, that we know of, that can be set down in a garden, without an architectural basement or other preparation, are the hollow roots of trees; but, even in the case of these, we would have a place indicated for them by a raised or a depressed panel of turf. The only description of architectural or sculptural objects that can be set down in a garden, without any architectural base or plinth, are those hollow Chinese jars or vases

92

which are used as seats; and even these never please so well when on grass, as when on gravel or pavement. With respect to mineral or geological specimens, or shells, madrepores, &c., we would, on no account whatever, place them on the naked turf, but always on a basement (even if it consist of only one course of bricks half sunk), which should be seen along with the shell or other ornament, and to which the latter ought to appear to belong. To place open vases on grass, whether intended to remain open, or to be filled with plants, without either plinth or pedestal, however frequently this may be done in gardens of otherwise great merit, is, in our opinion, altogether at variance with the laws of art; which require that extremes should never meet, and that there should be an obvious and sufficient reason for every thing. In all architectural objects whatever, and especially such as a vase or any other open vessel, there should be an obvious security for its erect and firm standing, and for the brim remaining in a truly horizontal position; and this can only be effected by placing it on a foundation of masonry. When such vases are placed on grass, they are generally seen leaning on one side; and, instead of conveying the ideas of appropriate and elegant decoration, they excite those of affectation, and of superfluous and tawdry finery.

Placed on a level surface of firmly rolled gravel, they are not so offensive without an architectural basement, as when placed on turf or dug ground; but, to be unexceptionable in this respect, all ponderous objects, of the stone, earthenware, or metallic kind, ought to be on an obviously secure foundation. What can be more offensive to the eye than the manner in which tombstones in burial-grounds (even in Père la Chaise, and the Kensall Green Cemetery) are seen leaning in different directions?

Fig. 93. shows two designs for front gardens, each containing about seven perches of ground. The walks are supposed to be paved, and the space on each side, not covered by the beds, to be turf. In the left design are shown six oval or circular beds; two of which, next the house, may be devoted to bulbs to make a show in spring, to be succeeded by mignonette for fragrance during summer. The two circular beds facing each other, in the centre, may be devoted to flowers which bloom during June, July, and August; and the two oval beds next the entrance may be planted with autumn-blooming flowers, such as the dahlia. The borders next the separation fences are supposed to be planted with low evergreen and deciduous flowering shrubs, and bulbs, but with no fibrous-rooted biennials or annuals.

In the right-hand design, the centre bed should be planted with evergreen shrubs; or with one handsome evergreen, such as an arbutus, or a rhododendron, pruned into a regular conical form, so as to look gardenesque, and not straggling. The eight beds in this garden may be planted with perennials, so as to

93

have each bed a mass of flowers during the whole of the floral months, and equally assorted as to colours, as explained in p. 217.

A Design by Mr. Rutger. Fig. 94. shows two designs for front gardens to street houses, also by Mr. Rutger, and containing about seven perches each. The walks in both gardens are paved, and in both the groundwork for the beds is turf. In the left-hand design there are five beds, which may be thus planted: — The one next the house with herbaceous plants of low growth, an equal number for every floral month; the two adjoining beds with different kinds of perpetual-flowering China roses; the next bed for dahlias or hollyhocks; and the bed next the entrance for rhododendrons and azaleas. The circles shown in the four angles may contain broad tazza vases, elevated on pedestals, and filled with mignonette; a plant which never ought to be wanting in any garden, and which gives out its fragrance much better when exposed by elevation in a vase to the action of a gentle breeze, than when recumbent on rich soil, in a close sheltered situation. Wherever it is wished to enjoy the fragrance of mignonette, and the situation is not open and airy, it ought to be sown or planted in pots, boxes, or vases, and elevated on pedestals, on the tops of walls, on brackets projecting from walls, or on balconies projected from windows

94

In the right-hand design, there are eight beds on the lawn, and a narrow border round the boundary walls. These walls we would plant with ivy, interspersed with some deciduous creepers; and the border and the beds we would plant, or rather sow, with flowers, without any shrubs whatever, except standard roses. On each bed there might be two of these, and in the borders one at every 10 ft. The beds and the borders we would sow with those kinds of North American annuals which have beautiful flowers; will stand the open air during winter; and which will ripen seeds in this country, and sow themselves. These annuals have the great advantage over all others, that, without any care or protection whatever, they come into flower very early in spring, and, if left to themselves, the seeds will ripen, drop, and spring up again; so that a bed, once sown with one of these kinds of annuals, will continue covered with it for several years, without any other care than that of occasionally pulling up the dead plants. The sorts for the eight beds might be the following, one sort being confined to a bed: —

Leptosìphon densiflòrus, lilac.
Œnothèra álbicans, white.
Clárk*ia* pulchélla, purple.
Gíl*ia* *achilleæfòlia*, blue.
Màdia élegans, yellow.
Collòmia lineàris, red.
Chrysè*is* (Eschschóltz*ia*) cròcea, orange.
Collíns*ia* bícolor, pinkish.

On the border under the surrounding walls, we would sow the following sorts of North American annuals, equally mixed together, with the addition of a little mignonette, and leave the whole to reproduce themselves for several years, doing nothing more in the way of culture than pulling up the weeds, and the dead plants as they drop their seeds, watering in dry weather, and occasionally scratching the soil to cover the seeds; or to give vigour to the plants when they appear to want it, sprinkling over it, during winter, a little thoroughly rotten dung or soot. The sorts are, —

Œnothèra Lindley*àna*, purple.
  álbicans, white.
  decúmbens, purple.
  vimínea, purple.
Clárk*ia* pulchélla, purple.
  élegans, pinkish purple.
Gíl*la* trícolor, whitish.
  *a*chille*æfòlia*, blue.
Màdia élegans, yellow.
Collòmia grandiflòra, pink.
  lineàris, red.
Nemóphila insígnis, blue.
Collíns*ia* grandiflòra, blue and white.
  vérna, blue and purple.
Collíns*ia* bícolor, pinkish.
Leptosìphon densiflòrus, lilac.
  *a*ndrosàceus, lilac.
Chrysèi*s* (Eschschóltz*ia*) cròcea, orange.
  califórnica, yellow.
Calliópsis Atkinsoni*àna*, yellow and brown.
  bícolor, yellow.
  atrosanguínea, rich dark orange and brown.
*R*esèda odoràta (the mignonette), greenish white.

The beauty of a front garden so planted and managed would not be of the gardenesque, or neat and distinct, kind; but it would exhibit neatness combined with the picturesque. The neatness would be displayed in the smooth-kept turf; the delicate and correct edgings to the walk, and to the beds and borders; and the regularity of the position, and the symmetry of the heads, of the standard roses. The picturesque would be exemplified in the annual flowers, which would exhibit an assemblage of many brilliant colours, with a perpetual fragrance arising from the mignonette. If these beds were altogether neglected, they would occasionally get bare in some places, and become too much crowded in others; but, if a little attention were paid to thin out the plants where they were too thick; to transplant some into places where they were too thin (previously stirring the soil); to cut down and water where it was wanted to produce a second crop of blossoms; to pull up dead, decaying, or diseased plants; to stir the soil, and occasionally, in winter, to top-dress it, or water it with liquid manure; such a garden would make a splendid show from March till December, at scarcely any cost. The mixture of so many fine plants in the borders, and the variegated masses of colour that they would produce there, would be quite original. The eight beds would each produce a brilliant mass of one colour; while the roses, if they were of the perpetual-flowering kinds, would make a fine display throughout the whole summer. Even in a botanical point of view, this garden would be interesting, since it would display 35 different kinds of standard roses, 16 kinds of creepers mixed with the ivy, and 12 genera and 24 species of annuals.

It is needless to observe, that this mode of planting may be adapted to any design, but not so well to those where dwarf shrubs are to be intermixed with herbaceous flowering plants; because, in that case, the latter are always apt to appear to be crowding and injuring the shrubs; and, in short, as weeds.

We are indebted for the idea of this mode of planting and managing a flower-garden to Mr. Caie, gardener to His Grace the Duke of Bedford at Camden Hill; the elegantly designed and admirably managed garden of which villa, His Grace has kindly permitted us to give a plan and views of in this work.

Designs for Front Gardens common to several Houses connected together in a Street or Row. The leading principle for laying out gardens of this kind is the same as that for laying out the gardens to individual houses; viz. that the character of the whole should determine the character of the parts. Hence the rule, that, when the space to be laid out is a square or a parallelogram with straight sides, the walk which surrounds it should be straight; and when one or more of the sides are curved, then the walk or walks in that part of the garden should coincide with the curvature. The conspicuousness of the regular boundary, and the formal geometrical character of the adjoining houses, will always supply an obvious and sufficient reason for adopting this mode of laying out a narrow strip of public garden. There may, however, be exceptions, as there are to all rules: such, for example, as the existence of some objects within the garden that it is desirable not to remove, and which the walk must deviate from the straight line or regular curve to pass round; or of others, either within or without its boundary, which it is desirable to make extraordinary exertions to conceal or to display, and which might require naked places or plantations that would interfere with regularity. Without these or some other anomalous circumstances, however, it does not appear to us in good or in reasonable taste, to introduce many winding walks, or walks irregular in their direction, in such gardens; though we acknowledge that this is the fashionable mode, and one which, on that account, and also, perhaps, as presenting a more violent contrast with the lines of the buildings, is at present more generally pleasing. There is another argument against the regular symmetrical mode recommended by us of laying out these gardens; and that is, that the public have not yet got rid of their violent antipathy to straight lines in ornamental gardens, as belonging to the geometric, or, as it is called, the formal, style; a style still as much disliked by many persons, as Gothic architecture was half a century ago. As soon, however, as reason shall have its due influence in matters of taste, and the geometrical style and the irregular style shall be estimated according to their respective merits, and their adaptation for particular purposes, it will be found that the geometrical style is alone adapted for all town gardens whatever, whether public or private. We shall now submit one example in what we consider to be the true style, and two or three by Mr. Rutger, in the style which is at present fashionable in London.

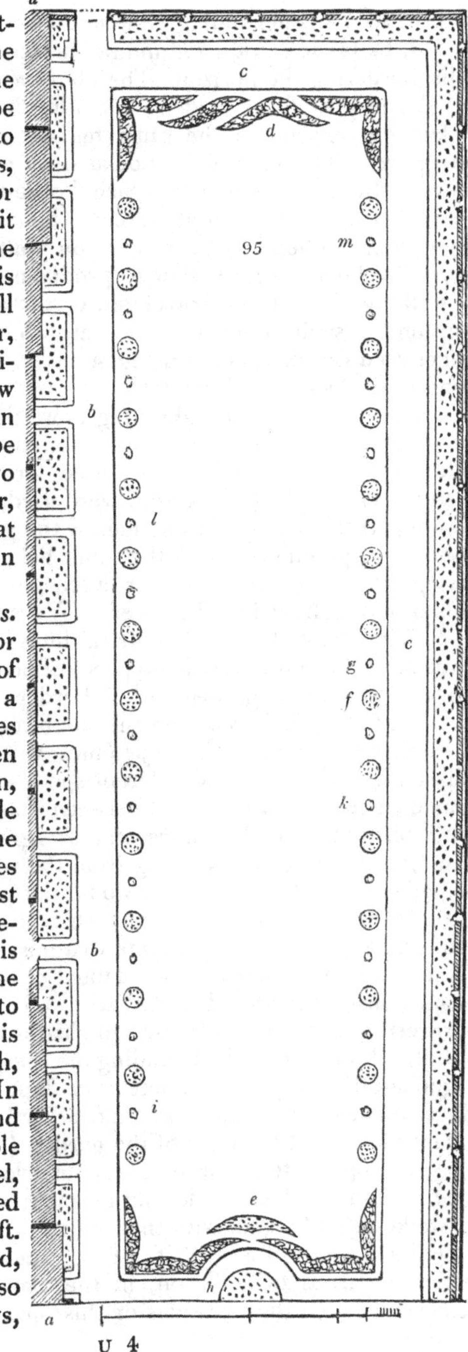

Hyde Park Gardens. Fig. 95. is a plan for laying out a piece of ground in front of a row of first-rate houses now building between Bayswater and London, and to be called Hyde Park Gardens. The direction of the houses and of the ground is east and west; and, consequently, the garden is finely exposed to the south. The ground to be laid out as garden is above 800 ft. in length, and 150 ft. in width. In front of the houses, and extending their whole length on one level, there is a terrace raised over the offices, 8 ft. high, and 50 ft. broad, from which, and also from all the windows, there will be a commanding view over the garden in the foreground, to Hyde Park, as a middle distance; and to the Surrey Hills, bordering the horizon. The plan (*fig.* 95.) shows only one half the length of the garden; the other half being supposed to be laid out exactly in the same manner as that exhibited in the figure. The terrace is indicated at *a a*; and *b* shows the descents from the terrace of each house to the garden. The portion of terrace in front of each house is separated from the portions adjoining by separation walls. The garden we propose to lay out with a view to producing long straight spaces for walking on, both of gravel and of turf; and to contrive the planting in such a manner as that, while all the walks are shaded to a certain extent during summer, the sun will shine on them during winter. The style of planting we propose to be strictly gardenesque, not admitting more than two plants of any species or variety of tree or shrub throughout the entire garden, with the exception of the shrubs used for the boundary hedge. In so far as two plants of a kind were used, we would not place these together, but one in one half of the garden, and the other in the corresponding part of the other half, for the sake of preserving symmetry; or, in other words, in order that one half the garden might reflect the other. We are aware that this is rendering it liable to Pope's satirical lines; but satire, in this as in various other instances, is no proof of error. We propose to surround the whole garden with a border (*c c*) 12 ft. wide; the sides next the public road, on the south, east, and west, having an evergreen hedge of *Aúcuba* japónica, with standard variegated hollies alternating with the different species of Cratæ`gus, at regular distances; and the space between that hedge and the walk being planted with deciduous and evergreen flowering shrubs, forming a sloping bank, rising from the walk to the summit of the hedge. In this border we would introduce bulbs, but no fibrous-rooted herbaceous plants, and no roses. The border next the terrace we would devote to a miscellaneous assemblage of the more choice flowers, all planted at regular distances, and conspicuously named. The terrace wall we would plant with evergreen and deciduous trees, and shrubs of the more rare and beautiful flowering kinds, including the evergreen and deciduous magnolias, the fragrant Chinese and Japan honeysuckles, jasmines, pomegranates, myrtles, and other beautiful and fragrant or rare plants. At the ends of the garden (*d*), and in the middle (*e*), we propose to place some beds of evergreen low trees and shrubs, to shelter benches for sitting on, and conceal them from the walks behind; and, with the side rows of trees and beds, to form the interior of each half of the garden into a compartment for large parties to walk on, in the manner of a public promenade; and on the two sides of this space next the walks, we would place alternately a circular bed of flowers (*f*) and a low tree (*g*). The circular beds are 12 ft. in diameter; and each is supposed to be planted with only one kind of flower, so as to produce a brilliant mass of colour. The trees are proposed to be alternately of deciduous and evergreen kinds, and of such sorts as do not exceed 30 ft. in height. The rays of the sun will penetrate through the heads of the deciduous trees in winter, so as to dry the walks; while in summer, the trees, being in leaf, will, in conjunction with the evergreen trees alternating with them, throw the greater part of the walks into shade. It will be observed, that the trees in the front hedge alternate with the spaces between the trees on the grass, in consequence of which, the view of the public road will be completely excluded from the terrace, and the trees in the garden joined in effect with the trees in Hyde Park. In the plan, the trees make a very insignificant appearance, from the circumstance of the situation of their stems only being indicated: but, if the soil be properly prepared, their heads, when of ten years' growth, will project as far as the line of the walk; and so far inwards, on the compartment of turf, as to reduce the width of the space for the parade to 25 ft. or 30 ft. In the centre of each of the circles of flowers we would plant a tall standard rose; and in the border next the terrace there might be a row of standard roses, with here and there a cone of climbing roses. The central circle (*h*) may either be an open temple for taking shelter in during rain, or a basin of water and a fountain. On the lawn side of the masses of plantation at *d* and *e*, there should be benches on which parties might sit and look along the promenade. Under most of the single trees on the lawn, there ought to be one or two light bamboo chairs or China-ware seats, to be moved about at pleasure, so that the parties wishing to sit on them might either look towards the broad gravel walk, or the central promenade. We have shown no garden ornaments, such as vases, &c., considering that these, with the exception of the Chinese porcelain seats, which will serve as ornaments as well as useful objects, ought to be confined to the parapet wall of the terrace, and to the marginal wall of the basin of water.

We have said nothing respecting draining the ground, preparing the soil, &c., having already shown the importance of these and other preliminary preparations; but it may be necessary to suggest, that, in the keeping of the garden, the expense will be considerably lessened by having pipes of water conveyed under ground, with cocks and branch pipes rising to the surface at moderate distances, so as to afford a ready supply for the purposes of watering both the plants and the turf. The branch pipes which come up to the surface should be concealed by small caps, fitted into stones, on a level with the surface of the grass; and these pipes should have stopcocks close under the caps, to which delivery pipes may be screwed on by the gardener, when he wants a supply of water for filling his watering-pot, or barrow-engine, and taken off when he has done watering; or, if the water be supplied direct from the main source, and that is sufficiently elevated to occasion a considerable pressure, the watering may be performed at once by means of flexible hose screwed on to the branch service-pipes. If these modes should be considered too expensive, pumps may either be placed at certain distances, so as to be partially concealed in the boundary border of shrubs or in the terrace wall; or on the lawn, in the place of two or three of the trees, and covered with some description of weeping tree. Thus, four pumps would be sufficient for each half of the garden, and they might be placed at *i*, *k*, *l*, and *m*, and concealed by a weeping lentiscus-leaved ash, a weeping *Sophòra* péndula, a weeping Queen Mary's thorn, a weeping Turkey oak, a parasol acacia, and a weeping purple beech; the branches of the trees being trained to framework, so as to form a regular canopy. Or, in lieu of any of these trees, artificial weeping trees might be formed of giant ivy and variegated common ivy, trained over hemispherical framework, which would have an admirable effect, especially in winter.

Expense and Mode of Management. If this garden were planted with the finer kinds of trees and shrubs, and these, as they ought to be, of considerable size, the expense, for plants alone, would amount to nearly 500*l*. This, by some, may be considered a large sum; but, surely, no description of garden in London so well merits being planted in the very best manner as that in question, situated as it is on a good soil and subsoil, open to the south, and sheltered by high houses to the north, east, and west. It has also the great advantage of having Hyde Park in front, and Kensington Gardens on the south-west; so that, as Hyde Park and Kensington Gardens are never likely to be built on, Hyde Park Gardens will be less likely to be injured by suburban smoke than any other London garden whatever. If this piece of ground were planted as we propose, with no triplicates, except in the case of the hedge plants, and of some of the finer trees and shrubs for the terrace wall, such as Magnòl*ia* grandiflòra, M. conspícua, M. Soulange*àna*, and M. purpùrea, camellias, myrtles, Eriobótrya, Photínia, Stransvæ`*sia*, Chinese and Japan honeysuckles, &c., of each of which there might be four plants, it would contain all the *élite* of the hardy trees and shrubs that live in the climate of London in the open garden; as well as of the somewhat delicate or half-hardy trees and shrubs on the terrace wall. These will amount to about 500 sorts; and, if they be named, and the herbaceous plants be named also, this would add considerably to the interest of the garden, especially for young people. With respect to the management, it ought either to be under the care of a commercial gardener, who has a nursery, with hot-houses or pits; or under the care of a head gardener, who ought to have a small piece of ground as a reserve garden, somewhere in the neighbourhood, with pits. The reason why the person having charge of this garden should have pits and a reserve ground is, that the flower-beds will require to be renewed two or three times a year, so as to keep up a continual succession of flowers; and, as many of these flowers will be pelargoniums, fuchsias, calceolarias, and other green-house plants, the necessity of pits, for preserving these during winter, and bringing them forward in spring, is obvious. One of the objects of interest connected with every garden is, seeing the operations performed in it; and, during hot weather in summer, when it is found useful to water the heads of the trees and shrubs to free them from dust and insects, and sometimes also the grass, to preserve its verdure, the playing of the engines, and the refreshing influence of the artificial showers, will be gratifying and amusing to spectators in the houses and on the terrace. Every one who has seen the watering of the garden of the Palais Royal in Paris will readily understand what we mean. If, instead of a temple, a basin and fountain were formed in the centre, it need only play during so many hours in the heat of the day, and till sunset; but, unless it be liberally supplied with water, so as to make a powerful jet, it would be much better dispensed with.

A fountain which does not throw the water up to the height of 50 ft., in a column 3 in. or 4 in. in diameter, would be unworthy of this situation. A drooping fountain of 10 ft. or 12 ft. in height, would be the most economical kind; but this kind of fountain, depending for its beauty mainly on its architecture and sculpture, is much less grand than a single jet. The fountain consisting of a single jet may be worked by steam, the engine being placed in the mews behind the houses, or in some concealed situation, which may be several hundred yards distant; and one argument in favour of such a fountain is, that the same water being forced up, and falling down alternately, scarcely any fresh supply is required; while a reason in favour of a drooping architectural fountain is, that, as its beauty depends more on its architecture than its display of water, it need not be played, except when water is wanted for the supply of the wells for the different pumps; each of these being understood to have a communication with the basin.

Management. The principal objects to be attended to in the management of this garden, besides keeping up a succession of flowers as already mentioned, will be training the trees against the terrace wall, and adjusting their branches, so that the strong-growing kinds do not injure the weaker; and that the branches of kinds which flower before the leaves come out, such as Magnòl*ia* conspícua, Chimonánthus fràgrans, &c., be more or less intermixed with the branches of evergreen trees, in order to display the blossoms, during winter and in early spring, with greater effect. The trees on the lawn should be trained to form well-balanced heads; and the shrubs in the boundary border should be kept symmetrical, and pruned or thinned out, so as to allow them to be covered with leaves and blossoms on every side; and, though close together, should never be allowed to touch each other. To insure this result, and to preserve a proper gradation from the front to the back of the border, as well as to make certain of a vigorous growth and an abundant show of blossoms, the greater part of the shrubs, particularly the rhododendrons, azaleas, and other American kinds, ought to be taken up and replaced in fresh soil every six or seven years. The standard roses ought to be kept perfectly symmetrical; and the climbing roses, intended to cover cones, pyramids, poles, or parasols, should be trained equally over them, and either often pruned, or scarcely pruned at all, according to the nature of the different kinds. The boundary hedge of *Aúcuba* japónica (or, if this shrub should be objected to, of *Laú*rus nóbilis or myrtle-leaved box) should not be clipped, on account of injuring the leaves; but carefully pruned with the knife, so as to keep it in the proper shape to form a compact hedge. The heads of the trees in this hedgerow ought to have none of their branches nearer the ground than 10 ft. or 12 ft., as well to keep them above the heads of passengers along the public path, as to prevent them from injuring the shrubs in the border by too much shade. If, owing to some of the trees being only to be purchased of a small size, and to others not being properly trained in the nursery, they should have short or crooked heads, it may become necessary to bestow some pains on their training. In general, tying the main stem to a single iron rod or Kyanised wooden stake will suffice; but, in difficult cases, the object is to be effected by four perpendicular wooden posts or iron rods being placed near the trunk, and two horizontal arms of the tree being nailed or tied to them at the height of 10 ft. or 12 ft.; while, placed on these, there may be three rings formed by iron rods, of, say, 3 ft., 6 ft., and 9 ft. in diameter. The stem may be trained to one of the four upright stakes, and the lower branches proceeding from it to the rings. A very handsome symmetrical-headed tree will thus be formed in the course of six or eight years; after which the posts and rings may be removed. If this be thought too much trouble, the trees in the hedge may be all trained into fastigiate shapes, like that of the common cypress or the Cratæ gus stricta, when they will neither be within the reach of the foot passengers, nor injure the shrubbery inside by their shade.

The grass of the lawn ought to be kept short and smooth, by frequent mowing and rolling; and of a dark green colour, by constant watering during dry weather; and the gravel should be often rolled, so as to be as firm and smooth as a pavement, and of a rich yellow; and the grass, gravel, and dug ground, ought all to be kept quite free from worm-casts, by watering them with pellucid lime-water. No insects ought to be allowed to appear on any of the plants; all remains of flowers should be cut off the moment they begin to decay; and not a single dead leaf or twig ought to be seen throughout the whole garden. To effect all these objects, a first-rate gardener ought to be employed; and such a gardener will reasonably require a salary greater than that given to a gardener in a gentleman's establishment in the country; because the situation in the public garden is out of the common routine, and, should he lose it, he would find a difficulty in getting another in private service. Suppose, however, that the total expense of this garden, including the salary of the head gardener and his assistants, amounted to 300*l.* a year; as this expense would be divided among thirty families, each, for the moderate annual sum of 10*l.*, will enjoy one of the finest gardens of the kind that can be created in the vicinity of London.

Remarks. We do not give the plan *fig.* 95. p. 295., as being better in its details than many which we or others could devise: we do not, for example, say the borders and walks should be exactly of the breadth which we have exhibited, or that the circle is the only proper form for the flower-beds; that there may not be a greater space covered with flowers, or a greater number of trees; or that the trees may not be planted in groups or rows, instead of being planted singly. On the contrary, we have no objection to make to the flower-beds being pentagons, squares, or parallelograms, or, in short, any simple regular figure, provided that figure be repeated at regular distances. We have no objection to the trees being placed in threes instead of singly, provided each of the three is placed in the angle of an equilateral triangle, and these triangles are regularly repeated. All that we contend for as to principle is, that regularity and symmetry ought to pervade every part of the design; and all that we insist on in regard to utility is, that in all public gardens the walks ought to be broad, and either straight or in very large curves, in order that several people may walk on them abreast; and that no individual or party may ever meet another suddenly, which they are likely to do in walks making quick turns. Another principle of utility is, that there ought to be long and broad spaces of lawn for walking upon; because some prefer lawn to gravel, and

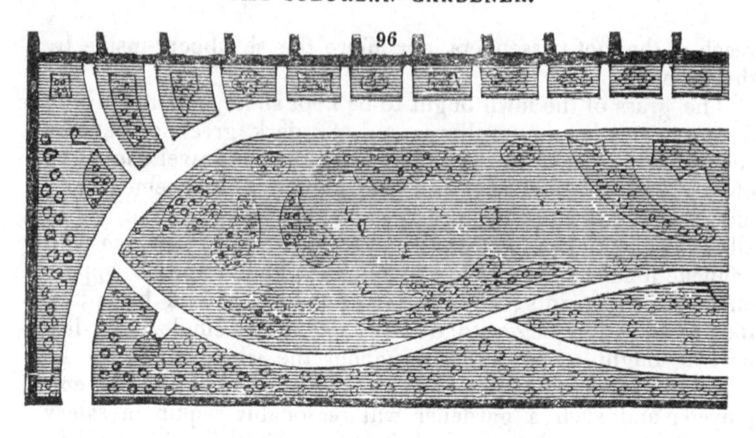

others can walk on the former, but not on the latter; and that these broad spaces of lawn ought to be continuous, whether in ample curves or straight lines, and never short, narrow, devious, or tortuous. Broad walks are always understood to be for the purpose of admitting two or more persons to walk together; and as, in walking, they will necessarily be engaged in conversation, sudden turns in the walk must have a tendency to divert the attention from the subject of discourse. Narrow winding walks are suited for persons who like to walk alone, for solitary thinkers, and for those who shun public view; they are therefore well adapted for private gardens, or for the more secluded and private parts of public gardens, but, in the latter, by no means of general application. A third principle of utility which we contend for is, that, in all gardens in towns, as these are necessarily in a confined atmosphere from the surrounding buildings, the trees and shrubs should be fewer in proportion to the extent of surface, than in open airy gardens in the country; and that trees or shrubs with clear stems to the height of 6 ft. or 8 ft. ought to be preferred to immense bushes covering a great breadth of surface, as allowing a more free circulation of air to carry off damps. An open airy public garden is also better adapted for a place of exercise for children; because the surface dries sooner after rain, and is of greater extent for them to play on. It is evident that these principles admit of being followed out on a small scale as well as on a large one, and with the commonest kinds of trees and shrubs, as well as with the most expensive.

The garden referred to is about to be laid out with winding walks of different breadths, and with clumps and scattered trees, in the manner of modern pleasure-grounds. The plan, of which Mr. Forrest, the author, has kindly permitted us to take a copy, we shall probably give as supplementary to this work; because, though it is for a public garden, yet, as the chief aim which we

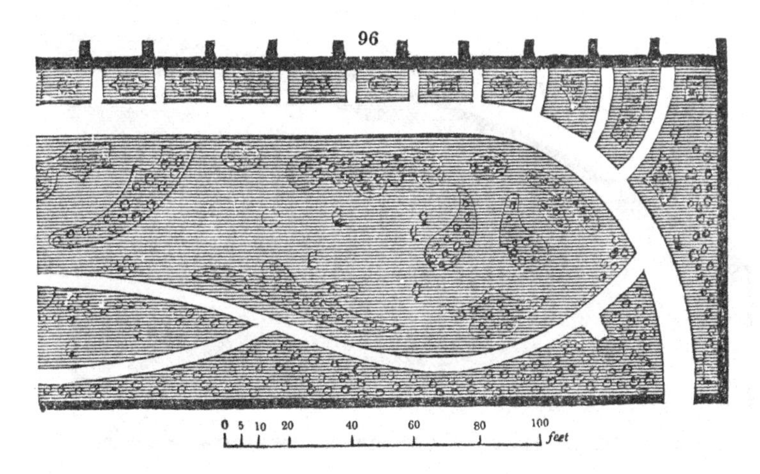

propose to ourselves is that of popularising sound principles of taste, discussions on a public garden may be rendered as instructive to the readers of the *Suburban Gardener* as those on a private one.

We shall now give two or three designs for gardens common to several houses, by Mr. Rutger.

A Front Garden common to 21 *Houses*. *Fig*. 96. is a design for a garden common to 21 houses, forming a connected row or street. The houses are approached by a carriage drive, with a small entrance-lodge at each end. The carriage road is 10 ft. broad; and from this there are paved pathways, 4 ft. broad, to the front door of each house. "The front," Mr. Rutger observes, "is laid out in the style of a modern flower-garden, with shaded walks towards the boundary fence. An alcove stands in the centre; and there is a seat at each angle where the two walks join. The small circles at the two extreme corners may either represent rotundas for sitting in, or situations for statues, vases, and other embellishments. The small walks are recommended to be laid down with flagstones." Nothing can be more judiciously contrived than the carriage road in this design, and the convenient way in which the occupier of every house may be set down before his own door; but, in point of utility, it may be questioned whether a gravel walk, perfectly straight from the one rotunda to the other, for children and their maids to walk on, would not have been preferable to the winding walk shown. In this case, the space between the walk and the boundary fence would remain as it is, viz. planted with trees and shrubs to form a screen from the street. The pavilions might be turned into alcoves facing each other. The walk which joins the terrace to the carriage approach should be narrower than the terrace; and, instead of communicating

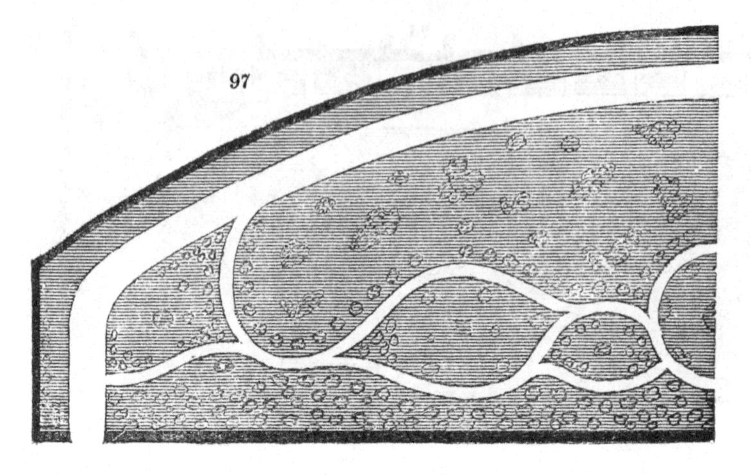

with it at the ends, which might endanger children leaving the garden and entering on the carriageroad when a carriage was coming in or going out, it might join it directly in the centre, where the point of junction would be attended with least danger. The flower-beds on the space between the terrace and the road would, in this case, require to be differently arranged, especially towards the middle, which might easily be done by any one who understands the general principle of connecting the beds which we have laid down, so as to preserve the appearance of design and of breadth of effect; the latter being attained by broad spaces of turf without beds. While we state this, it is but due to Mr. Rutger, to add that his design would have a more varied appearance, and be richer to look on, than ours.

Fig. 97., also by Mr. Rutger, represents the plan of a garden common to a crescent of houses, in which the separate entrances to each house are not shown; because the space, 8 ft. broad, between the carriage-drive and the houses is one continued pavement, which will form an excellent promenade for children when the gravel and the turf are too wet for their feet. The space between the common approach road and the front fence is laid out as a lawn and shrubbery, with the groups of shrubs, and winding rows of single shrubs and low trees, interspersed with beds of flowers. The centre of the circular space is intended for a fountain, or any other appropriate embellishment; and at the back of this circle is an alcove, with a bench for sitting on. The other small circles and squares are intended to represent the foundations for statues, and for vases to be filled with flowers; and the whole may be interspersed with movable seats. The walks may be of gravel, as more suitable for such as are of a winding direction than pavement; and they should be kept smooth and firm by being frequently rolled with a cast-iron roller, which may

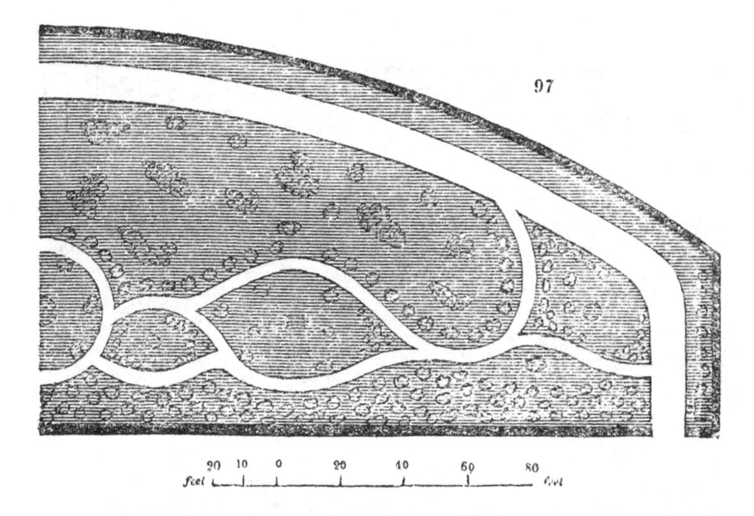

be concealed when not in use, among the screen of shrubs in front. At each of the entrance-gates to the approach road there is a lodge, the occupiers of which may be employed to look after the garden.

Fig. 98. is a design by Mr. Rutger for a front garden to a group of six houses. The carriage entrance is at each end, from the adjoining streets; and there is a broad terrace-walk between the approach road and the boundary fence. The groups on the turf at the two extreme ends are of evergreen and deciduous shrubs. The largest group in the centre consists of dwarf and standard roses; and the remaining five groups are of red, white, blue, yellow, and variegated flowers; the latter filling the central group. The boundary fence is concealed by a hedge of variegated hollies. The single specimens of shrubs and low trees on the lawn are chiefly evergreens; such as arbutus, aucuba, mahonia, photinia, phillyrea, laurustinus, and alaternus, with some deciduous flowering sorts; such as althæa frutex, Persian lilac, *Cydònia japónica*, &c. In the variegated holly hedge there may be standard thorns alternating with evergreen fastigiate-growing low trees, such as thuja, cupressus, juniperus, &c. Close to the houses there are beds for flowers, which may either be of low-growing perennial kinds, mixed with bulbs, calculated to make a display as much as practicable every month during the year; or it may be devoted to bulbs and annuals only. The advantage of this kind of garden common to several houses, Mr. Rutger observes, consists in its displaying a greater variety of trees, shrubs, and flowers, than could be done in any separate front garden; for each of the six families will enjoy the appearance of these, as much as if the entire garden were their own. It

98

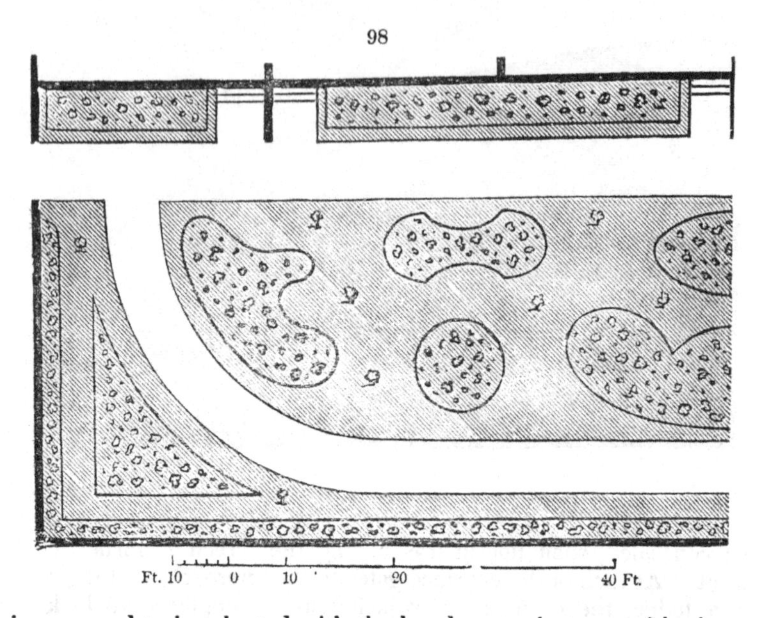

is supposed to be planted with the handsomer, but not with the most costly, kinds of trees and shrubs, and would, therefore, not be expensive at the outset, and would be easily kept in order afterwards. The first cost of preparing the ground, forming the walks, and planting, may amount to 60*l.*; and the annual cost to keep it in the very highest order, to 18*l.*; the saving, therefore, to each individual family is obvious. The only drawback to this garden is, that it cannot be considered private; but to each house it may be supposed that there is a back garden, to supply this desideratum, which, in our opinion, is essential to the comfort of every residence.

Remarks. From gardens common to several street houses, the natural transition is to those of public squares; but, as the same principles which we have laid down (p. 294.) for arranging public gardens generally are applicable to gardens for squares, as well as to those for crescents or rows of houses, we think it unnecessary to dwell longer on the subject.

Renovating Fourth-rate Suburban Gardens. Whoever does not build or take possession of a new house, so as to have the garden to lay out himself, will, on changing his residence, probably find that the garden of his new abode requires renovating. To ascertain how far this is necessary, he has only to test every part of his garden by the principles and rules for laying out and planting which we have already laid down; and we shall therefore confine our remarks here to directing his attention to those points in which an old garden will generally be found defective.

98

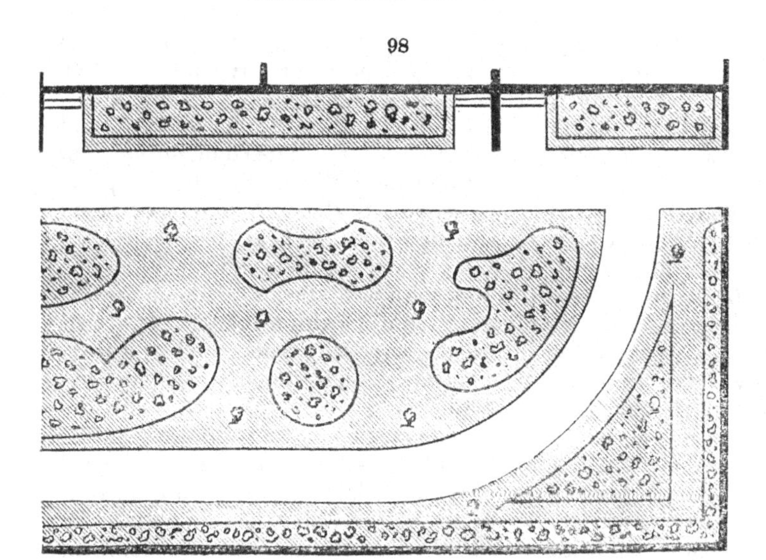

The underground drainage, in a garden which has been in cultivation for twenty years, will frequently be found more or less choked up; the indications of which are dampness and moss on certain parts of the walks, where the surface is lowest; a sodden, black, soft appearance in the soil, not easy to describe; and mossiness and canker in the fruit trees. When the drainage is defective, there is no remedy but digging out the drains, and forming others in their stead, in the same or in preferable directions, and with fresh materials. The surface drainage, also, will often be found defective, from the ground, when the garden was first formed, having settled unequally; and this evil having been aggravated during a series of years. In cases of this kind, the hollow places formed by the sinking of the soil will hold water in pools after every heavy shower; where the walk has sunk, the gravel will have become blackened or muddy on the surface, and the box or other plant edging will look pale and sickly. There is no remedy for this but releveling the surface; which implies taking up the gravel of the walk and its edgings, as well as such of the trees, shrubs, and plants as may grow in the hollow spaces; and, after raising the whole with fresh earth and gravel, to replace the plants. The sewerage, or drain, from the house to the public drain or sewer, and also the means of conveying water to the house, whether by pipes from a public company, or a well or tank, should be examined. The well will most probably require cleaning, and, possibly, its sides may want to be taken down and rebuilt. The same remark will apply to tanks.

The boundary walls, or fences, of the garden, of whatever kind these may be, can hardly fail to require some repairs. If they are brick walls, on which trees have been trained for many years, the bricks will be full of nail-holes, and their joints will be found open from the falling out of the mortar. The common mode of remedying this evil is to repoint the wall; that is, to fill up the interstices between the bricks with fresh mortar, replacing any greatly damaged bricks with new ones: but, as this is attended with considerable expense, a cheaper and more effectual mode is to brush over the entire wall with a mixture of Roman cement and water, as thick as it can be laid on with a brush. The wall having been previously brushed over with water to clear it from any loose materials, the mixture should be instantly applied; and it should be used as soon as made, or it will set in the vessel it was made in. If one coating of this liquid cement be not sufficient to fill up the holes, a second may be given, after the first has been two or three days laid on; or even a third, if necessary to render the surface of the wall tolerably even. An old dilapidated brick wall, in the Horticultural Society's Garden, was treated in this manner upwards of twelve years ago, and has ever since been almost as good as new. The coping of walls should be examined and rendered water-tight; otherwise, the water will be liable to get into the heart of the wall, and rot it. For the repair of copings, cement is admirably adapted; and, even if the coping should require to be entirely renewed, plates of artificial stone, formed of cement, will, in many situations, be found cheaper and better than any other. When a wall is in a bad state, and it is not considered desirable to bestow much expense in repairing it, the cheapest mode that we know of for keeping it standing for a great many years is, to plant ivy at its base, and allow it to run up; in which case the ivy will not only protect the sides of the wall but will form a mantled coping to its top. When the boundary fence is of wood, it should be examined, to ascertain whether the parts are rotten "between wind and water" (that is, at the surface of the soil), which is very frequently the case while the upper part of the fence is fresh and strong. In this case, the posts should either be entirely renewed, or the cheaper mode adopted of inserting fresh posts close to the rotten ones, of such a length as to reach two or three feet above the surface. By nailing these short new posts to the sound part of the old posts, the fence may be kept up during another lease. A full-grown hedge, with gaps in it, is one of the worst fences to repair; because young hedge plants will never thrive among old ones. The only mode, in this case, is to intertwine the adjoining branches, and to insert stakes for retaining these in their position. The hedge will often be found too broad at top; in consequence of which, the light and air are too much excluded from the bottom, which becomes naked; and here the only remedy is to cut in the sides of the hedge, so as to reduce the width at top; when, after two or three years, it will become equally thick from the ground upwards.

The walks, whether of gravel or pavement, will hardly fail requiring to be taken up and replaced or renewed; and, as the underground drains are generally along the sides of the walks, both may be renovated at the same time. The edgings to the walks, if of box, will probably be overgrown; in which case, they will require to be taken up and replaced by young plants from a nursery; unless the occupier be content to take up the old box, and plant it in rows in the interior of the garden, in the manner called by gardeners laying in, leaving only an inch or two of the plant above the surface, in which state it will commonly throw out fresh roots from the side shoots, and, in a year, be fit to plant out along the walks as edgings. Some gardeners plant old box at once as edgings, where it is finally to remain, burying the plants to within 2 in. of the ends of the shoots; but as, from the length of the old overgrown box, it becomes necessary, in this case, to bend the plants in planting them, and, consequently, many of them are broken, and thus are liable to die during the summer, it is better to make certain of plants that will live, by rooting the box first in nursery lines, as recommended above. In almost every case of renovation, of whatever material the edgings to the walks may have been made, whether grass, thrift, strawberries, brick, tile, slate, or stone, they will require to be taken up and replaced, if not renewed.

The turf, if there be any, if it does not require the surface on which it grows to be releveled, may be renewed by digging it down, forming and consolidating the surface by raking and rolling, and afterwards sowing grass seeds; or, if expense be not an object, fresh turf may be procured from an old pasture. In many cases, however, all the renovation that will be required for turf will be, the filling up of the inequalities of the surface with fine soil, rolling the whole firmly, and sowing grass seeds in the bare places after the first shower; then slightly raking it, and again rolling it after it becomes dry.

The general soil of a garden that has been twenty years in cultivation cannot fail to require to be renovated in some way or other. In general, it is impracticable to renew it with fresh soil from the country, even in part, on account of the expense; and, therefore, recourse must be had to more practicable means. The soil may have become exhausted by want of manure, or by being continually cropped with culinary vegetables of the same kinds (that is, those generally grown for small families);

or it may be soddened, and rendered worthless, by want of drainage, light, and air. In the first case, the remedy is easy, viz. that of supplying manure; in the second case, the evil may sometimes be remedied by a dressing of lime; but the more economical, and at the same time certain, mode is, to let the soil rest for two or three years. In many cases, however, it is only the surface soil that is exhausted by a succession of the same kind of crops; and in this case the remedy consists in trenching the ground, so as to bring the subsoil to the surface. If, however, the soil has been regularly trenched, and manured while it was cropped, and is nevertheless in an exhausted state, there is no other remedy than resting the ground, as already mentioned (by which is to be understood, cropping it with a different description of crop; such as laying down the whole of a garden, that has been occupied with culinary vegetables, in grass, &c.); or renewing a portion of the soil, say the entire surface, to the depth of from 6 in. to 9 in. In the suburbs of London, the soil of old fourth-rate gardens is seldom, if ever, exhausted from any other causes, than from want of drainage and ventilation, and from over-cropping; for which the remedies are, redrainage, thinning out the trees, and manure.

The most important consideration, however, with reference to renovating fourth-rate suburban gardens, is the state of the trees and shrubs. In almost all suburban gardens of twenty years' standing, this will be found to be most wretched. The ornamental shrubs will commonly be found to consist only of the coarser-growing kinds, which have remained and become vigorous, after having choked up and destroyed the weaker and more delicate kinds; and the ornamental trees will be found too large, overshading every thing, and rendering the ground damp and unproductive, and the atmosphere unhealthy. The fruit trees will generally be cankered, and producing little or no fruit; and the little which they do produce will have a bad flavour. The fruit shrubs, such as the gooseberry, the currant, and the raspberry, will be found overgrown, and crowded together for want of pruning and thinning. The only effectual remedy for these evils is, to root out the whole of the ligneous plants, and to introduce young healthy plants in their stead. There need be the less regret at doing this, because the kinds of fruit trees and shrubs, and of ornamental trees and shrubs, that were planted in suburban gardens twenty years ago, were very inferior to those which may now be purchased in the nurseries. There is another reason why it will almost always become necessary to root out the fruit trees in an old garden; which is, that their roots will, in most cases, be found too deeply buried in the soil. The main cause of this is, that they have been planted too deep at first; that is, the soil having been trenched perhaps 2 or 3 feet deep, previously to planting, the trees have been planted just as deep as they would have been on solid ground; in consequence of which, and of the watering and treading down the soil to the roots at the time of planting, the collar of the tree is, the very first year, 2 or 3 inches below the surface; and every year afterwards, from the additions made to the soil by manure, and from its swelling up in consequence of the pulverisation it receives by culture, the root of the tree becomes still more and more deeply buried, till, at the end of 20 years, the collar is perhaps 6 or 8 inches deeper in the soil than it ought to be. This burying of the collar is the grand cause of the unfruitfulness of fruit trees in small gardens; and, indeed, it is not too much to say, in all gardens whatever; and this deep burying of the roots is just as hurtful to a gooseberry, a currant, or a raspberry, as it is to an apple, a pear, or a plum. In the case of flowering trees and shrubs, it is equally injurious, by preventing them from flowering. If, therefore, old trees and shrubs of any kind, are to be retained in renovating the garden, they will require to be taken up, and replanted with their collars rather above the surface than under it, so as to allow for the sinking, of the ground, and to cause the tree, even when the ground is thoroughly settled, to have the appearance of growing out of a small mound. If we examine thriving trees in a state of nature, we shall always find that the collar (that is, the point of junction between the stem and the root) rises above the general surface, as shown in *fig.* 99. On the other hand, if we examine trees that have been planted by man in deeply trenched soils, we shall generally find that, though planted at first as shown in *fig.* 100., they will, after a certain number of years, have sunk, as shown in *fig.* 101.; or, if care has been taken to keep the ground about them level by adding fresh soil as it sinks, they will appear, as in *fig.*

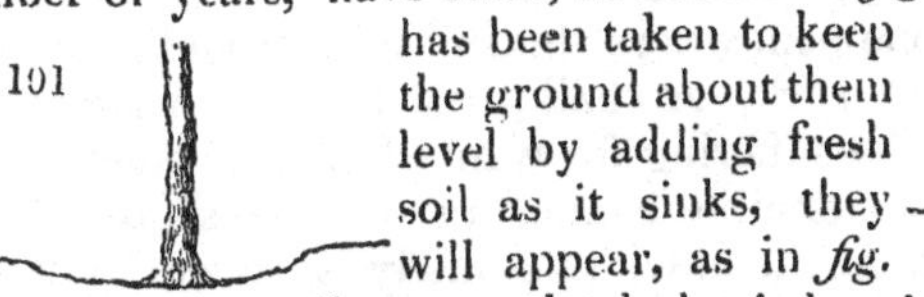

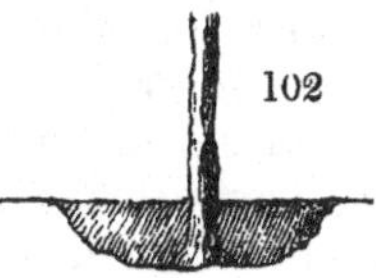

102., with their collars completely buried, and their trunks rising out of the soil like a post driven into it by art, instead of springing from a woody base formed by the roots, like trees and shrubs in an indigenous state, as shown in *fig.* 103. In replanting, either with trees already in the garden, and taken up in order to have their roots raised higher, or with young trees, they ought always to be planted on little hills, more or less in the manner

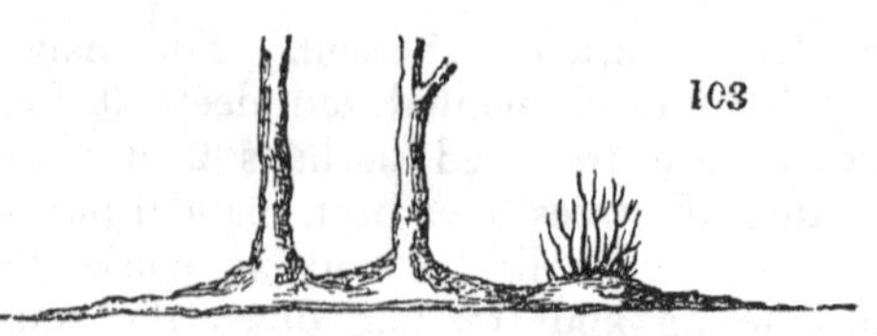

shown in *fig*. 104.; so as ultimately to settle down into flattened protuberances, as in *fig*. 99. It may be asked how the ground is to be cropped close to the roots of trees so treated; to which we answer, that we have already shown in p. 202. the utter impossibility of fruit trees planted in dug ground bearing well, unles an undug space be kept all around them. There is not a single point in the whole culture of trees and shrubs that is of equal importance to that of keeping the collar of the trunk above the surface; and there is not one which is more generally either neglected, or by some means or other counteracted, not only in small, but in large gardens.

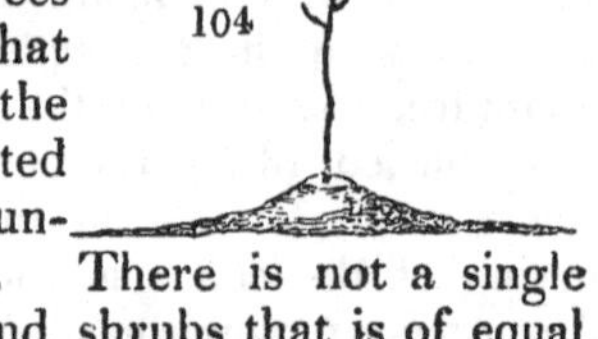

Where it is proposed to retain the ornamental shrubs or trees already existing in an old suburban garden, they will very generally require to be taken up and replanted, no less than the fruit trees, unless they should be so large as to be of a timber-like size; in which case, as in a fourth-rate suburban garden they will be disproportionate to every thing about them, they ought to be rooted up entirely. The only remedy for trees which have been too deeply planted, besides taking them up and replanting them, is removing the earth which covers their roots, as deep as the lower part of the collar, ; but, as this would very much disfigure any garden, whether large or small, it is a remedy which we cannot recommend any one to resort to. In the case of the smaller shrubs, whether fruit-bearing or ornamental, such as gooseberries, currants, roses, &c., there need never be the slightest hesitation in rooting up such as have been too deeply planted, throwing them away, and replacing them with young plants from the nurseries.

It may sometimes happen, where the soil is poor and shallow, and the subsoil gravel, rock, or chalk, that fruit or ornamental trees or shrubs may be suffering for want of nourishment, and may, from this cause, exhibit canker or mossiness on their branches. In this case, the obvious remedy is, to supply nourishment; which should be done, not by digging dung into the soil, but by top-dressing it with thoroughly rotten horse or cow dung, or with a composition of soil and some other animal or mixed manure. Where such trees stand in dug ground, the manure may be slightly forked in; but, where they stand on turf, spreading it on the surface in autumn, and allowing it to remain there during winter, will be sufficient. Renovating fruit trees by top-dressing the surface has the double advantage of rendering them more productive of fruit, and of making that fruit of higher flavour. Even when trees have been too deeply planted, if their collars are laid bare, and the surface soil removed for 2 or 3 feet all round, to within 2 or 3 inches of the main roots, and afterwards the surface top-dressed, from the trunk as far as the roots are supposed to extend, the evils of deep planting will be mitigated; it being understood that the surface is never to be afterwards dug. The fibrous roots are, by this treatment, tempted to come up to the surface in search of the nourishment afforded by the manure; and, whenever the greater number of the roots of a tree are near the surface, from the greater warmth in summer, and the influence of the air, it is certain to blossom and bear well, and, without being over-luxuriant, to be in a thriving state.

The only remaining subject connected with the renovation of fourth-rate suburban gardens is that of insects and other vermin. Where these have not been kept down by vigilant attention, the soil, owing to the state of closeness and dampness in which small gardens usually are kept, will be found teeming with snails, slugs, and worms; and the shoots of the trees with the eggs of moths and other insects. In winter, the snails and slugs will be found collected, for shelter and protection, under evergreens, among stones, or any similar materials that afford them shelter, and in the crevices of walls, &c. The most effectual mode, when they are in large quantities, is, to collect them by hand, and then destroy them; and, when they are less numerous, to water every part of the garden thoroughly and repeatedly with clear lime water, which, when it comes in contact with their bodies, effectually destroys snails, slugs, and worms. The eggs of insects attached to the trees should be rubbed off with a hard brush, and then collected by hand; and, early in spring, the progress of any that remain should be watched, and tobacco water, lime water, or soapsuds (which operate in the same manner as lime water, but are longer retained on the plant, in consequence of the greasy medium in which the alkali is enveloped), should be applied as soon as the insects are hatched.

In general, we may observe that, though there may be some advantage in taking possession of a second-rate, or even a third-rate, garden, which has been planted for some years, on account of the fruit trees being in a bearing state, and the ornamental trees being of a sufficient size to produce considerable effect, there is none whatever in occupying an old fourth-rate garden. With all such, the best mode will generally be found to be, to root up all the trees and shrubs, and renovate the garden by improving the soil, &c., laying out afresh, and replanting entirely.

§ 2. *Gardens of double detached Houses, or Houses in Pairs, which have Gardens on Three Sides of each, but which still form Part of a Row, as shown in the Diagram,* fig. 51. p. 178.—The chief difference between this kind of fourth-rate suburban garden and those of the preceding division consisting in the connexion of the front garden with the back garden by a piece of ground along one side of the house, we do not propose to enter much into the details of planting the back gardens; because, as these must almost always be parallelograms or squares, the same modes of surrounding them by walks, and planting them agreeably to the object or taste of the occupier, will be alike applicable; with only this difference, that, where the garden is larger, additional plants will be wanted, which may be either duplicates of those already named, or of allied kinds, selected by the planter from our lists. Nothing can be easier than to give plans which would show great variety in the disposition of the walks, and of the beds and borders, of even the smallest garden; and a number of such designs, placed side by side on paper, would, doubtless, possess greater attractions to the eye of the reader, than the repetition of a few plain designs. The case, however, is quite different in the execution, in which the space taken up by the walks of an intricate design would be so great, that the beds for culture would be too small to be of much use. Under these circumstances, the mere sight of a handsome design would soon pall on the feelings of the occupier. We can therefore, not only from reason, but from experience and observation, strongly recommend simple designs for laying out all small gardens whatever, in preference to complicated ones.

The objects in view, when double detached houses are built, are, to procure light and air on every side, both to the building and garden, and to give dignity and consequence to each dwelling by making it appear to have the magnitude of two houses. For this last reason, the entrances to double detached houses are generally so contrived that both cannot be seen at once; or, if seen at once, that they shall appear to be only one entrance, or that one of the entrances shall not appear to be such. Hence, though in this kind of suburban residence the back gardens are laid out and planted exactly in the same manner as in the preceding kind, yet some difference is required in laying out the front garden, and the piece of ground that connects it with that behind; and it is to this part of the subject that we shall chiefly direct our attention. In the designs which we shall give, the dark-shaded plants represent low trees, such as thorns, &c.; and the light-shaded ones and outlines, shrubs.

A double detached House with the ordinary Arrangements.—*Fig.* 105. exhibits one of the simplest of these modes; but, to give it effect, by preventing both entrances (*a a*) from being seen

105

at once, it requires a considerable projection between them, as shown at *b b*; or, instead of that projection, a mass of evergreen trees and shrubs. The projection may either be some additional conveniences to each house; or it may be a span-roofed green-house, with glass in front and on the sides, and separated by a wall to be covered with evergreens, such as camellias, myrtles, or oranges, as in the plan before us, at *c c*. Or, to avoid the risk of one of the occupants not paying the same attention to his green-house as the other, then, to insure a good effect from the exterior, and the proper cultivation of the plants, the green-house may be exclusively in the possession of the occupier of one of the houses, though equally looked into from the windows of both of them.

The principal object of the possessors of these houses, with reference to their gardens, is supposed to be, as the green-house would seem to indicate, ornamental display; for which reason, the ground is principally laid down in grass, and sprinkled with ornamental shrubs and low trees; the latter, as before observed, being indicated by a darker shade than the shrubs. On the supposition that the ground belonging to the adjoining houses is planted with a view to the same object, then the occupiers of the houses shown in *fig.* 105. should study to introduce species and varieties different from those in the adjoining gardens, but, at the same time, a sufficient number of the same kinds to preserve harmony in the general view of the whole, to a person passing along the street. If we suppose, for a moment, that, in one garden, small select trees, such as thorns, crabs, cypresses, &c., were planted, and in the adjoining garden large rapid-growing forest trees, such as poplars, limes, elms, &c., it must be evident that the contrast in appearance between the two gardens would destroy all unity of effect in the general view of the street; while the shade produced by the large trees of one garden, on the low and more choice kinds in the other, would so materially injure their growth as to prevent them from flowering, or being at all ornamental. "What, then, is to be done," it may be asked, "when a neighbour on the south side of our garden chooses to plant a row of poplars, so as to throw our garden completely into the shade?" All that can be done is, either to plant large coarse trees like our neighbour, or rather still larger and more rapidly growing ones; or to cultivate only such trees and shrubs as prosper best in the shade. Of these, one of the most valuable is the holly, of which there are twenty or thirty varieties, with very distinctly variegated leaves; all of which attain in time the size of trees, and, by the gaiety of their foliage, make a splendid appearance at every season of the year. There are also other varieties of holly which are curious; such as the hedgehog, the myrtle-leaved, and the recurved-leaved holly. Some foreign species of holly, which attain the size of low trees, might also be introduced; such as I'lex opàca and I. madeiréusis, both of which thrive better in the shade than in the sun. The yew, of which there are two very distinct sorts, the common and the Irish, will afford two other trees which, like the holly, will thrive better in the shade than in the sun; and to these may be added the box, than which, when planted singly in deep free soil somewhat calcareous, and allowed to take its own natural shape, there is not a more picturesque and beautiful low evergreen tree in the British arboretum. The tree box is picturesque, from its irregular outline, and the varied forms of the masses of its foliage; and it is beautiful from the smooth glossy green of the latter, and the graceful tendency of its lower branches, which recline on the ground. This tree, when grown in full sunshine, assumes a yellow hue, compared with the rich deep green which it has in the shade; and the same may be said, to a certain extent, both of the yew and the common green holly. Besides the common tree box, the broad-leaved and the myrtle-leaved varieties, and the variegated box, all which attain the size of low trees, there is the Minorca box, which forms a very handsome tree; and in addition to these trees, and others which will be found in our list of those that love the shade, there is the common laurel, which may be trained so as to assume the character of a tree, a shrub, or a hedge, at pleasure; and in each capacity there are a denseness and an appearance of comfort and gaiety in its foliage, which render this plant welcome almost every where. Few are better adapted for making evergreen screens, either between a front garden and the street, or between two adjoining front gardens. The privet, in like manner, may be trained so as to form a handsome small tree, an ornamental bush, or an evergreen hedge. The common arbutus, of which there are two very distinct varieties, also forms a very handsome low tree. For shrubs that delight in the shade, it is only necessary to mention the whole families of rhododendrons, azaleas, mezereons, and some of the evergreen daphnes. In the most shady part of the garden, against the walls, where nothing else will grow, the gold and silver ivy will produce a most brilliant effect; but, as both are weaker and more slowly growing plants than the common and the giant ivy, care must be taken that they are not overpowered by these kinds. With respect to rhododendrons, azaleas, mezereons, &c., the different species of these genera, with their varieties, are of themselves enough to fill a shady garden with a display of evergreens in winter, and splendid flowers in summer. We shall not enumerate any other plants adapted for shaded gardens, those above mentioned being enough to show that the possessor of a garden shaded by the trees of his neighbours, though unable to grow some of the finer kinds of trees, has still ample resources left for ornamenting his grounds. Among herbaceous flowers, there is a still more ample choice: for example, in annuals alone which will grow in the shade better than in the free air, we have only to point to the lists of Californian plants enumerated in p. 292. and p. 293.

To return to our plan (*fig.* 105. p. 315.), each house consists of a vestibule (*d*), hall and staircase (*e*), dining-room (*f*), and drawing-room (*g*), communicating with the conservatory (*c*). In the garden, at *h*, are shown flower-beds on each side of the entrance walk; and these, to harmonise with the green-house, may be symmetrically planted, the two central circles with fuchsias; the two beds *i i* with variegated pelargoniums (geraniums), and the remaining four beds with scarlet pelargoniums. These will continue in bloom during the whole summer; and, in order that the beds may not every year present the same appearance, heliotropes, rose-scented pelargoniums, celsias, and calceolarias may be employed in succession; or, instead of these kinds planted in the free ground, the green-house plants may be plunged in the beds in their pots; being symmetrically arranged with regard to size and foliage, so as still to maintain the rule laid down in p. 213.; viz. that symmetry of form in the beds requires symmetry in the form, colour, or arrangement, of the plants. It would, for example, be contrary to symmetry of arrangement, to plant two of these beds with green-house plants, and two with the common hardy kinds; though, by choosing plants of the same size, symmetry of form might be produced. If it were desired still further to vary these beds, they might be planted, for a year or two, with roses; a standard rose being placed in the central bed, surrounded by mignonette, and dwarf roses planted in the other beds; or the centre bed might be occupied by climbing roses trained over a cone formed by a framework of rods; or there might be an arch of iron rods thrown from one central bed to the other, across the walk, and covered with climbing roses, or with a cobœa, or some other ornamental climber. The beds may be edged with wire-work, to which mignonette may be trained; or with ivy, trained to an iron rod raised 6 in. above the surface; or they may be bordered with any very low-growing evergreen shrub, such as the evergreen iberis, thyme, &c.; or they may be edged with tiles or slates, or with cast-iron or wooden edgings. In short, these beds may be planted and edged in a different manner every year, during the whole period of a lease; and, in addition to the summer planting, they may be filled every autumn with bulbs, so as to produce a brilliant show in early spring.

The beds *k k* in the back gardens may be planted with a mixture of China roses, or of rhododendrons and azaleas, selected so as to present a bloom from April to August; with some clethras and *C*eanòthus azùreus, to continue to bloom till November; or they may be planted with a mixture of herbaceous plants and bulbs, so selected as to present some species in flower during every floral month in the year. Against the basement wall of the green-house, in this design, chrysanthemums may be planted, and carefully trained; in which situation they will flower beautifully: and against the lower part of the house *C*ydònia japónica may be planted, which, in that situation, would flower throughout the winter; while, in such other parts as did not interfere with the windows, Lonícer*a* japónica, *J*asmìnum officinàle, climbing roses, and other ornamental flowering creepers, might be planted, for their show and their fragrance. In the green-house, there ought to be vines or creepers trained on the piers between the windows, and under the rafters; and, in summer, when the plants in pots are taken out of the green-house, these creepers ought to be removed from the piers and rafters, and trained over the windows and sashes, so as to produce a sufficient degree of shade to admit of the interior being used as a sitting-room for the ladies in the morning, or as a banqueting-room for taking the dessert in after dinner, or as a room for taking tea in in the evening. The green-house may be heated in various ways, and, among others, by a fire-place or stove at the end, as shown in the plan at *l*; but, if only the more hardy green-house plants are grown, such as camellias against the wall, and heaths and Australian plants in pots, no other heating need be required than what may be given every night after the family go to bed, by leaving open the glass door communicating with the living-room, a fire being supposed to have been kept in that room during the day. The warm air from the living-room will raise the temperature of the green-house at least above the freezing point; and some degrees higher even in the most severe weather, provided the sashes and windows of the green-house fit tightly. It would be easy to heat this green-house in a most effectual manner, from the kitchen fire, or the parlour fire, by pipes of hot water; but the tendency of contrivances of this kind, after two or three years' use, to go out of order, renders us reluctant to recommend them if they can be avoided. Underneath the green-house, on the basement floor of the house, there may be a coal or beer cellar.

The part of the back garden which is not seen may be either wholly devoted to ornamental trees and flowers, on a lawn connected with that shown in the figure; or separated from it by a border of evergreens, and cropped with culinary vegetables. In either case, a portion of ground at the bottom of the garden must be fenced off as a reserve garden, in which there must be a pit or a hot-bed, to keep up the supply of young plants for the green-house, and for the beds at *h*; unless the cheaper mode is adopted, of having this done by contract with a commercial gardener.

Expense and Management. The gardens to these two houses might be laid out and planted for from 30*l.* to 60*l.* each, according to the kind of trees selected, exclusive of the reserve gardens, the green-houses, and the graveling or paving of the walks. From there being flower-beds, if the occupiers of the houses or their servants were not attached to gardening, the assistance of a man for each garden might be required, on an average, one day in a week throughout the year; which, with the requisite seeds, plants, &c., might bring the yearly expense of management to about 10*l.* This is supposing that the watering, and other attendance required by the plants in the conservatory, were chiefly done by some part of the family.

Remarks. These gardens are particularly well calculated for amateurs who are fond of performing the operations of gardening themselves; because there is no heavy, dirty, or disagreeable work required in them, and because the green-houses would afford recreation during winter.

106

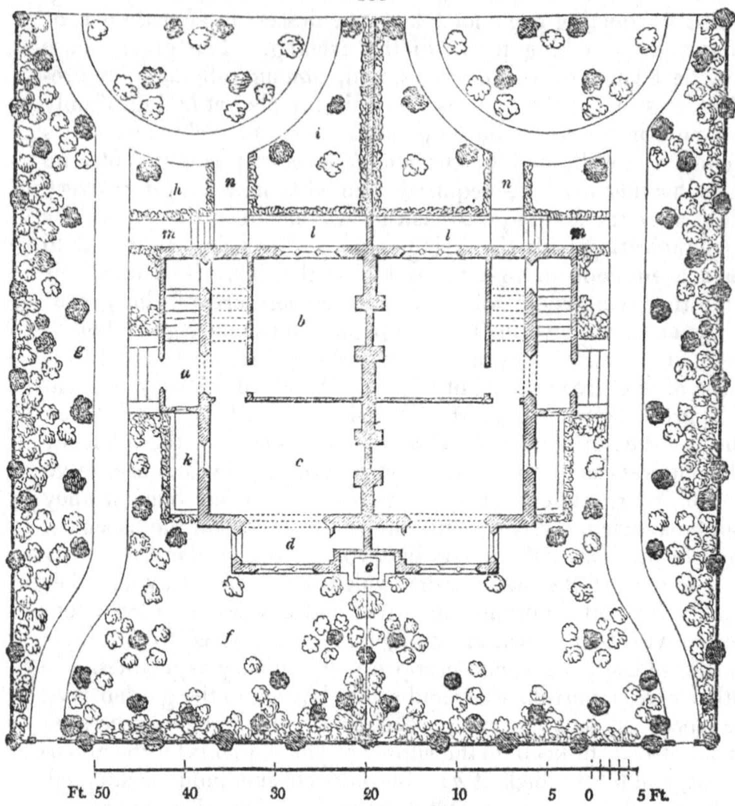

A double detached House, with the Entrance Porches at opposite Sides. Fig. 106. shows a common, and at the same time an effectual, mode of arranging and placing the entrances of a double detached house, so as to make it have the appearance of a single house. In this case, a square building, containing two houses, is entered by porches at opposite sides; and there are back entrances to each house, communicating with the area, and the garden behind each. The space in front of the houses is divided by a wire fence in the centre; so that a stranger entering from the street, and proceeding towards either house, sees across the whole width of the front garden; and both the houses and gardens appear to him to be one, and to be occupied by the same family. We have shown in this figure how shrubs and low trees may be distributed so as to aid this illusion. Each house contains an entrance porch and staircase (*a*), dining-room (*b*), and drawingroom (*c*), with a recess, which, in Elizabethan houses, is called a bay (*d*), communicating with a small closet. Between the two houses there is a pedestal and vase, as indicated at *e*; and on the lawns

107

(*f*, *g*, *h*, *i*, and *j*) there are no flower-beds, but only flowering shrubs and low trees: *k* and *l* are verandas; *m*, the sunk area, communicating with the walk by steps, and leading to the door of the back kitchen; and *n* shows the descent, by a few steps, from the veranda to the garden. The style of these houses, designed for us by E. B. Lamb, Esq., is supposed to be the Elizabethan; and *fig.* 107. is a perspective view, showing the front of both houses next the street, and the entrance front of one of them.

The object in laying out and planting these gardens we shall suppose to be a display of choice low trees and shrubs, but planted in such a manner as not to require much expense in keeping the garden in order. Flowering plants we shall imagine to be altogether dispensed with, except some in pots (which may be grown in a reserve ground, or supplied for a fixed yearly sum by a commercial gardener), for placing in the balconies over the bays, and under the verandas. The adjoining gardens we shall suppose to be planted much in the same manner; or, indeed, in any manner, provided a few trees, either fruit-bearing or ornamental, are sprinkled through them. Such gardens will not be offensive to look at, especially through a foreground of low trees; and, hence, it will not be necessary to proceed on the supposition that much requires to be planted out, that is hidden or partially concealed by trees. The ground being drained and levelled, and properly trenched and manured, the walks may be blocked out; but the gravel or the pavement should not be laid for a year; unless, indeed, the walks are formed of pavement laid on stone piers.

We would first plant common ivy against the exterior boundary wall, but not against the wall which separates the two back gardens. This ivy, if of the common kind, will require no farther attention whatever; but, if giant ivy be planted, it will require nailing, especially when it reaches the coping of the wall, as its claspers do not adhere nearly so well to brickwork as the common ivy. Against the wall which separates the two gardens, we would plant evergreen and deciduous magnolias, *C*eanòthus azùreus, Photínia serrulàta, Stransvæ`s*ia* glaúca, Eriobótrya japónica, Mahòn*ia* fasciculàris and M. *A*quifòlium, &c.; *J*asmìnum revolùtum; *C*ydònia japónica (not because it requires a wall, but because, if the blossom buds are picked off before expanding during summer, the plant will continue in bloom all the winter); Chimonánthus fràgrans; the Banksian, Macartney, and other evergreen roses; and various climbers and twiners, such as the fragrant Chinese and Japan honeysuckles, perpetual-flowering roses, wistaria, tecoma, passiflora, &c. If the direction of the wall be north and south, then these kinds may be placed on both sides of it; but, if it should be east and west, then the more tender plants should be placed only on the south side. It may be observed here, that many trees and shrubs, which are considered somewhat tender, thrive better on the north side of a wall than on the south side, provided they are planted in dry soil, not too rich. The reason is, they are not so soon excited by the action of the sun in spring, and, consequently, not so liable to have their buds and young shoots injured by spring frosts. On the other hand, from the want of sun to ripen their wood in autumn, they are more liable to be injured by the frosts of that season; but this evil may be greatly lessened by keeping the soil poor and dry, which will prevent the plants from making long succulent shoots, which can only be ripened by the full influence of the sun. All the plants above named, with the exception of ceanothus and eriobotrya, will succeed perfectly in a northern exposure, if the border be properly prepared with sand and lime rubbish, and the subsoil rendered thoroughly dry. Immediately within the exterior boundary walls of both gardens, we would plant a row of low-growing evergreen trees, at regular distances, among which we would include all the variegated hollies (unless these were abundant in the adjoining gardens), the box, the arbor vitæ, the common cypress, the red cedar, and, in some places, the evergreen and Lucombe oaks, and the cork tree. These trees, after they have grown 2 or 3 feet higher than the wall, will form, from the windows of both houses, a foreground to the exterior scenery; and, therefore, in planting them, the tall and broad-headed kinds, such as the evergreen and Lucombe oaks, ought only to be planted where there is something to conceal; and the tall narrow-growing kinds, such as the common cypress, the arbor vitæ, &c., where the exterior objects are required to be only slightly disguised. Where the objects beyond the boundary are such as may be looked on at all seasons with pleasure, such as trees, shrubs, or turf, then trees of low and slow growth may be employed instead of the larger ones, such as the arbutus, the myrtle-leaved box, &c. Within this outer boundary of trees, the ground may be sprinkled with evergreen and deciduous low trees and shrubs, in the manner indicated in the plan; none of these shrubs being planted in large masses, so as to require large breadths of dug soil; but in small groups, and nothing done to the soil round them, except preventing the grass from growing on a space above 1 ft. in diameter round the stem of each tree or shrub. This will prevent the grass from injuring the plants while young; and, what is of great practical importance, prevent the risk of injury to the tree or shrub, from the scythe of the mower. In a few years, when the shrubs have acquired strength, the grass may be allowed to spread as near to their stems as the closeness and shade produced by their branches will admit. It will be observed, that, though we recommend keeping a naked space round each plant for several years, till it has acquired sufficient strength, we by no means recommend digging this space, except in the case of standard roses, for the more tender kinds of which digging is necessary for the sake of applying manure.

After all the trees and shrubs are planted, the whole garden may be laid down in grass; the surface having been previously consolidated and raked quite smooth. If the trees and shrubs have been planted in the manner which we have already directed in p. 312., each plant will stand on a small hillock, which, for the lowest-growing shrubs, should not be less than 6 in. high, and 2 ft. broad at the base; and, for the larger shrubs and trees, not less than 1 ft. high, and 4 ft. wide at the base. On these hillocks no grass seeds should be sown; and this, besides being better for the trees, by preventing their roots from being exhausted by the grass, will save much trouble in their after-management, by rendering all clipping unnecessary. When trees and shrubs are thickly scattered over a lawn, only the glades of grass can be mown with the scythe, while that in the crowded parts is obliged to be clipped, either by a person using hedge-shears and kneeling while he performs the operation, or by using grass-shears and standing. Either of these operations is very hard and disagreeable work; and they are rendered quite unnecessary by the hillock manner of planting, and by not sowing the hillocks with grass. This hillock manner of planting will not, probably, be approved of by many practical gardeners, and especially jobbing gardeners, who have been accustomed to finish their planting by rendering the surface round the plants quite level. Our readers, however, may rely on its being by far the best mode for the plants: since both nature and experience teach us that nothing is more injurious to vegetation than having the collar, or base, of the stem, which ought to be above ground, placed under it, and covered with soil to the depth of 2 or 3 inches. Independently altogether of the use of these hillocks (which, in form, ought to be broad and flattened, and have concave sides uniting imperceptibly with the level surface all round), their effect in giving height and consequence to the plants on them, and the variety which is produced in the play of lines on the surface of the turf, ought to be an additional recommendation. The expense of this mode of planting can be no greater than that of the ordinary mode, except that, in some cases, stakes may be required to tie the plant to, for two or three years; which, in the ordinary mode of sinking the stem in the soil, may be dispensed with. We shall not obtrude on the reader a list of the trees and shrubs for this garden, as from those we have already given he must be able, in some degree, to make a selection for himself.

Expense and Management. Supposing the length of the back garden to be 150 ft., the soil moderately good, and the subsoil such as not to require much drainage, the expense of laying out, and blocking out the walks, may be 20*l.* or 25*l.* The number of trees and shrubs required, exclusive of the ivy, may be 250, at the average price of 2*s.* each. The ivy, the grass seeds, and other expenses, may amount to 5*l.*; so that the total expense of laying out and planting each garden, exclusive of purchasing and laying in the gravel in the walks, may amount to between 55*l.* and 60*l.* If trees and shrubs were purchased which averaged 1*s.* each, the sum would be reduced to from 40*l.* to 50*l.*; or, if the average of the trees and shrubs were 6*d.* each, then the total would be reduced to from 35*l.* to 40*l.* The expense of management, exclusive of taking care of the plants in pots, would be very trifling. If the walks were paved, nothing more would be required than mowing the lawn, clipping the edges of the grass along the walks, sweeping up leaves, and cutting off decayed flowers or dead twigs, all which need not cost more than 5*l.* a year; and for a similar sum a commercial gardener would keep the veranda stocked throughout the year with boxes of mignonette, and supply a succession of plants in flower, during the summer months, for the balcony.

Remarks. These gardens are well adapted for persons who take no pleasure in seeing the operations of gardening going forward, and who are rather annoyed than otherwise at meeting a gardener in walking round their grounds. Some, also, object to having a gardener, or any strange man, coming about the house; and in this case there would be as little of that as there could possibly be in any garden where a handsome display was to be made. The mowing, and all the operations required for these gardens, might be performed in the mornings between six and eight o'clock.

A double detached Suburban Villa, in Porchester Terrace, Bayswater. In the ground plan *fig.* 108. the entrances to the two houses are on opposite sides, on the same principle as in *fig.* 106.; but here, the porches not being so conspicuous, and it being utterly impossible to see any considerable part of both at the same moment, from any point of view, the illusion is more complete. As this figure represents the ground plan of a double detached house and garden, which we designed, built, and laid out for ourselves, we shall, as a detailed illustration of the mode of laying out and planting this division of fourth-rate suburban gardens, describe it at some length.

The object was to build two small houses, which should appear as one, and have some pretensions to architectural design; being, at the same time, calculated for invalids, and, therefore, furnished with verandas extending nearly round the whole building, for taking exercise in during inclement weather. The houses form part of a row of detached dwellings lying parallel to a street (Porchester Terrace, Bayswater) running north and south. According to the principle we have laid down, the diagonal of the square ought to have been parallel to the street, instead of one of its sides; and we should have placed the building in this manner: but, on stating our intentions to the surveyor of the estate, from whom we took the ground on a ninety-nine years' lease, he objected to it, as it did not appear probable that it would be generally followed in the other buildings in the same row, and, therefore, was, in his opinion, likely to disfigure the street.

Drainage, Service-Pipes, &c. The soil being a loam on a subsoil of gravel, no under-draining was required, except the drains for the water-closets, back kitchen, and rain-water pipes. For these purposes, each house has a barrel-drain, communicating with the sewer in the street. The general surface, and all the walks in the back garden, incline from one point at *a*, to another at *b*, where there is a drain to a small sewer in the back lane. From the point *a*, to *c* near the steps up to the front door, the surface of the walk and the adjoining ground is nearly level, but slightly inclining to *c*; and from *c* to the street entrance (*d*) there is a gradual slope of above a foot. There is also a very gradual slope of the general surface of the garden (say of 3 in.), from the margin of the walk *a c d*, to the party wall which forms the southern boundary of the garden; and between *c* and *d* there is also a gradual slope (say of 6 in.) towards the point *e*. In consequence of these inclinations of the ground and walks, no rain ever stands on the surface, however abundantly it may have fallen. The service water-pipes of both houses are laid down along one side of the walks,

108

from *d* to *a*, and are thence conducted to main cisterns for the supply of the houses, to others jointly for the houses and the gardens, and to other cisterns for water-closets.

Soil, Walks, &c. The whole of the ground, except the part on which the house stands, was trenched 4 ft. deep; the surface soil of what was excavated for the foundation and basement story of the house, and the half of the surface soil of the road, being previously distributed over the garden in such a manner, as to raise the ground at the house 2 ft. higher than the footpath in the street. Round the house the ground was kept nearly on a level for some feet distant; after which it was made to decline equally on every side, till at the front entrance it was higher than the gravel of the public path by the depth of the sill of the gate; and at the back entrance it was on a level with the path of the public lane. Before trenching, the ground was also limed, and thickly coated over with the best London stable dung. The lime was introduced, not only for the sake of adding calcareous matter to the earth, but for forming a comparatively insoluble compound with the dung, in order to prevent it from being all employed by the roots of the trees at once. By a part of it being rendered comparatively insoluble, there will be, as it were, a reserve of nourishment in the soil for many years to come; because it is well known that time and the soil gradually dissolve such a compound. The trenching was performed in autumn, and in the following spring the walks were hollowed out, the edgings firmly beaten, and planted with box, and the walks laid with gravel, and immediately after very heavily rolled.

The Houses. The general appearance of this double detached house, as seen from the street, is shown in *fig.* 109., which is a portrait taken by E. B. Lamb, Esq., in September, 1837. The two street entrances, or gates, are shown in the front wall, each having two small niches for the scrapers. The gates are of open iron rods, hinged to stone, without the intervention of wood, which always gives a temporary appearance when joined with iron-work, especially in a door or gateway exposed to the weather on all sides. Consistency, with reference to durability in architecture, is as much a fundamental principle, as consistency in point of effect; and to hang an open iron gate to a wooden door-

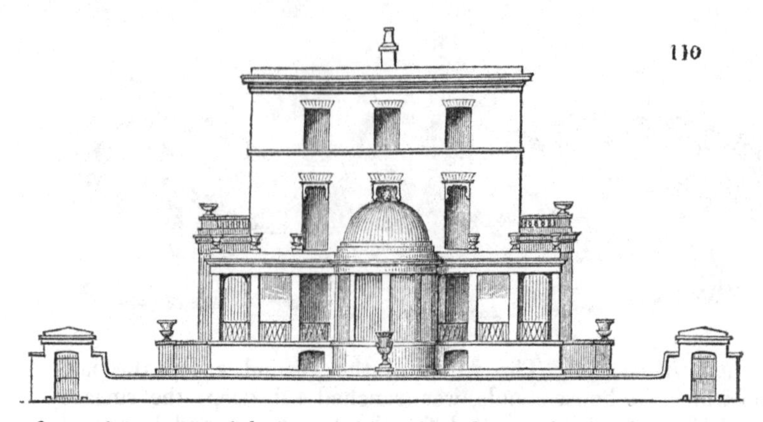

frame is a gross violation of this principle; and one which, when it occurs on the very threshold of a residence, gives but a poor example of the taste which may be expected within. Wooden lintels to the entrance gates of the front gardens of suburban houses are offensive on the same principle, whether the gates are of wood or iron. An error of a different kind, very frequently committed in street gates and railings, is that of having them too much ornamented for the style of the house. The gates of these two houses, and the railing on the low wall between them, are composed simply of straight iron rods, pointed at top; and, instead of being painted of a pea-green, as is very frequently done, they were coated over with boiling gas tar, when first put up, and nothing has been done to them since. This preserves a proper gradation from the front railing to the railing between the pillars of the veranda, which is of a more ornamental description, and is painted of a stone colour. The different sides of the house are shown in the three accompanying elevations.

Fig. 110. is the elevation of the front, or side next the street; showing the domical conservatory (*g* in the ground plan *fig.* 108.) in the centre; and in front of it the sculptured vase, the situation of which is shown at *o* in the same plan. The iron railing over the low front wall is omitted in this elevation, lest, from the smallness of the scale, it should create confusion.

Fig. 111. is the elevation of the entrance front to the house on the south side; in which the domical conservatory and the vase (*o*) are shown on the left hand; and a portion of the greenhouse (*r*) in the sunk area is shown on the right hand.

Fig. 112. is an elevation of the back front of the whole building; in which the projection *n* in the ground plan (*fig.* 108. in p. 326.), within which is the office of the *Gardener's Magazine*, is shown in the centre; and on each side of it, in the sunk floor, are the two arched openings to the two kitchens. The two square-headed doors, immediately under the windows of the office of the

111

Gardener's Magazine, are the doors to the dust-holes and garden cisterns of both houses; and the dark perpendicular line which separates these doors is intended to show the section of a door in the party wall, which has been introduced in that situation as being thinner than a brick wall, in order to admit of the doors to the dust-holes being of greater width. The elevation of this door and the party wall, and a portion of the wall and coping over the door, may be seen on the right side of the elevation of the south entrance front (*fig.* 111.). In the back elevation

112

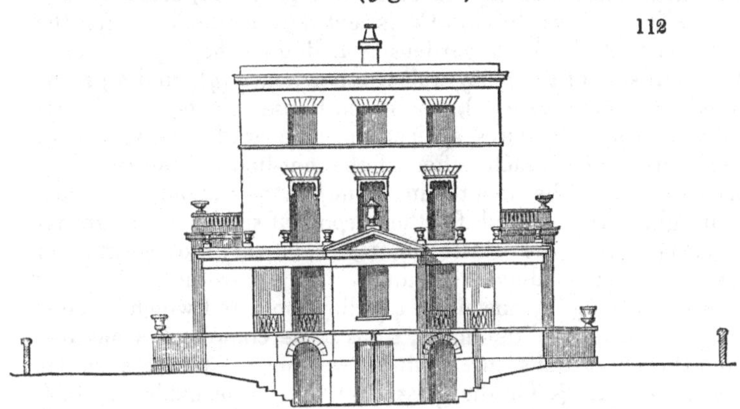

may be seen, at the extremes of the right and left, sections of the outer boundary walls to both gardens. The sections of the steps leading down to the sunk area in this elevation need no explanation.

The Ground Floor of each House consists of a porch and veranda, terminating at one end in a water-closet (*fig.* 108. *f*), and at the other in a small circular conservatory with a domical roof (*g*), a view of the plants in which is enjoyed equally from both verandas, as well as from both front rooms: *h* is the dining-room, with a recess for the side-board; *i* is the library, with two large, and three small, recesses for book-cases; *k* is a grating over the sunk area into which the back-kitchen door opens, and which communicates, under an arched opening at *l*, with the ascending steps (*m*). The aperture, which might have been divided so as to form a light closet to each house, is appropriated to the house on the south side, and now forms the office of the *Gardener's Magazine*. The sunk floor consists of a front and back kitchen; wine, beer, and coal cellars; pantry, store closet, and two other closets; and two servants' bed-rooms. So many apartments are obtained by making use of the space under the veranda; in which space, also, the main water cistern is placed, with a sink under it, and also a fixed safe. The dust-hole, water cistern for the garden, lumber closet, and servants' water-closet, for each house, are under the apartment *n*.

Out-door Arrangements. The front gardens are divided by a light wire fence, in the centre of which is the pedestal *o*, surmounted by a handsome sculptured vase of Coade's artificial stone. The back gardens are separated by a flued wall, which, though only 10 in. thick, yet, being built with bricks set on edge, and not having the ends of the bricks quite flush with the wall on the north side, has a flue left in it 4 in. wide. The wall on the north side is chiefly covered with ivy; and, consequently, the surface of its brickwork is entirely concealed. At the lower end of the back gardens is a double shed (*p*); and on the south side of the party wall is a hot-house (*q*), and a green-house, or glass case (*r*), for green-house plants, which are trained to upright wires against the flued wall; one species to each wire. Over each rafter of the hot-house is an iron rod, placed so as to be about 4 in. from the glass, and connected with light horizontal rods for the purpose of supporting a canvass covering during the winter nights, or in very hot weather in summer, when shading is required. This covering, when not in use, rolls up by means of a pulley, on a rod which extends the whole length of the house, so as to be completely concealed from the sight, and secure from the weather. There are similar canvass coverings for the glass case (*r*). The mode in which these coverings are constructed will be hereafter described. *s* is a double wire trellis for training fruit shrubs, such as gooseberries, on both sides; and in the south garden there is an additional trellis (*t*), placed on a dwarf wall, for ornamental shrubs. These trellises, the dwarf wall, and other arrangements, will be better understood by the section *fig.* 113.; in which *a* is the green-house, with the flued party wall behind; *b b*, the two paved walks in the sunk area; *c*, the strip of grass between these paved walks; *d*, the shelf for alpines in pots; *e e e*, the double trellises; *f f*, the gravel walks; *g g g*, raised beds for herbaceous

113

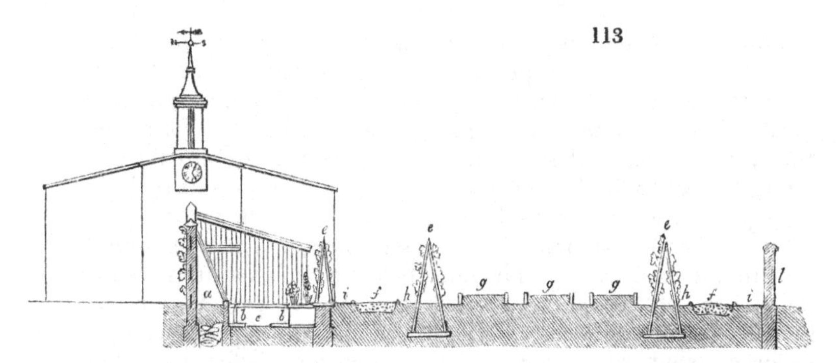

plants; *h h*, borders for the arboretum and fruticetum; *i i*, miscellaneous borders; and *l*, the southern party wall. Between the dwarf wall under the trellis and the glass case, or green-house, the ground is lowered, so as to form the sunk area (*b c b*), which is on the same level as the floor of the back kitchen, and that of the hot-house (*q* in *fig.* 108.). On the south side of this area there is a stone shelf (*u*) for alpines in pots; on the north side, the green-house, or glass case (*r*), before mentioned; and the space between has a paved walk on each side, and in the centre a strip of grass, as already shown in the section *fig.* 113. In the middle of this small grass-plot, at *w*, is a socket for a double clothes-post; that is, a post having two horizontal arms at top, each of which supports a clothes-line, which is fastened to a hook in the rafter of the hot-house at one end, and passes over a pulley fixed to the basement of the veranda wall at the other. When the clothes-post is not wanted, it is taken down, and laid in a place appropriated for it, under the stone shelf *u*. At the hot-house end of the grass-plot, the alpine shelf terminates in a shallow cistern for marsh plants, and a deep cistern at *v* for aquatics. Near the cistern *v* is a sun-dial, and at the opposite end of the grass-plot a vase, the plinth under which forms a cover to a liquid manure tank, supplied from the water-closets. The double shed *p* has a turret with a clock in the centre of the gable facing the houses (as shown in in the section *fig.* 113.), and a semicircular window in the centre of that towards the lane, which lights the lofts of both sheds; so that, from whichever side it is viewed, this double shed is symmetrical, and appears as completely a single one, as the double detached dwelling to which it belongs appears to be only one house. The shed belonging to the north house is fitted up as a wash-house on the ground floor, and has a loft over. The shed on the south side has three floors: the middle floor, which is shown in the plan, is one step above the level of the walks of the garden, and the ascent to it from the hot-house is by four steps. In it there is a potting bench, a pump, a carpenter's bench; a wooden safe for preserving fruit, bulbous roots, or large specimens of plants that will not lie flat between paper; and, against the walls, a small glass case for a garden library for the use of the gardener, and shelves for seeds, roots, tallies, &c. Beneath the potting bench are bins for different kinds of soils, broken pots, &c.: the empty pots being kept in regular order under the front shelf of the hot-house. There is also a wooden staircase to the floor above, and a stone staircase to the floor below. The floor below contains a fireplace for heating the hot-house and green-house, with a space for fuel or lumber, and the remaining space is used for growing mushrooms, or forcing rhubarb, chicory, &c. The loft floor is for mats; the sashes of the green-house, which are taken off during the summer time; and for onions, bulbs, and similar articles; with an opening to the machinery of the clock, for winding it up; and the half of the window, before mentioned, opening to the back lane. The smoke from the fires of both sheds comes out through the vertical openings between the columns of the turret, in order to avoid the incongruous appearance of a chimney top over a clock; the turret being for the purpose of supporting a vane to indicate the direction of the wind to both houses. At *x y*, in the south garden, are two pits for green-house plants, 4 ft. deep, with hollow walls and hollow bottoms; and with an iron rod over each rafter, and about 4 in. above it, for the purpose of keeping the canvass covering necessary in winter a few inches from the glass, so as to preserve a non-conducting vacuity between the canvass and the glass. The part of the walls of these pits that is above ground is covered with ivy, to keep them warm. The bed *z* is for a perpetual flora, a plan of which, with a list of plants, will be hereafter given in detail; and the border at 1 is a trough with a brick bottom and sides, for marine and bog plants. At & & & are brick pedestals, 9 in. square, and 1 ft. high, for supporting vases, or pots of choice plants when in flower; or which may be used for seats: the pedestal indicated at *m*, in the south garden, is surmounted by a statue in Portland stone; and that at *a*, by a crouching Venus in Austin's composition. At 2 2, in the front garden, are brick pedestals supporting bee-hives covered with earthenware covers, which are scored to imitate a straw hive; and at 3, in the back garden, is a Polish hive, fixed against the wall.

The Reserve Ground for the garden occupied by us is rented from Mr. Hopgood, a florist and nurseryman. It lies immediately on the other side of the lane 4 4, the door to it being exactly opposite our back door at 5, and is more than double the extent of the whole of our garden. We formed in it a range of temporary pits, the sides being of wickerwork, in which, for several years, we made use of the sashes of the green-house *r*, and of the two pits *x y*, in the summer season, for growing cucumbers and melons. We grew, also, in this garden, asparagus, sea-kale, tart rhubarb, strawberries, an excellent collection of Lancashire gooseberries kindly sent us by Mr. Saul of Lancaster, and other articles, so as to render it a useful kitchen-garden, as well as a reserve ground.

Fig. 114. is an isometrical view of the house and both gardens, in which will be observed most of the objects described, except the trellises, and the trees and shrubs; which are omitted to leave the ground plan more distinct.

General System of Planting both Gardens. The object being to make both gardens, as well as both houses, appear as one, whether when seen in front or from behind, the more conspicuous ornamental and fruit-bearing trees and shrubs chosen for both gardens were of the same kind; any tree placed in a particular position, at the angle of the house, or adjoining the back or front entrance, &c., or of a particular kind or form (such as the cedar of Lebanon), in one garden, having a corresponding tree of the same kind, in the other. For this reason, also, double trellises for gooseberries, &c., were formed in both gardens in the same situations (*s s* in *fig.* 108. p. 326.); and even, to a certain extent, similar ornaments were introduced. At the same time, as we intended to occupy the house and garden on the south side ourselves, we introduced a great many subordinate trees and shrubs of rare kinds into that garden, which were not planted in the other; and pursued a system of management and culture which, as it will show how much may be done in a small suburban garden, we shall give an account of under a distinct head, after noticing the mode of planting common to both gardens.

In the Front Gardens of both Houses, close by the entrance gate to each garden, there are the following trees and shrubs: — a cedar of Lebanon, a walnut, a sweet chestnut, a purple beech, a *P*ȳrus *S*órbus, a *P*ȳrus spectábilis, a *P*rùnus *Mahàleb*, a scarlet thorn, a laburnum, purple and white lilacs, a syringa, a mountain ash, a *L*ýcium bárbarum, a *C*lématis Vitálba, the common and giant ivy, and the Virginian creeper. These 18 kinds of trees and shrubs are in two groups, one on each side of the entrance, immediately within it; and they form a dense mass of shade over the walk, and project over the wall into the street. This produces a dark shade, both without and within the entrance gates, which acts like the case of a telescope to the distant glimpse caught of the steps which ascend to the front door of each house. If we imagine for a moment that there were no trees or shrubs immediately within the entrance gates, the total want of shade and foreground, and, consequently, of what artists call effect, would be felt in an instant by every man of taste. Embracing the angles of the verandas near *c*,

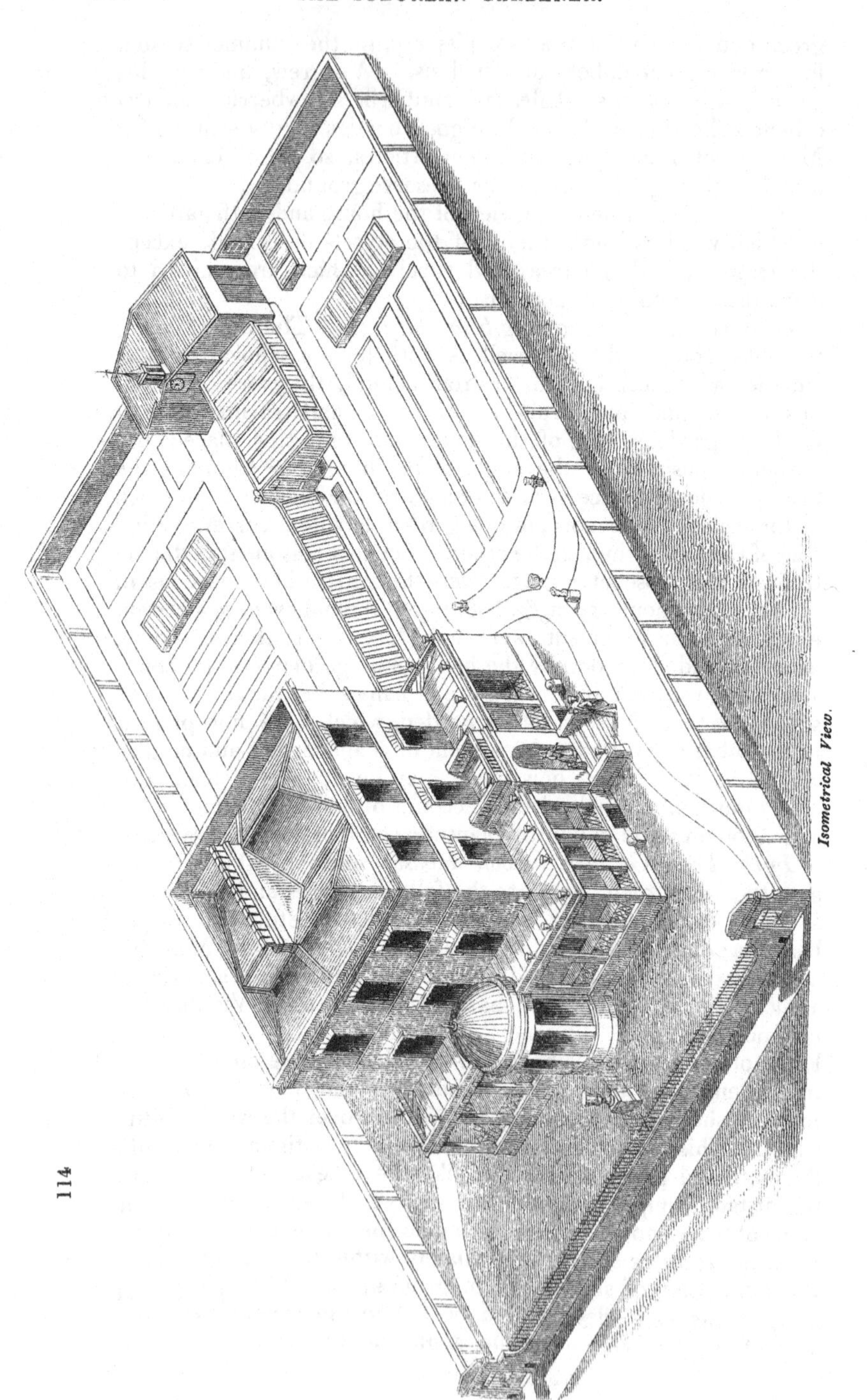

114

Isometrical View.

so as to connect both houses with the scenery as far as the boundary walls, are two groups, the same kinds of trees and shrubs being planted in each group; viz. a cedar of Lebanon, Robín*ia* viscòsa, *S*órbus hýbrida, *C*érasus *P*àdus, *C.* nìgra, *C.* semperflòrens, a variegated and a common holly, a Portugal laurel, pinaster, *P*rùnus myrobàlana, *Amelánchier* Botryàpium, a deciduous cypress, *P*hillýrea angustifòlia, *Aúcuba* japónica, evergreen and variegated hollies, Symphòria racemòsa, Persian lilac, rose acacia, and *H*ibíscus syrìacus. These form picturesque masses, and contain a sufficient number of evergreens to look well in the winter season; while, in spring, the myrobalan plum comes into flower at the end of February or the beginning of March, and immediately afterwards the amelanchier, and then the bird-cherry, next comes the *S*órbus pinnatífida, and then the Robín*ia* viscòsa. In the autumn, the purple berries of the bird-cherry, the red berries of the sorbus, and the white snow-berries have a fine effect; as have the coral berries of the holly throughout the winter and spring. There are two other main groups, the one to the north, and the other to the south, of the pedestal and vase at *o*, in the separation wire fence. These, in each garden, consist of a scarlet-flowered arbutus, rhododendron, azalea, kalmia, *C*ydònia (*P*ýrus) japónica, mezereon, *M*éspilus *a*rbutifòlia, and *C*ýtisus purpùreus and sessilifòlius, the two latter being grafted standard high. From the windows of the front rooms on the ground floor, this lengthened group makes an excellent middle distance between the pillars of the veranda, with the clustering foliage of the fig and the grape round the windows of the dining-rooms of the two houses as a foreground, and the trees and shrubs within the front fence as a distance. In winter, the fruit of the arbutus, and the flowers of the *C*ydònia japónica, make a fine appearance; and the latter shrub, which is indeed a truly valuable one, is more or less in flower during the whole year. Immediately within the front fence, which consists of a dwarf wall surmounted by an iron railing, is a row of variegated hollies, of as many kinds as there are plants; between each of which, when first planted in 1823, was a standard rose: but all these, as well as every other holly, have been since removed, in consequence of the vigorous growth of the hollies. Among the hollies are planted, as standards, nine thorns. Those next the entrance gates are the scarlet, that in the centre is the *C*ratæ`gus glandulòsa, and those between it and the scarlet thorn, on each side, are, the double-blossomed common thorn, *C.* tanacetifòlia, and *C.* orientàlis (odoratíssima). The double-blossomed common thorn comes into leaf a week or a fortnight before any other deciduous tree in the garden, and is profusely covered with its rich white blossoms, which die off of a beautiful pale pink, every year. Had it been a single-blossomed thorn of any kind, and, consequently, a fruit-bearing tree, in all probability it would only have been prolific in flowers every other year, as we have already stated, p. 191. *C.* glandulòsa comes first into flower, and produces every other year a profusion of scarlet berries; though, if the blossoms were thinned out, there would be a crop of fruit every year. *C.* tanacetifòlia is an upright, fastigiate-growing, rather singular-looking tree, with large yellow fruit; and *C.* orientàlis is a low spreading tree, with somewhat drooping branches and coral-coloured fruit. Had the variety *C.* orientàlis sanguínea (*Arb. Brit.*, p. 828.) been in British gardens at the time we planted this tree, we should have preferred it, its fruit being of a very deep port wine colour. The scarlet thorns, which, in both houses, are next the entrance gates, come into flower at the same time as the *P*ỳrus spectábilis, the laburnum, and purple and white lilacs; and, at that season, when these groups are looked down on from the drawingroom windows of both houses, they appear like gigantic nosegays. As scarlet thorns seldom set their fruit, they generally flower profusely every year. Between the entrance gates (*d*) and the points *&* *&*, in both gardens, there are, close to the boundary wall, common hollies, planted at regular distances, and between them pears and plums alternately as standards. The plums come early into blossom, and form a fine contrast with the dark green of the hollies. The pears were 20 ft. high when planted, and bore large crops of fruit for several years, till, with the plums, the greater part of them were obliged to be cut down, on account of their smothering the hollies and other plants. Behind the hollies, and immediately against the wall, common laurels were planted, and nailed against the wall so as to cover it; but plants of the giant ivy and of the Virginian creeper being planted there at the same time, they have since destroyed the laurels, and taken entire possession of the wall, forming a rich mantle along the coping.

Between the line of hollies and the walk there are various trees and shrubs, besides fruit trees; including the autumn-flowering mezereon, of which there is a large plant exactly opposite the steps which ascend to the porch, the purple laburnum, the weeping variety of the common oak, *S*órbus vestìta, *Q*uércus palústris and *Q. I*'lex, *I*'lex opàca and *I.* baleárica, the hemlock spruce, the common yew, the variegated common laurel, the scorpion senna, white and yellow broom, *B*úxus baleárica, *S*pártium virgàtum, *R*ùbus pauciflòrus, *L*aúrus nóbilis, *C*órnus sibírica, *J*uníperus virginiàna, *C*upréssus sempervìrens, *P*hiladélphus hirsùtus, *P*ỳrus nivàlis, *R*hús élegans, and a great many others. On the opposite side of the walk, between the points *c* and *d*, the kinds, being in a great measure concealed from the road, are more rare on the south side of the building than on the north side; and include Salisbùr*ia a*diantifòlia, a male plant, with the female grafted on it; *D*iospy̆ros *L*òtus, a very handsome tree, of which there is a portrait in our *Arboretum Britannicum;* Maclùr*a* aurantìaca, Magnòl*ia* purpùrea, Benthàm*ia* fragífera, Mahòn*ia* rèpens, *Pæ*òn*ia Moútan*, *Thèa* víridis, Illícium floridànum, Caméll*ia* japónica, *L*aúrus nóbilis, and several standard roses.

In the curvilinear triangles *&* *&* *&* of each garden are, a cedar of Lebanon, Siberian crab, *C*órylus *C*olúrna, a quince, a berberry, a filbert, laurustinus, rhododendrons, *C*ornús más, *J*uníperus suecica, and various others. In the triangular spot in the south garden, between *&*, *m*, and the trellis (*t*), are, a weeping birch, sweet briar, Halimodéndron argénteum, grafted standard high, the common single-blossomed furze, *C*ydònia japónica, *E*uónymus europæ̀us, *Syrínga* rothomagénsis, *R*hododéndron máximum, *R*úscus aculeàtus, *R.* hypophy̆llum, &c. In the plot in the north garden extending from *l* to the shed *p* are some dwarf apples, and various roses and evergreen shrubs; and the party wall is covered on that side with ivy, which, by protecting the wall from cold, contributes to the warmth of the green-house on the south side; and honeysuckle, which, with the shoots of the vines from the glass case in the south garden, form a rich and varied mantling along the coping; and the whole has a most agreeable effect from the back windows of both houses, and more especially from that of the office of the *Gardener's Magazine.* It is worthy of remark here, that one of the most ornamental shrubs on this piece of lawn in the north garden is the common *R*òsa canìna of the hedges. It has thrown out shoots 15 ft. in length, which bend over the party wall, and in other directions, in the most graceful manner. It continues flowering the greater part of the summer; and in autumn and winter the branches are borne down by the weight of the scarlet hips. The laurustinus, some standard Noisette roses, *C*ydònia japónica, and the double-blossomed furze, on this lawn, are also very ornamental. At the angle of each house, between *&* and *f*, there is an evergreen holly, and a swan's egg pear (the latter a handsome pyramidal-growing variety); and these, when the building is viewed from the lane behind, connect and harmonise it with the cedars at *&*, and with the hollies and pear trees ranged along the north and south party walls, from *n* to *d.*

Against the veranda on the side of both houses next the street, there is at each angle a giant ivy; next, *C*ratæ̀gus *P*yracántha and Magnòl*ia* grandiflòra; and, round the dome, Lonícer*a* flexuòsa, *Técoma* capreolàta, Magnòl*ia* cordàta, *R*òsa Boursaúlt*i*, Ayrshire rose, Wistàr*ia* sinénsis, and *C*ydònia japónica; and, in the centre, a claret grape, on account of its purple foliage in autumn. On one side of the wire fence, near the dome, is a double-blossomed furze, and on the other a *Rìbes* sanguíneum; and, in front of the kitchen window (the semicircular sunk area before which is shown in the plan), are *Vínca* màjor and mìnor, and *Yúcca* gloriòsa.

Against the veranda on the south side of the house are planted Escallòn*ia* rùbra, Lonícer*a* gràta; Magnòl*ia* acuminàta, M. conspícua, and M. grandiflòra; *Ròsa* moschàta, Passiflòra cærùlea, Wistàr*ia* sinénsis and frutéscens; China roses, budded on *Rò*sa arvénsis, *C*ydònia japónica, *J*asmìnum officinàle, Kérr*ia* japónica, and the double-flowered pomegranate. Against the veranda on the north side of the building are honeysuckles of different kinds, Virginian creeper, and ivy; and against the east side are honeysuckles, ivy, figs, and vines. The veranda all round the building is roofed with glass, and under it, on the south-west and east sides, are trained on upright iron rods, placed 8 in. apart, figs, grapes, roses, and Magnòl*ia* grandiflòra; the stems of all these being conducted, through the area under the paved floor of the veranda, to the ground outside of it.

Between the pillars, immediately within the panels of open iron-work, are placed narrow boxes, the exact length of each opening, which are filled with small pots of plants in flower, changed throughout the year as they go out of bloom. In autumn and winter, these are chiefly Russian violets, protected from frost by having a mat thrown over them; and, in spring, forced hyacinths, which make a brilliant display. At the bottom of the back wall of the veranda, a collection of chrysanthemums are placed when in flower, and their stems being tied to the rods which support the vines, they flower beautifully in the month of October and November, remaining in mild seasons till Christmas. We mention this for the sake of showing that a veranda of this kind is capable of affording a great deal of floricultural enjoyment. Under the veranda on the north side of the building no plants are trained, the occupier preferring to be without them. The conservatory (*g*) is planted with different varieties of Caméll*ia*, and the piers with Eriobótrya japónica, Lonícer*a* flexuòsa and japónica, Wistàr*ia* sinénsis, Passiflòra cærùlea, and vines: the latter are the royal muscadine, the muscat of Zante, muscat of Alexandria, Money's West's St. Peter's, and the black Hamburg. These vines are trained round the inside of the glass dome, and produce a good crop yearly. The door shown in the back part of the veranda has the panes of looking-glass; and, before the camellias grew large, it reflected them, as seen from the front garden, and from the road, in a very striking manner.

We have now completed the planting of the front gardens; and all that remains is, to state that the surface of the soil among these plants, composing much the larger half of both gardens was next sown down with grass seeds, in order to be kept in turf; the cropping of the grass among the stems of the shrubs being done by hedge-shears. As a finish to the turfed part of the gardens, and also as a definitive line of demarcation between it and the dug part, the dwarf pedestals and vases at *&* *&* were introduced. These vases also harmonise with the vases which form the crowning termination to the pillars of the veranda.

The Planting common to the Back Gardens of both houses will not require many details. In each of the two small angles between the back entrance and the shed a walnut tree is planted, which, having been 20 ft. high in 1823, soon overshadowed, not only the entrance, but even the roof of the shed. There are also common ivy, Virginian creeper, a China rose, and *L*ýcium bárbarum planted in these angles, from which they are trained over the shed and the boundary wall, mantling over and greatly enriching both from the lane. The other trees immediately within the eastern boundary wall are pears, a golden pippin apple, and a mulberry. The pears are the Chaumontelle, glout morceau, Duchesse d'Angoulême, Marie-Louise, and beurré Spence. On the west side of the north shed a giant ivy and a vine are trained; and on the south side of the south shed Lonícer*a* japónica and *C*ydònia japónica. The fragrance of Lonícer*a* japónica (the Lonícer*a* flexuòsa of most nurseries) is so great, when in flower, that, when coming home from London late in the evenings, when the wind has been in the west, we have felt its sweetness at the distance of nearly a quarter of a mile. The wall of the north garden which faces the south is planted with peaches and nectarines, and the east wall with apricots. Down the centre of the spaces enclosed by the trellises *f f*, a row of standard apple trees was planted, chiefly the Hawthornden and other early-bearing sorts, as it was intended to take them away as soon as they produced too much shade on the ground below. The north side of the party wall of the south garden was planted with cherries and plums, and with standard pear trees, at regular distances, so as to produce shade in the summer time on the walk, and to admit the sun's rays during winter. Ivy, honeysuckle of different sorts, and climbing roses, were subsequently planted against this wall; but the ivy has now taken entire possession of it, and forms a mantling covering to the coping from one end of it to another. The eastern boundary wall of the south garden is planted with Chimonánthus frágrans, Magnòl*ia* Soulange*àna*, *J*asmìnum revolùtum, and *R*òsa Boursaúlt*i*, and other roses. In the centre is a plant of ivy, which is trained with a single stem as high as the coping, on the top of which, under a chevaux-de-frise, it spreads both ways, and forms a fine mantling canopy, The footpath in the lane being formed over an old, deep, dry ditch, by filling it in with the soil of an old grubbed up hedgerow, it occurred to us, when building the wall, that it would form an excellent place for the root of a vine. We accordingly had a small hole left in the wall, about a foot under the surface, and, in due time, thrust through it a strong three-years-old cutting of the black Hamburg 4 ft. in length, which has since grown with extraordinary vigour, mounting to the tops of the pear trees, and producing abundance of fruit, which, in fine seasons, are coloured (though not ripened), and which give the whole a character of Italian scenery, admirably in keeping with the veranda round the house.

The edgings to the walks of both back gardens are of box. That of the front gardens, from *d* to *&*, was originally of turf; but the grass not growing well, in consequence of the crowded state and greatly increased size of the shrubs, we found it advisable, some years afterwards, to substitute an edging of brick, laid flatwise, without mortar, alternately header and stretcher; and which, joining in with the turf on each side, soon became covered with weather stains, and now forms altogether an admirable edging for this description of garden: it may be considered a kind of architectural string-course or band, which, expanded into a parallelogram or square basement, as at *a* or *&*, forms an excellent foundation for erecting a pedestal for a statue or a vase. It also harmonises with the band of brick laid on edge which forms an apparent basement round the veranda. Having now given a general view of the laying out and planting of both gardens, the remaining part of what we have to say respecting this design will be limited to the south garden, or that which is in our own occupation. The north house is let to parties who care little about gardening; but, as they are precluded in the lease from either taking down or planting trees, or adding to or taking from the house, without our permission, the general appearance of both gardens, as seen from the road in front and the lane behind, is the same; though the greater part of the back garden of the north house is now under grass.

A main object that we had in view in binding down our tenant neither to add to, nor take anything from, the exterior of the house, without our permission, was, to preserve the unity of effect of the two dwellings; and to prevent that appearance so common about London, of one half of a double detached house being painted, whitewashed, or otherwise renovated, and the other half not so. We also wished to prevent the placing on the chimney tops of any of those unsightly contrivances for curing smoky chimneys which are so common in and about London; undertaking, however, that, if the chimneys should smoke, we would cure them, or grant permission for this being attempted in any way the occupier chose. A certain number of the trees in both the front and back gardens being also necessary to preserve unity of appearance, a list of these was given to the tenant, who was bound not to cut them down without leave. We mention these particulars, to afford hints to others intending to build double detached houses, which every person of observation will allow are frequently disfigured for want of some similar arrangement.

Culture and Management of the south Garden. The following particulars respecting the planting, management, and changes which have taken place in the garden in our occupation, from the autumn of 1823, when the exterior of the house and the boundary walls of the garden were completed, till the autumn of 1837, when this account was drawn up, may not be without their use, in showing how much may be effected, and how much interest and enjoyment produced, on a very limited spot, and in a moderate space of time. We do not offer this gardening history as an example to be generally imitated; but rather as to show what may be done, where there is the inclination and the means. We have already described the manner in which both gardens were laid out, and planted for general effect, whether seen from the public road, or street, in front (Porchester Terrace), or from the road at the back (Craven Lane). While this general effect was studied with reference to both gardens, the south garden, which we intended for our own occupation, was planted with specimens of almost all the kinds of trees and shrubs that could, in 1823 and 1824, be procured in the London nurseries; our object being to allow the whole to grow up together as long as they could, in order to study the species and varieties; and then to thin out, by degrees, all the finer kinds rendered weak by the stronger kinds, when it was necessary to retain the stronger kinds for general effect; and to thin out the stronger kinds as soon as they began so far to injure one another as to become unsightly. In this way, we had the satisfaction of, as it were, passing in array before us a living specimen of almost every shrub and tree propagated in the London nurseries; the only exceptions being those kinds of which plants could not be procured, and those of which the varieties were too numerous for admission in our limited space. Among these were the pines and firs, the willows, the oaks, and the roses; though of these last we had 100 select sorts, chiefly standards and budded dwarfs. With respect to fruit trees, we planted above fifty kinds of apples, nearly as many kinds of pears, and a corresponding number of plums and cherries, with some peaches, nectarines, apricots, figs, and vines. We planted all the best gooseberries on both sides of the double trellis *f f;* and from these, and all the fruit trees, even to the walnut, filbert, mulberry, medlar, quince, true service, and almond, with the single exception of the sweet chestnut (three successive trees of that species having died), we have had fruit. Many of the trees, as soon as they began to bear, we were obliged to cut down to make room for the others; and there are not now above six sorts of apples, fifteen of pears, three of cherries, and three of plums, remaining. The peaches, nectarines, and apricots, all of which bore fruit, were rooted out, about seven years ago; and also a number of the vines, which bore most abundant crops.

In the hot-house, for the first year, we grew melons; and, for the next two years, pine-apples; though we did not grow the latter very well, as, from the house being heated by a flue, and the air being dry, the plants were stunted and thrown prematurely into fruit, which, however, though small, contained a larger quantity of the saccharine juice of the plant, than if it had been properly grown, and consequently swelled to a large size. We mention this for the sake of showing that highly flavoured fruit is not always a proof of good culture on the part of the gardener; and also that, when large fruit is the object, the pine plant should be kept in a growing state during the whole course of its culture. Whenever a pine plant, a palm, a bulb, a grass, or any monocotyledonous plant of a certain age, and in good health, has received a check, either by the withdrawal of heat or of moisture for a short time, when these stimuli are reapplied, the plant will, in all probability, produce flowers; or, as the technical phrase is, when speaking of the pine, "start into fruit." About this period, we also fruited Cattley's guava, and had the cherimoyer, and other tropical fruit trees, on the back wall, grapes under the rafters, and cucumbers in boxes. In 1830, this house was entirely devoted to a representative system of tropical plants in pots; but it is now (1837) only occupied with grapes and figs. It has been heated successively by flues and tan, by flues without tan, by hot water in the common method, and by Perkins's system of hermetically sealed hot-water pipes. In altering the interior of the house to fit it for a representative system, we excavated the floor 5 ft. in depth, so as to form a large tank for water, which collected there from the gravelly subsoil, and added considerably to the quantity supplied by the pump in the shed. Over this tank there was a paved floor supported on piers; and on this floor a stone shelf in front and at the ends, and a broad stone platform in the centre, in lieu of the pit, were erected; and on these the pots for the representative system were placed, and earthed up nearly to the brim. The number of plants required to represent all the orders and tribes of hot-house plants in cultivation in Britain in 1830 was eighty-two; and this number of pots was arranged on the shelf and the platform. For such of the orders as we could not procure plants, we had pots merely filled with sand, to distinguish them from the others, with labels bearing the names of the orders and tribes for which they were intended. Through the kindness of Messrs. Loddiges and Mr. Aiton of Kew, we were enabled to procure plants to represent the greater number of orders; and these were inserted in their proper places, with the name of the order and tribe on one label and that of the plant on another. In the centre of the front shelf there was a cistern for the orders represented by aquatics; in which we had *Papy̆rus*, *Nelúmbium*, Pontedèria crássipes, &c. The labels were of porcelain, written on with Indian ink, and varnished; those for the orders and tribes being a size larger than those for the specific names of the plants. The plants and the labels were so placed as to be seen and read from the path. The great heat and constant attention required to keep these plants in order being wanting after the first winter, and in the second winter the pipes having burst with the frost, they all died, and we have never had the courage, nor, indeed, could spare the time, to renew them.

The green-house (*r*) was originally planted with peaches and nectarines against the back wall; and with cucumbers, melons, and gourds in the border. In the second year, we divided the border into compartments, by cross walls of brick on edge laid in cement; and in every alternate compartment we laid a layer of rotten stable dung, powdered bones, and sugar-maker's scum, filling up the compartment with very rich sandy loam and lime rubbish. In this we planted two vines close against the back wall, and two between the back wall and the front; training the latter on two horizontal iron rods, fixed at about 14 in. from the surface of the ground. The vines had been five or six years in large pots before they were turned out, and began to bear immediately. In 1829, they bore an enormous crop, as did those trained against the back wall, and outside the glass frames on the coping. In 1830, we removed all the peaches and most of the vines, and planted against the back wall a representative system of green-house plants; that is, one plant of every order and tribe requiring the protection of a green-house in the climate of London. Most of these plants still remain, and are in a thriving state, though they have had no fire heat for the three last winters. Among them is the dwarf fan palm, which has grown vigorously. The border we planted with Cape bulbs, especially ixias and their allied genera.

The pit *x* was for two years devoted to the culture of melons, and in it we fruited the water-melon in 1825. In the pit *y*, cucumbers were chiefly grown. In 1826, both pits were filled with green-house plants, chiefly natives of Australia and the Cape; and a good collection was kept in them till 1835, when we presented the sashes to a friend, and filled up the pits with soil; planting the one with a collection of hyacinths, and the other with a collection of pæonies. The hyacinths flowered beautifully the first year, but during the second winter, from neglect, the bulbs were entirely eaten by the snails. This bed has now some varieties of China roses planted in it; and on both beds the ivy is allowed to make advances from the side walls; so that both will soon be evergreen masses, and require no attention whatever.

The bed *z* was, for the first year, devoted to culinary vegetables; but, in 1825, an arrangement of herbaceous plants was made in it, in such a manner, as that an equal number of each of the most common colours among flowers was in bloom every month in the year, except in December and January. In 1830, the trees surrounding this bed having completely overshadowed it, as well as the pit *y*, the herbaceous plants were removed, and only bulbs and a few shrubs were allowed to remain. As the trees are rapidly increasing in size, the next change will be to turf.

In the small compartment under the south boundary wall, marked 1 in *fig.* 108. p. 326., we had at one time a tank of salt water, in which we grew, or rather kept alive for a year, several marine fuci. These were collected for us on the sea shore at Leith, by Mr. Barnet of the Experimental Garden, Edinburgh, and sent by him, attached to the stones on which they grew, in a cask of salt water. We also grew in this compartment, for two or three years, *R*ùbus Chamæmòrus, received from Mr. Barnet, it having been dug up by him on the Portland Hills, near Edinburgh, and sent to us in masses of turf (divetts, *Scotch*). There were in the same compartment, *R*ùbus árcticus, some American cypripediums, and British *O*rchídeæ. In a few years, these gave way to vacciniums of different kinds, Oxycóccos palústris and macrocárpus, Trientàlis grandiflòrus and europæ`us, and some other peat-earth American plants, received from Col. Carr of Bartram's Botanic Garden, Philadelphia. Subsequently, these beds were filled with a collection of hardy ferns (of which family we had at one time 60 species, partly in the open air, but chiefly in the hot-house), received from Mr. Barclay of Bury Hill, Messrs. Loddiges, and Mr. Aiton.

The plot enclosed by the double trellis *f f* was originally laid out in beds, in the same manner as the corresponding plot in the north garden; and, in the beds of both gardens (for the north garden was two years in our possession before the house was let), we passed in review specimens of all the culinary vegetables grown in British gardens, including salsify, skirret, cardoons, asparagus, artichokes, sea-kale, rocambole, and even the roots of *Œ*nothèra biénnis, and of *S*tàchys palústris; and the Bath asparagus, that is, the flower-stalks of *O*rnithógalum pyrenàicum: the stachys roots having been brought into our notice, about that time, by Mr. Houlton, a surgeon, at Paddington; and the ornithogalum flower-stems by Mr. Capper of Bath. In 1826, the cross beds in the south garden were destroyed, and the ground laid out in three longitudinal beds, with side walls of brick on edge, laid in cement, so as to raise them to the height of 18 in. above the alleys; and, in order to cover these brick walls, an edging of box was planted outside of them, which, in three years, grew as high as the bricks, and has ever since completely concealed them. For three years, these beds were planted with florist's flowers; and in them we had above 150 sorts of tulips, nearly half that number of ranunculuses, anemones, pinks, carnations, primroses, polyanthuses, and even dahlias. In 1831, the florist's flowers gave way to a representative system of hardy herbaceous plants, with which, as far as the plants are alive, they are still occupied.

The bed 2, at the bottom of the garden, from the Polish beehive to the corner, was devoted to low-growing herbaceous plants of the more rare and beautiful kinds, intermixed with select bulbs; and against the wall were varieties of China roses, fuchsias, Aloÿsia citriodòra, and other half-hardy shrubs, with Chimonánthus, and the other kinds enumerated in p. 339. Besides these, there is a plant of ivy, trained to a single stem till it reaches the top of the wall, when it branches to the right and left, so as to cover it with an evergreen mantle. There is also a vine, the root of which, as already observed, is in the lane, and the stem brought through a hole in the wall under ground; and, as the roots run along in an old road-side ditch that was filled up when the wall was built, it grows with great vigour, and produces a fine effect among the foliage of the ivy and the pear trees, some of its shoots reaching their summits, and reminding the spectator of the south of Italy.

The border between the trellis *ff* and the walk in the south back garden was originally planted with a complete collection of strawberries, named agreeably to the Horticultural Society's *Catalogue of Fruits;* but, in 1826, these were removed, and the border was divided by flints and vitrified bricks into 464 small divisions, to contain that number of orders and tribes; and, consequently, representing the whole vegetable kingdom, hardy or tender, indigenous or exotic, cultivated in Britain. All the hardy species, ligneous and herbaceous, that we could procure, were inserted in their proper places; and as many of the green-house and hot-house species necessary to represent orders, as we could obtain, were planted out in the summer, and taken up again in autumn, and kept under glass till the following spring. In 1830, we removed all the flints and vitrified bricks, and took up all the plants, substituting for a universal representative system one limited to the representatives of the hardy trees and shrubs cultivated in Britain, amounting to 58 species. In the same border was planted a complete collection of hardy bulbs, especially narcissi. The stronger-growing trees and shrubs in this miniature arboretum are taken up every other year, and their roots reduced, in order to keep them of moderate dimensions. The border between the trellis *t* and the walk is planted with a selection of dwarf roses, all budded. They were procured from the Hammersmith Nursery in 1824; and, though they have never been taken up and replanted, they still continue to grow and flower with considerable vigour. Among them is planted a miscellaneous collection of the most ornamental bulbs. Against the side next the walk of the trellis *t* are planted showy flowering shrubs, such as Cydònia japónica, *R*ìbes sanguíneum, *H*ibíscus syrìacus, double-flowering peach, and a variety of others, including several kinds of honeysuckle, and upwards of twenty varieties of China and climbing roses; among which are the Rose de Lisle, *R*òsa Boursaúlt*i*, *R.* Drummónd*i*, *R.* rùga, and *R.* Russell*iàna.*

The north side of this trellis is covered entirely with the red Warrington gooseberry, which bears abundant crops every year; and the fruit, being in the shade, hangs on, in a dry season, for a month or six weeks after it is ripe; but in a wet season, or when flies are very abundant, it bursts, or is rendered unfit for use, in less than half that time.

The border under the south boundary wall, marked *i* in the section *fig.* 113. in p. 331., was originally planted with fruit trees to be trained against the wall, and subsequently with ivy and standard fruit trees, and with different species of *D*áphne placed at regular distances. Between the daphnes were pæonies and bulbs; and, next the box edging, a collection of different sorts of crocuses, mixed with snowdrops and the early-flowering species of *S*cílla. At one period, we attempted to grow a collection of mosses in this border, and procured a cart-load of common live moss, from Mr. M'Intosh of Claremont, as a nidus to grow the finer sorts in.

The stone shelf for alpines (*u* in *fig.* 108. p. 326.) contained, at one time, 600 species in small pots, all named, including *Cúscuta* europæ`a and *O*robánche màjor: and the cistern at the end next the hot-house contained, in pots, *N*ymphæ`a álba, *N*ùphar ádvena, *S*círpus lacústris, *T*ỳpha latifòlia, Sagittària sagittifòlia, and other aquatics, all of which, though crowded together, flowered luxuriantly; the bullrush growing several feet higher than the end of the hot-house. The shallow cistern, in which the water stood at only the depth of 1 in., contained, also in pots, *C*ómarum palústre, *B*ùtomus umbellàtus, *C*álla europæ`a, *C*áltha palústris, and a number of others, all of which flowered. In this substitute for a marsh, we also kept Drósera, Lycopòdium, *S*phágnum, Marchánt*ia*, and several other mosses and lichens.

That portion of the garden which is under turf was interspersed with crocuses, snowdrops, narcissi, scillas, hyacinths, and other bulbs, in great quantities, near the roots of the trees and shrubs. Besides these, broad irregular patches of cowslips, with sweet-scented violets, lilies of the valley, and daisies, were introduced; and these, particularly the cowslips, produced a fine effect in the spring months. About 1829, it became necessary to cut down several of the trees and shrubs, which had already, in many places, destroyed the bulbs and other plants; and, when the shrubs were removed, the ground appeared naked. These naked places we planted with cuttings of ivy; and we introduced among them some hundreds of plants of the commoner kinds of fern, collected by the hedge sides, chiefly in the neighbourhood of Edgeware. All the parts of the surface so planted were covered with flints, to retain the moisture during summer, and to give a rough picturesque character to those parts of the scene. Not liking the appearance of this, we soon had all the flints and ivy removed, and substituted grass seeds. We still retained the primroses, violets, bulbs, and many other plants which come up among the grass in spring, and created an intensity of botanical interest in every square yard of the surface. At one time, there were between 300 and 400 different species distributed over the lawn and among the shrubs (many of them named); insomuch that there was scarcely any room for mowing, and the grass was obliged to be kept short by clipping.

We have omitted to mention that we endeavoured to cultivate the truffle on a portion of this lawn, but without success, though we had tubers and mould sent to us by Mr. Pearson, from Wyre Forest, near Bewdley.

We might say a great deal more on the subject of the numerous experiments we have tried in this small piece of ground, in the course of twelve years; our great object being, as far as we could, to try every thing that could be tried in so small a space. Owing to the thorough preparation which we gave the soil, and the liberal supply of water during the growing months (viz. May, June, and July), the growth of the trees and shrubs was so rapid, that it would hardly be credited, were we to detail it; and, as we procured the fruit trees of as large a size as they could be found in the London nurseries, and planted them in the autumn, in the "watering-in manner," to be hereafter described, they produced some fruit the first year, a tolerable crop the second and third, and in 1828-29 and 1830, many bushels of apples and pears. So large had these trees become, that there is scarcely one of them now remaining. The greatest number of species that we ever had at one time (exclusive of varieties) was about 2000; but these are now reduced to a very few hundreds, not only from the growth of the trees, but from our having ceased, two or three years ago, to keep a gardener. Notwithstanding this, the general effect of our garden, not only from the street in front and the lane behind, but in walking through it, remains the same;

because that effect depends on the disposition of the trees and shrubs: and the great advantage of these over herbaceous plants is, that, after they are grown up, they require very little care; and that this care, which consists principally in reducing them by pruning and thinning, can be performed at any time of the year that is most convenient. The results of our experiments that we consider as likely to be most useful to the occupiers of similar gardens are, the general disposition of the trees and shrubs, which has been already given in the plan *fig.* 108. in p. 326.; a list of the collection of plants which will form a representative system of the whole vegetable system, as far as respects the plants indigenous to, or introduced into, Britain; and a list of the hardy herbaceous plants calculated to form a perpetual flora. The first of these lists, which is chiefly interesting to a botanist, will be found in the *Gardener's Magazine*, vol. vii. p. 150—176.; and the second we shall give on some future occasion.

Expense and Management. The two houses, including the sheds at the bottom of the garden, but not the hot-houses, cost upwards of 5000*l.* The trees and shrubs for both gardens, exclusive of those more rare kinds which were planted in the south garden, being the largest plants that could be procured, cost upwards of 100*l.*; and the glass structures about 300*l.* The whole was obtained at the trade price, or, at least, at 20 per cent less than what it would have cost, had we not been considered as belonging to the professions both of architect and gardener, from our having practised as a rural architect and a landscape-gardener.

This large amount for two small dwellings, not worth more to rent than 100*l.* a year each, is not to be taken as any guide for the general cost of such buildings; because the price of bricks and timber, at the time we built, was nearly one third higher than at present; and because the expenses attendant on the glazed-roof veranda which surrounds both houses, and the domical conservatory, may be considered as having been incurred for a species of luxury rather adapted to the taste of the individual than to the general taste of persons who occupy similar houses. If we deduct 1500*l.* for extra expenses on the house and garden, we shall have about the cost which such houses may now be built for; with gardens laid out and planted in the way which we would recommend in ordinary cases. With respect to the expense of managing these gardens, if committed to the care of a jobbing gardener, each garden would be kept in order, and all the requisite seeds, plants, manure, &c., supplied, for 20*l.* or 25*l.* a year for each garden. It is needless to say, after what we have stated of the experiments we made, that our expenses of management were more than ten times either of these sums, till within the last three or four years, when we greatly reduced our establishment, devoting all our resources (pecuniary and otherwise) to the *Arboretum et Fruticetum Britannicum*, the last manuscript of which we have this day (Nov. 25.) just sent to press.

Remarks. It will, perhaps, be difficult to find any person but ourselves, to whom a garden treated as ours has been would be suitable; because the end that we had in view was scientific knowledge, rather than ordinary enjoyment. We allude to the great number of changes which we made, in order not only to have a great many plants in the garden under culture at the same time, but to prove the culture of the different plants, and to familiarise our minds, as much as possible, with the aspect of all the plants, useful and ornamental, of the hardy kinds, which can be grown in British gardens. In particular, we were desirous of becoming, as it were, intimately acquainted with all kinds of hardy trees and shrubs adapted for enriching pleasure-grounds; and in this we certainly succeeded in an eminent degree. Had we confined ourselves to herbaceous plants, instead of growing 2000 species at one time, we might have had 10,000 in our limited space; as may be seen by turning to p. 234., where a calculation is made, by which it is shown that 1800 species may be grown on the tops of the boundary walls of a garden very little larger than ours.

Additions and Alterations. In a convenient and prosperous country, where the circumstances of individuals vary considerably in the course of their lives, it seldom happens that the same house will suit the same individual, from the time he begins housekeeping to the end of his life. He may require to change his locality, to have a smaller house, or to remove into a larger one. For the first two of these cases, no provision can be made in building a house, as they necessarily involve a change of residence; but this is not the case with the third; and, consequently, it is not unusual, in building a house, to keep in view probable additions. We mention the last circumstance here for the first time, because it is not applicable in the case of houses which form a connected street or row; except, indeed, by adding to their height. In detached houses, however, whether single or double, additions may be made on one or more sides; more especially if the idea of the probability of such additions being required entered into the mind of the architect when he was forming his plan. In all double detached houses which are strictly symmetrical, it is impossible to add a room to one of the houses without destroying the symmetry of the entire building; and, even if this point were given up, if the rooms on the ground floor are lighted on both sides, an additonal room cannot well be added. In the case of the double house before us, the rooms on the ground floor are each lighted by one large window on one side; and hence, supposing the occupiers of each house to agree in a desire to add one or two additional rooms, this might be done without the slightest deviation from perfect symmetry. Both houses might also be raised by means of a curbed roof, with dormer windows; though this does not look well, unless the greater part of it is hidden by the parapet.

Fig. 115. shows how one room may be added to the south house, the plan of which is shown in *fig.* 108. in p. 326. A portion of the veranda is sacrificed from the entrance porch in *fig.* 108., and the water-closet *f*; the latter is removed to the situation *b*, in *fig.* 115.; the porch is enclosed, as shown at *c*; a room is formed at *a*, 12 ft. by 16 ft., and joined to the library (*k*) by a door through the situation in which a window would have been placed, had there been two in the room, and which was arched over when the house was built, in case a window or a door should afterwards be required in that situation. There is a similar arch in the wall, in the same position, in the room *i*, from which a door might be opened to another room of the same size as the last. In that case, the water-closet *b* would be removed to a situation at *e*; or, what would be better, by increasing the length of the porch (*c*), a water-closet might be got in the situation of *b*, and a bath room of the same size on the opposite side. Both rooms, in this case, might be projected 3 or 4 feet at *f* and *g*; which projections would harmonise very well with those of the domical conservatory (*g* in *fig.* 108.), and the office of the *Gardener's Magazine* (*n*). We by no means say that these alterations are desirable; we merely speak of them as possible, and for the sake of directing attention to the advantage of keeping in view, when a house is built, the possibility of additions or alterations being at some future time required. For this purpose, wherever openings, such as windows or doors, might have been placed, it is desirable to throw arches over such places, in order that, should the alteration be required, it may be effected at any future time, with ease and perfect security.

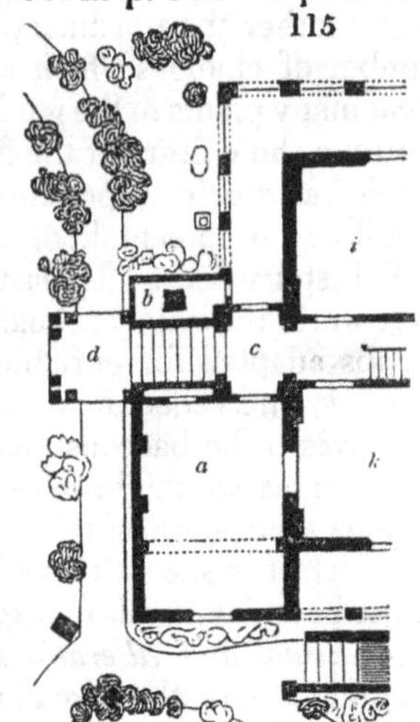

A double detached House with both Entrances under one Porch. In this design (*fig.* 116.) there is a double street entrance, with double doors under a porch, with a walk between the porch and the street for each, separated by a wire fence. In this plan, the houses are, in effect, brought one half nearer to the front entrance than in the preceding plan; partly from the porches being projected from the main body of the house, but chiefly from the only entrance to each house being in front, and immediately opposite to the street gate.

In small houses like those before us, this is a considerable advantage, in saving the time of servants in opening the street entrance, and one which will be particularly felt, if the occupier of the house should be a professional man; such as a medical practitioner, a solicitor, or the holder of any parish or public office, &c. The present design is also well adapted for a person wishing the one house to be occupied as his office, and as the residence of his principal clerk, and the other to be retained as his own residence. By such an arrangement, he would have all the advantages of a large house; while he would only pay the government and parochial taxes for two small ones. Thus, there need not be more than nine windows in each house; which, for the two houses, would cost two guineas per annum; while, for eighteen windows in one house, he would pay nearly 5*l.* per annum. In *fig.* 116. (which is to a scale of 20 ft. to an inch, as

116

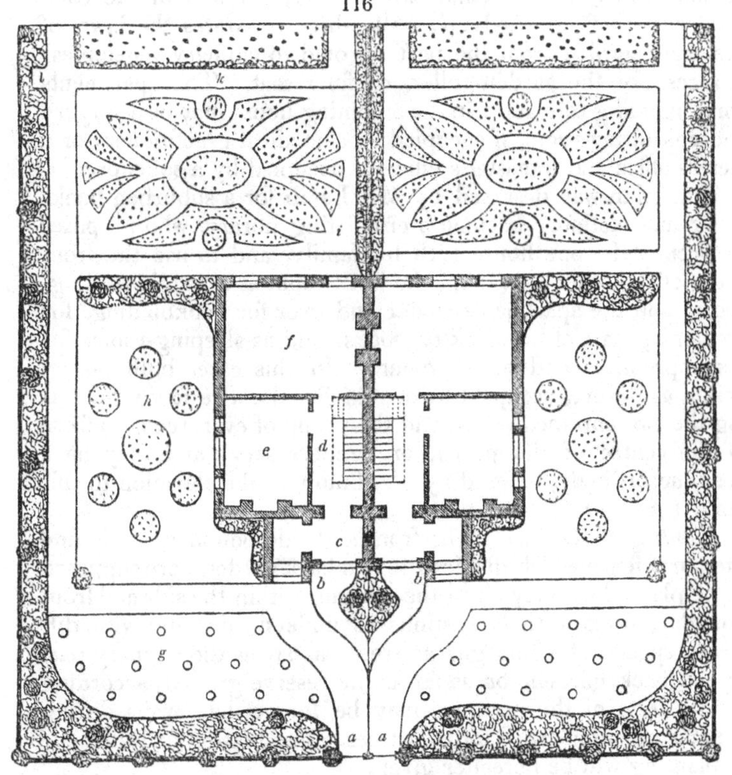

in fig. 106. p. 320.), *a a* are the front entrances; *b b*, the entrances to the double porch; *c*, the entrance to one of the houses; *d*, the hall and staircase; *e*, living-parlour; *f*, drawing-room; *g*, front lawn, with small circular beds for growing dahlias;

117

h, side lawns, with circular flower-beds; *i*, lawns in the back garden, with flower-beds; *k*, walk which separates the lawn of the back garden from the part devoted to culinary vegetables; *l*, recess for the garden roller, or for a seat. The apartments corresponding to *e* and *f* in the adjoining house may be a surgery and consulting-room, if occupied by a medical practitioner; or a clerk's office and a master's office, if occupied by a solicitor.

This plan was designed by Mr. Lamb for a suburban bookseller and stationer keeping a circulating library, who proposes to occupy the one house with his family, and to use the front room of the other as a shop, the back room as a public reading-room, and the apartments under and over for bookbinding, for containing part of his stock of books, and as sleeping-rooms for his apprentices and his shopman. In this case, both houses being, as it were, occupied by one family, the wire fence separating the two entrance walks, and the group of evergreens in front of the centre of the porch, are unnecessary; and they have been, accordingly, omitted by Mr. Lamb, in the elevation to this plan. (*fig.* 117.)

Planting the Garden. The front and side boundary walls, and the party fence which divides the two back gardens, are supposed to be planted with ivy; and the margins within the side and front boundary walls with laurustinus as undershrubs, and with different species of *Cratæ'gus* as trees; as far as the culinary part of the back garden, or as far as the reserve ground, according as the taste of the occupier may be for verdant walls during winter and summer, or for fruit trees. The reason for this mode of planting will be hereafter given.

In the front garden, *g* in *fig.* 116. shows small circular beds (which may be from 1 ft. to 1 ft. 6 in. in diameter) on turf, at regular distances, for containing a few select dahlias, neatly trained to stakes. Instead of dahlias, any other tall-growing plant with showy flowers might be substituted; and the best of these would be chrysanthemums, provided the situation were exposed to the south, the soil warm and dry, and the chrysanthemums brought forward in pots, so as to commence flowering in August. In situations not adapted for chrysanthemums, the Lavâtera arbòrea, *L*upìnus mutábilis, *Œ*nothèra biénnis, Persicària orientàlis, and various other tall-growing flowering plants might be substituted. The object of this mode of planting is novelty; that is, to produce a kind of display which is at present seldom, if ever, to be met with in front gardens. For the same reason, standard roses might be substituted for tall-growing herbaceous plants; or standard rhododendrons, mixed with standard azaleas, might be used; or, where there was a large pit in the reserve garden, fuchsias, brugmansias, or pelargoniums might be preserved in it through the winter, for filling the beds during summer; or a green-house or hardy twiner or climber might be planted in each bed every spring. Even the scarlet runner and the common nasturtium, when neatly trained to stakes, will form most ornamental plants; continuing to flower for a whole season, while their fruit, also, is of use. Lophospermums, maurandyas, eccremocarpus, *C*onvólvulus màjor, &c., are most ornamental climbers or twiners, and very suitable for the purpose in view.

The groundwork of the side garden is also turf, varied by circular beds, of greater diameter than those of the front garden, and symmetrically disposed. These beds consist of a centre bed, 6 ft. in diameter; and six smaller beds, each 4 ft. in diameter. In each of the latter, one kind of annual or low-growing perennial may be planted, so as to cover the entire surface of the bed; and in the centre bed there may either be a collection of herbaceous plants of low growth, and with an equal number of the common colours, for flowering in every floral month, or China roses may be substituted: or, to give a greater latitude to choice, and to include some of the cheaper modes with others that are more expensive, one or other of the following methods of planting may be adopted: —

1. Sow each of the six small beds with one of the Californian annuals mentioned in p. 293., leaving it to perpetuate itself with the occasional aid mentioned in the page referred to; and sow the centre bed with a mixture of all of these annuals, and some mignonette, in the manner described in p. 293.

2. Plant all the beds with low-growing perennials, from the list given in p. 217., arranging the colours according to the taste of the party.

3. Plant the beds with low-growing perennials, alternating with bulbs, agreeably to the list in p. 220.

4. Plant the beds with annuals every spring, one kind in

a bed; the kinds to be raised in the reserve garden, or procured from a commercial gardener.

5. Plant all the beds with evergreen perennials; each plant to be kept perfectly distinct. Those for the centre bed may be *S*axífraga crassifòlia, which is as valuable among evergreen herbaceous plants as the holly is among evergreen shrubs. The other beds may be filled with different species of evergreen saxifrages; or, what would look better in winter, with carnations and picotees in two opposite beds, and pinks and sweetwilliams in the four others, disposed symmetrically; that is, the two beds of sweetwilliams being placed opposite each other, and the pinks in the two other beds; or the centre bed might be filled with wallflowers, and the others with different kinds of stocks.

6. The beds might be planted with low evergreen undershrubs, which would look well at every season of the year. For example, the centre bed might be filled with rosemary; and the surrounding beds with lavender, sage, hyssop, winter savory, and thyme. These plants would serve, at the same time, as sweet herbs. Or the centre bed might be filled with *C*ístus créticus, or any low-growing cistus, and the surrounding beds with helianthemums of different colours: or the centre bed might be filled with *I*bèris sempervìrens, and the surrounding beds with alyssum, arabis, aubrietia, vesicaria, cardamine, and barbarea.

7. The centre bed may be occupied by a cone of climbing roses; the cone to which the plants are trained being formed by a circle 3 ft. in diameter, composed of eight rods of Kyanised wood, or cast iron, placed about a foot apart at the bottom, and brought to a point at top, about 10 or 12 feet from the ground, and there tied together by wire. One kind of rose may be planted at the foot of each rod, and trained to it; and some of the kinds ought to be evergreens, so as to prevent the cone appearing bare in winter. The roses may be selected from the lists which will be given hereafter; but, in the mean time, we may suggest that some of the most free-flowering sorts are the Boursault rose, the rose de Lille, and the Queen of the Belgians; the last being a beautiful white. *R*òsa rùga is also a free-flowering and most beautiful rose. There are several varieties of evergreen roses, one of the finest of which is the triomphe de Bollwyller, which has large and fragrant flowers. The surrounding beds may be planted with dwarf China roses; or with the rose des quatre saisons; or entirely with Lee's crimson perpetual, one of the most beautiful roses grown.

8. The surrounding beds may be planted with Fúchs*ia* globòsa, which, though it dies down to the ground every winter, comes up again the following spring, and flowers most luxuriantly all summer; and the centre bed with Fúchs*ia* microphýlla, which is a hardy evergreen, and does not die down during winter. The

flowers of Fúchs*ia* microphýlla, though small and inconspicuous, will also afford a striking contrast to those of Fúchs*ia* globòsa, which are large, and produced in such abundance as to make the whole plant appear one mass of brilliant scarlet.

9., and lastly. The centre bed may be planted with the tree ivy, all horizontal runners being cut off as they appear; and the surrounding beds covered with vitrified bricks, or fragments of stone, and planted with different kinds of variegated ivy. The principal advantage of this last mode is its cheapness, as not a shilling of expense per annum would be required for keeping the beds in order, while the variegated ivy has a remarkably gay appearance in winter; and the common ivy, being covered with flowers and fruit at that season, is also beautiful, and, every fine day, when the bees are abroad, is covered by them. The beds at *i*, in the back garden, may either be filled with annual or perennial flowers, with or without bulbs; and we shall leave the reader to exercise his taste in doing this, referring him to the rules already given in p. 213., and the hints in various pages; and to the lists from p. 217. to p. 230., as well as to those at the end of this Volume.

The boundary wall, and also the wall which separates the gardens of the two houses, as already mentioned in p. 352. are supposed to be planted with ivy, at all events as far as the walk (*k*) which separates the ornamental part of the back garden from that devoted to culinary crops. The border next the boundary wall is planted with low trees, indicated by a darker shade in the plan, and the shrubs under them, instead of being of different kinds, are supposed to be wholly of laurustinus, than which no shrub forms more picturesque masses in every stage of its growth, or is more beautiful when in flower. The small group in front of the porch is supposed to contain two scarlet arbutuses, and beneath them laurustinus. The undergrowth of the two borders which separate the side gardens from the back flower-gardens is also supposed to be of laurustinus; so that there will be a powerful contrast between the general mass of verdant scenery in this design, and the numerous small parts formed by the beds of flowers. If the trees consisted only of thorns of different kinds, forty-five sorts might be got into the two gardens; so that the planting of this villa would be quite unique.

The part of the back garden which is devoted to the culture of culinary vegetables is supposed to be enclosed by a dwarf hedge; which we should prefer of ivy trained to wire-work, but which may be of box, juniper, laurustinus, common laurel, privet, furze, or of the Lucombe oak. The reserve ground, at the farther extremity of the plot, may either be separated from the gardens by a wall, or an evergreen hedge; or by a row of evergreen or variegated hollies. If, however, the occupiers should be much attached to the culture of fruit trees, a wall or espalier will naturally be preferred in that situation. We may observe, however, that, fruit trees being deciduous, the walls, in small suburban gardens, which are covered by them have always a dreary town-like appearance throughout the winter season; particularly in the neighbourhood of London, where they remind us of those interminable masses of bricks, unvaried by either architectural ornament or vegetation, which meet the eye in every direction in the streets of the metropolis. Hence, if fruit trees are grown on the side walls, ivy ought to be planted against the end wall, or against whichever part of the side walls directly meets the eye from the windows; or such walls ought to be disguised by standard hollies, or other evergreens, being planted in front of them. For our own part, we greatly prefer ivy to fruit trees, for every part of the walls of a street suburban garden; and, next to ivy, a mixture of evergreen and deciduous trees or shrubs, such as magnolias, honeysuckles, roses, &c.

Expense. The cost of a double detached house, like that shown in fig. 117. p. 352., of two stories high, with the kitchens half sunk, would be about 1200*l.*; and each garden might cost 30*l.* more in forming the drains and walks, and in planting the trees and shrubs. This is supposing the house to have very little architectural ornament, either externally or within; and, also, that the roof has projecting eaves, so as not to require lead gutters, and consequently not to be liable to have these frequently out of repair.

Management. There being no green-house attached to these dwellings, so as to require attention during summer, for giving air, &c., and during winter for both giving air and supplying heat, the gardens are well adapted for being managed by persons who have an hour or two to spare in the mornings and evenings; such as clerks in public offices, banking-houses, &c. If the occupier lived all day on the premises, and had a small pit in the reserve ground, heated by a flue, he might preserve green-house plants through the winter, and bring forward seedlings in spring; and, in another pit or frame, in the same ground, if the occupier were disposed, he might keep a collection of chrysanthemums or auriculas in pots; or grow alpines, or, indeed, any kind of plants which require a little protection during winter, but no fire heat. The beds at *i* would afford a fine exercise for his skill and taste, to keep them constantly covered with flowers, symmetrically displayed in regard to colour, and to vary the colours every year. He might also change the kinds of plants every year: for example, one year he might have the beds filled entirely with bulbs, to be succeeded by mignonette alone; another year he might grow only perennials; a third and fourth, biennials; in the first of which years the beds would be green, but would not exhibit a single flower; and a fifth, he might sow annuals, &c. Chrysanthemums very rarely flower freely, even in the climate of London, in the open garden; but, if grown in pots, and brought up when just coming into bloom, and set under the entrance porch, or neatly trained to the walls of the house, they will flower beautifully, and make a fine display till the middle or end of December. If a collection of auriculas were grown in pots, the grower might exhibit them under the porch in spring.

Remarks. A design, planted in the manner suggested, with scarcely anything but laurustinus as undergrowth, nothing but thorns as trees, and with a collection of choice dahlias, distributed at regular distances all over the front lawn, is adapted for a row, or street, in which there is a great deal of botanical variety and richness in the planting of the adjoining gardens. The object is to form a garden which shall contrast with these, become distinguished from that circumstance, and, at the same time, add to the beauty of the verdant scenery of the street. It may be thought that there would be great sameness in having nothing but thorns as trees, and nothing but laurustinus as shrubs; and this would certainly be the case as far as this garden only was concerned; but it must be recollected that they would never be seen by the occupiers, or by the public, but as contrasted with the adjoining gardens, which are supposed to be full of arboricultural variety. The grove of dahlias in the front garden, would also be a new feature in such gardens, and would cooperate with the other causes mentioned, in distinguishing it from the others in the same street. The house, as we have already observed, is adapted for two small families; or the one house for a professional man or tradesman, to reside in with his family, while he carries on his business in the other.

A double detached House in the Italian Style, with the Entrances from a common Terrace. This house, of which *fig.* 118. is the plan, and *fig.* 119. a perspective view of the elevation, may either be occupied by a professional man, as in the preceding design, or by two private families. In the first case, the main entrance would be from the terrace (*d*); but, in the second, though this would be the ostensible entrance, the real entrances would be at the sides, through the green-houses, or plant lobbies (*k*).

The Ground Plan (*fig.* 118.) shows the main walk (*a*) from the entrance gate to the terrace (*d*); two servants', or side, entrances, in the event of the house being occupied by two different families (*b*); stairs down to the sunk area and the kitchen floor (*c*); terrace common to both houses (*d*); open porch, also common to both (*e*); vestibule, leading to the staircase (*f*); vestibule to the green-house (*g*); dining-room (*h*); drawing-

118

room (*i*); green-house, or plant lobby (*k*); library (*l*); balconies to the dining-room and drawingroom windows, with steps descending to the lawn in the back garden (*m*); flower-beds on the lawn in the back garden (*n*); lawn in the front garden, with two arabesque beds for flowers (*o*).

Laying out and Planting. The object, in laying out and planting these two gardens, we shall suppose to be picturesque effect, so as to harmonise with the broken outline, and numerous parts which compose the elevation of the house. The disposition of both trees and shrubs is consequently irregular, and by no means gardenesque. The kinds we shall suppose to be partly evergreen, and partly deciduous; and the prevailing species to be such as are common in the gardens or general scenery of Italy. A list of these will be found at the end of this Volume; but the principal kinds, which are to be had in every nursery, may be here mentioned, and are as follows: —

Evergreen Trees. Pìnus Larício, *P.* Pináster, *P.* Pínea, and

119

P. halepénsis; *Q*uércus *I*'lex and gramúntia; and *C*upréssus sempervìrens, and *J*uníperus phœnícea.

Deciduous Trees. Among these may be mentioned, the Lombardy poplar (which, however, must be very sparingly introduced); the Turkey oak, and the Neapolitan and other acers, various species of thorns, the catalpa, tulip tree, Judas tree, laburnum, almond, *P*ỳrus spectábilis, &c.

Evergreen Shrubs. These include the holly, box, sweet bay, common and Portugal laurels, arbutus, phillyrea, rhododendron, daphne, yucca, and many others.

Deciduous Shrubs. These are very numerous, and include the *P*aliùrus aculeàtus, various species of *R*hámnus and *R*hús, azaleas, lilacs, cistuses, almonds, Coronílla E'merus, colutea, and many others.

As characteristic of Italian scenery, the vine ought to be planted, and allowed to climb up the trees, not for the sake of its fruit, but for effect; and one of the best kinds for this purpose is the claret grape, on account of the colour of its leaves in autumn. The pomegranate, the phillyrea, and the ilex are highly characteristic of Italian gardens; the pine and the *A*rúndo *D*ònax, of Italian scenery; and the agave and the orange tree, both in tubs and vases, of Italian villas. The most characteristic shrubs of the flowering kind belonging to Italy are, the cistus and the cytisus. We do not here mention the myrtle, the olive, or any other trees or shrubs which will not thrive in the open air in Britain, because they could not be introduced with effect in British imitations of Italian scenery. The *R*úscus hypophỳllum, and the shrubby species of asparagus, are also found more frequently in Italy than in any other part of Europe, unless we except Greece. It fortunately happens for the imitator of an

Italian villa in the suburbs of a great city, that the *Pìnus Larício*, the most common pine in the open scenery of Italy; the pinaster, the next common; and the stone pine, which is most generally found near Italian houses, and in their gardens, will all grow remarkably well in the smoke of London; as may be seen by examining the trees of these species in the arboretum of the Messrs. Loddiges, and comparing their appearance with that of the American pines, and even the Scotch pine, growing beside them.

The Flower-Beds. The two arabesque beds in the front garden may be planted with carnations and pinks, which are peculiarly the flowers of Italy; or with wallflowers and stocks, which are also very much prized there; or with evergreen saxifragas, which are common on the Italian alps. Or these beds might be wholly planted with the indigenous bulbs of Italy, which include some sorts of tulips, narcissi, crocuses, scilla, &c., interspersed with Neapolitan violets at regular distances. The beds *n* will have the best effect relatively to the other beds and to the trees and shrubs, if planted with China roses, which were first improved by cross impregnation in the Royal Gardens at Monza. At the roots of the deciduous shrubs, in the groups on the lawn, may be planted some of the hardier bulbs of Italy, along with primroses, violets, &c., to come up through the grass; and, more especially, the colchicum and the *Cýclamen europæ'um*, which are highly characteristic of Italian scenery in autumn, as the crocus, the *Scílla itálica*, and the poet's narcissus are in spring.

The Italian Terrace and the Green-House. For ornamenting these in summer, we would cultivate, in a green-house or orangery in the reserve garden, some orange trees, oleanders, pomegranates, olives, myrtles, and jasmines, in large pots or boxes, to place on the terrace and in the green-house about the middle of May. For training against the walls of the terrace in the autumn, we would recommend a collection of chrysanthemums to be kept in pots in the reserve ground till the beginning or middle of August, when they may be brought out, and the more delicate and late-flowering kinds placed against the walls of the house within the terrace, and the other kinds placed against the walls all round the house; some of the most select being arranged in the plant lobby. The kinds may either be chosen from the old Chinese varieties, of which there are above thirty in cultivation in the nurseries, or from the new British varieties, which have been raised from seed in Jersey and other places, and of which the following is a selection out of above one hundred kinds, which we saw in flower in the autumn of 1837, in the nursery of Messrs. Chandler and Son, Vauxhall Road, London. (See *Gard. Mag.*, vol. xiv. p. 36.)

Selection of Jersey Chrysanthemums for a small suburban garden: —

Early-flowering.

WHITE.

Lùcidum. Petals incurved, flowers in clusters.
Spectábile. Flowers large and quilled.
Vésta. Flowers round, with cupped petals.
Conqueror. French white; flowers large, flat petals.
Virginia. Flowers finely quilled.
Imperial. French white; flowers large, and incurved.
Diadem. Flowers quilled.
Surprise. Paper white; flowers quilled.
Coronet. Flowers round, and in clusters.
Victory. Large flowers, with reflexed flat petals; dwarf.
Marchioness. Flowers large.

PINK.

Queen. Light rose-coloured; flowers double, a little quilled.
Adonis. Flowers in clusters.
Marquess. Pale rose; flowers double, and in clusters.
Exímium. Rose; flowers rather small, round.
Gem. White, with pink tips.
Rosamond. Pink, quilled.
Unique. Flowers twisted, and in clusters.
Celestial. Rosy white; flowers very double; dwarf plant.

DARK ROSE AND CRIMSON.

Calypso. Flowers small.
Decòra. Flowers large.
Grándis. Flesh colour; large flat petals.
Venus. Dark eye, and becoming paler on the outer rim.
Insígne. Pink and red; flowers in clusters.
Conspícuum. Crimson; flowers very large, semi-double.

PALE YELLOW AND LEMON-COLOURED.

Bícolor. White and yellow; flowers small, very double; dwarf plant.
Formòsum. White, with lemon-coloured centre, incurved.
Goliath. Yellowish white; flowers very large, incurved.
Sulphùreum. Pale yellow; flowers in clusters.
Flavéscens. Lemon colour; flowers small, in clusters.

DEEP YELLOW AND ORANGE.

Magnet. Yellow; flowers rather small, very regular.
Adventure. Bright yellow; flowers large.
Aurántium. Orange, quilled.

BUFF.

Triumphant. Pink and buff; flowers large.
Cleopatra. Pink and buff; flowers in clusters.
Mirábile. White, with buff tint; flowers rather small; dwarf plant.
Penelope. Buff; flowers round, semi-double.

Late-flowering.

WHITE.

Enchantress. Creamy white; flowers large and incurved.
Compáctum. Flowers small, very double.
Diana. Quilled.
Invincible. Creamy white; flowers large and reflexed.
Countess. Creamy white; flowers large, semi-double.
Defiance. Silvery white; flowers quilled, incurved.
Eclipse. Paper white; flowers incurved.
Glory. Paper white; flowers large, flat petals.
Grandíssimum. Incurved.

ROSE.

Eminent. Light pink; incurved.
Perspícuum. Pink; fl. quilled, large.
Pulchérrimum. Rose, with white tips, very double.
Empress. Pink; flowers large and tasseled.
Hero. Rose; flowers large.
Rival. Light rose; flowers incurved, semi-double.
King. Pale rose; flowers large, incurved.

Striàtum. Light rose, and red-striped.
Chancellor. White and pink; flowers large, quilled, and incurved.

LILAC AND PURPLE.

Elegance. Flowers very small, very regular.
Perfection. Flowers large, incurved.
Aurora. Purple; flowers large, quilled.

LEMON-COLOUR AND YELLOW.

Angelina. Quilled.
Champion. Lemon colour; flowers large and reflexed.

The Back Garden may either be wholly or in part under turf, and varied by trees and shrubs planted for picturesque effect, as in the front garden; or, it may be in part laid out in beds for culinary vegetables, as in fig. 55. p. 201. The walls should, we think, either be covered with ivy, or with evergreen and deciduous shrubs, and especially the flowering and odoriferous kinds. Among these, and also among the trees and shrubs planted in the back garden, may be some dwarf and standard fruit trees, of the more hardy free-bearing kinds; such as the Hawthornden apple, the glout morceau pear, the Orleans and magnum bonum plums, the morello cherry, the green gage and Warrington gooseberries, Wilmot's red currant, the champagne or striped currant, the Dutch white currant, the Naples black currant, and the cane and Antwerp raspberries. Even if there are no beds for culinary crops, there may still be a few circular beds, distributed in open places, for a few strawberries of different kinds: or, the strawberries may be grown on a cone of earth faced on every side with bricks, flints, or stones; the strawberries being placed in the joints between them. By such an arrangement, the strawberries are obtained a week or fortnight earlier than they would be on flat beds; particularly on the south side of the cone. The advantages of this mode of growing strawberries are, that the fruit may be gathered without stooping; it is certain of being always clean; and, if water be supplied liberally during the flowering and swelling seasons, it will attain a large size. (See *Gard. Mag.*, vol. v. p. 438.) The alpine, or common wood, strawberry, treated in this manner, and supplied with water (which can be done by pouring it into an opening made on purpose in the apex of the cone), will continue in bearing all the summer.

The Reserve Garden should contain a house 10 or 12 feet high, to preserve the myrtle, olive, and orange trees from frost during the winter. This house, provided that no plants are kept in it but evergreens, which, like those above-mentioned, make their young wood during the summer, need not have a glass roof, but only a glass front; in consequence of which, the frost will be very easily excluded by a very little fire heat. The chrysanthemums and other plants may be kept in a pit without flues. All the heat required for the orangery may be supplied by one of Joyce's portable smoke-consuming stoves; and the same means may be adopted for supplying what heat is wanting to the two small green-houses, *k k* in *fig.*118. If tender annuals or green-house plants are to be turned out into the beds *n n* during summer, they, and any other articles of the kind, may be brought forward in a pit heated by flues, by hot water, or by one of Joyce's stoves.

Expense. It is evident, that a house of this kind, covering a large surface, and containing such a great extent of external walling, cannot be considered as economical, even though the greater part of it be only two stories high. It may cost, including the green-houses, between 4500*l.* and 5000*l.*; and the expenses of the garden, including the orangery, pits, and a shed in the back garden, may cost 200*l.*

Management. Notwithstanding the rich appearance which such a garden would have, yet, from the small number of flower-beds, and from the manner in which the arabesque beds (*o*) are planted, it would require less care than the garden of the preceding design (*fig.* 116. in p. 351.); in which the flower-beds alone, to produce their full effect, would require more care than the whole of this Italian garden. This garden might either be managed by the occupier and his servants, in the mornings and evenings, with the occasional assistance of a man to mow the grass, assist in moving the orange trees, &c.; or, what would be more complete, a man might be engaged at 60*l.* a year, with a house to live in, who would do the entire work of both gardens, and keep them both at all times in the highest order. The annual expense of seeds and manure, and of nails, lists, ties, stakes, tools, and fuel, would not exceed 5*l.*

Remarks. It must be observed that this cannot be considered an economical house; but it would afford high gratification to lovers of architectural effect, to an architect, or to an artist of any kind; to whom, also, the garden, with its Italian trees and shrubs, would be an appropriate accompaniment. The house would have an excellent effect in the suburbs of a town among hilly scenery; such as near Newcastle, Edinburgh, Perth, &c. By turning two of the rooms in one house into a picture-gallery, the double house would form a very complete and appropriate dwelling for a landscape-painter.

A double detached House, with the Gardens laid out principally with a view to the Culture of Fruit Trees and Culinary Vegetables. In *fig.* 120., *a* is the entrance from the street; *b*, the steps to the front door; *c*, the hall and staircase; *d*, a library or business-room; *e*, breakfast parlour, school-room, or bed-room; *f*, dining-room; and *g*, drawingroom. This house is supposed to be only two stories high; the upper story containing two best bed-rooms, a servant's bed-room, a nursery containing a bath, and a water-closet; and the basement floor containing a kitchen, back kitchen, and other offices, with a servants' water-

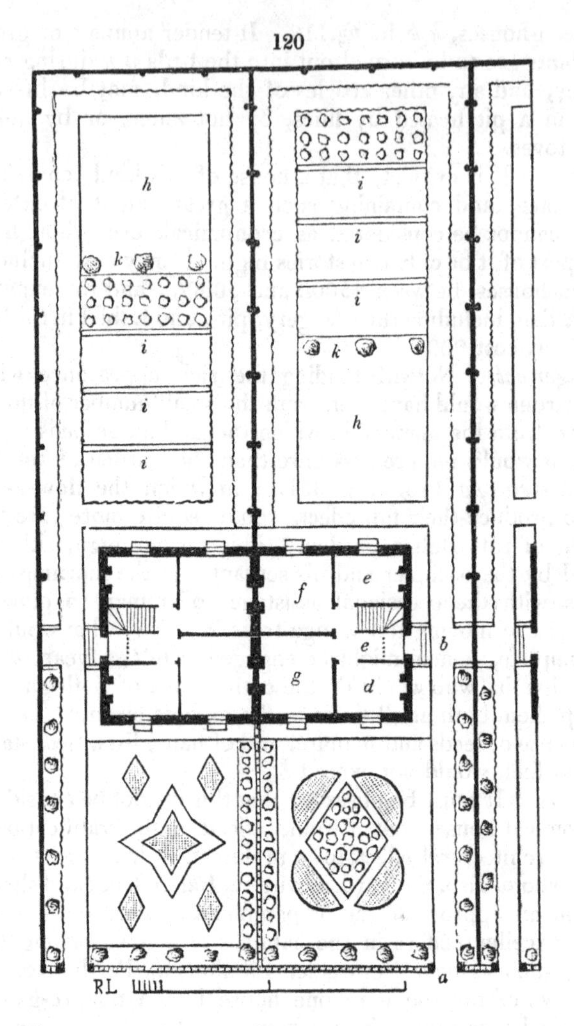

closet. The entrance to each house is protected by a porch, which is extended as far as the boundary wall on each side; thus forming a low horizontal roof, which will contrast effectively with the higher roof of the house. Beneath each porch, there is a bench for sitting on. Both houses are surrounded by a platform, or terrace of pavement, 3 ft. broad, and 9 in. higher than the surface of the walk, and which covers a vacuity all round the house as deep as the foundations; thereby keeping it perfectly dry. This platform also adds much to the dignity of the edifice, and forms a walk which is at all times firm, and comparatively dry and sheltered.

The Front Gardens are of turf, varied by flower-beds. That on the right has a rhomboidal figure in the centre, planted with gooseberries, with four semicircular beds round the central one, planted with strawberries. That on the left has a raised bed in the centre, with a border of turf, and four small diamond-shaped beds to harmonise with the rhomboidal bed in the right-hand garden. The centre bed may be filled with peat earth, and planted with *V*accínium madeirénse, or some other species of *V*accínium which bears fruit freely; and the diamond-shaped beds may be planted with the wood strawberry. The fruit of the common bilberry (*V*accínium Myrtíllus) arrives at maturity at the same time as that of the alpine, or wood, strawberry; and the two, when mixed together, and eaten with sugar and cream, are delightfully refreshing, as every one who has travelled in Poland or Sweden in the summer season must have experienced. In open airy situations, the central bed might be planted with *R*ùbus árcticus, which is also a most delicious fruit, with a flavour resembling that of the raspberry, but far superior. If none of these fruits will thrive in the situation, or if peat soil to grow them in be thought too expensive, then the twice-bearing raspberry may be planted, which requires no particular soil, but only to be taken up and have the soil renewed every four or five years; which, indeed, is the case with all travelling-rooted plants, which it is desired should continue to fruit freely. The row of trees adjoining the street fence is supposed to be damsons, which come early into flower; and the fruit, being small and thickly set among spines, is not easily stolen. The plants

121

on each side of the party railing may be gooseberries, currants, or raspberries, according to the taste of the occupier; or they may be dwarf apples. The row of detached standard trees between the entrance gate and the porch (*b*) may be pears, trained *en quenouille;* that is, distaff fashion (see *fig.* 121.); or *en pyramide* (see *fig.* 122.); by either of which modes they will occupy little space, and look well: some kinds also bear well when trained in these manners. Behind these trees against

122

the wall, we would train common laurels, laurustinus, or ivy; because, though currants or gooseberries would be more appropriate there, yet under the shade of the pear trees they would not thrive; besides, if they would thrive, we think they ought to be sacrificed, in this particular spot, to the evergreens mentioned. We may observe, here, that the beds in which the strawberries are grown will be improved in effect if they are either surrounded with a marginal line of brick or tiles, or if the surface is covered by some description of stones, pebbles, flints, or vitrified bricks. The cause of this improved effect is, the contrast between rocky or stony materials and the turf.

The Back Garden has fruit trees against all the walls, and some standards and dwarfs in the central beds. The spaces *h h*, in both gardens, are supposed to be devoted to annual crops, such as peas, spinach, salading, &c. The beds marked *i*, in both gardens, may be planted with asparagus, sea-kale, and tart rhubarb. The pot herbs are supposed to be grown in the borders next the walls. The three fruit trees at *k* are Hawthornden apples (standards), for immediate effect in bearing, and to be cut out in a few years after the three rows of dwarf trees, shown in each garden, have come into full bearing. If the front gardens supply a sufficient number of gooseberries and currants, these dwarf trees may be apples, pears, cherries, and plums; but gooseberries and currants may be substituted for these, if those grown in the front garden are not considered sufficient. In order to insure good crops, the ground among the dwarf fruit trees should not be cropped with vegetables, but only occasionally manured on the surface, and forked over to the depth of a few inches. The trees against the walls, on the supposition that the direction of the gardens is east and west, may be peaches and nectarines on the south aspect, plums and cherries on that facing the north, and apricots on the end wall; but, if the garden should lie in the direction of north and south, the peaches, nectarines, and apricots may be planted against the side walls, and figs and vines against the end wall that faces the south; or morello or other cherries or plums, if it faces the north. The edgings to the walks of the back garden we have supposed to be box; but strawberries might be substituted, more especially if the walks were formed of flagstone, as shown in the section, *fig.* 123. In this figure the pavement is supposed to be about 4 or 5 inches below the general level of the garden; and about a foot in width on each side of the pavement is excavated to the depth of about 2 in., in the centre of which is to be planted the row of strawberries which are to form the edging. The pavement of course should be supported on brick or stone piers, founded on the solid subsoil, below the depth to which the surface soil is trenched.

123

Expense, &c. The houses, being only two stories high, might be built for 500*l.* each; and the gardens laid out and planted for 30*l.* each, on the supposition that the soil and other circumstances were favourable. This garden is suited for an open airy situation, not in the immediate smoke of a city; because, under the latter circumstances, neither culinary vegetables nor fruits can be grown to any degree of perfection, except under glass. The culture of such a garden might be undertaken by any person, at any period of life; because the points of culture and management are of the ordinary kind; and those in which he could receive aid or instruction both from local gardeners, and common gardening books. On the other hand, where the culture of many different kinds of flowers or forcing was to be carried on, the occupier would require to begin at an earlier period of life, to have some slight knowledge of botany, or to have the occasional advice and assistance of a scientific, or professional, gardener.

§ 3. *Single detached Houses, which have Gardens all round them, but which still form Part of a Street or Row.*—The reader will bear in mind what we have stated respecting gardens of this division in p. 181.; viz. that these gardens differ only from those of the preceding division in entirely surrounding the house, instead of embracing it only on three sides. We shall therefore pass briefly over this part of our subject, and, instead of our own designs, shall commence by presenting some by different contributors.

A single detached House, or Suburban Villa, with a Fourth-rate Garden, designed by R. Varden, Esq., Architect, Worcester. We shall give the description of this villa in Mr. Varden's own words; premising that it stands on half an acre of ground, in the neighbourhood of Worcester: — "As, in giving designs for villas, it is desirable that plans should be given for every variety of situation, I have chosen the following, in order to show the manner in which natural difficulties may sometimes be surmounted, and those points that are usually considered unfavourable be rendered unimportant, or be converted into interesting features. The land on which it is proposed to lay out this villa consists of somewhat less than half an acre, and is one of a series of allotments now (1836) on sale in the neighbourhood of Worcester. My object in selecting it is, that the advantages or disadvantages to which I may allude shall be such as really exist in nature, and not merely the creations of my fancy; as that may probably suggest only such circumstances as can be readily turned to account, or made subservient to the beauty of the whole, and, therefore, not be so suitable for your purpose.

"This piece of ground is situated between two roads, and is

90 ft. wide and 225 ft. deep, sloping gently to the west, in which direction it commands a pleasing view of the city of Worcester; the towers, spires, and roofs of the cathedral, and the other churches and public buildings, being alone visible, as the common structures are concealed by trees planted lower down the hill, and in the intervening meadows. The whole is backed by the Malvern Hills, a granitic range 1500 ft. high, the outline of which is sharp and bold, and of a mountainous character. In the north and north-west direction, the Abberley Hills are visible over a considerable expanse of rich and highly cultivated country, consisting of luxuriant meadows and tillages, thickly interspersed with those beautiful apple and pear orchards that are the pride of the county. To the south, a distant view, over a comparatively flat country, may be obtained, in which there is not discernible any particularly striking feature.

"One great disadvantage in this locality is, the necessity of placing the house in the upper part of the enclosure, with its principal elevations east and west ranging with the buildings already erected; thus giving to the garden and the garden front a westerly aspect; one of the worst that can be named, and which is unfit for the windows of the principal rooms. This has caused me to adopt a somewhat novel arrangement, for the purpose of obviating these defects, which I will describe.

"The villa is small; and is to be rendered suitable for the residence of a genteel family who have a fondness for the fine arts, and who wish economically to possess, and be surrounded by, the usual comforts of modern life, with the addition of some of those elegancies that are now seldom seen but in the mansions of the wealthy, and not always even there. The front is to be formed of real or imitation stone, and it is intended to be somewhat more ornamental than is usually seen in the neighbourhood of a provincial city; but the whole must be characterised more by elegance and neatness, than by splendour or magnificence.

"The front garden (see *fig.* 183.) is rectangular; and it is bounded by the house on the west, a conservatory (A) on the north, stable offices (B) on the south, and a dwarf wall and iron railing on the east.

"On entering the gates, a broad gravelled walk leads direct to the portico (C), which is raised from the ground four steps, and is formed of two columns and two antæ, in front of a curved-backed cell (D), from which it is separated by three arches that support the front wall of the house. I imagine there can be no objection to columns and arches being employed together in such a case as this; for they each have their distinctive uses; columns could not, with propriety, be placed for the support of so large an incumbent weight of masonry as the whole story of a house; and I know of nothing so evidently suitable for the purpose as arches, both on account of their real and their apparent strength; and the effect of arches seen between columns is good, and generally meets with the approval of painters, as I think it is likely to do in the present case, particularly as the deep curved-backed cell and niches will show them out to advantage. The cell is to be raised one step above the portico, and is to be paved with ornamental tiles or tessellæ, if they can be obtained. The entrance door is in the centre of the curved back; and has, right and left of it, a niche for the reception of a plaster cast of some beautiful antique statue the size of life. Against the side walls may be placed stands for such rare exotic plants in flower, as may be thought most deserving of attentive examination.

"The columns of the portico are to be but one story high; and they are to be surmounted by an architrave cornice, that is to pass all round the house, arcades, conservatory, and enclosure wall of the stable offices. The roof is to be finished with an eaves-cornice and gutters, and to be covered with artificial stone tiles, similar in form to those of the Temple of Diana Propylea, the eaves-cornice and ridge being ornamented with antefixæ. The chimneys are to be visible, and to form a part of the design.

"The kitchen offices, water-closet, &c., are to be in the basement floor, and are to receive light from a broad area, having a wall round it 2 ft. high, above the front garden; on the corners of which, against the steps and portico, are to be placed slightly enriched vases of artificial stone. The side intercolumniations of the portico are to be fenced in by a neat metal railing; framed, not into the antæ or the columns, but into stiff standards placed close against the bases.

"The area wall is to be continued all round the front garden, except the doorways, forming at the sides a basement to the conservatory, and to the stable enclosure wall; and forming in front a foundation for the iron railing. Against this wall earth is to be thrown, and formed to the angle that flower-stands are usually arranged at; and in this are to be planted choice flowering exotic evergreens of small growth, placed so as to produce the effect of ranges of stands. The piers, which the gates and front iron railing are framed into, must be massive, and have enriched heads, or be surmounted by short, thick, enriched bronze shafts and lamps, according to circumstances. There is to be a broad walk, crossing at right angles that leading to the porch, which will conduct to the stable-yard and to the conservatory.

"The conservatory must be made ornamental without any great outlay of money, and must form a part of the general design. This, I think, may be done by carrying the architrave entablature, and also the dado, of the house, round it. The uprights and sashes may be framed into them much in the usual manner; but the front lights I would have contrived to slide down into this dado wall, and be out of the way entirely when air is admitted to the conservatory, in preference to being opened outwards, as in the usual manner. The rafters are to be tolerably close together, and be two to every upright in the front framing; and they are to be ornamented along the ridge and cornice with antefixæ similar to those to the roof of the house. Should any flue chimney be shown, it must be introduced as an acrota. The stoke-hole is in the area, and is approached from the stable-yard by steps, by the door into the arcade. On each side of the steps into the conservatory, and of the yard gates, are to be piers the height of the area wall, with vases placed thereon.

"The entrance to the house from the curved cell (D) is into the vestibule (E), the ends of which have strong segment-arched ceilings; but the centre is to be open to the upper story, and the aperture to be surrounded by Corinthian columns thick set, and supporting a lantern light of stained glass. These columns, with a connecting metal railing framed into upright standards, part off a gallery that likewise surrounds the aperture to give access to the sleeping apartments. The doorways into the cell, and the staircase, are to have simple draperies, hanging partly over the doors. The side walls are to have, in the centre of each, a door into the dining or drawingroom; and each corner is to have placed in it a pedestal and a statue. The effect of this little vestibule will be striking; and, although small, it will, I think, possess an air of considerable grandeur; and, if the colours of the ceiling, walls, floor, &c., are properly chosen and arranged, I think it cannot fail of pleasing. I should recommend the walls being of the colour of the feathers on a jay's breast; and the glass in the lantern light the same, with ornaments, and a border of a somewhat darker tint of the same colour, relieved with blue black, and dulled white. The ceiling I would have of a similar colour to the walls, but much darker; and the columns, architectural features, and statues of the same, but so much lighter as to be nothing more than white tinged with the predominating colour. The ground of the carpet must be light blue, with the pattern made out by white and dark grey or black. Should there be no carpet, chairs with blue covers must be introduced. The draperies before the doors are to correspond with the colour of the walls, and to be lined with blue; and are to hang from red bronze rods and rings, in large vertical folds.

The dining-room (F) is an oblong, having in the centre of each side, parted off by Ionic columns, a curved recess; in one of which is the door, with a niche for a side table and lamp on each side; and in the other a curved sideboard, and three windows, looking northward into the small strip of garden. There will be a flower-bank in front, and over that an oblique view of the distant landscape. Behind the columns should be plain simple draperies, that may usually be somewhat spread at the top, so as to back and show out the enriched capitals of the columns, but be drawn aside below, to display the plate, &c., on the sideboard; which is never seen to greater advantage than in such a situation, especially if it be placed on black marble, as it may be with great propriety in this case. On the north side of the room, but more towards the ends than the curved recess, are small glazed doors; one of which is to be fixed, and the other made to open and give access to an arcade (z) of three arches, that has a groined roof, a niche and statue at the end, and two others opposite the side arches, with a doorway into the conservatory, opposite the centre arch and the flight of steps down to the terrace. Opposite each of these sash doors is a niche; and there are two others at each end of the room; the whole of which are for the reception of plaster casts from antique Bacchanals, or similar figures. The fire-places are in the centre of the ends; and each is to have over it a small window, commanding the front or back view, with a looking-glass sliding shutter, that may be opened or shut by a concealed rack, and be made to fit exactly into a gilt margin, that shall form the window-dressing at one time, and the glass-frame at another. I describe these shutters thus particularly, as, during a considerable portion of the day, that to the west window must necessarily be closed, to exclude the unpleasant glare and heat of the afternoon sun. Similar sashes may with great propriety be placed to the glazed doors; where mirrors will produce a good effect, as they will reflect the opposite niches and statues. The dining-table is represented by the dotted lines in the plan; and is sufficiently large to accommodate, when drawn out to its whole length, twenty-six persons. Were this room for my own habitation, I should prefer having the colouring deep and rich. I would have the chimney-pieces, sideboard tops, columns, and antæ, of real or imitation black marble; and the entablature, architraves, dado, and skirtings painted and polished to match. The compartments of the ceiling should be of a deep purple; and the cornice, ceiling beams, doors, window-sashes, &c., painted in representation of a reddish-coloured bronze. The curtains should be crimson damask, lined on the edge and bottom with one width of purple sik, and be looped up with thick purple cords. They should be merely suspended from plain bronze rods and rings, and should hang in large vertical folds, and be without any valances, or draperies, at top. There should be inner sun curtains of figured muslin hanging down, with worked

Vandyck edgings, and drawn back with crimson silk cords. The walls I would have of very dark oak, or covered with a purple paper. The carpet green, patterned with light and dark tints of the same hue. The chair-covers crimson; and the table-covers the same, lined with white, and bound with gold, with a large tassel of the same hanging at each corner.

In this room, especially when thus decorated, I think there would be some very pictorial effects. The curved recesses, and the sideboard behind the columns and curtains; the columns backed by curtains; the view up the arcade, with the statue and niche at the end as a central object; and the vista seen through the vestibule and drawingroom, when the doors are opened, looking between the columns, &c.; would all be sources of interest, and, I trust, of beauty. The dark rich colouring I have described, I consider to be appropriate to the destination of the room, and by no means likely to produce a gloomy effect; as the great number of windows will admit ample light during the day, and at night any quantity may be artificially commanded. The colours are powerful, and some of the contrasts great; but I think they would harmonise well. The crimson and purple would produce an effect of splendour; and the many dark compound tints surrounding the furniture would impart to each article a brilliancy unattainable in an apartment coloured in a lighter tone.

The drawingroom (G) is of similar form and size to the dining-room; and the door from the arcade is an entrance to the stable-yard. If it were not for the particular aspect of this plot of ground, I would change the situation of the two rooms: then the drawingroom would be in immediate connexion with the conservatory, and the dining-room with the stable offices; a much better arrangement than the present; but which, if carried into effect here, would assign the worst aspect to the best room, and the reverse. However, if such an alteration were considered desirable by a proposed occupant, it could be carried into effect without disarranging the design. The drawingroom is to have windows over the fire-places at the ends; with looking-glass shutters, for excluding the afternoon sun when it becomes troublesome. In addition to the columns parting off the curved recesses described in the dining-room, there are to be other columns parting off the ends of the room, and forming a square compartment in the centre, which is to have a groined ceiling, with a large flat dome in the middle.

In this room the columns are to be Corinthian, of scagliola marble, in imitation of verd antique; with the entablature, ceiling, doors, skirtings, dado, architraves, &c., painted to resemble bronze; and the chimney-piece really of that metal, made to assimilate, as to height and decoration, with the dado and skirting; which it might do, without being inconveniently low, were the fire to be made on the hearth, with air admitted from below, as in Sylvester's stove. The walls are to be covered with apple-green silk, either stretched plainly over the whole surface, or introduced as fluted panels in bronze frames. The curtains are to be of the same colour, lined (those to the windows partially, and those behind the columns wholly) with rose-coloured silk: they are to be full, but very plain, merely hanging in large downright folds when in use, being at other times looped back with gold or rose-coloured silk cords, and large tassels. The sun curtains are to be of light figured muslin, either quite white, or of a delicate pink colour, looped up with rose silk cords. The rods will be best of bronze; but, if a large quantity of gilding, by way of chandeliers, lamps, &c., be introduced, then they may be of gilt brass: but this I do not recommend, as bronze appears to me far preferable. These articles, with gilding or glass, and other showy materials, scattered about the ceiling and upper part of the walls, always have the effect of destroying the repose of the apartment, and diminishing the brilliancy of the furniture, which ought to be preserved in its greatest strength and splendour, as it would, in the absence of living beings, unquestionably form in a painting the chief and most prominent object; and, for the apartment to be pleasing, each article must have the same repose, strength, or prominence, as would be requisite to form a good picture; as every view we take of it ought to present, in reality, to the eye, exactly the selfsame scene which the painter produces artificially on his canvass. They differ only in so far as one is reality, and the other its representation; the same laws of position (viz. grouping, colour, chiaro-scuro) applying equally to both. The chairs are to be covered with mohair plush (a newly introduced article, of great beauty and richness, and likely to stand much wear,) of a rose colour. The carpet may be a maroon; but I should prefer a buff or white ground, with red, yellow, and green flowers. The table-cover, for grand occasions, to be rose colour and white, lined with white satin, edged with a broad gold band, and decorated with massive bullion tassels at the corners; but, for common use, one with yellow binding and tassels, and white kerseymere lining, might be substituted.

" The curtains introduced behind the columns will greatly enhance the beauty of the apartment, and will, when wholly or partially spread, form a backing to the columns that will display their enriched capitals, whether gilt or not, to very great advantage. At the same time, they will be far from useless; as, by drawing them, either of the ends may be converted into a very comfortable apartment for a small party, that will have most of the advantages of one of a suite of rooms connected by

124
A
B
L
C
D
Y
F
E
G
N
H
I
Q
K
M
R
P
S
T
U
W
X
V

folding-doors, without its unsightliness or liability to draughts of cold air. The decoration of this room would not be particularly expensive; as the effect depends more on the arrangement of the objects and colours, than on the costliness of the materials composing them.

"H is the staircase, the only one in the house; for, in so small a villa as this, where few servants would be kept, a second is hardly called for. It is of ample size and elliptical form; which readily admits the introduction of a stone staircase, indestructible by fire. There is a lower flight of steps to the basement offices. The floor should be of Wright's ornamental tiles, or tessellated pavement; and the walls and ceiling should be painted in the cream and ochre colours, and with ornaments similar to those found on the ancient terracotta vases.

"I is a veranda. Architects avoid introducing these appendages to their designs, from the idea that they are always unsightly. Surely, there is a possibility of their being made ornamental, let the difficulties be what they may. At any rate, one placed here will be found by the inhabitants of the house a great luxury; as it will partly shade the western wall from the afternoon sun, and preserve the principal rooms, towards the close of the day, from the great heat to which they would otherwise be liable. I propose having it to appear as a temporary awning, and the architecture to be made complete without it; and thus, like Alexander, to cut the knot I cannot untie. No person, let his taste be ever so fastidious, can, I imagine, object to those delightful awnings, spread in the streets of every Continental town, before the principal shops; adding, by their brilliant-coloured stripes, to that gaiety in which our dwellings, both in town and country, are so deficient. These are not displeasing, even to artists; as witness the prevalence of them in the works of Prout, Stanfield, Nash, and our other topographical painters; and, therefore, I cannot see why we should hesitate to adopt them as accessories to our villas. The columns which support this awning are to be very light, and slightly ornamented: and, as well as a wire railing between them, painted stone colour. The roof, or awning, is to be of such a curve as canvass would naturally take, were it fastened to the wall along one side, and the other allowed to hang loosely over the plate that passes from the top of one column to another. This plate is to be of copper, having the joints concealed; and it is to lap over the front, and to hang down 18 in. or 20 in.; the end of it being cut into pointed pendants, one short and one long, alternately; while hanging from each point, and suspended by a wire hook, may be a turned wooden tassel, 4 in. or 5 in. long, that may be moved by the slightest breeze. The inside of the roof of the veranda is to be painted a cream colour; and the outside the same, but with vertical stripes of light bright red, one broad and two narrow ones to form a pattern, the centre of which is to correspond with the point of each pendant. The tassels are to be of the same colour; and there is to be an edging of it round the pendants, as though it were a binding. The floor is to be paved with stone, ornamented tiles, or tesselæ. The niches under the veranda are to be for statues, and the remaining space may be occupied by flower-stands, glass globes containing gold and silver fishes, bird-cages, &c.

"From the veranda, opposite the house door, are broad steps down to the terrace, having flank walls surmounted by sculptured vases. K K are stone platforms, somewhat lower than the floor of the veranda, on which are to be placed stands to receive pots of the most choice exotic plants in flower. L L are the areas to the basement story; all of which have plain breast-walls to fence them from the garden. The area is continued under the veranda, that being supported on piers.

"M M are flower-borders, edged with an ornamental kerb of artificial stone, similar to Austin's mignonette troughs, that may, perhaps, have a pipe within it, pierced so, that, when connected with a head of water, each hole may be converted into a miniature fountain, jetting in different directions, so as to cover the whole surface of the border with an artificial shower.

"N N are flower banks; that is, earth thrown against the wall at the angle which artificial flower stands are usually arranged at, into which pots of hardy plants may be plunged while in flower, and the space behind them filled up with small, flowering, evergreen shrubs. The walls (O O) are supposed to be 2 ft. or 3 ft. higher than the floor of the principal rooms of the house, which will not be sufficient to interfere with the view of the distant landscape; but, from the height it is above the natural surface of the ground, will be sufficient for every purpose of protection. P P are the parapet walls of the gravelled terrace, the same height as those round the area; and Q Q Q are vases.

"R R represent fountains; but, if these cannot be obtained, groups of statuary, or sun-dials and flower-stands, must be substituted.

"S S are steps from the terrace into the garden. The walk here is broad, and forms a regular or artificial figure, as the piece of ground is too small to attempt concealing the boundary fence. It is, therefore, made to run parallel with it; having on the outer side a range of flower-banks (T T) edged with turf, and backed by the enclosure wall, covered with evergreen and flowering creepers. Much effect may be produced by occasionally varying the outline of this wall by the introduction of vases, or other objects of art. U U communicate with sheds and other requisite garden buildings, formed under the terraces, that will be invisible from the principal walk. In the centre of the garden, but separated from the walk by a parterre, is another parallel flower-bank, having a pedestal and statue on the ridge of the circular end. This bank encloses the small space of ground V, that may be entered from the main walk, and is intended to be set apart for the female members of the family, who, if fond of botany or horticulture, may here tend and rear their favourite plants. It is well suited for this purpose; as it is easily accessible, and yet as private as the confined nature of the ground will admit; also from its having a portion of its surface inclining to each point of the compass, a convenient spot may be selected for every plant that will grow in our climate. If not used for this purpose, it may be made the nursery, or reserve-ground, from which the parterres and borders can be supplied with plants in flower.

"W W are flower borders; and X X are wicker or wire baskets, to be filled with climbing flowering plants, such as jasmine, clematis, maurandyas, roses, &c.

"Y Y are stands; on each of which may be placed, in a pot, any of those plants which display their blossoms to the greatest advantage when allowed to hang down over the sides.

[The house may be carried up three stories above the ground floor; the first floor containing two best bed-rooms, with dressing-rooms, a morning or breakfast room looking into the garden, a bath-room, and water-closet; the second floor, bedrooms for the family; and the third story, or attic, a nursery, servants' bed-rooms, lumber-room, &c.]

"In carrying such a design as this into execution, many ideas will naturally suggest themselves for enhancing the effect of the objects I have described, and for the embellishment of the whole. This sketch must be looked upon as the outline only of the design, the detail of which yet remains to be filled up; but this I shall not attempt, as it would swell this already long description to a very tedious length, without answering any important end. The finishing touches of this, as of all similar works of art, must be given by the hand of female taste, the superior elegance of which most writers acknowledge; and the distribution of the various kinds of shrubs and trees may be determined by the lady, and her nurseryman or gardener. My object has been to show that considerable variety and elegance, or even splendour, may be obtained in a confined space, at a moderate expense, by attending to the peculiarities of the site, and judiciously grouping the different useful or ornamental objects that enter into the composition. I need not enlarge upon the manner in which I have attempted to carry this into effect, as by glancing at the plan it may easily be perceived. Space I have endeavoured to give the appearance of, by preserving uninterrupted vistas in every possible direction; and splendour, by always having a number of objects of art along the sides of these vistas, and by terminating, always, with some interesting central objects, that may form a chief point to the particular view. Thus, on entering from the upper road, there will be seen, in succession, one beyond the other, the broad walk; the steps up to the portico, with the vases on the sides; the columns of the portico itself; the arches of the cell, and its curved back, with its niches, statues, and central opening; the vestibule; the staircase; the veranda; the steps; the terrace, &c.; the whole terminated by the raised statue in the garden. Only a part of these objects will be seen at the same time, on account of the alterations of the level of the floor; but this will rather add to than diminish the interest and the appearance of size; and a sufficiency of objects will always be visible together to form a very rich composition. On the veranda, the vases will form central objects at the ends. From the circular part of the terrace, past the fountain, will be seen the arcade, with its arches, niches, and statues, its ornamental tiled roof and enriched cornice, and its doorway opening to view the whole length of the conservatory. On looking the other way, from the end of the conservatory, past some of the same objects, the fountain will form the terminating object in the grounds; but beyond it will be visible the distant hills and landscape. These, and the many accidental groupings of the objects I have supposed to be introduced, will vary with every change of position, presenting a numerous succession of views, the greater part of which, I flatter myself, would possess interesting combinations of parts, fit for separate sketches. I feel convinced that there is nothing visionary or chimerical in this plan, or that might not be obtained by a moderate original outlay, and be kept in the neatest order at a very trifling expense."

A single detached House adjoining the preceding Design. This design (*fig.* 125. also by Mr. Varden) is supposed to be an adjoining allotment to the last, having precisely the same extent, aspect, and view. "This residence," Mr. Varden says, "will be suitable to a genteel family of but moderate fortune; as it will possess considerable effect of a simple, though elegant, description, produced by the arrangement and forms of its several useful parts. These may be left plain, or ornamented with a few artificial stone statues and vases (as I have shown), according to circumstances; but in either case it will be by no means an expensive dwelling.

"The principal floor of the house consists of a circular entrance hall, in which is the staircase; of a vestibule communicating with the principal rooms; of a dining and drawing room, each with circular ends, in which are the windows looking east and west, but others north and south may be introduced, if desired; and of a circular parlour or morning room, leaving a sash door into the garden for the ingress and egress of its occupants only; the communication from the other rooms to the back garden being round by the front door. I shall make no observations on the finishing or decorating of the rooms, farther than that their shape and arrangement will allow of much architectural effect being easily produced.

"The offices and common water-closet are in the basement. The chamber floor is to contain four sleeping-apartments, with bath-room and water-closet; and there may be an attic story above with the like, or a smaller, number of rooms, if required. There is no conservatory, and the stable offices are placed at the lowest end of the ground, where they need not be made unsightly, or to interfere with the view, from the house, of the distant scenery in that direction.

"The front garden is small, and is laid out in a simple manner. A walk leads direct to the front doors, and there is another crossing it at right angles, each portion of which leads to the back of the house; from one of these there is a flight of steps down to the basement, for the convenience of persons having business with the servants.

"In the plan, *fig.* 125., *a a a* are flower borders on the turf; *b b b* are flower stands raised upon podiums, that exactly range in height with the area enclosure walls; *c c* are the areas enclosed by walls (*d d*) the exact height of the dining-room floor, having at the corners plain vases; *e e* is the flight of steps down to the kitchens; and *f f* are paved stone paths, by which there is a communication between the back and front gardens, having, towards the former, steps to connect the different levels: *g g* are steps up to the door of the morning room; *h* is a pedestal for a statue; and *i i* are for large vases: *k k* are flower borders on the grass, and *l l* are baskets; *m m* are flower banks, and *n n*, are spaces for shrubs, backed by the boundary wall, which may be covered with ornamental climbing plants. The stable-yard is shown at *o*, and *p p* are the entrances to it; that from the garden is down a few steps, and may be through an openwork iron gate; for, if the yard should be seen from the garden walk, I think it will not be objectionable, and then there will be a certainty of its always being kept in a neat state. The stable and coach-house are at *q q*, each of which is to have a niche for the reception of a statue in the side next the garden, intended to form the terminating objects to the principal walks; and the roofs of both are to be concealed by parapets and cornices. The wall separating the yard from the garden must have a plain

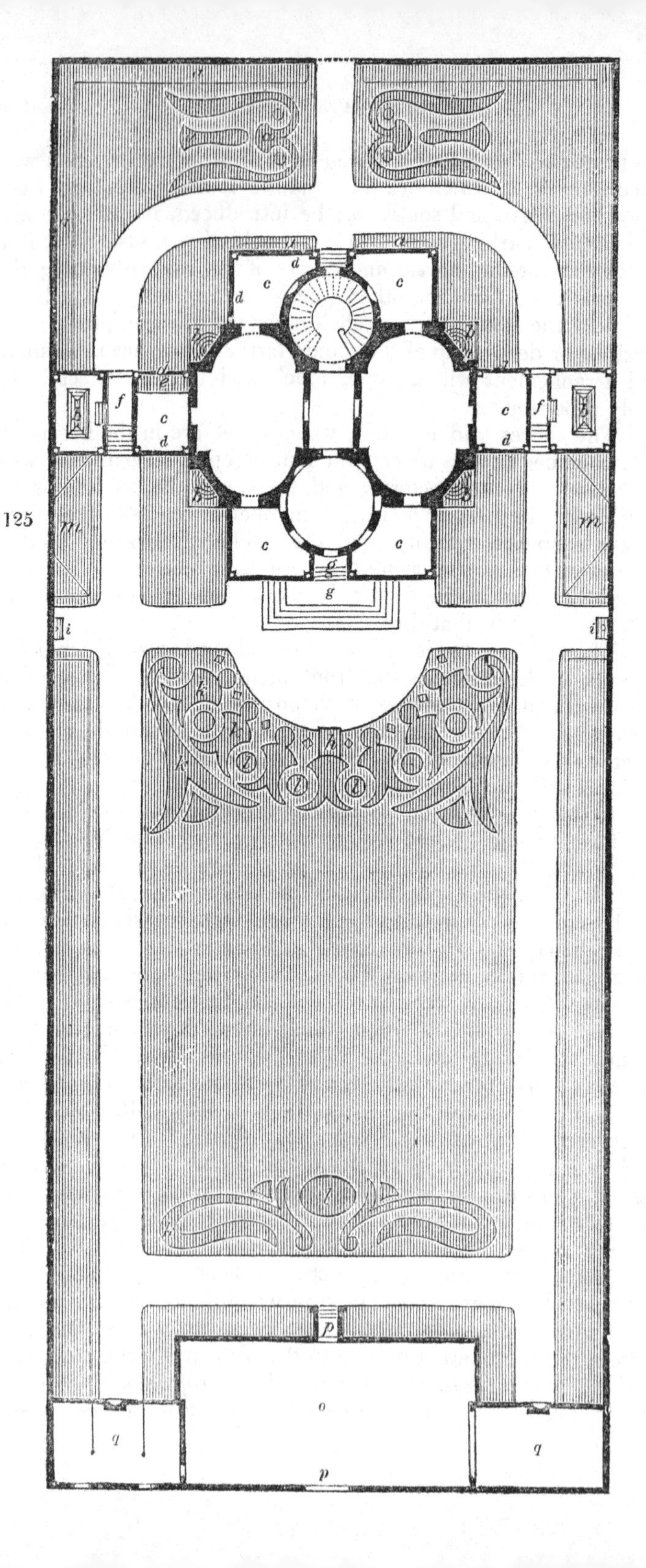

125

architectural character given it, and be made to appear as a natural connexion between the two buildings, and not as if it were merely stuck against them.

"I shall not descant on the effects that will be here produced, as they will be sufficiently obvious from the plan. I have been guided by the same principles as in the last, but have been less profuse of decoration; that which is introduced will, I think, tell strongly, as it is pretty much concentrated round the house, and none of the statues or vases, with the exception of that marked *h*, are isolated; for such objects seldom appear advantageously but when backed by, or in close connexion with, buildings, or other mural appendages. The transverse wall at the back of the house is so placed, that, if the adjoining design be carried into effect, a fountain may form its central terminating feature; and then the vase I have shown may be omitted."

A Design for a small Garden attached to a single Street House in a Country Town. This design was arranged, laid out, and planted under the direction of Mr. Varden, in the town of Godalming in Surrey, for H. Marshal, Esq. We saw it in 1834, six months after it was laid out and planted; and, being very much gratified by its appearance and suitableness for the peculiarities of the situation, we requested Mr. Varden to furnish us with a ground plan and other particulars, which he did as follows. It may be useful to premise, that the house fronts the main street of Godalming, and that the garden behind is on a steep ascent, which communicates at the top with a paddock belonging to the same gentleman.

"The front of the house (*fig.* 126.) is close to the street (*a*), and the back is towards the garden, which gradually ascends from it to a paddock, whence the ground rises rather sharply, till it terminates in a steep wooded hill, from the summit of which there is a pleasing view of the town, and of a grassy valley enclosed by steep copse-covered hills, with the river Wey winding through it; the whole terminating with the chalk hills and cliffs of Guildford.

"The garden was overrun and shaded by straggling trees of very little beauty, that gave it a damp and cheerless appearance. I had these removed, and the surface of the ground lowered, for the walk to cross the lawn without being seen from the drawing-room windows; the earth was shifted lower down the hill, and formed in such a manner as to give an appearance, when seen from the garden, of the house being on slightly rising ground. This alteration gave the whole surface of the lawn an undulation that greatly enlarged its apparent extent.

"In *fig.* 126. *a* is the street; *b*, the front garden; *c*, the steps to the front door; *e*, the carriage entrance. The house is at *f*,

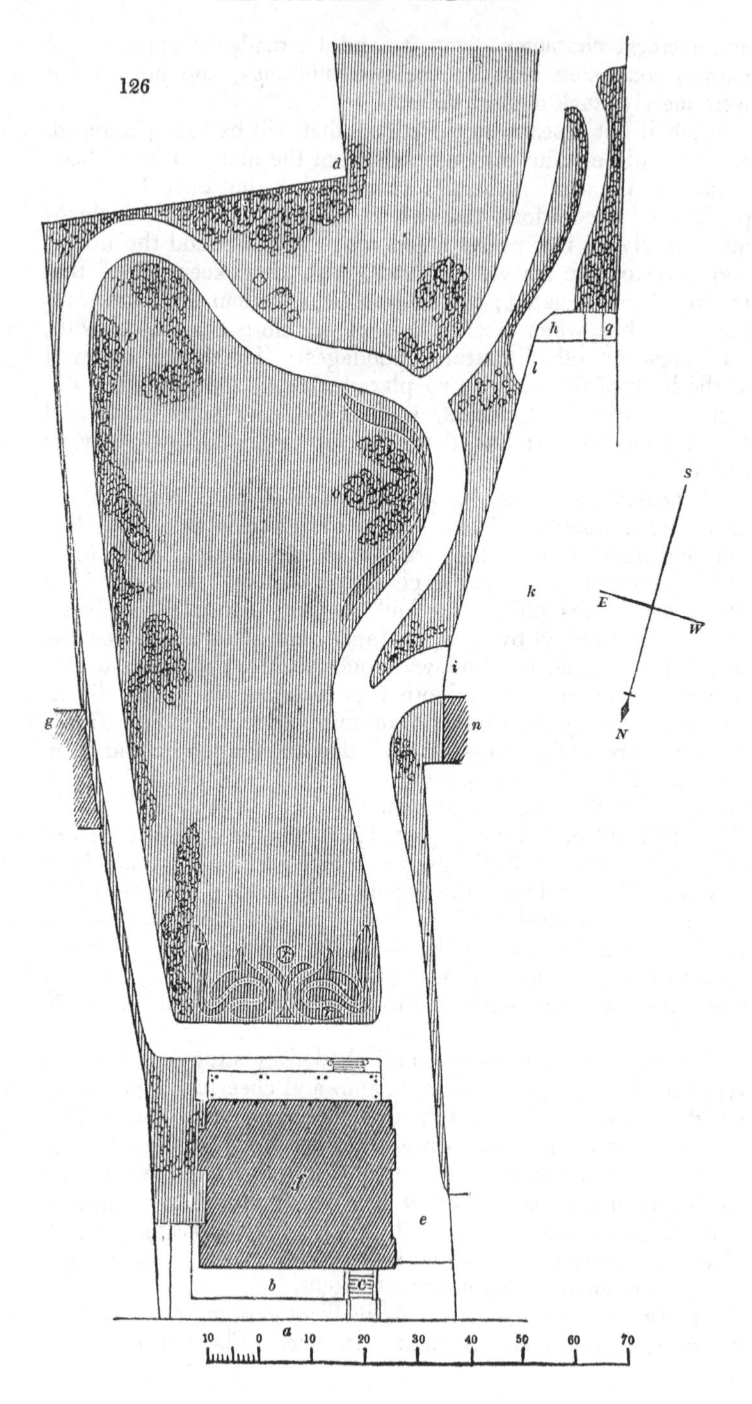
126
d
h
q
l
p
S
E
W
N
k
i
g
n
f
e
b
c
a
10 0 10 20 30 40 50 60 70

having against it, on the lawn, or south side, a gay-looking red-striped veranda, the stone floor of which is supported on piers so as to admit light to the rooms in the basement, by an area paved with white glazed tiles, and surrounded by sloping rough flint walls, planted with sun-roses (helianthemums), and other dwarf rock plants.

"The garden, on the east side, as far as *g*, is enclosed by different erections, the outline of which has been formed so as, when covered with creepers, to resemble a part of an extensive old English building. To this point the view is quite confined; but over the wall beyond, which is finished with buttresses and pinnacles, so as to display advantageously the climbing plants covering it, may be seen a part of the steep wooded hill before mentioned.

"Between the points *d h* are visible the paddock, and a garden sloping up the hill. *i* is the coach gate into the yard *k*; and *l* is a door for the horse to pass to and from the paddock, without going along the garden; the path used being enclosed by a thick yew hedge, clipped on the inner, but left rough on the outer, side.

"The stable is at *h*; and the end of the building may be covered with passion-flowers and trumpet-flowers (Passiflòra cærùlea and *Técoma* radìcans), to which may be added Caprifòlium japónicum, &c. The wall between this and the street is covered with ivy, and there is a row of good-sized horsechestnut trees to hide the neighbouring buildings. The walls and fences are covered with nearly every kind of creeping plant to be obtained, that will stand the winter without protection. The shrubs are nearly all evergreens, mostly of small growth, and consist of a great many different species. Those of the same colour of foliage are grouped together, in order that each mass may, by its depth or brightness of tone, form a satisfactory contrast with its neighbours. In some places they nearly approximate, but in others the difference of tint is very great: the points *o o* are the lightest, and *p p* the darkest; so that, when viewed from the house, the former will appear to stand strongly out from the latter, and will give (at least, I hope so) the effect of considerable space between.

"Wherever it was possible, the walks were hidden, so that the breadth of the lawn, and the repose of the whole scene, might not be disturbed by the sight of much glaring gravel. Thus, the left-hand walk is concealed by the planting, while that across the lawn is sunk out of sight from the drawing-room, and the path to the paddock is masked by a clump of American plants (*q*). The beds marked *r r r* are exclusively devoted to herbaceous flowers and roses. Unfortunately, I have not preserved a list of the shrubs, but they are all evergreen;

the common and Portugal laurels being, however, almost excluded: not from their being deficient in beauty, but because their preponderance in gardens is now so great, as to produce a monotonous effect wherever we go; and this I wished to avoid."

Design for laying out a Street Garden to a single detached House; the whole Plot being only one eighth of an Acre in extent. The objects in laying out this garden were utility and comfort, more than ornament. The situation is in a country town in the west of England, where, the situation being open, and the air good and free from smoke, there was no objection to the culture of culinary vegetables, and the growth of the finer fruits, both on espaliers and walls. The house is a square building, and contains three rooms and a hall and staircase on the ground floor. The entrance from the street *a*, in *fig.* 127, proceeds in a straight line to the hall door; on the left are the kitchens and offices, which are entered from the front court *b*; and on the right, the coach-house and stables, entered from the front court *c*. The walls which enclose these courts are covered with common ivy and Virginian creeper. The front garden is bordered by a shrubbery of low evergreens; the walls being clothed with variegated ivy; and on the grass plots are two oval clumps, which, to suit the peculiar taste of the proprietor, are planted with hydrangeas; the soil of those clumps being of a peculiar kind of peat, obtained in the neighbourhood, which changes the colour of the flowers from pink to blue. Among these plants are introduced crocuses, snowdrops, scillas, hyacinths, and narcissi of various kinds; which make a fine appearance in autumn. The shrubs in the surrounding borders are chiefly rhododendrons, laurustinus, azaleas, mezereons, one or two plants of *Rìbes* sanguíneum, *C*ydònia japónica, and the common myrtle, which, in that climate, stands without protection, and is only occasionally killed down to the ground. The flowers among these shrubs are bulbs of various kinds, including colchicums and the saffron crocus, both of which flower in autumn; and there are also some species of *L*ílium, which make a fine appearance during summer. Close to the door of the house, on each side, are two circular pedestals, surmounted by vases, which, in summer, are filled with pelargoniums.

On the garden front of the house are two plant cabinets (*d* and *e*.) A broad walk, or terrace, (*f*) stretches along the front of the house, from one side of the garden to the other, and terminates in handsome benches in alcoves, the whole breadth of the walk at *g*. From these alcoves a canvass awning is projected in the summer time, so as to cover a portion of the walk 10 ft. by 15 ft. Under one of these awnings, it is the custom of the family occasionally to dine, and under the other to drink tea. The main walk of the garden (*h*) has a broad border on

127

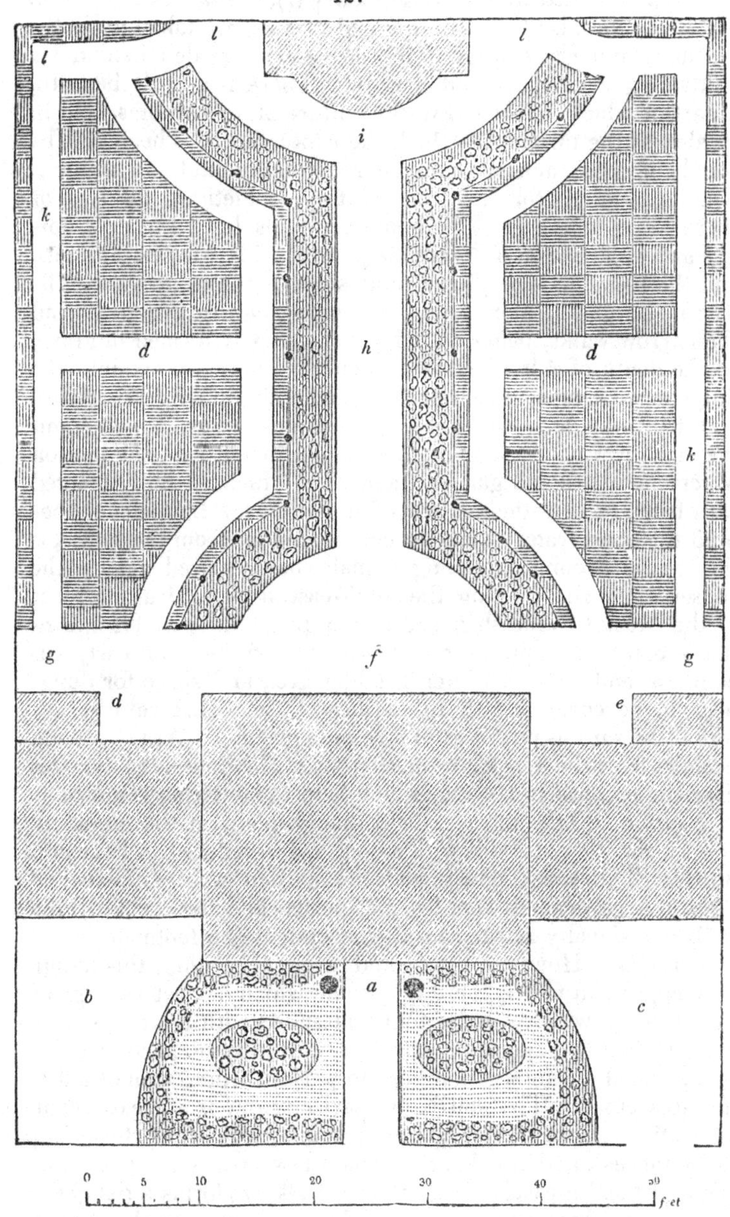

each side, devoted exclusively to herbaceous flowers, and to an espalier, which is covered with roses, and with standard apples at regular distances, as indicated on the plan. This walk is terminated by the conservatory (*i*), the sashes of the roof of which, and also the front and end glasses, are taken off in the summer, and are employed in the reserve garden behind for cultivating cucumbers and melons. There is a vault beneath, a part of which is devoted to the culture of mushrooms, and in it, also, is the furnace by which the conservatory is heated. The soil being dry, and the climate remarkably mild, no artificial heat is required for this conservatory, sometimes for two or three years together. The boundary walls have a narrow border and walk (*k*); and the wall is planted with peaches, apricots, and the finer kinds of plums, and some figs and grapes. The narrow borders to these walls are very slightly cropped; and the narrow walks, being on a dry soil, are only formed of gravel to the depth of 6 in., beneath which the roots stretch into the dug compartments, which are cropped with culinary vegetables, and fruit shrubs. At *l l* are open iron railings, which admit the view of a paddock behind, of equal extent with the plot on which the house and garden stand, and which is thinly planted with fruit trees, in the manner of an orchard. Immediately behind the conservatory, and, of course, hidden from the house, is the reserve ground, including a small vinery placed against the conservatory and fronting the south-east, a pit for fruiting pines or the Mùs*a* Cavendísh*ii*, succession pits, and pits for greenhouse plants to supply the plant cabinets, frames for early cucumbers and melons, winter salading, &c., and space for dungbeds to be covered with the sashes of the conservatory, and for all the various purposes of a reserve ground. As the ground declines from the terrace walk in front of the house, the eye sees over the orchard to the sea, which is in the extreme distance.

Expense. The house, which, though plain in the elevation, is very complete, is built with stone, and is reputed to have cost 3000*l.* The conservatory and the garden were estimated at 500*l.* A gardener is kept for the garden and the orchard, and he is occasionally assisted by the coachman and footman.

Remarks. However plain, and even uninviting, this design may appear on paper, there is not one which has yet been given in this work which will be found attended with more ease and comfort in the enjoyment. It is simple, all the parts are distinct, and the whole, when kept in perfect order, cannot fail to be satisfactory. The soil being perfectly dry is a great recommendation; as the walks and the court-yards were thus more easily formed, and are kept at much less expense. The house fronts the south-east; and the terrace walk (*f*) forms a delightful promenade, from the centre of which the vista to the conservatory, richly bordered with herbaceous flowers and roses in summer, to which are added the more beautiful kinds of apples in autumn, has a fine effect, more especially with the distant glimpse of the sea seen over the whole. Even in winter, the dryness of the walks and the glass of the conservatory give an air of comfort, warmth, and gardening enjoyment, which all feel who have been accustomed to indulge in the associations connected with these objects. When the proprietor has a large party, the company promenade on the terrace, and down the centre walk; and a band is frequently stationed in the reserve ground behind the conservatory, which has a very pleasing effect. The only defect which this place has is a want of water in the subsoil, which renders it necessary to collect all that falls on the roof of the dwelling-house, and to convey it to a tank in the kitchen-court, where it is kept cool, and drawn off by a pump as wanted. In the same court there is also an ice-house, which is a much more convenient situation for this luxury than in the grounds at a distance from the house; in the latter case the labour of finding the gardener, going with him to the ice-house, opening it, taking out the ice, shutting it, and securing it from the air, and carrying the ice to the house, consumes so much time of two persons, that it is generally neglected; whereas, when the ice-cellar is connected with the house, like any other cellar, the ice is obtained, when wanted, with almost as little trouble as coals or beer. The rain which falls on the conservatory is conducted to a cistern within, for watering the plants. Notwithstanding these measures for securing a supply, water is obliged to be carted from a distance of two miles in the summer season. The occupier of this house is a wealthy banker, but we are not at liberty to give his name.

A Residence in the Suburbs of a Country Village, containing a Quarter of an Acre. This design (*fig.* 128.), which is contributed by Mr. Rutger, is supposed to be situated on a piece of irregular ground, on the outskirts of a large village, or small country town. Its boundary is irregular, as is very commonly the case in such situations; and the front is to a road, which, being conducted over an uneven surface, is irregular also. For the same reason, there are scarcely any straight walks. The whole is surrounded by a fence of pales, except on the north side, at *h*, where there is a wall. The street entrance to the house is shown at *a*; and the space *e* is supposed to be a covered way, not so much as a luxury, as to afford shelter to persons waiting for the occupant; *b* is the carriage entrance to the kitchen-court, which contains a gig-house and a small stable; *c* is a walk that conducts from the kitchen-court, by *i* and *g*, to the lawn front of the house, and to the kitchen-garden (*d* and *f*). Within the boundary fence, on the south side, at *k*, are fruit trees, disposed in the form of a shrubbery; those immediately adjoining the fence being intermixed with hollies. The two beds to the right and left of the front entrance are supposed to be devoted to low evergreen shrubs; those at *i i*,

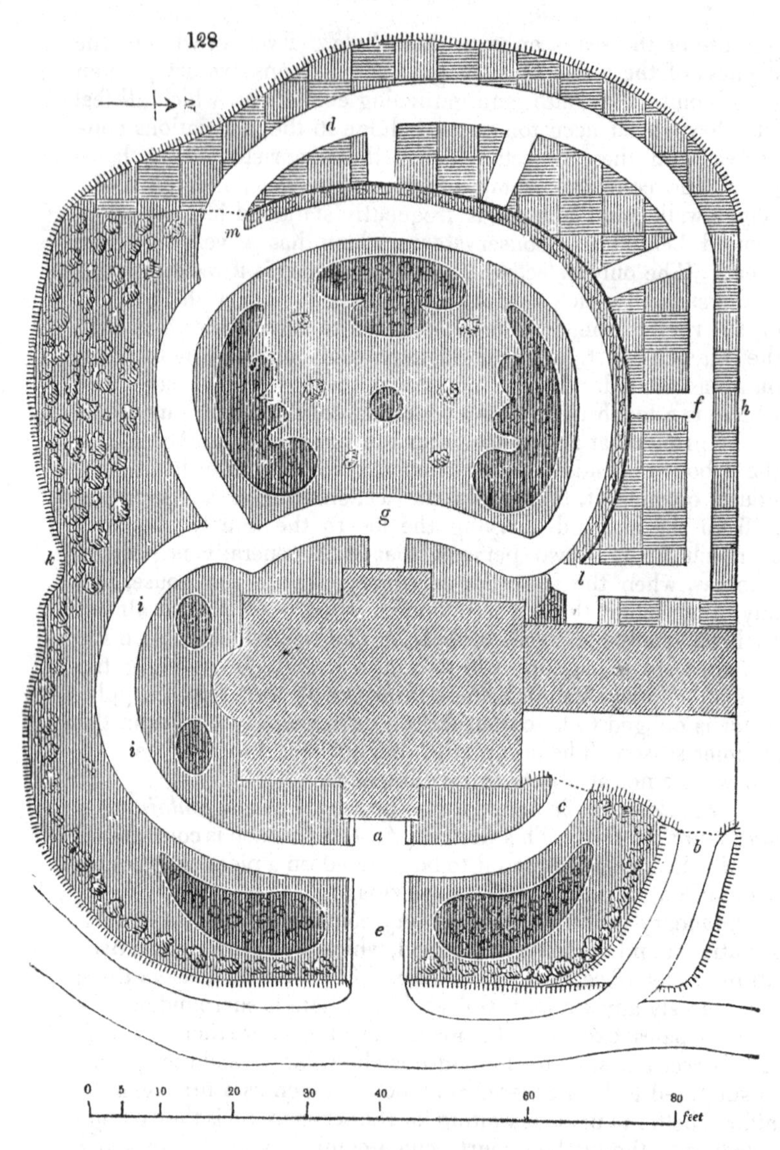

entirely to flowers; and those on the front lawn also to herbaceous flowers, with some dwarf and standard roses. There is an espalier rail for fruit trees from *l* to *m;* at which points are two openwork iron gates, to shut in the kitchen-garden.

Expense, &c. The house and offices in this design are supposed to cost 750*l.*, and the laying out and planting of the garden, 100*l.*; no draining being supposed requisite, and the principal part of the expense being incurred in forming the boundary fence and the espalier rail. Such a garden may be kept in order by a common country gardener, who will, at the same time, look after the horse and chaise. Perhaps there is rather too large a space devoted to walks in this design: but it must be considered that these are always a great advantage where there are children; and that, where the soil is dry, and the walks properly made at first, they may be kept at very little expense. This design may be considered as suitable for a country clergyman, who takes delight in working in his garden himself, or for a country attorney or medical man.

A Suburban Villa of Four Acres in extent. This design (*fig.* 129.), which was contributed by Mr. Rutger, from its parallelogram outline, obviously belongs to a series of villas fronting a street, and with a lane behind. The house (*a*) is approached by a semicircular road (*b b*). To the right are the stables and stable-court (*c*); and to the left, at *d*, are the washhouse, laundry, poultry-court, ice-house, and such other appendages of this kind as may be considered requisite. Both these courts of offices have carriage entrances from the street (*e e*); and foot entrances (*f f*), communicating, by the stairs shown at the two ends of the house, with the basement story, and with the approach road; and also with the walks in the garden, by the side paths (*g*). The ground, which is supposed to be perfectly flat, is laid out in lawns and shrubbery, with a pond (*h*), in the centre of which is a fountain. There is a straight central walk, which leads from this pond to the kitchen-garden, in the centre of which is another smaller basin of water, also with a fountain; and, beyond it, a green-house and forcing-house at *k*. To the right is a pine-stove (*l*), and the forcing-pits (*m m*), with a space to serve as the reserve ground at *n*. To the left are the gardener's house and court (*o*), and the orchard (*p*). It will be seen that, in the kitchen-garden, there are two long walls with southern exposures, for fruit trees, besides the end walls; and also a path (*q q*) from the stable offices to the reserve-ground, kitchen-garden, and gardener's house, which joins the paths *g g*, immediately within the boundary fence, and thus communicates with the poultry yard and laundry court.

Remarks. The arrangement of this design, as far as use and convenience are concerned, we consider as complete as it can possibly be made. The gardener and his men have a free, but concealed, communication with the stable-court, poultry-court, &c., for manure; and with the basement story of the house, to supply the kitchen, without once crossing a single walk of the pleasure-ground, which the family may walk in and enjoy in perfect retirement. Objections may be taken by some to the form of two or three of the clumps; but some allowance must be made for the difference of taste of individuals; and, be-

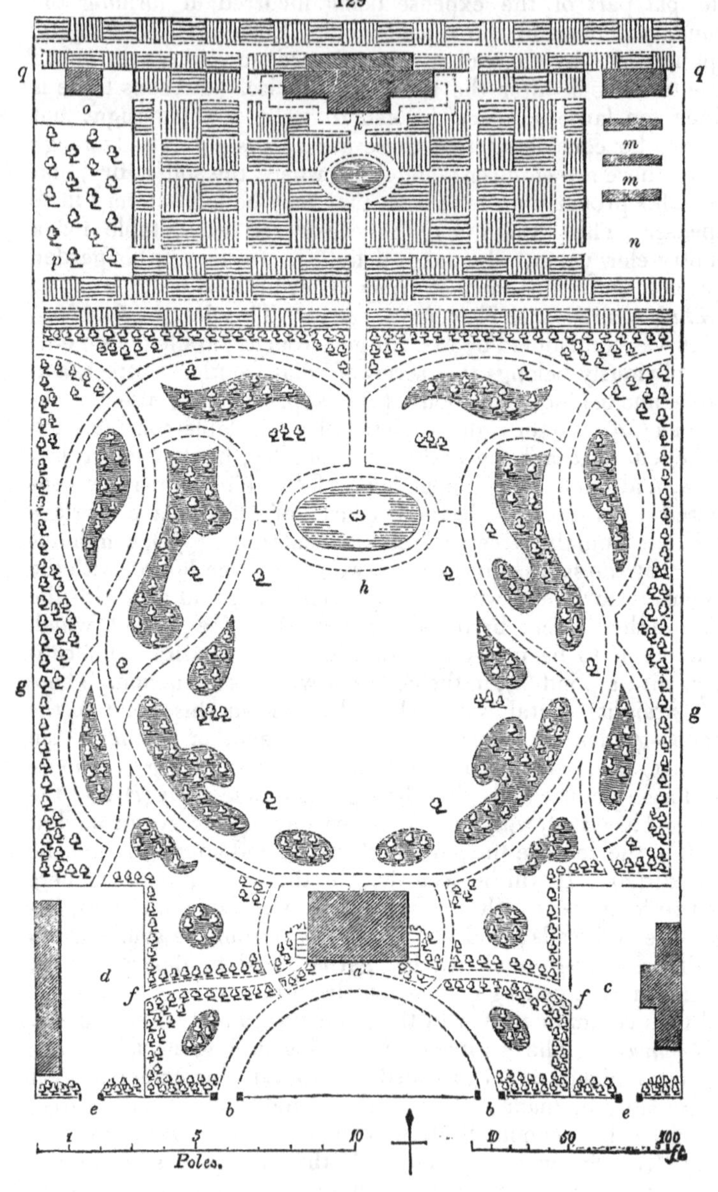

sides, as these clumps are for the most part large, and can never, by any possibility, be actually looked down upon, and seen as a composite figure, as they are in the plan on paper, their form is, in reality, a matter of no great consequence. From the trees indicated in these clumps, as well as those on the lawn, being isolated, they are evidently intended to be planted in the gardenesque manner.

If the plan were to be drawn over again more in detail, it would be advisable to distinguish between those clumps which are to consist of trees and shrubs, and those which are to be planted solely with flowers; but this we shall leave to the reader, who, if he has paid sufficient attention to the remarks given in preceding pages of this work, will be able to supply all that is wanting for himself. We shall only farther remark, that, from the extent of the pleasure-ground in this design, it may, if not more than two or three plants of each species or variety are introduced, contain all the finer trees and shrubs purchasable in British nurseries; as the kitchen-garden and orchard may contain all the most select fruits. The expense of forming such a residence will vary from 5000*l.* to 7000*l.*, or upwards, according to the style of house; and the annual keeping of the gardens will amount to 300*l.* The whole is suitable for a person having a clear income of 2000*l.* a year.

A Design for a Suburban Villa, the House of which forms part of a Row, the whole containing about One Acre. In this design (*fig.* 130.), which is also by Mr. Rutger, the carriage entrance forms a sweep, which is carried up directly to the porch of the house (*a* in *fig.* 130.). The two wings to the main body of the house are curvilinear in the ground plan; that to the left is a conservatory, and that to the right a billiard room. It may be objected to these wings, that, being curved, they do not harmonise with the straight lines of the main body of the house; but the reason for their curvature is as follows:—Had the conservatory been a parallelogram, placed at right angles to the house, it would have fronted the west; whereas, by the curvature given to it, it is made to front the south-west, which is doubtless a considerable advantage to the plants. The curvilinear plan of these wings also harmonises with the plan of the approach road. The stable offices are shown at *b*: the carriage entrance to them is from the street, and they communicate with the melon-ground and reserve garden (*c*), and with the kitchen offices, which are in the basement story, under the billiard-room. They also communicate with the kitchen-garden (*d*), as the latter does with the pleasure-ground by the walk *e*. The stable offices may also include a poultry-house and yard, and an ice-cellar, which we greatly prefer, when the subsoil will admit, to have placed near the kitchen (as we have before observed, p. 387.), instead of having it in the grounds. The pleasure-ground is laid out in a very neat manner; the walks are 5 ft. broad; and the verge of turf on each side of them is nowhere less than 4 ft. wide, in order that those who do not choose to walk on the gravel may walk on the soft

130

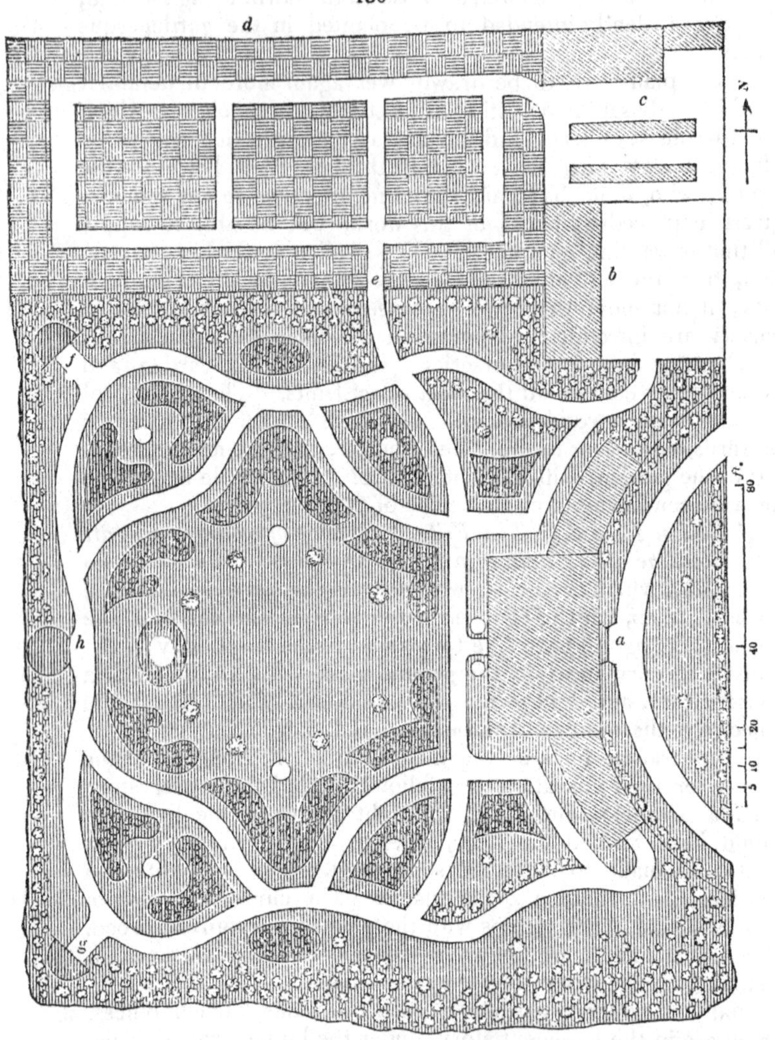

grass. This, in our opinion, is an arrangement which ought to be attended to, more or less, in every pleasure-ground of any extent; because there are many infirm persons who greatly prefer walking on grass to gravel; and because the ample space thus left on each side of the walks takes away from that confined appearance common to walks having shrubbery on both sides, and turf edges not more than 18 in. or 2 ft. broad. On a clayey soil and flat surface, we should say that the turf verges should be nearly as broad as the gravel walk. The groups of shrubs in this design, it will be observed, fit in very well to their different situations among the walks; and they are supposed to be planted in the gardenesque manner, as are the trees forming the boundary plantation. The flower-beds, which are not numerous, are indicated by circles; and those circles in the centre of the two clumps near the house are intended for pyramids of roses. There are two alcoves at *f* and *g*, and a circular building at *h*, which may either be a summer-house, an aviary, or, as Mr. Rutger suggests, a summer reading-room. Instead of a circular summer-house, there may be a polygonal rustic structure, as shown in the ground plan *fig.* 131., and the

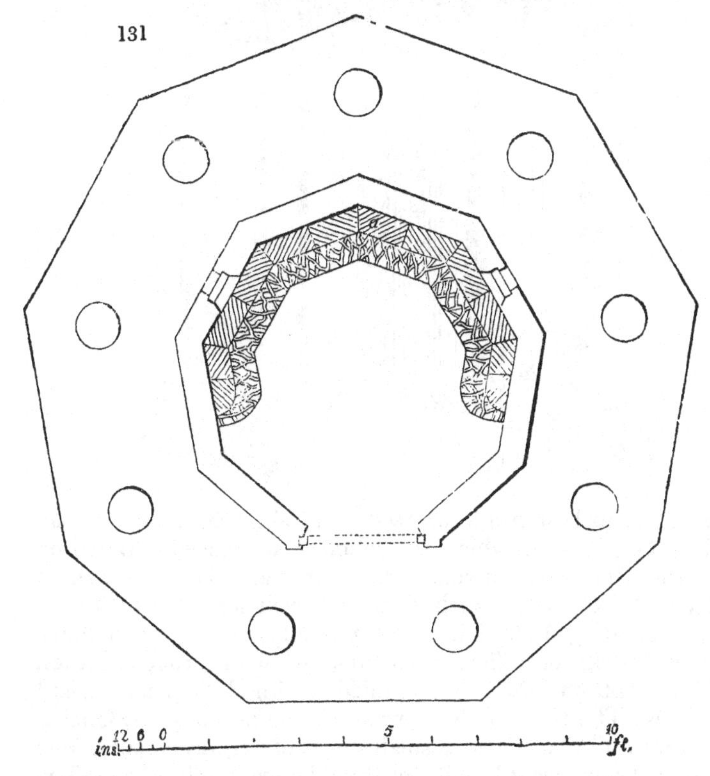

elevation *fig.* 132., which will have an exceedingly good effect from the house. The ground plan of this rustic structure has nine equal sides, with a portico all round, supported on nine pillars, each of a young fir tree, 1 ft. in diameter, with capitals 1 ft. 2 in. square, formed of square boards 4 in. thick. On these caps rest four courses of rustic planks, with the bark on each, 6 in. thick, and perfectly horizontal, which connect the whole of the columns, and support the rafters of the projecting roof, which

132

may be formed of heath or reeds, or of larch, birch, or oak bark. The floor on which the columns are placed is raised on a base 1 ft. high, and surrounded by a gravel walk. The position and form of the seat, which is made of well-seasoned wood, are shown at *a* in *fig.* 131. The floor of the cell may be laid with flints, pebbles, bricks of different colours, or with broken bottles, with their bottoms upwards, and filled in with Roman cement and sea shells. The floor of the portico may be paved in any fanciful manner, with bricks, or with small stones of various colours, from the nearest convenient locality, or from the sea beach. The rafters are 4 in. broad, and 5 in. deep; and, to secure them from wet, they may be first covered with a layer of slates, and above that with a coat of heath or of reeds, 9 in. thick. If heath or reeds cannot be conveniently procured, chips of wood, or of birch or hazel spray, may be used; and covered, or not, according to the taste of the party, with plates of larch, birch, or oak bark, laid in the manner of slates, with the outer surface upwards. If bark is not used as an outer covering, the materials may be Kyanised.

The wall of the cell is 9 in. thick, and consists of nine upright posts, on the inside of which are nailed horizontally (the upper edge standing out about 1 in. from the post) fillets of wood ¾ in. square, and 1 in. apart. In the interstices between the strips, moss is rammed in with a wedge-shaped piece of wood. The ceiling is formed in the same manner, except that in the centre there is a polygonal star, formed by moss of a different colour. On the supposition that the general surface is clothed by the common ground moss, the star may be formed of the same moss, dyed by steeping it in a decoction of logwood; or of the white moss found on trees. The cornice in the inside is made of the cones of the common Scotch pine, or of the pinaster; three rows being laid horizontally, and one row projecting a little outward. The door is made of rustic-work, and has the two upper panels filled in with latticework and stained glass, as are the two other windows in the side walls, shown in the plan. The seat is formed of different kinds of wood with the bark taken off, as shown in the plan. The best kinds of wood to use for the columns, and the horizontal planks over them which form the frieze, are spruce fir and larch; because these retain their bark a longer time without its decaying, than any other common British wood. If these kinds cannot be procured, then it is better to remove the bark. Structures of this kind are not uncommon in pleasure-grounds: that now described was erected, some years ago, at Murtle, in Aberdeenshire, by Mr. James Alexander, then gardener there. (See *Gard. Mag.*, vol. xi. p. 467.) The cost is chiefly labour, the whole timber and other materials not exceding in value 5*l*.

In front of the building there is a basin of water, with a fountain at *h* in *fig.* 130. p. 392.; which, if the building be rustic, may be rustic also. The fountain may be supplied from a well, from which the water may be pumped up to a cistern on the roof of the house, and conveyed thence, not only to this fountain, but to the conservatory, the frame-ground, and the kitchen-garden. The conservatory might by this means be watered by a shower from the roof, in the manner alluded to in p. 111.; and, indeed, the whole garden, if there were an abundant supply, might be occasionally watered, during the hot summer evenings, by a leathern portable tube, to fit on stopcocks concealed in different parts of the grounds; and connected with a main communicating with the supply cistern. It is true that this mode involves considerable expense; but, to persons fond of having a garden in first-rate order, and having the turf and everything else fresh, green, and vigorous, during the hottest months of summer, this would not be an insuperable objection. In some of the suburban gardens in the neighbourhood of Paris, this mode of watering is carried to an extent that many persons can have no idea of; partly because it is carried on early in the morning, or late in the evening, when most Englishmen visiting Paris are either in bed or at dinner; and partly because, as far as we have observed, the persons who have the finest small gardens in the neighbourhood of Paris do not mix with the English society in that city. Though watering is not nearly so essential to a London suburban garden as to one in Paris, yet still it adds wonderfully to the beauty and vigour even of the former, more especially in dry seasons, and on a dry subsoil. In the case of trees and shrubs, the effect of watering is almost incredible, if applied freely every evening that it does not rain, during the months of May and June, and until the leaves and young shoots are fully developed. Whenever the buds on these are considerably swelled, which takes place with most ligneous plants in the course of the month of August, artificial watering should be entirely withdrawn, in order that the wood may be thoroughly ripened. In some seasons, indeed, the weather is so dry in September and October, that the grass, in ordinary gardens, becomes burnt up, and the trees prematurely lose their leaves; but we have never known this to take place where artificial watering has been continued till the buds were fully swelled. After this swelling has taken place, the supply required by the leaves is much less than before; as they become comparatively mature, and, consequently, the evaporation from their surface does them less injury.

Irrigation of Lawns. However much we may approve of the watering of the grass or trees of a garden from the rose of a watering-pot, or a barrow engine, or a flexible tube, there is nothing less suitable to our ideas of an English pleasure-ground, than irrigating the lawn by small open channels conducted along its surface. There can be no comfort or enjoyment from turf kept green in this way in any country. Fortunately, in England, such gutters are not required; but, in English gardens in Italy, as for example in that of the Royal Palace at Caserta, near Naples, this system of irrigation is practised throughout the summer; and, besides the unsightly appearance of the gutters, the wetness of the ground, and the rankness of the grass immediately adjoining them, are most disagreeable. In English gardens, where the soil is exceedingly dry, and it is desirable to supply water during summer, both to the turf and trees, a mode of subterraneous irrigation may be adopted by means of brick drains. These may either be about 6 in. deep, and 2½ in. wide, formed of bricks, as shown in *fig.* 133.; or of tiles, as shown in *fig.* 134. The appearance of the top of the drain *fig.* 133. would be as represented in the vertical profile *fig.* 135. The upper surface of the bricks, being exposed to the air and weather, would soon assume a dingy green colour, so as to present no eyesore as contrasted with the green turf; and the surface of these bricks, being as perfectly even and smooth as the lawn on each side, would offer no obstruction to walking, rolling, or mowing: in short, they would not be recognised by any one but the gardener. The drains should be formed without mortar, in order to let the water escape at the bottom and sides, and to admit of taking off the top bricks to clear out any roots of grasses or trees, or other obstructions, which may be formed in them. The chief difficulty in carrying into execution such a system of under-ground irrigation is, to arrange the direction of the drain in such a manner as that every part of the subsoil shall be equally watered. This can only be done by adopting the principle employed in irrigating pasture ground where the surface is uneven; viz. beginning on the highest ground, and leading the drain about, always descending as nearly as possible at the same degree of slope, till the lowest or level ground is reached. By this means, the water will be equally distributed through every part of the subsoil, and never any where to such an extent as to render the surface soil wet, or to make the grass ranker in one part than another. The distance between the drains may be greater or less, according to the porosity or compactness of the subsoil; but, in general, they need not be nearer together than from 30 ft. to 40 ft. At this rate, about 250 yards of drain would suffice for an acre: and, as every yard would contain 16 bricks, the number of bricks for an acre would not much exceed 4000; which, at the present price, in the neighbourhood of London, would be under 8*l.*; say, with the labour, 12*l.* This expense, in laying out a pleasure-ground intended to be kept in the highest order, is a mere trifle in comparison with the immense advantage that would result from it. The supply of water to a drain of this kind should be given from a well, or other source, at the highest point; and, at the lowest point, the waste water, if any, might run into a basin or ornamental pond. If a brick drain on a level with the surface, or of the thickness of a turf under it, were considered too expensive, a drain might be formed, as in *fig.* 134., of ridge-tiles, covering the tiles with six or eight inches of soil. This mode, however, would have the disadvantage of not being

133

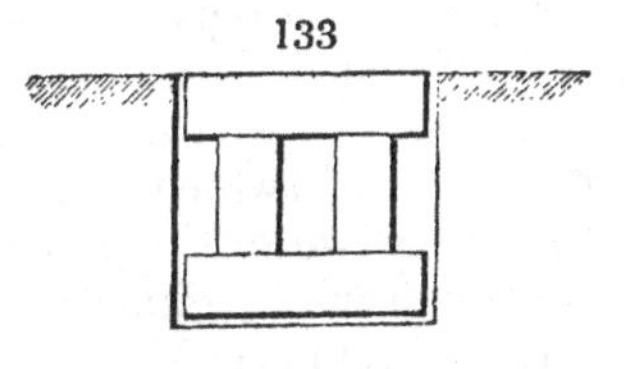

134

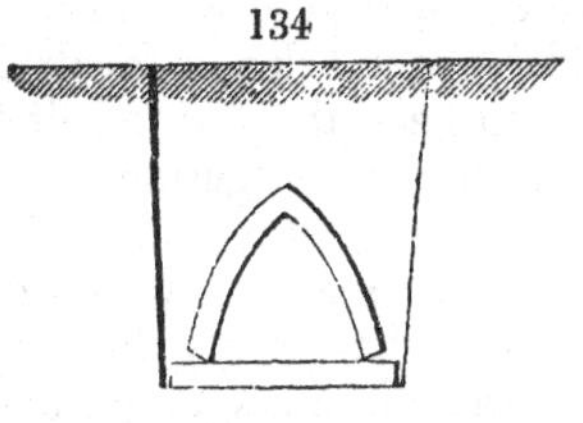

135

so readily kept in repair as the other; because it could not be examined without disturbing the roof; and it must be remembered that such drains would form a powerful attraction to the roots of trees and shrubs, which would be very apt to choke them up, unless cut off every two or three years. A system of drains of this kind, for under-ground irrigation, does not imply that no drains for drying the wet parts of the subsoil are requisite: on the contrary, these are just as necessary where irrigation drains are to be introduced as where they are not contemplated. Indeed, strong retentive clays, which, when they occur in a pleasure-ground, require under-ground drains for drying, at the distance of every 15 or 20 feet, to keep the ground fit for walking on in winter and spring, require, also, irrigating drains for use in the hottest weather in summer, almost as much as sandy or gravelly soils. Those soils which are least in want of irrigating drains are free unctuous loams, rather sandy than clayey, which, from their texture, retain water like a sponge, but never to an injurious extent.

Forcing-Ground. In the forcing-ground are shown, a range of pits to be heated by dung (*i*); another range, to be heated by hot water or a smoke-flue (*k*); a broader pit, for fruiting pines (*l*); and a house for general purposes (*m*).

A glazed House adapted for the Culture of Peaches, Grapes, and ornamental Plants. Fig. 136. shows a section of a house adapted for these purposes: it is 40 ft. long, 16 ft. wide, and heated by one fire: the details will be found immediately under the section. Provided that very early crops of peaches and grapes are not required, and that only pelargoniums and the more common showy green-house plants are grown, there can be no doubt that this house will answer its end. If, however, any attempt were made to procure early crops, as the progress of the peach tree is quite different from that of the vine, the result would certainly be a failure. The peach, as every one must have observed, produces its blossoms before its leaves; while the vine does not produce a single blossom till it has brought forth shoots of such a length as to have from three to five leaves fully developed on each.

Both the peach and the grape will ripen in the same climate; but, in this case, the latter is always much later in the season before it comes to maturity, than the former. Hence, to force the vine to bear fruit much earlier in the season than is natural to it, (say, for example, three months earlier,) it requires a much higher temperature than is necessary to mature the peach proportionately early. As the peach will not bear excess of heat, whenever it and the grape are grown in the same house, the temperature ought to be regulated as if for the peach alone: in which case, the grapes will not ripen near so early, but they will come in in succession to the peaches, and will, on that account, be better adapted for a small family.

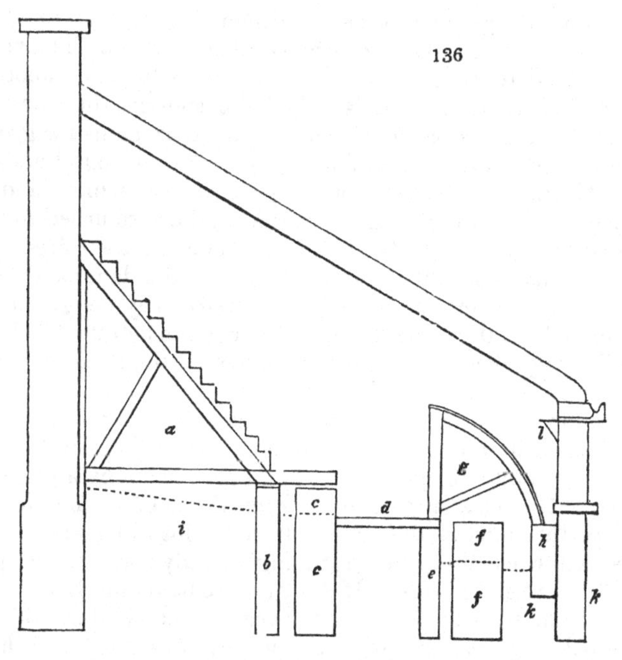

a, Stage for plants. *b*, Arched wall for support of stage. *c*, Arched wall and back flue. *d*, Raised walk, or gangway, in front of stage. *e*, Arched wall for support of walk. *f*, Arched wall and front flue. *g*, Peach trellis. *h*, Stone for support of peach trellis. *i*, Made border for vines, which is 14 ft. wide, 5 ft. deep at back, and 3 ft. at front. Here the vines are planted against the back wall, and trained down the rafters, one branch to each; the spur mode of pruning being adopted. *k*, Peach border within and without the front wall. *l*, Front shelf, for forcing strawberries, &c.

If no more heat is generated in a house containing peaches and vines, than wha is necessary to grow green-house plants, all the three will thrive very well together; but if, in order to force on the vines, a greater heat be applied than the green-house plants and peaches require, they will be spoiled; that is, the green-house plants will be drawn up with weak slender stems, and will only produce insignificant flowers; and the blossoms and incipient fruit of the peaches will drop off. The house before us is the invention of Mr. Livingstone, and was published some years ago in the *Caledonian Horticultural Society's Transactions*. It is arranged, as the description of the section shows, so as to keep the border for the roots of the vines quite distinct from the peach border; the latter being within the house, and the former exterior to it. The peaches are trained upon a low front trellis, sufficiently near both the front and roof glass, to derive every benefit from the light; and yet so far distant from it as not to throw any shadow on the stage of green-house plants behind. The vines it is proposed to train against the rafters only, one shoot to a rafter, and treat in what is called the spurring-in manner; which, though not so favourable as some other plans for growing very large bunches, is eminently so for producing abundant crops. On the supposition that the green-house plants stand in the open garden in the summer season, and are returned to the stage of the house about the middle of October; then fires may be made, so as to do little more than exclude the damp and frost till Christmas; after which the temperature may be gradually raised to 48° or 50°; which, in the course of the month of February, will bring the peaches, and a number of the pelargoniums, and other Cape plants, and also the Australian plants, into flower; a much more desirable period than June, which is the common season for the flowering of pelargoniums, when they are kept in a mere green-house temperature. In April, the temperature may be increased in clear sunshine to from 55° to 65°; and, in the course of this month, the vines will come into blossom. The peaches will ripen about the beginning of July; and the grapes about the middle of August. If this practice be regularly pursued with the same trees for a number of years together, they will acquire a habit of coming early into flower, with much less heat than will be required for the first two or three years. The same thing may be observed with respect to the green-house plants; provided the same individual plants are used, and that they are not allowed to come into flower in the summer season, after they are set out in the open air. This remark, it is obvious, will apply to forced plants generally; and it would be even worth bearing in mind when propagating peaches, vines, and pelargoniums, which are to be forced. Cuttings, grafts, and buds, in this case, ought to be taken only from plants which have acquired a habit of earliness. It is much to be wished that this principle were generally attended to in the nurseries.

In the fruiting pine-pit, which may be heated by flues or hot water, from fifty to a hundred plants may be fruited every year; these plants being brought forward for from ten to twelve months in the other flued pits: melons, cucumbers, potatoes, salading, and various forced articles, may be grown, or forced in the other pits. A cellar may be formed under the coach-house and stable, in which mushrooms may be grown, and tart rhubarb, chiccory, &c., forced all the winter. In one corner of the reserve ground, if the soil be dry, a pit may be formed for keeping ice; or, if the subsoil be moist, a mound may be raised for this purpose in one of the clumps in the pleasure-ground, and covered with trees and shrubs.

Expense and Management. A place of this extent may be

built on, laid out, and planted for from 4000*l.* to 5000*l.* It will require to be managed by a regular gardener, and a man and woman as assistants, which, with the ordinary garden expenses, may amount to 300*l.* a year. The whole may be considered as calculated for a person living at a general expenditure of from 1000*l.* to 1200*l.* a year. If, however, the forcing department were given up, and the conservatory and billiard-room omitted, the first cost would be less by a 1000*l.* and upwards; and the whole might suit an expenditure of 500*l.* a year.

Remarks. This place, as we have described it, is suitable for the residence of a mercantile or professional man, living at a few miles' distance from town; but, if it were to be occupied by a person of leisure attached to gardening, he might, with the assistance of a couple of labourers, manage the whole himself. Every thing, in short, depends on the annual sum which the occupier can afford to lay out; because the rate of expenditure may be regulated accordingly. In like manner, the rate of enjoyment which any place will afford to the occupier depends much less on its extent, and on the money annually expended, than on his knowledge and taste. It is impossible for any man to derive much enjoyment from a garden, who does not take an interest in the operations going on in it; and who does not know something of the various plants cultivated, such as their names, native countries, comparative rarity in this country, and the comparative magnitude, or degrees of perfection, to which they attain in different parts of it, &c. In short, in this case, as in every other, knowledge is pleasure as well as power.

A Street Villa, laid out in the Village of Great Malvern. Of this design (*fig.* 137.), which is by Mr. Varden, only the plan of the house and of the front garden is shown. "This little front garden," Mr. Varden observes, "is part of my design for a villa now being erected in the village of Great Malvern, in Worcestershire.

"The situation is the third part of the way up the east side of an abrupt range of hills, 1500 ft. high. The ground on which the house stands falls very quickly from the back, as much as 24 ft. in 60 ft.; and that allows of two very good stories being formed in the house, below the level of the turnpike-road, which is in front of it. This little front garden is raised partly on arched cellarage, and partly with earth brought from a distance, till the surface has a slight descent towards the road, sufficient to insure in that direction the drainage of fallen rain; as otherwise the water would be forced into the offices and lower rooms, and it would, by saturating the mass of loose earth, endanger the safety of the structure. In front, the view is confined to the steep hill side; but from the back may be seen the broad, rich, well-wooded valley of the Se vern, including a city

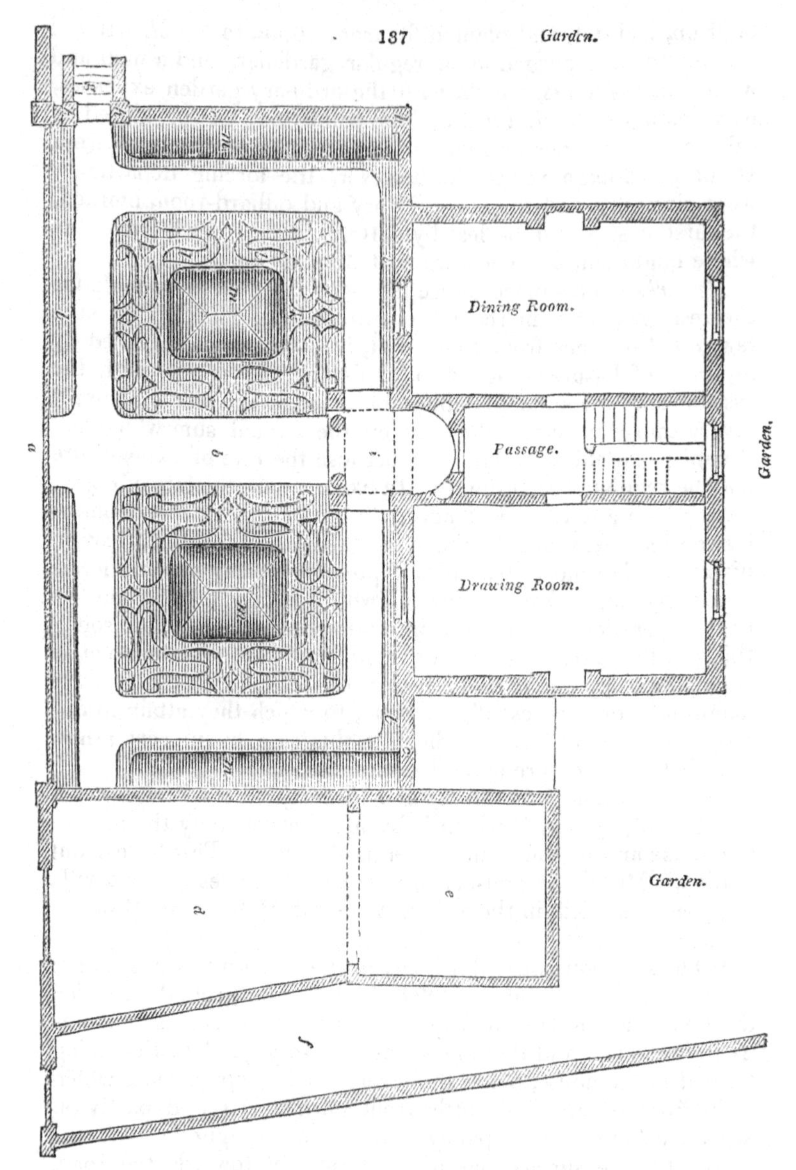

towns, villages, hamlets, and detached dwellings, and forming a scene of great beauty and extent.

"In *fig.* 137., *a* is the turnpike-road; *b*, the pathway from it to the house; *c c*, the parapets enclosing the garden; *d*, the yard; *e*, the coach-house, with a stable underneath approached by the inclined plane *f*; *g* is a flight of steps to the winding path leading to the back gardens, which are lower down the hill;

and *h h* are vases. A portico, supported on columns and antæ, is shown at *i;* and behind it, within the body of the house, is a curved recess, in which are two niches for statues, and the hall door. This recess is for the purpose of giving increased effect to the entrance; which it will, in some measure, do, by determining the place for, and showing the use of, the portico; and, by its great depth, giving additional relief to the columns and antæ. Its great variety of light and shade will also command attention. In the front garden, *k* is turf; *l l* are flower-borders; and *m m,* raised banks, also for flowers. The small size of this garden will allow of its being highly kept, which, with its regularity and its mural boundaries, will give it a decided character of art, and make it form a strong, but not improper, contrast with the wild and almost mountainous scenery surrounding it."

Remarks. We have introduced this design chiefly for the sake of showing what may be done in an extremely steep situation. The idea of having the stable under the coach-house is not one likely to be familiar to the general reader; nevertheless, there is a detached house in this street (Porchester Terrace), where, from the declivity of the surface being in the direction of the road, the line of frontage is several feet lower at one end than it is at the other; and at this lower end an entrance is made to the stable and coach-house, which, by excavating the ground a little, are obtained under the principal floor of the house. To render this arrangement more clear to the reader, we refer to the longitudinal section (*fig.* 138.), in which the line *n n* shows

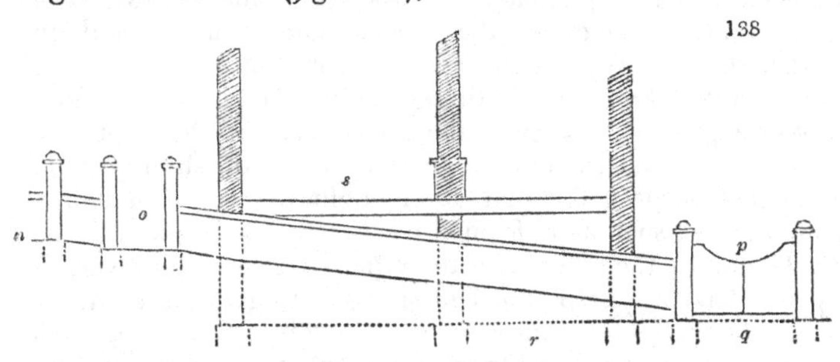

the declivity of the street; *o*, the principal entrance which is at one end; *p*, the entrance to the stables and the garden, which is at the opposite end; *q*, lines showing the depth to which the ground is excavated opposite the doors of the stable and coach-house, and to which there is a gradual slope from the street entrance; *r*, dotted lines showing the level of the floor of the coach-house and stables; and *s*, the level of the principal floor of the house.

A single detached House and Grounds, occupying about an Acre and Three Quarters. The general form is that of a parallelogram, as shown in *fig.* 139. In this plan the street entrance is by the veranda (*a*) to the porch (*b*), which leads to the staircase (*c*), dining-room (*d*), library (*e*), drawingroom (*f*), and green-house, heated from the back of the drawingroom and library fires (*g*). The kitchen-court is shown at *h*, and steps from the drawingroom to the lawn at *i*. At *k*, are steps down to the kitchen area, for servants; and at *l*, a flight of steps up to the green-house, for the gardener. The green-house has one glass door to the drawingroom, and another to the library; and, where the waste heat is not sufficient to keep out the frost, recourse is supposed to be had to one of Joyce's stoves. This last resource, however, will seldom be necessary, if, every night during the most severe weather, the family, immediately before quitting the drawingroom and library, throw open the green-house doors; which, as we have before stated (see p. 319.), will tend to equalise the temperature of the three apartments. As the furniture, books, walls, &c., in the drawingroom and library, must necessarily be heated to a temperature of about 60°, it will be several hours before the demand for heat by the green-house will reduce these rooms 20°, which would give a temperature common to the three of 40°; at which or even at 35°, green-house plants will take no harm. Two detached pavilions, supported on four latticed pillars, form a break in the walks at *m* and *n*, and serve as a foreground to the back garden, and *vice versâ*; and between these and the boundary walls there are covered seats at *n n*: *o o* are flower-beds; *p* is a plantation of low trees and shrubs, each plant standing distinct, in the gardenesque manner, and, in the plan, the trees being distinguished from the shrubs by their darker shade. For two or three years after these trees and shrubs are planted, the ground about them may be kept free from grass and weeds, and occasionally hoed or slightly dug; and for this purpose the plantation ought to be included in a definite outline, such as is formed by the edging of the walk on one side, and by the line *q r* on the other. As soon, however, as the plants have acquired sufficient strength to grow on a grassy surface, the outline on the lawn side may be obliterated, such of the trees and shrubs as touch one another thinned out, and the surface sown down with the finer grass seeds; the whole uniting and harmonising with the lawn, as indicated at *p* and *s*. We may observe here, as a general rule, that, in the gardenesque manner, wherever the ground is to be dug among trees or shrubs, the boundary should be definite; because the principle is, that a definite outline is most convenient for culture, and for the display of individual beauty. On the other hand, in a picturesque plantation, where the surface is to be dug, the outline should be indefinite, or consist of a ragged line; because indefiniteness and irregularity are properties of the picturesque.

139

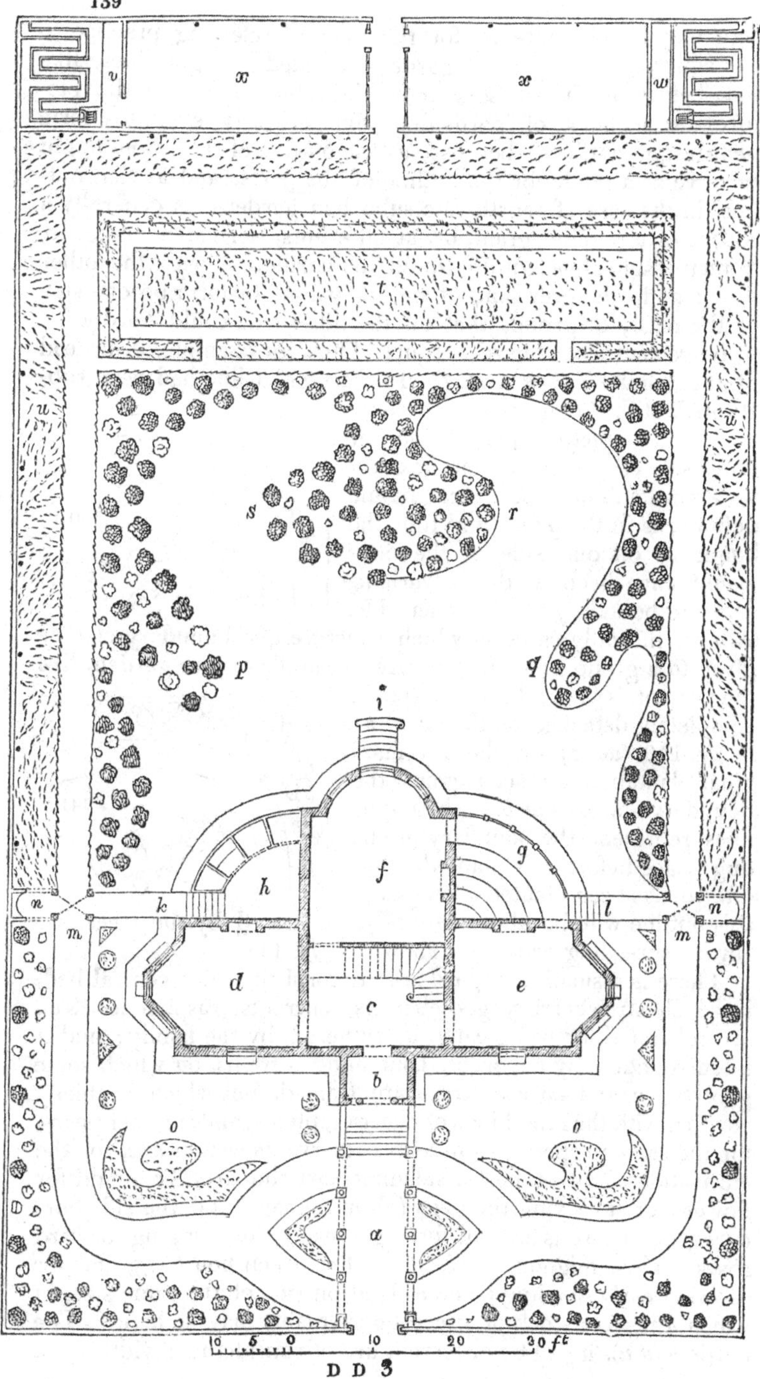

We may farther observe, for the sake of referring practice to principles, that, in a small garden bounded by right lines, like that before us, it will seldom be desirable to imitate the picturesque manner of gardening, and scarcely ever to form picturesque outlines; because, as the outline of the whole ought to serve as a guide for the outline of the parts, and as that outline, in the case of fourth-rate suburban gardens, is generally a square or a parallelogram, or, at all events, a right-lined figure, a picturesque line within would ill harmonise with the other lines; and, whatever kind of outline we might form on one side of the mass, or group, that on the other side could hardly fail to be geometrical or gardenesque. Thus, for example, if, instead of the definite line *q r*, in *fig.* 139. a ragged line had been substituted, still, though that side would have had a picturesque outline, the other side of the plantation next the walk must still have been straight and definite, as in the plan; and thus the boundary on one side of the mass would have been at direct variance with the boundary on the other side.

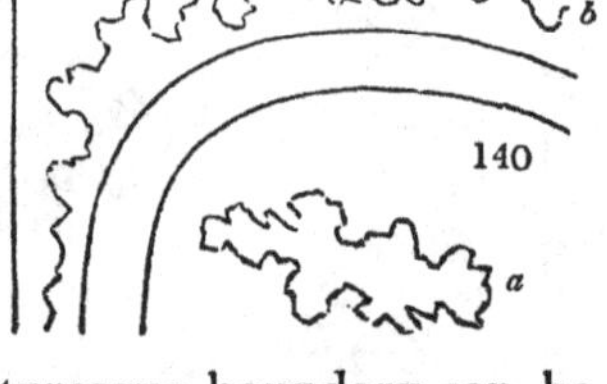

140

Almost the only cases in which a picturesque boundary can be given to a group in a small square or parallelogram garden, like the present, is where the group stands completely detached on the lawn, as in *fig.* 140. at *a*; or where a marginal plantation is placed against the boundary fence, as at *b*. These outlines represent the boundary of the dug space before it is planted; but, after it has been planted a few years, this outline will be almost entirely obliterated by the trees and shrubs spreading over it, as shown in *fig.* 141.

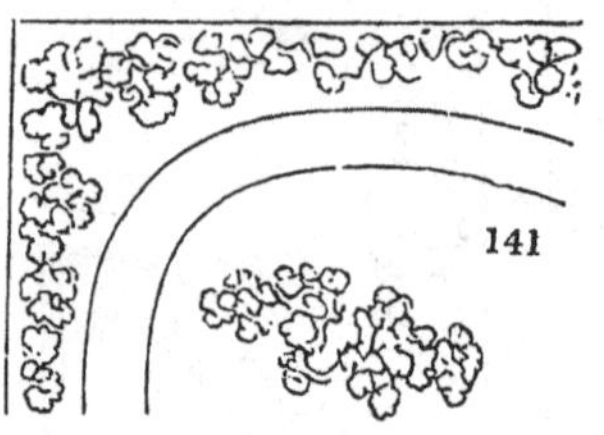
141

There is a small fruit-garden at *t*, consisting of a central bed, in which strawberries, gooseberries, currants, raspberries, &c., are cultivated for being eaten as gathered, by the family; and it is surrounded by a border, containing a trellis, on which some gooseberry and currant trees are trained, but which is chiefly covered with the finer kinds of cherries, plums, and summer pears, for the same purpose. There is a reserve garden at *x x*, for the cultivation of sweet herbs, salading, tart rhubarb, &c.; and for flowers for supplying the beds (*o*) on the lawn and the side borders (*u*). There is a small forcing-stove (*v*) for bringing forward roses, bulbs, mignonette, &c., for the green house (*g*), and for growing early salading; and a flued pit (*w*), of the same size, in which a reserve stock of green-house plants may be kept. The borders (*u u*), it will be observed, are of considerable width for a small garden; and it is proposed to devote them entirely to flowers. The walls may either be planted with fruit trees, or with the higher class of ornamental low trees and shrubs, according to the taste of the owner. If the latter plan were adopted, and no duplicates introduced into the plantation *q r s p*, or into the boundary plantation from *m* by *a* to *m*, then in this garden there might be included all the finer low trees and shrubs of the British arboretum. The display of roses, Japan, Chinese, and American honeysuckles, magnolias, wistarias, passion flowers, pomegranates, *C*lématis, *C*ydònia japónica, Chimonánthus, and a host of other articles of the same kind, which either have been or will be enumerated, would be delightful at every season of the year; while in the borders might be passed in review all the finer herbaceous plants, annuals, perennials or bulbous. The green-house (*g*), though small, yet being supplied by the small forcing-stove (*v*) and the reserve-pit (*w*), would make a fine display throughout the year; and, if it were thought desirable, there might be a vault under each of these pits, for growing mushrooms, and forcing sea-kale, tart rhubarb or chicory. The arabesque beds (*o*), and the small circular and triangular beds which accompany them, will contain a very fine display of half-hardy annuals during summer, and masses of crocuses, snowdrops, tulips, and other bulbs, in spring. In short, this garden, of an acre and three quarters, under the management of a master fond of gardening, and with the assistance of a single labourer, might contain almost every thing that is desirable in a suburban garden. The general appearance of the house, veranda, walls, and walks, is shown in the isometrical view, *fig.* 142.

Expense and Management. Such a house as we have shown need not cost more than than 750*l.*; and 250*l.* more will lay out and plant the garden, including the erection of the two pits. As so considerable a portion of the surface is under grass, the cost of yearly management will be much less than if the whole, or even the greater part, were under the spade. The grassy surface, from *o* to the small fruit-garden (*t*), occupies three quarters of an acre; the house, walks, and other buildings, fully one quarter of an acre; leaving three quarters for flowers, fruits, and vegetables. This space might either be cultivated by the master, with the assistance of a labourer and a mowing-machine; or by a head gardener, with a labourer, or the occasional assistance of the house servant. In the case of this villa, however, as in most others, very little can be said with certainty as to the expense of management, unless the style in which the garden is to be kept be given as one of the data on which the calculation is to be made. High keeping may always be set down at double the expense of ordinary keeping.

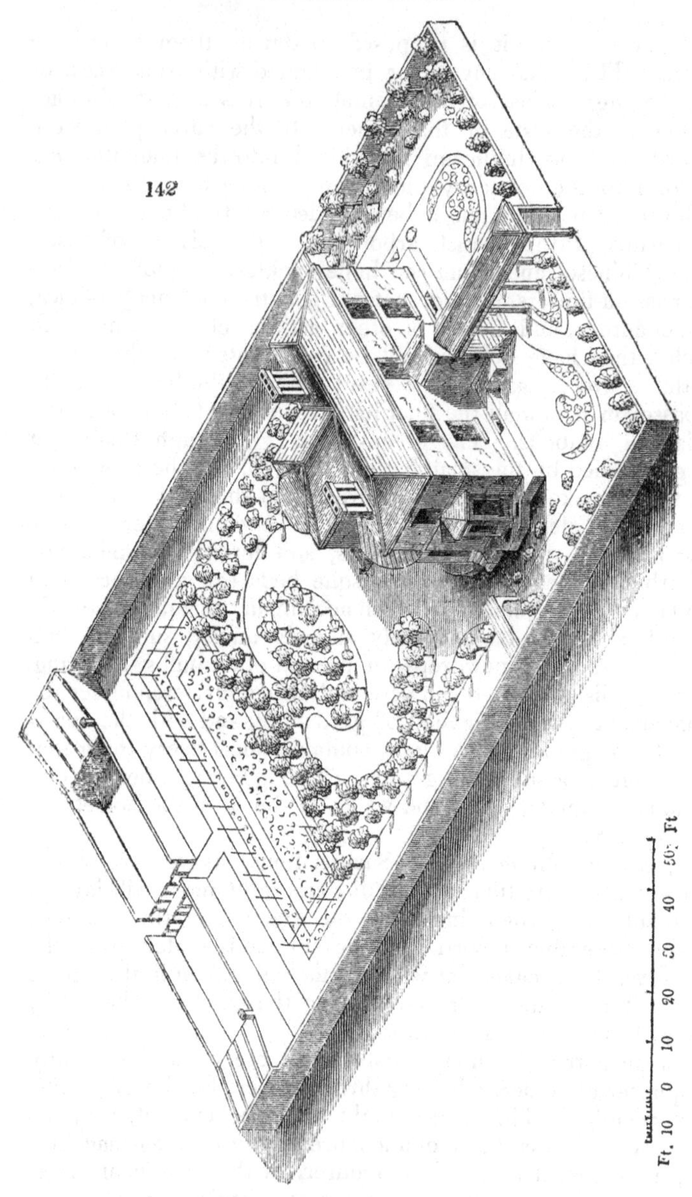

Remarks. This residence, it is thought, would suit a citizen with a large family of children, or with a number of sisters, or grown up daughters; as it would contain a fine display of flowers, and also abundant space for amusement on the lawn, on which a tent might be placed during summer. If the occupier were his own head gardener, he would find something to do every month in the year; more especially with the two forcing-pits, the principal green-house, and the two forcing-vaults. Instead of ornamental trees, on the lawn, at *p*, *q*, *r*, &c., in *fig.* 139., fruit trees might be planted, so as to form an ornamental orchard, which to many persons would be a source of greater enjoyment and interest, than a collection of merely ornamental trees and shrubs. The walls might also be planted with fruit trees, the green-house with vines, one of the pits with pine-apples, and the other devoted to forcing culinary vegetables, &c.

We had prepared a number of other designs for fourth-rate gardens; but, lest we should swell this volume beyond a reasonable size, we shall pass on.

SUBSECT. 2. *On laying out, planting, and managing Third-rate Suburban Gardens.*

THIRD-RATE gardens, it will be recollected (see p. 171.), are such, as are not regulated in the situation of the house by any street or row. Hitherto, we have been limited, by the necessity of keeping the front of the house in a line with the fronts of other houses belonging to the same row or street; and also by the right-lined boundary of the entire plot, as well as its comparatively limited extent; but now we are released from these trammels, unless we except some cases where a plot of a few acres may be bordered by straight lines, which is no great disadvantage in this comparatively large space, and more especially where the ground has an irregular surface. We shall, therefore, in these designs, be enabled to introduce greater variety, as well as a greater number of objects, in each design. If to irregularity of outline, and the extent of a few acres, irregularity of surface be added, and the choice left as to the situation of the house, all the leading desiderata necessary to make a fine place are present; and, whether such a place be produced or not, will, all other circumstances (such as climate, soil, water, &c.) being favourable, depend on the taste of the designer, and the liberality of his employer in regard to the sum to be laid out.

In giving designs for fourth-rate suburban gardens, we have supposed, in almost every case, that the surface is flat, and the space limited; and, consequently, that the beauties to be produced must necessarily be included within the enclosure. In some cases, indeed, we have suggested that adjoining trees, and other objects, might be admitted to view; and, in others, that unsightly objects might be excluded by trees: but, as there is in general, very little scope for exercising this kind of art in laying out street gardens, we have hitherto not said much on the subject. In the gardens of which we are now about to treat, however, and especially such as have an uneven surface, the exterior landscape necessarily forms a constant object of attention, in laying out the surface of the area within the boundary. Without a nice attention to the adjoining grounds, and the manner in which they are planted, built on, or laid out; and adapting the foregrounds and middle distances to them, as the case may require; the effect produced will not be entitled to be considered as landscape, in the artistical sense of that word.

What constitutes an artistical Landscape. It may be necessary to remind the general reader that by a landscape, in the sense in which it is understood by artists, is meant a scene, verdant, architectural, or aquatic, which can be seen by the eye looking horizontally and directly in front, without moving the head; which consists of strongly marked parts, within a few feet or yards of the eye, forming what is called the foreground; which contains a number of parts in the horizon, sufficiently distant to appear greatly diminished in size, which is called the distance; and which is completed by an intermediate portion of scenery, neither so indistinct as the distance, nor so large and bold, nor so much in detail, as the foreground, which is called the middle distance. Such landscapes admit of infinite variety: 1. as to the extent of the space within which the three component parts of them, viz. the foreground, middle distance, and third distance, are contained; and 2. as to the nature, number, and magnitude of the objects which form each of these component parts. One property, however, is common to every landscape that is satisfactory to the eye; viz. that it is composed of parts or distances, and that these are not less than three. The principle on which this is founded is, that no landscape can please that does not form a whole; and that three are the smallest number of parts in any object by which a whole can be produced. In addition to constituting a whole, a landscape may be expressive of different other kinds of beauty: it may be gay or gloomy, simple or intricate, and so on; and though the landscape-gardener may not always have it in his power to confer much expression of this kind, yet he is always able to form a whole, and, if he does not do this, he is without excuse.

Having endeavoured to give the general reader a clear idea of what constitutes an artistical landscape, our next object is to impress on his mind the proposition, that in every well laid out place there will be a considerable number of views, consisting of at least three distances; all more or less striking or captivating, and all more or less forcing themselves on the attention of the spectator. These views are to be obtained from the windows of the principal rooms of the house, from the walks of the pleasure-grounds, or from seats placed in different parts of them. If, in moving along the walks, one part of the scenery be found just as interesting as another, and none capable of attracting much attention, then the artist has failed in his attempt to create a series of landscapes. It does not follow from this that he has not created beauties of another kind; for there may be handsome beds of flowers, elegant sculpturesque objects, and fine specimens of trees and shrubs: but these beauties belong to floriculture, architectural decoration, and arboriculture, and not to landscape-gardening. Neither does it follow that every step in a pleasure-ground ought to produce a new and striking landscape, though this is practicable under very favourable circumstances; and we shall show, from a garden of less than an acre in extent, how upwards of a score of striking landscapes may be produced, in almost immediate succession. Nor is it necessary that every landscape should have its third distance beyond the limits of the area under the control of the landscape-gardener: on the contrary, this distance may always be obtained within the grounds when it cannot be found beyond them; though the views, under such circumstances, will be confined rather than grand. It is principally to effect this sort of pleasing, gardenesque, or picturesque views, that sculptural and architectural objects are introduced into gardens.

Before proceeding to give examples of the manner of laying out the grounds of third-rate suburban gardens, we shall introduce some remarks on the principles and rules by which the artist ought to be guided in laying out such gardens. These remarks will be of greater length than may, at first sight, appear necessary; but, as most of them are also applicable to second-rate and first-rate suburban gardens, they will supersede the necessity of our again recurring to the subject of first principles, when those classes of gardens are treated of.

The House and Offices. In proceeding to lay out the grounds of a country residence, the first point that requires to be determined is, the situation of the house and domestic offices; the latter including the kitchen-court and the stable-court. We have already said so much on the advantages of a situation for the house so far elevated as to be commanding, and on a soil so dry, and a surface so open, as to be healthy, that we shall not enlarge further upon these subjects here; but the site of the house, relatively to the public road and the boundary fence, remains to be considered Where there are offensive objects in or near the boundary, there is a motive for placing the house at a distance from it; but, where this is not the case, the house may be set down on the spot containing the best views, provided that spot allows of properly placing and disposing of the offices and kitchen-garden. Where it is intended that the establishment shall be small, there will often be found great convenience in placing the house near the public road and main entrance, in order to lessen the labour of servants in attending the gates. On the other hand, in the case of complete establishments, a situation at some distance from the public road has the advantage of concentrating the business of the establishment, and thus bringing all the servants more immediately under the eye and control of the master and mistress. Where it is desired to make a place look larger than it is, an advantage is gained by placing the house at a great distance from the entrance-lodge; because the approach road is thus lengthened, and the impression on the stranger, in regard to the extent of the place, is heightened in proportion to the length of space which he has to travel over between the entrance-lodge and the front door of the mansion. This idea of extent is greatly increased when the approach road separates into two branches at a short distance within the entrance-gates, and where one road is devoted to going to the mansion, and the other to returning from it; provided, however, that the one road is not seen from the other.

It is almost needless to observe that the kitchen-court should always be immediately connected with the mansion; and that in some cases, as in small houses, the kitchen and other offices should be, in a great measure, under the principal floor of the house. The drying-ground is most conveniently placed between the kitchen-court and the stable-court; and between the latter and the drying-ground, is a suitable situation for a poultry-yard, which, in our opinion, is an essential addition to every suburban residence which contains an acre or two of land; and, also, for the cow-house and piggery, where only one or two cows are kept, and no more pigs than are necessary to eat up the refuse. The stable offices should be as near as possible to the kitchen offices, in order that the servants employed in them may be within hearing of the bells by which they are called; that as little time as possible should be lost after ordering a carriage, till it appears at the door; and that the master may inspect his horses, carriages, and dogs, with little trouble, during severe weather.

The extent of the accommodation of the house and offices will vary with the style of expenditure of the family; and the style of architecture will depend on the taste of the proprietor; and, therefore, neither require to be here enlarged on. In order that access may be readily obtained to both the kitchen and stable offices by servants and tradesmen, without crossing before the entrance-front of the house, these offices should always be placed on that side of the house which is next the entrance-lodge.

The Kitchen-Garden should be near the stable offices, for the supply of manure; and so placed, that it may be at no great distance from the house, for the convenience of daily bringing to the kitchen the vegetables which it produces. It is also desirable to have the kitchen-garden so situated that ready access may be had to it by the master and mistress; because, in all small places, it is a part of the out-door establishment which is more frequently visited by the family than any other; being always interesting from the variety of operations carrying on in it, and the constant succession of productions in which the mistress of the family, more particularly, takes a daily interest. If the kitchen-garden be placed near the offices, it will necessarily be situated between the house and the entrance-lodge; and, consequently, be readily accessible to carts bringing manure, or soils, or other materials for the garden, from a distance, and also to the visits of neighbouring gardeners. In many small places, it is, for this reason, convenient to have the kitchen-garden near that part of the boundary which lies along the public road, in order to have a carriage entrance directly from the latter. This is also very frequently the case with the stable offices, in order to admit of bringing in hay, straw, corn, &c., without carting these along the approach road to the mansion. Where, however, the house is placed at a distance from the public road, this can neither be the case with the kitchen-garden nor the stable offices; for, in residences of limited extent, nothing will compensate for having these essential parts of an establishment at a distance from the centre, to which they are to contribute their quota of comfort and enjoyment. The kitchen-garden should consist of a wall for the growth of the finer hardy fruits, which wall is generally made to surround the enclosure; borders, protected by the wall, or by hedges or temporary screens, for growing early crops; open compartments, unencumbered by standard trees, for the culture of the main crops of culinary vegetables; a reserve-ground, or nursery, for bringing forward plants in their young state, till they are fit to transplant where they are finally to remain, as well as for keeping a stock to supply blanks occasioned by accidental deaths; and a forcing-ground. The borders, and the compartments of the main garden, are generally laid out in the form of a parallelogram or square, and enclosed by a wall with a border on the outside, in order that fruit trees may be trained on that side of the wall, as well as on the inside. The reserve-garden and forcing-ground are usually included in a small square or parallelogram; at one end, or on one side, of the large square, and in or near this small square, is generally placed the gardener's house. The forcing-houses may consist of pits of different dimensions, heated by flues or hot water; of wooden frames, placed on open brickwork, and heated by outside linings of dung, tan, or leaves; and of walled pits, not heated by artificial means, but simply protected by a covering of glass. Forcing-houses, in the general sense in which that expression is understood, may now be altogether dispensed with in moderate-sized gardens; and thus the great expense both of their original construction, and annual repairs of glass, &c., may be avoided. A forcing-house, such as was constructed thirty or forty years ago, consists of a glass roof, more or less lofty, and placed against a high wall; a glass front, and glass ends; with the interior laid out with at least one walk, which is generally paved; or with a raised bed in the centre, surrounded by parapet walls, flues, and a paved walk. A pit, on the other hand, such as is now built in the best forcing-grounds, whether large or small, has the side and end walls of masonry; the back wall seldom above 7 or 8 feet high; no part glass but the roof, and commonly with only one narrow walk at the back. The saving, both in original construction and in future repairs, is obvious; and every good gardener knows that in such houses the forcing is not only conducted with less labour and expense, but in general produces better results, than in large houses.

The Conservatory and Flower-Garden. Though neither of these luxuries can be considered essential to a suburban residence, they are yet additions which few persons, who can afford the expense will like to be without. The conservatory has been already treated of, as an appendage to the mansion, in p. 108. to 114. inclusive; to be truly enjoyed as a luxury, it ought to be connected with the house, and entered from one or more of the rooms as there shown. The flower-garden ought to be near the conservatory, in order that it may be at all times reached from the house in a few minutes, and also that it may, by its richness and high keeping, add to the appearance of art and refinement in the immediate vicinity of the mansion. Conservatories may either be of glass on all sides, and connected with the mansion by one end; or they may be placed lengthwise against the mansion, or against a wall projected from it, or against the offices; in either of which cases, only the front and the roof need to be of glass. The latter kind of conservatories, or green-houses, are always heated and managed at less expense in the winter season, than those which are of glass on all sides; and the plants in all conservatories always look well in an inverse proportion to the artificial heat required. The main reason why a conservatory with glass only on one side requires so much less artificial heat in the winter time is, that the glazed side is generally contrived so as to front the south; and, while this and the glass roof admit all the rays that the sun gives at that season, the thick wall on the north side protects the conservatory from the cold north winds. A very convenient arrangement, where the aspect is suitable, is, to place the conservatory against the exterior wall of the buildings of the kitchen-court; and, between this and the kitchen-garden, to form a flued conservative wall against the drying-ground, poultry-yard, and stables, and to continue it on, either in straight or in curved lines, or in a combination of these, as far as the wall of the kitchen-garden. The flower-garden might, in this case, be placed in some part of the lawn in front of the conservative wall, that is to say, the general flower-garden; because, where there is a great attachment to plants, and the place consists of two or three acres, there may be an American garden, an ericetum, an alpine garden, a marsh garden, an aquarium, &c., (which may all be considered as belonging to the class of flower-gardens,) in various parts of the grounds.

The Drainage of the house and offices, and of the entire place, may next be taken into consideration. In the country, in general, there are no public sewers, as there are near large towns; but, nevertheless, there are road-side, or field, ditches, and natural brooks, which receive the surface water produced by the rains on the given tract of country; and into these all drains, whether for drying the soil, or carrying off the waste water from the dwelling-house and offices, must be made to empty themselves. In general, there ought to be one main drain from the house, through the centre of the courts of offices, to a large liquid manure tank; or, what is preferable, there may be two tanks, that the one may be filling while the other is being emptied. If the manure produced in these tanks be properly valued, and, consequently, carried away as it is duly fermented, there will seldom be any overflow; but, nevertheless, to provide against the possibility of this occurring, there ought to be a drain from the tanks continued to the boundary of the property, and thence into the nearest ditch or brook. This drain, between its origin at the house and the manure tanks, ought to have all the communications with it furnished with proper traps, to prevent the rising of smells; and, where it communicates with the tanks, and also at its commencement and at its exit into the public drain, there ought to be either cast-iron gratings, or water traps, to exclude rats.

Traps to prevent smells are made of cast iron, and they are generally from 6 in. to 1 ft. square on the upper surface, according to the quantity of water that is supposed likely to pass through them. The construction and action of such a trap will readily be understood by the section, *fig.* 143.; in which *x* is the grated covering to the basin of the trap, and *y* the funnel through which the water escapes to the drain. Water falling into this trap through the cover (*x*) escapes by the funnel (*y*); but the bell-shaped cover placed over this funnel, by dipping into the water all round it, prevents the escape of air upwards. In order that such a trap may be protected from the effects of frost, it is convenient to have it sunk 1 ft. or 1 ft. 6 in. in the ground, and covered with an ordinary cast-iron grating, on

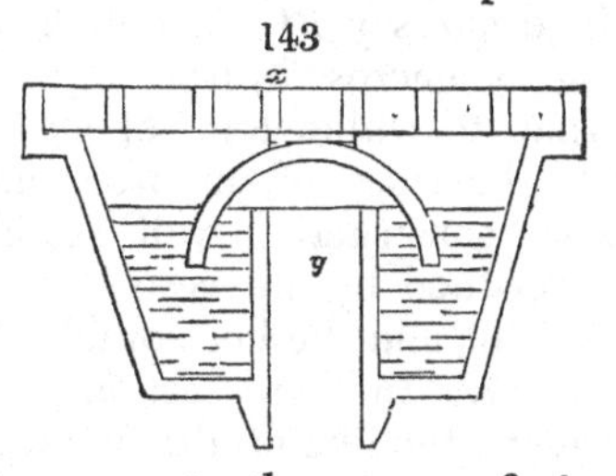

a level with the surrounding surface. The funnel may communicate with the drain, either through the top or sides. As the basin of the trap will be liable, after being a certain time in use, to become filled up with sand or other earthy matters, the grated cover to which the bell is attached is not fixed to the basin, but may be lifted out at pleasure, to admit of emptying the latter. *Figs.* 144. and 145. show the general view and section of a very strong cast

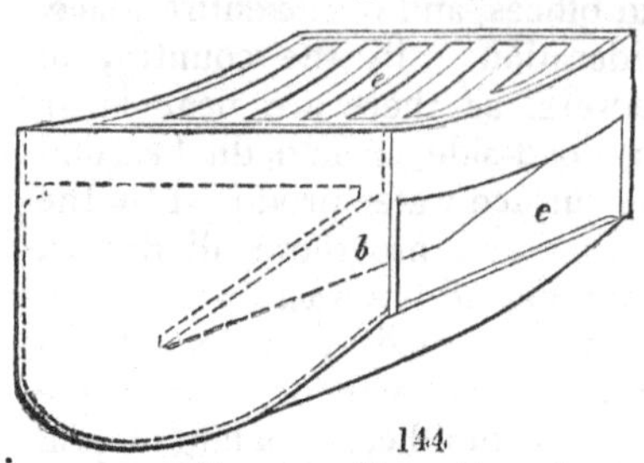

144

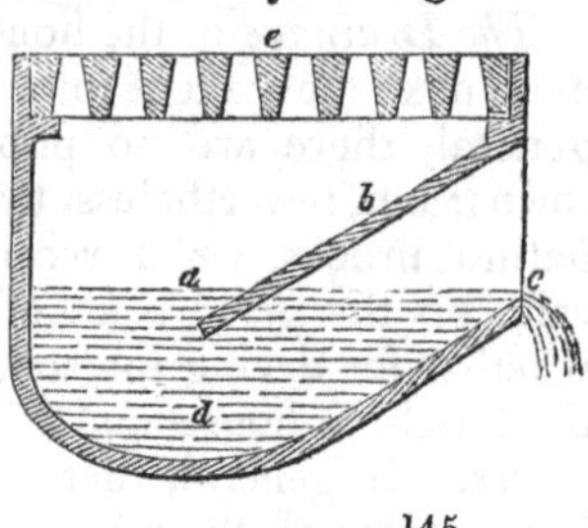

145

iron trap, for the sides of roads and the gutters of stable or farm yards, where it will be liable to be passed over by carts and waggons. In these figures, *a* shows the level of the water on both sides of the trap (*b*); *c* is the opening by which the water escapes; and *d* the place where the sediment is deposited. The grating at top (*e*) lifts off, in order to admit of the removal of this sediment. The grating, being hollow in the middle, indicates that the direction of the bars ought to be across the line of direction of the gutter in which it is placed; by which it will more readily receive the water of the gutter, and at the same time offer no obstruction to carriage-wheels. The weight of this trap, when executed of sufficient strength, will exceed 4 cwt. The first-described trap may be considered as suitable for the kitchen-court, and this last for the stable-court, the approach road, or the farm-yard.

A water trap, for the purpose of preventing the progress of rats along a drain, is formed by sinking a pit in the bottom of the drain, say 1 ft. 6 in. or 2 ft. deep, and suspending a piece of flagstone across it, from within half an inch of the cover of the drain, to within 6 in. of its bottom; the flagstone being built into the sides of the drain and pit. This construction will readily be understood by the vertical section, *fig.* 146., and the horizontal section, *fig.* 147. In these sections, *a a* represent the line of direction of the drain; *b* the flagstone which forms the dip or trap; and *c* the flagstone which serves as a cover, and which admits of being taken off without

146

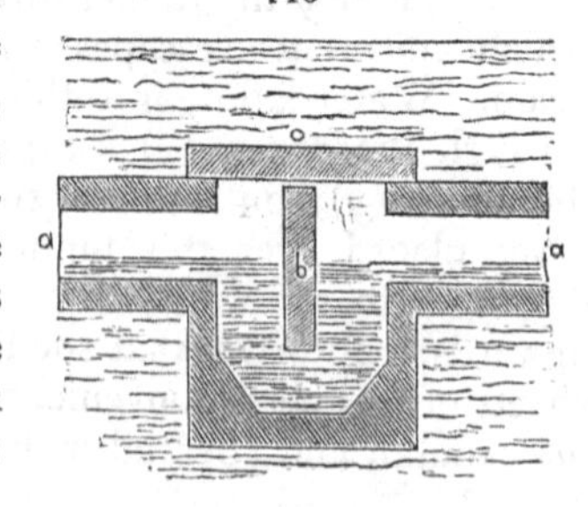

disturbing the dip-stone, in order to clear out the well, or pit, of the trap. The action of such a trap is sufficiently obvious. No animal can pass it without diving under the dip-stone; and this, it is ascertained from experience, rats will not do, unless where the water is clear, and both sides of the trap are fully exposed to the day, in which case they will dive under the dip-stone and ascend on the other side, though the stone should descend several feet into the pit. It should be observed, that only a small space of about half an inch is left between the dip-stone and the cover; this space is essential, in order to admit the passage of air; for though the water would escape without it, yet it would be at a much slower rate.

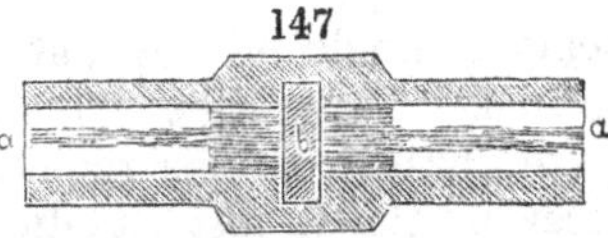

It may be useful to observe here, that all underground drains from offices, or from any part of an estate, which are intended to carry away night-soil, or any earthy matter, ought to have a certain slope or inclination; with the bottom of the drain not flat, but concave, and generally so much so as to form the section of a semicircle. A semicircular bottom, and a slope of 1½ in. in 10 ft., it has been ascertained by builders, will enable any drain, with a tolerable supply of water, to carry off all the night-soil suspended in it; but, where there are sand or small gravel, pieces of bones, broken crockery, and other matters of that description, to be carried away, the slope requires to be at least 3 in. in 10 ft.

If the subsoil of the property be dry, very few drains in the grounds, for the purpose of drying the soil, will be required; but, if the soil be loamy, or on a retentive bottom, every part of it will require under-draining. The kitchen-garden will require to be more thoroughly drained than the general surface which is to be laid down under grass, or planted with trees, because the excellence of culinary produce depends chiefly on the culture of the soil; and an extra attention to drainage must be paid in the flower-garden, and in all those parts of the lawn (especially such as are near the gravel walks) which are liable to be much walked on in autumn or spring. Whether the subsoil be naturally dry or wet, every walk and road, whether in the kitchen-garden, the flower-garden, the pleasure-ground, or forming the approach roads to the mansion, and the branches to the offices, will require to be drained, for the sake of absorbing as completely as possible the water that falls on them during heavy rains, or is produced by the thawing of snow. For this purpose, a drain should be conducted along one side of the walk, under the gravel in the kitchen-garden, and either under the gravel, or under the turf, in the pleasure-grounds; and to this drain small grated openings should be made, at regular distances, along one side of the walk. Where the walks in the pleasure-ground are of great length, branch drains require to be introduced at certain distances in the lowest situations, in order to conduct the water collected in the drains along the walks to the nearest public drain. The drains along the walks should be what are called box or barrel drains; the former having a bottom and square sides, with a flagstone cover, as in *fig.* 148.; and the latter being a circular cylinder of brickwork, a section of which is shown in *fig.* 149. Small iron or stone gratings on the surface communicate with the drains by short upright shafts, 9 in. square in the clear within; and, in the bottom of the drain, immediately under each shaft, a pit is formed, and walled on the sides, a foot square within, and from 1 ft. to 2 ft. or more in depth, as in the section *fig.* 150. The use of this pit is to receive the sand that is carried through the grating with the water, immediately after very heavy showers in summer; and thus to prevent the sand from being carried along the drain, and in a few years choking it up. Once a year, the grating at top is taken off, and the deposit of sand taken out with a spade having a kneed blade; and by this means not only may the drains be kept quite clear and effective for many years, but much smaller and less expensive drains may be made at first. Where the subsoil is sandy, gravelly, or rocky, and where it is not considered essential to use the walks immediately after rain, drains of the box or barrel kind may be dispensed with, and common rubble drains formed by filling up a trench with round stones, to within a few inches of the surface, as in *fig.* 151., and terminating it by a somewhat coarser portion of the common gravel used in forming the walk, through which the water will percolate, throughout the whole length of the drain.

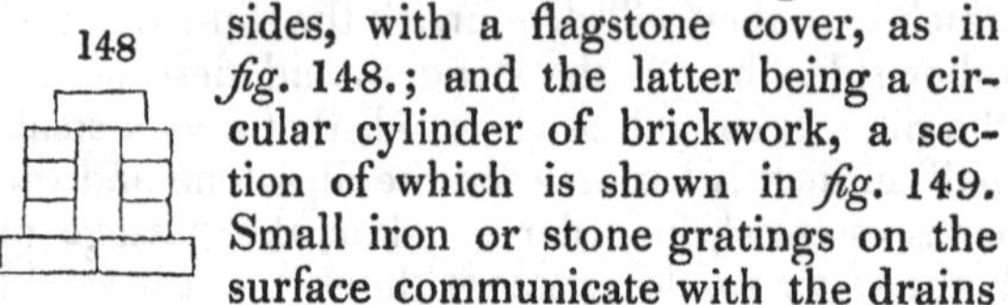
148

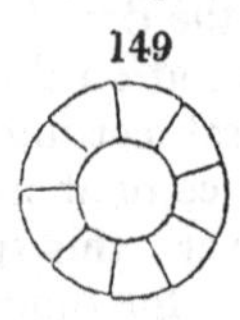
149

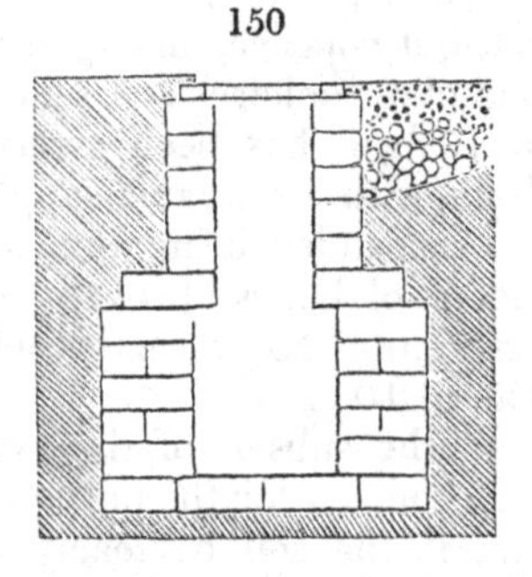
150

151

In flower-gardens, and indeed in extensive shrubberies, brick traps of the kind described above may be dispensed with, and the bell-trap represented in *fig.* 143., or some other trap of the kind, may be made use of; but, in this case, these traps must be cleaned out after every shower of rain.

Should it be considered necessary to form underground drains for subterraneous irrigation, the manner of doing this has been already described in p. 397.

In all extensive establishments, and even in small ones, where the drainage is in any degree intricate, it is desirable to have a map exclusively devoted to the under-drains, which should be accurately laid down on it, and accompanied by sections and dimensions, taken in different parts of the course of the drains, and projected on the plan alongside of the points where they were taken. Every successive addition and alteration should be introduced in the plan when made, and accompanied by the date. The use of such a plan is twofold: 1. to show the precise situation of the drains already existing, when any new drain is to be made; and, 2., as a guide when any drain becomes stopped up, and the point of stoppage is uncertain. By penetrating into the suspected drains at different distances, though only with a boring auger, or even in some cases with a crowbar, an immense deal of labour may be saved in detecting the seat of the evil; especially in cases where the principal servants have been changed since the drains were last opened.

The Pleasure-Ground, in places of only two or three acres in extent, may generally be considered as including the whole of the grounds, with the exception of the space occupied by the house and offices, and by the kitchen-garden. The number and the direction of the walks through this space will depend on various circumstances, but chiefly on the taste of the family for flowers and shrubs, and the annual expense which they are willing to incur in keeping the grounds in order. In general, the walks should be so far apart, or so hidden or disguised by undulations of the surface or by planting, that more than one walk shall never be seen at a time. In irregular surfaces, therefore, it is obvious that the walks may be much more numerous than in such as are even or flat; and, in surfaces richly varied by groups of trees or shrubs, the walks may be closer together than where the surface is chiefly in naked glades of lawn. The cheapest-kept pleasure-ground is, generally, that in which there are fewest walks, and broadest glades of lawn; while the most expensive are those where the surface is full of inequalities, the groups of trees and shrubs numerous, and the glades of lawn narrow. In general, in proportion as the surface of the ground is varied, so will be the beauty produced in any given space; more pictorial effect being found in one acre of undulations, than in three of level surface, the art and skill displayed being in both cases alike.

The direction of the walks should never appear to be forced, or to consist of bends made obviously for the sake of rendering the walk longer, or of avoiding a straight line. The direction may be straight or curvilinear, according as the one or the other may be required on account of the objects at which the walks are to touch; or apparently occasioned by obstructions on the ground, either natural or artificial. In curvilineal walks, no bend should ever be made that has not an obvious cause in the disposition of the flower-beds, or of the groups of trees and shrubs placed along its margin, or in the inequalities of the surface of the ground. Straight walks, where they are introduced, should have an obvious reason visible for their being straight; such as an accompanying wall, a row of trees at regular distances, or a covering of trellis-work, &c. In large places, the pleasure-ground, and consequently the walks belonging to it, are generally confined to grounds on one or on two sides of the mansion; while the grounds on the opposite side are considered in the nature of park scenery; but the grounds of third-rate residences, such as those under discussion, are, as already observed, generally considered as consisting wholly of pleasure-ground; and, consequently, the walks are carried through the scenery on the entrance front, as well as on the opposite side of the house, or what is called the lawn front. The grounds on the side next the entrance front, however, being more exposed to strangers coming along the approach road, have generally fewer walks, and these are accompanied by groups, more frequently of trees and shrubs, than of flowers; and by scenery, generally in a subordinate style of decoration. As the main walk through the pleasure-grounds generally requires, on the side of the entrance front, to cross the approach road, if the surface be much varied, or expense be not an object, the walk may cross the approach on an archway thrown over it, or by a tunnel carried under it: in either case, the road being concealed from the spectator on the walk, and the walk from the spectator on the road, by planting. The effect of this arrangement is greatly to increase the apparent extent of the grounds; and, indeed, in places where the surface is naturally varied, and the subsoil dry, the spectator may be led three or four times over the whole of the grounds, while, as he every moment enters on new scenery, he fancies he is traversing a place of three or four times its real extent. The great art of managing this description of scenery consists in the judicious use of tunnels, bridges, and raised surfaces in the form of wavy ridges, having their sides clothed with evergreens, and walks along their summits, and in the narrow winding valleys between them. Those who have never seen this kind of art applied to garden scenery will scarcely credit how easy it is by it to conceal one walk from another closely adjoining it; and what a magical effect may be produced in a very small space by this kind of landscape-gardening.

On varied surfaces, the general principles which we have laid down for the direction of walks must be taken in connexion with the principle of displaying to advantage such distant views as are considered good, and disguising or concealing such as are indifferent or bad. The object is, to show as many handsome landscapes as possible, and to conceal all disagreeable objects. Thus, the walks which are conducted round any place should, as already observed, p. 410., conduct to all the fine and striking points of view; so that, while none of these escape the attention of the spectator, no inharmonious or offensive object obtrudes itself on his notice.

The breadth of pleasure-ground walks may bear some proportion to the size of the place; though they should seldom be narrower than 6 ft., and, except in the immediate neighbourhood of the house, should rarely be broader than 12 ft. In general, there ought to be a main walk proceeding from the house, and making, as it were, the tour of the place; and this walk, even in third-rate residences, may commence at the house in a terrace parallel to the walks, and 10 ft. or 12 ft. or more in breath; and branch out to the right and left into main walks, commencing of the width of the terrace, or nearly so, and gradually diminishing, till they are at such a distance from the house as not to be seen by the eye of the spectator from the terrace, where they need not exceed the width of 6 ft. or 7 ft.; which width may be continued throughout the remainder of their length. The effect of this is at once to increase the grandeur of the scenery about the house, and to add to the apparent grandeur and length of the walks proceeding from it. The groups of trees and shrubs, or beds which are to be dug and kept planted with flowers, ought always to be separated from the walk by a verge or margin of turf; which, for the convenience of mowing, and of always presenting a healthy green surface, should not be less than 2 ft. in width. This applies to all the open parts of the pleasure-ground scenery; but, where the walk proceeds through dark shady woods, the trees and shrubs may be brought close up to it; and, the ground being clothed with the branches of the latter, their appearance will leave no room to regret the absence of turf. Even in open pleasure-ground scenery, a tree or a shrub may sometimes be planted close to the edge of the walk, and may occasionally project partially over it, for the sake of variety, and in order to produce particular effects. In this and every other case of the application of general principles and rules, such deviations may be made as are justified by the striking nature of the results; and it often happens that those features which principally distinguish one place from another are either the results of fortunate accidents, or unavoidable deviations from ordinary rules.

Besides gravel walks, there ought, in every case where there is a lawn of an acre or two in extent, to be either broad margins of turf to the main walk, running parallel to the gravel, on which persons may walk abreast with those on the hard surface; or glades in the interior of the scenery, so contrived as to admit of two or three persons making the tour of the place on them, instead of doing so on the gravel. In short, it ought never to be forgotten, that the enjoyment of walking on turf is to most persons much greater than that of walking on gravel; and that the preference given to the latter material, for what may be called the every day walks of a place, is owing to its firmness and dryness in moist weather, or after rain.

The Approach Road, next to the house and offices, is one of the most important features of a place, not only on account of its uses, but because it is that by which an impression, favourable or unfavourable, is first made on a stranger. It is, or ought to be, a walk or road fit for horses or carriages, from the entrance-gate of the premises, to the entrance-door of the mansion; formed and kept in a high style of art, corresponding with that displayed in the other component parts of the residence. It should be as different from a turnpike-road or public lane as park scenery is from that of a common grass-field or meadow. As it may be advisable to consider this important feature somewhat in detail, we shall endeavour to show how a high character of art may be conferred upon it, in its direction, in the slope or inclination of its surface, and in its material, margin, and accompanying scenery.

The line of direction between the entrance-gate and the mansion, in places laid out in the geometrical style, is in general straight, in the form of an avenue; but, in places laid out in the modern manner, and more especially in small residences like those under consideration, it is always more or less curvilineal. Now, in order to give these curves a character of art, they ought to have a certain uniformity in their degree of curvature; and the number of curves ought to be such as to give a character of regularity, or symmetry, to the whole line of road. To have a character of art, each separate curve ought to proceed, from its commencement to its termination, in a uniform uninterrupted degree of curvature; as opposed to the curves of natural paths, carried across meadows or commons, in which curves are often to be found of so undecided a character as to present tremulous-looking lines instead of uniform bends or curvatures. To constitute regularity among the curves, they ought all to be nearly of the same length; to preserve unity, they ought to be so united as not readily to discover where the one curve begins and the other ends; and, to constitute symmetry in the entire road, the curvatures which form one half of it ought to be balanced by those of the other half. This may, perhaps, be more readily explained by lines than by description. In *fig.* 152., the lower extremity of each road, indicated by double lines, is supposed to be the entrance-gate, and its upper extremity the entrance-door of the mansion. Supposing the space between the gate and the mansion a common grass-field, with no interruption of trees or ditches, but with some very slight inequalities; then the track of road which would be formed between these two points by carts or carriages would probably be somewhat in the manner of the line *a*, which is not straight, and yet can scarcely be called curved. To render this line simply artistical, each curve may be made uniform in its degree of curvature, as shown at *b*. To produce regularity in the line of curves, they ought to be all of the same length and curvature, as shown at *c*; and this

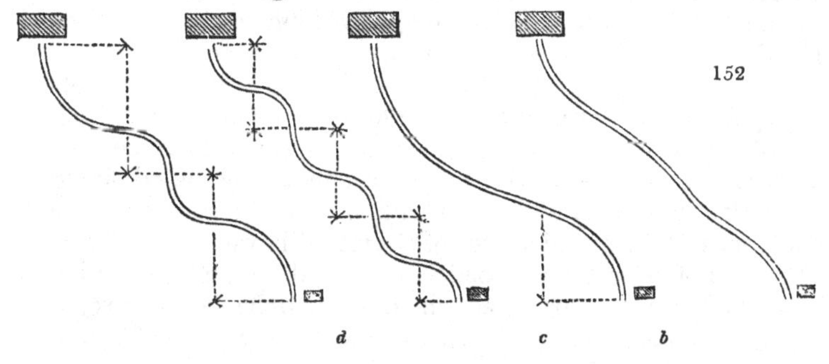

line may be raised to a higher character, and rendered symmetrical, by forming the two extremities of two curves of the same size, and by uniting them with smaller curves, as at *d*. The proof that this last line is symmetrical is, that if separated into halves, neither half would form a whole; whereas this would be the case with the line *c*. It may be useful to remark with reference to the first line (*a*), that all roads or lines formed by the traction of horses, or other draught animals guided by man, commence, say, after a gate has been gone through, by being inclined somewhat to the right; and terminate a little before the next gate, or other obstruction, by being turned somewhat to the left. This is supposed to be produced by the driver exercising his authority, by means of the whip and reins, at the beginning and ending of a line more than he does in the middle of it. Be that as it may, natural roads across fields, from one gate to another, will generally be found to consist of a short curve to the right immediately within the entrance-gate, a short turn to the left immediately within the gate of exit, and a nearly straight or indefinite line between the two. In ploughed fields, also, in former times, where an absolutely straight furrow was not as at present considered a desideratum, the direction of the furrow was serpentine, for the same reasons as those just given. We state these facts here, to showthat a curvilinealapproach road is quite natural; though the imitation of it by man, in laying out a place, must, as in imitating nature artistically in other cases, not be a mere fac-simile repetition, but a resemblance according to art.

In addition to the beauties of regularity and symmetry in the line of direction of an approach road, expression, or character, may be added. Thus, the expression of grandeur may be given by the increased size of the curves, and the general simplicity of the whole line (as shown at *b*); that of picturesque beauty, by very sudden changes in the direction of the curves, and by the partial introduction of irregularity; while a certain degree of elegance will be produced by gentle curves, some of which are very much prolonged, so as to be approaching to straight lines. The kind of artistical beauty least adapted to a curvilineal approach road, is regularity, as shown in the line *c*; which consists of a repetition of curves of the same form and magnitude, from one end to the other; and which, viewed merely as a curved line, and without reference to what might be effected by the adjoining scenery, if it were carried into execution, must be allowed to be monotonous. The grand source of character, however, in the line of direction of an approach road, is produced by the intervention of objects, natural or artificial, which set at defiance both regularity and symmetry; such as a rock, the base or talus of a hill, a building, a group of old trees, a stream, a pond, or water in some other form; all of which most commonly occasion sudden and picturesque deviations from regularity in direction.

The inclination of the surface of an approach road is subject to the same general principles as its line of direction. In a common road across a field, its surface follows every inequality in that of the pasture, and is, perhaps, nowhere either regularly sloping or regularly level; but, in the artistical imitation of such a road, the surface will proceed in a series of regular slopes, joined with spaces approximating to levels, on exactly the same principle as the regular curves, or nearly straight lines, constituting the line of direction. It will be taken into account, that the general surface of the ground over which such an approach is to be formed will have previously been rendered artistical, by the smoothing down of all minor protuberances, the filling up of small inequalities, and the obliteration of all minor elevations and depressions which interfere with the regular flow of outline, over the entire surface of the park or pleasure-grounds.

The inclination of the surface of an approach, ought also to be considered with reference to the effect which it has on the character, or expression, of the house. Every one feels that a house which is approached by an ascending road appears a far more dignified object, than one the road to which is level or descending. In order to obtain ascent in the direction of the road between the entrance-lodge and the front of the mansion, the former ought to be placed on a lower level than the latter; and the ascent should be either regularly distributed over the whole length of the road, or, what is preferable, it may be very gradual at first near the entrance-lodge, and increase as it proceeds towards the mansion. In residences of limited extent, where the approach is necessarily short, the road can seldom, with propriety, descend from the lodge, and afterwards reascend to the mansion; nevertheless, there are exceptions: such as when a brook or lake is to be crossed at no great distance within the lodge, to which the road may descend, and, having crossed it, may reascend immediately afterwards, and continue rising till it reaches the house. In approaches of considerable extent, there may be various ascents and descents between the entrance-lodge and the mansion, provided none of the points of ascent are as high as the ground on which the mansion stands; because, in that case, they would interfere with its dignity. In general, where there is great variety in the surface of the grounds in the limited space over which the approach road is to pass, the rises and falls in the inclination of the road will be numerous; though care must always be taken that the last variation before arriving at the house is one of ascent, and not of descent. On the other hand, where the variations on the surface are very gentle or very few, and where, as in all small places, the approach is not very long, there may be an ascent from the entrance-gate to the front of the mansion, either regularly distributed throughout, at the rate of so many inches of rise in a yard or pole of length, or the rate of inclination may vary in different parts of the road. One of the finest descriptions of approach road that we can imagine is, where a road of several miles in extent is made to wind its way through hilly or mountainous scenery at one uniform rate of ascent, till at last it arrives at an open level area containing the mansion.

The materials of which approach roads are formed, to be artistical, ought not to be those used in the common roads of the country: for example, if the common roads are Macadamised with granite, then the approach road ought to be gravel; or, if the common roads are graveled, as in most parts of the neighbourhood of London, then the material of the approach road ought to be gravel of a finer kind, such as that of Kensington or Bayswater; or the approach may be Macadamised with granite, flints, blue limestone, or sandstone. The artistical effect of the materials, whatever they may be, of approach roads, may be heightened by the use of a heavy roller, so as to render them perfectly smooth and even; and by keeping them at all times free from weeds, horse-droppings, and other offensive objects.

The margins of an approach road, to be artistical, ought not to be rough like those of a common road through a field; and, above all things, they should not have footpaths running parallel to them, with a gutter between, as in the case of public roads. The surface of the road ought to be on a level, or nearly so, with that of the lawn or grass on each side: it ought never to be raised above it, or to be rounded in the middle; nor ought it to be more than one inch sunk beneath it. The reason why the gravel ought not to be raised above the grass is, that the appearance produced gives the idea of the gravel in the road having been carted there, and laid down on the natural surface; whereas, when it is kept level with, or an inch beneath, the surface, the expression is conveyed of a proper foundation, or preparation, having been made for it. The reason why the margin ought to be one smooth, definite, delicate line is, that such a line is gardenesque, which is more likely to be the style employed in planting along an approach road through an open lawn, or through the park scenery of a small place, than the picturesque. If, however, the picturesque style of planting be employed, then the line of demarcation between the grass and the gravel may be indefinite, broken, and irregular, as in the case of walks with picturesque margins. (See p. 165.)

The scenery accompanying an approach road will, in all ordinary cases, and especially in small places, consist chiefly of groups of trees. Shrubs are seldom admissible along such a road, because, as its sides are generally supposed to be pastured by sheep or cattle, the shrubs would require fences for their protection. In planting along an approach, two objects ought to be chiefly kept in view: to dispose of the trees in such a way as to account for the curves of the road, and to arrange them so as to form suitable foregrounds to the scenery beyond. Subordinate objects are, to prevent two turns of the road from being seen at the same time; to conceal the house from the approach road till the spectator is near enough to see it in a bold and striking point of view; and, above all, to prevent it from being seen from the entrance-lodge, or from the road immediately within it, as that implies a limited extent of grounds, destroys the idea of seclusion, and shows the house without due preparation.

The width of approach roads, in very small places, where only a one-horse carriage is kept by the proprietor, need not be more than 8 ft.; because, when two such carriages meet, each driver can draw his horse to one side, so as to have one wheel on the turf, and may thus pass without difficulty. But where a two-horse carriage is kept, and, indeed, in every case where the mansion is of a respectable size, however small the grounds may be, the approach road should not be less in width than 12 ft. Indeed, much of the dignity and grandeur of every place depends on the walks and roads being of a good breadth, and always being highly artistical in their material and keeping.

The general Surface of the Ground, exclusive of what is occupied by the buildings and the kitchen-garden, requires to be rendered artistical, either in the geometrical style, by forming it into regular levels or slopes; or in the modern manner, by merely polishing the surface, by reducing roughnesses, protuberances, and petty inequalities, so as to form it everywhere into flowing lines, consisting of curves of various degrees of curvature and extent, insensibly gliding into each other, or into levels or slopes. The operations of levelling and smoothing are most conveniently performed before beginning to lay out the walks or roads, or to mark off the places for planting, or to trench the ground.

Trenching. The use of trenching ground which is afterwards to be covered permanently with grass, or trees and shrubs, is, to enable the roots of the grass and other plants to reach a greater depth, and thus to be more out of the reach of drought in summer; also, to increase the capacity of the soil for retaining water, in order to keep the grass, as well as the trees and shrubs, green during the warm season. By trenching, also, in connexion with manuring, draining, irrigating, &c., as already described, the growth of the trees and shrubs, immediately after being planted, will be much more rapid, and the final effect desired by the planter much sooner attained. It must not be forgotten, also, that the effect will be equally astonishing on the growth of the grass as on that of the trees and shrubs; and thus, not only a more intense green turf will be produced, but a greater quantity of hay and pasture for milch cows or sheep. We mention hay and pasture, because in many places of only two or three acres in extent, and in all places of five acres and upwards, that are not over-planted, a proportion of the lawn, or grassy surface, may be mown for hay every year; and portions of surface, also, may be hurdled off, for being pastured by cows or sheep. By some, no doubt, this kind of management may be disapproved of, as interfering with the smoothness and high polish of the lawn; but by others it will be considered as adding greatly to the value of a country residence, and to the interest and variety of the operations carried on in one.

The Boundary Fence to small places should, in most cases, be a wall, as being that which is most impervious to the smaller animals. Close palings of oak or larch, or of some other kind of wood, Kyanised, may be used; or a wall about 3 ft. high may be built, and a hedge placed immediately within it, and cut as it advances in growth, so as to form a continuation of the height of the wall; which, in this case, would be sufficient to keep out vermin at the base of the hedge, while the upper part of the fence would answer all the other purposes of protection and defence. A deep ditch on the outside, and a thorn hedge inside, will also form a good boundary fence; but, where the soil is a deep sandy loam, a hedge of holly is the best of all.

Planting. Trees and shrubs are distributed over the grounds of a residence, because these objects are considered the most effective in producing the kind of beauty which is desirable in the country. To trees and shrubs are added flowers, as minor ornaments. The question that we are now to consider is, the principles to be followed in the distribution of the trees and shrubs, and of those finishing ornaments, the flowers. The two sources from which we naturally seek instruction in this matter are, the practice of planters and landscape-gardeners hitherto; and the mode in which trees, shrubs, and flowers are distributed in nature. Till within the last two or three centuries, the number of trees, shrubs, and common garden flowers, in cultivation in any country, did not amount, in the number of kinds, to one hundredth part of what are now common to the gardens, not only of Britain, but of the temperate regions of the globe in both hemispheres. The distribution of this small number of plants, by gardeners, was formerly as simple as their number was few. All the woods of a place were formed of two or three kinds of trees, planted in rows or in masses, in the geometrical style, which we need not here describe; and the flowers were placed in a compartment near the house, walled or hedged round. The lesson, therefore, to be learned from the geometrical style is abundantly easy.

In every country, the plants which are indigenous, and found there when it is rescued from an uncultivated state, form but a very small number of those which will grow in it. Hence, with the progress of civilisation, an immense accession has been made, both to the useful and ornamental plants of every country. This, in Britain, has taken place more especially within the last three centuries; and the mode in which the newly introduced plants have been distributed by landscape-gardeners, since the introduction of the modern style of art about the beginning of the eighteenth century, is as follows:— The different kinds of trees are distributed over the parks in scattered groups or clumps, as single trees, or in a belt or strip forming its boundary. The order in which the kinds are placed with reference to one another, is considered of little importance; but, till lately, it has generally been attempted to mix the foreign and the indigenous sorts indiscriminately together throughout every part of the park. In more conspicuous or favourable situations, such as near the house, or along the approach road, some of the more choice trees were planted singly, and protected with more care than the others. The single trees and scattered groups, or clumps, were all guarded from the cattle by fences; and the boundary belt, commonly by a wall or close paling on the outside, and a hedge and ditch within. The ground was commonly dug or trenched before planting; and sometimes it was dug in the clumps and belts for two or three years afterwards. After this, the trees were left to themselves; thinning and pruning being more or less attended to in some cases, and altogether neglected in others. In consequence of the mixture of indigenous and foreign trees, and their after neglect, the indigenous and more vigorous-growing trees choked up, weakened, and ultimately destroyed the foreign kinds; so that, when the timber in such plantations arrived at an age to be cut down, there was seldom much which was good for anything that was not produced by the native trees of the country.

The shrubs, since the modern style of gardening has commenced, have been planted in the pleasure-ground, in the same general style as the trees were in the park; that is to say, they have been placed singly, or in groups or clumps, over a lawn two or three acres or more in extent; and around it, or along one or more of its sides, in strips, which were called shrubberies. Here, as in the case of the trees, the indigenous and foreign shrubs were mixed up together in every part of the clumps and shrubbery; and the result, thirteen or fourteen years after planting, was similar to that which took place in the plantations in the park; viz., the more delicate and foreign plants were choked up and destroyed by the vigorous-growing trees; and old pleasure-grounds, which had, perhaps, been originally planted with above a hundred kinds of shrubs, fifty years afterwards displayed only huge overgrown bushes, or low trees, consisting of not more than twenty or thirty indigenous species.

The Flowers were distributed in the front of the shrubbery, and in front of the clumps in the pleasure-ground. Like the shrubs, they consisted of foreign and indigenous kinds indiscriminately mixed together; and, like them, the former were destroyed by the latter and by the shrubs. It is only lately that beds wholly planted with flowers have been introduced on lawns; and, though what are called flower-gardens (that is, assemblages of beds wholly devoted to flowers) were to be met with in first-rate places during the latter half of the eighteenth century, yet, during that period, the principal places where they were planted were in the shrubbery or in the borders of the kitchen-garden. With the commencement of the present century, the practice of forming flower-gardens has increased; and, within the last twenty years, that of forming beds exclusively devoted to flowers on grass lawns, either in groups among the scattered shrubs, or by themselves, here and there along the walks, has become general. The flowers planted in these flower-gardens and beds were, till lately, mixed together indiscriminately in the same manner as in planting the shrubs in the shrubbery, or the trees in the clumps and belts; and, wherever the plants were not taken up and replanted every two or three years, the same results took place, of the stronger destroying the weaker. A great improvement has been made in the planting of flower-beds within the last twenty or thirty years. This consists in planting each bed with only one kind of flower, by which means a brilliant display of colour is produced; and in selecting for the flowers to be so planted those from warm climates, such as pelargoniums and fuchsias from the Cape of Good Hope and South America, &c., by which means a more brilliant display of colour is produced; the plants of warm countries far excelling, in this respect, those of colder climates.

Such, in a few words, has been the practice of planters, landscape-gardeners, and flower-gardeners, in British gardens, up to the present time. We shall now briefly enquire into the mode in which trees, shrubs, and flowers are distributed by nature.

In the natural scenery of every country, a certain number of plants will always be found congregated together, to which the soil and local situation are favourable. The number of species in these assemblages depends partly on the suitableness of the soil and climate for a great variety of species, and partly on the number of species naturally inhabiting that locality. In general, the greater number of species are found in alpine countries, on the sheltered sides of hills, where the soil is exceedingly various, and also the temperature, in consequence of the surface of the soil being exposed to the sun's rays at very different angles; as well as from other causes. The smallest number will generally be found in plains where the soil is wholly alluvial, and, perhaps, occasionally overflowed by a river; or in wastes, where the soil is a drifting sand. Even in the most favourable situations for a considerable number of species, that number, in a state of nature, and without the care of man, can never be great; because the stronger species, whether these acquire their strength from their nature, or from being placed in more favourable circumstances than other individuals of the same species, will soon weaken or destroy the others. Hence it is, that, in a natural forest, the number of kinds of trees, shrubs, and other plants, present in any one place, is comparatively few. These few will generally be found to consist of a prevailing species of tree, with perhaps one, two, or three, in a hundred, of some other kind of tree; a prevailing kind of shrub, with some subordinate undergrowths; and as generally, a prevailing kind of flowering herbaceous plant, fern, grass, moss, or even fungus or lichen. It is to be remarked of this mode in which Nature distributes her plants, that it is much more simple, and at the same time better calculated to be productive of variety, than the mode which we have described as being adopted by gardeners. In consequence of one kind of tree, shrub, plant, &c., always prevailing in one place, the aspect and interest of that place must necessarily be different from another where the species which prevail are different. In traversing a natural forest, with a view to studying the different kinds of trees and shrubs produced in different localities within it, we shall find the following causes are principally in operation:—1. Difference in the nature of the soil and the subsoil, or rocks. 2. Difference in the capacity of the soil for retaining water. 3. Inclination of the surface, by which the water of rains is more or less thrown off. 4. The aspect of the surface, from which it is more or less heated, according to the direct or indirect influence on it of the sun. 5. The elevation of the surface above the level of the sea. Hence, the greatest variety of trees, shrubs, and flowers, in any one country, will be found around the base and on the sides of its highest mountains.

Having seen the mode of distributing trees and plants adopted by the majority of gardeners, in the present and in former times, and their natural distribution in this country, and in every other, it may now be asked what mode we recommend to be adopted. To which we answer, briefly, an improvement on the present system, with some innovations for which a sufficient reason will be given. Improvements in gardening, as in every other art, can only be effected by degrees; and it rarely happens that any individual can do more than make a slight advance before his contemporaries. The principles on which we propose to distribute trees, shrubs, and flowers, in parks and pleasure-grounds, are the following:—

I. *Trees and Shrubs.—Choice of Kinds.*

1. We would introduce generally in every place, whether large or small, as many different species and varieties as would thrive in it; in small places, having no more than a single plant of each species or variety; but, in larger ones, introducing duplicates, triplicates, or a greater number, according to the extent of the place, but keeping each kind by itself.

2. Where the object was ornament, novelty, or variety, we would generally prefer kinds foreign to the country, or, at all events, to the locality; but, where the object was chiefly timber, shelter from high winds, coppice wood, hedges, &c., we would prefer indigenous species.

3. In places so small as not to admit of even one plant of each species and variety, we would give the preference to foreign kinds, and would generally exclude such as are indigenous, except greatly improved varieties.

4. In general, even where the space would allow of the introduction of the indigenous species, we would only plant them in small quantities, and chiefly for the purpose of rendering the collection of sorts complete, or for harmonising the woods of a residence among themselves, or with those of the adjoining residences, or for preserving artistical effect; unless, indeed, the object were timber, or other useful products, in which case we would plant these to the exclusion of the others, agreeably to Rule 2.

5. We would give the preference to such kinds as were likely to thrive best in the given soil and situation.

Trees and Shrubs. — Distribution.

6. We would adopt various modes of distribution, according to the object in view, either in the entire place, or in different parts of it.

7. In the park scenery, and in larger plantations, trees should prevail; and, in the pleasure-grounds, and in the lawn near the house, shrubs : because the former are more grand, and the latter more beautiful; and because those in the park require to resist horses, cattle, &c., while all domestic quadrupeds are excluded from the pleasure-grounds.

8. Near the house, or in what may be considered the more select parts of the grounds, we would make choice of the more rare and beautiful species or varieties, without reference to the countries where they are indigenous.

9. Near the house and offices, the prevailing kinds should be evergreens; and these should be introduced to a greater or less extent, in almost all the plantations of the place.

10. Marshy situations are those where evergreens are least natural; because there are few kinds that grow in marshes. On the other hand, most evergreens either grow naturally or will thrive in sand. Hence, in artificial scenery the presence of evergreens always implies a dry soil and healthy situation.

11. In general, a plantation where the trees are deciduous should have a portion of the shrubs introduced among them evergeens; and, on the contrary, where the trees are chiefly evergreens, a considerable portion of the undergrowth ought to be deciduous.

12. Throughout the greater part of the grounds, where any distribution might be adopted at pleasure, we would place such together as require the same soil and situation, or are generally found growing in the same locality in their native country, or are obviously allied by nature. This would bring together, in some places, trees and shrubs requiring moist soils, such as the willow, poplar, alder, &c.; in others, such as required peat or bog, or as thrive best in these soils, as is the case with all the *Ericàceæ*, and with most of the American trees and shrubs: it would also bring together such natural families as the *Quércinæ*, the oaks all requiring good soil; and the *Coníferæ*, as all requiring a soil comparatively dry and sandy.

13. While we arranged the general masses of the trees and shrubs of a place, so as to produce broad distinctive features, we would introduce certain species throughout the greater part of the plantations (though only sparingly), for the sake of harmonising the scenery of the particular residence with the scenery of the residences or of the country lying round it. The harmonising plants, in both cases, will be most effective when they are evergreens; but there ought, also, to be some of deciduous kinds, in order to admit of uniting an evergreen mass with an adjoining deciduous one.

14. As the species for harmonising the plantations of a residence within itself should not be of large size; and should be such as are acknowledged to be beautiful, and also known to general observers, in order that the harmony may be felt by all, the holly, the box, the laurustinus, and the American thorns are very suitable for this purpose.

15. To harmonise the plantations of a residence with those of the residences around it, the more conspicuous kinds of trees in the latter require to be introduced (though only sparingly) in the former. Thus, if in one part of the plantations of the adjoining residences, near the boundary fence, pines should prevail; then, two or three, or more, of the same sort of pines should be introduced within the boundary of the central residence; and farther in the interior there may also be one or two of this species of pine. The same remarks will apply in the case of exterior plantations of oak or other trees, or of coppice wood; but much of the good effect of carrying this idea into execution will depend on the taste and judgment of the planter.

16. In general, we would adopt the gardenesque manner of planting trees and shrubs, especially in the pleasure-ground, and near the house; but, in more distant parts of the ground, we might adopt the picturesque mode, for the sake

of variety and of saving labour; and this mode might also become necessary, in some cases, for harmonising the scenery within with the external scenery.

17. On no account whatever would we introduce nurse plants, as they are called, or common sorts, for the sake of covering the surface, in a plantation where ornament was the principal object in view. Such nurse plants, and common kinds put in to fill up, exhaust the soil by their roots ; and, by the shade and shelter afforded by their tops, draw up the plants to be nursed in such a slender sickly state, that, when the nurses are removed, the principal plants that remain are weak and unsightly; whereas, had they grown up without the nurses, they would have been bushy and handsome on every side. After all that has been experienced and written in favour of introducing nurse plants in plantations, we are very much inclined to question whether anything is gained by it in the end. It must be recollected, that, at the time when nurses were most strongly recommended by planters, the importance of exposing as large a surface as possible of the foliage of every plant to the sun and air was not understood ; and the circumstance that the nurses, from being the stronger plants, must necessarily draw the greater proportion of the nourishment from the soil, and in that respect retard the progress of the plants to be nursed, seems to have been overlooked. It is not denied that nurses will draw up young plants : but it is a question, whether a plantation, say of oaks, for example, drawn up according to the best practice by nurses, and another plantation of oaks alone, not drawn up either by nurses or by one another, would not arrive at the same height in 60 or 80 years after planting; and, if so, there can be no doubt that the latter would be of greater bulk.

18. The gardenesque manner of planting and managing includes the application of pruning and thinning, at all future periods of the growth of the trees and shrubs, so as to keep each plant perfectly distinct from those around it. The picturesque manner of planting also employs thinning and pruning ; but the object of these, in this style of gardening, is less definite ; being for the purpose of throwing the plantation into groups, or to increase the growth of underwood at some places, or of tall trees at others; and for similar objects connected with the production of picturesque beauty, and shelter, shade, or timber.

19. The proportion of surface, in a park or pleasure-ground, which we would cover with wood, will depend on the degree of variation in that surface. In flat grounds, a very small quantity of trees and shrubs is required in proportion to the extent of turf; and the contrary is the case in very irregular surfaces.

II. *Flowers. — Choice of Kinds.*

1. As the kinds of flowers are exceedingly numerous, unless there were a decided taste or desire for botanical knowledge, we would introduce only the more showy and vigorous-growing sorts, increasing the number of kinds according to the size of the place.

2. Where it is intended to form a collection, every kind may be procured that will grow in the given climate.

3. We would select some of each of the different colours, so as to have an equal number of these colours in bloom for each of the floral months. We would also select these from as many of the natural orders as practicable, in order to show greater variety in the forms, and in the botanical differences of the flowers.

Flowers. — Distribution.

4. We would seldom plant flowers among trees and large-growing shrubs; but either in beds or groups by themselves, or with an admixture of low-growing flowering shrubs ; such as roses, hydrangeas, helianthemums, &c.

5. Almost the only kinds of flowers that we would introduce among large-growing shrubs are spring-flowering bulbs.

6. In distributing beds of flowers over a residence, we would limit their range to a portion of the pleasure-ground adjoining the house; and we would place them along the walks, in order that they might be near the eye of the spectator.

7. Besides these scattered beds, we would form, in each place, one or more flower-gardens.

8. In distributing the kinds in the beds, we would aim sometimes at producing splendid masses of one colour; and, at others, of variegated masses of colours; sometimes we would have in view a fine display for only one month, to be succeeded by some other flower for the following month; while in other cases we would plant such a selection as should exhibit an equal number of plants in flower every month throughout the season.

9. Flowers requiring particular soils, aspects, &c., we would plant together, in the same group or bed.

10. Flowers characteristic of particular countries we would frequently plant together; such as Swiss plants, Highland plants, American plants, or the characteristic flowers of the flora of any particular country.

11. For the sake of a more brilliant display, we would introduce, in groups by themselves, beds of exotic flowers, or flowering shrubs, requiring to be brought forward under glass, &c.; such as pelargoniums, fuchsias, salvias, celsias, &c.

12. Where there is a taste for botany, and a desire to possess an extensive collection of hardy herbaceous plants, we would contrive to introduce as many species as might be thought desirable in a botanic flower-garden, arranged either geographically, that is, each country by itself; physically, that is, according to soil, aspect, &c., such as alpines, meadow plants, marsh plants, or aquatics; or systematically, that is, according to some system of botanical classification.

13. When the object of a bed of flowers is to present one unbroken mass of colour, we would plant and manage it in the picturesque manner; but, where the object is to exhibit each plant separately and distinctly, then we would plant and manage the flower-beds according to the gardenesque mode.

14. We would take up, and replant in fresh soil, all perennial flowers whatever, at the end of one, two, or three, or at most four, years; in order to prevent the stronger from overpowering the weaker, and the soil of the bed from becoming exhausted by the abstraction of nourishment, or contaminated by the addition of excretions.

15. Where the picturesque plan of planting the trees and shrubs is adopted, perennial flowers and bulbs may be planted among them, on the first formation of the plantation, and left to be choked up and destroyed as the woody plants gradually spread over the surface.

16. The proportion of space, in the pleasure-ground, which should be covered with flowers, will depend on the taste of the proprietor. If shrubs and breadth of lawn are preferred, or if there is a separate flower-garden, then the flower-beds need not be numerous, and may consist of a few near the house; but, if the taste for flowers is greater than that for trees and shrubs, then the flower-beds may be increased accordingly.

Architectural Ornaments. The number of these about a place depends on various circumstances, but chiefly on whether the pleasure-ground is plain, or much decorated in the immediate vicinity of the mansion. If there be a terrace-walk bordering the house, and leading to the flower-garden, the former may be enriched with statues, and the latter with vases, agreeably to principles already laid down in the preceding pages of this work. The walks in the pleasure-ground should, at all events, have seats placed in situations displaying the best views; and there may be covered seats, in the form of different descriptions of rustic buildings, constructed chiefly of wood, introduced where they would be useful as places for resting, and desirable as objects in the landscape. In general, garden buildings of a purely ornamental description, should either be very few (in which case they may be built of substantial materials, such as brick or stone); or, if of frequent occurrence, the materials of which they are constructed should be temporary, interest and value being given to each by its design. Expensive garden buildings, such as classical temples, porticoes, colonnades, &c., containing statues, busts, and sculptures of marble, can seldom be indulged in in small places, and, indeed, are better adapted for the grounds of hereditary residences. In residences liable to be sold on the death of the proprietor, ornamental buildings, whether of the permanent or temporary kind, add little or nothing to the price of the estate; and had Stowe, where the garden buildings are magnificent, been sold at the death of the Earl Temple during whose occupation of that estate they were built, they would have brought little more than the root-houses at the Leasowes, a contemporary ornamental residence, did on the death of their architect and proprietor, Shenstone. Ornamental buildings, therefore, are chiefly to be valued for the enjoyment they afford the proprietor during his lifetime; and, unless a part of this enjoyment consists in knowing that these buildings will descend to his posterity, prudence dictates that they should be built so as not to occasion great expense.

In all that respects the introduction of ornamental buildings, however, much must be left to the particular taste of the proprietor; and, as most proprietors take pleasure in having at all times some structure alteration, or addition, going forward, this is an additional argument in favour of structures of temporary materials, erected for picturesque effect, and as occasional resting-places; or for affording shelter, and not intended to last longer than the verdant scenery by which they are surrounded; or, in other words, to be equally subject to change as the rest of the garden.

Rustic vases, baskets of rustic work, boxes or beds of flowers cased with fantastic roots, hollow stumps of trees, rustic arcades, dead trees, and other places for forming receptacles for low plants, or supports for climbing ones, may all be introduced occasionally; care being taken that they are never displayed in such numbers, or of such magnitude, as to attract more attention than the flowers to which they are meant to be subservient. To keep within the happy medium, however, in this kind of garden ornament is extremely difficult; since the proprietor, who has had them constructed under his direction one after another, is apt to lose sight of the general effect, in the interest which each separate object has created in his mind, and which he looks upon with the fond indulgence of a parent; forgetting that a stranger judges of them solely by their effects as connected with the scenery around them.

Wirework, in the form of the rims and handles of baskets, edgings, and various kinds of props for climbers, may also be occasionally introduced; always, however, remembering that they are to be subordinate objects to the plants which they enclose, protect, or sustain. It is customary to paint wirework of this description green; but, in our opinion, this colour is the very worst that can be adopted for any kind of structure or utensil to be placed among the green of nature. A stone colour, or dull white, or greyish black, we think greatly preferable.

The Use of Vases, and other Garden Ornaments of artificial Stone, or of Earthenware, is now very fashionable in pleasure-grounds, and more especially in flower-gardens; but this kind of ornament is very frequently overdone or misplaced.

Whoever understands the phrase "unity of expression," and can examine any scene presented to him by the test of its being, or not being, "a harmonious whole," will be able to determine what is right and what is wrong in the disposition of sculptural ornaments in gardens. All architectural objects and statuary, being ponderous and intended for great duration, should be placed on bases obviously secure and durable. A vase or a statue should never be set down on grass, or on dug ground, without a decided pedestal, resting, or appearing to rest, on a secure foundation; and it should never be set on anything less obviously durable than masonry. Where such objects form the predominating features in a scene, they should always be connected with some kind of building, such as a parapet or terrace wall, or even a stone border to a walk, a bed, or a pond; and, in default of these, even a paved walk between a row of statues, the pedestals standing on a flagstone, projected from the pavement into the adjoining turf or dug ground, will tend to preserve unity of expression. Even an area of gravel projected from a gravel walk, and extending an inch or two all round the pedestal, will have a tendency to maintain the secure architectural character which ought always to accompany architectural and sculptural objects. Rootwork, rustic baskets, and other temporary objects or structures of this kind, should seldom or never be introduced in the same scene with vases, statues, or other ornaments of worked stone.

Rockwork, or a collection of fragments of rock, stones, flints, vitrified bricks, scoriæ, and similar objects, to serve as a nidus for plants, is allowable in particular situations; but no description of garden ornament, unless, perhaps, we except rustic work of the kind mentioned in the preceding paragraph, is more frequently misplaced and mismanaged. Collections of stones should never be heaped up about the roots of trees, at the base of walls, or against a mansion, hot-houses, &c.; nor should accumulations of fragments of stone ever be piled up on a level surface of turf, without some preparation or accompaniment indicating that they might have been there naturally. Two of the most remarkable pieces of rockwork in England, that at the Hoole, near Chester, constructed from the designs, and under the immediate inspection, of Lady Broughton, and that in front of the magnificent botanical conservatory at Syon, formed by Mr. Forrest, are both raised on level surfaces; but in both they are very well united by the principle of gradation. At the Hoole, there is a direct imitation of the glaciers of Switzerland, with a level valley between; and on this valley the mountain scenery projects and retires, forming a great variety of prominences, recesses, and sinuosities, aided by scattered fragments of rock, of different dimensions, and by shrubs and herbaceous plants. At Syon, no particular description of rocky scenery is imitated, and the scene can only be described as a ridge formed by piling up huge masses of stone, of different kinds; but the base of this ridge is so well united with the turf, and the whole of the ridge is so disguised by trees, shrubs, and plants, that the want of natural character scarcely ever occurs to the mind. Where the base of the ridge joins the level ground, there are at first seen here and there some slight protuberances of turf, each of which seems to indicate that there is a stone below, and quite near the surface. A little farther apart, portions of the stones seem to have burst through some of the protuberances; and here and there the upper parts of some stones appear quite bare. Near these, are blocks of stone raised in great part above the surface, and occasionally some entirely so; and, farther on, the grouping becomes conspicuous, and two or three stones are seen piled on one another. To groups of different sizes, so formed, is joined the great ridge, rising to the height of upwards of 50 ft.; and, on climbing up its sides, they are found to be varied by natural-looking paths among the stones, and a profusion of curious little rock plants, unseen from below.

Rockwork or stones, so disposed as to convey the idea that they form a part of, or protrude from below, the foundation of a building, produce the expression of stability and propriety; but, when they are heaped up on the surface of the ground against the walls, the idea of incongruity is excited, by seeing that which ought to be the foundation, and consequently under the building, piled up against it, and above the surface of the ground. Not only are the ideas of propriety and stability thus totally destroyed, but those of disorder and insecurity are produced. Neither, as we have observed in a preceding page, should roots, decayed trunks, stumps, or branches of trees, ever be mixed up with stones, on account of their obvious incongruity in point of durability. Each of these kinds of materials ought to be kept by itself; and thus we might have receptacles for plants formed entirely of decayed wood in one place; in another, of fragments of rock; in a third, of land stones or of flints; and, in others, of vitrified bricks, of scoriæ, of pieces of hewn stone, of fragments of sculpture, of shells, of corals, of spars, of petrifactions, &c. When there is nothing in the nature of the surface that indicates the presence of rock or stone in the soil, and when it is determined, at all events, to have some rockwork, one of the three following modes may be adopted to render it natural: —

1. On a level surface, a preparation may be made, beginning at a short distance from where the main body of rockwork is to be placed, by sinking some fragments of stone into the earth, so deep as to show only their edges, or angles, rising above the turf. As the main body is approached, these stones may become more numerous; larger portions of them may be shown; and they may be connected in lines, or ridges, in such a manner as to indicate, partly by protruding stones, and partly by raised places in the turf, something like the "cropping out," or rising to the surface of natural strata. A few half-sunk stones, of different sizes, may then appear in groups, as if they had been accidentally separated from these strata; and, immediately before the mass of rockwork, there may be some loose stones, with flat sides, piled irregularly on one another, so as to form subordinate masses to the large mass. This large or principal mass must be in imitation of some natural character of rock; and, whatever that character may be, the manner of the preparation for it which we have been just describing must be of the same kind. The rock to be imitated may be stratified in various ways, as we see sandstone, limestone, slatestone, &c., in nature; or it may be in masses, with no appearance of regular strata, but with cracks and fissures, sometimes horizontal or oblique, and at other times perpendicular; and differing both in the magnitude of the clefts or fissures, and also in their numbers, as we often see in masses of granite, trapstone, &c. In short, having thrown out the idea of imitating nature, both in the main mass of rockwork, and in the preparations for it, that alone will be sufficient to guide the artistical gardener, who has lived in a hilly or rocky country.

It will be seen, from these remarks, that the kind of rockwork which displays a heap or heaps of stones, however large some of these may be, all showing themselves above the surface, and to the same extent over the whole heap, as if it had been merely a mound of earth, dotted over with stones, has no claim whatever to be considered as rockwork in our sense of the word. It may represent a commonplace or a curious heap of stones, which may be more or less convenient for the culture of plants; but it is altogether unfit to be introduced into garden scenery, as an artistical object. In general, rockwork, to be truly natural, can only show the rock on one side, or, at most, on two sides; as scars, cliffs, precipices, &c., are seen in rocky districts. The upper part of the rock should be covered with turf, and trees and bushes, and the inclination of the turf should follow the supposed continuation of the rocky strata. This mode is not only natural, but has the advantage of being convenient; because, by making the angle of elevation of the strata more or less, according to the extent of the ground, the covering of turf can, at a greater or less distance, be made to unite with the level turf of the lawn. Where ferns, or plants requiring shade, are chiefly to be cultivated as rock plants, the abrupt side, or face, of the strata may face the north, and the talus, or sloping side, the south. Where early-flowering plants are to be cultivated, such as the Californian annuals, the abrupt side may face the south or south-east. It is almost needless to state that on the sloping side, which may be called the back part of the rock or hill, there ought to be no rocks or stones cropping out; but, on the contrary, the appearance ought to be such as to indicate depth of soil, where the slope joins the level surface; that being always the case in nature at the base of a declivity. The covering of the rock or hill, including the slope, and also a portion of the adjoining lawn, may be planted with trees and shrubs, chiefly in the picturesque manner; as their appearance in such a situation, and disposed in such a manner, is perfectly natural, powerfully supports the idea of the truth of the imitation, and serves artistically to unite the hill with the level surface. To render the imitation of the abrupt side, or face, of the stratified rock artistical, all that is necessary is, to let the flowering plants introduced in the clefts, fissures, or shelves (produced by one stratum projecting farther out than another), be of foreign kinds.

Where granite or basalt is the material used, the stratification, or lines of separation, may be chiefly vertical; but, in the case of sandstone or limestone, they should be chiefly horizontal. The most intractable materials for forming rockwork which is intended to have any grandeur of effect are, land stones, pebbles from the sea shore, flints, and chalk-stones; which, indeed, are only fit for facing an abrupt irregular bank, to be planted with creepers or alpines.

Fig. 153. may be described as a mechanical representation of a piece of artificial rockwork, in the form of a scar, or precipice, rising from a flat surface; and consisting of the cropping out of strata that have a considerable dip, or inclination. In this figure, *a b c d* show the face of the rock on two sides, in which the lines of the strata, viewed in front, appear nearly horizontal,

153

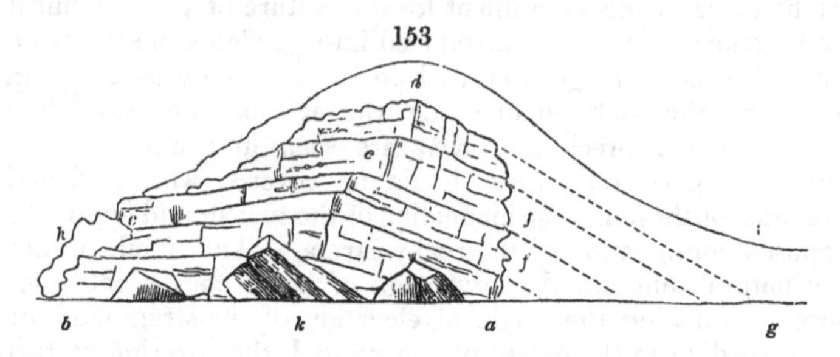

in different beds; each bed projecting somewhat beyond the one which is over it, in order to form ledges (*e*) for plants. The dotted lines *f f* show the supposed continuation and dip of the strata under the turf. The space *d i g* is covered with earth and turf, as is the rock on the opposite side at *h*; *i* shows the talus at the back of the scar or cliff, where the soil is always deepest and best; and *k* preparatory fragments in the foreground. It will be recollected that this sketch is made purposely plain and formal, for the sake of illustration; we shall hereafter give others, combining illustration with effect.

2. An excavation may be made in a flat surface, and in the bottom of it water may be introduced; from one or more of the sides of this small pond, stratified rockwork may be carried up to the surface, and considerably above it, so as to give some distant resemblance to an old stone quarry, which may be supposed to have existed there before the spot was turned into pleasure-grounds. This idea every gardener of taste will know how to improve and carry into execution.

3. A walk, or a glade of turf, may be sunk in a level surface, or carried through a knoll or raised surface; and the sides of the glade or walk may display scars or precipices of rockwork, more or less stratified, according to circumstances. On the surface above such rockwork trees may be planted, which would give the whole an air of truth, stability, and durability; the very reverse of the effect produced by heaping up fragments of stone about the roots of trees.

In general, no rockwork of any kind whatever can be put together in a manner satisfactory to the man of taste, except by a workman who has the eye of an artist, who can conceive beforehand the effect which he wishes to produce, who has some idea of connexion and grouping, and who knows the difference between peculiar and general nature. Those who attempt rockwork, without possessing one or more of these qualities of mind, can hardly fail to be unsuccessful; or, if they produce anything good, it must be by mere accident. They may put together heaps of stones, larger or smaller, according to the abundance of the material, or the sum expended; the heaps may be curious, from the variety of stones, spars, &c., brought together; or ludicrous, from the fantastic shapes of some of them; or childish, from the position of others; but nothing to affect the imagination can ever be the result of such accumulations. After all that has been done, they will still be only heaps of stones. Hence it is, that all the rockworks in Britain, worth looking at, have been constructed by workmen who have had a natural genius for this kind of work; or under the immediate direction of artists. For example, those at Pain's Hill, Wimbledon House, and Oatlands, were put up by a stonemason, who devoted himself entirely to this kind of production, and who was eagerly sought for in every part of the country; that at Hoole, as we have seen, was designed and executed under the eye of the proprietor, Lady Broughton; that at Drayton Green, which will be hereafter figured, by Mrs. Lawrence; and the interesting grottoes and cascades at Wardour Castle, by a mason who was much employed in that way throughout the country, and who, though he received nearly a pound a day, when employed, died a few years ago, as we were informed in 1833, in the parish workhouse. The best professional director of rockworks, that we are acquainted with, at the present time, is Mr. Gray. Under his direction were constructed the grotto-work at the Colosseum in the Regent's Park, and that at Clumber in Nottinghamshire; and other similar works, at a great number of places, both in the neighbourhood of London, and in the north of England, particularly in Yorkshire.

Water. To determine the extent to which water should be introduced, and the character which it ought to assume in small places, is a point requiring some consideration. In the case of a spring or a running stream the difficulty is not great; but, where there is only just a sufficient supply of water to maintain a pond or small lake during the summer season, the skill required is greater. The difficulty arises, not from any doubt of the effect of the water, in a picturesque point of view, for that is easily determined; but with regard to its influence on the salubrity of the atmosphere of the place. The exhalations, even from pure water, when they are taken up by the atmosphere of any given space to such an extent as to render it moister than that of the surrounding country, must be considered injurious; and much more so are exhalations from water rendered impure by the decay of vegetables along its banks, or by the admixture of impurities from the drains of offices, &c. Attentive observation, and some experience on this subject, have enabled us to arrive at the following conclusions, which may be considered as principles: —

I. With a view to health, water ought never to be introduced where it will increase, in any sensible degree, the quantity of moisture that would otherwise be taken up by the atmosphere of the locality, either in summer or winter. From this principle the following rules may be deduced: —

1. That the surface to be covered with an artificial piece of water should be small, in proportion as the general surface of the ground is flat; the soil retentive; the park, pleasure-ground, or enclosure, small; and the trees and shrubs numerous.

2. That the most wholesome situations in which artificial water can be introduced, are those where the general surface of the ground is elevated, and the soil naturally dry, and not thickly covered with trees and shrubs.

3. That, in flat situations with retentive soils, where the surface is thickly planted with trees or shrubs, water, even on the most limited scale, should never be introduced, without, at the same time, rendering the surface of the surrounding ground perfectly dry by the frequent drain system.

4. That in no description of artificial water ought the decay of vegetables to be allowed to take place, and more especially along the margin.

5. That the margin of all artificial pieces of water ought to be formed of a considerable thickness of gravel or small stones, or of blocks of stone, in imitation of rockwork; in order to diminish the quantity of spongy or marshy matter, by which evaporation never ceases; and substitute for it a smooth hard surface, from which evaporation will go on with rapidity, and which will thus soon become quite dry.

Gardeners are in the habit, when they plant out pelargoniums, and other spongy-wooded green-house plants, on lawns, to reduce them before planting out to single stems, and to free these from leaves to the height of 3 or 4 inches. After planting, the surface (which forms a very gentle knoll, about 1 ft. in diameter, and, say, from 1 in. to 2 in. high in the centre, where the stem is placed) is bedded over with small pebbles, about the size of pigeons' eggs; and these, by drying rapidly after rains, prevent the damp from lodging about the collars of the plants, which, without this precaution would, as it is technically called, damp off. Now, if gardeners would apply the rationale of this practice to the margins of basins, ponds, lakes, rivers, and all pieces of water whatever, natural or artificial, in pleasure-grounds, they would render the evaporation from these pieces of water wholly innoxious. The evaporation from clear water is simply injurious by increasing the quantity of moisture held in suspension by the atmosphere; but the evaporation from water containing a mixture of decaying vegetables from the park above it, contains, in addition to water, those deleterious gases known as malaria. As all pieces of water are liable to rise or fall with rains or great droughts, the breadth and height of the space along their margins, which is graveled or covered with stones, ought to be such as that, in the greatest drought of summer, when the water is sunk to the lowest point, it should not expose any of its earthy bed to the air; and that, in spring and autumn, when it is raised to its greatest height by rains, it should not touch the leaves of the plants along its margin. This is the beau ideal mode of treating artificial water, where the object is to render it wholesome; and it is gratifying to find that it is as superior to the common mode of treating the margins of pieces of water, in its accordance with the principles of picturesque beauty, as it is with the principles of health.

6. That, in distributing the trees and shrubs over a park or pleasure-ground containing a piece of water, provision should always be made for the exit from the grounds of the vapours which arise from the watery surface. This is to be done by broad spaces of lawn or turf without trees, extending from one or

from both ends of the piece of water, through the park or pleasure-ground, to ground on a lower level; to which lower level the air charged with vapour will find its way by its own gravity. Where this is neglected, parks in low moist situations, with trees scattered regularly over their surface, become covered with what may be described as one general pond of malarian vapour; or with a number of ponds, in which the vapour is dammed up by trees crossing the course which it would naturally take along the lowest level. The gardens of the New Palace at Pimlico may be considered as one immense pond of malarian vapour, confined by the palace and other buildings at the lower end, and by high walls along the sides; and which is prevented from being dispersed by winds or the sun, by the groups of trees, mounds, grassy banks, &c., contained in the interior.

II. With a view to beauty or effect, the distinctive properties of water, as contrasted with those of ground are, clearness and brilliancy, the power of multiplying objects, coolness, and motion. These are the principal properties which concern the landscape-gardener; and it is difficult to imagine a situation in which one or other of them may not be exhibited, and turned to excellent account. The following rules refer to this principle:—

1. Clearness in pieces of water is displayed by allowing them chiefly to reflect the sky; and brilliancy, by having the situation somewhat elevated, or otherwise so prominent as that the water may catch the rays of the sun, and reflect them directly in the eye of the spectator. This is a kind of beauty not often desirable in landscape; but it may occasionally be wanted to contrast with others; and it has this advantage, that it may be introduced in a very limited space, to which it adds great cheerfulness.

2. The power of multiplying objects by reflection is by far the most valuable property of water, relatively to landscape-gardening. Forms, shades, and colours may thus be doubled; and the most forcible contrasts may be produced between water surrounded by trees, buildings, or other objects, and open lawn or pleasure-ground. Hence, it is seldom desirable to form pieces of water without planting trees or shrubs, or both, along its banks. Water, without wood, may display clearness, and convey the idea of coolness; but it cannot have that intricacy, variety, richness, and force of effect, which are produced in greater perfection by the shade, colouring, and forms of trees and shrubs, with the occasional aid of buildings and animated nature, than by any other means at the command of the gardener.

3. Coolness, as a property of water, is actually produced by evaporation; but its cooling effect in landscape may be considered as depending on the idea of coolness which we associate with water, and which has been originally derived from experience. This property in water renders it a desirable material for contrasting with gravel, buildings, rocks, or other objects, which, with reference to their effect in landscape, are considered warm.

4. The motion of water, in the imitation of its more common forms in nature, is displayed in artificial brooks, rills, springs, &c.; and in the imitation of more extraordinary natural appearances, in artificial cascades, fountains, and jets. As the motion of water may thus be displayed on a very small scale; the property of motion, and, as already mentioned, that of clearness, are, perhaps, of the most universal use in gardening; there being scarcely any ground plot, however small, that will not admit of a fragment of rock, with a drooping fountain, or of a marble basin, or stone vase, of clear water, to reflect the rays of the sun, and tints of the sky.

III. With a view to picturesque effect, water may be introduced into the grounds of every residence, however limited it may be in extent; and the effect will depend, not on the magnitude of the piece of water, but on the character or expression which it is made to assume. Hence the following rules: —

1. That, in very small places of even a few perches in extent, a surface of water, of not more than a few square feet, sunk in the ground, and with a gravelly, rocky, or stony margin, by reflecting the objects near it, and the sky, may give an extraordinary interest to the landscape; the water being kept clear, and perfectly free from mud and decaying objects.

2. That a brook or streamlet, however small, can never be successfully imitated, unless there be a current of water equal to that which is found in brooks of the same magnitude in a state of nature.

3. That the character of a brook being briskness, it should never be imitated by art, except where there is not only an abundance of water, but a surface having a considerable slope, in order that the water may run rapidly.

4. That, where a natural brook passes through a park or pleasure-ground, it may be improved in effect by expanding some parts of it into pools; and by the distribution of gravel, stones, and other rocky materials, along its banks.

5. That a river, which, to be natural, ought to indicate on its banks the progress and action of water, can never be imitated in a park or pleasure-ground, except where there exists naturally a considerable stream.

6. That a lake may be imitated wherever there is a sufficient extent of nearly level surface to excavate; or where the water may be made to cover an extensive surface, principally by throwing a dam across a shallow valley.

7. That a prolonged lake, tame river, or even a canal of traffic, is occasionally admissible in artificial scenery, under particular circumstances; and more especially when it is at such a distance from the eye, and so far below it, as to render it impossible to detect whether the water is in motion or stagnant; the water, in this supposed case, being perfectly clear. Thus a pleasure-ground or park, which occupies the upper part and the sides of a hill, may have a zone of canal, or tame river, intermediate between the upper part of the hill and the valley at its bottom; by which means this canal will form an interesting foreground from the park above it to the country beyond. This is beautifully exemplified in many places bordering canals; for example, at Offchurch Rectory, near Leamington, in Warwickshire; and at Wentworth Castle, in Yorkshire.

The Union of all the Parts which compose a country Residence, so as to form a harmonious and expressive Whole. In the preceding remarks, we neither pretend to have embraced every part of which a place is composed, nor to have given all the details which belong to any one of the parts treated of. Our object has been to give the reader materials for thinking on the different subjects connected with the formation of a residence where there are a few acres of ground, in order that he may endeavour to find a reason for every thing that we may in future propose; and, in short, that he may feel a greater interest in this work than if it were merely a series of arbitrary directions. In pursuance of the same object, we shall next consider the residence as a whole.

After all the different parts which enter into the composition of a country residence have been duly weighed and considered by the proprietor and his landscape-gardener, and the proportionate extent of each agreed on, the next step is for the latter to put them together. It is chiefly in doing this that the artist has an opportunity of showing to what extent he is entitled to be considered as a man of genius and taste. It is easy to conceive that all the different component parts of a piece of music, a picture, or a piece of architecture, may be correctly executed; and yet that the want of due proportion between these parts may be so great, and the whole may be put together with so little connexion and harmony, as to form an object wholly without sentiment or expression, a body unanimated by a soul. It is this expression, formed by the due proportion, connexion, and cooperation of all the parts, that constitutes the main difference between a work of art, and one of mere mechanical skill. Two grand qualities in the artist, with reference to creating expression, are, the power of viewing every part of the scene which he is to create, with reference to the effect of the whole; and the power of foreseeing future effects. It is only by the union of these two qualities in the mind of the landscape-gardener, that the grounds of a residence can be formed into a composition, as perfect as a piece of architecture; every moulding of which, as well as every column and larger member, has reference to the elevation of the different sides of the building, forms a sort of index to it, and could not be removed without injury to its effect.

Perhaps these remarks may be better understood by noticing a few of the most common defects, or causes of defects, in country residences, than by describing a comparatively perfect model.

The first fault of a place that meets the eye of a stranger is generally at the entrance lodge. Here the building and gates are very frequently either too mean, or too much ornamented; too large and substantial, or too small, for the mansion and its accompaniments. What the happy medium is, it may be difficult to say, unless a particular case were before us: but, in every age, there is a sort of conventional agreement among men of taste, as to what is proper, and what exceeds the bounds of propriety. If every part of a place should give the spectator some idea of the style of art employed in every other part, then it is clear that the lodge and gate should be in the same architectural style as that of the mansion; and, hence, where the latter is Grecian or Italian, the former should neither be Gothic, nor in the ornamented English cottage style. The boundary fence connected with the lodge should be modern, where the art employed is modern; but, in the case of Gothic, or Elizabethan, lodges, walls only are admissible; hedges, as boundary fences, not being chronologically correct when connected with buildings in either of these styles. Where the lodge and gates have patches of plantation connecting them with the boundary fences, but no scattered trees or groups to unite them to the mansion and the scenery in the interior, the principle of connexion is grossly violated, and one of the finest sources of variety in the views along the approach road neglected. A mansion set down in a park or lawn, without any scenery of an intermediate character, such as an architectural basement, surrounding terrace, &c., is in opposition to the harmonising principle of uniting objects so powerfully contrasted as a house and a field, by scenery of an intermediate kind.

The mansion is often seen enclosed in a patch of plantation, which also includes the domestic offices; but it very frequently wants the connexion, which groups and small masses of trees would give it, with the general surface of the park. Sometimes the latter is too much crowded, by the trees being uniformly distributed over every part of it; and sometimes it presents a bald appearance, from the want of trees. Indeed, to adjust the proportion of trees and pasture, or clothed spaces and naked spaces, about a place, is one of those points of art, on which more of the future effect depends, than on almost any other. A residence, of which it may be said that it has just enough of trees and shrubs, and not too many, will generally be found a highly satisfactory one.

There are certain defects in grounds and buildings, which owe their existence to errors and omissions on the part of the first builder or planter; and certain other sins, perhaps of a more heinous nature, which are committed by the occupiers after the place is finished. The remote cause of these last errors is, the desire inherent in almost every body who is in possession of a house and grounds which he can call his own, of doing something to it; and the immediate cause is, that this something is usually done solely with reference to itself, and without any regard to its general effect on the house or grounds, considered as a whole. Perhaps a new flower-garden is to be formed, and it must have beds in it, or statues to ornament it, like those at B or C: without considering that B is a castle, and C a palace, while the scene where the flower-garden is to be formed is, perhaps, a plain modern villa. The desire of imitating those above us is thus at once the cause of the spread of improvement, and of the introduction of much absurdity. The purchase of articles at sales, because they are good and cheap, or, perhaps, beautiful in themselves, is also often the means of spoiling the general effect of a residence. The young family of A., who are growing up, have acquired a taste for plants, and are desirous of having a green-house, which A. kindly purchases for them at the sale of the first neighbouring nurseryman who becomes bankrupt. This shed-like structure is placed against one end of the house, in a conspicuous situation; and the entrance front has thus ever after a mean appearance. We have seen a handsome lawn spoiled by the desire of the lady of the house to have a piece of rockwork; and we have known the foundations of a house rendered damp by the occupier having purchased the flints and scoriæ of a rockwork at a sale, and, for want of any better situation, banking up the lawn front of his house with them. Nothing is more common than for gentlemen of leisure, who have small country residences, to attend nurserymen's sales, and purchase articles they do not want, merely because they are cheap. These cheap purchases are often fatal to the general effect of a small place. Room must be found for the trees and shrubs which have been bought; and, wherever there is an open space on the lawn, one or more are put down in the middle of that space. "Surely, Mr. L.," we have been told a hundred times, "there can be no harm in putting down a single tree, more especially as we always plant them in the middle of open spaces, where there are no others near?" Persons arguing thus, little know that a very few single trees, put down on this principle of "always placing them in the middle of an open space, where there are no others near," would destroy the effect of the finest place in existence. Single trees, in a park or pleasure-ground, are like the last touches of an artist in painting a landscape. The bold striking effects of light, shade, and character are given by the masses; but the expression of these masses, and the attractions by which the eye is led to enjoy them, are produced by the last touches. The great general effect of the grounds of a residence depends on there being open spaces where there are no trees, to contrast with other spaces which are entirely covered with trees; and the finish to these grand features of wood and lawn consists in the single trees and small groups which are distributed along their margins.

Soon after Sir Henry Steuart's *Planter's Guide* was published, and an unmerited importance given to it by the Report of the Committee of the Highland Society of Scotland, and a number of laudatory reviews in the periodicals, the idea of transplanting large trees by machinery took possession of many country gentlemen of leisure; and the breadth of effect of many lawns and parks, and the grouping along the margins of lawns in others, were destroyed, by always putting down these large trees in those open spaces where it was thought that they would not interfere with anything else. A gentleman who possesses one of the handsomest small places on the banks of the Thames, between London and Gravesend, covered almost the entire surface not occupied by old trees, with young single trees at similar distances, without even employing a gardener to guide his labourers in planting them. In such a situation as the very remarkable one to which we allude, before a single tree could be put down with propriety, its future effect should have been studied from various points of view; and there is no landscape-gardener who would not consider the putting down of even one tree, in such a case, a difficult and delicate task. But profound ignorance of any subject is favourable to courage, and our planter put them in by hundreds; but, fortunately, as they were planted three or four times too deep, in a soil consisting, in many places, of strong clay, almost the whole of them died the second year.

Professional Routine. We shall now go through what may be called the professional routine of fixing on a situation; making the plans for laying out a residence; and carrying these plans into execution.

Fixing on the Situation, Extent, &c. *Fig.* 154. may be supposed to exhibit a portion of country, the greater part of which

154

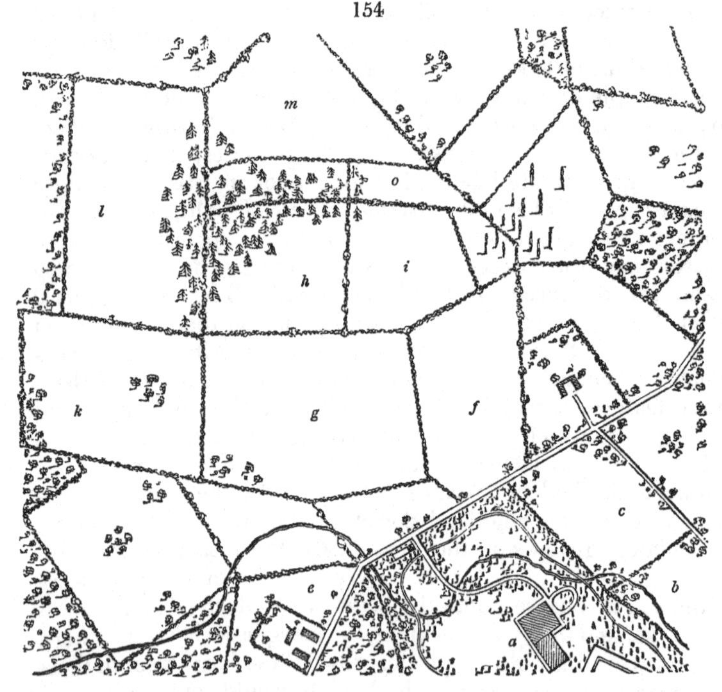

is to be sold by private contract, in lots as various in magnitude and form as the size and shape of the fields; or to be let on building leases. It may be supposed to have all belonged to the mansion and park *a*; but that, owing to death or other causes, the present owner of that park has determined to limit himself to it, and to the adjoining fields and farmyard (*b*, *c*, *d*, and *e*). All the other fields, therefore, are to be sold or let; and we shall suppose them to have been looked over by a gentleman desirous of forming a suburban residence, who has fixed on the fields *f*, *g*, *h*, *i*, and *j*, containing in all about ten acres. Here we must notice what is a common error, often committed in the first step of forming a country residence; viz. that the party (say a tradesman, a merchant, or a professional man) chooses the situation, and fixes the conditions of purchase, without consulting a surveyor; or, what is better, some friend, who has a practical knowledge of soils and situations, and of what are really essential in these to the formation of a comfortable residence. Perhaps the party who, in his want of knowledge, has rashly ventured to judge for himself, was too modest to let it be known to his friends that he intended to purchase land; or, perhaps, by keeping the matter quiet, he expected to get it somewhat cheaper; or, perhaps, one object was to prevent his friend and neighbour, Mr. A. B., from getting what he considered one of the most advantageous of the allotments. At all events, he has chosen these fields entirely on his own judgment, and, in one point of view, judiciously, because they lie very compactly in a ring fence. It is proper to mention, however, that there are situations, where, when the object is to form a country residence, compactness of outline possesses fewer advantages, and is in fact much less desirable, than an outline of great irregularity. This depends on the natural inequalities of the surface, and on the distant prospect: for example, in the plan before us, had there been a natural ridge proceeding from *f* to *g*, and thence by *k* to *l*, terminating in a considerable hill at *m*, then we should have preferred these five fields to the five which lie so compactly; notwithstanding the great extent of boundary line which in the latter case there would have been to keep up, and the circumstance of the fields *h*, *i*, *j*, *n*, and *o* lying, as it were, in the midst of the property. This, however, would be no disadvantage in an ornamental point of view; because we are supposing the ridge *f*, *g*, *k*, *l*, *m* to be far higher than the adjoining fields, and to overlook them entirely. The proprietor of such a ridge might build his house on the hill (*m*); have a going approach along one side of the ridge, and a returning approach along the other; the trees along both approaches being arranged so as to form foregrounds to the distant scenery, and to exclude near objects in the adjoining properties, which might be considered as not worth looking at. Along the centre of the ridge, there might be an irregular-margined avenue of turf, in the manner of the green drives at Fonthill, Goodwood, Stourhead, &c.

But to return to our choice. The five fields, *f* to *j*, are supposed to contain no great variety of surface; and the country around to be tame rather than otherwise, and in the same style as that which lies north of London, along the Edgware Road. The purchaser, we shall suppose, now employs a landscape-gardener, whose first business is to procure a plan to be made, such as *fig.* 155., in which the ring-fence of the five fields is shown,

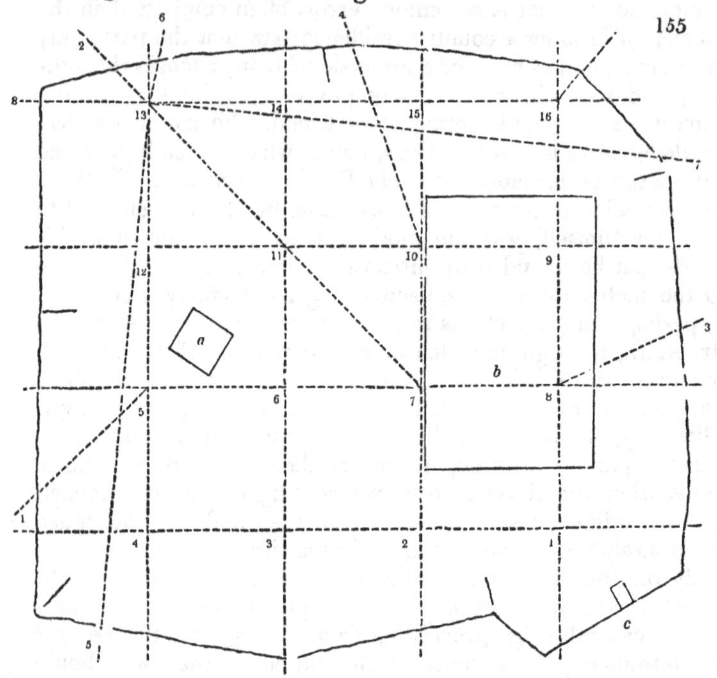

enclosing a space thrown into squares by dotted lines. These dotted lines are, as they ought to be in every working plan of this kind, in the exact direction of north and south, and east and west, for more convenient reference and description, and future use in marking out improvements on the ground. Before the squares are drawn on the plan, they ought previously to be marked out on the ground, and a small stake placed in every intersection of the lines; that is, at every corner of each square, as shown in the figure. The squares may be 50 ft., or 100 ft., or 200 ft., on the side, according to the extent of the plot, the inequalities of its surface, or the alterations which are to be made in it. In the case before us, they are sixteen in number, exclusive of the portions of squares round the boundary; each square is 150 ft. on the side, and each contains half an acre and 22 poles. On an estate where the surface is flat, the squares may be large; because, from the general sameness of the whole surface, the character of the ground included in one square must be very like that in all the others: but on a very irregular surface they must be small; because each square may have a different character of surface. By having a stake with a number on it, in the corner of each square; by having these numbers in regular series from one side of the plot of ground to the other; and then by having corresponding numbers on a plan, and a memorandum book for reference and description, it is evident that a tolerably correct idea may be conveyed of the soil, subsoil, surface, and distant views, even to a person who has not seen the estate.

The memorandum, or field, book may have the same numbers on two opposite pages, as in the specimen below: the one to describe the present state, and the other the intended alterations or improvements, thus: —

Present State of the Five Fields, purchased by Mr. C., and intended to be laid out as a Villa Residence.— The numbers preceding each paragraph refer to the numbers in the squares in the plan. (*fig.* 155.)

Remarks as to the Improvements which may be made on the Property purchased by Mr. C., &c.

1. Surface here only about 5 ft. higher than the public road, towards which it gradually slopes. Soil loamy.

1. This, being the lowest part of the ground, and also near the public road, would seem an eligible place for the entrance-lodge and the approach. On mentioning this to Mr. C., he said that Mrs. C. had determined on having a castellated lodge, like that at Cobham Hall, only smaller.

2. Surface still higher.

3. Continuing to rise.

4. About the same height as at 1. Surface apparently wet in winter, judging from the kinds of grasses growing on it; such as *C*àrex, *P*hlèum, &c.

2, 3, 4. Soil good either for trees or pasture. Little or no distant view; and, therefore, this part may, if thought desirable, be covered with wood.

5. One of the highest points within the ring fence, from which, in the direction of N. E., a tolerable view of the village church is obtained, backed by a green hill, as in sketch No. 1. (*fig.* 156.) The grasses chiefly rye-grass and meadow-fescue.

156

5. A few yards S. E. from this point promises to be the best situation for the house, as the ground falls from it on three sides. Considering that the country has an equal claim in point of beauty all round, a square house seems the most desirable; and a square is accordingly drawn on the plan, at *a*, in *fig.* 155.

6, 7. Ground flat, but fully as high as at 5. The view from 7 to the S. E. shows a viaduct thrown over a valley for a railroad. No. 2. (*fig.* 157.)

157

6, 7. The offices may very properly be placed in this direction, so that the view may be obtained from the pleasure-grounds and the conservatory, which may be placed against the kitchen-court. There is here a very fine opportunity of connecting the conservatory with the kitchen-garden, by means of an architectural conservative wall, with an open or covered walk.

N.B. Mr. C. objects to a conservatory on account of the expense, but would not object to a green-house. I told him that though a green-house would cost less at first, it would prove more expensive afterwards, on account of the plants being all in pots, and requiring much more attention in watering, repotting, &c. In the evening, Mrs. C. said she should prefer an orchideous house like Mrs. Lawrence's. Perhaps better that these things should be left for Mrs. C. to settle with her gardener.

8. Even surface, and soil loamy, with a view, in a direction W. by S., of farmer Swiney's house and farm-yard. See sketch No. 3. (*fig.* 158.)

158

8. A number of luxuriant docks growing here; on seeing which, Mr. C. said that the soil was undoubtedly bad. Endeavoured to convince him that docks are never found growing luxuriantly on bad soil; and, in short, that they are generally a proof of good, deep, loamy soil, as was the case here.

9, 10. Surface even and soil good. From No. 10., a distant view, in the direction of S. by E., of a viaduct for the railroad over the river Colne. See sketch No. 4. (*fig.* 159.)

159

9, 10. The ground included in the squares 8, 9. and 16. being good and even, and also the adjoining squares eastward and westward being of the same description, this part of the property appears to be a very favourable situation for the kitchen-garden, which is accordingly indicated by a parallelogram at *b*, in *fig.* 155.

11. Surface undulating; subsoil said to be stony. White clover.

11. This ground will require more smoothing, in order to produce an artistical surface, than any which we have yet gone over; but the effect will be satisfactory.

12, 13. Ground falls from this point to the boundary, both in a S. and E. direction. The view to the N. by E. is of the mansion belonging to the park (*a* in *fig.* 154. in p. 448.), shown in sketch 5. (*fig.* 160.), and

12, 13. Marks of ridges on the surface, the ground having been here under corn three or four years ago. At a short distance nettles growing; on which Mr. C. said, ironically, he supposed that these plants were also an indication of good soil, as well as the docks. Told him that the nettle was a domestic plant, and was seldom found any where, except about human habitations; and that, properly speaking, it could not be considered as an indication of the natural state of the soil, but only of the accidental

nearly directly S. to the suburban villa of Captain B. See sketch No. 6. (*fig.* 161.) The view W. by N. shows

161

the river Colne, with an old bridge, as per sketch No. 7. (*fig.* 162.); and

162

direct E., the windmill on the heath, as per sketch 8. (*fig.* 163.)

163

circumstance of lime rubbish having been placed there, in consequence of the proximity of buildings. Mr. C. laughed at this, as the field was at a considerable distance from any house. But, on enquiry, we found that a barn had stood there formerly; on which Mr. C. observed that the nettle was not only a domestic plant, but a historical plant, as it told the previous history of the ground on which it grew.

A good situation for a summer-house, on account of the fine views seen from it; and on the supposition that the pleasure-ground walk passes this way.

At 14, 15, and 16, the ground is low, and the views of the distant scenery not marked by any particular feature. Near 16., a view is obtained of an old barn turned into a cottage, in the direction of S. W. See sketch No. 9. (*fig.* 164.)

164

14, 15, and 16. Low, without much exterior view, and may therefore be planted if thought desirable. Being at one corner of the property, this would be a good situation for an archery-ground, or bowling-green. On mentioning this, the Misses C. appeared delighted at the thought of an archery-ground; but Mr. C. seemed to incline to a bowling-green. Mrs. C. thought it would be a good place for the younger children to learn to ride.

The left-hand column above gives a general idea of the surface, and sufficient indications of the soil and subsoil to enable the landscape-gardener to assist the architect in determining the best situation for the house and offices; and the remarks in the opposite column are for the same object, and also for the sake of indicating, as far as this can be done from incidental observations dropped from time to time, the taste and wishes of the proprietor and his family. The situation of the house we have indicated on the plan *fig.* 155. at *a*, that of the kitchen-garden at *b*, and that of the entrance lodge at *c*; and these three points, being fixed on, naturally determine the situation of the offices, which must be between the house and the kitchen-garden, and that of the approach, which must be between the lodge and the house; leaving the remainder of the place to be laid out as pleasure-ground.

Another mode, by which the most interesting views from different points within the ring-fence may be exhibited, is shown in *fig.* 165.; in which the estate, and the views from it, are drawn

165

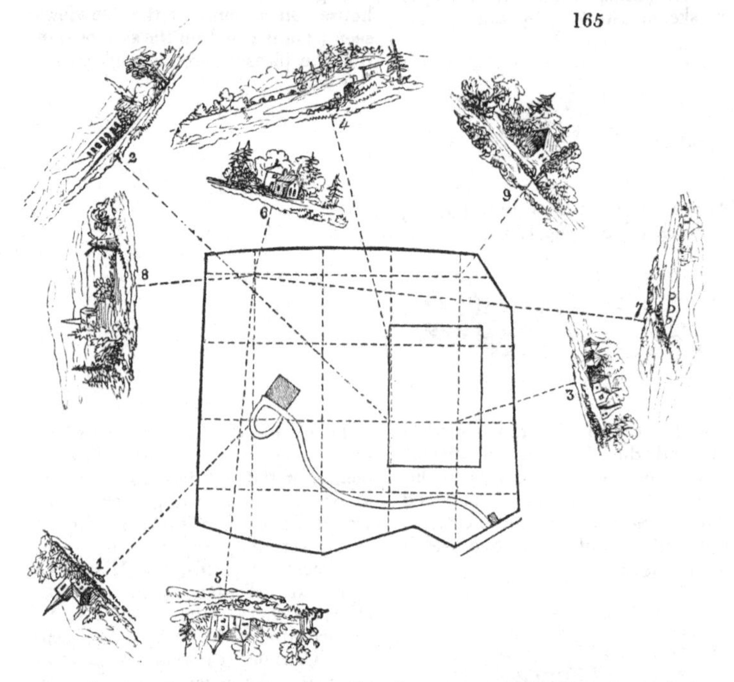

to a scale one half smaller than in *fig.* 155. and in the views (*figs.* 156. to 164.) given in the descriptive column. By the mode shown in *fig.* 165., all the different distant objects given in the sketches in the descriptive column, p. 451, 452, and 453., are shown in the exact position, and at the relative distances at which they lie from the different points; the scenery of the foreground, and that of the middle distance, being, in this case, as in the descriptive column, taken no notice of.

A third mode consists in giving a panoramic view of the entire country, exterior to the ring-fence of the part to be laid out, as in *fig.* 166. This panoramic view is supposed to be taken from a temporary platform, erected in the centre of the exact situation where the mansion is to be placed, and raised 20 or 30

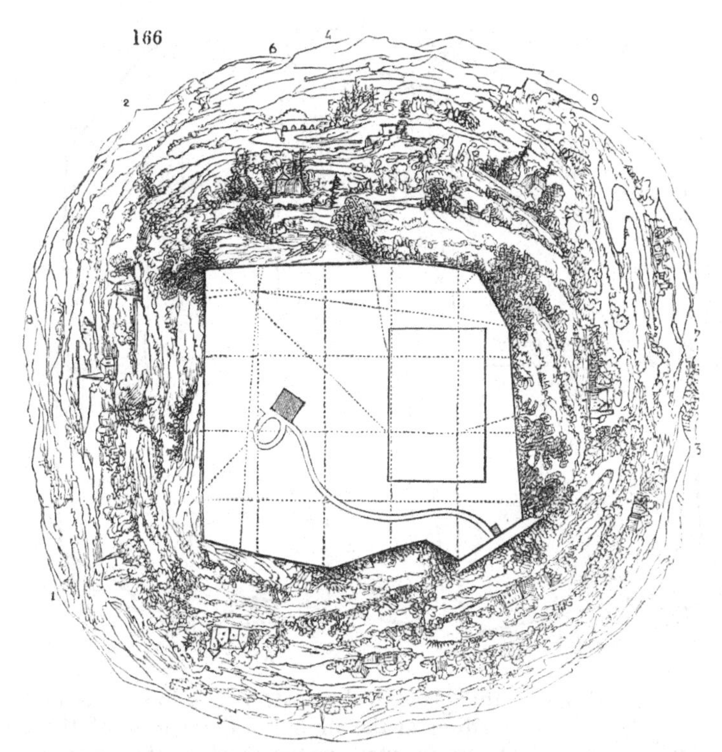
166

feet above the natural surface of the ground. The artist, in taking such a view, changes his position six times, so as to include in the panorama six connected views, each embracing an angle of 60°. This mode is calculated to give a very clear idea to a purchaser of the kind of distant scenery which will be viewed from the windows of his intended habitation; and it may be very conveniently adopted in a case which sometimes occurs; viz. of an English gentleman engaged in business in South America, or any distant country, who sends home orders to have grounds purchased, and a house built, preparatory to his return to England. The first step, in such a case, previously even to purchasing the land, would be to send out a ground plan, surrounded by a panoramic view, as in the figure just referred to.

Where the surface of the ground is considerably varied, and where it is also desired to form a correct idea of the subsoil, and, perhaps, of the mines and minerals, the kind of sectional plan exhibited in *fig.* 167. may be adopted. In this plan, the area is first thrown into squares, by N. and S. and E. and W. lines, as in giving surface plans. The lowest point on the ground is next fixed on, as the level of the base line of all the sections; and on

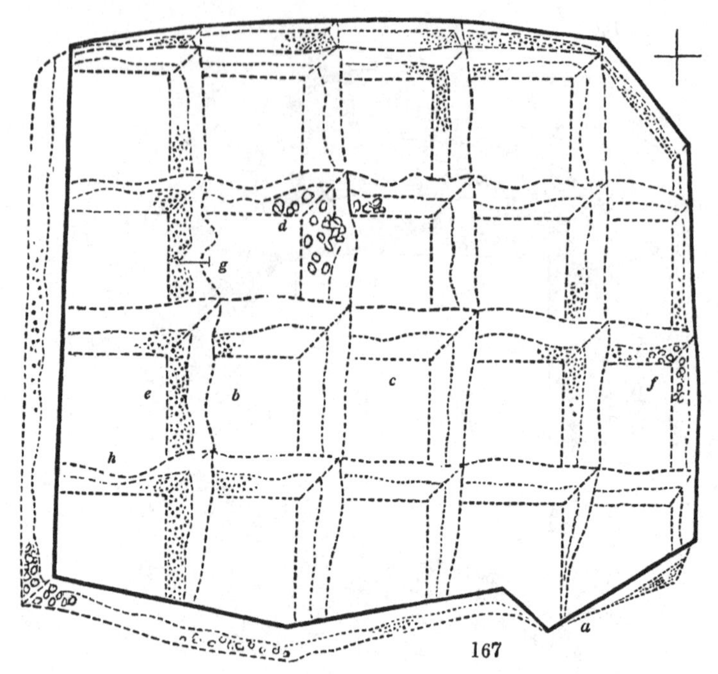

167

this line the rising of the ground above it, on the four sides of every square, is indicated, as in the figure. The nature of the subsoil may also be shown in such sectional plans (for example, whether it is stony, rocky, sandy, or gravelly); and, what is as useful as a knowledge of the subsoil, the depth and direction of under drains, the situation of springs, wells, pits, &c., may be accurately delineated. The levels are obtained in the manner well known to surveyors, along the lines forming the squares; and the nature of the subsoils, mines, minerals, &c., is ascertained by digging pits, boring, or trials with an iron probe. Where the object does not extend beyond those of the cultivator or the landscape-gardener, the nature of the subsoil, for 3 or 4 feet below the surface, is all that is required to be known; and this can be attained with very little trouble. In *fig.* 167., *a* is the lowest point in the ground, from which point the approach is made to enter; *b*, is a square including the highest part of the grounds, on the surface of which the house and offices are placed; *c*, the appearance of the section when the subsoil consists of the same earths as the surface soil; *d*, the appearance of loose stones; *e*, sand; *f*, gravel; *g*, a hollow in the surface where there is a pump-well; and *h*, a hollow from which brick earth has been dug.

The working Surface Plan. Having, by means of the plans and sections already described, and of conversations with the proprietor and his family, acquired a knowledge of what are technically called the data and desiderata of the place, the next business of the landscape-gardener is to form a working plan for laying out the proposed improvements; the situation of the house, the kitchen-garden, and the entrance lodge, being already fixed on, as before indicated. We shall suppose that the numbered sticks at the angles of the squares still remain on the ground; because it is more convenient to adopt squares of the same dimensions as those already marked on the ground in the plan on which we are to trace the roads, walks, kitchen-garden, plantations, &c., in detail. The sides of the squares, also, in this plan, must (in order to admit of readily indicating objects with reference to the points of the compass), be directly east and west, and north and south. It will often happen that the same plan which is used to indicate the levels of the surface, and the principal points of view, and to afford data for the description given in p. 451., will serve also for tracing the lines which constitute the working plan; but, in the present case, the plan (*fig.* 155. in p. 450.) is on so small a scale, that, if we were to trace the necessary lines upon it, it would become confused, and unfit for our purpose. *Fig.* 168., therefore, must be had recourse to; and we shall go

168

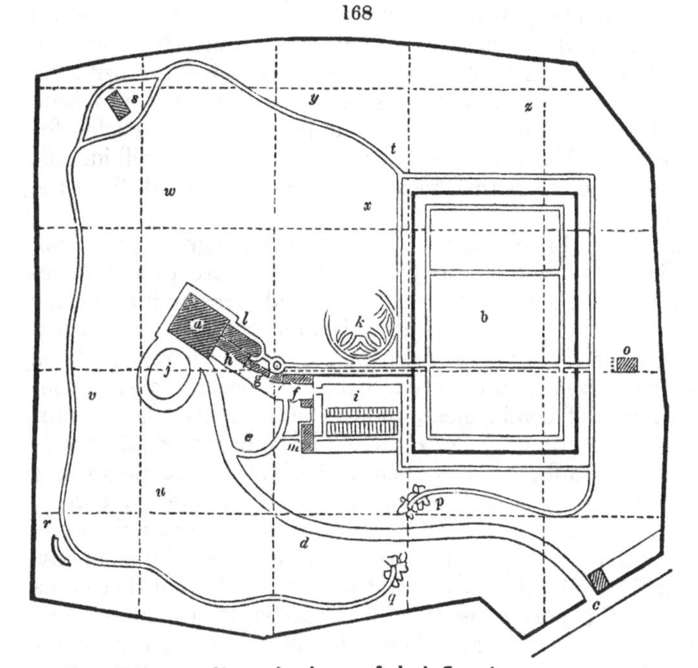

over the different lines in it, and briefly give our reasons for producing them.

The position of the house (*a*) and lodge (*c*) being fixed on, the road between them might either have been made straight or curved. If it had been straight, it would have been inconsistent with this style of art; and, if the grand sweep which it takes had been bent to the right instead of to the left, as at *d*, it would have interfered with the arrangements connected with the offices and kitchen-garden. Before arriving at the entrance front of the house, there is a branch road (*e*) to the stable offices (*f*), which are connected on the one hand with the poultry-yard (*g*) and the kitchen-court (*h*), and, on the other, with the reserve ground to the kitchen-garden (*i*). This branch road (*e*) is made narrower than the main approach, in order that it may never be mistaken for it: it is bent, so that persons either going to the front entrance of the house, or coming from it, may never see along it as far as the gates of the stable-court; and it is joined to the main approach in such a manner, that it may rather invite a person to enter on it, when returning from the house, than when going to it. The reason of this is, that this branch road will be principally used by the carriages of visiters going to the stable offices, after they have set down their company at the front door of the house; and by the carriage of the family going from the stables to the house, and back again, after the carriage has been used. For one person, therefore, who comes out of this branch road and turns towards the entrance lodge, there will be at least ten who come out of it, and turn towards the mansion, agreeably to the inclination of the road at its junction with the approach. In short, tradesmen's carts, and carts with provision for the horses and poultry, with manure and other articles for the garden, and with coal, &c., for the kitchen-court, will include every kind of carriage that goes along this branch road, from the entrance gate.

Instead of a broad mass of gravel for carriages to turn on, immediately before the entrance portico, we have preferred retaining the road there of its usual width, and carrying it round the large oval of turf (*j*). This is not only a much safer mode of turning a carriage, but gives an idea of ample space; whereas a dilated surface of gravel, immediately in front of a portico, always conveys the idea of confinement, and of the front or entrance court to a town mansion. The width of the approach road is supposed to be 15 ft., in consequence of which two carriages may pass each other with ease without going off the gravel; and, in the case of a large party, 50 or 100 carriages might wait round the oval, and any one of them be called out to take up company, without the slightest derangement to the others. All that is necessary for this purpose is, that all the carriages in waiting round the oval should stand along one side of the road, leaving the other side free for each carriage, as it is wanted, to turn out of the line and drive up to the door of the house.

The ground plan of the house and of the offices, of the conservatory (*l*), and gardener's house (*m*), we shall leave to be determined on by the architect, only stipulating that their precise position, as indicated in the plan, shall not be altered; and that there be the proper drainage to a main sewer, commenced at the house, and conducted in a right line through the kitchen and other courts, and through the reserve garden, to the boundary fence, with manure tanks in the reserve garden (see p. 415.); and that over the stables, or in some conspicuous part of the offices, there be a handsome turret clock, the dial of which may be seen from the windows of the kitchen, and from those of the gardener's living and sleeping rooms; and from the forcing-ground.

The position of the house (*a*) and kitchen-garden (*b*) being fixed on, the most convenient situation for the offices, as we have already observed, is between them; and to disguise these offices, or, at least, the courts belonging to them in which the business of each office is carried on, nothing can be more convenient than garden structures, or garden scenery. In the more ordinary cases, a shrubbery may thus serve to disguise the offices, and connect the house with the kitchen-garden; but, in cases where more art and expense are employed, like that before us, instead of the shrubbery, there may be a conservatory of an architectural character, to harmonise with the house; and, adjoining that, either a conservative wall, with or without a glazed veranda, or a wall with a common veranda. The conservative wall is supposed to be covered with the finer kinds of woody plants, and the supports of the veranda with climbers and creepers.

169

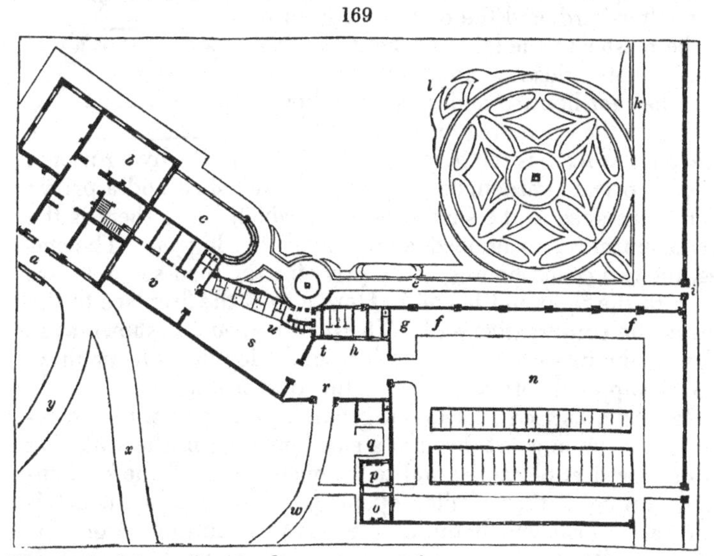

Through this wall, whether conservative or common, there may be private entrances for the master and mistress to the stable-court and poultry-yard, as indicated in *fig.* 169., which shows the plan of the offices on a larger scale. In this plan, *a* is the entrance-portico to the house; *b*, the drawingroom, with three windows at one end (opening down to the floor, and serving also as doors), which look into the conservatory. In this there is a broad walk down the middle (*c*), terminating with a door in the centre of a semicircular end, outside of which are steps descending to a circular basin and fountain, beyond which is the walk (*e*) in front of the conservative wall (*ff*). Instead of a basin and fountain, a circular stage of stone steps for containing greenhouse plants in the summer season, may terminate in the centre in a sundial, such as *fig.* 170.; and in that case the basin and fountain may be transferred to the centre of the flower-garden. The kind of fountain there used may resemble *fig.* 171. This walk terminates in an archway (*i*), which forms the main entrance to the kitchen-garden; and in the angle at the right is the flower-garden (*l*), which is also shown in the same position at *k*, in *fig.* 168. There is a walk at *k*, in *fig.* 169., communicating with the other parts of the pleasure-ground. There is an open loggia with a seat, opposite the fountain; and on each side of this loggia is a small door, the one forming an entrance for the mistress to the poultry-yard, and the other an entrance for the master to the stables: here are also summer water-closets. In the reserve garden, the hot-houses and pits are shown at *m*; and the open area for composts, manure, &c., at *n*: *o* is the gardener's kitchen; *p*, his living-room; and *q* his private garden, near which are a fuel shed and a privy: *r* is the entrance to the stable-court, in which, at *t*, there is the private entrance, mentioned above, from the loggia. The stables, the two coach-houses, and a privy for the men-servants, are shown to the right and left of *h*. Here also is the fireplace to the flues in the conservative wall. The poultry-court is shown at *s*; and at *u*, the private entrance to it from the loggia. The poultry-yard is supplied with water from the overflowing of the basin and fountain, carried to it under ground. The poultry have access to the stable-court through a small opening in the wall, that can be closed at pleasure; and to the open lawn and the kitchen-court, through other similar openings. The kitchen-court is shown at *i*, near which there is a servants' entrance from the approach. Part of the branch road leading to the stables is shown at *w*; part of the approach, at *x*; and part of the sweep round the oval, at *y*.

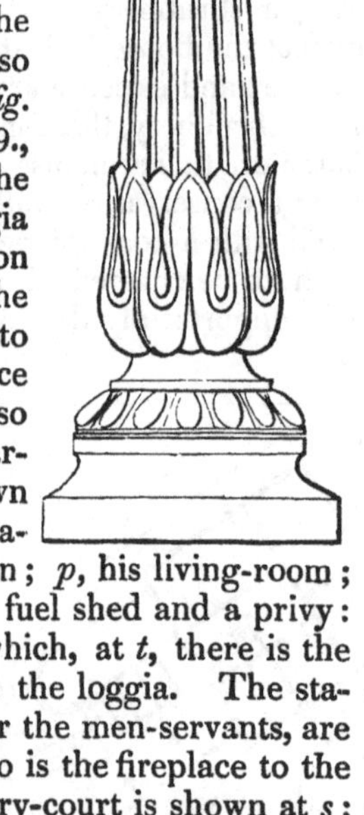
170

171

The conservative wall (*ff*) should not be a common erection, presenting only a flat perpendicular surface and a horizontal line at top: it may have piers at regular distances, terminating in caps surmounted by vases, above the height of the wall, but arranged in form and proportion, so as to harmonise with the conservatory and the house. In the case of a Gothic or Elizabethan building, these piers and their terminating ornaments should, of course, vary accordingly. Instead of piers, the face of the wall might be broken by arched recesses; and, while a more delicate kind of plant was trained against that part of the wall which formed the back of each recess, a more hardy sort might be trained against the projections between them. We have seen a wall of this sort at Genoa, on which all the recesses were covered with roses, and the piers with ivy; the effect of which was beautiful, as the roses continued in flower throughout the year. The same effect might be produced in England, by having the wall flued, and protected by matting during severe weather. Where the style was Gothic, the wall might be covered with a series of piers and intersecting arches; and, if the piers and imposts of the arches were covered with ivy, and the rest of the wall with deciduous plants, the effect, more particularly in winter, would be very striking. An excellent plan for varying such a wall is, to form the ground plan in a zigzag line, with piers at the angles; in which case, the length of each angle may be 10 ft., and the deviation from a straight line from 2 ft. to 3 ft. In going along the walk in front of such a wall, one series of angles would meet the eye; and in returning, another series. Another plan is, to have the wall straight, and a temporary or permanent roof projecting from it. In this case, if the roof were permanent, it ought to be composed of glazed sashes, which might be taken off in the summer season, and used for growing melons, leaving the pillars and rafters which supported the sashes, as fixtures; and these might be covered with rapidly growing climbing plants. Such a roof ought to extend over the walk, in order that the latter may be used during rainy weather in summer; and that, during the most severe frosts in winter, it may afford a somewhat more temperate place for taking exercise than in the open air. The most complete glazed veranda of this kind would be one where the whole of the skeleton framework, as well as the sashes, might be removed in summer, without leaving any marks to disfigure the scene, and replaced every autumn. A temporary veranda, in which the framework is to be covered with hurdles clothed with thatch, or with canvass fixed to framework or oiled paper, forms a very good protection for plants while in their dormant state; but requires to be removed much sooner in spring when they begin to grow, than a glass roof; because, when the plants begin to grow under an opaque roof, they become etiolated and blanched for want of light. In general, conservative walls should be flued, in order to give the gardener the power of assisting the ripening of the wood in autumn; and, in this case, the fireplace might be conveniently situated behind the wall, as indicated in the plan *fig.* 169., at *g*, where it is placed in the corner of the stable buildings. A conservative wall may often form one of the sides of a range of office buildings; and this is the case with a part of the wall we are now describing, which forms the side wall to the stable (*t*) and coach-house (*h*).

The conservatory, it will be observed, has a south-east aspect; while the conservative wall, and the forcing structures (*m*) in the reserve garden, front direct by south. By turning to the plans *figs.* 156. to 165., it will be seen that the distant views from the lawn front of the house, and from the walk in front of the conservative wall, are equal to any obtained from other parts of the grounds.

The situation of the circular flower-garden at *k* in *fig.* 168., and *l* in *fig.* 169., is good in every point of view. It is completely sheltered from the north and east by the conservative wall and the kitchen-garden wall; and the side next the south is open for the free admission of the sun and air throughout the year. The walk from the flower-garden along three sides of the kitchen-garden, and leading to the point *n*, is made straight, and parallel to the walls; being, as we consider it, governed in direction by the latter. The space between the wall and the walk is a border, which may either be devoted to ornamental shrubs and flowers, the wall being covered with half-hardy ligneous plants; or it may be planted with culinary vegetables, and the wall covered with fruit trees. In either case, the exterior space is supposed to be turf, varied with trees and shrubs. The subdivision of the kitchen-garden into compartments, by walks 6 ft. broad, is so obvious an arrangement as to require no remark. The walls we shall suppose to be of brick, 12 ft. high, and built hollow. The two walls which form the ends, having eaeh one side facing the south, may be flued. At *o* is a rustic building of an architectural character, so as to be ornamental, which serves as a shed in which various articles may be kept, that are required in the pleasure-ground, and partly also in the kitchen-garden. Among these are, hurdles for dividing the lawn, or hurdling off portions to be fed by sheep; portable racks, mangers, &c., for feeding sheep and lambs in winter and spring; trees which have been thinned out, and are to be cut up at a convenient season, for poles, or sticks for peas, and other uses; reed-hurdles, canvass and oil-paper frames, &c., for protecting fruit trees, and the plants on the conservative wall; and, in short, all articles required in the garden, which are too bulky or numerous to be kept in the reserve ground.

The pleasure-ground walk, from the point *n* to *p* in *fig.* 168., is to be lowered by forming an artificial winding hollow, or valley, and placing the walk in the bottom of it. At *p*, this hollow is supposed to be so much below the level of the approach road as to admit of the walk in it being carried under the road through a tunnel. Both sides of this artificial hollow are intended to be densely clothed with evergreens, to prevent any person on the walk from seeing any object on the approach, or discovering that he is at all near it; and also to prevent the sunk walk from being seen by persons driving along the approach. The direction of the tunnel under the road must be perfectly straight, in order that the light may penetrate freely through it; for, if bent so as to obstruct the direct passage of the light, it would appear dark and gloomy while entering from either end. The arch may be of brick or stone, set in cement; and it may be covered with a coating of clay, or of asphaltic mastic, so as to prevent the surface water from sinking into it. The interior of the arch, if built of stone, may have open irregular joints, in the rustic manner; and something of the same sort may be effected in brickwork, if the arch be first built in the usual manner, but with an extra half brick in thickness, and the soffit afterwards blocked out with the aid of cement so as to resemble large stones, and weather-stained so as to imitate an arch of great antiquity beginning to decay; or the bricks which form the soffit of the arch may remain of their natural colour, but be broken and stained in such a manner as to imitate great age. Both extremities of the arch may terminate in irregular masses of stone, or with vitrified brick, placed according to art, as a finish: but great taste and judgment are required, not to overdo this part of the structure. Above all things, in tunnels of this description, let no attempt ever be made to communicate a grotto-like character, by lining the arch with spars, shells, stalactites, &c., as if it were intended for a place to linger in, and contemplate these, and other grotto-like or hermitage-like, objects. In short, no attempt ought to be made to give the tunnel the appearance of being anything else than what it is (that is, an archway thrown over a sunk walk), except with reference to age. The older such an arch appears to be, provided there be no indication of insecurity or rapid decay, the less objection is likely to be made to its use, by a stranger, as a place for a walk to pass through. Whatever is very old, and at the same time very strong and secure, commands so much respect on account of these qualities, that other points are not thought of: we cannot hesitate to use a passage that has apparently been so often traversed by others before us; and that we may fancy has been only made for the convenience of exploring what lies beyond. Were such a tunnel to appear to be constructed of raw brick, or stone newly dug from the quarry, the question would arise in the mind of a stranger, why it had become necessary; and suspicions might arise as to some trick or deception about to be played off; or that some advantage had been taken in the way of making the place appear larger or better than it really is.

Supposing the subsoil to be very wet, and incapable of being drained to any great depth, then, instead of conducting the walk under the approach in a tunnel, we would carry it over the road on a bridge. In this case, the approach road should be placed in the bottom of a hollow, and the bridge should be sufficiently wide to allow of a plantation being made on each side of the road over it, so as to conceal the approach from the spectator on the walk. In short, the approach should appear rather to pass through a short tunnel than under a bridge; and the same remarks as to age, dryness, light, strength, &c., will apply, as in the case of the sunk tunnel for the walk. Many persons, in such a case as this, would propose to carry the walk over the approach road, on some kind of iron bridge; perhaps even one on the suspension principle; but such bridges, though excellent in themselves, and admirably adapted for useful purposes, are the very reverse of rural or picturesque. Their lightness does not at all harmonise with the forms of trees and shrubs, and of the rough surface of their trunks and stems; and the associations connected with iron and its manufacture are much more at variance with rural life, than those which arise from the works of the builder or carpenter, to which mankind have been accustomed from the most remote ages. Where it is desired that the grounds of a residence should appear truly rural and picturesque, we would allow of no iron gates, iron hurdles, or wire fences; much less of tree guards of iron, iron seats and benches, iron trelliswork, iron stakes for plants, &c. Where, however, a high degree of elegance and finish was to be joined to the picturesque, such as near the house, we would allow of a wire fence, to separate the lawn from the park, paddock, or sheep walk; and, for grand mansions, castles, and palaces, we would admit of highly wrought entrance gates; but this is as far as we could go. Of course, we do not object to iron in the construction of verandas, glazed structures, &c., where it assumes an architectural character, and is, besides, disguised, so as to appear like wood. Our principle is, that we object to iron, where, from the form of the article, the material of which it is composed is obvious to the eye of every spectator; and, consequently, its use interferes with rural and picturesque beauty.

From *q* in *fig.* 168. p. 457., where the walk emerges from the tunnel, it gradually rises till we arrive at the seat *r*. From this seat, which may be a rustic thatched structure, in the manner of *fig.* 172., with a wall behind, and open arches in front supported on rustic pillars, a good view of the entrance front of the house is obtained, which will have a somewhat striking effect upon a stranger, his last view of it having been the very opposite side, viz. the conservatory front.

The construction of this seat is as follows: — The whole of the ground on which it stands, being dug out to the depth of 2 ft., is filled in, to within 6 in. of the surface, with concrete; or with chalk and flints, or gravel, rammed hard, and finished with a level surface. On this are raised the bases of the rustic pillars which support the roof, which bases should each consist of a single stone; or, if it is wished to be economical, of brickwork, which will reach as high as the intended surface of the floor, and be there covered with a square or octagonal paving-tile. On these

172

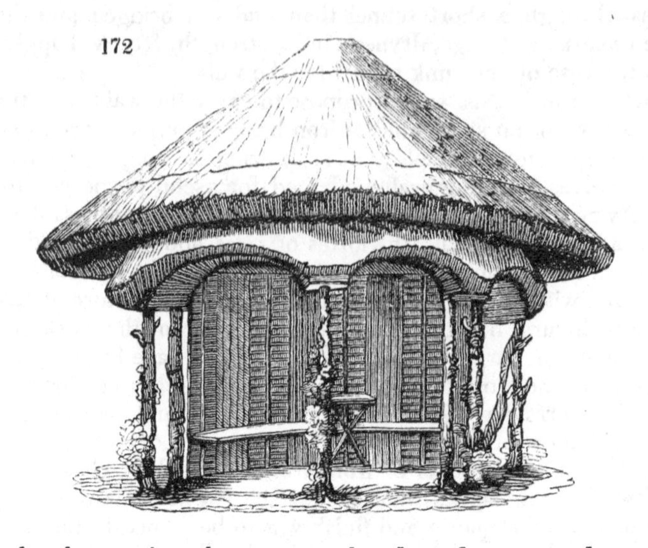

plinths the rustic columns are placed, and connected temporarily at top by strips of deal nailed to their sides. The columns, or props, for the back part of the structure, will be placed upon projecting plinths, from one continued plinth; because they are not intended to be seen on the inside, and to appear detached only an inch or two from the wall on the outside. Standards must now be raised, so as to form the skeleton for the solid part of the back wall; and this skeleton must first be covered with thin boards on both sides. This being done, the columns all round the structure are to have caps placed on them, consisting of squared blocks of wood, somewhat longer than the 9-inch tile which formed the plinth below, but of the same width. On these, skeleton arches are to be constructed, and afterwards a skeleton roof. The roof, on the inside, is to be clothed with thin boards, so as to form the interior of a cone; and on the outside with laths, so as to render the exterior of the same shape. The next step is to clothe the exterior of the cone with reeds or thatch, as in the figure. The interior being now secured from rain, the floor may be paved with pebbles, or with sections of the trunks of trees placed endwise, or with brickbats placed on end; and the interior of the back wall may be covered with strips of wood, of such kinds as have a thin bark, such as hazel, birch, &c., disposed as in the figure. The exterior of this wall may be clothed with rough bark, such as that of the oak, covered with lichens, moss, &c. The soffits of the arches should also be covered with strips of smooth-barked trees, with the bark on; as should the whole of the interior of the cone, which forms the ceiling. There remains only to fix the seat, which is a simple bench, attached to the wall behind, and with supports in front.

Fig. 173. is another design for a rustic seat of the same general character, but on a smaller scale, and more elaborately finished. The lower part of the bonnet roof, instead of being of thatch, is of strips of wood with the bark on, closely joined, so as to exclude rain. The seat is also more elaborately finished.

173

From *r*, in *fig.* 168. p. 457., the walk proceeds along a uniform surface to *s*; where there may be a Doric temple constructed in rustic work, or any other building that the proprietor may think desirable. We have mentioned a Doric temple, because it admits of the interior being fitted up as a good room. This room might be a billiard-room, or a place for occasionally drinking tea or dining in; or for a children's party, or a dance; or for stationing a band of music in, on extraordinary occasions; or it might be a place for statues, minerals, models, or other objects in which the proprietor was curious. Instead of one large room, there might be two open porticoes, back to back, with a small room between them communicating with both; in which case, the portico of the end next the house would make a good object to be seen from the windows; while from a seat in it would be obtained an excellent view of the conservatory front of the house; and the seat in the other portico, at the back, would command the interesting distant prospects indicated in *figs.* 160. to 163., p. 452, 453.

The walk is continued from *s*, in *fig.* 168. p. 457., till it joins that round the kitchen-garden at *t*, thus completing the tour of the place. It would have been easy to increase the number of walks, and to conceal them from one another by strips of evergreens in some places, and by raised mounds of turf in others; but we have preferred a comparatively simple style for laying out this place, and have indicated no more walks than what are likely, under common management, to be kept in good order. We have carefully avoided laying down any branch walks from the circumferential walk *p q r s t*, towards the approach, the house, or the flower-garden; not only because we wished to preserve the breadth of effect of the lawn (which we consider, when combined with the concealment of the boundaries by planting, as the best means for giving dignity and extent to a place), but also because we wished to keep an uninterrupted glade of turf from *d* to *x*, by *u*, *v*, and *w*, for those to take exercise in who prefer turf to gravel for walking on; and also as a place where children may learn to ride. (See the open spaces among the trees in the plan *fig.* 174.) We have shown no walk from *y* to *z*, leaving that situation as a fit place for an archery-ground, should one be thought desirable.

The width of all the walks in the pleasure-ground we propose to be 6 ft.; and we have already given directions for their execution in the best manner. (See p. 419.)

Planting. The general disposition of the trees and shrubs is exhibited in the plan *fig.* 174.; but it remains for us to indicate

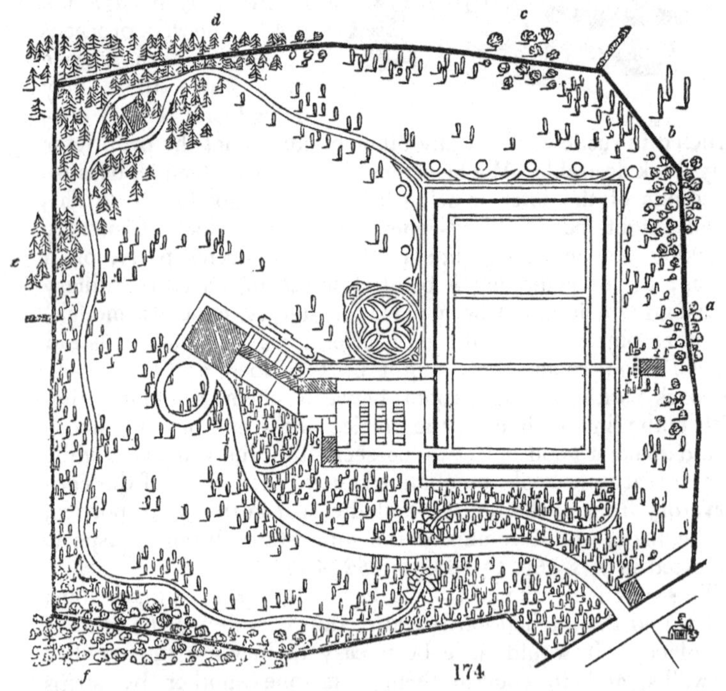

174

the distribution of the leading genera. The first point which deserves to be noticed is, that, in order to harmonise the wood within the ring-fence with the trees in the hedgerows or strips beyond it, a few trees should be planted within, of the same kinds as those without. For example, in *fig.* 174., there are elms in two narrow slips beyond the boundary at *a*; and, therefore, there must be a few elms of the same kind scattered among the other trees within. At *b*, there are some Lombardy poplars in the foreground of the exterior scene; and, therefore, two or three of the same species are planted within the boundary. At *c*, there are some sweet chestnuts without; and, consequently, some are planted in the interior. From *d* to *e*, there is an extensive plantation of larches in the exterior scenery; and, therefore, to appropriate this plantation to the newly formed residence, a good many larches must be planted in the enclosed grounds. At *f*, the exterior plantation is a natural oak wood; and, therefore, Quércus pedunculàta and Q. sessiliflòra are the prevailing trees in the interior. In some cases, it may not only be desirable to have the same sorts of trees on both sides of the ring-fence, but even to allow portions of the old hedgerows, with the trees in them, to remain in the improved grounds, in order to harmonise what is within with what is without; but this kind of treatment should never be attempted, except where its effect in appropriating the adjoining grounds, as well as in harmonising the general landscape as a whole, will be such as to silence all objections. All the trees introduced for the purpose of harmonising the interior of the ring-fence with the scenery without, in such a case as that before us, need not exceed 50 or 60; and it will most commonly happen that these may already exist in the condemned hedgerows, or in scattered groups in the fields, as is the case, to a considerable extent, in the plan *fig.* 154. p. 448. The trees in the old hedgerows which it may be thought advisable to save, will only want to be freed from the thorns and other hedge plants (except a few left to group with them), and to have the ground about them levelled down and smoothed. The groups already existing may either be thinned out or added to, as may be found requisite.

It may be here observed, that, in the geometric style of laying out grounds, the mode of harmonising a residence with the surrounding country was, by the projection into it of a continuation of those avenues and lines of trees which formed part of the residence. This implied the possession of the grounds beyond the park fence; and, when this is the case in the modern style, a few groups without the boundary fence, of the same kind of trees as those within, will effect the same object. This used to be frequently done when a park or parks were surrounded by a common: and Hounslow Heath was formerly clumped, for the sake of the adjoining country seats of Syon, Sion Hill, &c.; as Cobham Common was, to appropriate it, in the eye of exclusive taste, to Claremont and Esher.

All the foreign trees and shrubs that are to be introduced in order to complete this residence, we intend to plant without fences, and to place at such distances as that they will not require any thinning for at least ten years. The ground on which these are to be planted is supposed to be thoroughly trenched and drained, and even, where necessary, manured, before planting it; so that the trees and shrubs cannot fail to thrive rapidly, and, in the climate of London, to attain the size of the trees ten years planted of which portraits are given in our *Arboretum Britannicum*, and as will be indicated, also, in the catalogue of trees and shrubs which will be given at the end of this work. During the ten years' growth of these trees, we would not allow any grass or weeds to grow within several feet of their stems; but, at the same time, we would not dig the surface, but only hoe it, to destroy the weeds, and fork up the soil to the depth of 3 or 4 inches, to render it a non-conductor of heat and moisture, and to admit air to the fibrous roots which rise up near the surface. We would not, during the same period, prune off any of the side branches, except such as had begun to decay; leaving each tree and shrub to assume its natural shape.

Some will imagine that a plantation in which the trees are at such a distance apart will appear thin; and this, we allow, will be the case for two or three years at first; but in the fourth year, in consequence of the lateral branches extending themselves, there will be an appearance of richness and massiveness in such a plantation, which plantations as thick as they are generally made never acquire at any age. There is not a greater mistake, nor one more commonly made by practical gardeners, no less than by amateurs, than that of planting thick, with a view to producing a dense mass of foliage that shall not be seen through; and which shall thus become a screen to objects which it is desired to conceal. On the contrary, the true and only mode of producing a dense mass of trees or shrubs is, to plant so thinly as to admit of each tree or shrub becoming clothed with branches from the ground upwards. A plantation of this kind, of only two trees or shrubs in depth (the plants being inserted alternately), will, as soon as their branches touch, form an effectual screen; whereas a plantation of scores of trees in depth, if the plants are so close together as to draw one another up, and leave the stems without side branches, will be every where seen through. Any one may have a proof of this in the strips of plantations along road sides, made to shut out the public road from gentlemen's parks or pleasure-grounds. At the distance at which the house is placed, the road is concealed from its windows by the heads of the trees; but the spectator riding along the road sees through between their stems without any difficulty. The immense masses of wood in Kensington Gardens being without undergrowth, and never having been thinned, are seen through in every direction, though some of them are a thousand feet in depth; and thus (now that the old yew hedges planted by London and Wise have been cut down) there is not a single space on which the imagination can rest throughout the whole of these gardens. A plantation which is thinly planted has this other great advantage; viz. that, while it cannot be seen through, it can be seen into: its margin, instead of being a line of naked stems, forms a succession of prominences and recesses, each varied more or less in form, and in light and shade; and thus constituting a rich and varied boundary, instead of a meagre and monotonous one. There is scarcely any point which we are more anxious to impress on the minds of our readers, than the necessity of planting trees and shrubs thinly, and of thinning out afterwards, as the trees advance in growth, so as, in general, to keep them clothed from the ground upwards; and always to do so when the object is concealment.

With respect to the distribution of the kinds of trees, setting out from the natural oak wood (*f* in *fig.* 174. in p. 468.), we would introduce, from that point to the entrance lodge, and along both sides of the approach road from the entrance lodge to the house, all the different species and varieties of *Quércus* that would grow freely in the given locality. All the evergreen varieties of *Quércus Cérris* we would make use of over the tunnel, and on both sides of the hollow of which the tunnel walk forms the bottom; and all the other evergreen oaks, such as *Quércus I'lex*, the cork tree, &c., we would introduce next the house; partly because these species will not grow so high as to overpower it, and partly to vary the appearance of the domestic offices. As secondary trees, we would introduce along the approach the maples and sycamores, which will contrast well with the oaks with which they are mingled, from the earliness of their foliage and flowers, and from the interesting nature of both in many species. Along with the oaks and acers, we would also plant a few spring-flowering trees, such as the ornmental crab trees, laburnums, and amelanchiers, to add to the gaiety of the scene at that season; some summer-flowering trees, such as the different kinds of horsechestnut, pavias, and robinias, to give beauty during summer; and some of the autumn showy fruit-bearing kinds, such as the large-fruited thorns, *Sórbus hýbrida*, &c.; so as to produce a rich appearance at that season. We would introduce no shrubs along this approach, except such as the holly, the yew, and the box, which, as they grow up, take the character of low trees. The box trees should be most numerous near the house; and the hollies and yews over the tunnel and along the banks of the sunk walk. None of these trees and shrubs should be planted nearer to one another, or to the road, than from 10 ft. to 15 ft.; but the common and Turkey oaks may be as much as 20 or 30 feet apart at planting; and no nurse trees whatever ought to be planted along with them, however bleak the situation, for the following reasons, and others already given in p. 433. The ground being trenched and in good heart, and having neither weeds nor nurses to extract the nourishment from the soil, the trees will grow with extraordinary vigour; and, being exposed to the light, air, and wind, on every side, will become bushy vigorous plants, capable of resisting violent winds, from whatever direction they may come. If drawn up by nurses, they will attain a given height in a shorter period; but, being deprived by the nurse of a great part of the nutriment contained in the soil, and also precluded by the same false friend from enjoying the light and air on the sides, they will be without lateral strength or beauty. Hence, a long period of stagnation occurs in all plantations that have been forced forward by nurses, as soon as these are withdrawn, and the trees are left to be weaned by time and the weather.

In the south-east angle, where accident has placed a number of larches, we would introduce all the species of the pine and fir tribe that could be admitted in a place of this extent; distributing them along the southern boundary, among a few deciduous trees and shrubs, for the sake of harmony, and also along the eastern boundary, in the same manner, and for the same reason. The trees that are scattered on the lawn, we would have chiefly of the low or middling-sized kinds, in order that they might not interfere with the dignity of the house. Among these we would include all the species and varieties of tree thorns, all the crabs, a number of fruit trees of the kinds most ornamental in blossom and in fruit; and, in short, all the trees belonging to the order *Rosàceæ*. In this way, we would, in this residence of ten acres, include every species and variety of tree and shrub that would endure the open air in the given climate; being careful to admit only a few of such kinds as grow very rapidly, or to a great height; such as some species of poplar and willow, and some varieties of the English elm. Of these very rapid-growing lofty trees, there are about a score which it might, perhaps, be as well to omit altogether in a place like this, of small extent, and nearly flat. With respect to the distribution of all the other trees, we have already (p. 432.) laid down the general principles; and it ought to be the business of the landscape-gardener, in a case like the present, to mark the situation of every particular tree and shrub, with its name, on a ground plan; but such a plan (of which a specimen is given in fig. 75. in p. 254.) would obviously, in the present case, be much too large for our pages. We cannot, however, too strongly impress it on the mind of the reader, that a working-plan, indicating the precise situation of every tree and shrub, is essential in every small place, where the object is to lay it out to the greatest advantage. Even an experienced landscape-gardener, when laying out a small villa, without such a guide, will be liable to err, by admitting too many of one species or variety, and too few, perhaps, of another; by omitting some altogether; and by planting others in wrong situations. In short, the mature consideration which is required for the landscape-gardener, before he can make a proper working plan for the plantations, is one of the greatest securities, not only to the proprietor, but to the artist himself, for the maintenance of due proportion among the kinds of trees employed, and for the general taste of the place, as far as trees are concerned. Another advantage of having the situation of every particular tree and shrub marked on the plan, with a corresponding list of their names, is, that no nurseryman or jobbing gardener, who may be employed to execute the planting, can have any pretence for sending in more trees, or trees and shrubs of other sorts, than are indicated in the plan; unless, indeed, as is often the case, the nurseryman or jobbing gardener pretends to improve the plan of the landscape-gardener, and having, by dint of perseverance and talking, got the ear of his employer, the latter is prevailed on, for quiet's sake, to yield to the proposed alterations, and to admit trees and shrubs in such quantities as, in some cases, entirely to destroy the effect which the landscape-gardener intended to produce. It is well known that, in all new plantations, especially in those made in the grounds of small suburban residences, the plants are now generally put in so thick as almost to touch each other, or at the rate of six or eight thousand plants per acre, even of trees alone; whereas, according to the mode of planting which we recommend, the number per acre, when trees alone are employed, will scarcely ever amount to a thousand; and, when shrubs alone are planted, to not more than between two and three thousand, according to the small or large size which the kinds will attain when fully grown, or the rapidity of their growth when young. Another advantage of this mode of wide planting is, that no thinning will be required for several years afterwards: and every landscape-gardener knows that the effect of the plantations, in nine tenths of newly made places, is most materially injured by the neglect which generally takes place in thinning. When thinning is neglected after the branches of the trees touch each other, the plants are drawn up as if they were in a nursery. In a few years, the more tender kinds are choked; and the coarser kinds, filling up the space thus left, are, in their turn, drawn up; so that, at the end of fifteen or twenty years, the whole presents a mass of naked stems, with diminutive tops: and, if thinning is then had recourse to, the results are hideous, at first, in regard to effect; and, after one or two seasons, from the wind and weather being admitted where they never were before, they are destructive to the trees; which either gradually decay, and at last die standing, or are blown down by the first violent storm of wind.

The planting of the kitchen-garden with fruit trees and shrubs may be considered as having been in part treated of in preceding pages; and it will be given systematically when we come to the subject of suburban horticulture.

Execution. All that we have said hitherto may be considered as only committed to paper in the form of plans, and a report, for the consideration of the proprietor and his family. When these are approved of, the next step is to carry them into execution. This is sometimes done by contract, and sometimes by the proprietor employing his own workmen, under the direction of a competent manager; but, most frequently, partly in one way and partly in the other. The house, offices, lodge, garden walls, and, in general, all that belongs to architecture, may be done by contract, provided a respectable and responsible builder is engaged as the contractor, and not beaten down to the lowest price by competition. The architect who designed the buildings should, of course, have the general inspection of the work as it is going on; and there should be a clause in the agreement between the contractor and the proprietor, that alterations or deviations from the plan may be made according to separate and specific agreements, without invalidating the general contract entered into at first. Many persons, from good motives, or from a view to economy, employ a carpenter or builder in a small way, whom they wish to encourage, to carry buildings into execution; but there cannot be a greater error. Tradesmen in a small way of business are generally deficient in capital; and, not having a stock of seasoned materials by them, they never can do justice to the work. The charges of such persons are, also, very frequently higher than those of first-rate builders. We could give the history of many houses, in the neighbourhood of London, which have cost from 500*l.* to between 30,000*l.* and 40,000*l.*, and have been built in this manner; and for doing which the owners have bitterly repented ever afterwards. We know one gentleman who, on his own judgment, aided by that of his carpenter and bricklayer, laid out upwards of 40,000*l.* in endeavouring to execute a plan for a mansion, received from a first-rate architect; but, in doing this, not thoroughly understanding the construction of a certain dome over a saloon, it has become depressed at one part of the sides, and admits rain at the haunches of the arch. Another gentleman commenced the building of a house on the day-work system, which, if it be ever completed, will cost him three times more than he ever contemplated laying out. B., a wealthy merchant about to retire, employed, to build a country house, a very worthy carpenter, who had married his wife's maid, and also had become a master in a small way: but, whether from not having supplied him regularly with ready money, or from some other cause with which we are unacquainted, certain it is, that unseasoned timber was used in the partitions, roof, and floors; and a very warm summer, that of 1826, happening soon after the house was finished and taken possession of, the whole of the partitions shrank and twisted to such a degree, as to produce large rents in the plaster. The carpenter endeavoured to persuade his employer that the foundations of the walls had given way; but this was too palpable an absurdity to be credited by any one. The rents in the plaster of the partitions were filled up with putty in some places, and with stucco in others; but they are still conspicuous, and must necessarily remain so till the lath and plaster are stripped off, and the stud-work reclothed. The whole of the boarded floors in this house shrank so much, that they were obliged to be twice taken up and relaid; and all the ceilings are cracked. Another merchant in a smaller way, a few years ago, built a house in the country, which cost him 2000*l.*, and employed a very respectable jobbing carpenter that he had confidence in, from having been long accustomed to employ him in petty jobs in town: but confidence is often the result of habit, want of enquiry, or indolence; and this confidence may be deserved by an individual in one point, or in several, and yet not be applicable to all that that individual will, or is even entitled to, undertake. In this case, the London carpenter and joiner, who could procure whatever credit was wanting for the execution of little jobs that he executed from time to time, under the immediate eye, it may be said, of his timber-merchant, could not so readily do this in the case of a more extensive contract for an erection in the country, where he had never before been employed; and where, as his timber-merchant well knew, if his employer did not, he had to purchase his experience, and that necessarily at the expense of others, from having himself nothing to lose. Whatever may have been the cause or causes, chalk lime, instead of stone lime, was employed for the outside walls, and unseasoned timber for the carpentry of the house, as well as for the joinery. The joints of the brickwork were soon obliged to be raked out, and filled in with tuck and puck; and there is not a single door in the house that has not shrunk and twisted; nor a window-sash that does not rattle in the frames with the slightest breeze. It is true there is no want of ventilation in this house, and there is not a single chimney that does not draw well; but the causes which produce these effects render the rooms so cold in winter, that they cannot be thoroughly heated by the largest fires. In all the larger windows, the panes have been cracked by the twisting of the sash-bars; and it is needless to say that the partitions and the floors are warped and rent, as in the preceding case. To employ first-rate workmen, and supply them with ready money for purchasing all the materials, is a mode often resorted to, under the idea of economy; but, however excellent and honest the workmen employed may be, and however well they may be acquainted with the various details of house-building, they must necessarily be without the experience of the master-builder, who has built several houses of different sizes; and thus be liable to be wrong in judging of what will suit. The persons selling the materials will also naturally keep their best articles for their best customers; and thus the purchaser of small quantities will not be able to get his materials either so good or so cheap. We acknowledge that it seems hard, to lay down rules which are calculated to prevent the mechanic or tradesman who has no capital from ever bettering his condition; but we cannot help it. We merely state facts, and the opinions which we have formed on them. In another house, built under somewhat similar circumstances, in addition to these evils, more or less aggravated or alleviated, in consequence of employing an ignorant plumber, all the lead on the roof of the house, which was too thin, and was, consequently, raised, cracked, and twisted by the heat of the sun, was obliged to be removed, and replaced by lead of a thicker quality at the end of the third summer; and all the water-closets were obliged to be altered. One of these, a self-acting patent one, which cost double the usual expense, was rendered useless by the plumber's attempting to improve its construction, while setting it up. On enquiry into the history of this plumber, it was found that he had never been regularly brought up to the business, but that, having failed in a totally different trade, in another part of London, he had set up as a plumber on the faith of public ignorance. In this manner many persons about the metropolis set up as petty masters in businesses which they have never been regularly taught. Every one knows that the occupations of coal-merchant, wine-merchant, publican, and coffeehouse-keeper, are universal resources to persons not regularly brought up to any trade; but those of petty builders, plumbers, glaziers, bricklayers, upholsterers, and a great variety of other trades and professions, are also the resources of persons who have been unsuccessful in their original pursuits. To these examples of bungled houses, we could add hundreds of other cases, some of which would hardly be credited by our readers, which have come under our observation during a period of nearly forty years, during all which time our attention has been constantly directed to the subjects of building, planting, and gardening. We shall only add, that our decided opinion is, that, whether a shed or a mansion is to be erected, the cheapest mode, as well as the best, is to employ a first-rate builder.

The principal planting about the place, including the trenching and levelling of the ground, the formation of the approach and walks, and also the laying out of the kitchen-garden, may be executed by contract, by a respectable nurseryman. In general, however, whether in small or large places, the work will be more carefully done, and not at much more expense, by employing a first-rate head gardener, who is to be continued afterwards, to manage the garden; and allowing him to execute the work, partly by letting it out in small jobs, and partly by day-work. Trenching, digging drains, excavating for roads or walks, and even smoothing and levelling, may all be done by the job; and, wherever work can be thus executed, it is by far the cheapest and best mode for the proprietor, and the most agreeable to the workmen. Much, however, in this case, depends on the knowledge and experience of the head gardener in letting out the work; and his attention in seeing it properly executed.

Whether the buildings or the plantations ought to be commenced first, is a question which admits of being variously answered, according to circumstances. Where the place is extensive, and much planting, wholly detached from buildings, is required, all the plantations may be commenced two or three years before the buildings; but, in the case of a place like that before us, where the principal part of the planting is connected with the house and offices, or the kitchen-garden, or borders the approach road, all the buildings ought to be completed, at least exteriorly, before even the ground is levelled or trenched for planting: even the cartage of materials to the house along the approach road will endanger young trees bordering on it; and, therefore, it is much better not to plant them till the road is made.

The first actual step in executing the buildings, more particularly the house and offices, is to erect a temporary shed for the use of the carpenter and other workmen, both as a place for working in, and to enable them to lock up their tools, &c.; the next is to mark out the foundations, and to have them excavated; and the third, which ought to go on at the same time as the first and second, is the procuring and laying down materials. It may be worthy of notice, that the first and the last workman connected with erecting any building is the carpenter. He is required to construct skeleton squares or triangles, before the ground is broken for the foundations, or a single brick or stone can be laid, in order that the angles of the foundations may be truly correct; and, as the building proceeds, he is employed in preparing and placing the flooring, roofing, doors, windows, and interior fittings, to the last step in joinery, viz. that of putting the finger-plates on the doors, after the painting and papering have been finished. Hence it is that carpenters (including under this term joiners) know more about the detail of executing buildings, than any other person connected with them, not even excepting the architect; and hence, also, the carpenter, other circumstances being the same, will invariably be found the most intelligent mechanic on an estate, as connected with construction; as the gardener will generally be found the most intelligent, as connected with cultivation. If the proprietor of a country residence, therefore, find himself reduced so as to be only able to keep two men to look after it, these two should be the carpenter and the gardener. Remarks of this kind may, perhaps, be considered by some out of place; but, this book being intended for those who have little or no previous knowledge of the subjects of which it treats, we think them likely to be useful; and it is consistent with our plan to introduce them incidentally rather than systematically.

The walks, situation of the trees, &c., are easily marked out on the ground, by driving stakes or pins in the direction of the lines representing them in the plan. This is rendered easy by the squares already marked out and indicated on the ground, and by stakes, with numbers on them, having been driven in at the angles formed by the intersecting lines of the squares. If these squares should not have been formed, or if the stakes indicating them should have been taken up, or partly misplaced, then the squares ought to be laid out afresh, remembering to keep the imaginary lines which form their sides correctly north and south, and east and west.

Estimate of Expense. As we have given no detailed plan of the house and offices, no special estimate can be given of their cost; but we may state that a very capital house, offices, conservatory, &c., for such a property, may be erected in the neighbourhood of London, of brick, and covered with cement, for from 7000*l.* to 10,000*l.* The garden, including the walls, the structures in the forcing-ground, the gardener's house, &c., may cost 1000*l.* more; and the tunnel, the walks, the plantations, and the ornamental buildings, from 1000*l.* to 3000*l.* additional. In short, such a residence, in the neighbourhood of London, may be considered as costing from 10,000*l.* to 15,000*l.*; to which is to be added the purchase of the ground, which will vary from 200*l.* to 300*l.* per acre, if in the most eligible situation. In some places, the ground will not cost half this sum, from its not being in a fashionable neighbourhood, or not being near a good road, &c. It is almost unnecessary to observe that a family, to occupy such a residence with propriety and dignity, would require a clear income of at least 2000*l.* a year. But, though from 10,000*l.* to 15,000*l.* would be required to complete a very handsome residence, and an income of 2000*l.* a year to occupy and keep it up properly; yet a very good house, garden, &c., might be erected for 2000*l.*, in which case, the residence might be consistently occupied by a person having from 500*l.* to 1000*l.* a year.

As we have strongly recommended professional men to be consulted previously not only to building and planting, but even to purchasing a property for the purpose of forming a residence, it may be useful to give some idea of the charges which surveyors, architects, and landscape-gardeners make for giving their advice. An opinion, or a visit, may be obtained at from one to five guineas, according to the reputation of the party employed; exclusive of travelling expenses, if the artist should be required to leave his home to look at the premises. If, after seeing the grounds, or the house, or both, he finds it necessary to make a written report, the charge for that report would be one, two, or three fees more; that is, at the utmost, from three to fifteen guineas. If plans are to be made, the charge will depend on their nature and extent; always bearing in mind that the difference in the price charged for plans is as different as the price charged for single visits; and that the general rule by which the artist charges for his plans or his reports is, the time which he is occupied in making them. Hence, a plan or report, from an artist whose charge is five guineas a day, would cost five times as much as the same plan or report from an artist who charged a guinea a day. It was formerly the general custom for architects, and it is still so in a considerable degree, instead of charging for their plans, to charge a per-centage on the amount of the work done, as a remuneration both for the plans and for superintending their execution; but this may truly be called an absurd mode of remuneration, since it implies no available responsibility on the part of the architect, and it makes no distinction between the skill required for the designs and working-drawings of the most elaborate mansion or villa, and the merest accumulation of bricks and mortar, in a manufactory or barracks; and, at the same time that it overpays the architect for a palace or church, it does not pay him adequately for a cottage. It used also to be the custom formerly for the architect not only to receive a per-centage from his employer, but also another per-centage from the builder, or the different tradesmen employed in executing the work from his plans; so that, instead of 5 per cent, the nominal remuneration usually paid by the employer to the architect, he was often in the actual receipt of 10 or 15 per cent. These extravagant and anomalous charges formed one of the causes which have led to the establishment of large building companies, who at once give designs, and carry their plans into execution, by contract; not only without charging any per-centage, but at the risk of loss (which is serious, they having a real capital to lose), if they should have deceived themselves in forming their estimates. The large sums paid to architects, in consequence of the mode of remunerating them by a commission on the amount of the work, is one of the grand causes which has led to the retardation of architecture in Britain. It has deterred persons from employing regular architects, and either prevented them from building and improving at all, or compelled them to have recourse to builders, carpenters, and persons of inferior taste and knowledge. At the present time, when this kind of monopoly has been partly broken up, and architects, in many instances, charge by the day, or at so much per job, twenty are employed for one that was consulted thirty years ago, and a proportionate number of handsome buildings have been erected all over the country. We certainly would not employ a builder to make a design for anything we intended to erect, however eminent he might be; because the designs from such a source, being generally prepared in a routine manner in the builder's office, are frequently monotonous, and of little value with reference to the taste displayed. We would employ an architect whom we believed to be a man of genius; pay him handsomely for his designs; and have them carried into execution, under his occasional inspection, by a first-rate builder.

Management. As data for the management of a place such as that we have decribed, there will be the plan of the drains from the house and offices, left by the architect, as suggested in p. 419.; and the map of the plantations, with the drains through the lawn, and along the walks and approach, and the report, &c., made by the landscape-gardener. The latter will show when thinning must be commenced, and to what extent it ought to be carried in order to produce the effect contemplated by the artist. It will always be necessary, in such a place, to have a head gardener who understands his business well; and he ought to have two, three, or more men under him, according to what the proportion of the lawn which is to be mown is to that which is to be covered with trees (and, consequently, only hoed to keep down the weeds), and to that which is to be mown for hay, or hurdled off and fed by sheep.

Remarks. Let not the reader suppose that, in order to lay out a place of ten acres, it is necessary to have all the plans, sections, vignettes, reports, &c., which have been here given, or spoken of: on the contrary, a single working-plan, such as fig. 174. in p. 468., without any of the vignettes, or underground sections, will generally be enough; but, in that case, the plan must be the work of an artist, whose mind has been employed in studying the situation, and the kind of residence to be produced, under all the different points of view which we have treated of, and exhibited in the other plans; and the person employed to execute it must be a gardener well acquainted with his profession, and of some experience. Our object in going so much into detail has been, to show what a great variety of points require consideration in laying out grounds; and that, to do this as it ought to be done, not a single step ought to be taken without the artist being able to assign a reason for it: and, also, to show how a person who has a property, or has purchased ground in a remote part of the country, may have such plans and views of it and the surrounding scenery made by a local artist, as will enable him to get the advice of a landscape-gardener at a distance, without incurring the expense of his travelling to the property. If we can induce the reader to enter somewhat into our views on the subject, it will not only assist him in choosing a residence, and laying one out, or in improving one already existing, but it will enable him to derive a great deal of pleasure in studying the defects and beauties, and their causes, in whatever residence may come before him, and give him an intellectual interest in residences generally. One error we must guard the reader against; viz. that of supposing that a small place of ten acres, like that which we have been treating of, requires proportionately less skill, care, and attention than a place of a hundred or a thousand acres: the comparison does not hold in the slightest possible degree. Every essential component part, whether ornamental or useful, is required in a residence of ten acres, or even one acre, that is found in a residence of a hundred, or even a thousand, acres. The difference consists merely in the greater magnitude of the parts in the larger residence, and in some of the parts being carried more into detail. Hence, the professional charges for laying out a small place will always be much greater, in proportion to its extent, than those for laying out a large place.

Hendon Rectory. (*figs.* 175. to 179.) — This residence is selected in order to show what may be effected on a very small spot by the choice of trees and shrubs of a superior description, by the distribution of green-house plants in tubs and pots, and also by combining the gardenesque with the picturesque. There is nothing remarkable in the art or taste displayed in laying out this place; that having been done before the present occupier, the Rev. Theodore Williams, had acquired a taste for botany and gardening. On the other hand, the selection of the plants grown in pots, boxes, and vases, and their disposition on the lawn; the kinds of trees and shrubs planted in the masses and groups; and the manner in which these are managed; display the greatest taste, and a degree of care and high keeping in the management, which is very rarely to be met with in either small or large gardens. To give some idea of the extent to which this high keeping is carried, we may mention that, though the whole space occupied by the garden and pleasure-ground is only about an acre and a half, yet several gardeners are kept; the head gardener, Mr. Lawrence, is a man who ranks high in his profession, and his foreman is also a very superior cultivator.

Mr. Williams, considering that, in all works of art, and in all natural objects which are to be examined singly, one of the greatest beauties is symmetry, has those trees and shrubs which he manages in a gardenesque manner brought into the most perfectly symmetrical forms, by tying the branches up or down, inwards or outwards, as may be necessary, with small almost invisible copper wire; by which means, not only every plant in a tub or a pot is perfectly symmetrical, whatsoever be its form, but those trees and shrubs which stand singly on the lawn, or compose gardenesque masses, are individually so treated; and, standing as they do a few inches apart from each other, the separate shape of each plant is seen by the spectator. The same care is bestowed on the dahlias, which are here grown in large quantities, and of sorts most of which were raised under the direction of Mr. Williams, from seeds saved in his own garden.

That which renders Hendon Rectory altogether unique in a gardening point of view is, a collection of Coníferæ in pots; as remarkable in its way as the collection of fruit trees in pots which was maintained for so many years, for the purpose of proving and describing the kinds of fruit, by that celebrated pomologist, Dr. Diel of Nassau-Dietz, the author of *Obst-Orangerie in Scherben.* These Coníferæ are in part set out on the lawn in the summer season, and in part kept under glass; and all of them are trained into the most beautifully symmetrical shapes that are any where to be seen. As the pine and fir tribe is liable to be attacked by insects in the summer season, it is the business of one gardener to attend entirely to them and to the Cuprėssinæ; in other words, to the pines, firs, cedars, araucarias, dammaras, cypresses, dacrydiums, junipers, and arbor vitæs, in pots. They are thus kept regularly watered, accurately tied into shape, and perfectly free from insects. Some of the plants of this kind at Hendon Rectory are of great value; one, a dacrydium, in particular, is matchless for its size, beauty, and rarity. The same plan of dividing the labour of the place is adopted with reference to the dahlias, which, from the day they are planted out till the time the roots are taken out of the ground, are constantly under the care of one individual. Another man is solely occupied in propagating by cuttings or otherwise; and one is kept as a man of all work, to assist the others, and to look after the walks. The duty of the head gardener is to see that the rest perform the several works assigned to them, and, in general, to be careful that the whole is kept in perfect health and order.

The fence which separates the grounds of Hendon Rectory from the road which leads from Hendon to Mill Hill, is of oak pales; and the main entrance is through a door, also of oak. Besides the garden, the property consists of a grass field, of 3 or 4 acres, surrounded by an irregular hedge with oaks and elms, which harmonises so well with the adjoining fields similarly enclosed, that the limits of the property are no way discernible. The details of the plan (*fig.* 176.) are as under, the botanic names being those used by Mr. Lawrence: —

a, Principal entrance. To the left is shown a small path in an ogee direction leading to a door, opening into the court of offices (*m*).

b, A point, from which the view *fig.* 175. is obtained.

View at Hendon Rectory. 175

c, A point, where the spectator, having his back to the house, sees before him a narrow strip of lawn, with handsome symmetrical plants of the following kinds: — Next the entrance door, Taxòdium dístichum; then *Sophòra* japónica péndula; next, *P*ìnus *Mùghus*, Dáhl*ia*, Taxòdium dístichum nùtans, Dáhl*ia*, *P*ìnus rígida, Taxòdium dístichum pàtens. Beyond this, there is a row of dwarf hybrid rhododendrons, as a margin to a bank of common laurel, cut smooth above, with standard roses, and other trees, all cut into symmetrical roundish forms, rising through it, as seen in *fig.* 175., which forms a very singular phalanx of objects, and serves to occupy the mind of the spectator, and prevent his recollecting that he is so very near the boundary and the public road. Turning round, with the face towards the house, a number of rare and beautiful plants are displayed on a rockwork com-

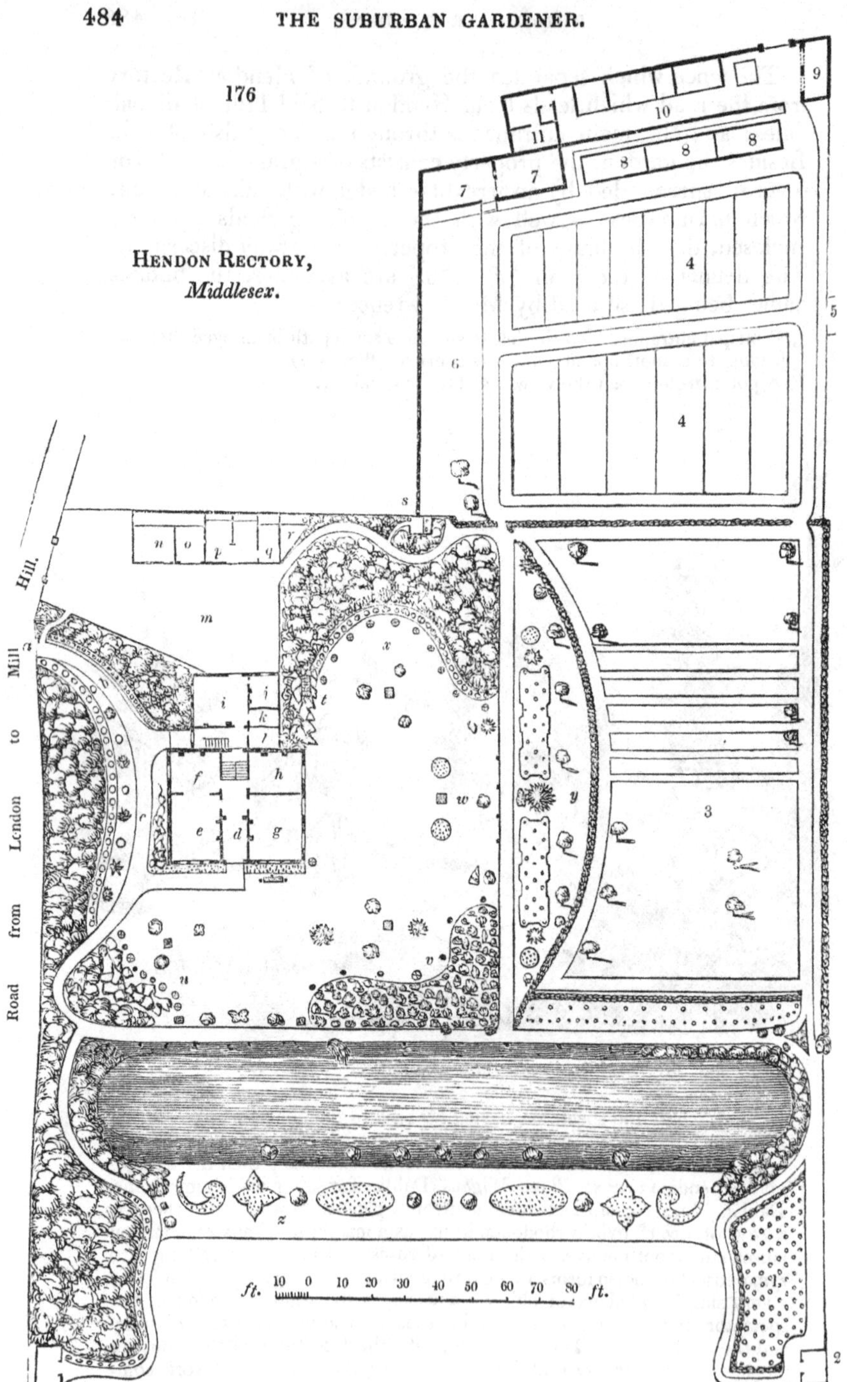
176
HENDON RECTORY,
Middlesex.
Road from London to Mill Hill.
ft. 10 0 10 20 30 40 50 60 70 80 ft.

posed entirely of crystallised spar. The walk turns round to the entrance to the house; which circumstance corresponds well with its cottage and unobtrusive character, and is, in reality, as we think, in better taste than if the walk had been conducted to the main entrance, with a sweep like that of a carriage approach.

d, The entrance lobby of the house. *e*, The dining-room.

f, The library, which opens into the dining-room with folding doors; both rooms being intended for books. *g*, The drawingroom.

h, Study. *i*, Kitchen. *j*, Back-kitchen. *k*, Pantry. *l*, Dairy.

m, Court, common to the kitchen and stables, with folding gates to the public road. *n*, Justice-room. *o*, Coach-house. *p*, Two-stalled stable.

q, Harness-room. *r*, Dust-hole.

s, Servants' privy, entering from a winding walk, which passes from the stable court to the garden.

t, Situation where there is a handsome Gothic aviary partially concealed by bushes, and containing a good collection of birds; Mr. Theodore Williams, jun., being much attached to the study of ornithology.

u, A point round which there is a constellation of rare and beautiful trees and shrubs in pots, besides various statuary and sculpturesque objects. Among the hardy trees are, Photínia serrulàta and *P*ìnus Pallas*iàna*; and among the green-house plants, in pots and vases, are, oranges, myrtles, fuchsias, tree rhododendrons, &c. The view from this spot, looking towards *v*, is indicated in *fig*. 177.

v, A gardenesque plantation, in which every tree and shrub is kept distinct, and every one trained into a symmetrical shape. The mass gradually rises from the height of 2 ft. round the margin, to the middle, which is 7 or 8 feet high. Among the plants in this gardenesque mass are the following:—

*Magnol*iàceæ.

Magnòl*ia* tripétala, 12 ft. high, and 6 ft. in diameter.

M. obovàta.

M. acuminàta, 9 ft. high, and 8 ft. in diameter; in bog soil and silver sand.

M. glaúca, 7 ft. high, and 3½ ft. in diameter; in bog and loam.

M. macrophýlla, 5½ ft. high, and 1½ ft. in diameter; in bog and loam.

M. Thompson*iàna*, 14 ft. high, and 6 ft. in diameter; in loam, bog, and silver sand.

M. exoniénsis, 5½ ft. high, and 3½ ft. in diameter; in bog, loam, and sand.

*Winter*àceæ.

Illícium floridànum, 3½ ft. high, and half a foot in diameter; in bog and silver sand.

Berberàceæ.

Bérberis fasciculàris, 3 ft. high, and 2 ft. wide; in loam, bog, and silver sand.

Pittosporàceæ.

Pittósporum *Tobìra*, 3 ft. high, and 3½ ft. wide; in loam, bog, and sand.

*Ternströmi*àceæ.

Camèll*ia* japónica quadrangulàris, 6½ ft. high, and 3½ ft. in diameter; soil half loam, and half bog and silver sand.

C. double red, 5½ ft. high, and 4 ft. wide; half loam, and half bog and silver sand.

C. double white, 3½ ft. high, and 2½ ft. wide; soil half loam, and half bog and silver sand.

C. single, 5 ft. high, and 4 ft. wide, in loam, dung, and grit.

Thèa víridis, 1½ ft. high and 2½ ft. wide.

Æsculàceæ.

Pàv*ia* rùbra, 15 ft. high, and 9 ft. wide; in loam, dung, and grit.

A*quifoliàceæ.*

I'lex, yellow painted holly, 5½ ft. high, and 3½ ft. wide; in loam, dung, and grit.

I. white-margined holly, 5 ft. high, and 5 ft. wide; in loam.

I. Cassìne, 5 ft. high, and 4 ft. wide; in loam and grit.

I. milkmaid holly, 6 ft. high, and 3 ft. wide; in loam, dung, and grit.

I. angustifòlia, 7 ft. high, and 3½ ft. wide; and *I.* a. scótica, 6½ ft. high, and 3½ ft. wide; in loam, bog, and silver sand.

177

View at Hendon Rectory.

I. Dahoón, 7 ft. high, and 3½ ft. wide; in loam, bog, and siver sand.
I. Peràdo 6½ ft. high, and 3 ft. wide; in loam, bog, and silver sand.

R*hamnàceæ.*

*R*hámnus *A*latérnus fòliis argénteis, 7 ft. high, and 4 ft. wide; in loam, bog, and silver sand.

Leguminòsæ.

*C*ýtisus pygmæ'a, 6½ ft. high, and 1½ ft. wide; in loam and bog.
Halimodéndron argénteum, 7½ ft. high, and 3½ ft. wide; in loam and bog.

R*osàceæ.*

Photínia serrulàta, 12½ ft. high, and 5½ ft. wide; in loam, bog, and silver sand.

Araliàceæ.

*H*édera arboréscens, 4½ ft. high, and 4 ft. wide; in loam and dung.

C*ornàceæ.*

*V*ibúrnum *T*ìnus, 6 ft. high, and 4 ft. wide.
V. T. lùcida, shining-leaved, 5½ ft. high, and 5 ft. wide.

Loranthàceæ.

Aúcuba japónica, 5½ ft. high, and 4 ft. wide; in loam, bog, and sand.

E*ricàceæ.*

A'rbutus hýbrida, 7 ft. high, and 4 ft. wide; in loam and silver sand.
A. Pince's hybrid, 4 ft. high, and 3 ft. wide; in loam, silver sand, and bog.
A. procèra, 4 ft. high, and 3 ft. wide; in loam, silver sand, and bog.
A. scarlet, 6 ft. high, and 5½ ft. wide; in loam, silver sand, and bog.
*R*hododéndron, scarlet hybrid, 4½ ft. high, and 4½ ft. wide; in bog and silver sand.
R. azaleoìdes, 6½ ft. high, and 4 ft. wide; in bog.
R. white, 4½ ft. high, and 4½ ft. wide; in bog.
R. altacleréns, 5 ft. high, and 5 ft. wide; in bog and silver sand.
R. caucásicum, 6½ ft. high, and 2 ft. wide; in bog and silver sand.
R. yellow variety 3½ ft. high, and 3 ft. wide, in ditto; and 16 other hybrid varieties, the greater part scarlet-flowered.

*V*accínium *A*rctostáphylos, 6 ft. high, and 4½ ft. wide; in bog.

Sapotàceæ.

*B*umèlia tènax, 4½ ft. high, and 3½ ft. wide.

L*auràceæ.*

*L*aúrus nóbilis, 3 varieties, 5½ ft. high, and 4 ft. wide; in loam, bog, and sand.

E*uphorb*iàceæ.

*B*úxus baleárica, 5½ ft. high, and 2 ft. wide; in loam and silver sand.

C*orylàceæ.*

*Q*uércus *S*ùber, 5½ ft. high, and 3 ft. wide; in loam, bog, and sand.
Q. "gramúntia, 6 ft. high, and 4 ft. wide; in loam, bog, and sand.
Q. I'lex, 2 varieties, 7 ft. high, and 2½ ft. wide; in loam and road grit.

O*leàceæ.*

O'lea europæ'a, 7½ ft. high, and 1½ ft. wide; in loam, bog, and sand.
Chionánthus virgínica, 11 ft. high, and 6 ft. wide; in loam, bog, and silver sand.

T*axàceæ.*

*T*áxus baccàta, 7½ ft. high, and 3½ ft. wide; in bog and silver sand.
T. b. hibérnica, 7½ ft. high, and 3½ ft. wide; in loam, bog, and silver sand.

Coníferæ, § Abiétinæ.

*P*ìnus *Cémbra*, 6½ ft. high, and 4 ft. wide; in bog and silver sand.
P. ponderòsa, 5½ ft. high, and 3 ft. wide; in bog and silver sand.
A'bies Clanbrasil*iàna*, 5 ft. high, and 2½ ft. wide; in bog and silver sand.
*P*ícea Fràseri, 5½ ft. high, and 4½ ft. wide; in loam, bog, and sand.
*C*èdrus Libàni, 2 ft. high, and 3 ft. wide, 20 years old, and shaped like a bee-hive.

Coníferæ, § Cupréssinæ.

*T*hùja plicàta, 6½ ft. high, and 3 ft. wide; in loam and bog.
Taxòdium péndulum, 10 ft. high, and 3 ft. wide; in loam, bog, and sand.
*J*uníperus chinénsis, 8 ft. high, and 3 ft. wide; in loam, bog, and sand.
J. suécica, 5½ ft. high, and 1½ ft. wide; in loam, bog, and sand.

178

Hendon Rectory.

The spectator, standing at *v*, with his back to the mass, sees the house on two sides to the greatest advantage, as in *fig.* 178.; looking towards *x*, he sees a fine display of plants in pots and vases, backed by a dense wood, of which some idea may be obtained from *fig.* 179.; and, looking towards *u*, he has also a great variety of rich and interesting objects. Among the plants in the lawn, in front of *v*, are, Brugmáns*ia* suavèolens, Magnòl*ia* conspícua, *A*'bies Douglàs*ii*, Taxòdium dístichum nùtans, and various others, which will be found in a list given below.

w. Here are two baskets of pelargoniums, with elaborately worked handles, and between them a vase, supported on an elegant shaft, as seen in *fig.* 179., and filled with *V*erbèna chamædrifòlia. Between this vase and the walk are, a fine specimen of Magnòl*ia* conspícua, and one of Robín*ia* Pseùd-*A*càcia tortuòsa.

x, A bay, ornamented with a series of standard fuchsias, with some choice plants, and elegant vases in front of them. Among these plants are, Kálm*ia* latifòlia, *R*hododéndron arbòreum hýbridum, and a standard *R*hododéndron pónticum. A handsome vase is filled with *Petùnia* phœnícea.

The fuchsias and myrtles grown in tubs and pots, and set out on the lawn during summer, are splendid specimens of the following kinds: —

Fúchs*ia* grácilis, 8½ ft. high, head 6½ ft. wide; tub 2 ft. wide, and 1 ft. 10 in. high.

Another specimen, 8 ft. 9 in. high, head 6½ ft. in diameter, with a clear stem of 4½ ft., and 6 in. in circumference; grown in a tub 2 ft. wide, and 1 ft. 10 in. high.

Another specimen, total height 10½ ft., with a clear stem of 7 ft., the head 5 ft. in diameter; grown in a tub 1 ft. 9 in. wide, and 1 ft. 4 in. high.

Another specimen, in all respects the same as the last, except that it has a clear stem of 7 ft. 10 in.

Fúchs*ia* cónica, total height 6½ ft., clear stem 5 ft., width of the head 5 ft. 9 in.; grown in a tub 1 ft. 5 in. wide, and 1 ft. 3 in. high.

Fúchs*ia* globòsa màjor, total height 6 ft., clear stem 3½ ft.; grown in a pot 1 ft. wide, and 1 ft. high.

Fúchs*ia* tenélla, grown in a pot 1 ft. 9 in. wide, and of the same height, forms a regular cone 10 ft. 6 in. high, and 5 ft. in diameter at the base.

*M*ýrtus commùnis flòre plèno, the common double-blossomed myrtle, total height 6½ ft., with a clear stem of 2½ ft., the head 4 ft. in diameter; grown in a tub 2 ft. wide, and 1 ft. 9 in. high.

Two handsome standard common myrtles, 3 ft. 3 in high; grown in pots 10 in. wide, and 10 in. high.

Two standard double-flowered myrtles, of the same dimensions; grown in pots of the same size.

179

View at Hendon Rectory.

The more remarkable of the trees and shrubs planted on the lawn are the following:—

Ranunculàceæ.

Clématis, several species.

Magnoliàceæ.

Magnòlia grandiflòra.
g. lanceolàta.
glaúca.
g. Thompsoniàna.
umbrélla.
acuminàta.
auriculàta.
macrophýlla.
conspícua.
obovàta.

Berberàceæ.

Bérberis fasciculàris.
vulgàris.

Ternströmiàceæ.

Caméllia víridis.
japónica semidùplex.
j. rùbro plèno.
j. álbo plèno.
j. quadrangulàris.
Stuártia Malachodéndron.

Aceràceæ.

A'cer créticum, 8 ft. high.

Æsculàceæ.

Æ'sculus rubicúnda.
cárnea.
pállida.
Pàvia discolor and P. flàva, 3 ft. high, and 3 ft. wide; in loam and bog.
P. hùmilis.

Sapindàceæ.

Kölreutèria paniculàta.

Vitàceæ.

Vìtis vulpìna.

Celastràceæ.

Curtísia fagínea.
Euónymus europæ'us fòliis variegàtis.

Aquifoliàceæ.

I'lex Aquifòlium.
I'lex A. heterophýllum.
A. crassifòlium.
A. fèrox.
A. echinàtum.
A. flàvum.
A. senéscens.
A. àtro-marginàtum.
A. álbo-píctum.
A. aúreo-píctum.
A. scóticum.
A. recúrvum.
baleárica.
Peràdo.
opàca.
Dahoón.
Cassìne.
myrtifòlia.
lævigàta.
a beautiful unknown kind.

Rhamnàceæ.

Rhámnus Alatérnus.
latifòlius.
baleáricus.
vulgàris.
glàber.
integrifòlius.
aúreus.

Leguminòsæ.

Sophòra japónica péndula, 6½ ft. high, and 5 ft. wide; in loam and bog.
S. japónica, 10 ft. high, and 6 ft. wide; in loam and grit.
Cércis canadénsis, 8 ft. high, and 3 ft. wide; in bog.
Genísta virgàta.
Pistàcia Terebínthus.
Cýtisus Labúrnum.
L. variegàtum álbum.
woolgáricus.
U'lex europæ'a flòre plèno.
Edwárdsia microphýlla.
grandiflòra.
Gledítschia hórrida.
Gymnócladus canadénsis.
Acàcia armàta.
A. Julibríssin.

Rosàceæ.

Cérasus àvium múltiplex.
Cotoneáster microphýlla.
Photínia serrulàta.
Pýrus spectábilis.
Cratæ'gus Oxyacántha.
O. præ'cox.
O. coccínea.
O. aúrea.
Ròsa, 100 dwarf standards of the very best sorts.

Calycanthàceæ.

Calycánthus flóridus.

Granatàceæ.

Pùnica Granàtum.

Philadelphàceæ.

Philadélphus coronàrius.

Myrtàceæ.

Mýrtus commùnis.
c. romàna.
c. múltiplex.
Eucalýptus, a beautiful pendulous species, with oval glaucous leaves, growing 2 or 3 feet in one season.

Passifloràceæ.

Passiflòra cærùlea.

Grossulàceæ.

Rìbes alpìnum, 3 ft. high, and 4 ft. wide; in dung, loam, and bog.
R. alpìnum var.
R. sanguíneum.

Araliàceæ.

Hédera arboréscens.
canariénsis.
Hèlix.
H. álbo variegàta.
H. flàvo variegàta.

Hamamelidàceæ.

Hamamèlis virgínica.

Loranthàceæ.

Aúcuba japónica.

Ericàceæ.
*R*hododéndron maximum.
álbum.
purpùreum.
variegàtum fòliis aúreis.
azaleöìdes.
hýbridum.
Smíth*ii*.
arbòreum.
campanulàtum.
catawbiénse.
barbàtum.
däùricum.
ferrugíneum.
hirsùtum.
caucásicum.
Russell*iànum*.
R. with larger foliage and flowers.
anticulénse (?).
magnoli*æfòlium*; and about 50 scarlet varieties, hybrids, &c.
Azàlea índica.
i. álba.
puníceа.
purpùrea.
calendulàcea.
c. chrysoléсta.
c. ignéscens.
Andrómed*a* floribúnda.
arbòrea.
A'rbutus *U*'nedo.
rùbra.
críspa.
canariénsis.
hýbrida.
*A*ndráchne.
procèra.
Pince's hybrid.
*V*accínium nítidum.
*V*acc. *A*rctostáphylos.

Myrsíneæ.

*M*ýrsine retùsa.

Oleàceæ.

*P*hillýrea, of sorts.
O'lea europæ'a.
*b*uxifòlia.
Syrínga vulgàris.
v. álba.
pérsica.
p. álba.
Chionánthus virgínica.
*F*ráxinus, variegated white-leaved.
F. curled-leaved.

Lauràceæ.

*L*aúrus *Benzòin*, 6 ft. high, and 4 ft. wide; in loam and bog.

Thymelàceæ.

*D*áphne Laurèola.
póntica.
collìna.
napolitàna.

Aristolochiàceæ.

*A*ristolòchia sìpho.

Euphorbiàceæ.

*B*úxus baleárica.
sempervìrens arboréscens.
s. a. argéntea.
s. a. aúrea.

Urticàceæ.

*F*ìcus Cárica.

Betulàceæ.

A'lnus incìsa.

Corylàceæ, or Cupulíferæ.

*Q*uércus gramúntia.
I'lex.
I. integrifòlia.
*Q*uércus *I*'lex serràta.
I. oblónga.
I. *f*agifòlia.
I. críspa.
*S*ùber.
coccífera.
Túrner*i*.
bícolor.
tinctòria.
coccínea.
palústris.
Æ'gilops.
pedunculàta.
p. fòliis variegàtis.
fastigiàta.
lyràta.
Cérris.
Lucombe*àna*.
Lucombe*àna* nòva.
fern-leaved.
víridis.
fulhaménsis, or dentàta.
*F*àgus sylvática *f*ilicifòlia, the fern-leaved beech.

Balsaminàceæ.

Liquidámbar Styracíflua.

Taxàceæ.

Podocárpus nucíferus.

Coníferæ.

*P*ìnus *P*ináster.
*P*ínea.
A'bies Douglàs*ii*, 10 ft. high, and 6 ft. in diameter; planted in a tub sunk into the ground and concealed.
Cèdrus Libàni, 20 ft. high.

Smilàceæ.

*R*úscus racemòsus.

y, A cedar of Lebanon; and to the right and left are two beds of select dwarf dahlias. Beyond the beds of dahlias are two rustic baskets of pelargoniums; and there is a shell filled with mesembryanthemums. The other plants are select species of pines, firs, and *C*upréssinæ.

z, A strip of lawn, bounded on the south by a wire fence, which separates it from a paddock, and varied with beds of dahlias, pelargoniums, and other showy flowers. On the north of this strip of lawn is an oblong pond, well stocked with water lilies and gold fish.

&, Shed for cows or horses.

1, Reserve ground for proving seedling dahlias.

2, House for hurdles for dividing the paddock.

3, Compartment entirely devoted to dahlias, which are planted in beds 3 ft. wide, with alleys 2 ft. wide between. The collection amounted, in the year 1837, to upwards of 400 sorts, which, according to Mr. Lawrence, are among the best in existence.

4, 4, Kitchen-garden, chiefly planted with gooseberries, currants, raspberries, strawberries, potherbs, asparagus, tart rhubarb, sea-kale, and several articles not so readily procured from the market-gardens in the neighbourhood.
5, Tool-house, including a potting-shed.
6, Border for choice flowers; the wall covered with half-hardy shrubs.
7, 7, Two green-houses entirely devoted to Coníferæ in pots. The following list includes those species which constituted the collection in November, 1837: —

Taxàceæ.

Dacrýdium cupréssinum, 6 ft. 9 in. high, and 5 ft. wide; in a pot 17 in. wide, and 18 in. deep: by far the largest and handsomest plant of this species ever seen in England.

Coníferæ, § Abiétinæ.

Pìnus sylvéstris.
P. s. rigénsis, 4 ft. high, and 2 ft. wide; in a pot 9 in. wide, and 9 in. deep.
P. pumílio. *P.* p. *Mùghus.*
P. Banks*iàna*, 3½ ft. high, and 2 ft. in diameter; in a pot 10 in. wide, and 10 in. deep.
P. ìnops, 5 ft. high, and 3½ ft. wide; in a pot 17 in. wide, and 17 in. deep.
P. mìtis. *P.* púngens.
P. Laricio, 4 ft. high, and 18 in. wide; in a pot 10 in. wide, and 10 in. deep.
P. austrìaca, 1 ft. high, and 9 in. wide; in a pot 6 in. wide, and 6 in. deep.
P. Pallas*iàna*, 2 ft. 3 in. high, and 2 ft. in diameter; in a pot 13 in. wide, and 13 in. deep.
P. resinòsa.
P. *P*ináster.
P. *P.* var. helénica, from St. Helena, 7 ft. high, and 4 ft. in diameter, grafted on *P*ìnus sylvéstris.
P. *P.* var. nòva zælándica *Lodd.*, from New Zealand.
P. *P.* escarèna, 6 ft. high, and 4 ft. wide; in a pot 13 in. wide, and 13 in. deep.
P. *P*ínea, 3½ ft. high, and 2 ft. in diameter; in a pot 12 in. wide, and 12 in. deep.
P. halepénsis, 4 ft. 3 in. high, and 2 ft. 3 in. in diameter; in a pot 10 in. wide, and 12 in. deep.
P. *Tæ*da, 3 ft. high, and 3 ft. wide; in a pot 8 in. wide, and 8 in. deep.
P. rígida, 6 ft. 2 in. high, and 4½ ft. in diameter; in a pot 2 ft. wide, and 2 ft. deep.
P. serótina. *P.* ponderòsa.
P. Sabin*iàna*, 4 ft. high, and 3 ft. wide; in a pot 19 in. wide, and 17 in. deep: and another, 7 ft. high, and 5 ft. wide; in a pot 15 in. wide, and 10 in. deep.
P. Coúlter*i* (macrocárpa *Lindl.*), 4 ft. high, and 3 ft. wide; in a pot 13 in. wide, and 13 in. deep.
P. longifòlia, 5½ ft. high, and 3 ft. wide; in a pot 13 in. wide, and 13 in. deep.
P. Gerard*iàna.*
P. austràlis, 3½ ft. high, and 18 in. in diameter; in a pot 17 in. wide, and 17 in. deep.
P. canariénsis.
P. sinénsis, 3½ ft. high, and 3 ft. in diameter; in a pot 12 in. wide, and 12 in. deep.
P. insígnis, 5½ ft. high, and 2 ft. wide; in a pot 13 in. wide, and 14 in. deep: another is 3 ft. high, and 1½ ft. wide.
P. *Teocòte*, 13 in. high, and 9 in. wide; in a pot 7 in. wide, and 7 in. deep.
P. leiophýlla.
P. Llave*àna*, 2 ft. high, and 1¼ in. wide, 6 years old.
P. *Cémbra*, 4 ft. high, and 4 ft. in diameter; in a pot 16 in. wide, and 16 in. deep.
P. *S*tròbus.
P. excélsa, 1 ft. 7 in. high, and 13 in. in diameter; in a pot 7 in. wide, 7 in. deep.
P. Lambert*iàna.*
P. montícola, 1 ft. high, and 9 in. in diameter; in a pot 9 in. wide, and 9 in. deep.
A'bies excélsa var., 2½ ft. high, and 1½ ft. wide; in a pot 10 in. wide, and 10 in. deep.
A. e. Clanbrasil*iàna*, 18 in. high, and 2½ ft. wide.
A. e. pygmæ'a, 13 in. high, 13 in. wide, and 20 years old.
A. nìgra, 5 ft. high, and 6 ft. in diameter; in a pot 18 in. wide, and 18 in. deep.
A. rùbra, 6½ ft. high, and 5½ ft. wide.
A. Smith*iàna*, 2 ft. high, and 1½ ft. wide; in a pot 10 in. wide, and 9 in. deep: another, 3½ ft. high, and 3½ ft. wide.
A. Douglàs*ii*, 4 ft. 3 in. high, and 3½ ft.

in diameter; in a pot 1 in. wide, and 12 in. deep. The plant on the lawn, which is 10 ft. high, is also in a tub, plunged, and the brim covered.
A. Menzièsii, 3 ft. high, and 2½ ft. wide; in a pot 13 in. wide by 13 in. deep.
A. canadénsis.
A. cephalónica, 3 ft. 9 in. high, and 5 ft. wide; in a pot 18 in. wide, and 21 in. deep: two plants of these dimensions, and three others of smaller size.
*Pí*cea pectinàta.
P. balsàmea.
P. b. Fràseri, 2 ft. high, and 20 in. wide; in a pot 9 in. wide, and 9 in. deep.
P. Píchta, 1½ ft. high, and 1½ ft. in diameter; in a pot 11 in. wide, and 11 in. deep.
P. Webb*iàna*, 3 ft. high, and 3 ft. wide; in a pot 18 in. wide, and 16 in. deep.
P. grándis, 3 ft. high, and 2 ft. wide; in a pot 12 in. wide, and 13 in. deep.
P. nóbilis, 1½ ft. high, and 2½ ft. wide; in a pot 13 in. wide, and 13 in. deep.
*L*àrix europæ'a.
L. microcárpa.
*C*èdrus Libàni.
C. Deodàra, 3½ ft. high, and 4 ft. wide; in a pot 14 in. wide, and 18 in. deep: another, 5 ft. high, and 4 ft. wide.
Araucària imbricàta, 3½ ft. high, and 3 ft. wide; in a pot 16 in. wide, and 17 in. deep.
A. brasiliàna, 6 ft. high, and 4 ft. in diameter; in a pot 18 in. wide, and 18 in. deep.
A. excélsa, 5½ ft. high, and 3½ ft. wide; in a pot 13 in. wide, and 10 in. deep.
A. Cunninghàm*ii*.
Cunninghàm*ia* sinénsis, 6 ft. high, and 5 ft. wide; in a pot 13 in. wide, and 13 in. deep.

Coníferæ, § *Cupréssinæ.*

*Th*ùja orientàlis.
T. plicàta, 6 ft. high, and 3 ft. wide.
T. articulàta, 7 ft. high, and 2½ ft. wide; in a pot 9 in. wide, and 10 in. deep.
Cállitris pyramidàlis, 2 ft. high, and 1 ft. wide; in a pot 7 in. wide, and 6 in. deep.
*C*upréssus Tournefórt*ii*, 17 in. high, and 8 in. wide; in a pot 5 in. wide, and 5 in. deep.
C. lusitánica, 3 ft. high, and 2 ft. wide; in a pot 8 in. wide, and 8 in. deep.
*J*uníperus phœnícea, 4 ft. high, and 18 in. wide; in a pot 8 in. wide, and 8 in. deep.
J. from Gossainthan, 2 ft. high, and 1 ft. wide; in a pot 8 in. wide, and 8 in. deep.
J. austràlis, 1 ft. 9 in. high, and 1½ ft. wide; in a pot 6 in. wide, and 6 in. deep.
J. excélsa, 3 ft. high, and 15 in. wide; in a pot 9 in. wide, and 10 in. deep.
J. bermudiàna, 2 ft. 10 in. high, and 15 in. wide; in a pot 9 in. wide, and 9 in. deep.
J. recúrva, 4½ ft. high, and 2½ ft. wide; in a pot 13 in. wide, and 13 in. deep.

8, 8, 8, Hot-beds for striking and bringing forward pelargoniums and other flowers, for the beds, baskets, vases, &c.
9, Potting-shed.
10, Large pits for preserving fuchsias, orange trees, brugmansias, camellias, and other large plants, in tubs, through the winter.
11, Working-shed, with store-room over, and containing, also, the furnaces for heating the green-houses and the pits.

Remarks. This residence is not given as one for general imitation; for there are but few persons whose taste for botany and gardening is so enthusiastic as that of Mr. Williams. It shows, however, what may be done by industry and taste, in the course of not more than four years, on a very small spot of ground, possessing few or no advantages, either natural or artificial. The amateur may see, in the lists which we have given, the names of many of the choicest trees and shrubs which can be procured in the London nurseries; and these names may serve as a guide to him in selecting plants that shall indicate a superior degree of botanical taste. Mr. Williams has not only selected the finest specimens and varieties, but the largest plants of these that could be procured in England.

The sum that the plants enumerated in our lists must have cost, we cannot tell, on account of the large size of the plants; but small plants of all the same species might be obtained for perhaps 200*l*.

The unique practice adopted by Mr. Williams, of growing in pots pines which will endure the open air (we make an exception in favour of those that will not, such as *P.* longifòlia, leiophýlla, &c.), and keeping them in a green-house both in summer and winter, cannot be recommended; because, though the plants, when taken so much care of as at Hendon Rectory, will look remarkably well for five or six years, yet, for want of room, they must ultimately become stunted and die; or, if they are turned out into the free soil, after being six or seven years in pots, even with all the care that can be bestowed in unwinding their roots from the balls, and spreading them out, their chance of living is very doubtful. If it were probable that keeping these pines and firs in pots would make them bear cones, like Dr. Diel's fruit trees, that might serve as an apology for this kind of taste; and, doubtless, if the trees can be kept alive in pots, till nearly the usual period at which they would bear cones in the free ground, this would be the case; but still, so little would be gained by it, that we cannot recommend the plan for imitation. The only plants, in our opinion, that can be legitimately grown in green-houses and hot-houses, are such as can be brought to as great a degree of perfection there, as they would attain in the open air in their native countries. Of these there are thousands of species which can be brought to greater perfection under glass in Britain, than they are ever seen to attain in their native countries, in the open air. This will apply to almost all the shrubs, and all the herbaceous plants, of warm climates. The trees of warm climates have, in general, a miserable appearance under glass, for want of room.

To return to Hendon Rectory, we wish not to be understood as denying the right of Mr. Williams to indulge in his own peculiar taste: we merely state that it is one which never can become general, on account of the expense and trouble with which it is attended, in proportion to the effect produced. Mr. Williams's taste for the gardenesque in the planting of his garden, and for the highest order and keeping in its management, is beyond all praise, and is equalled nowhere, that we know of, in the neighbourhood of London, but at Mrs. Lawrence's, at Drayton Green.

Since the above was written, Mr. Williams has made great

alterations and additions. He has added two hot-houses at 7 7, in the situation of the hot-beds 8 8. Towards the north end of the compartment marked 3, he has constructed an elegant curvilinear house, glass on all sides, for the Coníferæ; at each end of it he has placed large masses of rockwork, which are to contain collections of ferns and *Saxifrageæ*; and the whole of the ground in front he has laid out in beds on turf, as a flower-garden.

A Suburban Villa of Two Acres, in a rectangular Boundary. (*figs.* 180. and 181.) — Though this villa is bounded by straight lines, and, being rectangular, might form a portion of ground along any road, yet the manner in which the buildings are disposed prevents it from being considered as belonging to the class in which the house forms part of a street or row. The design is by Mr. Rutger, and, like all that gentleman's ground plans, is remarkable for convenience of arrangement.

"The ornamental part of this design," Mr. Rutger observes, "may be considered as in a mixed style, partly geometrical and partly irregular. The approach is by a lodge entrance; and it passes on from the entrance door of the house to the stable-yard, and out again into the public road. On the left of the house are a conservatory and its fireplace, &c., hidden at the back by a shrubbery. The wing on the right, also hidden by shrubs, is for the kitchen offices. There are two private entrances from the main road, that on the left (*a*) for the family, and that on the right (*b*) for the domestics, with a private walk to the latter from the kitchen offices. On the right of these offices is a poultry-court (*e*), with entrances for the family at *l*, and for the domestics at *g*, adjoining to which are the laundry and drying-ground (*f*), where a brewhouse or cider-house may also be constructed if required. The forcing garden and reserve ground lie at the back of the stables, and may be entered from the stable-yard, for the convenience of dung, &c., as well as from the kitchen-garden and the gardener's house. The erections at the north of the reserve ground are, a potting-shed, cellar for mushrooms, &c., with fruit-rooms over, and gardener's house. The range for two forcing-houses, with a green-house between, is placed against the wall on the south side of the kitchen-garden, and approached in front by a walk from the lawn or pleasure-ground. On the western front of the house is a colonnade, in front of which a terrace walk extends on each side to the extremity of the lawn; on the verge of which are small clumps and vases, or other ornamental devices, in alternate succession. On the farther side of the lawn are a pond with a fountain, and a summer-house or billiard-room behind it. The small circles in the shrubbery on each side of the summer-house, at *r r*, are intended for statues, or some kind of ornamental structure." See the isometrical view, *fig.* 181.

Mr. Rutger continues: "I beg here to observe, that, as a general principle, I would not admit of walks crossing each other at right angles, or nearly so, where it can possibly be avoided; and that, where it cannot, I should conceal the angles formed by their intersections with shrubs, planted as closely as possible to the walks. In the ground plan (*fig.* 180.), it will be perceived that, on three sides or corners of the walks which cross each other at *ss*, this principle is attended to; and that on the fourth side I have indicated a vase, or something of that kind, for the sake of variety, though I think that shrubs would be better."

180

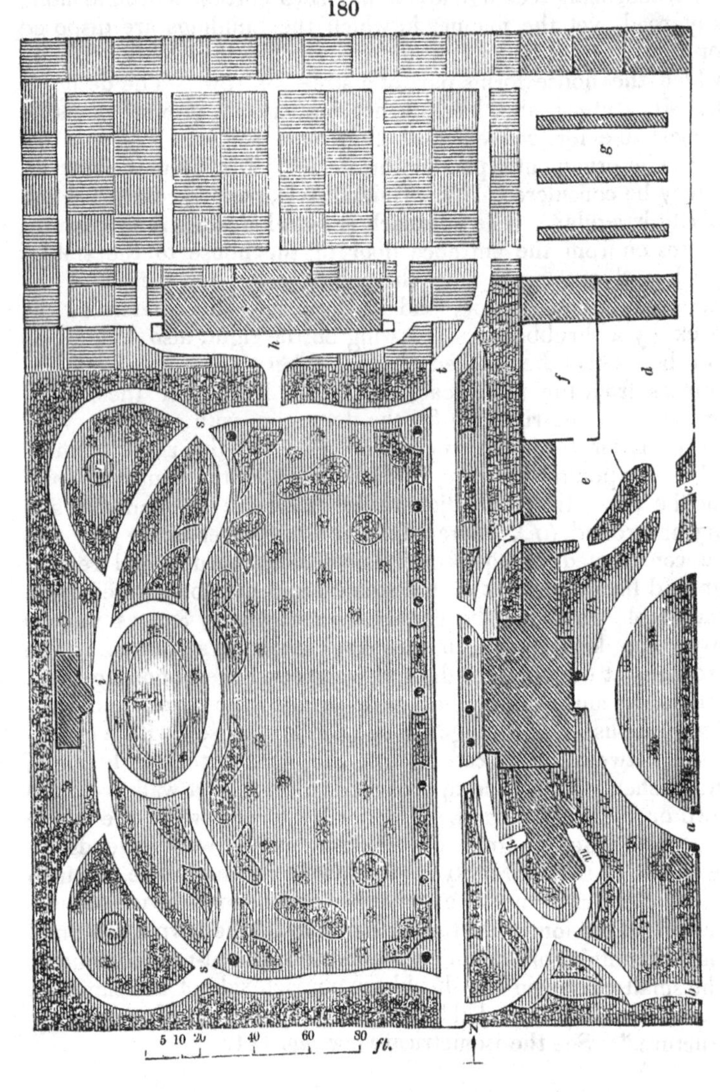

a, Carriage entrance from the Brighton road.
b, Private foot entrance for the family.
c, Private foot entrance for the domestics.
d, Stable-court, with coachman's lodgings over the harness-room; three-stalled stable, double coach-house, &c.
e, Poultry-yard, with a private entrance for the master and mistress at *l*, and an entrance for domestics.
f, Laundry and drying-ground, in which there may be a brewhouse and cider-mill, if thought requisite.
g, Forcing and reserve ground, the range of building on the north side including the gardener's house, mushroom-shed, and tool-house.
h, Peach-house and vinery, having a green-house between them, and behind the kitchen-garden, in which are the sheds and furnaces.
i, Summer-house. *k*, Conservatory.
l, The family's private entrance to the poultry-yard; and through it to the laundry (*f*), or to the stable-court (*d*).
m, Furnace-shed and potting-shed for the conservatory; behind which is a water-closet; the whole concealed from the entrance approach by evergreens. The water-closet is entered by a concealed door in the back wall of the conservatory, and it has also another door on the outside; so that no person need come out by the same door by which he went in.
n, Gardener's house, with an entrance from the public road, and another from the reserve ground.
o, Mushroom-cellar, with fruit-room over it. In this cellar may also be forced rhubarb, chicory, &c.
p, Tool-house and poultry shed.
The ice is supposed to be kept in one of the cellars under the right wing of the house.

Remarks. There is scarcely a single point in this design that we would wish to alter, with the exception of the termination of the terrace walk at *t* in *fig.* 180. We would have this walk terminate either in an alcove, and give up altogether the entrance in that quarter to the kitchen-garden; or we would carry it on to the kitchen-garden wall, and terminate it there in an archway, which should form an entrance to the kitchen-garden. The opposite end of the terrace walk, which, in the ground plan, is concealed by trees, is intended, we presume, to terminate in a covered seat of some kind, that has an architectural character harmonising with the house and conservatory. Wherever straight walks are used in laying out grounds, one of the main points to be attended to respecting them is, their finish, or terminations; as, when these are not of a decided description, there is a want of that character of art which is essential to this style of design. This rule is founded on the principle of a sufficient reason; for why should a walk be continued along in a straight line, unless it be for the purpose of arriving at some point or object that it is desirable to go and see? The windings of walks also depend, or ought to depend, on the same principle, viz., the recurrence of obstacles, either natural or artificial, which prevent them from being made straight. We go along a straight walk, in order to come at the object which is at the end of it; and along a winding one, for the sake of observing the objects that gradually develope

181

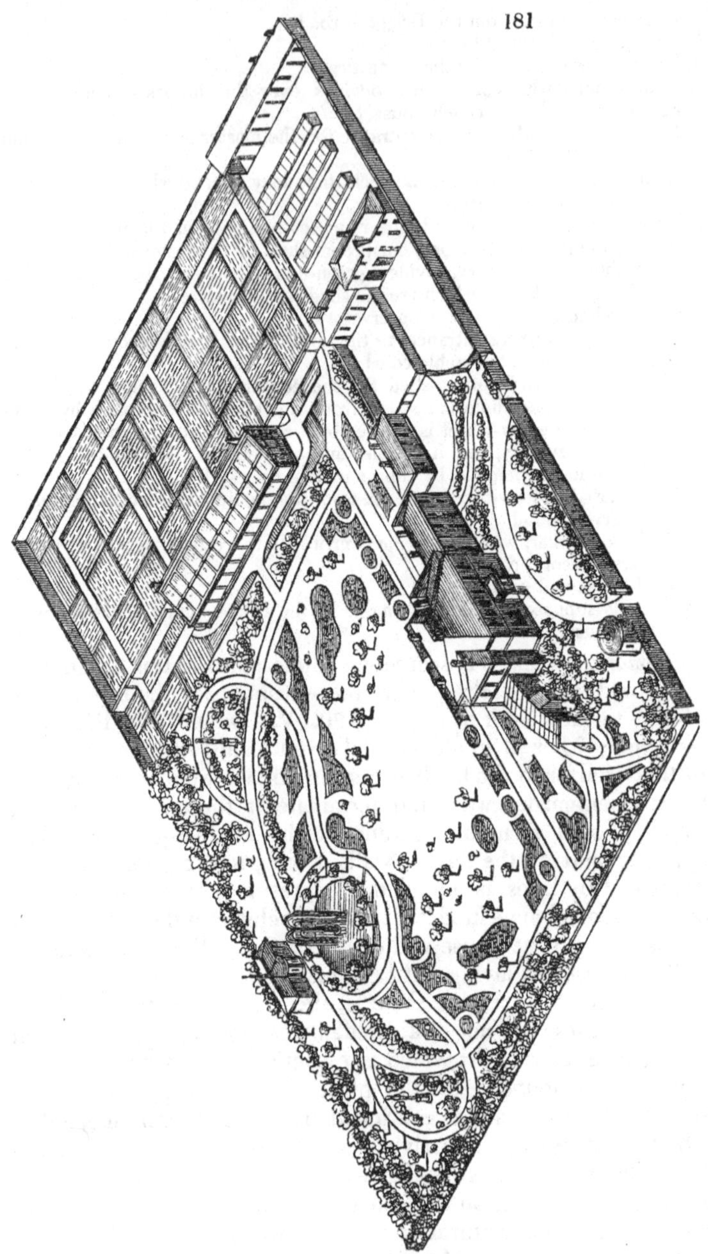

themselves as one turn succeeds another. The straight walk approaches its object in a direct and grand manner; the winding one in a manner which is varied and graceful. This place appears suitable to a gentleman possessing from 1000*l.* to 1500*l.* a year; but, by reducing the offices, it may be occupied by a person having 500*l.* a year.

A Suburban Villa, with the House in the Italian Style, and the Ground laid out in the picturesque Manner. (*figs.* 182. to 191.)— This design, by E. B. Lamb, Esq., is calculated for a piece of ground of two acres in extent, in the neighbourhood of Gravesend; and, with some trifling alterations, has been carried into execution. We have the greater pleasure in introducing it in this place, as, from the ample architectural details which Mr. Lamb has given, it will form some relief to the reader who is more fond of building than of botany and gardening.

"The general plan of the ground, and the situation of the house, are shown in *fig.* 182. In this plan, *a* is the entrance from the road; *b* is the lawn, *c*, the flower-garden and rosary; *d* and *f*, a walk all round the lawn, concealed in part by the shrubbery; *e*, a shaded walk, with grass paths leading to the other walks; *g* and *h*, garden seats of an architectural character; *i*, a grove of standard fruit trees, with here and there an evergreen, leading to the green-house (*k*); *l*, kitchen-garden; *m*, gardener's house; *n*, meadow; *o*, gate; *p*, flower-bed; *q*, garden seat under the drawingroom window; *r*, stable and yard. It will be observed, that the entrance from the road is placed near one side of the ground, in order to get more space in the situation open to the best and most agreeable aspect. A carriage drive and road to the stables are here shown. It will be better, perhaps, to begin my brief description from the steps at *d*. This path is not made the same width all the way round the lawn, but it is varied, to produce that agreeable effect which a parallel winding path never does. Proceeding to the north of the house, we arrive at the garden seat *g*, which is opposite some ornamental flower-beds, raised a little above the path *d*. Passing round these beds, we again enter the path *d*, at a more thickly planted spot, and gradually descend to *f*. This part is sunk gradually, so that it may not be seen from the house; and here is placed an architectural fence, which divides the lawn from the meadow. The lawn slopes from the house to this place in undulating lines, and the principal distant scenery is seen in this direction. We continue from *f*, gently ascending the path *d*, again to the house; but, before we leave this end of the ground, we will pass through the grass path leading to the shaded walk *e*: this walk leads to the flower-garden at one end, and the gate at the other. In the arrangement of this small plot of ground, I have been governed by the size of the house, and the supposed wants of the occupant. First, being small, it should never affect a concealment of art; but, at the same time, different degrees of art should

182

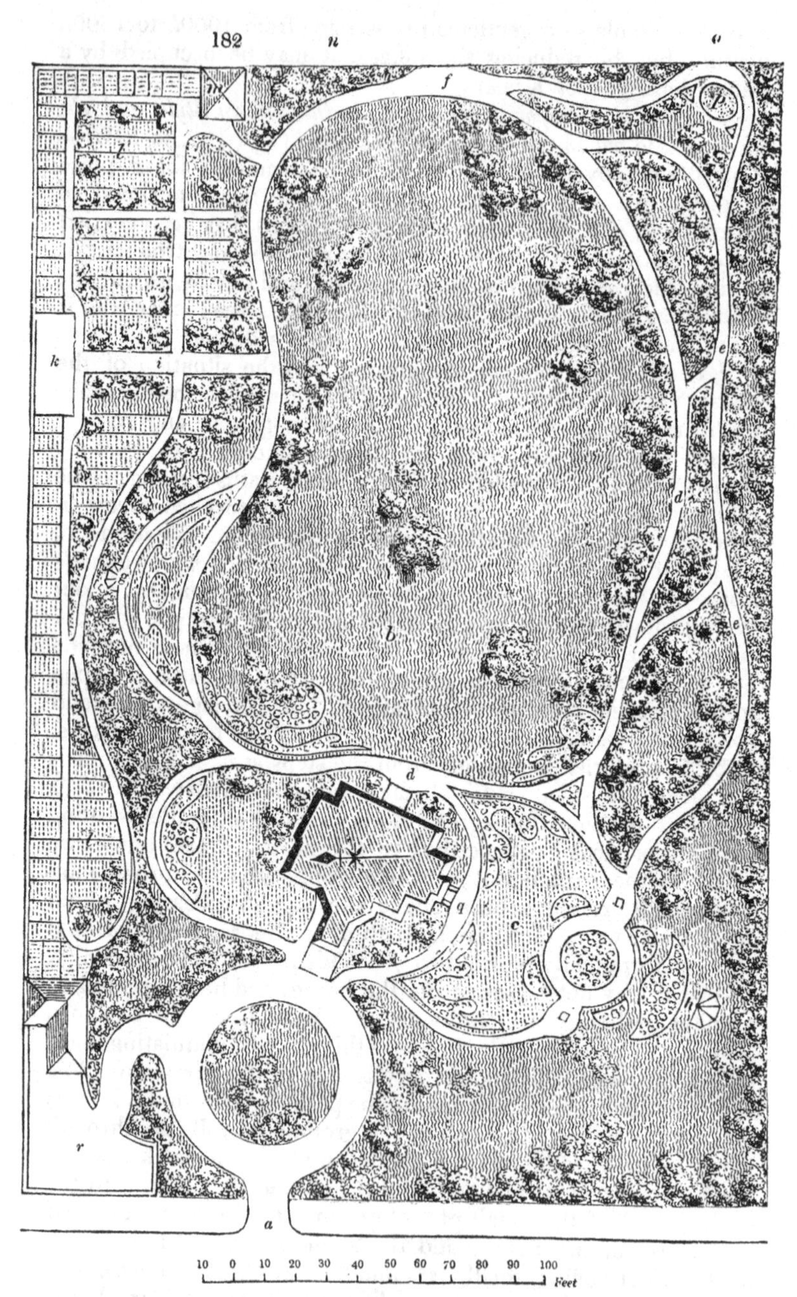

a, Entrance from the road. *b*, Lawn. *c*, Flower-garden and rosary. *d*, *f*, Walk round lawn. *e*, Shaded walk. G, *h*, Architectural garden seats. *i*, Grove of standard fruit trees. *k*, Green house. *l*, Kitchen-garden. *m*, Gardener's house. *n*, Meadow. *o*, Gate. *p*, Flower bed *q*, Garden seat under drawingroom window. *r*, Stable and yard.

be shown. Thus, near the house, every thing is of the most lively and cultivated kind: gradually proceeding towards the meadow, less dressed garden is shown, until, near the architectural fence and gate, an increase of cultivation would be consistent, and, without losing sight of art, the effects would be varied and pleasing. In the flower-garden, vases, figures, busts, &c., may be sparingly introduced; and, when good subjects, and well arranged, they have an excellent effect. Of the kitchen-garden very little need be said, except that the standard fruit trees, interspersed with evergreens, forming a grove to the green-house, would, I think, form an agreeable contrast to the waving lines of the other parts of the ground. In this description, I have merely given the general form, without mentioning the kind of trees, &c., for planting it; but these should be well studied for their form, colour, and changes through the different seasons; and, if judiciously planted, would, in a short time, produce a pleasure-ground of a picturesque nature.

The ground plan of the house is shown in *fig.* 183.; in which *a a* are steps leading to the porch (*b*). The porch is paved; and its roof is covered with large slates, or slabs of stone, upon fir joists, wrought, and open to the view. These slabs should be laid at such an inclination as to carry off the water; and the joints should be filled in with cement or white lead; or, more effectually to keep out the wet, one course of plain tiles might be laid over the slates, and bedded in cement; taking care to cover the joints of the slates; and, if expense were not an object, ornamental paving tiles might be used instead of the ordinary plain tiles; or two courses of slates alone, which last would certainly be the best. Of course it will be necessary to run a cement flashing or fillet round, to prevent the weather from getting into the brickwork. Round the under side of this covering, inside the porch, instead of running a many-moulded cornice, I would merely project a course of bricks, herring-bone-wise, with another course under it, only projecting half an inch from the face of the wall. The porch should not be plastered at all; but the brickwork should be well and neatly jointed, and the arches, and other decorative parts throughout the building, should be executed with white bricks (except where stone or other materials are mentioned); and the plain walls should be built with what are called seconds. The verandas, balconies, and all other weather projections, might be executed in the same way as the porch.

From the porch we enter the hall (*fig.* 183. *c.*) through folding-doors, glazed with large squares of glass, with rich stained glass borders; and the window on the right of the entrance should have stained or figured glass. The hall ceiling should have a

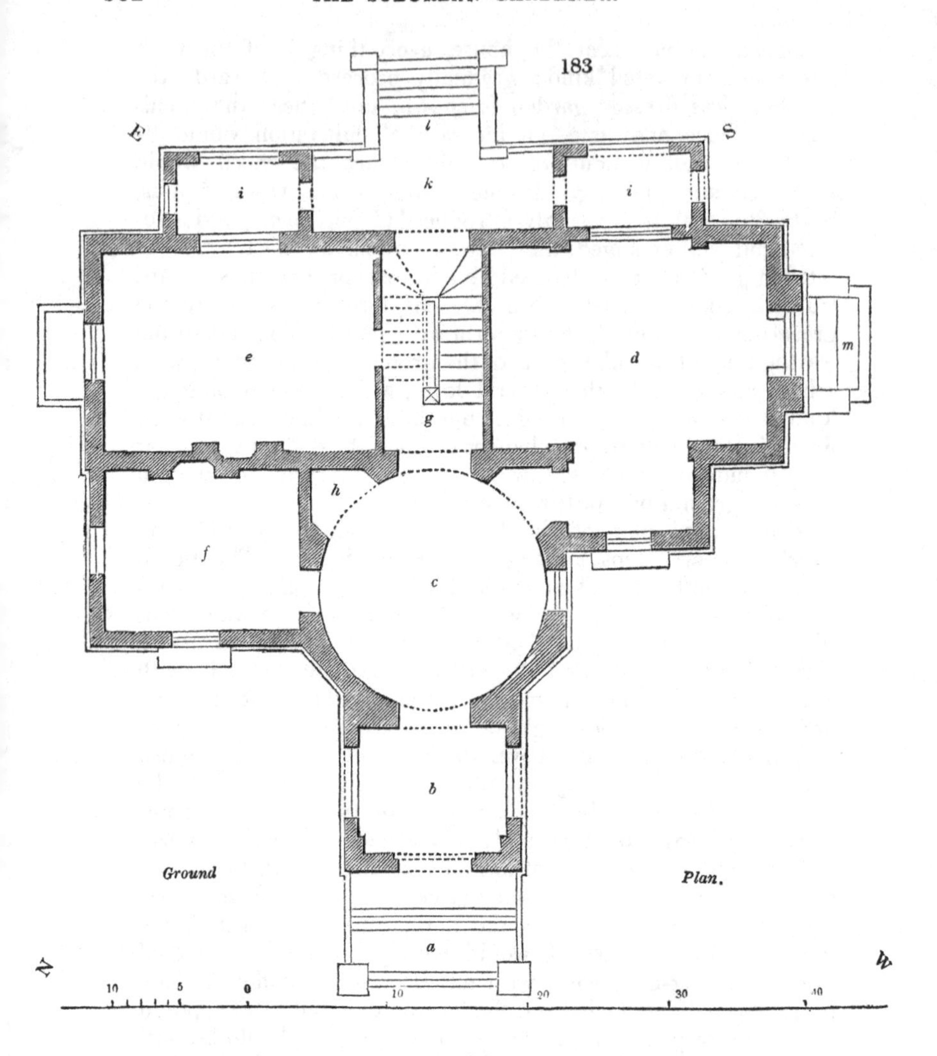

deep cove, with a very few mouldings under it, of rather a bold character, neither Gothic nor Grecian, but of a form well suited to receive the full benefit of the light and shade; the ceiling might be paneled, and adorned with many appropriate ornaments, without adopting the Grecian honeysuckle, or stripping the genealogical tree for heraldic devices. The colour of the ceiling should be rather dark, the walls of a lighter hue, and the floor still lighter; but these tones should be so graduated as to soften off, as it were, into each other by almost imperceptible stages. When I say the tones should be thus, I mean the general tones of the whole apartment; for, as the mouldings in the ceiling and walls will produce dark shadows, and, occasionally, strong lights, the floor must also partake of the same character; for which purpose, oil-cloth, or a pavement of two-coloured stones, or inlaid woods, when not too dark, would be proper. I am aware that this method will not be approved by many, nay, I may almost say most, people: they have been so accustomed to see white ceilings, dark walls, and darker floors, that it will be very difficult to turn the long-cherished custom upside down: but let us do it by degrees, and, if the dark colour is very objectionable, make it somewhat lighter — any thing but white. A large white ceiling, with an equally large portion of light in a room, completely destroys the effect of any piece of furniture which has only small portions of white in it. Every one must have observed how cold and cheerless a room looks when there are no curtains; and the cause of this is the glaring light and white ceiling. But to return to the hall; on the right of the entrance door, and adjoining the window, is an arched entrance to the drawingroom; and here would be a little triangular bit of ceiling, into which I would not forget to put its proper ornaments. How many and diversified might these be! A few minutes' thought would create so many pleasing designs to attract the eye and satisfy the mind at the same time, that I wonder antiques should be ransacked for them. Opposite this entrance is a corresponding recess, well suited for a candelabrum, a figure, or an elegant pedestal and glass vase for flowers; or it might be used for an umbrella-stand, or for hats and coats: the only objection to this would be its being opposite to the drawingroom entrance; on this account, perhaps, the best situations for hat and coat stands would be on each side of the entrance door.

We will now leave the hall for the drawingroom (*fig.* 183. *d*). As this room is appropriated to all the elegant refinements of the present day, it should be rather light in its general character; at the same time preserving the principles of light and shade, and gradation of tones, before mentioned. There are three windows in this room, one of which is over the fireplace, and from it is seen the flower-garden; that near the entrance commands a view of the entrance drive, and scenery in that direction; the other opens to the balcony, which is well stocked with plants, diffusing delightful perfumes and brilliant colours, which cannot fail to produce agreeable sensations in the mind. The windows of this room should all have some stained and ornamental glass; and the ceiling should be paneled and ornamented to as great an extent as would be consistent with the size of the house. The styles of the ceiling should be coloured oak, or painted in imitation of some ornamental English wood; the panels might be coloured of any lighter colour; and, if there were flowers, they might be picked out in two colours. I particularly mention that the ceiling should be in imitation of wood, as there would be a flat space of considerable extent; and an imitation of stone in it, or any colour which could be mistaken for stone, would be contrary to the principles of architecture, and would, consequently, produce a disagreeable effect. The cornice should be rather deep, and the large beams might terminate on an ornamental bracket: this would add to the appearance of strength, and produce variety without heaviness. The pilasters, which are shown on the plan, should go right up to the cornice, and would appear to support the largest beams, which should run in this direction across the room. The walls might be papered with paper of a lively pattern, but not too light; and a gold moulding would add very much to the effect of the paper. Opposite the window over the fireplace should be a large mirror, so as to reflect the flower-garden, &c. It will be quite unnecessary to say anything about the furniture, except the curtains, which are so connected with the architecture of the room, that their arrangement is of some importance. Between the pilasters, by the large window, there would be very little difficulty in hanging curtains, as these pilasters form a termination to a cornice, a pole, or piece of straight fluted drapery with a deep fringe: there would be, also, room for curtain pins or hooks. The opposite window could be managed much in the same way; carrying the cornice or drapery to the return walls. Over the fireplace there could be no drapery except at the top; but the shutters should be lined with plates of looking-glass, so that, when closed, they would reflect the objects in the room; and, if the boxings were splayed, and glass fixed in, the variety would be extremely pleasing.

We will now enter the staircase (*fig.* 183. *g*). The newel of the stairs should be a candelabrum, if one is not used in the hall; the stairs might have rounded nosings; the balusters should be ornamental cast iron bronzed, and the handrail should be capped with mahogany; but, if this should be thought too expensive, bold turned balusters would cost considerably less.

From the stairs we enter the dining-room (*fig.* 183. *e*). The entrance is placed here to be near the kitchen stairs, and to be away from the drawingroom entrance; which entrance is sufficiently far from it to prevent any disagreeable smells from the kitchen. The dining-room has two windows; one opening to the veranda (*i*), and the other looking to the lawn, &c.; these windows should have some stained glass. The general colour of this room should be of a rich tone; and the chimney-piece should be of such a marble as would, in colour, harmonise with the rest of the room; as, in fact, all the chimney-pieces in the principal rooms should do. It is a very common thing for persons to order a white marble chimney-piece for the drawing-room, a black or dark marble for the dining-room, without considering how it will look when in contrast with the walls and other parts of the room: this is a point which architects have not sufficiently considered; and marble is too often chosen for its rarity or purity, but not for its effect. Another thing respecting chimney-pieces I will mention: veined marble should only be used for plain designs, and plain marble for enriched ones. I was very much surprised to see, at the Marble Company's warehouses, some designs of Sir Robert Smirke's, of a rich Gothic pattern, executed in variegated marble, so that when one was standing opposite the light, the design was entirely lost; and even with a side light the design was very little better seen.

From the dining-room we reenter the hall to the library, or study (*f*), which might, if required, be used for a breakfast-room. This is a small room, but as much attention should be bestowed upon its effect as upon that of the others; and the ceiling should show some of the beams. Perhaps, by using the term beams, I shall be understood to mean those large unsightly projections commonly seen in houses of a century back; but this is not the case: the projections I mean are merely intended to represent the principal timbers of the floor above; and are to be moulded, and connected with each other by small paneling. I think the whole of this ceiling might be painted of a light oak; the walls might be of any plain colour; and some stained glass might be used in the windows. I will just go back to the porch, to show the relative degrees of richness by which I conceive a house of this size should be regulated. The porch should be the simplest; the hall the next step, at the same time possessing some of the characteristics of each room, so that the contrast is not too violent when entering those rooms; the breakfast-room, or library, the next; the dining-room the next; and the drawingroom the most decorated and most cheerful. These degrees are best marked by the furniture; but they ought to be studied also by the architect.

The basement plan requires very few remarks. First, all drains should be outside the house, and in such situations as to be easily accessible; the area round the house, for the purpose of keeping the basement dry, should be at least 1 ft. below the kitchen floor, and should be covered with Yorkshire paving, except where light was required; and the walls should be built battering, to prevent the earth from pressing them in. The kitchen floor should be boarded, and all the others paved with stone, tiles, or bricks. Stone is best for the floor of the wash-house and pantry, and for the passage; and it should be laid

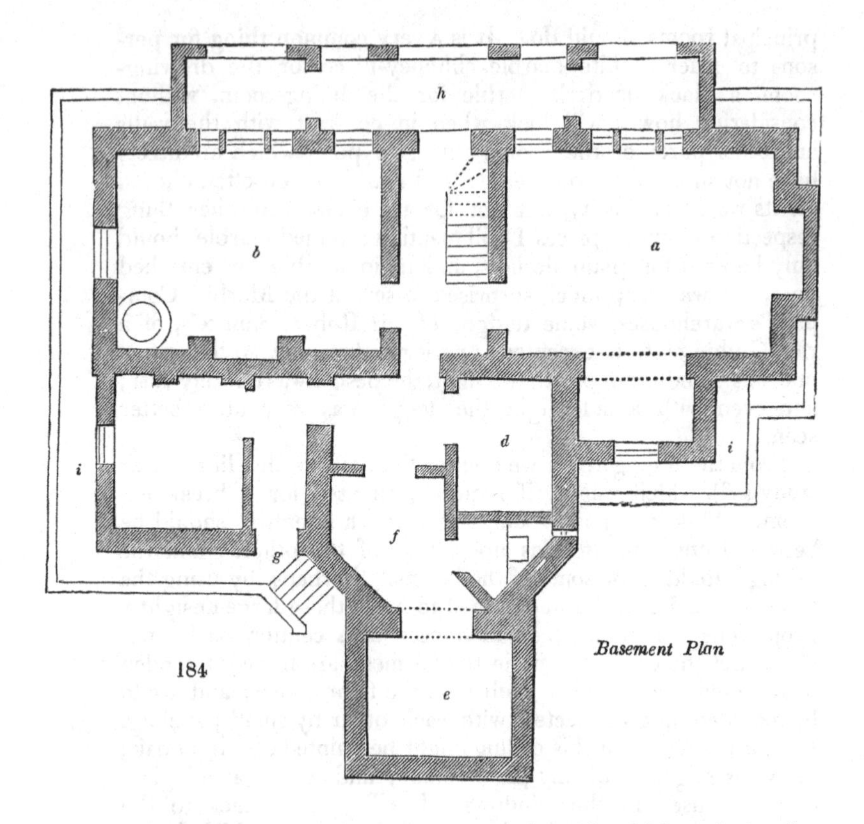

upon brick withs. A free circulation of air should be admitted under the floor and paving, by means of air-bricks, which would prevent what is called dry rot in the timbers, and damp from rising in the walls.

The details of the basement plan (*fig.* 184.) are as follows:— *a* is the kitchen; *b*, the washhouse and scullery; *c*, the pantry; *d*, the wine-cellar; *e*, the coal-cellar; *f*, the wood-cellar; *g*, the back entrance; *h*, the area under the veranda; *i i i*, the area round the house.

The chamber plan (*fig.* 185.) shows four bed-rooms, &c. The windows to these rooms are hung folding, and open on balconies, which are seen in the elevations.

Balconies are delightful additions to a bed-room: they also give a cheerful effect to the exterior of the house; and, when furnished with plants, nothing is more delightful, I think, than opening the window directly you get out of bed, and enjoying the fresh air and the sweet scent of the flowers at the same time. These balconies may either be formed in the usual manner, with stone balustrades, as shown in the eleva-

185

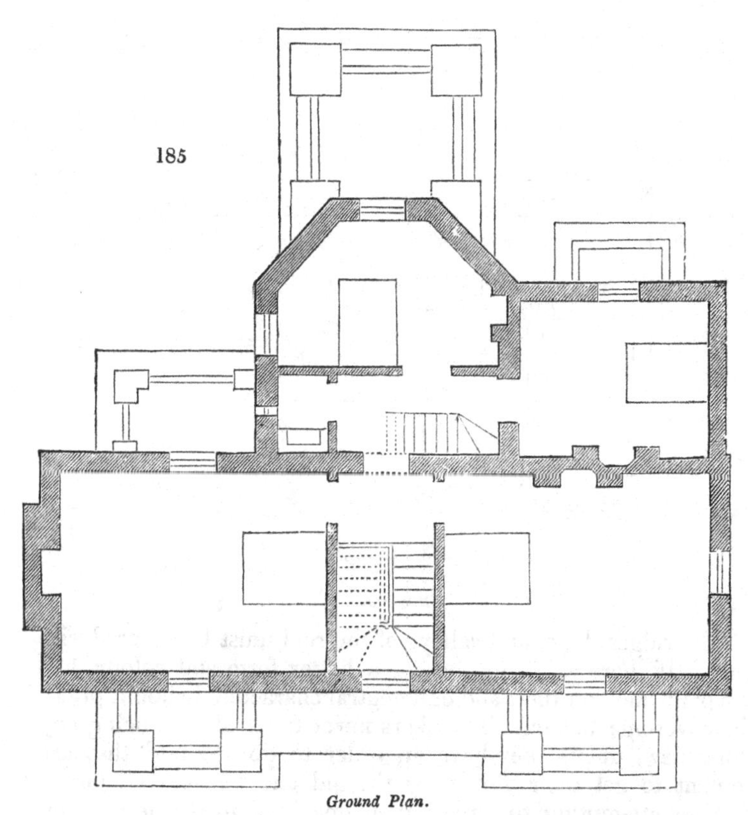

Ground Plan.

tions; or they may have stone pedestals at the angles, and the space between filled in with iron railing, supporting a cap of stone, as shown in *fig.* 186. The pedestals would be proper places for vases, flower-pots, or even figures if they were in sitting postures: but, as many stone or artificial stone figures would not agree with this brick building, great care should be taken in their introduction.

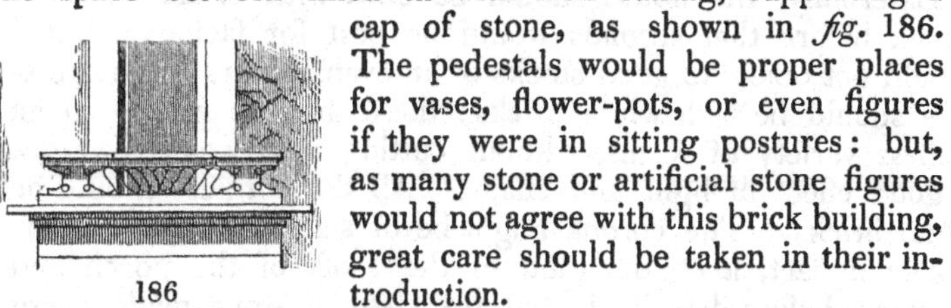

186

The attic plan (*fig.* 187.) shows two rooms in the tower, and the roofs, gutters, and chimney-stacks. The roofs should be covered with large slates; and, when they are new, if they were stained with oil lime and lampblack so as to take off the rawness of their material, they might be made to harmonise with the walls. Time is the best stainer; but, as harmony is as requisite to architecture as it is to music, we must study to obtain it artificially, if we cannot get it naturally.

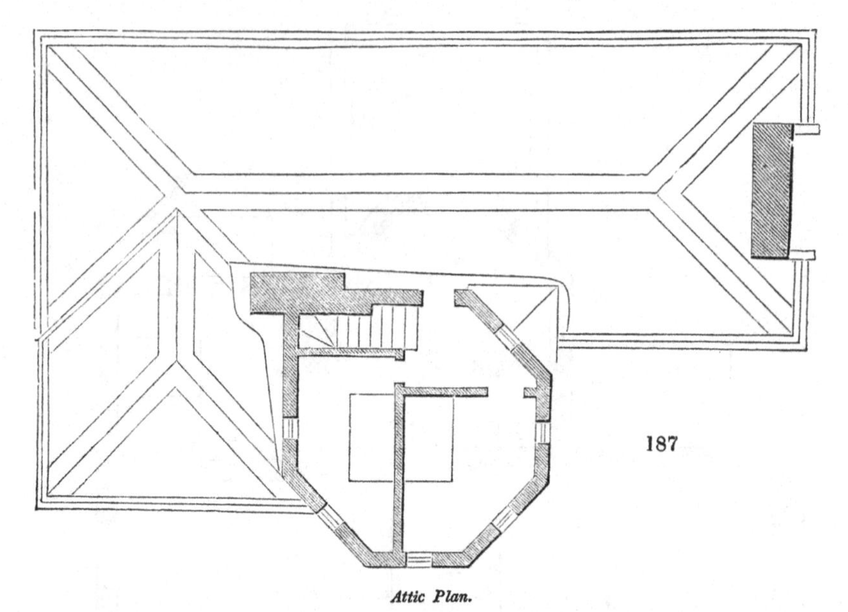

Attic Plan.

The ridges, hips, and valleys of the roof must be covered with lead. If there were any tiles of a better form and colour than the plain and pan tiles, and of a neutral character, I should prefer tile covering; but our tile-makers never think of producing any thing new, unless they have an order to do so; and they are content to get their money by the old and well-known forms, without attempting to stretch their imaginations or their purse-strings for new ones.

Externally, the house should be built with brick. I mentioned before that seconds would be best for facings; but I should not object to good stocks of an even colour. The dressings should be with white bricks: these may be placed in an endless variety of forms, without cutting, and would produce a good effect of light and shade. All the cornices should be white bricks. The corona might be of slate, or York paving-stone; in fact, a continuation of the roofs of the porch and balconies before described: but the slate or stone made use of must be throated, so that the water may drop off. The under side of the corona should be coloured, if it is slate, to resemble the white bricks; the top would not be of so much consequence. A great projection is the principal thing to be sought for, to produce a good effect: the means of obtaining it are not always of consequence; therefore, slate or stone coronæ in thin slabs, if not too finely tooled, or any other material sufficiently durable, would produce broad masses of light and shade at a

188

North-west Elevation.

small expense, and equal in pictorial effect to the richest and finest wrought cornice in expensive marble. Bricks, for the same reason, when properly managed, even without cutting, can be made ornamental; and they are our own material, and most durable; that is, the best bricks of the present day: then why should we reject them in ordinary buildings, or, rather, disguise them with cement, which, in a very few years, frequently cracks and peels off? Cement is excellent for using instead of mortar; but it does not adhere well to brickwork, unless it is the very best, and then it is almost as dear as stone: but for ornamental works nothing could be better; in some instances it is even better than stone. Wyatt's, for chimney moulds, balusters, vases, fountains, &c., and paving, has been found better than Yorkshire stone. Austin's cement, also, is capable of producing the same articles.

It will be unnecessary to make many observations on the elevations: I think they tell their own tale; and a close examination will clearly show the manner of the construction and effect.

Fig. 188. shows the north-west elevation, or the entrance front,

189

South-west Elevation.

with the porch, &c. The arched opening of this porch is in two half-brick rims; the outer arch projecting beyond the face of the wall three fourths of an inch. At the springing of the arches, both of the doors and windows, one or two bricks may be placed herring-bone-wise, if thought advisable. It may be proper to remark, that the shrubbery is not shown in these elevations, as it would be according to the plan of the ground, the object being, here, to show the building divested of all ornamental planting.

In this and the other elevations, the balconies are represented with stone balustrades; but, if it should be thought preferable, they may have light iron railings, &c., as shown in *fig.* 186. p. 507.

Fig. 189. shows the south-west elevation. In this front is shown the chimney-stack from the kitchen, drawingroom, and bedroom; also the window over the drawingroom fireplace, and a garden seat. This front is seen from the flower-garden.

Fig. 190. is the south-east elevation. In this front is shown the corona, continued through to form a covering to the landing: it is supported by cast-iron brackets. The entrance door has a plate of looking-glass in the place of a fan-light: this should be

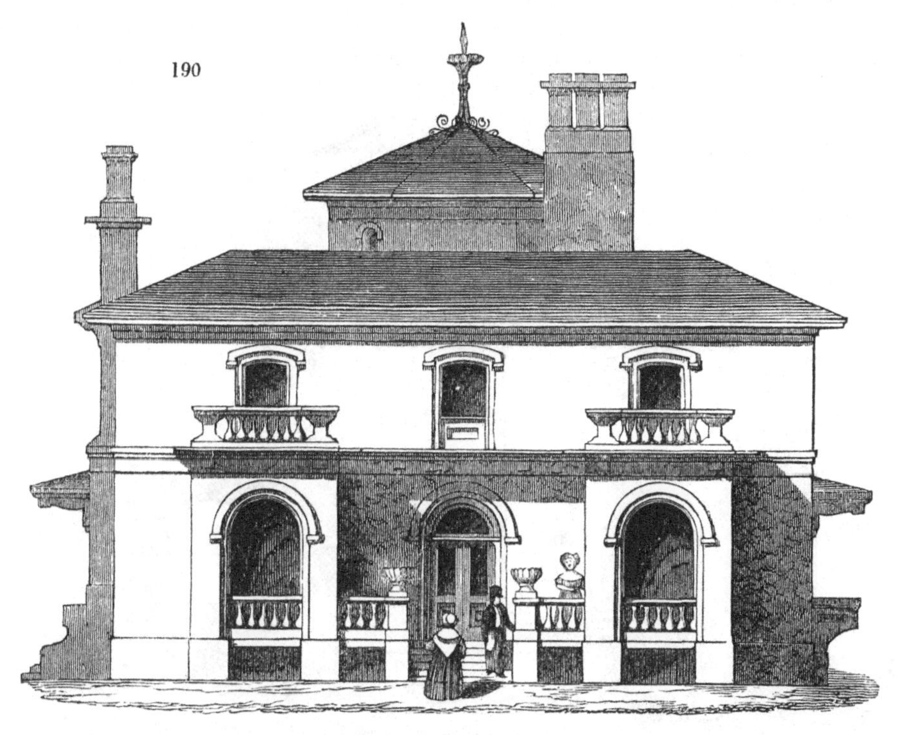

South-east Elevation.

fixed at such an angle as to reflect the lawn and distant scenery, and to be seen when ascending the steps. This would always be an interesting object, as it would be confined within a frame, and thus form a living picture more perfect than when the eye is left to range from place to place.

Fig. 191. is the north-east front. In this front, over the study window, is a tablet, in which an appropriate inscription might be placed, or the founder's arms. This tablet is instead of a window, which would have been most inconvenient here, as may be seen by referring to the chamber plan, besides admitting too much light into the room. In this elevation I have omitted the chimney shafts at the other end, as they would be seen only at a great distance, and in a particular situation.

To make the building fire-proof, or, at least, nearly so, the divisions of the rooms should be in brickwork, the stairs should be of stone, and the balusters and handrail iron: there would then be very little for fire to communicate with, as the principal conductors are generally found to be the hollow wooden partitions and the wooden stairs. Ceilings, if well plastered, are not likely

North-east Elevation.

to burn; and, as it is the nature of fire to ascend, it would not be a very easy task to make a floor burn: boxings to shutters, sashes, doors, and other wooden linings, only, are to be feared; but, with careful building, and tolerably careful servants, no very serious accidents could occur. Houses, to be made perfectly fire-proof, would be attended with very considerable expense. — *E. B. Lamb. Henrietta Street, Brunswick Square.*"

A Villa of an Acre and a Half, with an irregular Outline. — This design, of which we give only a ground plan (*fig.* 192.) is by Mr. Rutger, and illustrates the manner in which the kitchen-garden may be disposed, where there is not room to lay it out as a regular figure. "The wing," Mr. Rutger observes, "on the right of the house (*a*), supposed to be hidden by tall shrubs, is a part of the offices, and may be entered from the stable-yard at *b*. Adjoining the stable and yard (*c*), at the north-east corner, are the laundry and yard (*d*), a part of the building of which may be fitted up for a small brewhouse. The whole of these offices is supposed to be hidden from the house, by the shrubbery (*e e*) in front of the yard. The approach to the house is

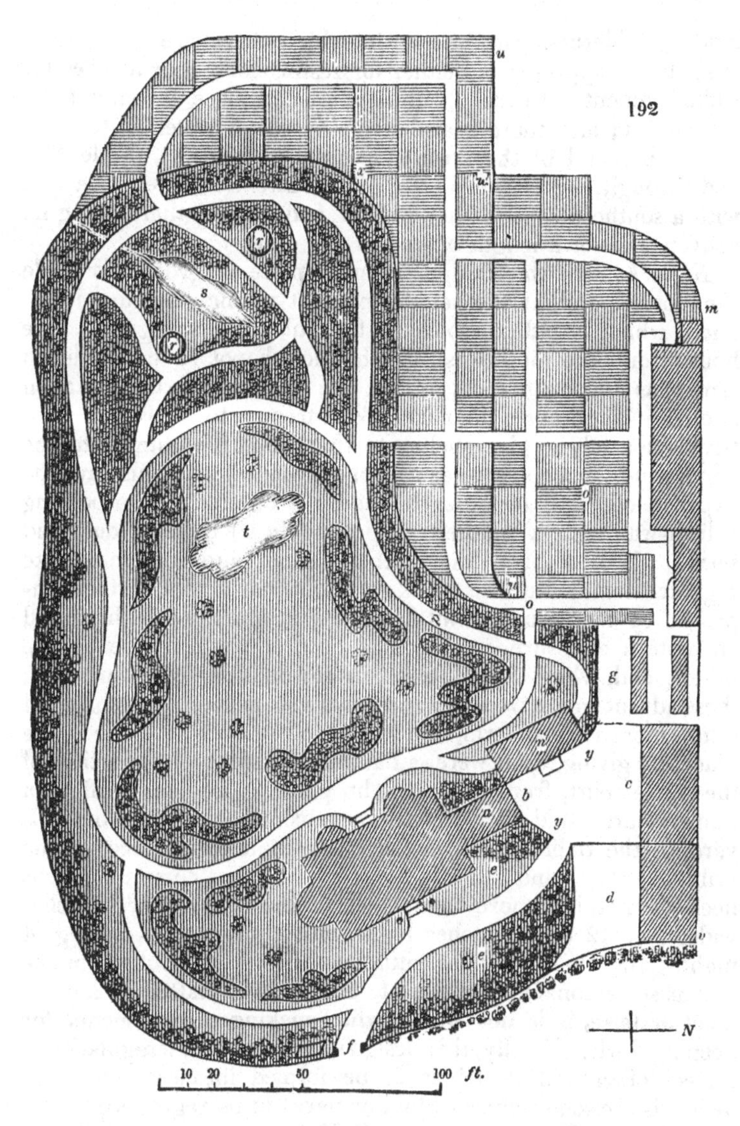

by a lodge entrance (*f*), which passes on to the stable-yard (*c*). Adjoining to the stables is the melon-ground (*g*); and adjoining the melon-ground are, the gardener's room (*h*), potting-shed (*i*), and stock-hole (*k*) to the forcing-house (*l*) beyond, at the west end of which is another stock-hole (*m*). On turning to the right from the south-west front, there is a green-house (*n*); and, on passing through it, you may enter the kitchen-garden (*o*); or, skirting the lawn by a walk (*p*), you may proceed to a kind of wilderness shrubbery, in which there are two circles (*r r*), to be appropriated either for resting-places, or articles for embellishment. In this shrubbery, a small piece of water (*s*) is introduced; and there is also a small piece of water (*t*) towards the farther end of the lawn. The dark line (*u u*), the length, and through nearly the centre, of the kitchen-garden, is a wall with a southern aspect, intended for fruit trees; there being no convenience for a length of wall in any other way."

Remarks. In this design, a small piece of ground is made the most of, and, at the same time, not a single feature introduced which is either arbitrary, fanciful, or extravagant. The house fronts the south-west, which, though not so desirable an aspect as the south-east, is yet the best which the situation affords. The range of offices, &c., from *v* to *m*, is well contrived, not only to give shelter to the grounds within from the north winds, but to serve as a boundary fence, and thus save the expense of wall, or other boundary. The mode of procuring a fruit-wall with a southern exposure (*u u*) is ingenious, and serves, also, to throw the kitchen-garden into something like regular forms; but we think it might have been farther improved by continuing a wall from *w* to *x*, which would have had an eastern exposure on one side, and a western one on the other; and, what is equal to, or, in our opinion, even surpasses, these advantages, it would have thrown a shapeless piece of ground into a more regular form, and added to the variety of the whole place, by giving, as it were, an additional scene. A portion of the stable-court, from *y* to *y*, might, perhaps, be fenced off as a poultry-yard; and there might be a communication from this yard to the public road, through the boundary fence of the stable-court (*c*) and laundry-court (*d*). These communications need be nothing more than holes in the walls, 6 or 8 inches wide, and 12 or 15 inches high, with shutters for closing at night. In some suburban situations, however, poultry can be had at so reasonable a rate, that, unless as a matter of amusement or taste, it is not worth while making arrangements for keeping fowls. Finally, this design shows that the irregularity of a piece of ground, which is to be formed into a country residence, is a circumstance seldom or never to be regretted.

A Villa of Two Acres and a Half, in an irregular Outline.— This design (*fig.* 193.), also by Mr. Rutger, is of the same general character as the last. "The house is approached by a lodge entrance (*a*), which passes on to the stable-yard (*b*); and the house has a colonnade at the carriage entrance. There is a private entrance (*c*) from the road, for the domestics, which communicates with the office yards and the back entrance. The yards and buildings at the east of the stables are intended for the poultry-court (*d*), laundry, brewhouse, &c. (*e*), which are

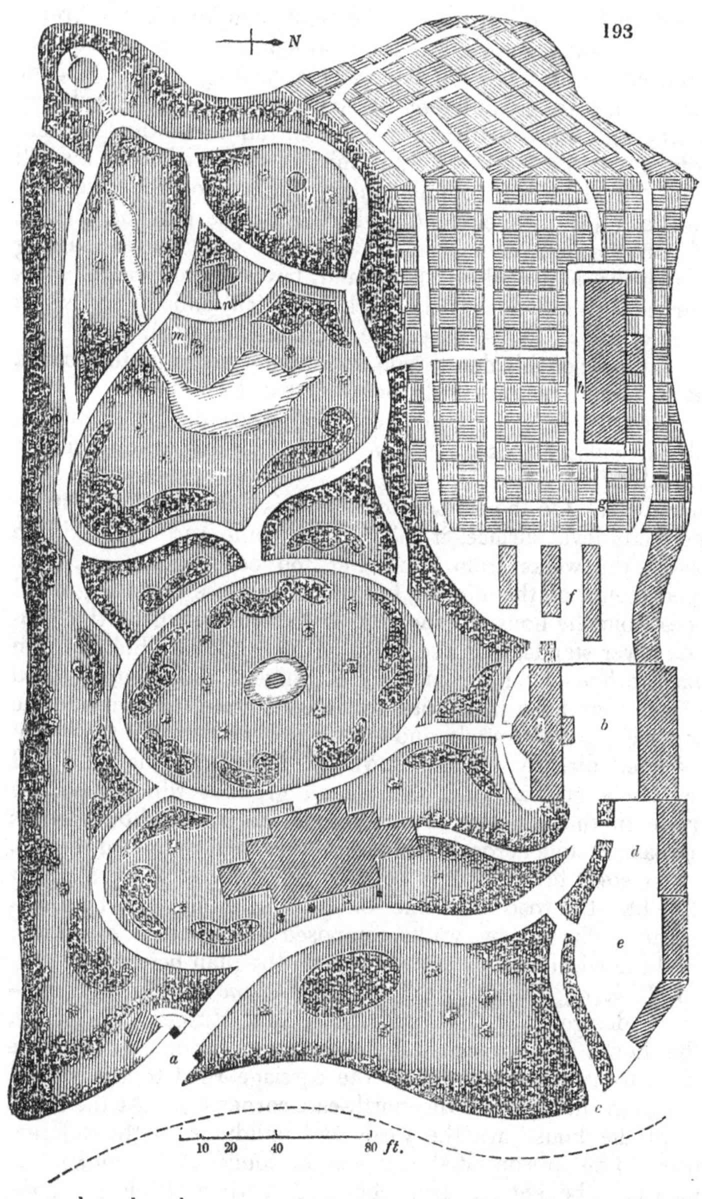

supposed to be shut out of view from the house by the shrubbery. Adjoining to the stable-yard, on the west, is the frame-ground (*f*), with erections at the back for the gardener's room, potting-shed, &c.; from this you enter the kitchen-garden (*g*), in which stands a range of forcing-houses (*h*). From the south-west front, on turning to the right, you approach a green-house, or conservatory (*i*), from which may be gained admittance to the frame-ground; or, by returning to the lawn, the walks will lead to different compartments of the shrubbery garden, and also to the principal entrance to the kitchen-garden. At the south-west extremity is supposed to be a small mount, on which is a rotunda, meant for a prospect tower (*k*). On the right is a circle (*l*), where a statue may be placed; and on the left is a rivulet, which is widened in passing through the first glade of lawn, and, beyond the bridge (*m*), is enlarged into a pond; at the back of which, and on the verge of one of the walks, is placed the summer-house (*n*).

"In places of this description, it may not be amiss to remark, that it should be the study of the artist to hide the view, as much as possible, of one walk from another, by shrubs or clumps, and particularly those that run nearly parallel or opposite to each other."

Remarks. On the supposition that this design is laid out on a perfectly level surface, it will require some care and skill to prevent the walks from appearing too conspicuous from the upper rooms of the house; but, if we suppose that the ground slopes from the house to the farther extremity, where the prospect tower stands, then the walks will not only be wholly unseen from the house, but, from their curvilinear direction, it would be more easy for a pedestrian to descend from the house to the lower part of the grounds, and also to reascend, than if they were conducted directly up and down the slope. It may be laid down as a general rule, that, in all grounds which have the surface in the slightest degree steep, the walks should be as much as possible across the slopes, rather than in their direction. A very steep hill can only be ascended with ease by a zigzag walk, like the road over the Simplon in miniature; by a serpentine walk; or by walks composed of segments of circles, like those laid down by Mr. Rutger in the plan before us.

A Villa of Three Acres, with an Outline nearly regular.— In this design (*figs.* 194. and 195.), Mr. Rutger informs us, "the house is approached by a lodge entrance (*a*), and the area in front presents an oval: the carriage road to the stables (*b*) is from the road at the north-east corner (*c*). At the north end of the house are the yards and buildings for the different offices, to be appropriated as it may be found expedient; all of which may be entered from the road *c*, which leads from the main road to the stables. On the west of the stables is a range of forcing-houses (*d*), with potting-shed, &c., and gardener's room. At the back, and farther on, is the frame-ground (*e*), from which the kitchen-garden (*f*) is entered; where there is a pond, and also a cross wall, to give more room for fruit trees.

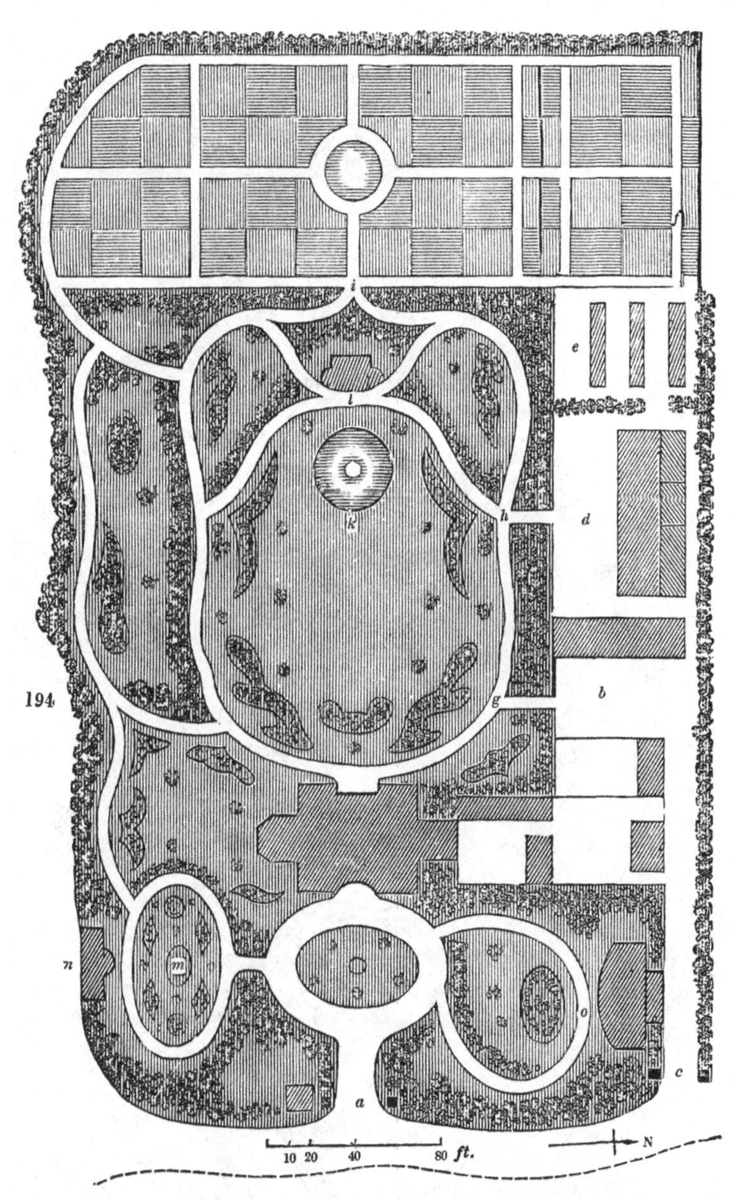

It will be observed that there is access to all these departments, from the stable-yard (*b*), for the convenience of carting in dung &c.; and, also, that the offices and yards appertaining to them are all shut out of view from the house by shrubs. On turning to the right from the west front of the house, a walk (*g*) is opened to the stable-yard; and, farther on, a walk (*h*) leads into

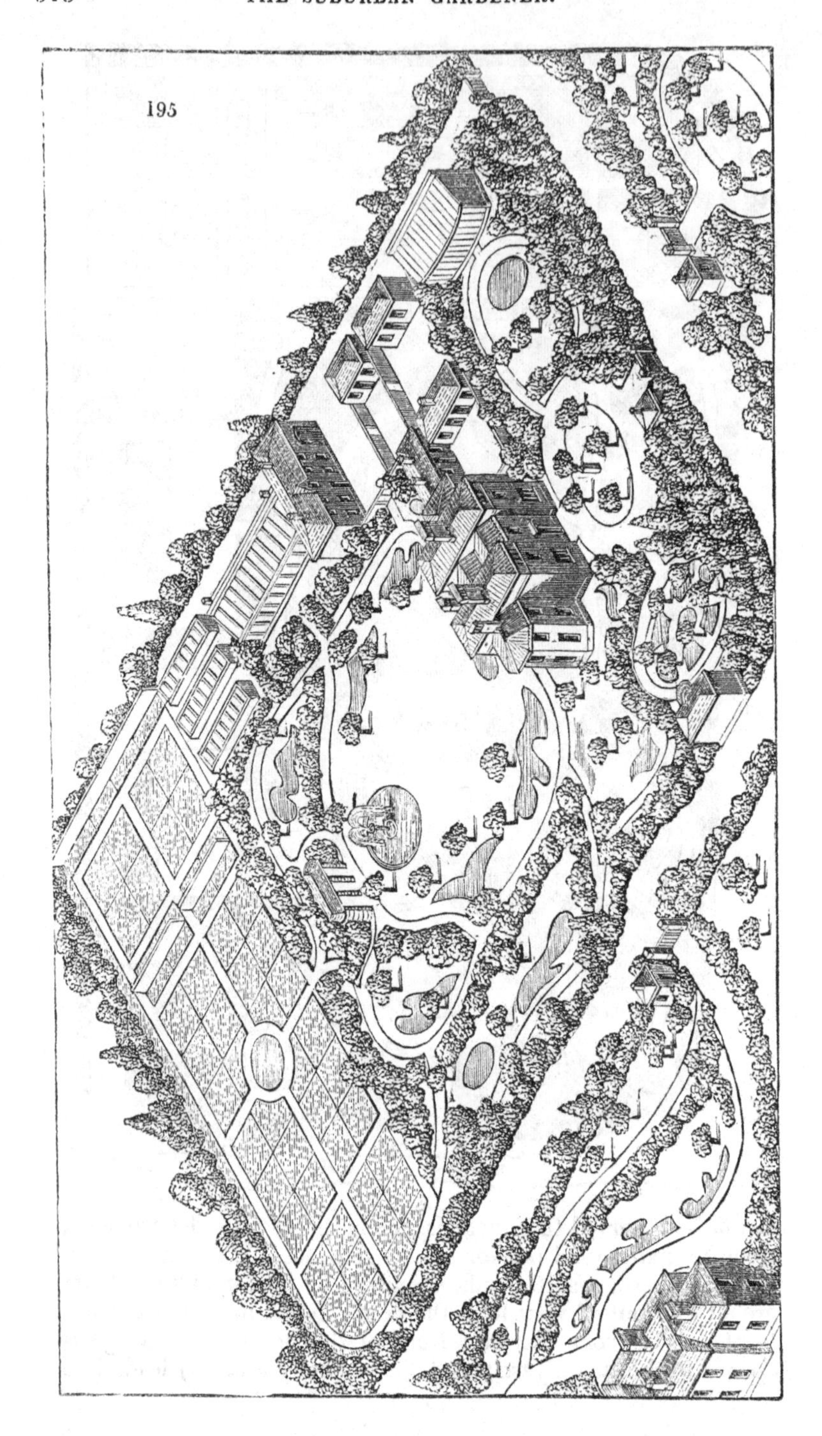
195

the forcing-house department (*d*). The principal entrance to the kitchen-garden (*i*) is in the centre, at the west end of the shrubbery. There is a pond with a fountain (*k*) on the lawn, and, opposite to it, a summer-house (*l*). On the south there is a side walk, which leads to a small flower-garden (*m*), enclosed by a shrubbery, with a reading-room (*n*); and, on the north, nearly opposite, is a conservatory (*o*), and an oval clump at its front on the lawn, with a walk to communicate from the carriage entrance."

Remarks. A very good idea of the general effect of this residence is given by the isometrical view (*fig.* 195.), by Mr. Robertson; in which also a portion of a villa on the opposite side of the road is shown. In this view, and in others of the same kind, only a few of the trees and shrubs are exhibited, because, to have put in as many as actually should exist in such a place, would have created confusion, and defeated the object in view, viz. that of giving a palpable idea of the general effect of the plan. It may be observed, that, though a first-rate house, and the requisite offices to such a house, are shown in this design, yet the disposition of the grounds is equally adapted for a house of a much more humble description. That before us is calculated for a person having from 1000*l.* to 1500*l.* a year; but, by substituting a cottage, with only a poultry-yard, stable, and gig-house, for the mansion and offices, it would be rendered fit for a person having a clear income of from 500*l.* to 750*l.* a year, who would act as his own head gardener.

A Villa of Three Acres and a Half, on a level Surface, with a First-rate House. — For this design (*figs.* 196, 197.) we are also indebted to Mr. Rutger. "The house is approached from the south-west by a lodge entrance (*a*), and, as the road indicates, the carriage returns to the stables on the north-east of the premises. The coach-house (*b*), in this design, is detached from the stables, at the back of which is a yard (*c*), which may be made available for the dung, and, also, for compost for the frame and forcing departments. In the frame-ground (*d*), adjoining to the stable-yard, is a row of sheds, for the gardener's room, and other purposes. The forcing-houses (*e*) are at the back of the frames, separated by a yew hedge (*f*), which also divides the frame-ground from the kitchen-garden (*g*), which is walled on three sides, with an irregular slip (*h*) round as far as the walls and forcing-houses extend. The offices (*i*) are placed at the south end of the house, and hidden by a shrubbery, through which there are a back entrance (*k*), and also a walk (*l*) to communicate with the out-door offices, such as the laundry (*m*), poultry-court (*n*), brewhouse, &c. (*o*), of all of which the view is supposed to be hidden from the house by the shrubs which form the boundary of the oval area at the east front. The north end of the

196

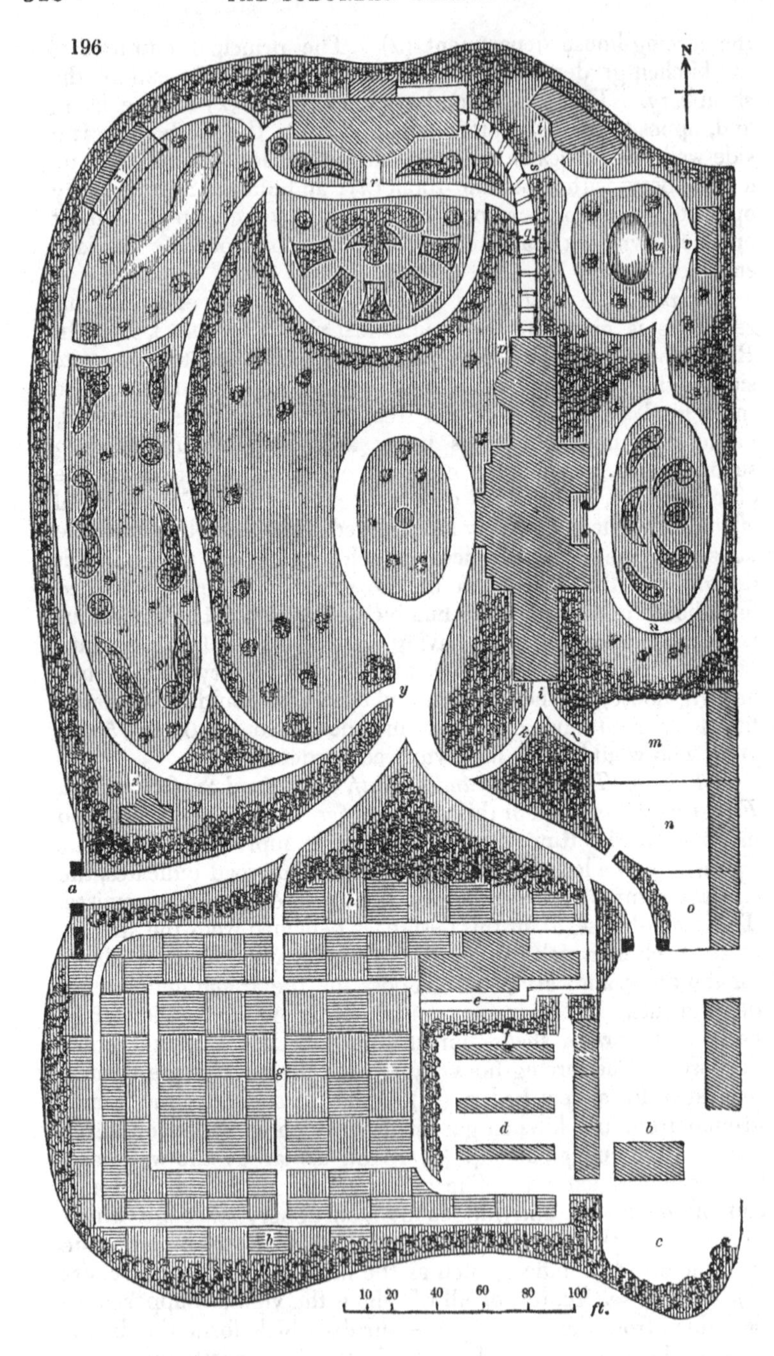

building (*p*) is supposed to be the drawingroom, from which a covered way (*q*) leads to a conservatory (*r*), in front of which is a flower-garden, which may be entered either from the covered way, or from the conservatory. From the covered way, on the right, a walk (*s*) leads out to an aviary (*t*), and to a compartment for fancy fowls, ducks, &c. (*u*), in which there is a building (*v*) to be appropriated according to the wishes of the proprietor. On leaving the flower-garden, and turning to the right, the walk leads through a covered way (*w*), wherein is a seat, and opposite to it a piece of water. Proceeding onward, a summer-house (*x*) is approached, situated so as to have a view of the whole length of the narrow lawn at its front."

The Conservatory. The ingenious manner in which the conservatory is made to front the south, and is, at the same time, connected with the house by the covered way, is worthy of notice. This covered way might be greatly increased in interest, if the roof were glazed, and movable sashes were adapted to the spaces between the supports in front. These sashes might be taken out, and those of the roof taken off, in the summer time, and used for covering beds of cucumbers and melons. A part of the floor of this covered way, next the back wall, should then be formed into a bed for plants, and might contain a very fine assemblage of such of the Australian and Chinese or Japan shrubs as bloom during winter or in early spring, or Cape heaths. The pathway need not, in such a narrow building, be above 3 ft. broad; and it should be carried along the front of the structure, in order that the eye may look on the best side of the plants, which is always the side next the light. Against each support, in front, a creeper may be planted, and trained up it and across the rafter to the back wall. A long, narrow, and not very high, conservatory of this kind is, doubtless, less grand and imposing than one of larger dimensions, a view down the middle of which is seen from one of the living-rooms; but it is calculated to afford much more enjoyment to the lover of plants, as well as much more variety to the general observer; because each plant is, as it were, passed before him in succession. The difference in the enjoyment produced by a long, narrow, and comparatively low, winding green-house or conservatory, like the covered way shown in the isometrical view (*fig.* 197.), and that produced by a rectangular one of large dimensions, such as those commonly appended to first-rate mansions, does not appear to us to be at all understood; or, at least, not to be sufficiently impressed on the minds of proprietors of suburban residences. The object with such persons, when they propose to construct a green-house or conservatory, is to have one which shall be considered large; and, hence, the fear of the expense often deters them from enjoying this innocent luxury. Now, a large conservatory, if it is architectural, forms a very handsome appendage to a house; but, if it is not archi-

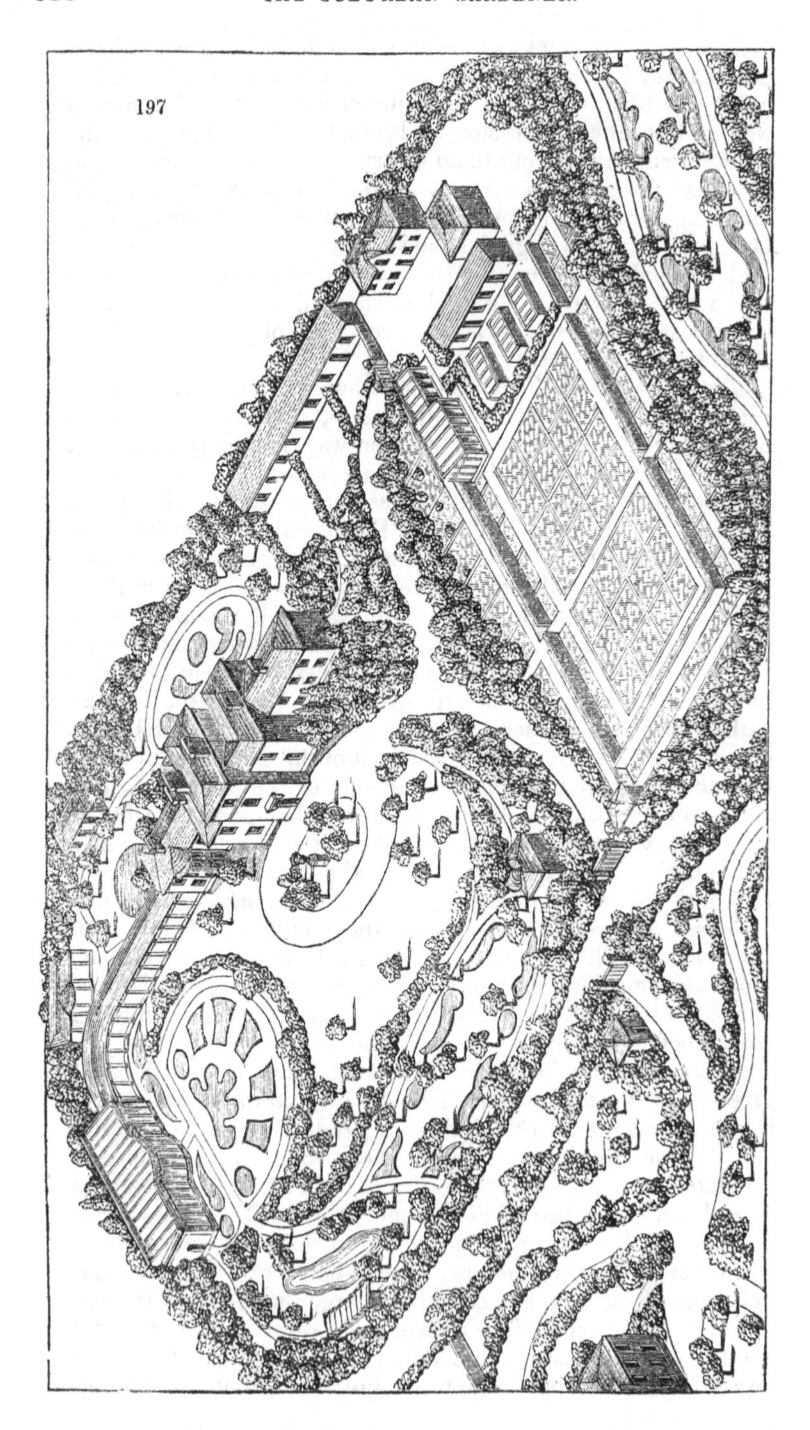
197

tectural, it injures the effect of the house, by its resemblance to the common shed-like forcing-houses of the kitchen-garden or of a nursery. This last kind of conservatory or green-house attached to a house is, therefore, a deformity, rather than a beauty; while the other, though it is a grand architectural appendage, is only adapted to first-rate mansions. On the other hand, a long narrow conservatory, such as we have described, is equally adapted to the cottage and to the mansion; and it costs much less in proportion to the space enclosed than the large structure; while the plants, being nearer to the light, arrive at greater perfection, and, being nearer to the eye of the spectator, are seen to greater advantage. Another advantage of a long narrow conservatory is, that it does not interfere with the windows of the house; because, being narrow, it requires but a small portion of wall for it to abut against, and may proceed from any door or living-room window that may be convenient.

Remarks. This is a very handsome and well-arranged design, suitable for a person having an income of from 1500*l.* to 2000*l.* a year; or with a cottage dwelling, and appropriate appendages, for a lover of quiet and retirement, having from 750*l.* to 1000*l.* a year. The apparent extent of this place might be greatly increased by connecting the walk at *y* with that at *z* by a tunnel; but the practicability and expediency of this will depend very much on the dryness of the subsoil. The extent of the walks might, also, be considerably increased, by conducting one from *x*, over or under the approach road near the entrance, and leading it round the kitchen-garden on a glade of turf properly furnished with trees and shrubs, and so in a tunnel, under the road to the offices, to *z*. Of course, this arrangement would diminish the extent, and vary the position, of the kitchen-garden; and it would require such an arrangement to be made at *i*, *k*, and *l*, as not to have a tunnel longer than three or four yards. All this, and many other arrangements of the kind, might be effected with the greatest ease, and produce a very striking result, provided the subsoil were thoroughly dry, and a little extra expense not objected to. In short, it may be held as an axiom, in laying out suburban gardens of from a quarter of an acre to 20 acres and upwards, that their apparent extent to the stranger walking through them may be doubled or trebled by judicious tunneling in some places, and carrying the road or walk over bridges, through a ruin or rockwork, or under an arcade or trellis-work, in others. Improvements of this kind cannot be shown with much effect on paper; but in reality, when judiciously executed, they have the effect of enchantment.

A Villa of Four Acres, with a regular Outline. — This villa (*figs.* 198. and 199.) is also by Mr. Rutger. "The lodge entrance

198

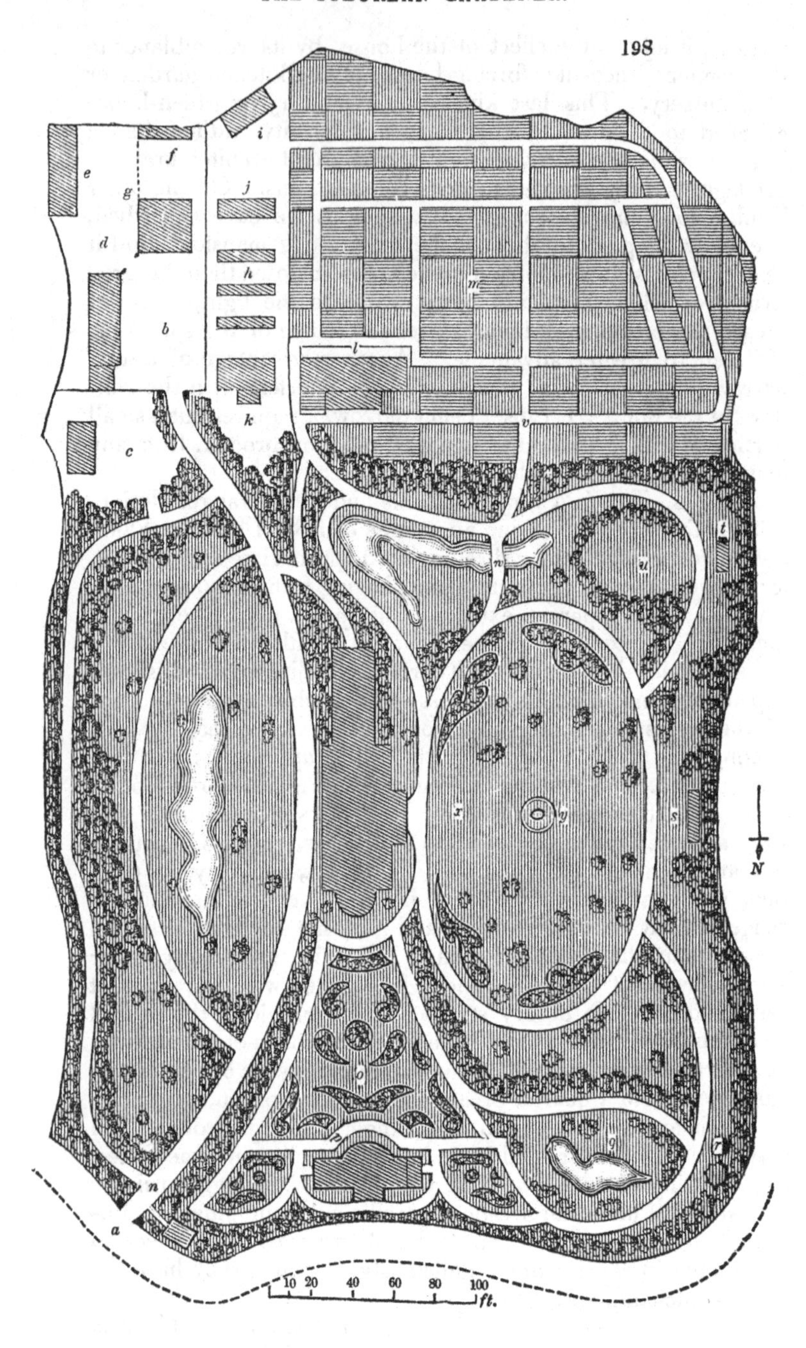

(*a*) is at the north-east corner: the road forms a sweep to the house, and passes on to the stables, at the south-east corner. The stable-yard (*b*) contains the stables, coach-house, and brew-house; at the north of which is placed the laundry, with its drying-yard (*c*). At the back of the stable-court is a yard, in which are a place for the dung (*d*), a piggery (*e*), and a mushroom-shed (*f*). The portion of the yard which contains the mushroom-shed may be parted off at pleasure in a line with the coach-house, as indicated by the dotted line *g*, and may, at the same time, form a place for compost, or for mixing and turning dung. In the yard adjoining, south of the frame-ground (*h*), is the gardener's room (*i*), which may be available, also, if made large enough, for a fruit, seed, and onion-room. The melon-ground (*h*) is supposed to contain three ranges of frames and a pit (*j*), and also a forcing-house (*k*) at the back. There is another forcing-house in the kitchen-garden, contiguous to the frame-ground, at *l*. The kitchen-garden (*m*) has a wall all round it, and slips on both sides and at one end. The walk at *n*, on entering the premises, immediately on the left of the lodge entrance, is intended for the domestics, and communicates with the laundry and stables, and it is also meant as a back entrance to the house.

"On the lawn of the entrance front is a piece of water, with a walk round, communicating at each end with the coach road. Proceeding from the western front to the right, you may enter the flower-garden (*o*), in which there is a conservatory (*p*); or, by leaving the flower-garden on the right, you are led on by different walks, as represented in the figure. At the north-west corner are a piece of water (*q*) and a summer-house (*r*). In the centre, on the west, is a seat (*s*); and at the south-west angle of the pleasure-ground is another structure (*t*), in front of which there is an oval (*u*), intended either for a small bowling-green, or for any other purpose. Proceeding onwards, you come to the principal entrance to the kitchen-garden on the right (*v*); and on the left is a bridge (*w*), leading over a piece of water to the oval (*x*) at the western front, in the centre of which is a circle (*y*), for a fountain, statue, or large vase.

"In a place of this extent, a good collection of choice trees and shrubs might be introduced in such a way as that the whole might almost be considered as an arboretum; and the line of shrubs, which is placed to hide the kitchen-garden, might be converted into an American border."

Remarks applicable to the Six preceding Designs.—The surface of the ground, in all these designs, is supposed to be even, and for the most part flat; in consequence of which, there can be no difficulty in laying out the roads and walks in any direction which the designer may think suitable. It may be asked, then, by what leading circumstance the designer is influenced in

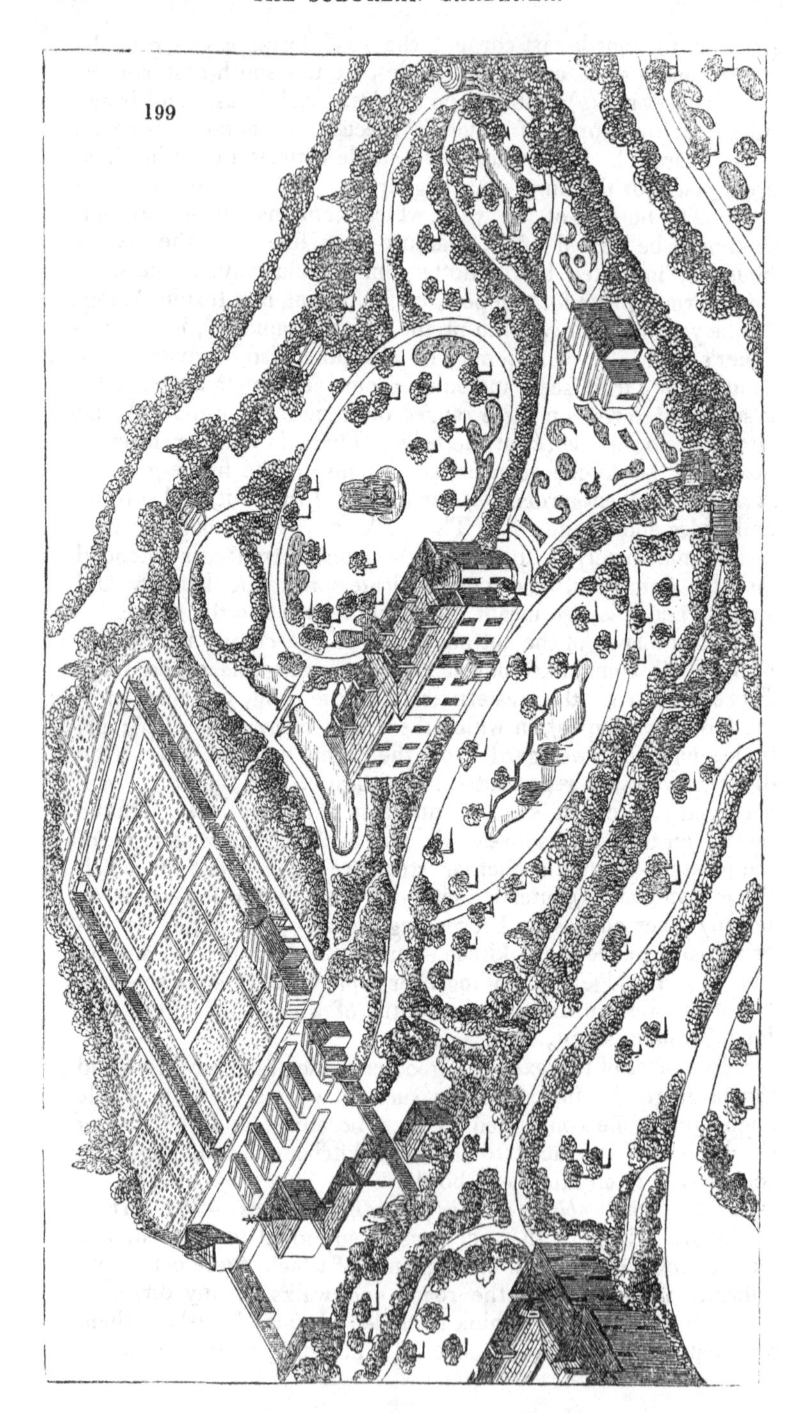
199

tracing the lines for the roads and walks, as shown in these plans? Is it arbitrary, depending on his particular taste; or is it guided by some circumstance in the situation, the wants of the proprietor, and the means which he places at the disposal of the artist?

To these questions we answer, that all the main features, in laying out a place, are indicated by certain circumstances, either peculiar to the situation, or peculiar to the wants and means of the intended occupier. For example, the situation of the entrance lodge is determined by the point or side of the estate on which it touches the public road. If the road touches on only one point, there is, of course, no choice; but, if it borders the estate on one or more sides, then a point is chosen for the entrance which it is considered may be most convenient for use, and may show the place off to visiters entering by it to the greatest advantage. The distance at which the house is placed from the public road, and its position relatively to the boundary fence on every side, depend chiefly, where the surface is even, on the extent of the ground, but partly also on its form; whether it extends in length or breadth, and whether the boundary is very irregular, or tolerably regular. Something, also, depends on the style in which the occupant intends to live. Where there are no horses or carriages, and but few servants are kept, labour is saved by having the house near the road; but, even in this case, if the family had but few visiters, and were devoted to their garden, placing the house in the interior of the property would be more suitable. The situation of the house is also influenced by the room which requires to be found for the offices and kitchen-garden; and, also, by the proper aspect which the fronts of the house ought to have; that for the garden or living-room front being generally the south-east or south-west. The number and direction of the walks are in a great measure determined by the outline of the ground. In all small places, it is desirable to have as long a walk as can be got, for the sake of recreation; and the longest walk is necessarily that which follows the boundary. Hence, almost every place, whether large or small, has a walk round it. The walks in the interior depend on the extent of the place. In the smallest residence, one is required from the house to the kitchen-garden; and a walk round the outside of the latter is found convenient for culture, as well as an agreeable place for recreation. Any other walks than one round the boundary, and another round the kitchen-garden, and connected with the out-door offices, must depend on circumstances peculiar to the situation or the occupier. If there are great natural inequalities of surface, a piece of natural rock or natural water, a walk may be conducted so as to show them to advantage; and, if the occupier is wealthy, and disposed to make the most of the situation, he may create hills, hollows, and other inequalities, form rocks and water, and erect buildings at pleasure; to all which objects walks are either required, or may be legitimately introduced.

The plantations necessary to form a small residence are, in the same manner, guided by reason in all that relates to general effect; and in detail they depend on the taste and means of the intended occupier. Trees are wanted throughout the grounds to connect one object with another; to unite the house with the offices, and partially to conceal the latter; and to unite the place as a whole with other places in the neighbourhood, or with the adjoining scenery. Trees are also required for shade, and for shelter. Shrubs are, in this sense, to be considered as included under trees. They are wanted for thickening masses and screens so that they may not be seen through, and also for sheltering and, in some cases, shading, herbaceous plants; and, in the form of hedges, they serve for subdividing compartments. So far the use of trees and shrubs is guided by common sense; the next point is, to determine the choice of species and varieties, which is in part determined by common sense, and in part by the particular taste and means of the occupier. It is natural that the finer or more choice kinds of trees and shrubs should be placed near the house, as the centre of art and refinement; and, also, that evergreens should abound there, as being in our climate indicative of culture and taste. In a small place, trees of small size will naturally be preferred to those which soon attain a very large size, such as some of the elms and poplars. Trees with showy flowers or fruit will be selected, in preference to those which have the flowers and fruit inconspicuous. In exposed situations, hardy trees which endure the blast will be preferred to such as are more tender; and, where objects are to be concealed, evergreens will always be preferred to trees and shrubs which are deciduous. These general principles are applicable to all small places. The farther pursuance of the subject belongs to the particular taste and means of the individual. One person may choose to render his place a complete arboretum; another may limit himself to a few species of the more showy trees and shrubs; a third may prefer large rapid-growing trees, that he may soon have the pleasure of pruning or thinning them for profit; while a fourth may give a preference to fruit trees, and so on. There is, in short, no end to the variations that may be introduced in the planting of trees and shrubs, independently altogether of those beauties which all trees and shrubs produce, viz. individual expression and character as pictorial objects, variety and intricacy in combination, and botanical interest.

The pieces of water, in all these designs, are liable to objections; but these are partly owing to the particular manner in which the ground plans are drawn, and their very small scale. If these pieces of water and the adjoining scenery are considered as being in the picturesque style, then the pieces of water would require a considerable addition of trees and shrubs to conceal and vary their boundaries, to connect them with the general scenery, and to disguise their poverty of shape. If, on the other hand, they are considered as gardenesque pieces of water, to be planted in the gardenesque manner, then the shapes of the ponds or miniature lakes ought to be handsomer in themselves than they now are, so as to produce an agreeable effect, even without any trees and shrubs whatever as accompaniments. At the same time we do not say that gardenesque pieces of water ought to be without trees and shrubs; on the contrary, they require this addition to display them to advantage, no less than picturesque pieces of water; all the difference being, that in the former case they must be added in a gardenesque manner. The pieces of water in these designs are not sufficiently handsome to be admired as gardenesque, and not sufficiently disguised by trees and shrubs to be decidedly picturesque; but this, as we have before observed, arises in a great measure from the smallness of the scale, and the fear of crowding the places with too many trees and shrubs.

In a future page we shall give plans of pieces of water by themselves, on a larger scale, to illustrate, in a more definite manner, both the gardenesque and picturesque styles, as they have reference to water.

A Villa Residence of Two Acres, within a regular Boundary, laid out in the Geometrical Style. — The object in this case is to produce a splendid effect at a moderate expense of annual keeping, but with no regard to profit. The general form of the ground is that of a parallelogram, and its disposition is so clearly shown in the isometrical view (*fig.* 200.), that it will require little or no description. The entrance is through a straight avenue to a flight of steps, which leads to a raised platform on which the house stands. To the right and left of the avenue are double rows of trees, which may be fruit-bearing kinds, such as the apple, pear, cherry, and plum. Beyond these, on each side, are two small kitchen-gardens, intended for gooseberries, strawberries, and other small fruits, and for pot-herbs, tart rhubarb, spinach, kidneybeans, and a few such vegetables as are desirable to have always at hand. The house and these kitchen-gardens occupy about half the entire residence. The other half is laid out in the form of a sunk flower-garden, consisting of a variety of curvilinear beds, bordered by a kerb of stone, and surrounded by turf. From the terrace walks there are four descents to this garden, each consisting of a double flight of steps. Each bed is supposed to be planted with one kind of herbaceous plant, so as to produce

200

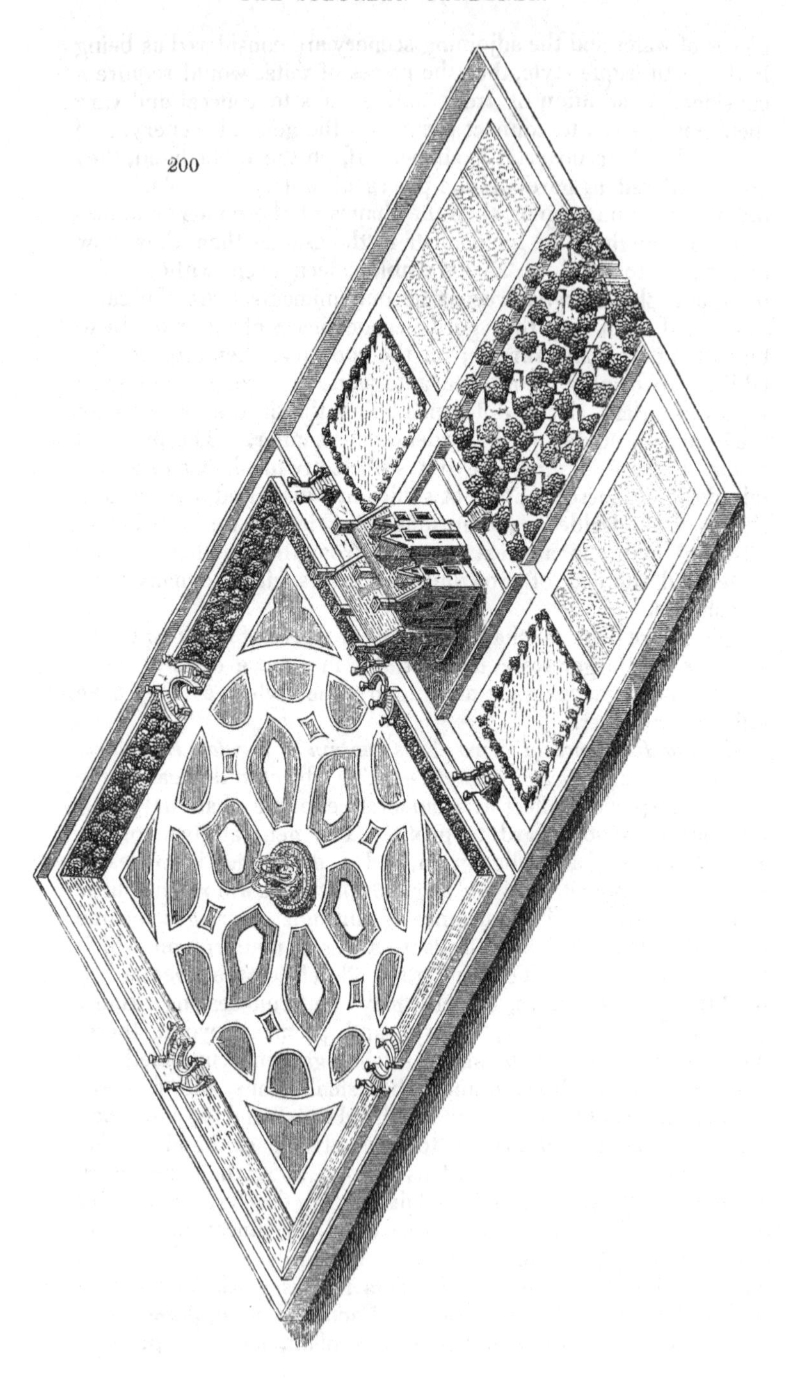

large masses of colour. The mode of selecting plants for this purpose, as well as lists of suitable plants, have been already given (p. 217. to p. 226.), and further resources will be found in our catalogue. The sloping border between the sunk area and the flower-garden may either be planted with low evergreen shrubs, with roses kept low, or it may be in turf, or in rockwork: in the latter case, it may be covered with a collection of rock plants. Perhaps the most appropriate disposition of this sloping border would be to vary it with ornaments of box, on a ground of turf, so as to give it the appearance of an architectural moulding. In the centre there is a fountain. In situations where so much turf was not desirable, the walks between the beds might be of gravel or paved; but they will produce the best effect in turf.

Instead of a raised terrace-walk surrounding this flower-garden, there might be in its place a narrow conservatory, such as that remarked on in p. 526; in front of this conservatory, a narrow border for Cape and other half-hardy bulbs; then a terrace-walk, two steps lower than the level of that within the conservatory, with another narrow border, and beyond that a low parapet wall, ornamented with vases. This terrace-walk, though not so high as that represented in the figure, should still be at least 5 ft. above the level of the flower-garden, in order that the spectator may look down on it in such a manner as to see the shapes of the beds. We have observed, in a preceding page, that wherever the figures forming a flower-garden are regular or symmetrical, to be seen in their full beauty, the eye should be so elevated, as, when looking down on them, to be able to comprehend the entire shape of each bed. What the height of the eye ought to be, to do this, may always be determined beforehand, by ascertaining the width of the flower-garden or symmetrical figure, the distance between the beds, and the position of the spectator. In general, the angle made by a line drawn from the farther edge of the most distant bed to the eye of the spectator, should not be less than 15°; and hence, taking the height of the human eye at 5 ft., a square or circular symmetrical flower-garden, of 40 ft. in diameter, ought to be surrounded by a walk raised to the height of at least 4½ ft. above its level; while a flower-garden of double the size ought to be surrounded by a walk 9 ft. high. When a parterre of symmetrical beds is to be planted with low shrubs, such as rhododendrons, azaleas, kalmias, &c.; the surrounding walk ought to be made higher, in proportion to the anticipated growth of the plants. These rules, the correctness of which no person of any experience will, we think, dispute, show that, when very large spaces are to be laid out as flower-gardens, symmetrical figures need not be attempted, except near the walk, where they will be immediately under the eye, unless they are so situated

as to be seen from different heights. Hence, in very extensive designs, such as *fig.* 201., two systems of symmetrical figures are adopted; one near the margin of the walk, to be seen from the terrace immediately adjoining; and the other in the centre, to be seen from more elevated and distant points of view. If the reader will always bear in mind, that, to see any figure distinctly, the rays of light reflected from the most distant points of it must meet the eye of the spectator at an angle of 15°; and that this angle will not be obtained by an individual, whose eye is 5 ft. from the ground, at a greater distance than 20 ft., he will never be at a loss.

Fig. 202. shows the different heights which a terrace-walk requires to be above the level of the flower-garden, according to the size of the garden.

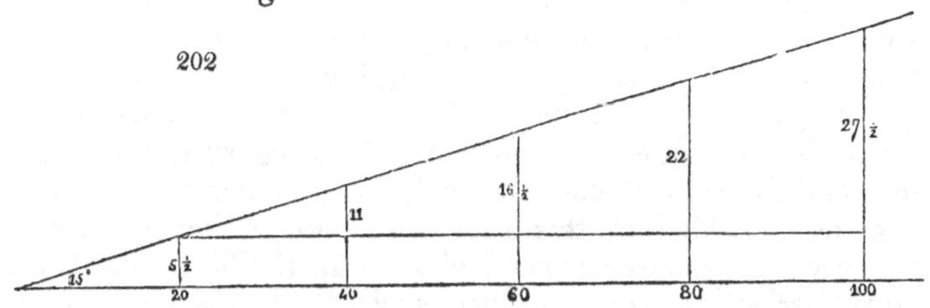

A Suburban Villa of Four Acres, with a regular Outline.—We introduce this design (*fig.* 203.), from the fertile pencil of Mr. Rutger, on account of the valuable discussion which he has connected with it. "The following hints," Mr. Rutger observes, "may be acceptable to such as have little experience in gardening. I shall commence by saying a few words concerning gardeners.

"*Gardeners employed in Suburban Gardens.* In many instances the gardens around the metropolis are too small for the employment of a gardener; hence, in gardens of this description, such as are termed jobbing gardeners are employed; among whom, perhaps, too few are to be found of sufficient judgment to crop to advantage; and, what is still worse, all the interest that they feel on the occasion is to get a day's work or so a week. In some instances, the care of small gardens is confided to a man of all work, who may have to act as groom, wait at table, &c. In cases of this kind, much cannot be expected in the way of cropping to advantage. An instance or two have come under my notice, where a servant out of livery has had the care of the garden, with permission to hire in a labourer occasionally when wanted. When this is the case, should such person have a taste for gardening, considerable advantages may accrue to the master. However, it must be understood that what is so far stated has nothing to do with places which are of sufficient scope to employ one or more persons to keep them in order. Where only one gardener is kept, or, as it is technically termed, where it is a single-handed place, the man must necessarily be chosen from the lowest order of regular gardeners, as all of a higher grade will expect to have labourers under them to do the rougher work. But as, even among the gardeners who will take a single-handed place, there are many very far superior to others, it will be worth while to give good wages to get one of the best, as it will be much to the advantage of the master to obtain a person of a tolerable degree of intelligence. A very few pounds a year extra will be well disposed of in this way, and go far towards securing the best interests of the master; and here, at the commencement of the subject, it may be proper to remark, that a considerable share of intelligence is necessary to be possessed by a gardener, even of the lowest grade. It is true that there are some who, in a sort of mechanical way, from what they have seen done by others, may go on tolerably well in producing crops in due season; but these are not the persons from whom there can be expected anything out of the general routine of gardening; and they will be found incapable to compete with the difficulties that may occur with regard to soil, the seasons, and the various enemies which the gardener has to contend against. Added to this, they will be found incompetent, with respect to taste, to regulate and keep up the flower-garden and shrubberies, and to introduce such plants and flowers as would have a good effect. In short, a gardener, to be one at all in the true acceptation of the term, must be a man of intelligence.

"Next to the lowest class of gardeners, those may be considered as a step in advance, who have a man or two under their control; and these should be men possessed of a tolerable education, as well as of a superior taste and judgment in gardening in general; as, in many instances, it will be found that a gardener in a small place, who has but one man under him, will be expected to plan and execute, or, at all events, to execute, alterations and improvements which will require a considerable degree of knowledge in his profession. In cases where there is only one man allowed, the gardener, in justice to himself, should be permitted to make choice of a person who is civil, active, and industrious, and of sober habits; which, with corresponding qualities on the part of the gardener, will cause things to go on pleasantly, and at the same time greatly facilitate the work.

"I might thus go on to show that, in proportion to the number of hands employed, and the nature and quantity of the ground to be kept up, so will it be necessary for the gardener to possess a taste for his profession, education, experience, &c.; but the above must suffice, and I will only add, that, without these acquirements (especially in the present day), a gardener will be thought but little of, either by his employer, or by his brethren in the profession.

203

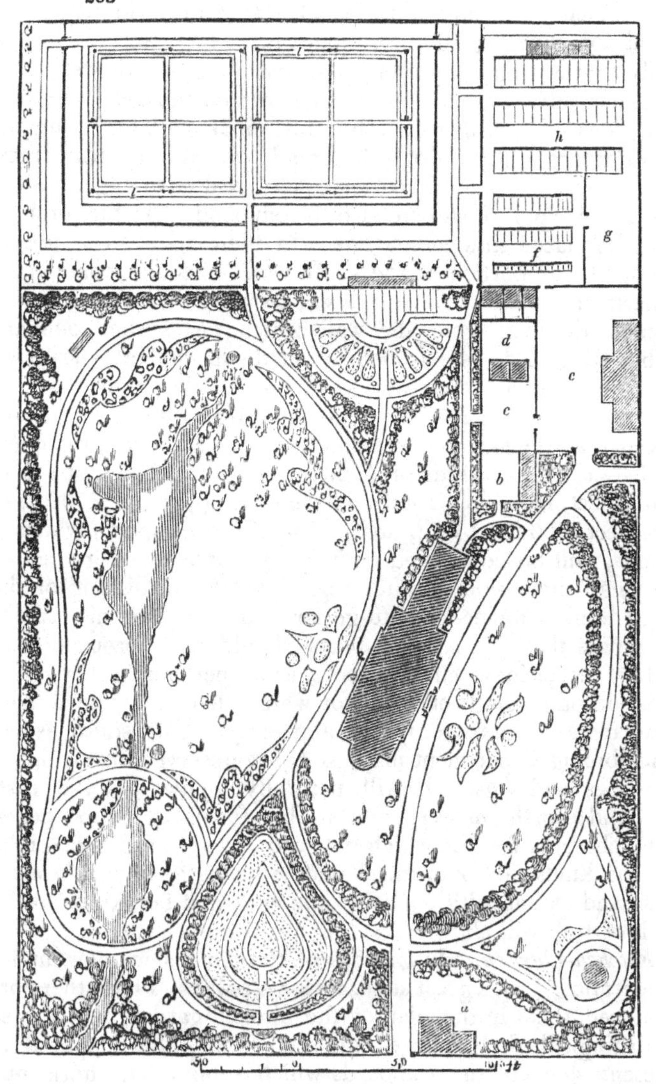

a, Entrance lodge. *b*, Poultry-court. *c*, Laundry, brewhouse, and yard. *d*, Piggeries. *e*, Stables and yard. *f*, Melon-ground and sheds. *g*, Dung-yard and shed. *h*, Forcing department. *i*, Rosary or dahlia ground. *k*, Conservatory and flower-garden. *l*, Espalier rail in the kitchen-garden.

" *Disposition and Appropriation of the Grounds.* I will now suppose a place, the extent of which shall be four acres, three acres of which are to be appropriated for the house, yards, pleasure-ground, flower-gardens, &c., and the remaining acre for the kitchen-garden, including the forcing and framing departments, which, according to the sketch (*fig.* 203.,) will reduce the ground for cropping to three quarters of an acre. It will be readily acknowledged by every practical gardener, that slips are at all times advantageous, and particularly on the east, west, and south sides of a garden; as, by this mode of laying out the ground, a larger portion of wall is made available for fruit trees, which, especially in small gardens, is a great advantage. As a general principle, I should also recommend that the forcing-houses be placed in a compartment by themselves, so as not to swallow up any portion of the south wall; and, also, that, in the compartment where they are placed, they should be erected at a sufficient distance from the wall which forms the boundary to the north side, so as not to shade it, and at the same time to give space for a walk and border, the latter of which may be made available for vines and figs. Thus far it will be seen that the sketch corresponds with the above principles, and that it gives upwards of 1100 ft. of wall; namely, 508 ft. of south aspect, including 108 ft. in the forcing-ground; 200 ft. of east aspect, 200 ft. of west, and 200 ft. of north aspect; which, I presume, will be considered sufficiently ample for a garden of such small dimensions. With regard to the furnishing of the walls, I think it unnecessary to go far into detail. It may suffice to intimate that the south aspects should be appropriated to peaches, nectarines, and apricots, with, perhaps, a few early plums, or other kinds of fruits for which there may be a partiality, and a desire to get in fine and early. The south aspect in the forcing department may, as above observed, be furnished with vines and figs. It will naturally follow that the east, west, and north aspects are to be furnished with pears, plums, cherries, &c., of which as good a selection as possible, from the kinds now known, should be made, and in as great a variety and extent, with regard to the early and late kinds, as the length of wall will allow.

"*Kitchen-Garden.* In gardens of the dimensions now under consideration, I am a great advocate for espaliers. I have therefore introduced them in the plan of the kitchen-garden, where they are distinguished by large dots, which may be supposed to represent some of the standards which support the horizontal wires. It is desirable that every garden, however small, should produce apples both for the kitchen and for the table; and, consequently, apple trees are frequently grown in the quarters, where they not only injure the quality of the culinary crops, but prevent regularity in cropping. The length of espaliers in the sketch is about 580 ft., which, if judiciously planted, will give room for a tolerable supply. A proper selection of sorts is, of course, necessary, in order to have a succession of table fruit, and also of fruit for the kitchen. I would here observe, that, in making a selection of the best sorts of apples for espaliers, preference should be given to those kinds that are most inclined to throw out spurs, and require least skill in pruning. Among the table fruits, the Ribston pippin and the Margil are excellent in this way, and highly to be recommended, both with respect to their flavour, and the crops they will produce. It is particularly requisite for espalier apple trees, that they be kept clean from the aphis, or American blight, as otherwise in a very few years they will receive serious injury; not only with regard to the health of the trees, but also to their appearance, as gouty-looking excrescences will be formed round the spurs, which will render them extremely unsightly. Where the trees are young, and the bark smooth, the blight, whenever it is observed (and its cottony appearance renders it immediately conspicuous), may be easily removed by a painter's brush and common water; and where the bark is rough, and there is reason to believe that insects are concealed in its crevices, they may be suffocated by painting such places over once or twice with a mixture of clay and water, about as thick as cream. Two coats, Mr. Waterton, the inventor of this process, says, will fill up every crack, and soon smother every bug and every egg which has been deposited. With the expansion of the tree, and the weather, the clay will drop off of itself. (*Waterton's Essays*, p. 276.)

"In small gardens," Mr. Rutger continues, "I should be inclined to place the espaliers near to the walks, and only leave room for a row of strawberries by way of an edging, and thus secure to the quarters as much ground as possible; and, for the same reason, narrow walks, not exceeding 4 ft. 6 in. in breadth, should be laid down. In case the walls should be found not sufficiently extensive to have enough of cherries and plums, a few standards of each, and also of damsons, may be planted in lines along the back of the south slip; and in advance of these a line of gooseberries, currants, and raspberries, fronted by as many strawberries of different kinds as may be required, leaving the remainder of the ground for garden crops. At the back of the western slip there may be a row of filberts and berberries, and in front as many pot-herbs as may be wanted. The border against the fence in the eastern slip may be devoted to rhubarb, sea-kale, and artichokes; as, being near to the frame-ground, it will be found convenient for bringing in dung or leaves, for forcing and protection.

"Should the subsoil of the garden be not congenial to the health of the wall-fruit trees, I am inclined to advocate the chambering system for the south borders. To do this in perfection, it will be necessary to excavate the soil to at least 3 ft. deep, and to pave with flagstone at the bottom; or, if chalk can be had in sufficient quantity, a layer of about 6 in. thick, well beaten down, may suffice. This pavement, or layer, should form an inclined plane, descending a little from the wall to the walk in front, where there should be a drain constructed to run parallel with the wall. The bottom of the drain should be about 3 in. under the level of the pavement, and constructed with pigeon-holes about 3 ft. distant from each other, at the side next to the border. On the pavement I should recommend a layer of brickbats or lime rubbish of about 8 in. thick, the roughest parts of which should be laid near the pigeon-holes, in order to facilitate the escape of an excess of moisture. The chambering, &c., thus prepared, the border may be filled with compost, to such a height, as, allowing for the chambering and the thickness of the lime rubbish, will leave the soil about 2 ft. 6 in. deep, which should remain two or three months to settle before planting the trees. The annexed diagram (*fig.* 204.) may, perhaps, explain

204

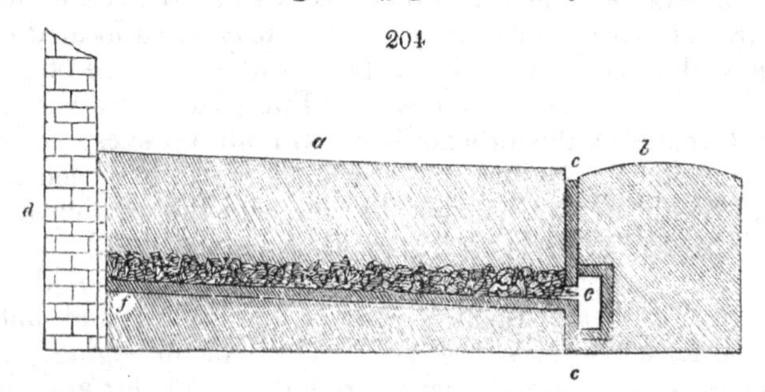

more clearly what is meant. In this diagram, *a* shows the surface of the border; *b*, the surface of the walk, separated from the border by the chambering wall (*c c*); *d* is the back wall; *e* is the drain; and *e f*, the paved bottom of the chamber, on which the brickbats, &c., are laid.

"Experienced gardeners, in general, consider that a good loam, of rather an adhesive quality, is the best soil for peach and nectarine trees; and, if taken from the surface of a field with the turf, all the better. In this case, it should be laid by for a few months, and receive two or three turnings, during which the turf may be chopped into small pieces, so as to cause it to be regularly distributed throughout the soil. Some gardeners mix a small portion of rotten dung or vegetable mould with the loam; but, if the loam be really good, the less that any manuring substances are introduced the better, as they only tend to produce a luxuriancy of growth beyond what is wanted, and to retard the ripening of the wood in the autumn.

"*System of Cropping.* I will now endeavour to say a few words on the subject of cropping: but, although a datum has been fixed in the sketch (namely, three quarters of an acre for crops), yet, as in this department much must depend upon the kinds of vegetables which are held in the highest estimation by different proprietors, some wishing to have more of this, and others more of that, and also on other contingencies which must be taken into the account, I doubt if much can be satisfactorily written on the subject; rather, therefore, than attempt to limit the precise quantity of ground for each crop as it may be called for in succession, I shall give a few hints upon general principles, which must suffice.

" In small gardens, the greater the variety and perfection that the vegetables are brought to, the greater the merit of the gardener; but, in order to do this, he should always have manure at command, as double cropping must be continually going on, in order to have a good supply. The gardener, on his part, should mark well the quantity of ground he has to work upon, so as to apportion to each crop so much, as not to have a redundancy on the one hand, or a scarcity on the other. Some inexperienced gardeners fall into sad mistakes in this way, with those kinds of vegetables the crops of which are inclined to come in nearly all at once: among these may be named cauliflowers, broccoli, and lettuces, with some others; consequently, these should be sown in quick succession, or small crops should be planted rather than large crops, half of which, however fine when brought to perfection, cannot be available for the kitchen. This rule is particularly applicable to spinach, it being a plant that runs quickly to seed. It should be the gardener's study, also, not to throw away a crop, by sowing at an improper season. For instance, if scorzonera and salsify are wanted for a winter supply, and the seeds are sown in March or April, the object will be defeated, by the plants running to seed in the autumn: the best crops of these for a winter supply I ever saw were sown in the middle of June. Much might, perhaps, be written upon this part of the subject, relating to the peculiarities of the growth and coming in of the different kinds of vegetables necessary to be cultivated, in order to obtain a constant supply from a small garden, but the above must suffice. It, however, will be seen from these brief remarks, that special care ought to be taken, in small gardens, to husband the ground well, in order to obtain as large and as constant a supply of vegetables as the space allotted will produce. During the latter years of my own practice, I entirely abandoned the broad-cast system of sowing, and adopted that of drilling for all my standing crops; and, for the extra time that it took to get them in, I found myself amply repaid in the after-work of thinning and keeping them clean from weeds; while it added a neatness to the appearance of the crops when growing; and, I think, I may also safely add that the crops were of a superior quality.

"*Kinds of Vegetables to be grown.* In order to bring under one view the various kinds of vegetables which are grown in the present day, it may not be amiss to transcribe the list as it stands in your new edition of the *Encyclopædia of Gardening*, p. 812., which is as follows: —

'*The cabbage tribe;* comprehending the white and red cabbage, cabbage-colewort, Savoy, Brussels sprouts, borecoles or winter greens, cauliflowers, broccoli, Portugal cabbage, &c.

'*Leguminous vegetables;* comprehending the pea, bean, and kidneybean.

'*Esculent roots;* comprehending the potato, Jerusalem artichoke, turnip, carrot, parsnep, red beet, skirret, scorzonera, salsify, and radish.

'*Spinaceous plants;* comprehending the garden spinach, white beet, orache, wild spinach, New Zealand spinach, sorrel, and herb-patience.

'*Alliaceous plants;* comprehending the onion, leek, chives, garlic, shallot, and rocambole.

'*Asparaginous plants;* comprehending asparagus, sea-kale, artichoke, cardoon, rampion, alisanders, hop, bladder-campion, cotton-thistle, and milk-thistle.

'*Acetarious plants or salads;* comprehending small salads, lettuce, endive, chicory, dandelion, celery, mustard, rape, corn-salad, garden cress, American cress, winter cress, water-cress, brook-lime, scurvy-grass, garden rocket, burnet, buckshorn plantain, ox-eye daisy, and some of those included in other sections, as the sorrel, tarragon, Indian cress, &c.

'*Pot-herbs and garnishings;* comprehending parsley, purslane, tarragon, fennel, dill, chervil, horseradish, Indian cress, marigold, borage, and some others included in other sections.

'*Sweet herbs;* comprehending thyme, sage, clary, mint, balm, marjoram, savory, basil, rosemary, lavender, tansy, costmary, and some of those in the preceding section.

'*Plants used in tarts, confectionery, domestic medicine, and for destroying insects;* comprehending rhubarb, gourd, angelica, anise, coriander, caraway, rue, hyssop, chamomile, elecampane, liquorice, blessed thistle, wormwood, tobacco, and some others.

'*Plants used as preserves and pickles;* comprehending love-apple, egg-plant, capsicum, caper, samphire; and the red cabbage, Indian cress, radish, kidney-bean, marsh marigold, &c., included in other sections.

'*Edible indigenous plants, neglected or not in cultivation;* comprehending the sea-beet, nettle, sea-peas, and a variety of other natives.

'*Foreign hardy herbaceous vegetables, little used as such in Britain;* comprehending the Claytònia and *Basélla* spinach, the maize, the tuberous-rooted nasturtium, the battatas, or sweet potato, Chenopòdium *Quinòa*, *O'*xalis crenàta, &c.

'*Edible British fungi;* comprehending the mushroom, truffle, and morel.'

"Although the above list comprehends many articles that may not be required by the proprietors of small gardens, it may still be of use, both to gentlemen and gardeners, to select from it such vegetables, &c., as may be thought the most useful; and in such quantities as the extent of ground they have for cropping will allow.

"If it be enquired, what proportion of ground should be set apart for the kitchen-garden, out of any quantity given to lay out as a place, the question will be difficult to answer; as it must greatly depend upon the wishes of the person who may be in possession of the ground which he wants to have laid out.

Should a good kitchen-garden, with all its appendages, such as forcing-houses, framing, &c., be preferred to the flower-garden and shrubbery department, then, of course, ample room must be devoted to the former purpose; and *vice versâ*, if preference should be given to the latter. Should the question be asked as to the smallest quantity of ground that is necessary to give a successional supply throughout the season, of all the vegetables that are in general use, perhaps the minimum of ground absolutely necessary for that purpose may be stated at half an acre; but in this case the crops must be very small, and the ground well attended to in regard to manure, double cropping, &c. But when a kitchen-garden is small in proportion to the supplies that it is expected to yield, a gardener will always be in difficulties, even under the best management, to find ground to carry on his operations, so as to give satisfaction either to his employer or to himself.

" *Forcing Department.* In turning to the forcing department in the plan (*fig.* 203.), there will be found a range of glass of about 70 ft. in length, and a pit in front of the range. It is not here intended to point out to what purposes these shall be applied; as pines, grapes, peaches, &c., may be grown in them in any proportion agreeable to the proprietor. For heating them, the hot-water system will be found decidedly the best; and, if the apparatus be properly constructed, one boiler will probably be found sufficient for both the range of houses and the pit

" In the frame-ground, there are two ranges of frames, about 40 ft. each in length, and in front of these a cucumber ridge. Brick walls, for supporting the frames, may be constructed on M'Phail's system, or in any other approved mode.

As an appendage to the frame-ground, two or three dozen hand-lights will be necessary for ridge cucumbers, striking cuttings, cauliflowers, &c. The frame-ground has in it two sheds at the back; and, for convenience, it is placed contiguous to the dung-yard and compost-ground, in which there is also a shed. These sheds will be found useful for growing mushrooms, forcing rhubarb, sea-kale, and for various other purposes which it is unnecessary to enter upon here in detail.

" *Pleasure-Ground Department.* In glancing over the ornamental parts of the sketch, it will be seen that the whole is thrown into a kind of shrubbery-garden, with the exception of a flower-garden in front of the conservatory, and a rosary or dahlia-ground on the left. However, it is not intended that the laying out of the clumps should be strictly confined to the mode adopted in the design; some of them, and particularly the two which are placed on the verges between the house and the water, may be displaced, and a group or two of small clumps on each side may be introduced in their stead, to be furnished with flowers in mass. The same mode may also be adopted in laying out the lawn at the entrance front, and on the small lawn at the bow end of the house; added to which, the whole may be heightened in beauty and variety by the introduction of ornamental objects, such as vases, &c., if placed in proper situations; and also by floral devices, such as pillars of roses, cones of geraniums, and other ingenious modes of training such plants and flowers as are applicable for the purposes of embellishment. The shrubs and trees for the lawns of a place of this extent should be of the most choice kinds that will endure our climate throughout the winter; of which, in the present day, a sufficient number may be easily procured, to give all the variety wanted to keep up a perpetual interest throughout the year.

" *Expense of Keeping.* The expenses necessary to keep up a place of the kind here described must depend, like all others, upon the state in which it is kept: high keeping will always make a suitable return to every admirer of a neat garden; and this will hold good with regard to the kitchen-garden, as well as to every other department. A gardener, with two men and a boy throughout the spring, summer, and autumn, and with one man less during winter, would, I think, be adequate to the labour necessary for this place, with its forcing department, framing, conservatory, flower-garden, and grounds; and with this assistance it might be preserved in what is termed high keeping. With regard to the number of hands requisite to keep up a place of any given size, it must be considered that it is not merely upon the extent of the ground that a calculation can be made; but rather upon the quantity of forcing and framing that is to be carried on, together with the labour that will be required to keep the shrubberies, flower-gardens, and green-houses or conservatories, in the highest order. In reference to the sketch here given, were it divested of the forcing-houses, conservatory, and flower-gardens, and the shrubberies, lawns, and walks but meanly kept up, perhaps a man and a boy would be sufficient for the purpose; but in this case the enjoyment would be proportionately small.

" In places of small extent, the principal art necessary to make them in the highest degree enjoyable is, to introduce as much variety as possible, void of confusion, turning into account all that nature may supply, or that art can produce. If a streamlet should offer itself, it may be made use of in a way somewhat similar to what is done in the sketch; and it may be made available, also, by a conduit, for the use of the kitchen-garden, conservatory, forcing department, stables, &c. Undulated ground may be assisted by raising a mound or two, in appropriate situations, from the parts which it would be desirable to have level; and on these mounds suitable erections may be placed, particularly when a distant and interesting view is to be obtained. The mounds should be at least partially covered with shrubs and trees, with a winding walk leading to their summits. A grotto will be always considered an agreeable companion with water; and, where it can be erected with advantage, it should not be lost sight of. Rockwork, also, in the vicinity of water, is agreeable, and, if constructed with taste, will add a pleasing feature to the variety which in sundry other ways may be introduced; such as a trellis arcade covered with creepers, a covered rustic seat, an alcove, vases for flowers, together with other devices, which fancy, combined with taste, may suggest; but, in all this, harmony of design should be studied in the laying out and planting, and also in regard to the embellishments that may be introduced afterwards."

Remarks. These observations by Mr. Rutger are exceedingly judicious; and, when our readers are informed that they are the result of the extensive observation and experience of a person of great intelligence and sound sense, who possesses a thorough knowledge of gardening in all its branches, as well as of all the wants of a gentleman's family, their value will be duly appreciated.

Renovating Third-rate Suburban Gardens.—We have little on this subject to add to what we have said respecting renovating fourth-rate gardens, from p. 306. to p. 313. Where a third-rate garden has been neglected, almost all the trees and shrubs will have become overgrown and shapeless, and will require to be either cut in or cut down, or even, perhaps, rooted out.

In cutting down trees and shrubs, regard must always be had to whether the plants so cut have the property of springing up again from the stool or root; or, as the technical term is, whether they stole. It often happens that trees and shrubs which have become naked below, and very unsightly, when cut down within an inch of the ground, will spring up with great vigour; and, in an incredibly short time, will form splendid bushes, or, if trained to a single stem, handsome trees. This most sorts of trees and shrubs will do when cut down, in less than one fourth of the time that would be requisite for the attainment of the same result from young trees. Before, however, the operator ventures to cut down trees or shrubs to the ground, he ought to inform himself of two particulars respecting them. First, whether they are of kinds that stole, and, secondly, whether or not they have been grafted. To cut down a tree or shrub that does not stole, or that stoles imperfectly, such as the spruce fir, the holly, or the juniper, is, in effect, to remove it altogether; and to cut down a grafted tree or shrub below the graft, even if it should stole, is to substitute one kind for another. For example, if a medlar, which is generally grafted on a thorn, and an almond, which is commonly grafted on a plum, were cut down, there would shoot up in their stead a common hawthorn and a wild plum; while, if they were cut above the graft, we should have the medlar and almond reproduced. Some of the finest trees that are introduced into pleasure-grounds are different species and varieties of crabs, cherries, thorns, plums, maples, variegated sycamores, oaks, elms, to which we might add a host of others, all of which are grafted on the common and free-growing species of their respective genera; some a few inches above the surface of the ground, and many at heights varying from 5 or 6 inches, to 5 or 6 feet. If, then, these trees were cut down close to the ground, we should have nothing but the commonest kinds coming up to succeed them. In some cases, indeed, where a stock has been used which does not stole, as in the case of all the resinous tribe which are grafted, there would be no succession at all, which is still worse. There are also a number of trees and shrubs which stole freely when young, but which scarcely stole at all if not cut down till they have attained their full size. Among these are the beech, hornbeam, Scotch elm, birch, privet, phillyrea, arbutus, and a number of others, all of which will be found indicated in our catalogue.

By neglecting to attend to cutting above the graft, we have known a tolerable collection of trees and shrubs reduced to a mass of the commonest kinds, which it was necessary to have regrafted, or rooted out, to give place to fresh plants; and we know a large camellia-house in Kent, in which a collection of the finest kinds of camellias having lost their leaves by insects and disease, and having been ordered by the head gardener to be cut down, in order to renovate them, were cut by an ignorant journeyman so low, that all the shoots from the stools proved to be the single red. When a gardener of skill, and some taste, has the renovating of an old place, he will generally be able to produce, by cutting in and cutting down, very striking effects in a short time, from the great rapidity of the growth of shoots from stools, the number which are thrown up, and the immense bush, or tree, which is formed by them. We have known the stool of a common laurel throw up shoots 6 ft. high in one season, and a tree 20 ft. high raised from the stool of an *A*cer Pseùdo-*P*látanus in five years. The stools of the locust (Robín*ia* Pseùd-*A*càcia), in suitable soils, will produce a considerable tree in three years, all the suckers being removed but one.

If the person who has purchased or taken a lease of a residence which requires renovating be much attached to gardening pursuits, and can afford the expense, his best mode will often be to root out every tree and shrub on the premises, except specimens of decided beauty, variety, singularity, or usefulness. By specimens of decided usefulness are meant such trees as shelter the house, or some object within the boundaries of the residence; such as conceal objects which it is not desirable to see either within the boundaries or beyond them; and such as produce shade, say for a seat or walk, during the heat of summer. The condemned trees and shrubs having been rooted out, the ground should be thoroughly trenched to the depth of 3 or 4 feet, and mixed with manure; the drains, walks, walls, and all the buildings, being attended to, and put in repair, according to the mode suggested in p. 306., &c. If the ground be trenched 3 or 4 feet deep, it should remain at least six months before it is planted, in order that it may be consolidated by rains. For this reason, the best time for deep trenching is early in autumn, in order that the rains of that season may equalise the sinking of the soil. In many, and perhaps in most, cases of renovating a third-rate garden, the situation of some of the main features, such as the flower-garden, the kitchen-garden, the conservatory, some of the ornamental structures, &c., will require to be changed; and this is another argument in favour of the rooting out of the greater number of the trees and shrubs. It may also be thought advisable to make some undulations or other inequalities on the surface of the lawn; or to enlarge the apparent extent of the place, by carrying some of the walks under others in tunnels, or over them on bridges; and this cannot be done without the removal of all the trees and shrubs in such situations. In changing the situation of the kitchen-garden, regard must be had to a proper communication between it and the stable-yard, for dung; because, if the garden has been properly placed at first, it will, in general, be immediately adjoining the stable-court; and, consequently, if its situation be changed, it can only be to some spot more or less distant from it, which will hence be more or less inconvenient.

In some cases, where the occupier only contemplates a temporary residence, it will be advisable to give only a temporary repair; and, when this happens, thinning and pruning applied to the trees and shrubs, and washing over with cement to the walls and out-buildings, will be the leading points of renovation. The lawn, in such places, will generally be more or less worn out, and particularly on those spots which have been covered with the trees or bushes removed. The whole of the lawn ought to have the coarser weeds, such as dandelion, &c., removed, and the inequalities levelled with soil of the same kind as that on which it is laid, in order that, when it is consolidated, it may form a similar surface in point of texture; after which, the whole ought to have a top-dressing of rich compost; and all the places not completely covered with grass ought to be sown thickly with the following mixture:—*A*grós-tis vulgàris var. tenuifòlia, *Festùca* duriúscula, *F.* ovìna, Cy-nosùrus cristàtus, *P*òa praténsis, *A*vèna flavéscens, and *T*rifòlium minus. These seeds should be mixed together in equal portions, and sown at the rate of from 4 to 6 bushels per acre. The gravel walks, instead of being taken up and relaid entirely with fresh gravel, may be stirred, and a thin layer of gravel laid on the top, and afterwards firmly rolled. Where the gravel is loose, it ought to be mixed with gravel of an adhesive nature, newly taken from the pit; or, if this cannot be procured, with Roman cement in a state of powder. The proportion of cement should be very small; not more than at the rate of half a pint of cement to one bushel of gravel. The gravel and cement ought to be intimately and rapidly mixed before laying them on the walk, and heavily rolled as soon after as possible. Where a yellow ferruginous clay can be procured, and where it is taken fresh from the pit, and instantly mixed with the gravel, it will answer the same purpose as the cement; but, if it be exposed to the air for a day or two, it will be of little or no use. A mixture of clay and sand, or of brick-dust and lime, is sometimes used instead of ferruginous clay, but it is much less effective. In general, ferruginous gravels (such as the Kensington), when laid down fresh from the pit, and heavily rolled, become almost immediately as solid as a mass of plum-pudding stone; but if exposed to the air for a few days, so as to lose their moisture before they are laid down, the oxidised surfaces become so dry, that they will no longer form a conglomerate union, and consequently such gravel can only be held together mechanically, viz., by rolling. This, indeed, is the case with most gravels as at present treated.

To render a place comfortable, even for a temporary residence, three things are essential: first, that the house, and every building connected with it, should be put in thorough repair; secondly, that the drainage of the grounds, including the walks, lawns, and plantations, both under and above the surface, should be effective; and, thirdly, that the trees and shrubs should neither be so thick, or in such quantities, as to prevent sufficient ventilation; nor so thin, or so few, as not to produce sufficient shelter and shade. These points kept in view will serve as a guide at once to the intending purchaser of an old suburban residence, and to the improver of one already in his possession.

Where the occupier has only a temporary interest in a residence, say of seven or eight years, though he may not think it worth his while to undertake substantial repairs, he will naturally require a supply of fruits and vegetables from the kitchen-garden. Where the soil of the part devoted to vegetables is worn out, it will not answer his purpose to renew it, by taking out the old, and carting in fresh, soil; but he may restore it, by allowing it to rest for a year, or for two years in succession, without any kind of crop whatever: and, though it would be unreasonable to expect that he should purchase new fruit trees, yet he may graft the newest and best sorts on the branches of such as are already there. Fruit shrubs, such as the gooseberry, currant, raspberry, &c., he may renew either by cuttings, or by procuring young plants from the nurseries. It may be useful here to introduce, as a general remark, that the simplest, most effectual, and least expensive mode of renovating worn out soil of every kind is, to allow it to remain a certain period without any plants growing on it. This period may vary from six months to three years, according to the state in which the soil is; that in which there is most inert vegetable matter, and most seeds of weeds and eggs of insects, requiring the longest period of rest. During the resting period, all weeds should be destroyed while they are in the seed-leaf; and, in order to bring the greater number of the seeds of weeds into a vegetative state, as well as to admit air to the interior of the soil, it ought to be dug over, or trenched, three or four times a year, each digging being of a different depth from that which preceded it, in order to expose a fresh surface to the influence of the atmosphere, and thus occasion the germination of the seeds which the soil may contain. The manner in which worn out soil is benefited by resting is, by the admission of air to the interior of the soil; by the decomposition of the roots and other organised matters which it contains, from heat and the alternate action of dryness and moisture; and by the germination of the seeds of weeds, and the destruction of the larvæ and eggs of insects, snails, and worms, or their complete developement, and migration from the spot, because no kind of animal can live where there are no vegetables to support it.

Manure will always repay an occupant, even if he should have only a single crop after it; but in this case he should take care that the manure is thoroughly decomposed, and equally and thinly distributed over the ground, so that the greater part of its nutriment may be imbibed by the plants the first season. Liquid manure is the best for this purpose; and next soot, and those manures or composts which are sold in a state of powder. Wherever a part of a residence is under grass which is to be made into hay, or even pastured by sheep or cattle, there is, in general, no mode in which the possessor can lay out money on his land to such advantage as in thickly coating the surface with stable manure. It rarely happens that the culture of arable land by a proprietor will pay its own expenses; but we know various instances, in the neighbourhood of London, where, by richly manuring grass land, and selling the growing crop every year by auction, the proprietor has obtained a good profit.

SUBSECT. 3. *On laying out, planting, and managing Second-rate Suburban Gardens.*

SECOND-RATE suburban gardens differ from those of the third-rate, in having a paddock and dairy; and from those of the first-rate, in not having a park and farm. The extent of second-rate residences may vary from four acres (three being, in most situations, the least quantity that will serve for keeping a cow, and one the least that will suffice for a suitable house, pleasure-ground, and kitchen-garden) to eight or ten acres, and upwards. The characteristics of this kind of residence being the paddock and dairy, we shall confine our introductory observations chiefly to them.

The Dairy. A cow, to a person with a family, is one of the principal sources of comfort derivable from a country residence. A cow, it is true, may be kept in town as well as in the country, and may occupy a stall in a stable, in the same manner as a horse; hay and straw being purchased for feeding and littering the one as well as the other. The cow, however, not being worked in the saddle, or in harness, like the horse, and not having either a large yard or a field to take exercise in, soon suffers in her health, and must, in that state, produce unwholesome milk. It is true there are some exceptions, where cows kept in gentlemen's stables in the metropolis are regularly exercised by driving them to some public park, where the pasturage is let out (such as Hyde Park, or the Regent's Park), and bringing them back again after they have remained there an hour or two; but this mode, besides being expensive, is too troublesome ever to become general; not to mention the injury which the cow sustains in being driven through crowded streets. Notwithstanding the evils attending want of exercise, it is a fact, though not generally known, that cows in some of the London dairies are kept stall-fed, and so treated as to give milk for two years in succession, without having a second calf. There are instances of such cows never having been once untied, from the day they were put up, till the day two years afterwards, when they were sold to be fattened for the butcher. (See our account of the principal London dairies, from personal inspection in the year 1830; published in the *Encyclopædia of Agriculture*, ed. 2., p. 1028.) The confined places (frequently dark cellars) in which cows are lodged, and the state of filth from want of litter and drainage, and of closeness from want of ventilation, in which they are kept in the crowded parts of the metropolis, such as St. Giles's, Saffron Hill, Liquorpond Street, &c., are disgusting in idea; and, in reality, must be highly injurious to the health of those who use the milk as an article of food. Even the milk from cows kept a year or two without exercise, however cleanly and suitable the treatment may be for their condition, cannot, we should imagine, be so wholesome as that produced by cows that have abundance of exercise and air. Hence it is that the better-informed inhabitants of the metropolis, and all the higher classes, have their milk from suburban establishments to which grass fields are attached; and where the cows, though highly fed in the house, are yet turned out into these fields for a certain portion of every day throughout the year, except when the weather is of an extraordinary degree of severity.

The cow-house should be of ample size, with complete drainage and thorough ventilation; and the cow should be carefully cleaned every day with a currycomb and brush, in the same manner as a horse. When there is only one cow, some other animal, the food for which will not be expensive, such as a pet lamb or a goat, ought to be kept in the cow-house; and this lamb, or a donkey, should be turned out with the cow into the field. The cow soon becomes attached to her companion, and this attachment is useful in keeping her quiet: for it is well known that restlessness and anxiety not only impair the quality of her milk, but considerably decrease its quantity. If two cows have been kept together for some time in the same pasture, and one is sold or taken away, the other will immediately begin to fall off in her milk (which will become thin and poor), and her uneasiness will be so obvious, as to occasion pain to every one who sees her in the field. The dairy should be cool and well ventilated; but the construction of this building, and also that of the cow-house, will be found in a succeeding chapter.

The breed of cows used for large public dairies is generally unsuitable for private dairies; the object in the former case being quantity of milk, and in the latter, chiefly quality. The Ayrshire breed is, perhaps, on the whole, the best adapted for a private dairy; but, where the main objects are butter and cream, the Guernsey or Alderney cows are preferable. In order to have a supply of milk and butter from one's own dairy throughout the year, it is necessary to have at least two cows; that one may continue to give milk during the month or six weeks which the other is dry before calving; but, in situations where dairy produce can be purchased for that period, then one cow may suffice for a small family. Whoever keeps either cows or horses, must occasionally send to market, either as a purchaser or a seller; and this is commonly the part of the business in which there is most risk of being deceived. To endeavour to instruct the reader on the subject, is out of the question in a work like the present; nor would it be worth any man's while to attend to instructions of this kind, even if we were to give them, unless he contemplated becoming a professional dealer or a farmer. All that we can say, that is likely to be useful, is, consult your neighbours, and employ such professional dealers to buy and sell, as you find are generally considered in the neighbourhood to be respectable. In all cases, the well-known maxim should be kept in view, viz. always to employ a man who has a character to lose.

The Paddock in which the cow is to pasture, and take her exercise, may either be barely sufficient for the latter purpose, or it may be of such an extent as not only to afford pasture for the summer, but hay for winter. In the neighbourhood of London, a small cow, of either the Guernsey or Ayrshire breeds, may be kept on two acres of ground; on the supposition that the soil is good, and that for every crop of hay taken off a dressing of manure is put on. In ordinary soils, and with less careful treatment, three acres of meadow are necessary. Whatever may be the quantity devoted to the cow, one half of it requires to be separated from the other by a hurdle, or other fence, and mown for hay; and the portion mown, and the portion pastured, should alternate with each other. Two tons of hay, with little or no pasture or other food, will, it is calculated, keep an ordinary-sized cow a year; and, therefore, if as much mowing ground be set apart as will produce one ton and a half of hay, there will be no want of food, even if the pasture should not be very abundant. Straw for litter, of course, must be purchased. In the case of residences where a part of the ground is under the plough, or where field crops, such as clover, lucerne, mangold wurtzel, carrots, &c., are raised by spade culture, a small cow may be kept very well on one acre of highly cultivated and thoroughly manured ground; a part of the clover, lucerne, or saintfoin, grown on it being used green, and the remainder made into hay. In this case, also, litter must, in general, be purchased, unless the cow be accustomed to do without it; which is practicable, as was proved in the Harleyan dairy at Glasgow, though it is not, in our opinion, congenial to the animal. A substitute for straw litter may sometimes be found in gardens, in the haulm of peas and beans, and other stems and leaves which have borne crops; and in some kinds of prunings, such as the clippings of hedges, which may be dried in the summer time, and stacked for the purpose. In many gentlemen's places in the country, the leaves of the trees in the plantations are swept up when dry, and kept in an open shed for use as litter; and they not only afford a soft and elastic bed for the cow, but the best of all manure for the flower-garden. Market-gardeners and others, who are obliged to be careful of every thing, save from the rot-heap all that can be dried and turned into litter, for both their horse and cow, and often do not require to purchase straw at any period of the year. It sometimes happens that the kitchen-garden belonging to a residence is too large for the present occupant; in which case the ground to spare cannot be more profitably occupied than with lucerne, to be cut green as summer food for the cow, and with carrots or mangold wurtzel for her winter food. Whichever way the food and litter for the cow may be produced, a paddock for her exercise may be considered essential. The size of this enclosure, when exercise is the main object, will depend chiefly on its shape. A square, roundish, or compact form, of limited extent, affords little temptation to the cow to reach the boundary; because it is everywhere near to her, and comparatively equidistant. If the paddock is to be square or round therefore, it must be large. A long narrow paddock, on the contrary, in which there is a turn, or in which trees are scattered in such a manner as to conceal the farther end, may be small, as the concealment of its termination will induce the cow, every time she is turned out, to find her way thither; and this she will be the more inclined to do, if there should be a pond or a shed at the farther extremity.

The soil of the paddock, if not naturally dry, ought to be rendered thoroughly so, by both surface and under draining. Where the surface is flat, and the soil is a strong clay, such as that common in the north of London, underground drains should be introduced at regular distances all over the field; and, however extravagant it may seem, they ought not to be farther apart, in many cases, than 5 or 6 feet. These drains need not be either broad or deep; and, if stones or bricks cannot be conveniently procured, they may be filled with faggot-wood from the thorn hedges. No money laid out on a wet clayey soil will pay better than that expended in forming drains. Where the soil is a clay, and not thoroughly drained, the feet of the cow will sink into it during winter, which they do much more deeply than the feet of the horse; not only because they are much smaller in proportion to the size of the animal, but from their being cloven; while the feet of the horse, being entire and larger, rest upon a greater breadth of surface. This treading on a wet surface produces what is called poaching; and, when this is the case, all that part of the surface which sinks down under the animal's feet is rendered unproductive, and the remaining part has its productiveness greatly diminished. Wherever a cow, therefore, is kept, and the field in which she is turned out to take exercise is not dry, either naturally or artificially, during winter; it is better at that season to exercise the cow by driving her gently along a lane or quiet road; and, in warm weather, by allowing her, in addition to this, to remain for some hours every day in the open yard attached to the cowhouse.

The paddock may always be harmonised with the pleasure-ground of the place, and rendered ornamental, by scattering a few trees over it; by introducing a shelter for the cow in the form of an open shed, in a proper situation; and by the judicious disposal and planting of the drinking-pond. It sometimes happens that a pond cannot be conveniently formed in a paddock, from the want of springs, or from the porosity of the surface soil being such that the rain-water cannot be collected by gutters in sufficient quantities to fill it. In this case, the usual substitute for a pond is a pump with a cistern, in some convenient part of the paddock, which may be partially concealed by bushes, or rendered ornamental as an architectural object.

The trees for a paddock may either be of the ornamental kind, or they may be the more robust-growing fruit-trees. If ornamental trees, they ought to be of such sorts as will grow freely, and soon be in a state not to require fencing; and, whether merely ornamental trees, or fruit-trees, or both are employed, they should neither be introduced in such numbers as to injure the pasture by their shade and drip, nor crowded round the drinking-pond in such a manner as to discolour the water in autumn by their falling leaves. A few trees may be considered as advantageous, for affording the cow shelter and shade during rains or very hot sunshine; but, beyond a certain point, they must be considered injurious to the grass, and only allowable on account of their ornamental effect, or of the fruit that they produce. Where good healthy pasturage is the object, from half a dozen to a dozen trees per acre ought on no account to be exceeded; because grasses and clovers, like all other plants, to attain their greatest nutrimental value, require to have their foliage fully exposed to the direct influence of the sun, and to the free circulation of the air.

Ornamental Trees. The trees for a paddock ought to be of small size, even when fully grown; but they ought to be such as grow rapidly when young, so as to be soon out of the reach of cattle. The following kinds possess these advantages, and are, besides, ornamental in appearance: —

A'cer Pseùdo-*Plá*tanus var. flàva variegàta, the Costorphine plane.
A. obtusàtum, the Neapolitan maple.
A. eriocárpum, Sir Charles Wager's maple.
*Negúndo f*raxinifòlium, the ash-leaved box elder.
Æ'sculus rubicúnda, the scarlet horsechestnut.
*C*ýtisus alpìnus, the Scotch laburnum.
Robín*ia* Pseùd-*A*càcia, the common acacia, or American locust.
Gledítsch*ia* triacánthos, the honey locust, or thorny acacia.
*C*ratæ'gus coccínea, punctàta, Crús-gálli, Arònia, tanacetifòlia, and orientàlis, different kinds of thorns.
Amelánchier vulgàris, the common amelanchier.
A. Botryàpium, the snowy mespilus.
*M*éspilus germánica and Smíth*ii*, medlars.
*P*ýrus spectábilis, Chinese crab.
P. A'ria, *P.* vestìta, *P.* pinnatífida, white beam trees.
P. aucupària, the mountain ash.
*P. S*órbus, the true service.
*T*ília europæ'a álba, the Hungarian lime.
O'rnus europæ'a, the flowering ash.
Quércus *C*érris Lucombe*àna*, the Lucombe oak.

Q. I'lex, the evergreen oak.
Sàlix álba, Russelliàna, vitellìna, amygdálina, and fragílis, tree willows.
Pópulus balsamífera, balsam poplar.
A'lnus glutinòsa laciniàta, cut-leaved alder.
A. cordifòlia, heart-leaved alder.
A. incàna, the hoary-leaved alder.
Taxòdium dístichum, the deciduous cypress.
Bétula álba and nìgra, birches, &c.

The last twelve trees, to which may be added the weeping willow, are adapted for planting in the neighbourhood of the drinking pond.

Fruit Trees. The following selection has been made with the approbation of Mr. Thompson of the Horticultural Society's Garden, as ornamental in the shape of the tree, and of vigorous growth as standards :—

Apples and Crabs.

The Red Astrachan. The tree is middle-sized, with a branchy head; the fruit is of a bright red, with a fine bloom, like that of a plum.

The White Astrachan, or Transparent Crab of Moscow.—The tree resembles the preceding sort, but has the branches tending upwards when young, and afterwards becoming pendulous. The fruit is of a wax colour, with a fine bloom on it, and is almost transparent. It is known in English nurseries under the name of transparent crab.

The Lincolnshire Holland Pippin is remarkable for the large size and high colour of its blossoms. Its fruit keeps till February.

The Tulip Apple is a great bearer of very bright red fruit.

The Violet Apple has fruit of a violet colour, covered with a bloom, like that of the plum.

The Cherry Crab is a subvariety of the Siberian crab. The tree is spreading, with drooping branches; and the fruit is numerous, and about the size and colour of a cherry.

The Supreme Crab has larger fruit than the cherry crab. The tree is of robust growth, and the branches are somewhat erect.

Biggs's Everlasting Crab is a vigorous-growing tree, with pendulous branches, and abundance of fruit, which, in form and character, are intermediate between the plum-leaved Siberian crab, and the common Siberian crab, and which remain on the trees long after Christmas. In sheltered situations, and mild winters, this tree appears almost a subevergreen.

Pears.

Beurré Diel. Leaves large and flowers very large. A hardy tree, somewhat fastigiate in its shape; a great bearer, and deserving of extensive cultivation on account of its fruit, independently altogether of its handsome shape and large flowers.

Beurré de Rans (not Beurré rance, as commonly written, which means rank or rancid). Branches spreading or pendulous. The best very late pear yet known. It bears very well as a standard.

Nelis d'Hiver. Branches slender and somewhat wavy. Flowers very abundant. Leaves narrow-elliptic, more so perhaps than in any other cultivated variety of the pear; and they make so little appearance when unfolding, that the whiteness of the blossom is almost untinged by them.

Glout Morceau. Branches spreading. Head pyramidal. A hardy tree, and a great bearer; the fruit of most excellent flavour, and hanging late on the tree. A tree in our garden at Bayswater, the trunk of which is covered with ivy, is loaded with fruit almost every year, without any care or attention whatever being bestowed upon it.

Napoleon. Leaves broad and shining. Blossoms large. The tree vigorous, and a good bearer. The fruit excellent.

Swan's Egg. A handsome pyramidal tree, and an excellent bearer; the fruit roundish or obovate. This is one of the commonest pear trees in the market-gardens about London; and we have introduced the name here from having ourselves observed the handsome shapes taken by the trees. The fruit, however, as compared with that of the sorts above recommended by Mr. Thompson, is not worth cultivating; though, in the months of November and December, it is more abundant in the London markets than that of any other variety.

The following Scotch Pears are recommended by Mr. Gorrie, as forms adapted for landscape scenery; but little can be said in favour of their fruit, as compared with that of the new Flemish varieties:—

The Benvie, the Golden Knap, and the Elcho take fastigiate forms; the latter, more especially, Mr. Gorrie says, may be called the Lombardy poplar of the pear tribe. These trees generally attain the height of from 45 ft. to 50 ft. in as many years, in the Carse of Gowrie, in Perthshire.

The Busked Lady and Pow Meg take spreading orbiculate forms, such as will assort with the A'cer Pseùdo-Plátanus, and may be called the oaks and elms of the pear family.

Cherries. Those recommended by Mr. Thompson are:—

The Bigarreau, a tree of vigorous growth, with large pale green leaves, and stout divergent branches.

Büttner's Yellow, a vigorous-growing tree, like the preceding, but with golden-coloured fruit.

The Kentish Cherry is a round-headed tree, with slender shoots, somewhat pendulous.

The May Duke is a middle-sized tree, with an erect fastigiate head.

The Morello is a low tree, with a spreading head, somewhat pendulous; most prolific in flowers and fruit, the latter ripens very late; and, from not being so readily eaten by birds as most other sorts, hangs on the trees a long time. In this last respect, however, it is now exceeded by Büttner's October morello.

Plums. The handsomest-growing trees, Mr. Thompson considers to be:—

The Red Magnum Bonum, which has a fastigiate habit of growth, and a trunk sufficiently vigorous to bear the rubbing of cattle without injury.

The Washington, a vigorous-growing tree, with a pyramidal head; and a great bearer of fruit of excellent quality.

The Green Gage, in the climate of London, forms a handsome tree; and, at a certain age, becomes a great bearer.

The Orleans is a vigorous-growing handsome, spreading tree, producing very excellent fruit.

The Wheat Plum is a very ornamental tree when in fruit, that being of a bright fiery red colour.

The Damson and Wine Sour Plums are vigorous-growing trees, quite hardy, very prolific both of flowers and fruit, which fruit is particularly well adapted for all culinary purposes.

Walnuts and Sweet Chestnuts.

The paddock would seem a very appropriate place for walnut trees; and every residence ought to contain one or two of these trees, for the sake of their green fruit for pickling, and their ripe fruit for the dessert; but more than two or three should not be planted in a paddock, unless it be very large; because their leaves, when eaten by cows, give a bad taste to the milk, and, when they drop into the pond, make the water bitter. The same objection does not apply to chestnut trees, one or two of which may be introduced for their fruit. The Downton and the Madeira chestnuts are recommended by Mr. Thompson; the latter has a large oblong nut.

Staking and protecting single Trees in Paddocks. Plants should be selected which have clear stems, at least 6 ft. in height: they should be planted on little hillocks, for the reasons given in p. 311.; and staked and protected from the rubbing of the cow. One of the best modes of doing this is that invented by Charles Lawrence, Esq., represented in *fig.* 205., and explained in *figs.* 206. to 208. The advantages which this guard has over all others, Mr. Lawrence finds to be, "free motion to the tree, without producing any friction on the bark; perfect security against the attacks of cattle; durability, cheapness, and neatness in appearance." Those guards which confine the tree are objectionable from preventing the motion of its stem, which is essential to its health and growth; and those which stand at a distance from the tree, consisting of three or four posts connected by horizontal spars, are very injurious to the bark in high winds, and are, besides, very expensive and unsightly. The following are Mr. Lawrence's directions for preparing and putting up his tree-guard: —

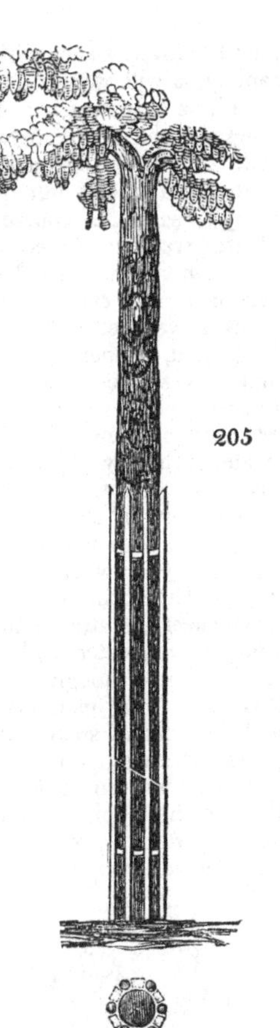

205

Mr. Lawrence's Tree-Guard against horned Cattle. "Procure stakes of ash or larch, or, in default of these, of any other straight-growing tree, when thinning young plantations, or cutting down coppice. These stakes should be 6 ft. in length, or more if requisite, and about 2 in. in diameter at the thickest end; and they should have holes drilled through them at the top and bottom, about 1 ft. from each end. Get a similar hole drilled 2 or 3 inches up the centre of a stake, and then saw off the length which has had the hole drilled through it, and which will give a piece that, when the string or wire is drawn through it, will resemble *b* in *fig.* 206. Repeat the operation till as many pieces are drilled and sawn off as may be wanted. Pass a strong piece of copper, or Rowland's metallic wire, or thick tarred string, through one stake by the hole at the top, and then through one of the 2-inch pieces, then through another stake, and so on, separating each stake at top and bottom by one of the 2-inch pieces of wood, until you have enough to surround your tree loosely, leaving plenty of space for growth. When this is done the appearance of the guard, before being put on, will be as in *fig.* 207. Place the guard thus formed round the tree, and fasten the ends of the wire or string. The guard is much the same as the cradle put round the neck of a blistered horse, to prevent his gnawing the irritated part. The ends of the stakes merely rest on the ground, and they should be cut quite flat at the bottom, to prevent their sticking in it. At the upper end, they should have a sharp slanting cut with a bill-hook, to throw off the rain. The motion of the tree will not be in any degree impeded; and the bark cannot be injured, let the wind blow as it may, for the

b

206

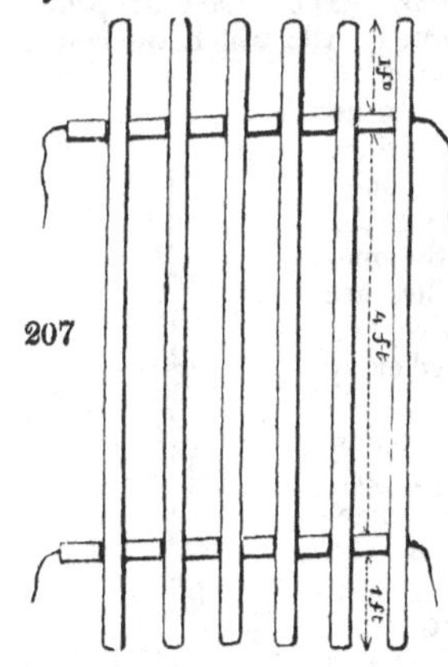

207

guard moves freely with the tree in every direction. If a tree is growing rapidly, it will want room before the guard requires renewing; in which case it is only necessary to unite the string or wire at the top and bottom, lengthen the string or wire by tying a piece to it, and introduce an extra rod, and two extra separating pieces. As a principal feature in this guard is that the tree is left quite at liberty to be blown about by the wind in every direction, of course it does not obviate the necessity of staking a newly planted tree until it becomes fairly rooted.

"I can vouch," says Mr. Lawrence, "for this guard affording perfect protection against cattle; for I had two cows, the most determined barkers of trees I ever met with, in a paddock in which there were several single trees. They had, sooner or later, contrived to get at my trees, and ruin them; and I was about to fatten them off, and condemn them to death for their offences, when I hit upon the mode of protection just described. My men and myself were curious to see the impression that would be made on the cows by the new guards when they were first turned into the paddock. They very soon proceeded to the trees, examined the cradles round them, and made several attempts with their mouths. Finding these unsuccessful, they made an attack with their horns; but, as the fence yielded with every blow, and merely turned round, they made no progress that way. Thereupon they began stamping with their feet, and, as we thought, from sheer vexation and disappointment. Be this as it may, they were completely defeated, and my trees have now (1837) continued 4 years in perfect security, though these identical cows have remained amongst them till this day. One man can fence in this way a great many trees in a day; and the cost of stakes, if purchased, would not exceed that of one of the posts necessary upon the ordinary plan." (*Gard. Mag.*, xiii. p. 167).

Having written to Mr. Lawrence, in April, 1838, to enquire if he had still the same high opinion of this guard, he says, "I find it perfect in all respects. It is a perfect protection against damage by any animal, however ingenious or determined it may be, and it never injures a tree in any respect."

Fig. 208. shows, on a larger scale, the ground plan, or rather horizontal section 1 ft. from the ground, and a portion of the elevation of a tree so fenced. In this figure the wire or string is shown passing through the upright rods and horizontal short pieces, from *c* by *d* to *e*; but, from *c* by *f* to *e*, the wires are only shown passing through the upright rods; the short pieces being seen in vertical profile, as they are in nature.

Protecting by Thorns. The mode of protecting trees from horned cattle which we consider next best to that of Mr. Lawrence is, to clothe their stems with thorn bushes, as shown in *fig.* 209.; a mode very generally adopted in the public parks about the metropolis. The branches are tied on with copper wire, and are loosened or renewed every two or three years.

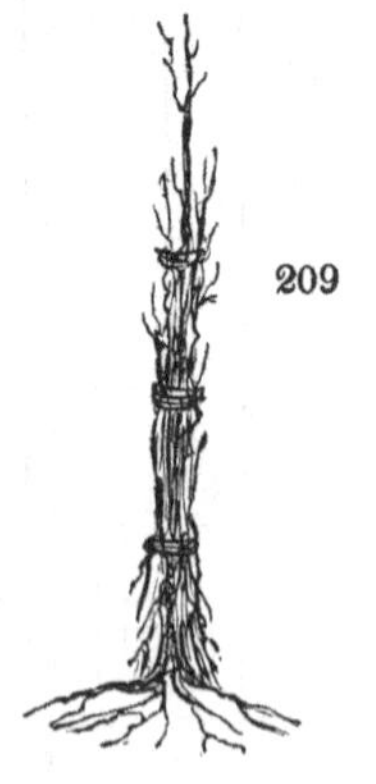

Protecting Trees against Sheep. A mode of protecting the stems of trees somewhat similar to that of Mr. Lawrence, and which has been long practised in Scotland, consists in tying plasterer's laths, set on end, and touching each other at their edges, round the trunks or stems; but this mode is better adapted for protecting trees from sheep than from horned cattle. Where sheep only are grazed, no mode can be cheaper, or, after the laths have been exposed to the air for a year, less conspicuous, unless we were to substitute for the laths, oak, birch, or willow bark, with the epidermis outwards. Where laths or strips of thin boards are used, they, and also the string by which they are tied on, might be rendered of great durability by being previously Kyanised, or steeped in Margary's composition, which is said to cost only one tenth of Kyan's process, and to be equally efficacious both on wood and cordage. (See *Brit. and For. Rev*, April, 1838, p. 695.)

210

Fig. 210. shows the horizontal section, and elevation of a tree thus protected, in which *a* represents the stem of the tree, *b* the wire which ties on the laths, and *c* the lower ends of the laths.

Staking Trees to protect them against the Wind. When single trees are planted which have large heads in proportion to their roots; or, when the situation in which they are placed is very much exposed, they are in danger of being blown off the perpendicular, or even thrown down, by high winds. This is more especially the case when the trees are planted, as we recommend, on little hillocks; and this is one reason why planters have been led to deep planting, or, at all events, to prefer a flat surface to a raised one. To retain a tree fast in its place that has but few ramose roots, fix previously in an upright position, in the bottom of the pit, a stake of such a length that it may be rendered firm by ramming, &c., below the level of the soil which is to contain the roots of the tree, and that it may reach 3 or 4 feet up the stem above ground; afterwards tie the stem of the tree to this stake, as shown in *fig.* 211.; or, notch two short pieces of wood into each other, so as to form a cross; and at one of the angles of intersection fix a stake, to which the tree is to be tied, as in *fig.* 212. No wind whatever will blow down a tree thus supported, provided the arms of the cross are long and strong. In order to give additional strength, the arms of the cross may be pegged down with hooked sticks, agreeably to the practice

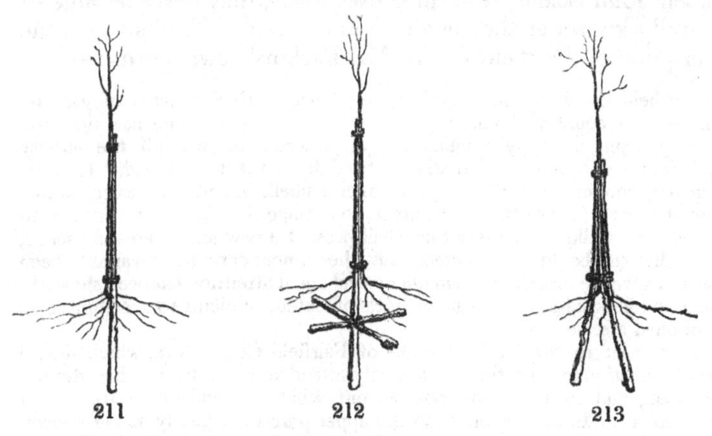

211 212 213

of Mr. Thom of Annan, as detailed in *Gard. Mag.*, vol. vii. p. 452. Another mode consists in tying the tree to two stakes in contact with the trunk, and deeply inserted in the soil, as shown in *fig.* 213.; and the last mode we shall here mention, consists in strengthening the ramose roots of a tree by tying them to pieces of branches laid on them, and in close contact with them, throughout their length, as shown in *fig.* 214.; thus producing, in effect, large, strong, and powerful roots, to act as levers to keep the trunk in its place. After one of these methods of fixing the tree has been adopted, it may be protected from cattle or sheep by Mr. Lawrence's mode, by clothing with thorns, or by a covering of laths, as above described.

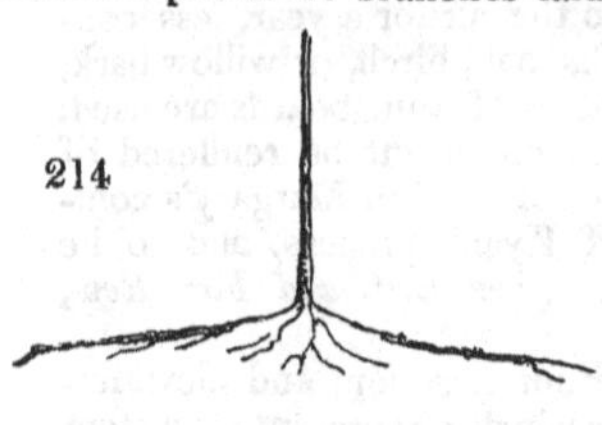
214

It is almost unnecessary to mention that the introductory paragraphs to our subsection on laying out third-rate gardens, which occupy from page 409. to page 448., are equally applicable to second-rate gardens, and, indeed, to all others which contain more space than fourth-rate gardens. We shall, therefore, without farther preface, now give our examples, and shall commence with a design which shows some variations in what we have called professional routine.

Fairfield Cot, of which *fig.* 217. is a plan (with some additional single trees), and *figs.* 215, 216. 218, and 219. sections and sectional geometrical views, is taken from a work entitled *Six Designs for Laying out Grounds*, by G. J. Parkins, Esq., published in 1793. Mr. Parkins was connected with the army; and, being fond of rural scenery, and of considering how it might be improved, he employed much of his leisure time in studying natural situations, and imagining how he could adapt them for residences of different degrees of extent, from 5 acres upwards. Mr. Parkins's designs were published in an expensive form, and in connexion with a large work on architecture by the late Sir John Soane; and, in consequence, they never became so generally known as they ought to have been. We shall give the description of Fairfield Cot in Mr. Parkins's own words:—

"Fairfield Cot is a small secluded residence, with five acres of pasture, situated in a country abounding with numberless picturesque beauties, and affording, from a happy combination of external objects, all the variety requisite to enliven the mind when retired from the busy world. Its size, being very confined, will not admit of much embellishment; therefore, to distribute the parts to the utmost advantage, to arrange the shrubberies, and so to conduct the walks that every turn shall present a new and interesting scene, are all that can be done. In doing this, the utmost care has invariably been observed never to lose sight of simplicity, and equal attention has been shown to induce the imagination to credit an idea of extent, difficult to accomplish on so confined a scale.

"In respect to the local situation of Fairfield Cot, nature, which should never be overlooked, or too incautiously intruded upon by art, has done a great deal, and in a measure pointed out what is requisite for its utmost improvement. In consequence, on the upper part of a gently rising ascent, commanding over the adjacent country most luxuriant views of hill, dale, wood, and water, the cottage is erected. Immediately adjoining are the necessary offices and kitchen-garden; these by a shrubbery are concealed. The orchard, placed at an extremity, gives variety, and hides the bounds, where otherwise they would, by being seen, defeat the deception of extent; and likewise confines the view, which, but for such management, would be considerably too extensive. The remainder of the ground is again subdivided by a hedge, decked with roses, honeysuckles, and other wild shrubs, and irregularly planted, partly to allow the walk to take an easy bend betwixt the two enclosures, and partly to give internal variety; a circumstance which never fails to cheer the imagination, and relieve the eye.

215

"Having thus noticed the general disposition, it will be necessary to observe, that the walk, in no instance, has been suffered to approach too near the bounds. From the shrubbery, after passing a seat under a few trees (*e fig.* 217), it leads down the side of the hill to a copse overhanging a purling stream. A bridge, adjoining to a root-house (*f*), crosses the rill, the path accompanying its meandering course, till a rustic plank (*g*), thrown over the same, again unites it with the lawn. Every thing here is simple and unadorned: to load with ornament a scene dedicated to contemplation and repose, would destroy the effect which a sequestered situation ever has upon a congenial mind. Pursuing the walk, a sunk fence on the right admits the country. Clumps of trees in the adjacent pastures unite it with the distance. A little gate, on gaining the summit, leads to the temple of Concord (*h*). From the window is seen to peculiar advantage the view, purposely hid by the orchard from the house. This circumstance renders the temple much more interesting than it otherwise would be, and increases the variety of the whole. From this place the walk waves to the left. Entering a small shrubbery, with a seat in a sequestered situation (*i*), it soon opens on the green, and terminates at the cottage.

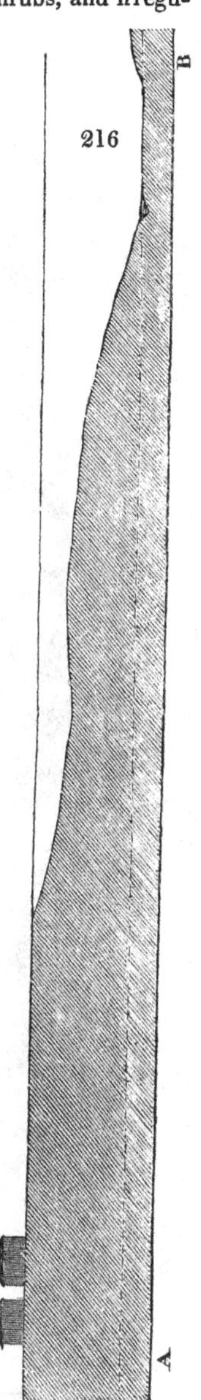
216

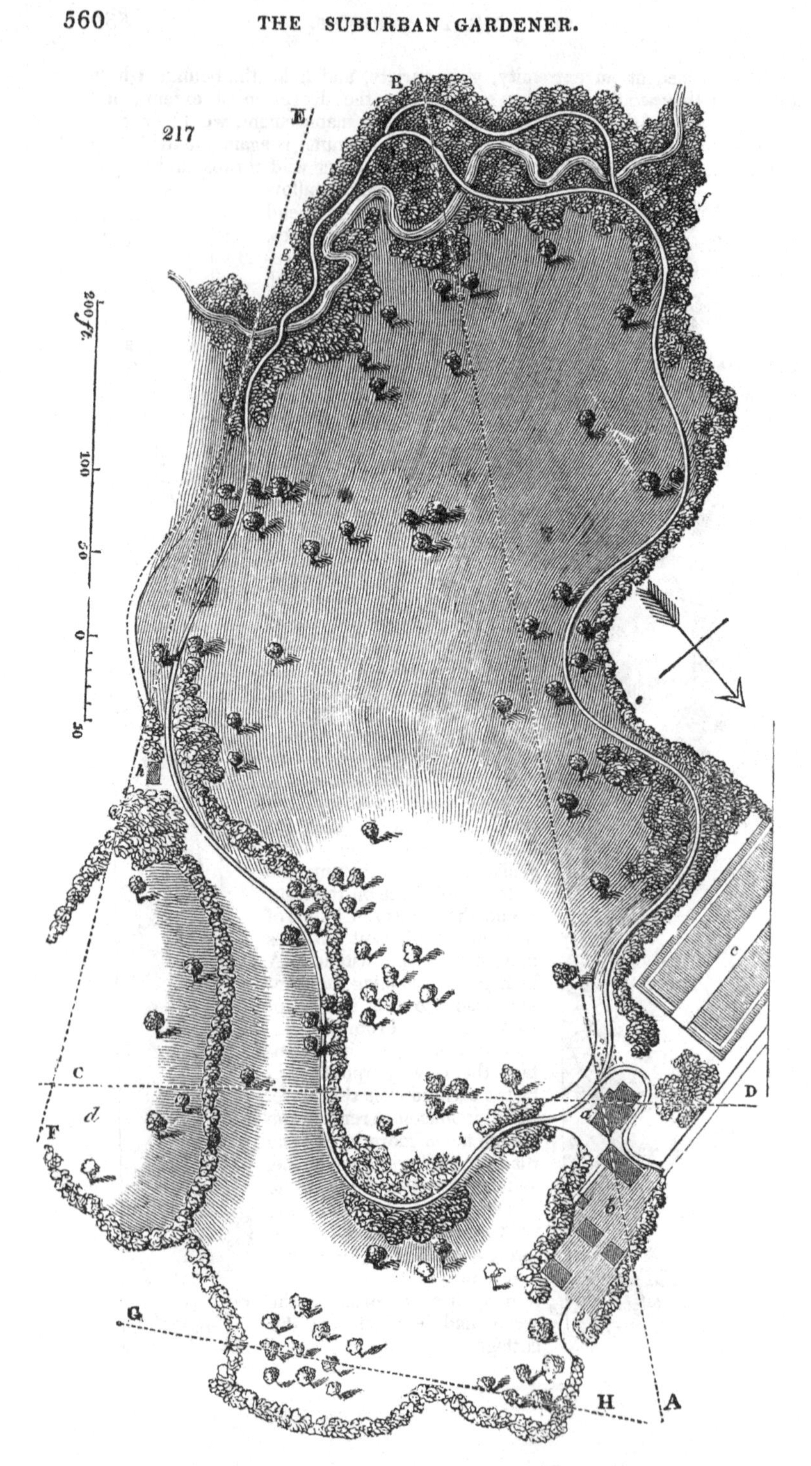
217
B
E
g
f
200 ft.
100
50
0
50
h
c
C
D
d
F
a
i
b
G
H
A

a, House.
b, Stable, cow-house, piggeries, &c., hid from the grounds by a plantation, and approached by a road overshadowed with tall trees.
c, Kitchen-garden, screened in the same manner. An opening in the shrubbery, however, admits a view down the principal walk, on each side of which the beds are arranged.
d, Orchard. This, from the cottage, has an interesting appearance: it was placed there to hide the extremity, and to confine the eye to a ruined tower, to a river meandering through the vale below, and to distant mountains, seen from the house across the lawn.
e, A seat, composed of rude materials, situated under trees. From this spot is seen an extensive distant country, adorned with water, hanging woods, &c.
f, Root-house, built of roots of trees, and thatched; the inside lined with moss. Ivy creeps over the door, along with the honeysuckle and jessamine. A table and two rustic benches constitute its furniture; on the former an appropriate inscription.
g, A bridge. A few large stones, supporting a plank or two, with a rail on one side, will generally be found sufficient for such a situation. It accords with simplicity, and is therefore infinitely more attractive than a formal structure.
h, Temple of Concord. A small square building, the walls emblematically painted in fresco. From the windows, a most extensive view, particularly of objects in the distance, screened from the house by the orchard.
i, A seat in a sequestered situation. On the opposite side of the walk, under cypress and flowering shrubs, an urn, dedicated to Friendship.
A B, First sectional line. See *fig.* 215. C D, Second line. See *fig.* 218.

218

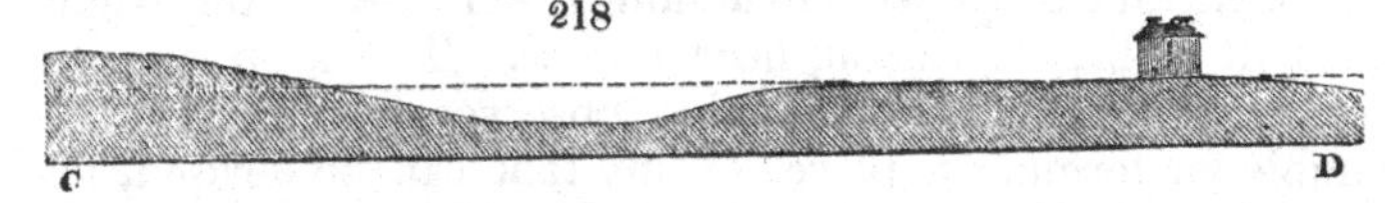

E F, Line for first sectional geometrical view, giving the general appearance of the lawn rising to its summit; the woods, the house, Temple of Concord, and distant country. See *fig.* 216.
G H, Line for second sectional geometrical view. In this are shown the cottage, surrounded with trees; the lawn; the fence bounding the same; a bend of the serpentine walk, with part of the orchard, and the distant country." See *fig.* 219. (*Six Designs for Improving and Embellishing Gardens*, p. 2.)

219

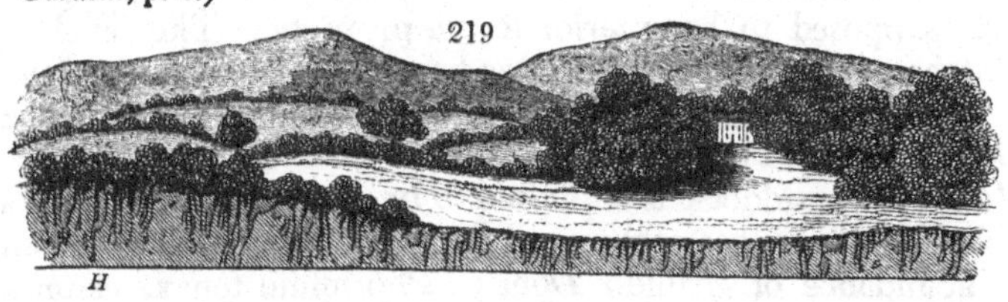

Remarks. The sections and sectional geometrical views in this design would answer exceedingly well for giving a general idea of the effect of improvements; but they would not serve instead of a working plan. Such designs were commonly given, when modern landscape-gardening was in its infancy, by Kent, Brown, Wright, and others, and executed by contract with alterations almost at pleasure, by a contractor under the name of a new ground workman; or sometimes in the ordinary routine, under the direction of the gardener. All the details of execution, and the choice of the trees and shrubs planted, were passed over in the general design of the artist, and left to be supplied according to the taste, knowledge, or means, of the contractor or the gardener. It does not appear that, in the infancy of landscape-gardening, any great value was set upon having a variety of trees and shrubs in plantations; and, accordingly, in the first laid out places in the modern style, with the exception of Pain's Hill, and one or two others, the trees and shrubs are all of the common kind. At present, however, the taste is decidedly different, and there is a laudable desire on the part of proprietors, and especially on that of the females of their families, to render garden scenery botanically as well as pictorially interesting. The subject of the kinds of trees is scarcely at all mentioned by Mr. Parkins in the description of any one of his designs: he looks on garden scenery entirely with the eye of a painter and a poet, while the modern artist adds to these the eye of the botanist and the cultivator.

A Second-rate Suburban Villa of Seven Acres, contained in a Space nearly square. — This design, of which *fig.* 220. is a ground plan, is contributed by Mr. Glendinning, gardener to the Right Hon. Lord Rolle, at Bicton, near Exeter. The form or outline of the ground, Mr. Glendinning observes, is the most unfavourable for forming a place, of any that can be devised, for producing effect; but it is chosen as being the one which most frequently occurs in the neighbourhood of towns, and particularly in the neighbourhood of new towns, such as those of America and Australia. The surface is supposed to be even, but with an inclination from the house towards the pond and the paddock. This pond is supplied with water from certain fountains in the pleasure-ground, and these are supplied from a spring or brook, supposed to be exterior to the property. The level of the kitchen-garden is considered as being about 20 ft. above the level of the pond in the paddock; and the house stands on a platform, supposed to be 6 or 8 feet above the level of the kitchen-garden. The paddock is supposed to be grazed by cows and a few sheep. The edges of the pond Mr. Glendinning would plant with abundance of *Arúndo Dònax*, *Phórmium tènax*, cannas, and hedychiums. The pond should also, he says, swarm with fish; and should have some wild ducks, a couple of swans, the bernacle goose (*A'nas erýthropus L.*), the Egyptian goose (*A'nas ægyptìaca L.*), &c. The boundary fence to this residence, whether wooden pales or a stone or brick wall, should be such as to resist every description of cattle; and within it a row of evergreen Lucombe oaks are supposed to be planted at 20 ft. distance from the boundary, and the same distance from one another. Within these, there should be a row of the ilex oak, and a third row of hollies, yews, and Portugal laurels alternately. This will completely shelter the grounds within, and exclude all objects in the immediate vicinity of the residence. In the pleasure-ground there is a terrace-wall (*s s*), within which there is a border sufficient for 500 herbaceous plants; and the rockwork (*f f*) will afford room for 500 additional species, including the cistus and helianthemum families. The narrow border in front of the botanic hot-houses (*l*, *m*, *n*) will afford ample room for a collection of ixias, watsonias, gladioluses, amaryllises, oxalises, and similar Cape bulbs and tubers; so that a very creditable collection of every kind of garden plant, whether ligneous or herbaceous, may be found in even a small place of this extent. This will be rendered more obvious after perusing the list of trees and shrubs given for planting the boundary plantation. The following explanations refer to *fig.* 220.

a, The house. *b*, Entrance from the public road. *c*, Stable-court.
d, Drying-ground; a part of the circular building enclosing the stable-court being a laundry, opening to the drying-ground; a part of it a brewhouse, opening to the west; and a part of it a poultry-house, opening to the south-east.
e, Pit for the stable dung, with a movable roof to prevent evaporation.
f, Place for hay-ricks, stacks of wood, and various other matters required for the stable-court and kitchen-garden.
g g, Two marble fountains, rising from octagonal basins.
h h, Sarcophagi, elevated on angular balls of stone, filled with pelargoniums, salvias, &c., in summer; and with rhododendrons in winter.
i i, Beds for florist's flowers, with stone kerbing 8 in. high, and with an awning to fit over them when requisite. After the florist's flowers, such as tulips, hyacinths, &c., have been removed, their place is supplied with showy annual flowers, brought forward in the reserve-ground.
k k k k, Circular beds, lined inside with brick, with stone kerbing, for bulbs in spring, and the best kinds of dahlias in summer and autumn.
l, Green-house for a miscellaneous collection. *m*, Stove.
n, House for Cape heaths. *o*, Back sheds to the range of botanical hot-houses.
p p, Conservative walls. *q*, Aviary.
r, Flower-beds, with glass frames to fit over them; being chiefly used for growing hyacinths.
s, Terrace wall, with ornamental vases, immediately within which is a border for choice herbaceous plants.
t, Terrace walk. *u*, Lawn. *v*, Scattered shrubs.
w, Groups of variegated plants.
x, Lawn, varied by groups of shrubs, baskets of flowers, standard roses, &c.
y, Flower-baskets. *z*, A sun-dial, placed against a conservative wall.
A, Rosarium. B, Covered seat, with a basket of flowers in front.
C, D, E, F, Rockwork, rustic seat, grotto, and rocky fountain; the whole varied by rock-plants and climbers; the grotto consisting of two stories, the upper one serving as a prospect tower.
G, A pond, or artificial lake, the shape being such as to be wholly seen from every point of view, and therefore not calculated for picturesque effect so much as brilliancy, and the display of aquatic plants along its margin.
H, Wire fence, separating the paddock, which is to be pastured, from the lawn, shrubbery, and arboretum, which are to be mown.
I, A covered seat, and in front of it a basket of Californian annuals; beyond which are a pedestal and vase, to the memory of Douglas the botanist.
L, Reserve-ground to the kitchen-garden, and liquid manure tank, with a pump, for receiving the drainage from the stables, &c., communicating with

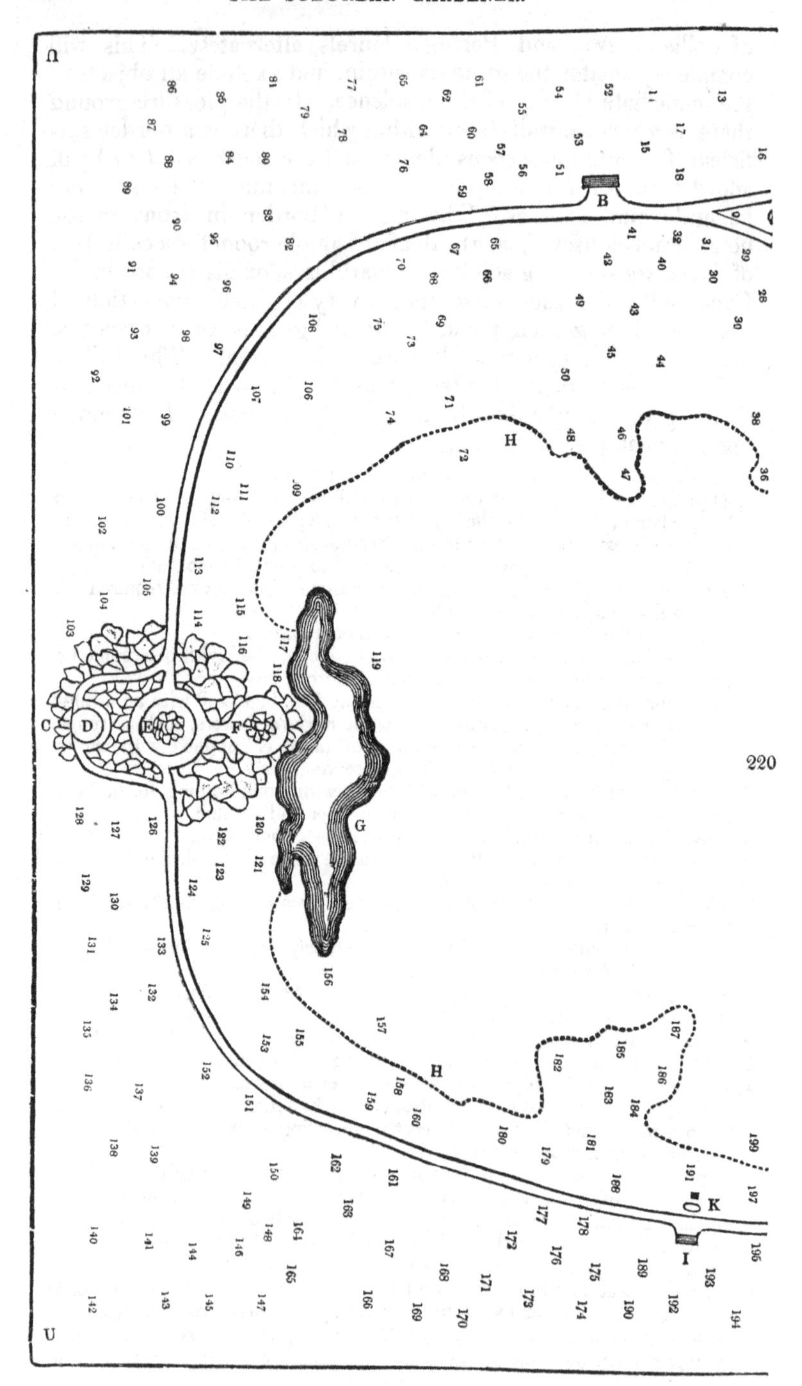

220

220

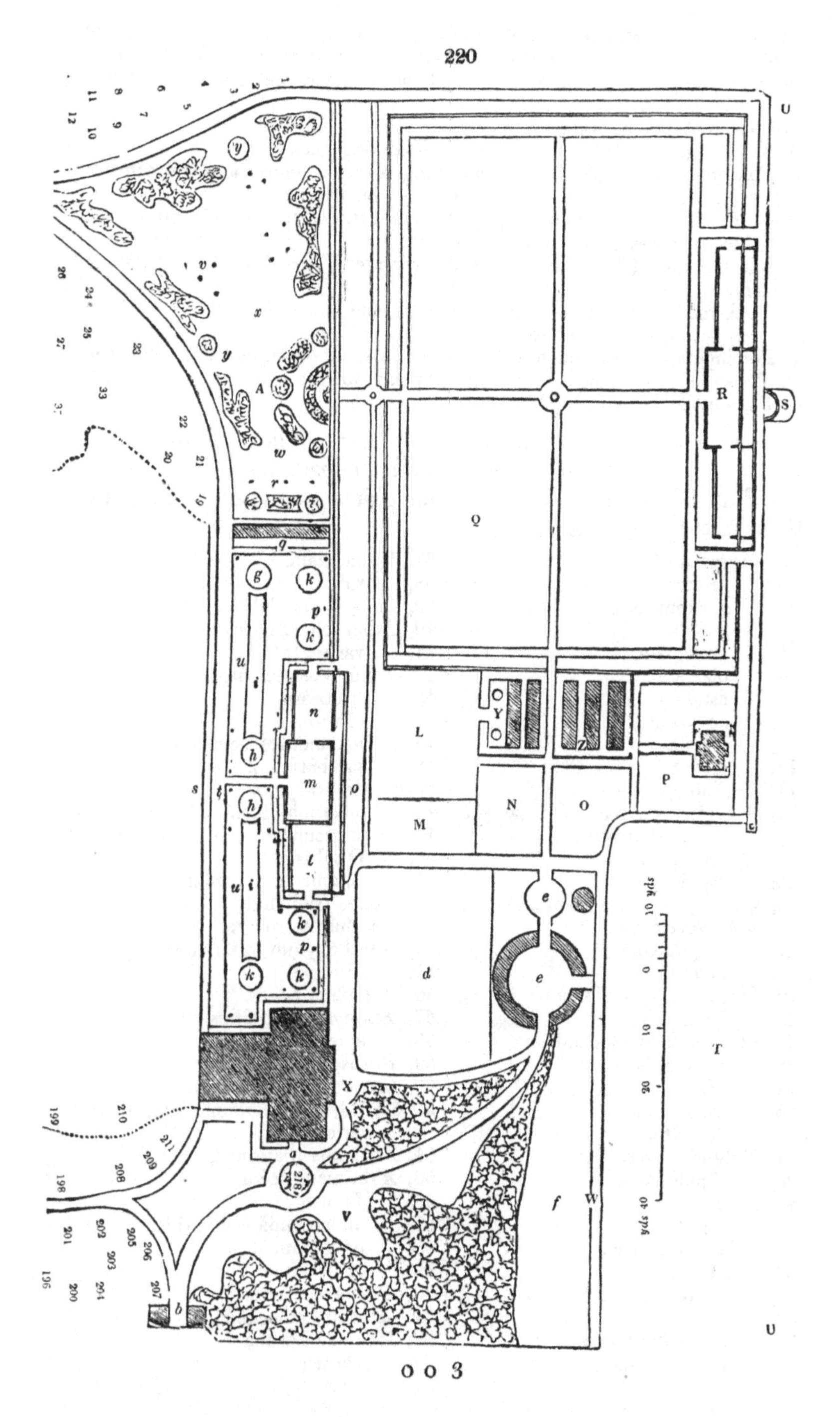

O O 3

M, The botanic frame-ground, placed in this situation to be near *e*, for stable dung. N, Reserve-ground, and rot heap for the botanic hot-houses.
O, Compost-ground, liquid manure tank with cover, and rot heap for the kitchen-garden.
P, Gardener's house, and private court.
Q, Kitchen-garden, of which *fig.* 221. is a working-plan.
R, Pine-stove in the centre; the one wing being a vinery, and the other a peach-house. S, Rubbish-heap. T, Orchard.
U U U U, Situation of the four rows of trees intended to surround the whole, for the purposes of shelter and seclusion.
V, Lawn, bordered by shrubbery, chiefly evergreens, treated in the picturesque manner.
W, Back road to the stables, kitchen-garden, and house.
X, Back entrance to the house.
Y, Forcing-pits to the botanic stoves, with two small basins of water, as a place of reserve for bringing forward aquatic plants for the pond (G).
Z, Pine-pits and forcing-pits for the kitchen-garden.

References to the numbers in *fig.* 220., indicating the kinds of trees that are to be planted round the paddock. The names are Mr. Glendinning's, and all the plants may be obtained by them in the nurseries.

1, Magnòl*ia* grandiflòra.
2, fuscàta.
3, g. exoniénsis.
4, pùmila.
5, auriculàta.
6, obovàta.
7, caudàta.
8, acuminàta.
9, grácilis.
10, glaúca.
11, Thompson*iàna*.
12, Araucària excélsa (to be protected in winter).
13, braziliàna.
14, imbricàta.
15, Cunninghàm*ia* lanceolàta.
16, *Ribes* speciòsum.
17, sanguíneum.
18, níveum.
19, Liriodéndron Tulipífera.
20, integrifòlia.
21, 22, Hybrid rhododendrons.
23, *Bérberis* aristàta.
24, dúlcis.
25, floribúnda.
26, dealbàta.
27, Mahòn*ia* fasciculàris.
28, *A*quifòlium.
29, nervòsa.
30, rèpens.
31, *Ribes* *m*alvàceum.
32, glutinòsum.
33, Kölreutèr*ia* paniculàta.
34, *T*ília argéntea.
35, parviflòra aurea.
36, heterophýlla.
37, *T.* laciniàta.
38, rùbra.
39, *Ailántus* glandulòsa.
40, Aristotèl*ia* *Mácqui.*
41, variegàta
42, *P*aliùrus aculeàtus.
43, *A*'cer striàtum.
44, rùbrum.
45, *p*latanöìdes laciniàtum.
46, sacchárinum.
47, dasycárpum.
48, macrophýllum.
49, créticum.
50, O'palus.
51, japónicum péndulum.
52, *P*tèlea trifoliàta.
53, Photínia serrulàta.
54, Xanthóxylum *f*raxíneum.
55, nítidum.
56, *Z*ízyphus *L*òtus.
57, *E*uónymus latifòlius.
58, americànus.
59, Pittósporum *Tobìra.*
60, *C*eanòthus americànus.
61, 62, 63, *I*'lex *A*quifòlium, three varieties.
64, *R*hàmnus *A*latérnus.
65, *A*'rbutus procèra.
66, *U*'nedo.
67, *a*ndrachnöìdes hýbrida.
68, *a*ndrachnöìdes.
69, *A*ndráchne.
70, *B*úxus baleárica.
71, *Æ*'sculus rubicúnda.
72, Pàv*ia* americàna.
73, cárnea.

74, Pàv*ia* flàva.
75, macrostàchya.
76, Virgíl*ia* lùtea.
77, Gledítsch*ia* monospérma.
78, hórrida.
79, triacánthos.
80, brachycárpa.
81, macrántha.
82, *C*ýtisus *L*abúrnum péndulum.
83, purpùreus.
84, *D*iospýros virginiàna.
85, *Negúndo f*raxinifòlium.
86, Plánera Richárd*i*.
87, *F*ráxinus parviflòra.
88, víridis.
89, argéntea.
90, Gymnócladus canadénsis.
91, Robín*ia* Pseùd-*A*càcia tomentòsa.
92, inérmis.
93, viscòsa.
94, híspida *r*òsea.
95, *Caragàna* arboréscens.
96, Halimodéndron argénteum.
97, *Thèa Bohèa.*
98, víridis.
99, Caméll*ia*, white.
100, striped.
101, Warratah.
102, *C*èdrus Libàni.
103. *P*ópulus heterophýlla.
104, *P*ìnus halepénsis.
105, *Cémbra.*
106, *A*càcia *Julibríssin.*
107, dealbàta.
108, *C*ércis *S*iliquástrum.
109, *C*ratæ`gus *Azaròlus.*
110, *C*érasus lusitánica.
111, *P*àdus.
112, *P*rùnus insitítia.
113, cerasífera.
114, *S*pártium monospérmum.
115, *C*ratæ`gus nìgra.
116, *S*pártium *j*únceum.
117, *C*ratæ`gus grandiflòra.
118, *S*àlix vitellìna.
119, babylónica.
120, pentándra.
121, *C*ratæ`gus Douglàs*ii*.
122, *U*'lex europæ`a flòre plèno.
123, *C*ratæ`gus orientàlis.
124, tanacetifòlia.
125, *C*ratæ`gus coccínea indentàta.
126, punctàta aúrea.
127, *P*ìnus palústris.
128, *P*ópulus balsamífera.
129, *P*ìnus sylvéstris.
130, *A*'bies Clanbrasil*iàna*.
131, *P*ópulus álba.
132, *C*ratæ`gus *p*yrifolia.
133, macrántha.
134, Crús-gálli *p*yracánthifòlia.
135, *Q*uércus pedunculàta.
136, *P*héllos.
137, *S*ùber.
138, vìrens.
139, *I*'lex.
140, *Ballòta*
141, fastigiàta.
142, *R*òbur.
143, *Taúzin.*
144, *Æ*'gilops.
145, *C*érris.
146, *P*látanus cuneàta.
147, orientàlis.
148, Salisbùr*ia* *a*diantifòlia.
149, Liquidámbar Styracíflua.
150, *C*éltis orientàlis.
151, Tournefórt*ii*.
152, Escallòn*ia* rùbra.
153, bífida.
154, *F*àgus americàna purpùrea.
155, ferrugínea.
156, *S*àlix babylónica.
157, *B*étula papyrífera.
158, Broussonèt*ia* papyrífera.
159, Maclùr*a* aurantìaca.
160, *H*amamèlis virgínica.
161, *P*rìnos glàbra.
162, verticillàta.
163, *Amelánchier* Botryàpium.
164, sanguíneum.
165, *Catálpa syringæ*fòlia.
166, *U*'lmus críspa.
167, rùbra péndula.
168, *C*èdrus Libàni.
169, *A*'lnus cordifòlia.
170, *J*ùglans nìgra.
171, *O*'strya virgínica.
172, *L*aúrus nóbilis.
173, *C*astànea vésca.
174, *P*ópulus álba.
175, Nýss*a* tomentòsa.
176, *H*ippóphae Rhamnöìdes.
177, Halès*ia* tetráptera.
178, *P*hiladélphus grandiflòrus.
179, Leptospérmum emarginàtum.
180, Aràlia spinòsa.
181, *L*aúrus *Benzòin.*
182, Plánera Richárd*i*.
183, Gmèlin*i*.
184, *P*istàcia *T*erebínthus.
185, *M*òrus nìgra.
186, *C*ydònia sínensis.
187, Liriodéndron obtusíloba.
188, Edwárds*ia* microphýlla.

189, *P*ìnus uncinàta.
190, ponderòsa.
191, *A*'bies Doúglas*i*.
192, *P*ìnus Sabini*àna*.
193, *C*èdrus *Deodàra*.
194, *A*'bies Menziès*ii*.
195, Deútz*ia* scàbra.
196, Taxòdium dístichum.
197, Benthàm*ia* fragífera.
198, Eriobótrya japónica.
199, *L*àrix péndula.
200, *M*èlia *Azedarách*.
201, *C*upréssus sempervìrens horizontàlis
202, *T*hùja orientàlis.
203, *C*upréssus lusitánica.
204, *P*ìnus *P*ínea.
205, *J*uníperus repánda.
206, suécica
207, *P*ópulus trépida.
208, Edwárds*ia* chrysophýlla.
209, *J*uníperus recúrva.
210, *Q*uércus pedunculàta variegàta.
211, Chimonánthus fràgrans.
212, *C*upréssus sempervìrens stricta.
213, *F*ráxinus excélsior péndula.
214, *P*ópulus trémula.
215, *F*àgus sylvática péndula.
216, *P*hillýrea péndula.
217, *J*uníperus phœnícea péndula.
218, *C*èdrus Libàni.

A few of the species above enumerated will not stand in the open air without protection, in the neighbourhood of London, but in the south of England, where Mr. Glendinning at present resides, the case is different. Camellias there stand in the open air, and grow with the greatest luxuriance; and Araucària brasiliàna, Pittósporum *Tobìra*, *Thèa Bohèa* and *T.* víridis, *A*càcia *Julibríssin*, and several of the Australian species, the edwardsias, Eriobótrya japónica, and *M*èlia *Azedarách*, are nearly hardy, though they all require protection in the climate of London.

The Kitchen-Garden. (*fig.* 221.) Though the extent of this garden is but small, being only 1½ acres, Mr. Glendinning observes that the accompanying list will show its capacity for containing fruit trees. The borders are proposed to be uncropped, or, at the most, to have only a row of strawberries near the edge of the walk. For the ground lost in this way in the inside of the garden, the space contained in the slip in the outside will be an ample compensation. The apples and pears to be trained on the espalier rails should be worked on quince and paradise stocks.

The following are the references to the plan:—

A, Tank of pure water.
a, Entrance from the pleasure-ground.
b, Back sheds, fireplaces, coal-bins, working-shed (including a place for making baskets, preparing and painting labels, &c.), potting-shed, mushroom house, and fruit-room.
c, Pine-stove. *d*, Vinery. *e*, Peach-house.
f, Entrance from the frame ground.
g g, Departments for early vegetables.

1, Apple, Borovitsky.
2, Dutch mignonne.
3, Ribston pippin.
4, Golden pippin.
5, Summer golden pippin.
6, Pear, Bellissime d'hiver
7, Beurré d'Aremberg.
8, Pear, Beurré d'automne.
9, Beurré de Capiaumont.
10, Beurré Diel.
11, Bon Chrétien fondant.
12, Beurré, Easter.
13, Beurré de Rans.
14, Beurré Spence.

221

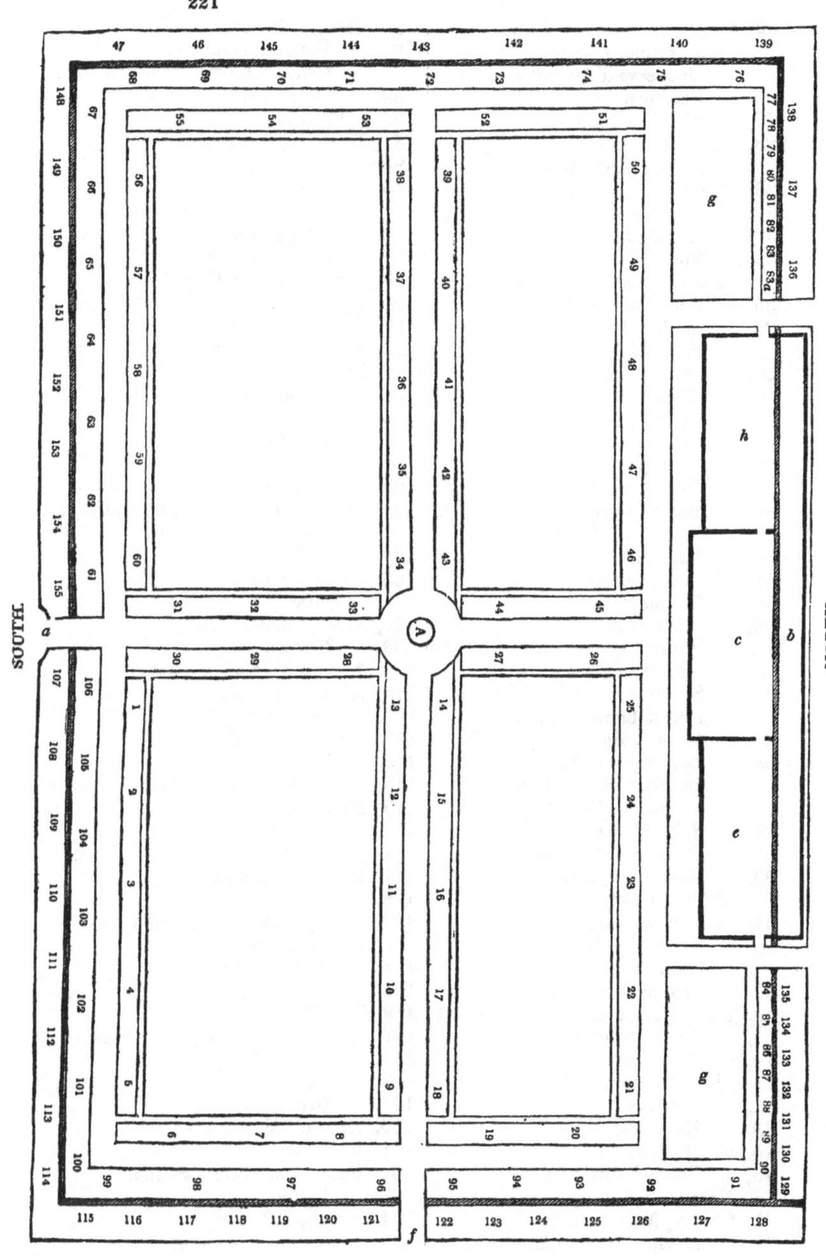

15, Pear, Catillac.
16, Chaptal.
17, Chaumontel.
18, Citron des Carmes.
19, Pear, Colmar, automne.
20, Comte de Lamy.
21, Crassane, Althorp.
22, Crassane, winter.

23, Pear, Délices d'Hardenpont.
24, Doyenne, white.
25, Duchesse d'Angoulême.
26, Famenga.
27, Figue de Naples.
28, Apple, Gloria mundi.
29, Coe's golden drop.
30, Golden Harvey.
31, Hawthornden.
32, Hicks's fancy.
33, Juneating, white.
34, Pear, Flemish beauty.
35, Fondante d'automne.
36, Fondante du bois.
37, Franc réal, summer.
38, Gendeseim.
39, Glout morceau.
40, Henri Quatre.
41, Hessel.
42. Incomparable, Hacon's.
43, Sucre vert.
44, Marie-Louise.
45, Monarch, Knight's.
46, Napoleon.
47, Nelis, winter.
48, Ne plus meuris.
49, Passe-Colmar.
50, Rouse Lench.
51, Seckle.
52, St. Germain, Uvedale's.
53, Louise bonne (Jersey).
54, Beurré, Easter.
55, Beurré de Rans.
56, Apple, Margaret, early red.
57, Nonpareil, Braddick's.
58, Syke House russet.
59, Reinette du Canada.
60, Russet, Boston.
61, 62, 63, Cherry, Morello.
64, Pear, Marie-Louise.
65, Beurré, brown.
66, Glout morceau.
67, Incomparable, Hacon's.
68, Plum, Morocco.
69, Coe's golden drop.
70, Drap d'or.
71, Cherry, Elton.
72, Downton.
73, May duke.
74, Peach, Late admirable.
75, Colonel Ansley's.
76, Nectarine, Murrey.
77, Cherry, May duke.
78, Grape-vine, Willmot's muscat.
79, Espérione.
80, Grape, Royal muscadine.
81, Paramata cluster.
82, Grove End sweetwater.
83, White cluster.
83 *a*, Pitmaston white cluster.
84, Figs, Figue blanche.
85, Genoa, large white.
86, Ischia, yellow.
87, Marseilles.
88, Pregussata.
89, Brown Turkey.
90, Apricot, Moor Park.
91, Nectarine, Elruge.
92, Violette hâtive.
93, Peach, Barrington.
94, Plum, Green gage.
95, Reine Claude violette.
96, Pear, Beurré Easter.
97, Beurré de Rans.
98, Gansel's bergamot.
99, Colmar.
100, Plum, Magnum bonum, white.
101, Magnum bonum, red.
102, Apricot, Turkey.
103, Cherry, Griotte de ratafia.
104, 105, 106, Morello.
107, Peach, Bellegarde.
108, Royal George.
109, Nectarine, Violette hâtive.
110, Apricot, Royal.
111, Hemskirke.
112, Cherry, May duke.
113, Elton.
114, Bigarreau Napoléon.
115, Pear, Easter beurré.
116, Beurré de Rans.
117 to 126, Currants, white and red
127, Pear, Chaumontel.
128, Passe-Colmar.
129 to 135, Currants, red and white.
136, Plum, Orleans.
137, Washington.
138, Mirabelle.
139, 140, 141, Green gage.
142, 143, 144, Apricot, Moor Park.
145, 146, 147, Cherry, Morello.
148, Plum, Drap d'or.
149, Green gage.
150, Apricot, Red masculine.
151, Royal.
152, Nectarine, Hunt's Fanny.
153, Elruge.
154, Peach, Noblesse.
155, Grosse mignonne.

Remarks. This design is very valuable on account of the list of trees with which it is accompanied, and the indications by figures in the plan of the kitchen-garden (*fig.* 221.), and in that of the shrubbery (*fig.* 220.), of the situations in which they are to be planted. All the fruit trees enumerated in the former list have been fruited by Mr. Glendinning, and he can therefore speak confidently of their merits. The arrangements immediately connected with the kitchen-garden, such as the melon-ground, &c., seem also very good; and those for saving all the liquid manure, excellent. Considering that the surface of the ground is nearly flat, or, at all events, that the difference of level between the highest and the lowest points is not more than 30 or 40 feet, and that the whole is surrounded by a belt of Lucombe oaks, which will completely exclude all exterior view, this residence will depend for its interest entirely on its interior beauties; and hence the propriety of introducing as great a variety of trees and shrubs as can be grown in the gardenesque manner in so limited a spot. The close plantations near the house, it will be observed, are planted thick in the picturesque style, which will serve by contrast to set off the gardenesque plantations to advantage.

A Suburban Villa of Four Acres. — The occupier of the villa, the plan of which is shown in *fig.* 222., took a long lease of seven or eight acres of ground in the parish of Hammersmith. He took about four acres into his own possession, and the remaining part, lying along the public road, he let off in portions of a quarter of an acre each, for the purpose of being laid out and built on. This gentleman's own portion extends along the public road, from *a* to *b* in the figure. On the left, it is bounded by a brook (*a c*), beyond which are the grounds of a handsome suburban villa, the two grounds serving mutually to set off each other. On the east, it is bordered by the boundary belt of a gentleman's park; and it extends behind the portion of ground allotted off, in the form of a paddock of upwards of two acres. This portion of the residence, which extends behind the frontage allotments, is kept in pasture, for the purpose of grazing a horse and cow; and, as the soil is rich, and moist rather than dry below, it produces a great abundance of nutritive herbage, and is a source of much comfort and enjoyment to the occupier. The quantity of grass produced is greatly increased by dividing the field by hurdles into three equal portions, and by only grazing one portion at a time. This practice might even be improved on, by dividing it into four equal portions, keeping the cow and the horse always separate, and making the one animal always follow the other. The reason why this is advantageous is, that the droppings from horses produce a rank growth, which is not objected to by cows, though it is by horses; and the same thing takes place in respect to cows.

The surface of the ground of this villa is perfectly flat, but fortunately elevated 2 or 3 feet above the level of the brook (*a c*). There is a carriage entrance near *a*, and a private door for domestics near *b*. The house consists of two parlours (*d e*), a kitchen (*f*), and the usual offices, including a cow-house, stable,

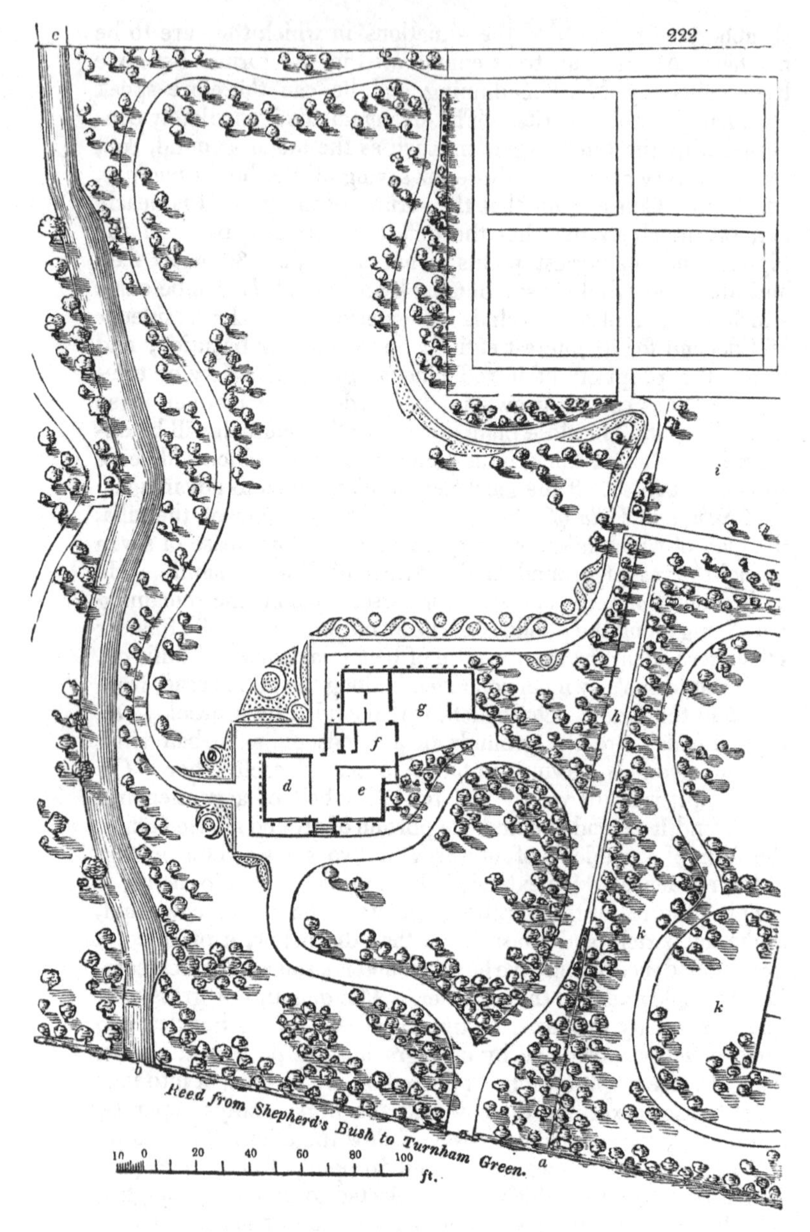

and gig-house. The kitchen and stable-court (*g*) are screened by plantations, as is the carriage road (*h*), which leads to the field (*i*) behind the range of front allotments, a portion of one of which is shown at *k*. The beds of flowers along the pleasure-ground walk, in this design, require no explanation; and the kitchen-garden is evident from its rectangular form. It is surrounded by a holly hedge, and, therefore, requires very little effort on the part of the planter of the pleasure-ground to conceal it. From the turn of the pleasure-ground walk at the principal entrance to the kitchen-garden, it is obvious that this garden is not intended as a place to walk in. In short, a kitchen-garden without walls is too like an arable field anywhere, and, in London, too like a market-garden, to be much resorted to as a place of recreation. This is more particularly the case where the plan is a square of limited extent, as in the enclosure before us. The square form and limited extent give the idea of confinement; whereas a long narrow slip, even though bounded by hedges, has more in it to amuse: on entering the strip at one end, something may be hoped for before we reach the other; but, in the case of a small square, the whole is seen at once the moment of entering; and, all the walks being equally short, and all the boundary fences equally exposed to the eye, there is no desire to proceed farther.

These remarks as to the kitchen-garden apply chiefly in the case of a visiter walking round the place for the first or second time: to the occupant, the crops and the cropping are sources of particular interest. The conclusion that we wish to be drawn in the way of principle is, that, for interesting the imagination, and for picturesque effect, a small spot of ground, whether a plot of two or three perches, or a residence of five or six acres, should extend in one of its directions much more than in another: it should be much longer than it is broad, and the direction of its length should be crooked rather than straight.

Remarks. The house, in this design, is placed rather too near one side; and it should have been elevated on a platform, on account of the dampness of the soil. The offices, also, project rather too much on the lawn front, which is thereby rendered much less imposing than, from the expense bestowed upon it, it ought to be. The design *fig.* 223., by Mr. Lamb, would

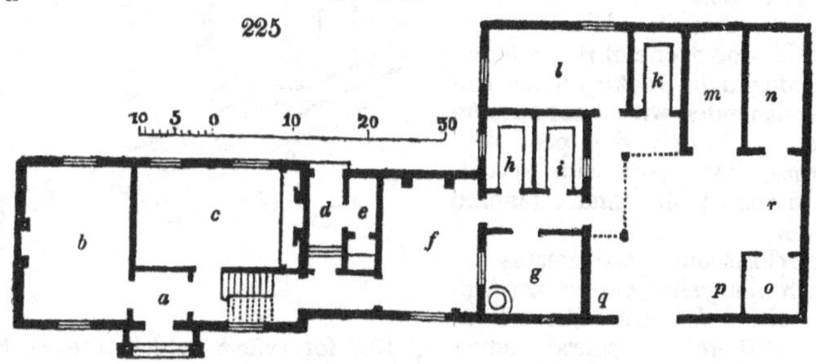

perhaps have suited the situation better. In this plan, *a* is the entrance lobby; *b*, the dining-room; *c*, the drawingroom; *d*, descent of three steps to the closet (*e*), and to the lawn front; *f* is the kitchen; *g*, the scullery; *h*, cellar; *i*, pantry; *k*, dairy; *l*, laundry; *m*, gig-house; *n*, stable; *o*, cow-house. At *p* is a place for ashes, and other refuse fit for manure; and at *q* is a bin for broken bottles,

224

and other matters unfit for manure. At *r* is a dung-pit, with a cover; and at one side of it a tank for liquid manure, with a pump. The front elevation of this house is shown at *fig.* 224.

The Lawrencian Villa, Drayton Green. (*figs.* 225. to 245.).— This villa, of which *fig.* 225. is the ground plan, is unquestion-

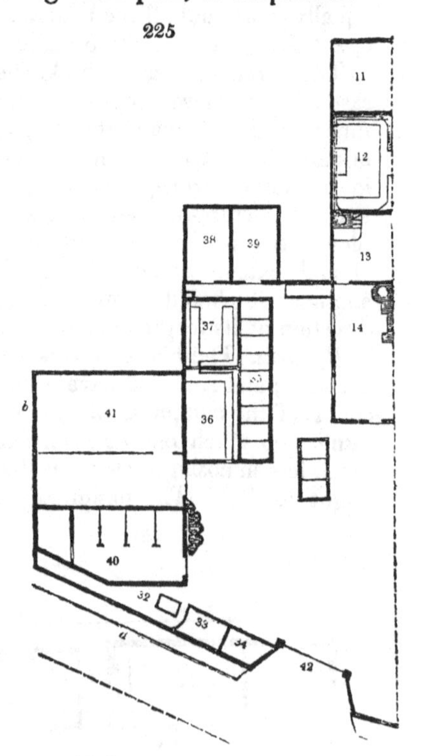

225

a, Grounds of an adjoining villa.
b b, Grass fields, occupied by a farmer.
c, Grass field, belonging to Colonel Sir James Limond, separated from the lawn by a sunk wall and ditch, surmounted by a slight fence formed of four horizontal rods of iron wire.
d d, Village lane, leading on the right to the London road, and on the left to Perrivale, Greenford, and Harrow.
e, Entrance to the house under a covered way; at the end of which, on each side of the hall door, is a niche, with a statue.
f, Entrance lobby.
g, Hall and staircase.
h, Drawingroom, opening under a veranda to the lawn.
i, Dining-room, opening into the garden walk.
k, Mrs. Lawrence's boudoir.
l, Breakfast-room, one of the windows opening to the front garden, which is ornamented with a border, and beds of low-growing peat-earth shrubs, intermixed with spring-flowering bulbs and standard roses.
m, Store closet under the staircase.
n, French wine cellar, entered through the ale and spirit cellar; from which there are stairs leading to the wine-cellar below.
o, Kitchen. *p*, Butler's pantry.
q, Back kitchen, serving also as a scullery to the dairy.
r, Dairy.
s, Housekeeper's room.
t, Dust-bin.
u, Cinder-bin.
v, Bin for refuse which cannot be burned or turned into manure.
w, Coal-house.
x, Lumber-house for bottles, hampers, &c.
y, Knife-house; adjoining which is a privy for the family.
z, Wood-house, adjoining which is a privy for the servants.

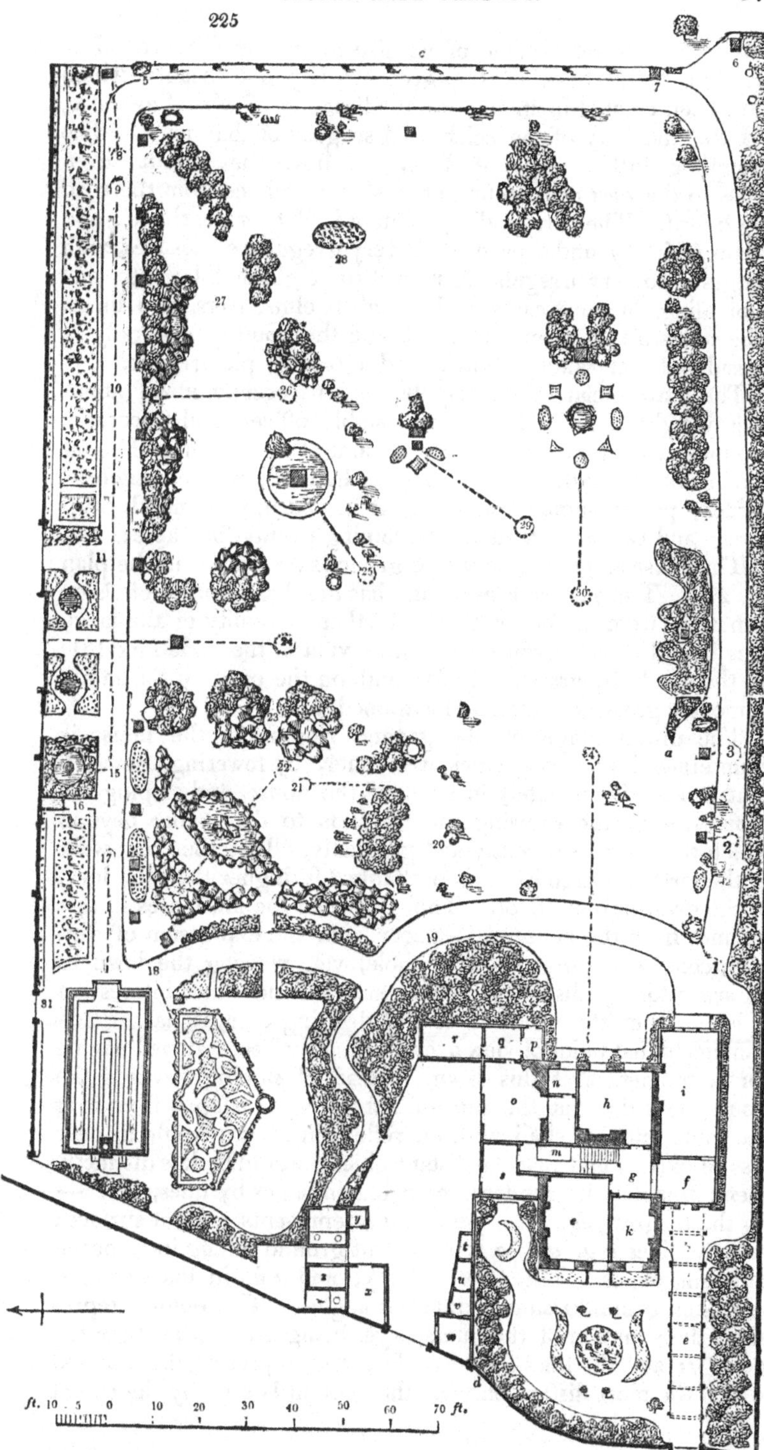
225
ft. 10 5 0 10 20 30 40 50 60 70 ft.

ably the most remarkable of its size in the neighbourhood of London, on account of the great variety and beauty which have been created in it, under the direction of Mrs. Lawrence, F.H.S., the lady of the celebrated surgeon of that name. The straggling little village of Drayton Green lies about seven miles to the west of London, at a short distance from the road to Oxford. The surrounding country is flat, or nearly so; and it is principally under pasture, in very irregular enclosures, with hedges, also very irregular in regard to height and breadth, and abounding, in most cases, with English elms, oaks, and limes. The soil is a thin loam on gravel, and the country generally is considered remarkably healthy, and agreeably picturesque.

The Lawrencian Villa may be said to occupy in all about twenty-eight acres. The house, stable offices, and decorated grounds stand on about two acres; and at the distance of about two or three hundred yards, across the road, are the kitchen-garden, poultry-houses, and piggeries, occupying nearly two acres; and two pasture fields, containing twenty-four acres.

The house and the ornamented grounds are shown in the plan *fig.* 225. The surface is even, and has the disadvantage of rising somewhat from the house to the further extremity of the lawn. It is bounded on the south by another villa of the same kind (*a*); on the north, by grass fields (*b*); and, on the east, by Sir James Limond's grass field already mentioned (*c*).

The disadvantage of the ground sloping to the house is counteracted in a very efficient manner, by lowering the walk that crosses immediately in front of the house, and sloping the ground from the drawingroom veranda to that walk; beyond which the lawn rises gently and gradually, till, at the cross walk at the farther extremity, it is probably 6 ft. higher than the level of the drawingroom floor. Though, when the lawn rises in this manner from the house, it detracts from the expression of dignity, considering the villa as a whole, yet, viewing the lawn as an arena for the display of plants, statues, and other interesting objects, from the windows of the drawingroom, it has an advantage in that point of view over a falling surface. If we imagine for a moment that this lawn, instead of sloping towards the house, as it does, at the rate of 1 ft. in 50 ft., sloped from it at the same rate, we shall find, on reflection, that it would appear less in extent, and that the distant objects would be less distinctly seen: this may be rendered palpable on paper by lines, thus:— In the diagram *fig.* 226., the line *a e* represents a level surface; and the lines *a d*, *a c*, *a b*, represent ground falling in slopes at different angles. The lines *a f*, *a g*, and *a h*, in like manner, represent ground rising at different angles. The point *k* represents the situation of the human eye, being 5 ft. higher than the point *a*; and the lines *k b*, *k c*, *k d*, &c., represent the angle at which the most distant part of the ground is seen by the eye at *k*. Now, the larger the angle at which this distant point of the ground is seen by the spectator at *k*, the more distinctly will he discern objects there; and, as these different angles are represented by the sines to each (*i i*,) it follows that, in rising ground, the most favourable slope for seeing objects from a fixed point is that represented by the line *a f*, or some slope near to that line; say a slope forming an angle between 20° and 30° with the horizon. In the case of falling ground, it will be observed that the most favourable slope lies between the same angles; though in falling ground the objects are not nearly so advantageously seen as in rising ground. A level surface, it will be observed, possesses exactly the same advantages, in point of seeing objects placed on it, as a surface rising at an angle of between 20° and 30°. Hence, for the display of flower-beds, a lawn which has a level surface, or one which rises at any angle under 30°, is much better adapted, than one which slopes from the eye at any angle, however small.

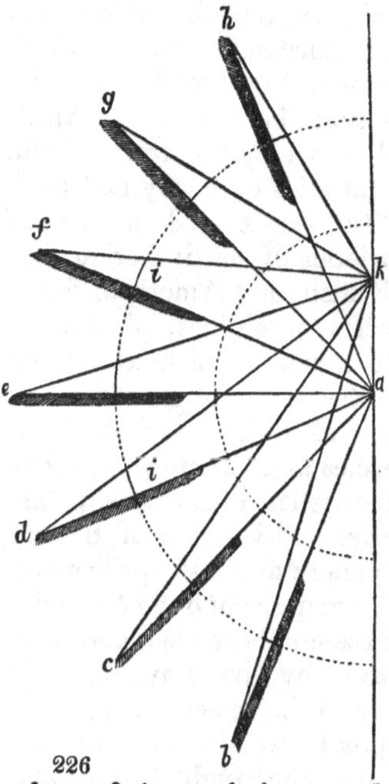

226

The decorated ground in the Lawrencian Villa is remarkable for the very great variety which it contains in a very limited space; and the secret of producing this variety consists in introducing numerous small groups of trees and shrubs, sometimes combined with flowers or climbers, at other times with rockwork, and with statues, fountains, basketwork, and so on. The trees and shrubs are of good kinds, though not remarkable in this respect, having been in part planted before the place was taken possession of by Mrs. Lawrence; but the flowering shrubs, including rhododendrons, azaleas, roses, &c., and the herbaceous flowers, are of the most rare and beautiful kinds. The collection of green-house and hot-house plants may be characterised as among the most select and valuable in the neighbourhood of London.

The numerous prizes which Mrs. Lawrence has received, for some years past, from the Horticultural Society of London, evince the excellence of the articles which she exhibits at their meetings. The first prize which Mrs. Lawrence received from the Horticultural Society was the silver medal for plants sent to the exhibition at the Chiswick Garden in May, 1833; and the last, previously to the moment at which we now write, the silver Knightian medal, for a collection of plants exhibited in Regent Street, May 1. 1838; making in all 53 medals. (See *Gard. Mag.*, art. "Horticultural Society and Garden," from 1833 to the present time.) As a general summary of the flora of the Lawrencian Villa, we may mention that there were, in April, 1838, exclusive of what were killed down by the frost of the preceding January, 212 species and varieties of hardy and half-hardy ornamental trees and shrubs; 130 species and varieties of hardy fruit trees; 600 species and varieties of hardy herbaceous plants; 30 species and varieties of British and American ferns, planted in the rockwork; 140 species of alpines, planted in the rockwork; 34 species of hardy aquatics, planted in the basins; 200 varieties of heartsease; 500 varieties of garden roses, creepers and standards; 12 varieties of ivy; 40 species and varieties of American plants; 9 species and varieties of hardy ligneous climbers; 140 species and varieties of florist's pelargoniums; 172 genera and 992 species and varieties of Botany Bay, China, and Cape shrubs; 134 genera and 340 species and varieties of hot-house plants; and 57 genera, and 227 species and varieties of stove Orchídeæ. These numbers are taken from a manuscript catalogue, kindly lent to us by Mrs. Lawrence.

Next to the grouping on the lawn, and the select collection in the green-houses and stoves, the points worthy of imitation in Mrs. Lawrence's management are, the high order and keeping which pervade every part of her residence, from the most obscure recesses of the offices, to the most brilliant scenes on the lawn. This is effected, also, by a smaller number of gardeners than might be expected: the number kept in the summer time being six, with one or two women for collecting insects and dead leaves, and during winter three. It is only farther necessary to add, that all the different scenes in these gardens, all the beds of flowers, pieces of rockwork, &c., as well as the green-houses and hot-houses, were designed by Mrs. Lawrence herself, and executed under her direction.

The dwelling-house of this villa has been much enlarged and added to at different times, in consequence of which there is a want of regularity and symmetry in the arrangement, and of proportion in the dimensions of the different apartments, which is unăvoidable in such cases. It is often, however, useful, to give the plan of such houses; because it shows how additions may be made according to the wants of the occupier. These additions show in a more forcible manner than a regular or symmetrical ground plan, the accommodations which cannot be dispensed with, as well as what may be considered as the minimum extent of these accommodations. In a regular plan, an apartment is sometimes added to complete the regularity of the figure; and the size of this apartment, as well as of that of some of the closets, &c., and some of the outbuildings, is often larger than there is any occasion for, and sometimes smaller than it ought to be, for the same reason. In the case of a house like the present, consisting originally of five rooms and a kitchen on the ground floor, being rendered, by additions, fit for the occupation of a family enjoying every comfort and luxury, we discover not only all the necessary supplementary rooms and offices, but the smallest size of each that will answer the end in view. We have not, in this dwelling, either a laundry or a brewhouse, because washing and brewing are not done at home; nor a large library or wine-cellar, because Mr. Lawrence residing principally in London, his principal stock of books and wines is kept there. We have, however, an ample pantry and dairy, and all the smaller outdoor offices which are required in the largest mansion. Baths and water-closets are not shown, because they are on the bed-room floor; and there is also a bathhouse in the garden.

Entering the lawn from the drawingroom (*h*), we find a gentle descent from the veranda to the walk. Turning to the right, at the angle at 1, we observe the foliated vase *fig.* 227., the base of which is concealed by a plant of tree ivy; proceeding onwards towards 2 and 4, we pass the pedestals and vases *figs.* 228. and 229. We are now at a sufficient distance from the garden front of the house, to see it to advantage by turning round; and, if we step on the lawn to the point 4, we shall find the view *fig.* 230., to the left of which will be observed the ivy vase, and a basket containing a pyramid of roses; and to the right an elevated rustic basket of pelargoniums. The large window on the left is that of the dining-room.

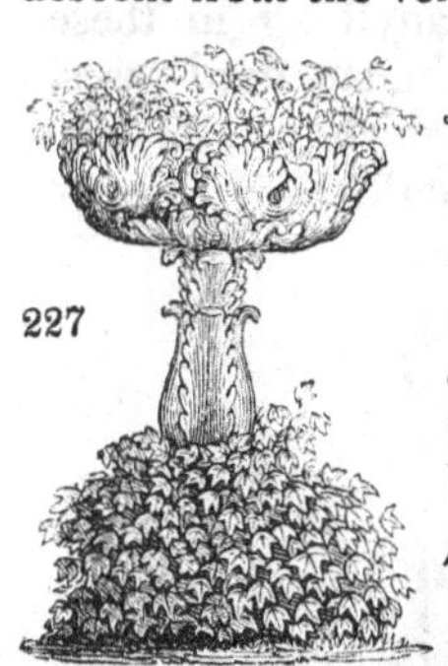
227

Looking from 1, 2, and 3, across the lawn, the eye observes an intricate maze of agreeable and beautiful objects, but sufficiently distant not to create the idea of being crowded or confused; the reason of which is, that there is always an ample surface of naked lawn in the foreground, or middle distance, to contrast with the ornamental groups, and to throw them to a sufficient distance from the eye.

The margin of plantation on the right is composed partly of evergreen trees and shrubs, and partly of deciduous flowering kinds. The groups on the left hand are, in part, of more rare sorts, and contain a great many fine hybrid rhododendrons and

228

229

azaleas. All these plantations and groups are treated in the picturesque manner; there being scarcely anything in these grounds, except the single plants, such as the standard roses, and some rhododendrons and other shrubs, which can be considered as treated in the gardenesque style of culture.

230

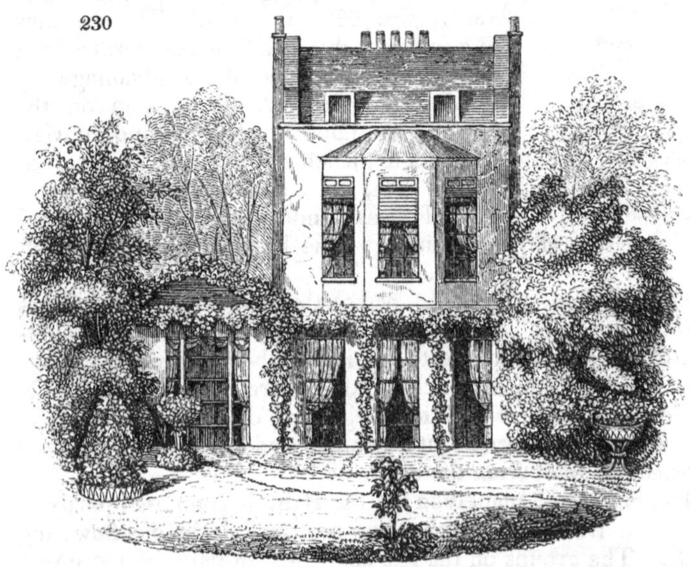

At 5, there is a fine specimen of double-blossomed furze, and two splendid vases on elevated pedestals; both combining to form an interesting termination to the comparatively straight walk from the house to this point. One of these vases at 6, when the spectator is at a distance, appears to him to be the terminating point, while that at 5 comes into his view afterwards. The walk from 5 to 7 is several feet higher than the floor of the veranda in the front of the drawingroom; and hence the views towards the house, being along a descent, are less interesting than, from the number of objects on the lawn, they otherwise would be. The view into the paddock, to the right, affords an agreeable relief from the excess of beauty and variety on the lawn, as it consists of a plain grass field, grazed by some fine Alderney cows, and planted with two or three scattered elms, oaks, and aspens and other poplars.

The next scene of interest is the Italian walk, arrived at the point 8, in which, and looking back towards the paddock, we

231

have, as a termination to one end of that walk, the rustic arch and vase *fig.* 231.

From the point 9, we have the view of the Italian walk *fig.* 232., with a span-roofed green-house as the termination at the farther end, and an elegant fountain on the right hand. The border on the left is planted with the most choice herbaceous flowers, interspersed with standard roses at regular distances; and the wall is devoted in part to the finer fruits, but principally to climbing roses, and other climbing or twining shrubs of fragrance or beauty. At the point 10, there is a rustic archway of rockwork on the right, from which an interesting view across the lawn is obtained. At the point 11, there is a walk across the border to the bath-house, adjoining which is a camellia-house (12); and beyond that two long sheds (13, 14), for tools, pots,

232

&c., with potting benches, and other places for garden materials and operations, and for the boilers to heat the bath, the camellia-house, &c. In the camellia-house there are sixty-seven of the finest species and varieties that can be procured.

At the point 15, we are immediately in front of the fountain *fig.* 233., supplied from a cistern which forms a small tower on the top of the tool-house; and beyond that is a walk to the stone cistern at 16, which supplies water for watering the garden. The water is raised to these cisterns by a forcing pump in the stable-yard.

233

Farther on in the Italian walk, at 17, we have the front view of the span-roofed green-house *fig.* 234.; and a little to the left, at 18, the view of the French parterre *fig.* 235.

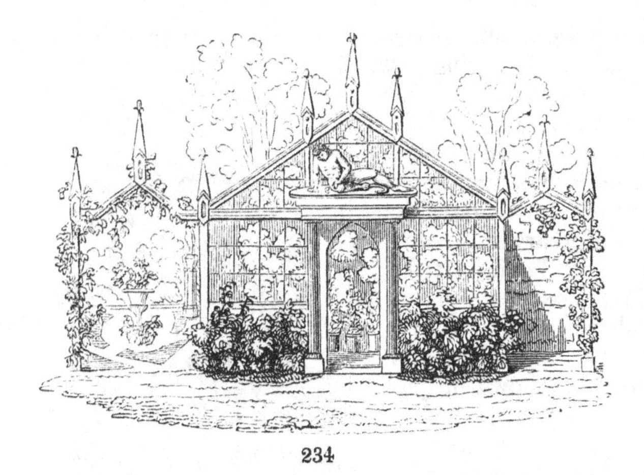

234

Proceeding towards the house, a view of a handsome weeping ash (20) is obtained from the point 19; and, at the farther extremity of the walk, the vases placed at 1, 2, 3 on the plan have an excellent effect, backed by the marginal plantation of evergreens. Leaving the walk at 19, and passing the weeping ash at 20, if we advance on the lawn to 21, and look towards the south, we have the pollard vista *fig.* 236.; and, changing the

235

position to 22, we have the view of the rockwork, statue of Fame, &c., shown in *fig.* 237.

On the right and left of 23 are two groups of rockwork, with concealed springs, which drop from rock to rock, and from stone to stone, and form curious little moist places for aquatic plants. Advancing to 24, and looking northwards, we have the statue of Mercury in the foreground, and behind it the camellia-house, the wall on each side of which is heightened with trelliswork for creepers, as shown in *fig.* 238.

236

At 25, we have the view of the fountain and arch behind, shown in *fig.* 239. In the basin are nymphæas and other aquatics; and on one side is a Napoleon willow.

At 26, we have the view of the rustic arch and Cupid, shown in *fig.* 240.; and, at 27, the tent seen in *fig.* 239. is frequently pitched in the summer time, which gives the idea of the warm season, and of the enjoyment of coolness and refreshing breezes in the midst of intense sunshine.

237

At 28, there is a bed of *R*òsa índica, in the centre of which is a large plant of *Yúcca* gloriòsa; and, proceeding across the-

238

lawn to 29, we have the view *fig.* 242.; and, at 30, we have the fountain, surrounded by baskets of flowers, with the two garden nymphs *fig.* 245.

239

240

We shall now suppose that the spectator walks across the lawn, and, passing the span-roofed green-house, enters the court of offices by the door at 30. In this court which forms a part of *fig.* 225, and which we here repeat (*fig.* 241.), he finds, —

241

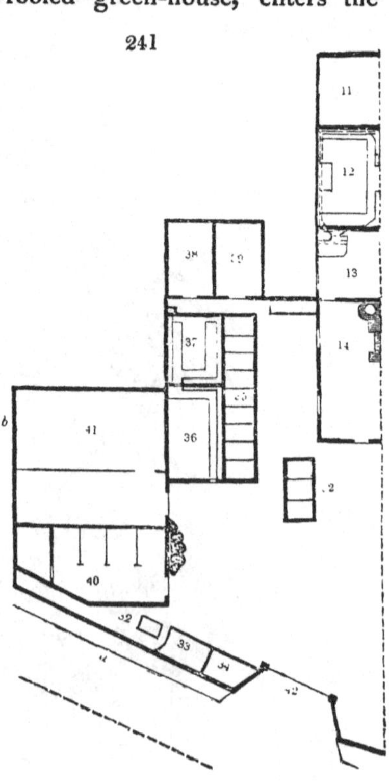

31, A large pit for plants.
32, A small pit.
33, Dung-pit for the stable.
34, Rubbish-pit, and rot-heap for the garden.
35, Pit for heaths.
36, Green-house.
37, Dry stove.
38, Shed for flower-pots.
39, Rubbish-shed.
40, Four-stalled stable, with hay-bin at the farther end.
41, Coach-houses, harness-room, and sleeping-room for coachman.
42, Place for plants in pots that have done flowering.
43, Carriage entrance from lane.

The kitchen-garden, the gardener's house, the stove, and the poultry-houses, grass fields, and cow-shed, are situated on the other side of the lane, and at the distance of 100 yards from it. These

242

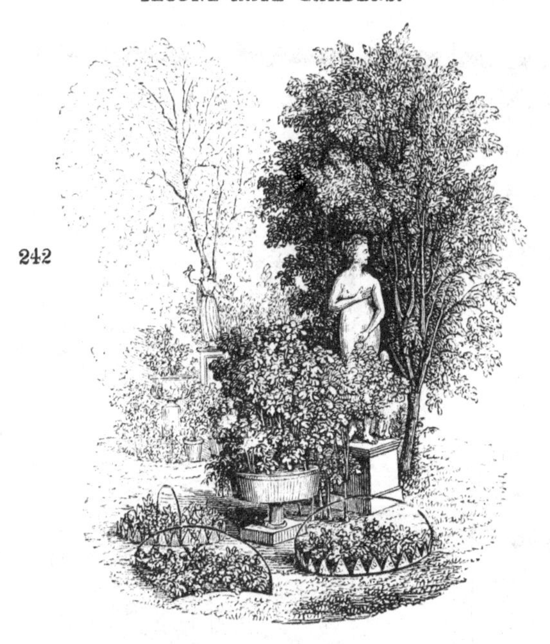

are shown in the plan *fig.* 244, which is accompanied by an explanation of the references.

Remarks. This villa may be considered as a model of its particular kind; and, though it may not be in the power of many to imitate it in every thing, yet the humblest and most economical possessor of a villa residence of two acres may take a lesson from Mrs. Lawrence's taste, as displayed in the manner in which the trees and shrubs are grouped on the lawn. Every one cannot have so many fountains, or form rockwork of spars, fossil organic remains, and other geological specimens brought from distant parts of the country; but every one may sink in the

243

ground a few small wooden cisterns lined with lead, and supply them with water by hand, as it evaporates in the summer season. Some of these may serve as brilliant spots to attract the eye, and others as habitats for aquatic plants

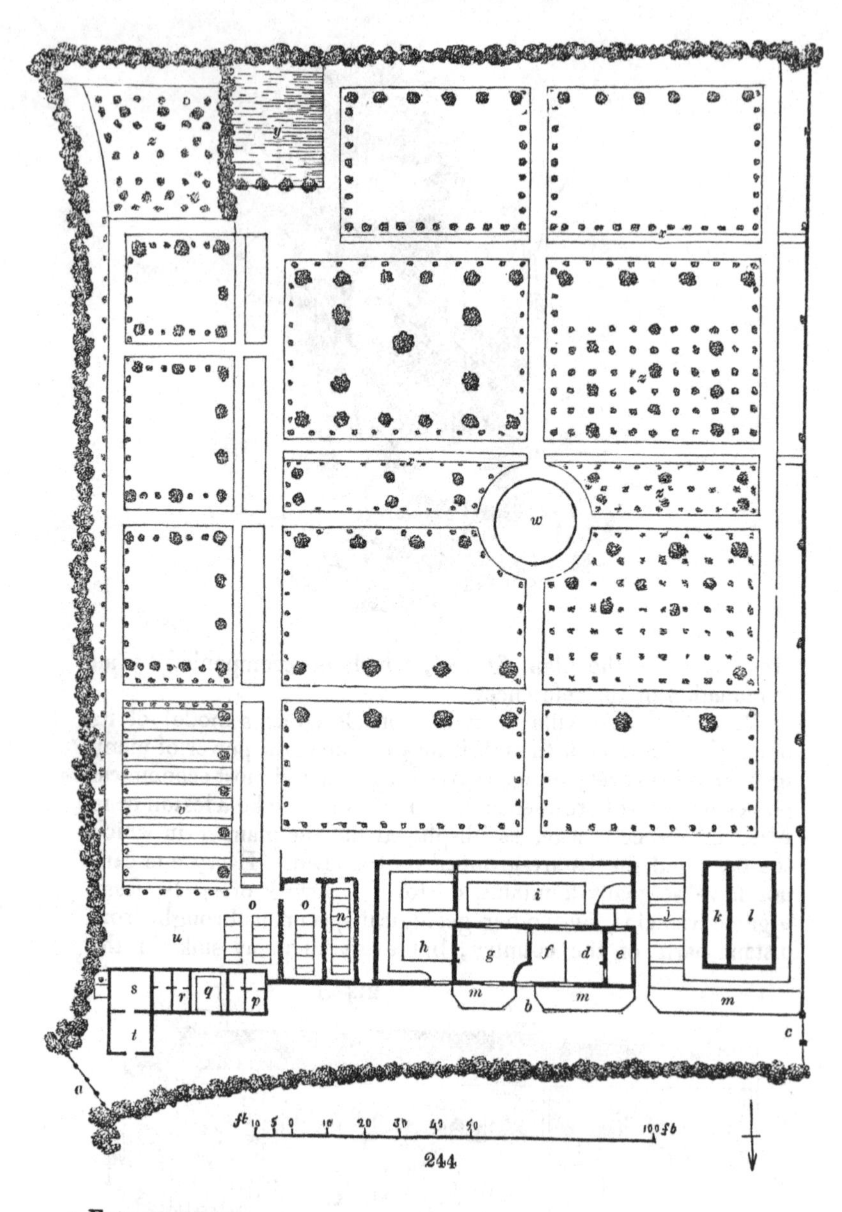

244

a, Entrance gates.
b, Entrance to the gardener's house, the elevation of which is shown in *fig.* 243.
c, Entrance to the cow-field, in which the cow-shed is placed.
d, Kitchen.
e, Wash-house, or back kitchen.
f, Gardener's sitting-room.
g, Apartment divided into two bedrooms.

245

h, Stove, heated by hot water, the side elevation of which is seen in the view of the gardener's house, *fig.* 243.
i, Orchidaceous house, with miniature rockworks and artificial hillocks, for terrestrial *Orchídeæ*; and small basins and fountains, formed of shellwork, for aquatics.
j, Beds of reserve flowers.
k, Cold-pit.
l, Span-roofed green-house.
m m, Children's gardens.
n, Situation for a hot-bed, surrounded by a privet hedge, 18 in. high.
o, Compost and frame ground.
p, Two pigsties.
q, Poultry-house, with pigeon-house over. This house has a span roof, with a gable end over the door; and the triangular part of the gable end has 4 rows of holes for the pigeons, the rows having narrow shelves in front for the pigeons to rest on, and an enclosed space behind, 3 ft. in depth, for the nests.
r, Two other pigsties.
s, Rabbit-house.
t, Tool-house, in which, also, the ducks are kept.
u, Frame-ground.
v, Asparagus-beds.
w, Circle of grass where a tent may be fixed, for eating fruit in during the summer season.
x x, Open drains, the soil of the garden being a retentive clay.
y, Pond.
z, Fruit trees and fruit shrubs.

The margins of basins of this sort can be effectually disguised with rockwork, and this can be procured from the nearest brick-field, stone-quarry, or, perhaps, from old houses. which are taking down, chalk-pits, ferruginous gravel-pits, &c. If suitable vitrified bricks cannot be procured, common bricks may be joined together, in masses of any size and shape, by cement; and there is no reason why blocks so formed, or any other materials to be substituted for rockwork, should not receive weather stains artificially, no less than the walls of a house, where the object is to imitate an ancient building. As to the wooden cisterns, they will last long enough: and we know, from experience, that it is cheaper, in the end, to form such cisterns of wood, lined with lead, than to build cisterns of brickwork and cement; for, unless these are of considerable size, the cost is as great as where lead is employed; and they are much more apt to leak and receive injury from frost.

It is worthy of remark, that a good deal of the interest attached to the groups on the lawn of the Lawrencian Villa depends on the plants which are planted in the rockwork. Now, though every one cannot procure American ferns, and other plants of such rarity and beauty as are there displayed, yet there are hundreds of alpines, and many British ferns, which may be easily procured from botanic gardens, or by one botanist from another; and, even if no perennials could be obtained suitable for rockwork, there are the Californian annuals, which alone are sufficient to clothe erections of this kind with great beauty and variety of colouring.

With regard to the statues, vases, &c., though some of these, at Drayton Green, are of bronze, marble, or stone, and have cost considerable sums, yet others of composition, equal in point of taste, though far inferior in pecuniary value, may readily be procured, at a moderate cost, of Austin's artificial stone, or of earthenware.

We are aware that there are many persons, of a simple and severe taste, who will think that the Lawrencian Villa is too highly ornamented with statues and sculptures; but allowance must be made for individual taste, for devotion to the subject, and for the limited extent of the place. Were Mrs. Lawrence in possession of a villa of 100 acres, there can be no doubt that she would display on her lawn a taste as appropriate to a residence of that extent, as the taste she has displayed at Drayton Green is suitable for that place.

Parsonages. — Assuming that every clergyman's house in the country ought to have ground enough attached to it to enable its occupier to keep a cow, this will bring parsonages into the second class of suburban residences; and, as the characteristic of parsonages is their proximity to, or connexion with, a church and churchyard, we shall here submit some observations on churchyards generally, on the choice of a situation for a church and churchyard, on laying out the churchyard, and on planting and managing it; including, under each head, the treatment of churchyards already existing, with a view to their improvement.

Churchyards, like every other description of yard or garden, ought to be laid out, planted, and managed, with reference to their use; and the scenery produced should, in its expression and general effect, indicate what that use is, or, at all events, be in accordance with it. A churchyard ought not to be laid out so as to be mistaken for a pleasure-ground, a shrubbery, or a flower-garden; neither, on the other hand, ought it to be left in a state of utter neglect, without regular walks, and overgrown with weeds and rank grass. The use of the churchyard is as a place of burial, as an enclosure and protection to the church, as a place sacred to the memory of the dead, as a place of weekly meeting for solemn purposes, and as an approach to the church. All its uses are of a serious and important nature; and it is therefore to be considered as a grave and solemn scene. Now, the question to be solved in laying out a churchyard is, what treatment of the trees, the surface of the ground, the grass, walks, graves, gravestones, and tombs, will be most conducive to solemnity of effect. The expression of the exterior of the church is grave and solemn, by its long-established association with our religious feelings; and it therefore may be considered as having a similar influence on the scenery around. The feeling of solemnity is one more of a passive, than of an active, nature: it neither needs to be much cultivated, nor much exercise of the imagination. Strong contrasts are not required to excite this feeling, nor varied and intricate scenery to prolong it. It has its origin in the uses of the place, and will only be interfered with, or weakened, by the introduction of such objects as interfere with these uses. Simplicity, therefore, ought to be a governing principle in every thing relating to churchyards; and, as the appearance of neglect or slovenliness always implies want of respect, order and neatness are next in importance. By order, we mean the avoiding of every thing like confusion in the disposition of the trees, or the placing of the tombs and gravestones; and by neatness, we allude more particularly to keeping the turf short and smooth, the walks firm, even, and free from weeds, the gravestones upright, and the tombs in a state of repair.

The character of a churchyard, as a place of burial, will always be more or less influenced by the character and manners of the people to whom it belongs. In Britain, churchyards have much less care bestowed upon them than in Central Germany, and in some parts of France, Belgium, and Holland. The sentiment of respect to the memory of deceased persons in these countries is shown by planting flowers over the graves, and frequently cultivating them there for some years afterwards. Among the Moravians, on the Continent, the churchyard is sometimes laid out in compartments, with walks between, like a garden; and the compartments are kept dug, and planted with flowers and ornamental plants. Two powerful arguments are advanced in favour of this practice: the first is, that a churchyard so managed costs less than if it were in turf, and kept short by mowing; the second, that the surface of the ground has always the same appearance, there being no gravestones or tombs, and the ground being left level, and replanted with the plants which stood on it before, after every interment; these having been carefully taken up, and placed on one side, before the grave was dug. It is evident that this mode of treating a churchyard, however consonant it may be to the ideas of those who adopt it, is not in accordance with our desiderata. It does not indicate its use, as it has neither raised graves, tombs, gravestones, nor any other appearance of its being a place of burial; and it is not calculated to excite solemn emotions, as it has all the gaiety of a flower-garden.

In Britain, respect for the dead is not generally shown by the introduction of flowers over their graves; but the practice prevails in some places throughout the country, more especially in Wales, and is not unfrequent in the metropolitan and other cemeteries. Perhaps it ought to be commended and encouraged, as the frequent recollection of deceased friends has a tendency to sober the mind and cultivate the affections of the living. In every part of Germany where the inhabitants are in the habit of cultivating flowers on the graves of their friends, or even of visiting these graves annually on a certain day and decorating them, the inhabitants are a reflective, and very humane and amiable, people; for example, at Munich. The introduction of flowers in churchyards, therefore, where they are planted over the graves by the relations of the deceased, is a very different thing from their introduction in the margins of plantations of trees and shrubs, as is done in some of our public cemeteries, in imitation of shrubberies; to the utter neglect, as we think, of appropriate character and expression. Bearing in mind, therefore, the three principles of simplicity, order, and neatness, as guides in laying out churchyards, we shall next proceed with the details.

Situation and Soil. It is almost unnecessary to observe that a country church ought either to be built adjoining the village for which it is intended, or, if it is to serve two or three villages, in a situation central to them. The surface of the ground ought to be an elevated knoll, in order that the church and the spire may be seen on every side, and, if possible, throughout the whole extent of the parish. The knoll should be sufficiently large to admit of its summit being reduced to a level, or, at all events, to a nearly level, platform, or piece of table land, about the size of the churchyard; a level surface being more convenient for the purpose of interment than a sloping one, for a reason that will be given hereafter. Besides which, the ground plan of a church being a parallelogram, to see it rising out of a round knoll would be contrary to every idea of a suitable and secure foundation. Where there is no want of room, or not many burials likely to take place, the surface of a churchyard, instead of being level, may be quite irregular; but, in this case, the places for graves, and the walks of communication to these places, must be rendered easily accessible, and, to a certain extent, level. This can always be effected by laying the ground out in terraces; a mode of disposition which may be as advantageously adopted in churchyard gardening, as it is in gardening as an art of culture. The soil should, if possible, be sandy or gravelly, as being most suitable for promoting animal decomposition; but any soil may be rendered fit for this purpose by deep trenching, and the addition of sand, gravel, and lime rubbish, so as to form an artificial stratum as deep as it is intended to dig the graves. The worst of all soils for a churchyard is a stiff wet clay; which, by its compactness and retention of water, prevents the natural decomposition of the body, and has even been known to change it into an adipose substance.

The Size of the Church, and the Extent of the Churchyard, will depend on the population for whose service they are intended; and on the probable slowness or rapidity of its increase. The form of the church may be considered as fixed, by precedent and immemorial usage, in that of a parallelogram, with or without projections at the sides, so as to give it the form of a Latin cross; and having a tower, steeple, or cupola, at one end, for the church bells, and a clock. There are some examples, however, of churches having been made semicircular, circular, or polygonal, in the plan, so as to suit them to particular situations.

The form of the churchyard is not fixed, like that of the church, but will naturally be determined jointly by the form and position of the church, and the form of the ground which surrounds it. If the ground be level, or nearly so, then the outline of the churchyard may coincide with that of the church; so as also to form a large parallelogram, in the direction of east and west, that being the prescribed bearing of all Christian churches. If the church be situated on the summit of a conspicuous conical hill, or dome-like knoll, then the outline of the churchyard will be determined solely by the ground, and may be circular, oval, or roundish; and we may here observe, that, when cases of this kind occur, as they are not very common, we think the ground plan of the church ought to be round, or roundish, also. In general, the position and form of the churchyard ought to be such as will have a good effect from all the different parts of the surrounding parish from which it is seen; while, at the same time, it should look well from its immediate vicinity, and also from the different doors and sides of the church.

The Site of the Church should be central to the natural shape of the ground which is to constitute the churchyard, when that shape is in any way remarkable; but, where the surface of the ground is level, the church may be placed nearer one end of the parallelogram, or other-shaped piece of ground, which forms the churchyard, than another; or even nearest to one side, provided this is not attended with injustice to the parishioners. In general, the exact position of the church within the churchyard, when not determined by natural circumstances, ought to be regulated by the number of sides on which it is approached. If the parish lie equally round the church on every side, there will be at least four gates to the churchyard, corresponding with the four cardinal points; and in that case the church ought to be in the centre of the churchyard: but, if there be only a gate at one end, or if there should be several gates, but all nearer one end than the other, the church ought to be placed accordingly.

The Ground Plan of the Church, its exact position in the churchyard, the boundary lines of the latter, and the different churchyard doors or gates being fixed on, before anything farther is done, the church ought to be built; and we shall suppose that its elevation is so designed as to appear to rise from a platform of gravel or pavement, of from 10 ft. to 20 ft. wide, according to the size of the church; this platform, or terrace, being supported by a sloping bank of turf, at an angle of 45°, and furnished with flights of steps opposite each of the churchyard gates. Underneath the surrounding platform, there ought to be a deep barrel-drain, or box-drain, for receiving the rain-water from the roof of the church, and thus keeping the foundations dry; and from this drain there ought to proceed others of the same kind, under each of the walks which lead from the church platform to the boundary wall. These last, besides carrying away the water collected in the drain which surrounds the church, will dry the subsoil of the churchyard generally; and receive the surface water from the walks, through gratings placed at regular distances.

The Boundary Fence of the churchyard should be such as to exclude every kind of domestic quadruped; but it is not, in general, necessary that it should be so high as to prove a barrier to man, because it may fairly be supposed that most persons will reverence the interior more or less, and that those who are without this reverence will have, in general, nothing to gain by breaking into such a scene. We here exclude altogether the consideration of body stealing, which can only be practised by a particular set; who, in the country at least, are rarely to be met with. As swine and rabbits are particularly offensive in churchyards, especially where the soil is sandy, the boundary fence should either be a low wall of 3 ft., surmounted by a holly or thorn hedge; or a wall of 6 ft. or 7 ft. in height, without any hedge. In the latter case, the inner face of the wall may be planted with common ivy. Where the churchyard is to be united with the adjoining lawn, garden, or pleasure-ground of the parsonage, the boundary fence on the side next the residence may be an open iron railing; and, where it is to be united with a pleasure-ground on a large scale, or a park, it may either be surrounded by an open iron railing, or by a deep and wide sunk fence. If a hedge is in any case determined on as the boundary to a churchyard, it ought to be kept much broader at bottom than at top, in order that it may grow quite thick and close there; and the only plants fit for such a hedge are the common white thorn and the holly.

The Walks of a Churchyard are of two kinds: those for proceeding from the different gates in the boundary fence to the church doors, for persons going to, or returning from, the church; and those which make the circuit of the churchyard, for the more conveniently viewing the tombs and graves, and for conducting funerals. The walks proceeding from the entrance gates in the boundary fence to the church doors should be always in straight lines, and of a width proportionate to the size of the church and churchyard, but never narrower than 6 ft.; because this is the least width which will allow two persons abreast, carrying a coffin between them, to pass solemnly along: the width, indeed, should be greater rather than less, because nothing can be more indecorous than to see a funeral procession crowded and huddled together for want of room. In every case, we would, if possible, place the entrance gates so that the walk from them to the church, whether to its sides or its ends, might always meet the building at a right angle.

With respect to the walk round the churchyard, it should in every case, and whether the churchyard were small or large, be at a distance of at least 10 ft. from the boundary wall, in order to leave a border sufficiently broad for a range of graves to be placed at right angles to the wall. This walk should be of the same breadth as the others; and, like them, in no case less than 6 ft., for the reasons already mentioned. In most churchyards this boundary walk, and the cross walks necessary as approaches to the church, will be sufficient; but, where this is not the case, cross walks from the boundary walk to the terrace round the church may be added; or a second surrounding walk may be formed, half-way between the terrace or walk round the church, and the circumferential walk.

The Walks of old Churchyards. The preceding remarks constitute our *beau idéal* of churchyard walks, supposing that a church were to be built, and the walks to be laid out on a flat surface; but, as this is not a case of every-day occurrence, the next point is to show what can be done in the case of old churches and churchyards, having, perhaps, only one gate, and one walk from that gate to the church; and where the churchyard is so thickly studded with graves and gravestones as scarcely to leave room for any additional walk, and, at all events, not for a straight one. The mode in which we should proceed in this case would be, to lay out the requisite walks on the same general principles as in a new churchyard, but to make them of double or treble the usual breadth, so that where a tombstone occurred in the middle, or in any part of the walk, there might be room on one side to pass it. In some cases, we might, perhaps, adopt a winding course for the circumferential walk; but, unless the curves coincided with those of the boundary fence, or were made very large, this direction would be so unsuitable in point both of use and expression, that we would avoid it as much as possible. A serpentine walk in a churchyard is a waste of ground, with reference to the placing of the graves; and it has too much of the pleasure-ground air, or even, perhaps, of an air of affectation, for the gravity of the scene.

In every old churchyard there are some of the gravestones, and perhaps even of the tombs, in which no person living has any interest. We are far from saying that any of these should be done away with, in order to allow a walk to be made straight; but we think that, in cases of this kind, there would not be any impropriety in reducing raised tombs, so as to leave the covering stone on a level with the walk, and forming part of its surface; or in laying down upright gravestones in a flat position for the same purpose. Some families, also, might consent to this being done with the tombs that they are interested in, as it is frequently done purposely; and, probably, with a very few alterations of this kind, the straight, or other suitable direction of the walks might be maintained. With respect to turf graves, without gravestones, it does not seem unreasonable to suppose, that, after a certain number of years, these may be levelled, and a walk carried over them; because, in the common course of things, such graves are reopened, often in a few years, for the purpose of interring persons who have no connexion whatever with those previously buried there. In a word, the interest of the relations of the deceased in the ground, in these cases, can only be considered as lasting for the ordinary period of rotation in the common parts of a burial-ground; and this period, which varies in different churchyards, according to the demand for ground, and the nature of the soil, seldom in any exceeds twelve or fourteen years.

The grassy Surface of a Churchyard, when it is newly laid out, should, of course, be even; and the nearer it is to level, the more convenient will it be for all the purposes of interment. Whether even or uneven, it should always have a descent from the church, rather than towards it, for the sake of throwing off the surface water; and in strong clayey soils, in moist climates, provision ought to be made by surface gutters, even in the turf, for conveying the water to underground drains, or directly along the surface to the boundary of the churchyard. In churchyards which have been long in use, the grassy surface is, in general, very uneven, on account of the greater accumulation of graves in one place than in another, and from certain parts being chiefly occupied by tombs, and others by turf graves without marks Perhaps the chief disadvantage of these inequalities is, that they occasion the production of rank coarse grass, and large weeds, than which we scarcely know anything more unsightly in a churchyard; because they give the idea of neglect, and this seems to derogate from the idea of that respect for the dead, which ought to be the prevailing sentiment raised in the mind by the appearance of a churchyard. If the graves were regularly distributed on some general plan, such accumulations of soil could never occur; and the grass would not be stronger, or the weeds more numerous, in one place than another. One great source of neatness in a churchyard, and consequently of apparent care and respect for the dead, is, the shortness of the grass; and, as the surface is seldom or never so even as to admit of effecting this by mowing, it can generally only be done by grazing it with sheep. In this case, however, flowers cannot be planted over the graves, unless the site, or grave, is enclosed with an iron fence; but, as this would entail an unreasonable expense on the poor man, who, perhaps, had no other means of evincing his respect for the deceased than by planting flowers on the grave, it follows that mowing or clipping with garden shears is preferable to grazing, for keeping the grass short, and the turf smooth, in churchyards. Another mode which contributes to the same end is, that of never raising the graves above the level of the surface, which may thus be easily mown; but this also appears unjust to the poor man, who, perhaps, can afford no other means than a raised mound to distinguish the grave of even his dearest friend; and, besides, it requires, after a certain period, from the sinking of the graves, to have the turf taken up, and soil introduced to raise their surface to the general level. In all cases, we repeat, it is best to have recourse to the scythe for such places as admit of using it, and in other parts to the hedge-shears. By the constant use of the hedge-shears, indeed, the roughest surface may be kept short and smooth; and, if clergymen were only to see that the leisure time of the sexton and his men was employed in keeping the grassy surface short, churchyards would, in general, be much improved in appearance. The scythe and the shears should always be used when the grass, or other herbage, is quite short, in order to weaken the root, and check the future luxuriance of the plants; in consequence of which, the labour of mowing or clipping, after the first two or three years, would be greatly reduced. On the other hand, if the herbage be allowed to perfect its leaves, and come into flower, as it too often is, before it is cut down, the roots, instead of being weakened by mowing, will be strengthened; the plants having received all the nourishment they require from the leaves, and being prevented from exhausting that nourishment by bearing seed, spring up more vigorously than before, and thus the labour of mowing, instead of being annually diminished, will be increased.

Trees in Churchyards. The number of trees which may be introduced into a churchyard depends on its situation and soil; the great object, next to that of leaving abundance of room for the graves, being to preserve dryness, in order to permit the escape of the mephitic effluvia, which can only be effected by the admission of abundance of light and air. Where the soil is clayey, and the situation low, very few trees are admissible; and these few should be small fastigiate-growing kinds, that neither cover a large space with their branches, nor give too much shade when the sun shines. In an elevated open situation, where the soil is sandy or gravelly, the trees in a churchyard may be comparatively numerous; because the shelter which they will afford in winter will produce warmth to persons crossing the churchyard to church; and, from the airiness of the situation, and dryness of the soil, they will not produce damp when their leaves are on in summer, but will freely admit of evaporation from the surface.

Supposing a new churchyard to be planted, we should place the trees chiefly at regular distances, in rows parallel to the walks. There are very few churchyards that would bear more trees than a row on each side of the circumferential walk, and also on each side of the walks leading from the entrance gates to the church doors; while, in cases of limited extent, and a clayey soil, a row of trees, planted at regular distances along the boundary fence, will, perhaps, be as many as can be introduced without producing damp; and, in others, a few trees along each side of the principal walk from the entrance gate of the churchyard to the church will, perhaps, be enough. It must not be forgotten, that the principal part of the area of a churchyard, in general, lies from east to west, and, consequently, that all trees planted in that direction will throw a shade upon the ground the greater part of every day that the sun shines, throughout the year. For this reason, where the soil is so damp, or the situation so confined, as to render it advisable to introduce but very few trees, these ought either to be in lines along such of the approaches to the church terrace as lie in the direction of north and south; or to be introduced as single trees, at the intersections of the cross walks with the boundary walk.

The kinds of trees to be planted in a churchyard form a subject of as great importance as their number; because a single tree of some species will produce more bulk of head, and consequently more shelter, shade, and damp, than half a dozen trees of some other kinds. As a guide in the choice of the kinds of trees, it may be adopted as a principle, that none ought to be planted which will grow higher than the side walls of the church; because to conceal the church by its appendages or ornaments is inconsistent, not only with good taste, but with common sense. By good taste, in this instance, we mean allowing the church to have its proper expression, as the principal and most dignified object in the landscape. Thorns, hollies, maples, sycamores, yews, mountain ash, wild service, &c., are suitable trees for the churchyards of very small churches; and the common maple, some species of oaks, such as the evergreen oak, the Italian oak, and some of the American oaks, with a host of other middle-sized trees, are suitable for the churchyards of churches of the ordinary size. There are very few country churches indeed which have even their towers, or spires, sufficiently high to admit of the stronger-growing elms or poplars to be planted in their churchyards. The Oriental plane (not the Occidental) may be especially recommended, on account of the stone-like hue of its bark and foliage, its finely cut leaves, and agreeable shade, for churches of both the largest and the middle size. The purple beech would harmonise well in churchyards with the dark yew; and the flowering ash is also a very suitable tree.

As all trees in churchyards must be liable to have their roots injured by the digging of graves, this is one grand argument for planting the trees alongside the walks; because in that case there will be always one side of the tree, the roots of which will remain untouched, viz. those which spread under the walk. For the same reason, trees with roots that spread near the surface, such as the pine and fir tribe, should seldom be made choice of. Were it not on this account, the cedar of Lebanon would be one of the most fitting of all trees for a churchyard, from the sombre hue of its foliage, and its grand, and yet picturesque, form; from the horizontal lines of its spreading branches contrasting strongly with the perpendicular lines of a Gothic church; and, above all, from the associations connected with it, on account of its frequent mention in Holy Writ. For all these reasons, it were much to be wished that, in all new churchyards, two or three spots (each of about 30 ft. in diameter) were set apart, not to be broken up for interments, and each planted with a cedar of Lebanon. In many old churchyards in the country, a spot sufficiently large for at least one cedar might easily be spared; and the clergyman or the churchwardens who might plant a cedar on such a spot, and fence it sufficiently while young, would confer a very grand and appropriate ornament on the church, and would deserve the gratitude of the parishioners.

No trees should be planted in a churchyard the natural habit of which is to grow near water, such as willows, alders, &c.; because the expression conveyed by such trees, being that of a moist situation, is, as we have seen, altogether unsuitable for a churchyard. On the whole, the different species of thorns, the common, Montpelier, mountain, and other maples, the wild service, the whitebeam tree and its hybrids, the holly, the yew, the Irish yew, the red cedar, the Oriental arbor vitæ, and a few others, are the most suitable low trees for churchyards; next, those which grow about the height of the Norway maple; and, lastly, those which rank in point of size with the Oriental plane.

In the case of old churchyards crowded with graves and gravestones, it may be difficult to introduce trees in regular lines, and at regular distances; in which case, a picturesque disposition may easily be effected, by scattering them irregularly, but very thinly, over the surface. It is also proper to observe, that, when a churchyard is to be united with a garden or pleasure-ground, or with a park or paddock, some of the trees characteristic of pleasure-grounds and parks, and already existing in the particular locality, will be required in the churchyard, in order to produce harmony, and to show that the one scene belongs to the other. On this principle, we would, where the churchyard joined a garden or pleasure-ground, occasionally introduce the gay laburnum, the showy Chinese crab, the perfumed cherry, and similar low trees; and, where a churchyard was to be harmonised with a paddock or a park, the horse and sweet chestnuts, or the oak, elm, or beech, might be admissible, according as the one or other of these trees prevailed in the park around it.

The System of Interments in Churchyards is, in general, very imperfect; and, indeed, in many cases, no system whatever is adopted. The obvious principle, we think, is, to place the tombs near the edge, and consequently near the walks; and to place the graves without marks in the interior of the compartments. For this reason, we would reserve a strip of ground, 10 or 12 feet in width, along both sides of all the walks (which would include the whole of the space between the boundary walk and the boundary wall, these strips should be devoted exclusively to family burial-places, whether merely indicated by corner stones, or railed in, or containing gravestones, or tombs. The whole of the compartments being thus bordered by strips for family burial-places, or purchased graves, the interior of each compartment might either be laid out in strips parallel to the borders, with gravel walks between; or devoted to graves without marks, laid out in the manner of a garden, with regular alleys of turf between.

The total space devoted to graves without marks should be divided into such a number of parts as will correspond with the years required, in the given soil and situation, to decompose a corpse and an ordinary coffin. If possible, there should not be less than fourteen divisions; that number of years being sufficient, in almost any soil, to insure decomposition. Then, commencing at one side of one of these fourteen portions, the markless graves may be placed parallel to each other, with rather more than the width of a grave between them; all the graves which may be required during a year being dug to the same depth. At the expiration of this period, the next division may be commenced, whether the one previously in use were filled up or not; and at the end of fourteen years, when the first rotation was completed, the first compartment would be begun on the second time, and graves only made in the intervals between what had previously been graves. In the third rotation, the graves may be made in the same places as the first, there having been an interval of twenty-eight years between the two interments in the same grave.

In using any division of this kind, more especially where the interments are numerous, the sexton should commence at the lower side; because in all crowded grave-yards the substratum is loaded with the moisture of decomposition; and, if the sexton were to work down a declivity, instead of up it, he would find his labours extremely offensive and unwholesome, while the surrounding atmosphere would be contaminated.

In Germany, it is customary, in some churchyards, to bury all the children under a certain age, who are not to have grave-marks, in a compartment by themselves; not only because the waste of ground occasioned by placing large and small graves together is thus avoided, but because it is found that, in the case of children, the ground may be used again much sooner than the ground in which adults have been buried. But we do not think it necessary to recommend such a practice for Britain, where churchyards are, or may be, increased in size with the increase of population.

Fig. 246. is the ground plan of a churchyard laid out agreeably to the foregoing principles; and *fig.* 247. is an isometrical view, supposing the trees to have been ten or twelve years planted, and some of the gravestones and tombs to have been erected. The churchyard is of small size, and is adapted for an agricultural parish, where the majority of the inhabitants are in moderately good circumstances, and whence it is supposed that the superfluous population will migrate to the towns, and leave the number of permanent inhabitants comparatively stationary. There is only one entrance to the churchyard, at *a*, over which there is an archway for the protection of persons waiting during rain or snow. The walk is 8 ft. broad, and proceeds direct to the steps (*b*), which ascend to the platform on which the church stands. The circumferential walk (*c*) is 6 ft. wide, with a border for tombs and gravestones on each side, 12 ft. wide. There is also an inner walk (*d*), of the same width, between which and the platform on which the church stands there is another 12 ft. border for tombs. The space for graves without marks lies on each side of the walk *e*, and is in 14 divisions, with room in each for 24 graves. Each of these divisions is separated by a grass path 2 ft. wide. The two surrounding borders, intended for tombs, are planted with trees 20 ft. apart. At the angles (*f f*), these trees are cedars of Lebanon; at the main entrance (*g g*), they are yew trees; and the remainder of the trees are different species of thorns (Cratæ`gus) (*h*), and evergreen cypresses (*i*), alternately; except opposite to the side entrances to the platform, and at the angles adjoining the cedars, where there are the yew trees, marked *k k k k*. Whatever tree is introduced on one side of the walk, the same sort is also planted on the other; for the sake of preserving uniformity in the perspective. The number of trees wanted for this churchyard will be 8 cedars of Lebanon, 20 yews, 28 cypresses, and 32 plants of Cratæ`gus. The latter may be of the following 16 species or varieties: —

C. coccínea.	*C.* Arònia.
C. c. corállina.	*C.* Oxyacántha *ròsea.*
C. punctàta.	*C.* O. múltiplex (flòre plèno).
C. Crús-gálli.	*C.* O. melanocárpa.
C. C. salicifòlia.	*C.* O. præ`cox.
C. orientàlis (odoratíssima).	*C.* glandulòsa.
C. tanacetifòlia.	*C.* heterophýlla.
C. t. Lee*àna.*	*C.* flàva.

Half the yews may be of the upright Irish variety; but the cypresses should be all of the common upright-growing kind. In many parts of England, and generally in Scotland, the climate is too severe for the cypress; but in all such places the Irish yew, the Oriental arbor vitæ, or the *P*ìnus *Cémbra*, may be substituted. The *P*ìnus *Cémbra*, from the slowness of its growth, and its narrow conical form, is admirably adapted for a churchyard

246

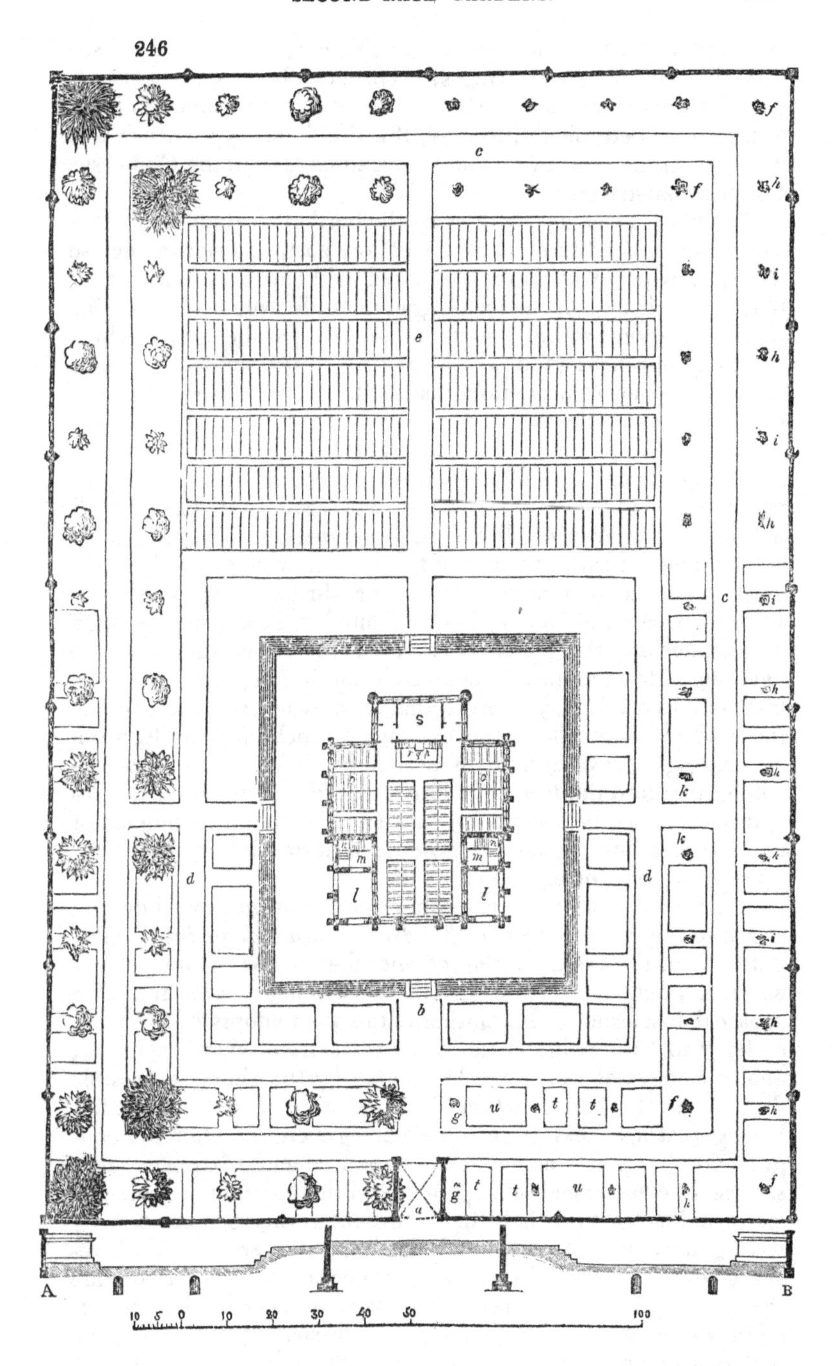

tree, and is perhaps, next to the Irish yew, the best of all substitutes for the evergreen cypress. The next best is the upright-growing variety of the Oriental arbor vitæ. The common holly is also not a bad substitute; and, if a deciduous cypress-like tree were required, we know of none more suitable than the *Cratæ`gus Oxyacántha stricta*.

The parties wishing to bury in the borders are not to be considered as obliged to erect tombs of any sort, or even to enclose the spot which they have purchased with an iron railing; all that they will be held under obligation to do will be, to confine their operations within the limits of the parallelogram which they may purchase (and which may be either single, as shown in the plan at *t*, or double, as at *u*), and the four corners of which will be indicated by four stones let into the soil at the expense of the parish. The party purchasing the ground may erect any description of gravestone, tomb, statue, or monument he chooses within it; or he may leave it in naked turf, which will be mown or clipped at the expense of the parish; or he may plant it with shrubs and flowers, in which case, he must keep it in repair himself. Trees, or shrubs which will grow 15 ft. high, cannot be allowed to be planted in these graves, as they would interfere with the effect of the cypresses and thorns. We have suggested the idea of not rendering it compulsory to erect tombs or iron railings, in order that we may not seem to exclude those who cannot afford the expense of such memorials, from purchasing a grave to hold in perpetuity. A poor man may be willing to afford the price of a grave, in order to preserve the remains of his family from being disturbed; though he might not be able to afford the farther expense of decorating it, by setting up a gravestone, or erecting a tomb.

The Church is on what is supposed to be an improved design, suggested by an architect in the *Architectural Magazine;* and it differs from the ordinary plan of churches in the manner of the entrances, and also in the general form being nearer that of a square than is usual. The author of this plan adopts it as a principle, "that the point in the outer walls from which each pew, and each class of pews, can be gained by the shortest possible distance, is the best situation for an entrance; and for the following reason: that a person entering a church after the congregation has partly assembled, or, as frequently happens, after service is commenced, may gain his sitting as soon as possible, and avoid at least one half the disturbance otherwise created, by having only half the length of an aile to traverse." With respect to the general form, this architect considers "that plan the best which concentrates the greatest number of benches or pews within a given distance of the preacher; and hence he prefers a square to a parallelogram." He adds: "Never let the

247

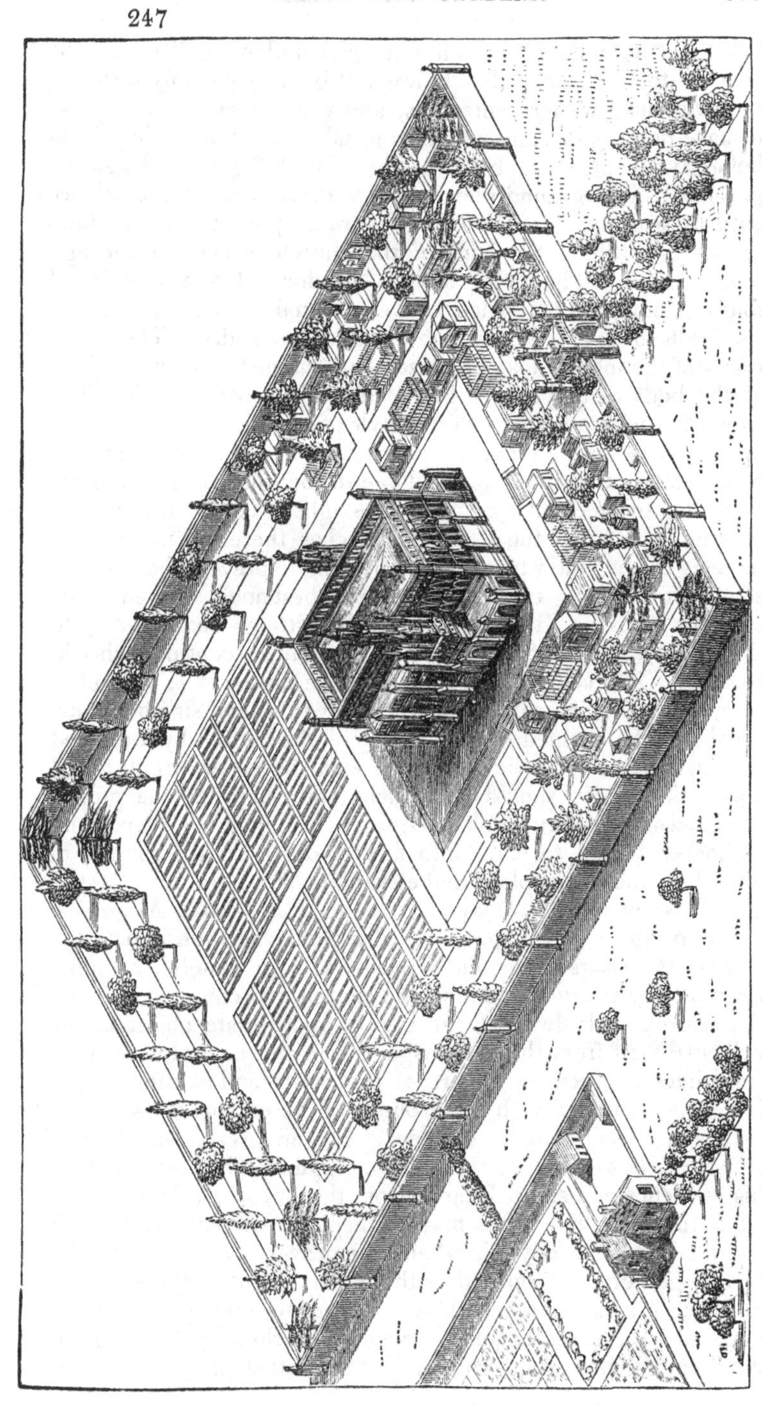

inner entrance door of a church open under a gallery, or the effect of the interior of the church will be irrecoverably lost. If you will have western entrances, and western galleries, contrive to have porches or cloisters, so as to take you to the gallery front before you enter the body of the church." (*Arch. Mag.*, iv. p. 568.) The ground plan in *fig.* 246. is made in accordance with these principles: *l l* are the entrance porches; *m m*, staircases, from which the body of the church is entered through lobbies at *n*. The inner lobbies are formed by two pairs of folding doors, with a space between, equal to the thickness of the walls of the towers which contain the stairs. The inner doors of the lobbies may be glazed with stained or painted glass. If the body of the church be fitted up with benches, the effect would harmonise better with this style of architecture; and, in the opinion of several clergymen with whom we are acquainted, this arrangement would be more suitable to the spirit of Christianity, according to which all are equal in the sight of God. It is worthy of remark, that in the Russian churches there are no benches or seats of any kind whatever, and nothing to prevent the meanest slave from standing by the side of the highest noble, or even of the emperor himself. The portion of the sittings marked *o o*, to the right and left of the pulpit, our architect considers should be free. The communion table is to be placed at *p*, the pulpit at *q*, and the reading desk at *r*. "The vestry, and singers' seats, (*s*) should be divided from the body of the church by a pierced screen, finished upon the same level with the gallery fronts; and above this screen should be a niche and canopy to the pulpit, designed as much as possible to improve the sound." (*Ib.*, p. 571.) Whoever wishes to enter into farther detail on the subject of churches, and to see plans and elevations on a large scale of the one shown in *fig.* 246., may consult the *Architectural Magazine*, vol. ii. p. 393.; vol. iv. p. 237. and p. 566.; and vol. v. p. 223.

The Parsonage House and Grounds will, in general, be most conveniently situated adjoining the church and churchyard; and the church will always form a most appropriate object in the principal view from the parsonage. The churchyard, also, may sometimes be seen as a part of the view; and at other times it may be so united with the grounds of the parsonage as almost to seem a continuation of them. In the greater number of situations, however, we believe the clergyman will prefer having his residence at a short distance from the churchyard; not only from the idea that there may be mephitic exhalations from it (especially in churchyards where the graves are crowded pell-mell together, and opened without any regular system), but also because familiarity with the interments taking place in it may lessen the sentiment of solemnity excited by them in his children and domestics, and may obtrude that expression more powerfully than is desirable upon the minds of strangers who may be his guests. Another, and a decisive, reason why the church and churchyard should generally stand alone is, that the expression of solemnity is heightened by this circumstance. Solitariness is unquestionably a powerful ingredient in all feelings which are the opposite to those of gaiety; and, on this account, the church and churchyard should stand completely isolated, and, as we have said before, they should, if possible, be so elevated as to be seen from all the surrounding country.

In laying out the grounds of a parsonage, it ought to be borne in mind that the profession of the occupier is one rather of a sedentary than of an active nature, and that it would be very desirable to infuse in him, not only a taste for gardening, but for working in his garden. Every person who has tried the effects of gardening operations and labours, as a recreation from sedentary or literary pursuits, will bear witness to their suitableness for this purpose. There are some judicious observations on this subject in the *Manse Garden*, a work which, though written with a view to the clergy of the church of Scotland, (by one of their number, the Rev. Nathaniel Paterson,) well merits the perusal of the clergy of every country. After showing that it is not enough for a studious man to trust to mere walking as sufficient bodily exercise, (which is too frequently the case, from the mere want of a taste for some more active and invigorating pursuit,) addressing himself to his brethren, he says: —

"Besides the walk taken purely for health, you have many out-of-door duties, to the performance of which you must travel no small distance; and hence you are apt to imagine that the inconveniency of a too sedentary mode of life will be sufficiently counteracted. A little attention, however, to the principles of physiology would correct this mistake. These duties, discharged amongst the distant members of your flock, are all of a solemn kind, and many of them deeply affecting, keeping the mind as intent as in the study; causing the heart and throat to swell, and tears to flow; and keeping in quick vibration all those untraceable chords that serve for a correspondence between the mind and the remotest material parts of our system. This mode of overworking and wearing, by only one sort of application, which is as inconsistent with the health of our frame, as it is inconsistent with the nature of our mind, soon destroys either the mind or the body; and indigestion, or bilious disorder, is frequently the first intimation that violence has been done to the laws of our constitution.

"Let it be a fixed thing in your mind, that temperance, air, exercise, with diversity of attention, are essential to a healthful and useful existence. This law holds on its even tenor, regular as the sun, and steadfast as the mind of the Eternal. Conformity or suffering is the only alternative; let the character of the transgressor be, in other respects, good or bad, the punishment is equally sure. God doth not suffer his law to be changed: he changes the countenance of the violator, and sendeth him away.

"To render your observance of the above law both cheerful and constant, nothing can be more efficacious than to betake yourself to the study and labour of your garden. In summer or in winter, you will always find there something to do, and something that will give pleasure when it is done. Your required exercise never wants an object; one, too, that sufficiently draws off attention from more serious things, and has that peculiar interest which arises from a work that is progressive. While the mind is refreshed by a continual variety, the exercise to which the body is called has not only the advantage of being in the open air, but of accommodating itself, by various degrees of activity, to every change of temperature. In the training of trees, the mind is agreeably occupied, whilst the free air and moderate exertion are admirably calculated for relieving, in the early part of the week, the languor and debility incident to the labours of the pulpit. When the air is colder, and the frame more energetic, the saw and the pruning-knife, the one toilsome and the other easy, are excellent companions; and the spade, in one half hour, will bring on a summer glow in the coldest days of winter. Here, then, you have a kind of exercise suited to all circumstances, ever at hand, and the motive to which is ever new, strengthened by the love of progress, and the grateful survey of the work you have accomplished. A mere walk, compared with this, is like the amusement which children take in writing their names on the sand of the sea shore: you derive advantage from the motion as you pass along, but you leave no abiding trace on the path that you have trodden.

"It is more important to observe, that, whilst the mind is invigorated by diversity of pursuit, there is this further benefit, that the reciprocity of mental and manual exertion creates for each an increase of relish and aptitude. The garden recreation quickens the appetite for study, and the quiescent posture of study renews the desire for garden activity. Whoever has maintained, for a sufficient length of time, a regular system of employment, in which bodily and mental application are upheld in due proportion, will be surprised by the spontaneous appearance of those energies which hitherto lay dormant in his frame: nor is this the discovery of a fact merely, it is a source of delight; for the healthful play of either muscular or mental power is as certainly a pleasure to the human creature, as is skipping to the lamb, or singing to the bird. A man used to this renovating process cannot become sluggish, and is a stranger to the sloth that eats into the bone. He keeps disease at a distance;

and duties, which to the sluggard are a load, are light and easy to him. Whatever he has in hand, he has also in heart; his movements are so impetuous, that it is almost dangerous, from the velocity with which he is carried, to meet him at the turn of a corner; and, when the bodily energies are for a time suspended, but not exhausted, and there is a return to study, he enjoys, in the exercise of the thinking faculties, an actual revelry in the flowing of thoughts, which amount to more in a brief space, than the most laborious efforts could produce by the longest application, in a more languid state of the system.

"To possess this efficiency, and to promote its continuance, it is necessary not only to alternate, as above stated, the muscular and mental activity, which, by a mutual reaction, improve each other, but it is necessary alike for both, to avoid either fatigue or too long rest. Do not continue in study till mental application be overstretched, or till the circulation of the material fluids has become clogged and stagnant; and do not labour with hands or feet till weariness come upon the body, whilst the mind has been too long inactive. The moment that the thinking powers begin to flag, hasten to your garden; and as soon as weariness affects your body, return to your books. Let rest and fatigue be your tropics, and you will travel with unabated vigour over the undulating line of your ecliptic: but let quiescence be too long indulged in, or labour too long sustained, and the consequence will be a long unfitness for any achievement: the one state terminates in leaden slumbers, the other in excessive lassitude; the one makes exertion seem appalling, the other makes it really impossible." (*Manse Garden*, p. 57.)

The reverend author next endeavours to show how a taste for horticultural pursuits may be acquired by one who has hitherto paid no great attention to them. An accident requiring immediate assistance is one of the commonest modes of exciting an interest in us for something which we have not cared for before. A branch of a wall tree laden with fruit has become detached by the wind, and we perceive that the fruit will be lost, if it be not immediately nailed up again. This leads to the use of the hammer, lists, and nails, and the progress of the branch and its fruit will be watched afterwards with an interest which was not taken in it before. Tying up and trimming flowers are operations generally undertaken by the ladies of the family; but the pruning of shrubs and trees, and tying up to stakes such as require it, invite the hand of the master. A taste for watching the progress of plants, and the effects of culture, once commenced, soon leads to raising improved varieties from seed, to the various modes of propagating, by striking cuttings, layering, grafting, budding, and to all the endless variety of garden operations. To the cares of the garden we might recommend adding those of the small farm; in which, if it should be only a pasture field, there will be the boundary fence to keep in order, hedgerow or scattered trees that will now and then require a little pruning, or, perhaps, protecting from the cattle turned to graze in the field by some kind of tree-guard; and there will be, besides, the draining of the ground, and the manuring and weeding of the grass. If the farm should contain some arable land, the operations would be still more numerous and varied; and they might even include the occasional holding of the plough, the most delightful and invigorating of all agricultural labours, and one which has been honoured by some of the greatest names of antiquity.

In laying out a parsonage residence, we should say that, as a *beau idéal*, it ought to include something of every thing that is contained in a complete residence; in order to produce some work to be done, something to be cared about, and some production to be hoped for, every day in the year. A clergyman, who knows something of every country pursuit will derive much more enjoyment from his situation in life, and be much more useful and agreeable to his parishioners, than if he were conversant with the pursuits of only one class of society; and, if to some knowledge of country matters be added a knowledge of general physiology and the constitution of man, and a taste for natural history, and if he were withal a liberal and benevolent man, what greater blessing could there possibly be to the parish in which he resided? The parsonage residence, therefore, should have a small farm attached, in which those products should be raised, which were most commonly grown in the parish or neighbourhood; and the reason why we mention these products is, that, where the arable land was, perhaps, only one or two acres, which in most cases would be quite enough for our purpose, persons within the parish might be more readily found to furnish the labour required for such arable land on hire. The pasture land should, in general, be not less than 3 acres, but 5 will be more convenient; and this quantity and 2 acres of arable land, will keep two cows, a horse, and a few sheep, throughout the year; besides pigs, poultry, rabbits, &c., which would receive a portion of their food from the kitchen-garden and the house. We do not say that a clergyman would gain anything by this quantity of ground, even if he paid no rent for it; but it would be a source of recreation to him, and give him a deeper interest in the pursuits of his neighbours. Adjoining or surrounding his house, he should have a lawn and pleasure-ground, even if they did not occupy more than half an acre; and, attached to the house, there ought to be a green-house or conservatory for the mistress of the family and her daughters; and near it there should be the ladies' private flower-garden, and also some small gardens, each two or three yards square, for the children. The pleasure-ground ought to be laid out and planted in the gardenesque rather than in the picturesque manner; because, by the former mode, there will be much more exercise for the master, in the way of pruning and thinning out the trees and shrubs. There ought to be a kitchen-garden, surrounded by a wall or other fence for the finer fruit trees, and there ought to be a piece of ground near it, kept under pasture, as an orchard; unless, indeed, the trees in the paddock or cow pasture supplied this desideratum. We consider it unnecessary to give a *beau idéal* plan of a parsonage residence; and, therefore, having already given that of Hendon Rectory, situated at some distance from the church, we shall here give that of Dunchurch Vicarage, which adjoins the churchyard.

Dunchurch Vicarage, Warwickshire, is the residence of the Rev. John Sandford, whose lady is the distinguished authoress of various works on female improvement. The vicarage grounds lie along the churchyard, from which they are separated by an open fence, so that the churchyard forms, as it were, a part of the pleasure-ground. The general surface, both of the vicarage and churchyard, is flat, but the soil is excellent, and there is an agreeable declivity from the lawn front of the house to the boundary of the property. The living was taken possession of by Mr. Sandford only two or three years ago, on which occasion that gentleman consulted us professionally, and sent us the ground plan of the place in the state it then was. We made some slight alterations in the position of the entrance gates, and in the direction of the approach; and suggested the idea of forming a circular kitchen-garden; the circle being the largest geometrical figure that could be introduced into the particular spot where it was placed. As the situation of this kitchen-garden is at no great distance from the front of the house, in order to prevent the intrusion of a garden wall so near the eye, we proposed having that half of the kitchen-garden boundary which is next to the lawn an open iron railing, which, being partially covered with fruit trees, and disguised by the shrubs which intervene between it and the house, would neither give the idea of nearness to the eye, which a wall would have done, nor of confinement. Some difficulty occurs in carrying a walk round the boundary of a churchyard, in consequence of the numerous gravestones which would come in its way; but the most likely way to get over it appears to be, the expanding of the walk on both sides of the regular line, on the principle recommended in a preceding page. The house has been improved under Mr. Sandford's own direction, and is now a most commodious and comfortable residence, as the ground plan will show; and a very handsome school has been erected adjoining the churchyard by subscription; the proprietor of the parish, Lord John Scott, having subscribed the largest amount, as well as given the ground.

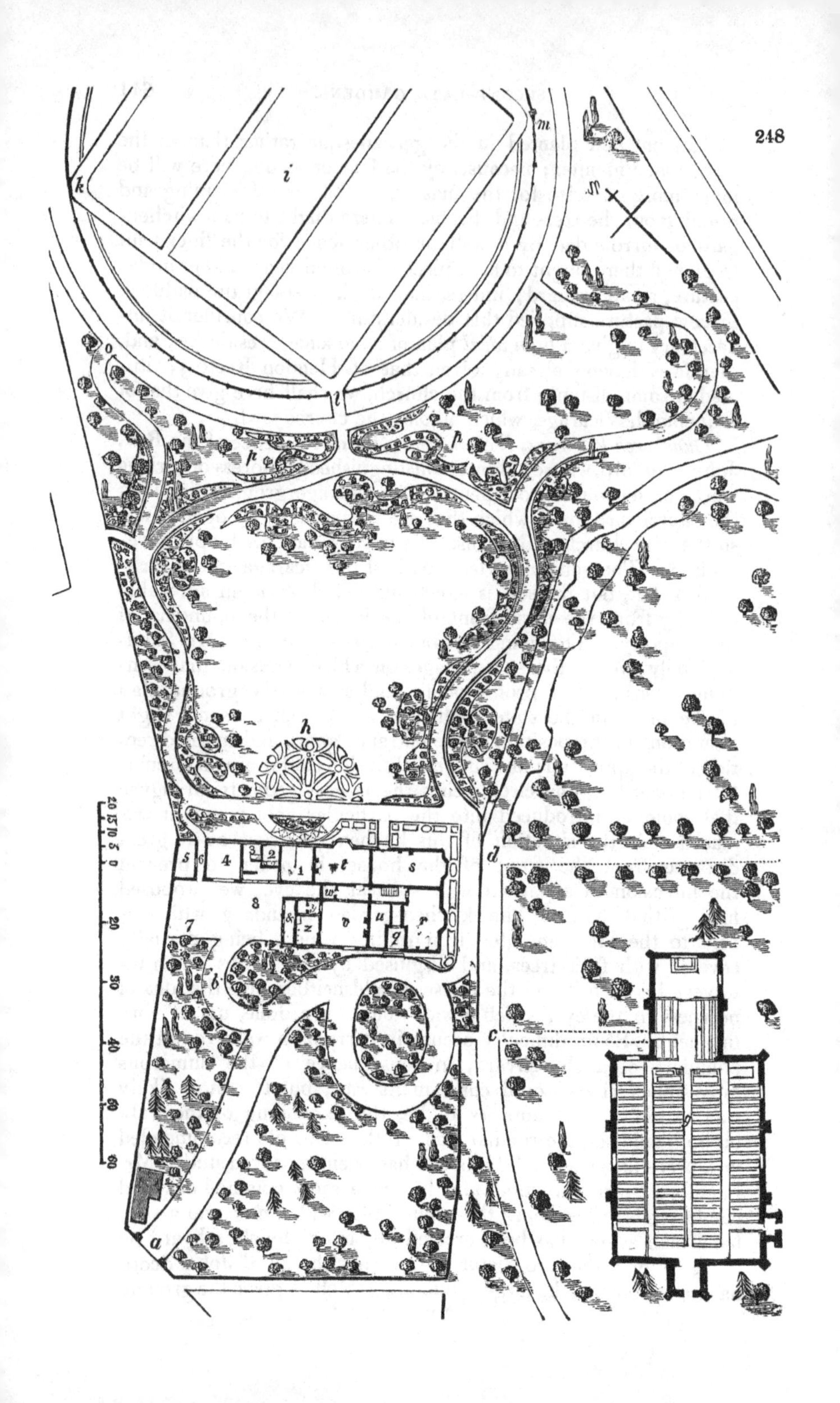

k
i
m
o
l
p
p
h
d
c
a
b
7
8
s
6
4
3
2
1
w t
s
z
v
u
r
q
0 5 10 15 20
20
30
40
50
60

Fig. 248. is the general ground plan of the buildings and grounds, and *fig.* 249. is an isometrical view. In the ground plan, it will be seen that all the effect produced by the trees is the result of single trees and small groups; that these are most abundant on each side of the approach on the entrance front of the house, and in the churchyard; and that the groups on the lawn are chiefly shrubs and flowers. With respect to the kinds of trees, those in the entrance front are in great part fruit trees; those in the churchyard, trees of the third rank, with regard to size, some of which have been already enumerated in a preceding page; and those on the lawn, and around the kitchen-garden, embrace so great a variety, both of trees and shrubs, as to constitute a tolerably good collection.

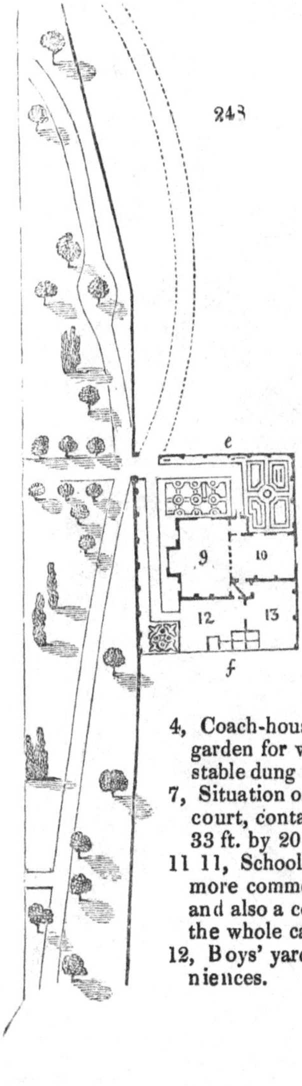

248

a, Carriage entrance and lodge; the trees on each side being chiefly fruit-bearing kinds, and ornamental trees, such as thorns, &c., of moderate size.

b, Entrance to the offices. *c*, Walk from the vicarage to the church. *d*, Walk across the churchyard to the school. *e*, School and school gardens. *f*, Paddock. *g*, Church; the details of the plan within the walls are imaginary. *h*, Flower-garden on the lawn.

i, Kitchen-garden.

k l m, Open espalier railing for fruit trees; the remainder of the boundary fence to the kitchen-garden is a fruit-wall of brick, which faces the south-east, and is, therefore, suitable for the finer fruits.

o, Circular walk bounding an outside border to the fence of the kitchen-garden.

p p, Beds of low flowering shrubs.

q, Entrance hall of the house.

r, Dining-room. *s*, Drawingroom, with a window opening to the lawn.

t, Library and study. *u*, Gentlemen's room.

v, Kitchen. *w*, Butler's pantry. *y*, Larder, and cold meat safe. *z*, Scullery. *&*, Dairy.

1, Stable. 2, Best privy, approached under trelliswork, from the lawn front. 3, Servants' privy, with a screen wall, which completely protects the entrance door, from being seen from any part of the kitchen-court.

4, Coach-house. 5, Coal-house. 6, Entrance to the garden for workmen, and for the convenience of wheeling the stable dung to the kitchen-garden by the side-road, *o*.

7, Situation of the dung-pit, and liquid manure tank. 8, Open court, containing wood-stack, pump, &c. 9, Boys' school, 33 ft. by 20 ft. 10, Girls' school, 28 ft. by 18 ft.

11 11, School gardens, intended to contain a collection of the more common English plants, especially those of the locality, and also a collection of the more dangerous poisonous plants; the whole carefully named.

12, Boys' yard with conveniences. 13, Girls' yard with conveniences.

249

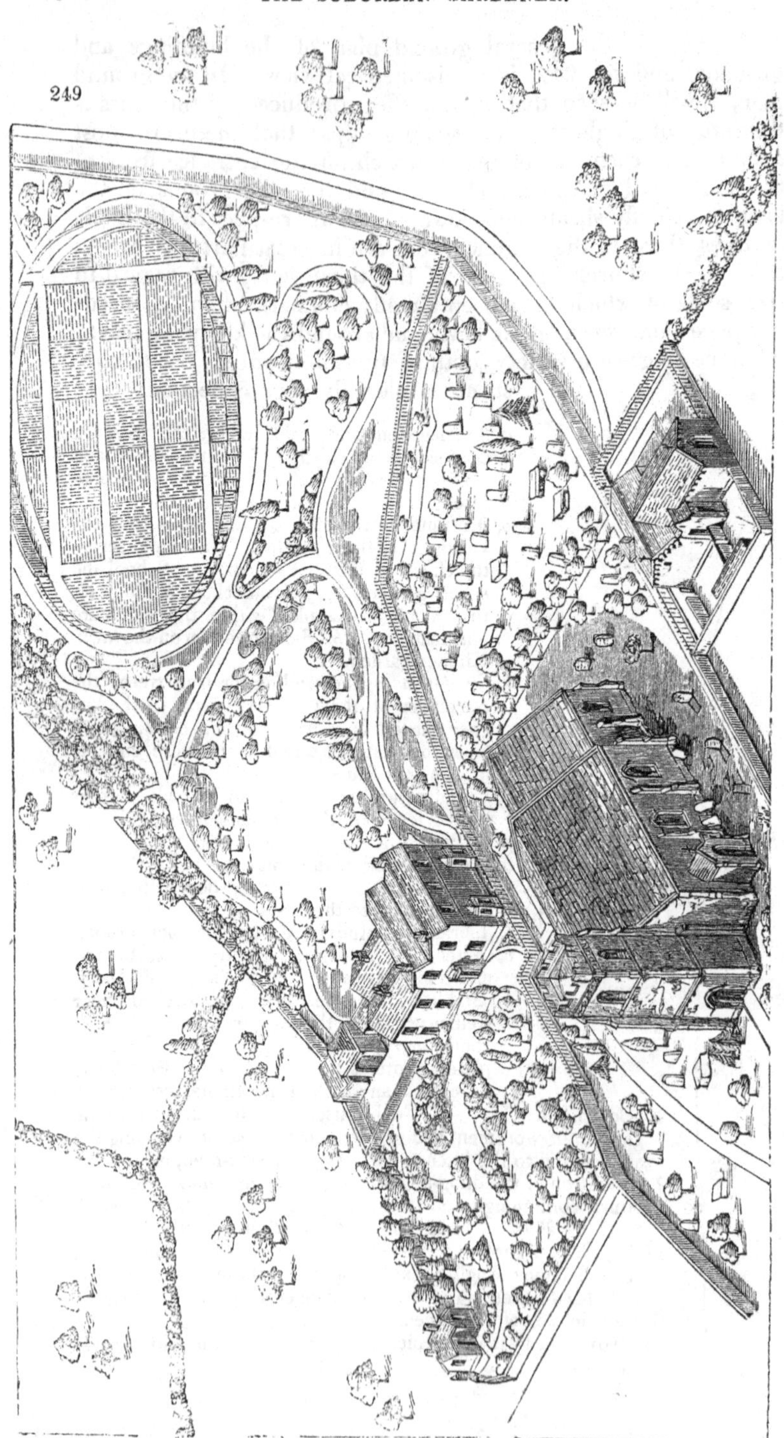

Fig. 249. is an isometrical view of Dunchurch Rectory, church, and school, as seen from the north-west. It is proper to observe respecting this view, that the gravestones in the churchyard, and all the scenery surrounding both it and the vicarage, are entirely imaginary, neither the draftsman, Mr. Robertson, nor ourselves, ever having seen the place.

Remarks. As a parsonage, Dunchurch Vicarage is chiefly valuable as showing how the grounds of such a residence may be united with the churchyard; and perhaps, also, some instruction may be derived from the form of the kitchen-garden, considered relatively to its position; and the adoption, in a part of its boundary, of an open iron railing instead of a conspicuous brick wall.

Renovating Second-rate Residences.—We have little to add, under this head, to what we have advanced in the preceding subsection, p. 543. to p. 547. It very frequently happens, in neglected residences, where there is only pasture sufficient to keep one or two cows, that the grass has become deteriorated in quality, in consequence of its being mixed with many of the grosser weeds, such as docks, thistles, crowfoot, &c.; and with the coarser grasses, such as carexes, cock's-foot, *A*vèna elàtior, &c. A pasture in this state will not be productive of one half, or even one third, of the quantity of milk and butter which it ought to yield; and, though we have no doubt the fact will be doubted by many, yet we can assert, from our own knowledge, that this is the case with the pastures of one half the suburban residences in the neighbourhood of London. The reason is, that persons who have not an intimate knowledge of country affairs are not aware that there are different qualities of grasses, and think that one green field, which can be mown for hay, is just as good as another. They are not aware that grasses may, and indeed should, be cultivated like any other plant; and that there are many kinds of grasses, even in our best pasture fields, which can only be considered as weeds, and which ought to be eradicated. The kinds of these weed-like grasses are different in different soils, and their number increases in proportion as the soil is neglected. Every grass field contains several different species of grass, some of which are deserving of culture, and others are mere weeds; and, in proportion as the field is kept in "good heart" and properly managed, or neglected and impoverished, will the one kind or the other acquire the ascendency. To the superficial observer, however, all grass-fields are alike; and, hence, the same routine, as far as respects management, is applied, whatever may be the nature of the soil, and whether the prevailing grasses be wholesome or injurious. The field is shut up in April, in order to produce a crop of hay; and when this is removed in July or August, the cows are turned in every day, except in the most severe weather, till the return of the shutting-up season. Not to speak of the neglect of draining and manuring, a positive evil resulting from turning the cows out on a clayey undrained soil during winter is, that the ground gets poached, in consequence of which weeds and bad grasses, such as thistles, docks, crowfoot, carexes, oat-grass, brome-grass, &c., make their appearance; and, by degrees, choke the clovers and good grasses, such as white clover, rye-grass, fescue, &c. Among the weeds, one of the most abundant is the common meadow crowfoot, or yellow buttercup (*Ranúnculus àcris L.*), remarkable for its acrid juice, which, it is alleged, blisters the mouths and stomachs of cows, and injures the quality of their milk and butter. To say nothing of its injurious effects, the proportionate space which this weed occupies in green herbage or in hay is a sufficient reason for wishing to get rid of it; though, from its abundance, not only in bad soils, but in good soils which have been neglected or mismanaged, this is a work requiring some time, and depending on some knowledge of the nature of plants. The crowfoot, as the name *Ranúnculus* (from *rana*, a frog, alluding to the moist places where most of the species grow) implies, naturally loves moist soil, which rye-grass and the other good grasses as naturally dislike. Hence, the first process to get rid of the crowfoot is thorough under-draining by shallow drains, which need not be, in general, more than 1 ft. 6 in. deep, and placed not farther apart than from 8 ft. to 15 ft. The next thing is to apply manure liberally; and the third (without which success would not be complete) is, to pasture the surface at least till the beginning of July, before shutting it up to be mown for hay. When the pasture is shut up for mowing, in April or the beginning of May, the leaves and flower-stems of the crowfoot shoot up uninjured along with the grass; and thus the roots of the crowfoot are nourished and invigorated for the following season: but, on the other hand, when the surface is pastured till the beginning of July, the leaves and flower-stems of the crowfoot are cropped by the pasturing animals; the root is weakened in consequence of not deriving as much nourishment from the leaves as it otherwise would do; and, as the crowfoot is one of those plants that scarcely produce any leaves after midsummer, it is in a great measure suffocated by the growth of the grasses in August and September. If this course be pursued for three years in succession with a field overrun with crowfoot (draining and manuring having been properly attended to the first season), the number of these weeds will be found to have greatly diminished, and the clover and good grasses to have increased. To increase the number of the latter plants, some white clover and rye-grass seeds may be scattered over the surface the first year, early in spring. It may be thought that the eating down of the herbage in the beginning of summer, instead of shutting it up for mowing, while it destroys the crowfoot, would also have a tendency to destroy the clover. This will, no doubt, be the case to a certain extent; but the clover has the advantage of being a much more vivacious plant, it having creeping stems, which throw up numerous leaves, and continue growing the whole summer. The meadow crowfoot, on the other hand, is a stationary plant, which increases but slowly except by seed, which throws up only one set of leaves in spring, and which does not renew these in the course of the season. Even the creeping crowfoot, which, however, is more commonly found in arable fields than in meadows, throws up but few leaves when compared with the clover; and is much more easily choked or killed by cropping these leaves early in the season.

The same practice as that recommended for getting rid of the crowfoot will apply in the case of all broad-leaved weeds; but, as docks and thistles are not so readily eaten by cattle, they may be either rooted up; or, what is a much more simple and economical mode, and yet quite as effectual, the incipient leaves may be kept constantly cut over close by the surface of the ground as soon as they appear. As the health, and even life, of the root of a plant depend upon the nourishment which it receives from the leaves, so, by removing the incipient leaves or buds from any plant the moment they appear, and continually doing this as long as any leaves or buds are produced, all plants whatever, and consequently all weeds, will ultimately be killed. In this way ferns, which are so injurious to pastures in some sandy soils, may with more certainty be destroyed than by any other mode; for their roots, or rather underground shoots, are so numerous and brittle, that it is scarcely possible to eradicate them by digging them out. It is a common practice to mow weeds and ferns for the purpose of destroying them; but, as this mowing never takes place till the leaves have expanded, a considerable portion of nourishment has been already sent down to the root, and thus the object of the operation is in a great measure defeated. Our readers, therefore, will bear in mind, as a general principle, that all plants, from the loftiest tree to the humblest herb, depend not only for their growth, but for their continued existence, on the leaves which they produce; and that, as we before observed, by removing these leaves before they are fully expanded, all plants whatever, and consequently all weeds, may be destroyed.

Mosses frequently abound in pastures, and occupy great part of the space that ought to contain plants of grass or clover. Where pastures are open and fully exposed to the influence of the sun and air, the appearance of moss in them is a symptom of extreme poverty in the soil; as in the case of lawns in the fronts of mansions, which have been mown for many years together without being pastured or manured. When moss is abundant in pastures which have been fed or mown, it indicates moisture, shade, and want of air and sunshine, accompanied, most generally, by want of manure. The remedy in this last case is sufficiently obvious: some of the trees must be removed, or the hedges cut in and kept low, while the soil is sufficiently drained, and manured and pastured one or two seasons, without being mown. The reason why we recommend pasturing without mowing, is founded on the fact that all other plants are more injured by being deprived of their leaves than grasses; consequently, pasturing a surface, by eating down all the leaves close to the ground, must be more injurious to all other plants, whether weeds or clovers, than it is to the grasses, and must have a tendency to give the latter an advantage over the former.

Sorrel is very abundant in some soils, and is almost always indicative of an excess of ferruginous earths. The remedy in these cases is the application of quicklime, which neutralises the ferruginous matter. The lime may be given as a top-dressing on the surface of the pasture; but it is most effective when applied after the soil has been broken up, and kept under corn crops a year or two, and then the lime applied when the soil is in a dry and powdery state like itself. Ant-hills are also very common in some pastures. They may be destroyed by breaking open the hills, and scattering them abroad, repeating the operation whenever the hills begin to reappear; because the principle on which success depends is, the interruption of the insect's labours in hatching its young, and in laying up its food for the winter. If the interruption be continued, the ants will in a short time be dispersed and die. The naked spaces from which the nests have been pared off should be sown down with grass and clover seeds. Ants seldom appear except in worn-out grass lands; and the most effective mode of getting rid of them in this case, at least on a large scale, is, to break up the land, and to keep it three or four years under the plough, before it is laid down again. We can hardly recommend this practice, however, in the case of a paddock of a few acres; because, among other reasons, the loss of two or three years' pasture would be a very serious drawback to the comforts of the family. For so small an extent, therefore, we recommend draining, frequent top-dressings with manure or compost; and scattering in the seeds of rye-grass, fescue, and white clover, wherever the grass appears thin. This ought to be done early in the spring; or in September, after a crop of hay is taken; and the ground ought afterwards to be thoroughly harrowed and rolled.

In order that the draining of permanent pasture may not appear a more formidable business than it really is, we shall here show some of the cheapest and best modes of performing it, premising that the object is not to drain the subsoil of water arising from springs, but to collect what rain-water would otherwise stagnate on the surface, or soak into the ground, and be retained there longer than is beneficial for the health of the plants, or to such an extent as to render the soil unfit for the tread of heavy animals. The first thing to determine is the direction of the drains; the next thing, their distance from one another; and the third, their form and the materials with which they are to be filled. If the surface of the land have been formed into ridges before it was laid down in grass, it will, in general, be found sufficient to make one drain in the bottom of each furrow; because we may fairly presume that the ridges were laid out in the direction best adapted for carrying off the water, and that they were formed of such a width as to leave their furrows sufficiently near one another for that purpose. If the land should not have been laid out in ridges, and the surface should be nearly level, then, after having discovered the lowest side of the field (because there is no such thing in nature as a perfectly level field), drains ought to be made at 10 ft. or 15 ft. apart, and parallel to each other, from the highest side of the field to the lowest. If the field be very steep, however, then the drains ought to be made more or less obliquely across the declivity, so as to intercept the surface water, and, at the same time, not to carry it off with such a degree of rapidity as to wear out the bottom of the drains. Where the surface slopes in various directions, and consists of a series of eminences and hollows, then drains must be conducted in such a manner round the eminences as to intercept the water every where, especially near the bottoms of the declivities, and to carry it off to the boundary of the field, or to some general outlet or public drain. In strong clayey soils, having a moderate declivity, the distance of 15 ft. between the drains may be sufficient; in loamy soils, it may be more; and where the soil is a strong clay, and the surface quite flat, it ought to be less. Under the last circumstances, we should not hesitate to recommend putting in drains every 6 ft. or 8 ft., and filling up these drains to within an inch of the surface with gravel if it could be obtained, or with small stones, or with turf and soil, in a manner which we shall describe hereafter.

With respect to the dimensions of the drains, as the surface is not to be ploughed, they need not be deep; and any greater width than may be required for attaining the depth fixed on is unnecessary. In general, from 15 in. to 18 in. in depth, and from 8 in. to 10 in. in width at the surface, and from 3 in. to 4 in. in width at the bottom, will be sufficient; except in particular parts of the field, where the drain may have to cross a slight rise in the surface, and where it should be deeper in order to maintain a sufficient slope at the bottom to give currency to the water. The drain, being dug out, may be filled two thirds of its depth with small stones, brick-bats, or very coarse gravel; or with thorn branches, straw ropes, exhausted tanner's bark, carpenter's shavings, or any refuse vegetable matter that will partially or wholly rot, and leave a vacuity for the passage of water. In Essex, the material most commonly employed is straw twisted into a rope 3 or 4 inches in diameter. Reeds tied in cylindrical bundles of a few inches in diameter, are also employed the same county, as willows are in Huntingdonshire, furze in Warwickshire, broom in Nottinghamshire, the spray of the larch in Wales, and heath in the north of Scotland. The most convenient materials in clayey districts, which are never productive of either gravel or rock, are the branches procured by cutting down thorn hedges. These branches, including the spray, may be laid lengthwise in the drain, and firmly trodden in; covered with soil, and again firmly trodden in; and, finally, the surface turf replaced. In order to make allowance for the sinking of the materials of the drain, this surface turf should be kept two or three inches higher than the adjoining surface, its edges being beaten down so as to form a small semicircular ridge in the direction of the drain. *Fig.* 251. is the section of a drain thus formed and covered. In the digging out of these drains, two kinds of spade are used: the first (*fig.* 250. *c*), 8 in. broad at the tread, is for taking out the surface-spit; and the second (*b*), which is 5 in. wide at the tread for digging out the remainder. Afterwards a sort of hoe, or scoop, is used (*fig.* 250. *a*), for clearing out the loose particles and small lumps of soil from the bottom of the drain.

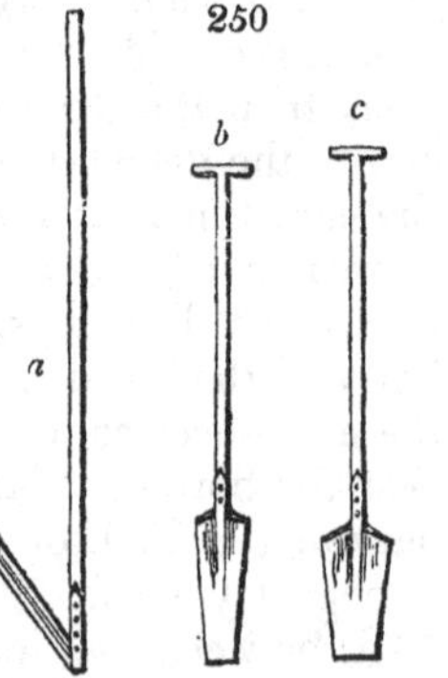

A simple and not expensive mode of forming drains in strong clayey pastures is shown in *fig.* 255., in which the surface-spit is taken off 9 or 10 inches in width; and, a gutter having been formed in the subsoil, 18 or 20 inches in depth, the surface-spit is replaced. This mode of draining is most effective where the surface is naturally tolerably even; but, where inequalities are to be passed over, the gutter beneath requires to be cut down to such a depth, that its sides are apt to crumble in.

A simple and economical mode of draining a grass field, where the surface is laid out in ridges, is shown in *fig.* 257., in which a furrow-slice, being thrown out by the plough from every furrow of the field, has its apex (*b*) taken off by a spade; and the remainder of the furrow-slice being replaced, a hollow is formed, as at *a*.

There are several kinds of draining ploughs, and one, the mole plough, forms a cylindrical furrow drain, as shown in *fig.* 256.; but none of these machines can be recommended for using in a suburban villa; and, indeed, from their original cost, and the great power required to draw them, it is very doubtful to us whether they pay any where.

Another mode of forming drains in grass land, and which will serve for draining the subsoil of springs, as well as the surface of rain-water, is explained by *fig.* 253., in which a gutter is formed in the bottom of a rectangular drain, with a shoulder on each side, for supporting the reversed surface-spit which was dug out when the drain was commenced: the rest of the drain is then filled up in the usual manner. A variation of this kind of drain is shown in *fig.* 252.; and another in *fig.* 254.; which latter is calculated for drawing the water from the subsoil at a considerable depth, as well as from the surface. These three modes are better adapted for free loamy soils than for retentive clay soils.

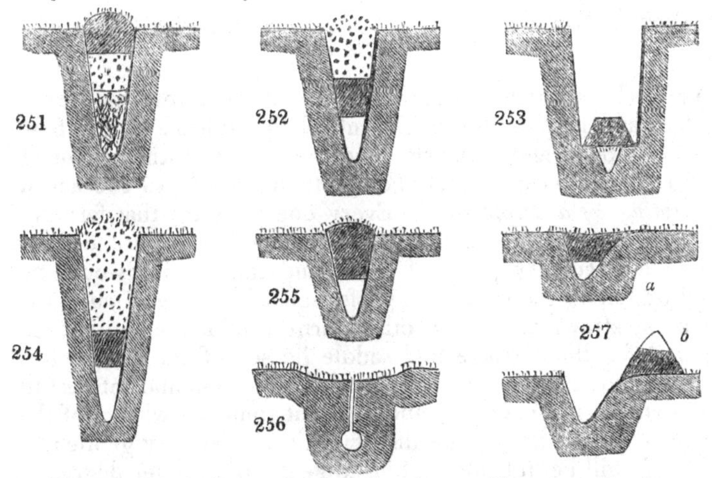

Wherever it is determined to break up old pasture land, on a wet or retentive soil, a different kind of underground drain is required to what was sufficient when it was under pasture. Though not quite so numerous as in grass lands, yet these drains require to be much deeper, and are consequently more expensive; nor will they altogether supersede the use of frequent shallow drains, after the ground is laid down again for permanent pasture.

We have dwelt longer on the subject of renovating the paddock than those will think necessary who are not aware how much the difference of produce depends on management. Our wish is to see the paddock in as high order and keeping, in its way, as any part of the pleasure-ground or the kitchen-garden; and, well knowing how much of the comfort of a suburban residence depends on the abundance and excellence of the milk and butter it produces, we feel that we can scarcely sufficiently impress our readers with the importance of bringing their grass fields into the highest state of cultivation, and afterwards keeping them in the highest order.

In renovating the kitchen-garden, the reader should bear in mind the importance of not having the roots of the trees too deep in the soil, and that, should this be the case, he must either take up and replant, or renew; unless, which is very rarely the case, he can reduce the surface of the garden so as to leave all the main roots of the trees at a proper distance beneath it. The soil of the kitchen-garden may, as before observed, be restored by rest, without any crop whatever; by the substitution of a large portion of fresh soil; or by the application of animal manure, such as good stable dung or night soil.

SUBSECT. 4. *On laying out, planting, and managing First-rate Suburban Gardens.*

As the characteristics of a second-rate suburban residence are a paddock and dairy, so those of a suburban residence of the first class are a park and farmery; and the extent of the whole of such a residence can scarcely be less than from 50 to 100 acres.

Farming by a Proprietor. Every one is aware that farming by a proprietor or an amateur is, for the most part, attended by loss in a pecuniary point of view, and that money would be saved by letting the lands to a professional farmer, and purchasing corn, straw, and such other farm produce as might be required for the carriage and saddle horses of the proprietor. Farming, however, is a great source of recreation and interest to a resident in the country; and, without some knowledge of the practices of the art in the district where a country gentleman resides, it will be difficult for him to keep up a proper degree of social intercourse with his neighbours. Farming and the weather, are topics which every countryman can discuss, from the humblest labourer to the most wealthy proprietor. If there are any exceptions, such persons must be out of the pale of general country society. Since, then, every person living in the country must, of necessity, take some interest in farming, it would seem worth while for those who can afford to do so, and have an opportunity, to increase that interest by farming themselves. If we enquire into the habits of our greatest landed proprietors, while they reside in the country, we shall find that a considerable portion of the time of each individual is devoted to the inspection of his farm, and to giving directions to his bailiff, hearing a relation of his market transactions, and auditing his accounts. When a landed proprietor has left his country residence to reside in town, or is gone on a visit to some distant part of the country or abroad, the weekly receipt of his bailiff's journal (of which there are regular printed forms, on separate sheets, to be sent by post, and which, when bound at the end of the year, form a volume,) constitutes no inconsiderable part of his enjoyment. During the late war, accounts of this kind, not only from bailiffs, but from head gardeners, foresters, and land stewards, were transmitted to such of the heads of families as held commissions in the army, even while they were on the field of action; and answers and directions were regularly returned by the greatest generals.

To a citizen who has altogether retired from a commercial or manufacturing concern, the cares of a farm must form an agreeable contrast to those of the counting-house, and will amuse and recreate, without fatiguing; for no mistake can be greater, than to suppose that there is any enjoyment in retiring to the country and doing nothing there. Every retired man of business, who wishes to be as happy in the country as he was in town, must betake himself, if his residence be on a small scale, to gardening, and, if it be on a large scale, to farming and planting. Some of the operations of farming may be recommended to a country gentleman, as a substitute for hunting and shooting. We particularly allude to the occasional holding of the plough; an operation which calls into moderate exercise every part of the body, and which also engages the mind in keeping the furrow straight. We speak from experience, when we say that we consider this the most agreeable of all farming operations, and one by which a maximum of exercise may be obtained with a minimum of fatigue. The handles of the plough draw, as it were, the operator after it, and the necessity of keeping his eye on two points, seen through between the pair of horses, occupies his attention. This attention, however, is only kept alive in ploughing with a plough in which the horses are yoked abreast; for, in those cases in which they are yoked in a line, the straightness of the furrow does not depend on the holder of the plough, but on the driver of the horses. The holder of the plough, in this latter case, is little better than a machine, and the operation, as he performs it, can no more be compared to holding a plough and pair, as practised in Northumberland, Berwickshire, &c., than the wooden plough in Middlesex can be compared to the iron plough of Mid-Lothian. In short, the occupation of ploughing with two horses is a fit exercise for a gentleman and a philosopher; and we can readily conceive the country gentlemen of Britain, at some future time, substituting this, and other agricultural labours, for the sports of the field. We have known several gentlemen in Scotland, of independent fortunes, follow the plough a portion of every day, when they were not otherwise engaged, and the weather would permit. There can be no doubt, also, that emigrants take pleasure in this exercise; and we can readily imagine that the sons of some of our landed proprietors, who now cultivate their own grounds in Australia or North America, are far happier in labouring in their fields with their own hands, than they would have been had they remained at home, and been compelled to seek for occupation in mere amusement.

Moral Influence of Farming. It has generally been thought that the habit of labouring with animals, or looking after them, has a tendency to brutalise, or at least to render coarse, rather than refined, farmers and their servants. This is, no doubt, to a certain extent true, where farming is pursued on the old system, and where all the animals of the farm are managed by main force; but, on the modern system of farming and managing animals, the whip and the goad are no longer employed; and horses, instead of being broken in by main force, undergo a similar treatment to human beings, commencing with their earliest years, on a system analogous to that of the infant schools. It is easy to conceive that this mode of managing animals must require quite a different kind of masters from what they have hitherto had; and that, by reaction, the gentleness and humane treatment which the man is compelled to show the animal, cannot fail to have a corresponding effect in humanising himself. This mode of treatment was first published by William Skirving, Esq., of Strathruddy, but it has only been lately adopted as a system, by some of the best Scotch agriculturists, and by the trainers of cavalry horses.* When we take this into consideration, and also the benefits which agriculture is capable of receiving from its operators possessing some knowledge of natural philosophy, natural history, and chemistry, our readers will, we think, readily allow that farming, in all its branches, is likely soon to become as little debasing a pursuit as gardening. The essential reason why the English ploughmen are, as a body, so very far inferior to the labouring mechanics is, that the ploughman, while he enjoys exuberant health and vigour of body, has comparatively no mental exercise, either during his work or during his hours of leisure: the mechanic, on the other hand, requires much more intellect even to exercise his business; and, in his leisure hours, from his mind being comparatively cultivated, he can amuse him-

* "The principal object in the treatment of young horses is, to render them docile; and the same gentle means are now used for that purpose, which are found to answer best in the treatment of children. They are rendered quiet and tractable by frequently patting, handling, and rubbing them, and by taking up their feet. They are led about the barrack-yard to accustom them to the sight of mounted horses, and to the glitter of arms; and, in the course of four or five months afterwards, they are transferred to the riding-school to be trained. The good old plan, like that preferred by our ancestors for teaching boys Latin and Greek, was, to whip all fear and shyness out of them; but kindness and common sense have at last gained the ascendency, to the great delight of the organs of sensation, both in boys and horses." (*Education, &c., of Cavalry Horses in Britain; by Mr. Marshall, Assistant Surgeon to the 7th Dragoon Guards:* in the *Edin. Med. and Surg. Journal*, for 1838.)

self with books. His mind, in fact, gets so strong from constant exercise, that it is continually craving for food.

We have made this digression in favour of farming, because the pursuit ranks high, in our estimation, as a source of useful occupation to emigrants, and to proprietors who farm their own lands; and also as a source of healthful recreation and rational interest to the retired citizen who possesses what we have called a first-rate suburban residence.

Situation of the Farm. The part of a first-rate suburban residence which is to be cultivated as a farm will be most conveniently situated when it adjoins the park on one side, and when the farm buildings are placed centrally to the lands, and, at the same time, not far from the mansion and offices. However, in amateur farming, when on a small scale, there is no great objection to the fields which are subjected to the plough lying round the park, or, perhaps, at a short distance from it. In many cases, particularly where the park is large, and the soil a loam that admits of being readily converted from arable land to pasture, and the contrary, a part of the park may be kept under the plough. This part should lie towards the confines of the park, and it should be changed from time to time, by breaking up a fresh surface, and laying down in pasture what had been for some few years under the plough. In short, all the park, except twenty or thirty acres adjoining the house, pleasure-grounds, domestic offices, kitchen-garden, &c., may be subjected, in turn, to the convertible husbandry; and, in our opinion, should be so wherever the soil will permit, and the object is to create as great an interest as possible within the park. Some inconvenience may result from the single trees and small groups scattered over the park, impeding the operations of ploughing, harrowing, &c., when that portion of the park is in aration; but these inconveniences are unavoidable: and it must be recollected that the object of this kind of farming is not mere profit, but recreation, combined with useful produce, as well as with picturesque beauty. In order neither to injure the trees, nor to cultivate any portion of the surface that will not produce an abundant crop, it is advisable to leave, untouched by the plough, a circle of turf round each tree or small group, of a diameter 3 or 4 feet greater than the spread of the branches. Round the margin of larger masses, and also along the margin of the boundary plantation, there ought to be a border of turf 10 or 12 feet wide, that is, sufficiently broad to admit of driving a carriage along it; which will at once be convenient for farm operations, such as carting out manure, carting home produce, &c., and serve as a drive, or a place for riding in, for the proprietor and his family, to survey the farm. It is the existence of green drives of this kind round the fields, and ornamental trees and shrubs in the hedges, and in groups and masses, which constitute what is called a *ferme ornée;* and, when these margins are mown, and the hedges and plantations that accompany them are planted or varied with rare or ornamental trees and shrubs, the interest and enjoyment produced by a farm may be considered as carried to the highest pitch.

The Kind of Farming, or of field culture, that will be commonly adopted in the suburban park farm, will, for the most part, be of the ordinary description, such as raising oats for the carriage horses, and straw for littering them and the cows; raising clover, lucerne, saintfoin, or tares, according to the soil and subsoil, for the cows and the farm horses; raising carrots, potatoes, and other root crops, for the cows and sheep, and partly for use in the house; and sometimes raising buckwheat, for the nourishment of game, &c. It may sometimes happen, however, where the object is chiefly recreation, that some of the kinds of farming peculiar to different districts of country may be adopted, to increase the interest of a *ferme ornée.* For example, where the soil and climate were suitable, the hop culture might be adopted; or apple, pear, cherry, walnut, or filbert orchards might be introduced. A willow-ground would be suitable for some situations; and a coppice of ash trees for walkingsticks, crate-ware, or hop-poles, for others; and so on. Near large towns, the raising of garden crops and small fruits, such as currants, gooseberries, raspberries, strawberries, &c., will form a source of interest for the cultivator of the *ferme ornée*, and one which, in some cases, may afford a little profit. In other parts of the country, peculiar crops may be grown; such as madder, woad, and other plants for dyeing; liquorice, rhubarb, poppy, and other plants for druggists and apothecaries; and a very general source of interest, and occasionally of profit, is the raising of seeds for the field, such as those of mangold wurtzel, turnip, &c.; or for the garden, such as those of peas and other pulse, carrots, onions, &c. We mention these crops, to show the endless source of interest, and the constant occupation, which a *ferme ornée* is calculated to afford to an amateur who will devote his chief energies to it; but, at the same time, we consider it proper to state that, where only a little attention can be given, and where farming is not wanted as a profitable occupation, the commonest kind of produce in the given neighbourhood is the best, as it can be raised with least trouble, and may be entirely directed by a local bailiff, or, if very small, by a single ploughman.

The Farm Buildings for a house or park farm, or for a *ferme ornée*, should, if possible, be situated at no great distance from the stable-court of the mansion; partly in order that the latter may be readily supplied with hay, corn, &c., from the former; and partly that they may be more readily inspected by the master.

The buildings which are to compose the farmery will depend on the kind of farming to be carried on; but as, in almost every case, there must be a portion of the farm devoted to the growth of corn, for the sake of the straw for litter; so, in every home farmery, a barn will be required. Where the extent of the lands under the plough at any one time does not exceed 20 or 30 acres, a threshing-machine is scarcely necessary; but, above that quantity, it will be found a source of both convenience and profit. The barn should generally form the main feature of the farm-yard; and, if threshing with the flail be adopted, it should be at least 20 ft. wide, and 40 ft. long, with the side walls 10 or 12 feet high, and open to the roof. Where a threshing-machine is introduced, the side walls require to be higher. The stables and cow-houses may be arranged to the right and left of the barn; and there may be a cart-shed on one side, with a granary over, and piggeries on the other, which, with a root-house, and a boiling-house for preparing food, may complete three sides of a square, or parallelogram, which should be open to the south. It is not, however, intended in this work to enter into the details of farm buildings; and we must therefore refer the reader, who may wish to know more on the subject, to our *Encyclopædia of Cottage, Farm, and Villa Architecture.*

It is only necessary farther to observe, that, when a proprietor intends to cultivate a farm on his own account, if he is not master of the subject himself, he must necessarily commit the management of it to a bailiff. Now, of this class of servants there are three kinds, very distinct: one is the local bailiff, who knows only the farming of the neighbourhood, to which he has been accustomed since his infancy, and who is, in general, wholly unwilling, and indeed incompetent from ignorance, to introduce any kind of improvement; another is the Scotch bailiff, who, in general, is so prejudiced in favour of Scotch farming, that he can see nothing good in that of any other country; and the third is what may be called the reasonable bailiff, who is generally an Englishman of some education and reading, who has seen and practised farming in different parts of the country, or who has only seen the farming of the neighbourhood, but who reads extensively on the subject. In most cases, unless the proprietor understand farming so far as to be able to become his own bailiff, it will be better for him to content himself with a bailiff of the first kind; because a better kind of bailiff must necessarily require a higher salary than the kind of small home farm which we contemplate can afford. Where there is a farm of 300 or 400 acres, then it may be worth while, according to the soil and situation, to employ a Scotch or Northumbrian bailiff, even in preference to what we have called a reasonable bailiff; because the north countryman will have had the advantages of experience among the experienced: whereas the reasonable bailiff may only have had the experience of a novice; that is, what he has experienced himself when he has endeavoured to carry into effect what he has learned from books. We repeat, however, that, in the case of small farms of 20 or 30 acres, attached to suburban residences not exceeding 100 or 150 acres in extent, the proprietors ought to become their own principal bailiffs, and to employ local working men under them.

The preliminary observations which we have given with the two preceding subsections are equally applicable in the case before us; and, as they render it unnecessary for us to say anything further on the subjects upon which they treat, we shall now proceed at once to our examples.

Wimbledon House, Surrey, the property of Mrs. Marryatt, F.H.S. (*figs.* 258. to 274.). —This estate, which once belonged to the celebrated Bond Hopkins, Esq., and was laid out for him about the middle of the last century, consists of about 100 acres of table-land, slightly varied on the surface, not by undulations, bold swells, deep valleys, or precipitous declivities, but by unconnected hollows, or large gullies, of little beauty in themselves, but capable of producing considerable effect when filled with water, as the principal one now is. In a word, the grounds at Wimbledon House have naturally little to recommend them, beyond their extent, and the circumstance of their falling in a gradual manner from the lawn front of the house, so as to display from that front an interesting view of the distant country. The chief merit of the place, as a suburban residence, consists in its completeness, the whole lying compactly within a ring fence, and there being a most commodious mansion, with complete domestic offices, a park, a farm (including a dairy and a poultry-yard), a kitchen-garden, and a flower-garden; the latter, perhaps, unrivalled in the neighbourhood of London, for the number of species and varieties of herbaceous plants that it contains.

Mrs. Marryatt has long been an enthusiastic admirer of flowers, and especially of such as are sufficiently hardy to make a display in the flower-garden. She also maintains a good collection of green-house and hot-house plants; and many of these, as well as hardy plants, have flowered at Wimbledon House for the first time in England, as the botanical periodicals for the last twenty years bear ample witness. In 1830, 1831, and 1832, Mrs. Marryatt's eldest daughter, Mrs. Palliser, made an extensive tour on the Continent, whence (and more especially from Italy and the Alps) she sent home a great many species which had never before been seen in a living state in Britain. The names of some of these plants, and also of new fruits introduced by Mr. and Mrs. Palliser, will be found in the *Gardener's Magazine*, vol. x., for 1834, p. 340. Among the more beautiful and remarkable green-house

258

View looking towards Wimbledon House.

plants which have flowered for the first time at Wimbledon, is the *Tacsònia* pinnatistípula, one of the most elegant, and at the same time singular, of climbing shrubs. Mrs. Marryatt is a liberal patron of the London Horticultural Society, assisting their exhibitions, at all seasons, by sending plants; and, indeed, there are few horticultural exhibitions that have taken place since the Society commenced this mode of encouraging gardening, in which Mrs. Marryatt, or her gardener, Mr. Redding, has not obtained a medal.

The park at Wimbledon House cannot be considered as having been planted with much taste; but there are some fine old trees in it, near the house, especially evergreen oaks. The pieces of water do not form agreeable shapes on paper, but, in the reality, at a distance from the eye, and with their outlines more or less disguised by trees, the effect is good, particularly that of the principal one, seven acres in extent, as seen from the house. *Fig.* 258. shows a portion of this piece of water, looking towards the house. Beyond a certain size, the form of a piece of water is of little consequence; but, when it is so small as to be readily comprehended as a whole by a spectator at no great distance from it, then, in order to please, it must be of an agreeable shape.

What constitutes an agreeable Shape in a Lake, Pond, or Piece of still Water? To this question we answer, that whatever constitutes an agreeable shape on level ground or lawn, or on any other flat surface, will look equally well in water.

A shape, whether of smooth lawn or still water, to be agreeable, must be such as to be readily comprehended by the eye and the mind at a single glance; and for this purpose the general form must be simple. It may be an oval or a circle, or it may be some form in which the length and breadth are more conspicuous elements than the sinuosities of the margin, and in which, consequently, the whole will always produce its full effect before the eye is attracted by the parts. These may be considered as fundamental principles; and, in applying them to the different styles in which water is formed, or surfaces of lawn surrounded by trees or shrubs, the characteristics of these styles readily come to our assistance. For example, in forming a piece of water in the geometric style, the outline must be composed of either straight or regularly curved lines. The general form may be a square, an octagon, or a parallelogram, and this general form is that which will produce the first impression on the mind of the spectator; but the margin of the geometrical figure may be composed of numerous segments or tangents, so as to form an intricate outline, which will afford amusement to the spectator when he approaches near enough to the figure to examine it in detail. If we form a piece of water in imitation of the ponds and lakes of nature, in what is called the picturesque manner, we must still begin by some simple form, and trust to this for the general impression. This being firmly established, the margin may be varied by prominences and recesses, and trees and shrubs, to any extent that does not interfere with the effect of the figure as a whole. If a piece of water is to be formed in the gardenesque manner, the same general principles are followed; but, as all the details of the margin require to be comparatively distinct, and all to appear decidedly artificial, greater care is requisite to combine regularity, uniformity, or symmetry, with variety.

In modern gardening, a long winding canal of still water is frequently the most effective form in which that element can be introduced; and it is sometimes so good an imitation of a river in a tame country, as to leave nothing to be desired. This, however, can only take place where long reaches of it are seen at a considerable distance from the eye, as at Oatlands; or where it is seen only in successional portions, by a spectator walking near it, and the walk being of considerable length, as at Esher. Most commonly such canals are near the eye, vary considerably in width, and are generally contracted towards both extremities, so as to leave no doubt in the mind of the spectator that they are only ponds. Pieces of water of this kind are frequently of such a length as not to be seen all at once, but in succession; and as, from the irregularity of their width, they can never, when so seen, be mistaken for a river, they ought to be treated in such a manner as to give them the appearance of a winding lake; which, being seen in successive portions, each of these ought to have an agreeable shape. These shapes are produced on the principles already laid down, viz., of displaying some obvious figure as a whole, and concealing the water altogether by wood wherever the desired form is interfered with by the opposite shores of the lake or canal approaching, or seeming to approach, each other too nearly. When a winding sheet of water is at such a distance from the eye as to be seen all at once, then all that can be done is, to cause it to wind agreeably, either by directing its bends in a manner more or less regular or symmetrical, or by creating an obvious allusion to some recognised figure. Thus, the general tendency of the winding may be to produce a crescent-like form; or it may be serpentine; or there may be one conspicuous bend in some regular form, and the subordinate bends may be comparatively indistinct. The safe guide in this case is, the principle already laid down, of producing some form which may take the lead, and be recognised at once; and trusting for variety to the subordinate forms, and to the details.

As there is no such thing in nature as a piece of water without an outlet, or as a lake from which there does not issue a brook; so, in landscape-gardening, no piece of still water ought to be formed, in which there is not some point indicated, which the brook might be supposed to flow from. In the geometrical style of forming pieces of water, the brook is represented by a formal cascade; and, in natural-looking shapes, there is commonly also a cascade, or something of the kind, in the way of outlet; or, if there is not, a portion of the margin, in what is decidedly the lowest part of the general surface of the ground, ought to be concealed by trees, or by an island, so as to leave the spectator room to suppose that there is an outlet in the proper place. Most lakes are also supplied by brooks; and, for this reason, in all imitations of them, whether gardenesque or picturesque, there ought to be an indication of the point at which the brook or rill expands into a lake, as well as of that where it contracts again into a brook. Where a rill or small stream runs through a narrow valley, it sometimes happens that a succession of dams are formed across the valley, in order to produce a succession of ponds; and when the dams are not disguised by wood, and the margins properly varied, this is one of the most formal and unartistical appearances which artificial water can be made to produce. In valleys which are nearly flat, ponds formed by throwing dams across are more easily managed; but even in cases of this kind, the dam is frequently left naked, and the margin of the water unbroken by trees or islands.

Fig. 259. shows two ponds which existed some years ago in a

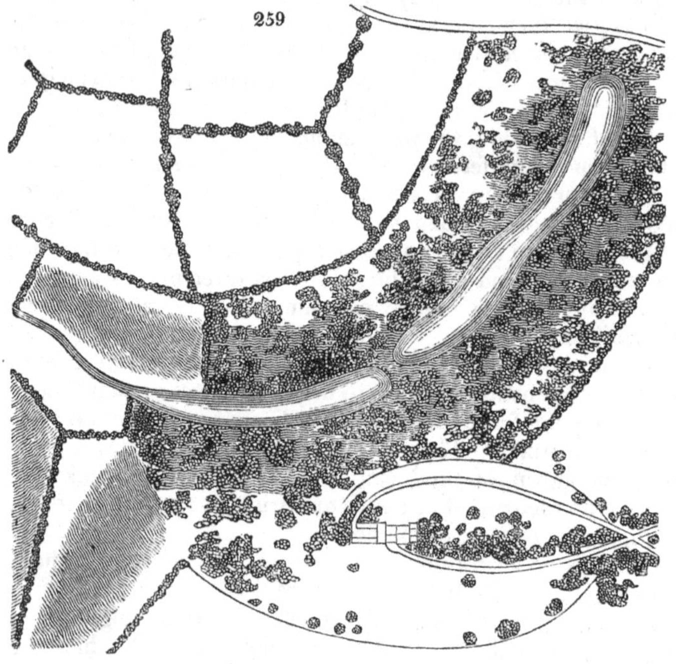
259

valley at Lyne Grove, near Chertsey, Surrey; and *fig.* 260. shows the manner in which we proposed to reduce both pieces to the same level, to vary the margin by islands, and to form a cascade at *a* in *fig*. 260.; where also a hydraulic ram might be erected for

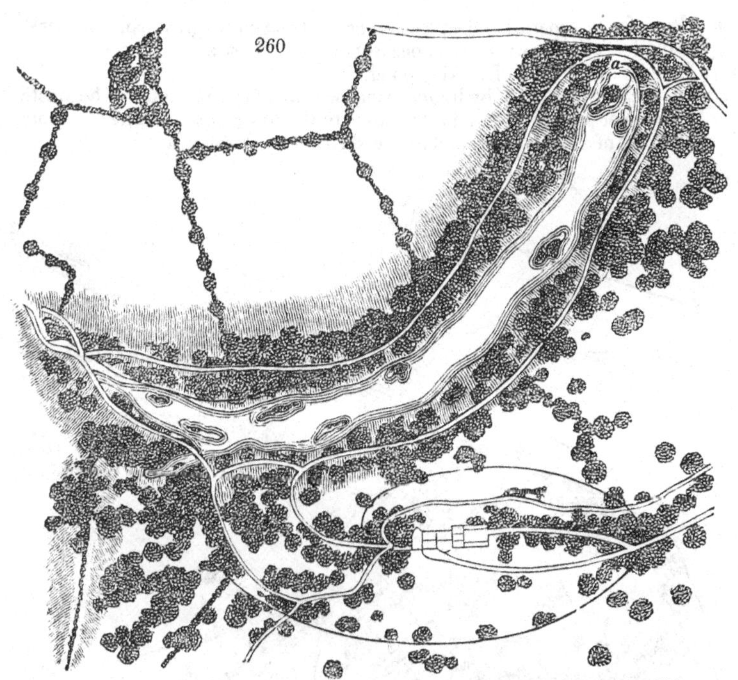

forcing up the water to a cistern on the top of the dwelling-house.

Notwithstanding this digression on the forms of artificial water, yet that element is so great an addition to landscape, that, where it has a clear surface, and is of a sufficient extent to receive a breadth of light, it is admissible in almost any shape; and this is exemplified by the effect produced by the pieces of water in Wimbledon Park, than which none were ever formed of less artistical shapes.

We shall now proceed to describe the plan, *fig*. 261.

a, The entrance lodge.
b, The entrance portico to the mansion.
c, The kitchen-court. *d*, The stable-court.
e, The laundry, and brewhouse, court. *f*, The drying-ground.
g, The green-house, with a summer-house at the back.
h, Paddock, containing an ancient building, formerly a mill for raising water, surrounded by a mass of plantation.
i, Flower-garden. *j*, Kitchen-garden. *k*, Poultry-yard.
l, Farm buildings. *m*, Rick-yard. *n*, Arable land.
o, Paddock, with a beautiful architectural grotto at *p* (formed by Bushell, who constructed the grottoes at Oatlands, Pain's Hill, and other places), and another ornamental building at *q*.
r, Osier-ground.
s, Island and ruin; the latter used as a Catholic chapel by the Prince de Condé, who resided at Wimbledon House before it was purchased by the late Joseph Marryatt, Esq.
t, Wilderness, and fish-stews. *u*, Open grove of ancient trees.

v v v, Wire fence, separating the pleasure-ground and mown grass from the park.
w, Hurdle fence, separating the upper from the lower park.
x x, Fish-ponds. *y*, Drinking-pond.
z, Well, near which is a hydraulic ram, which, put in motion by the waste of the upper piece of water in its course to the lower piece, forces a supply to the roof of the mansion, and to the sheds of the hot-houses

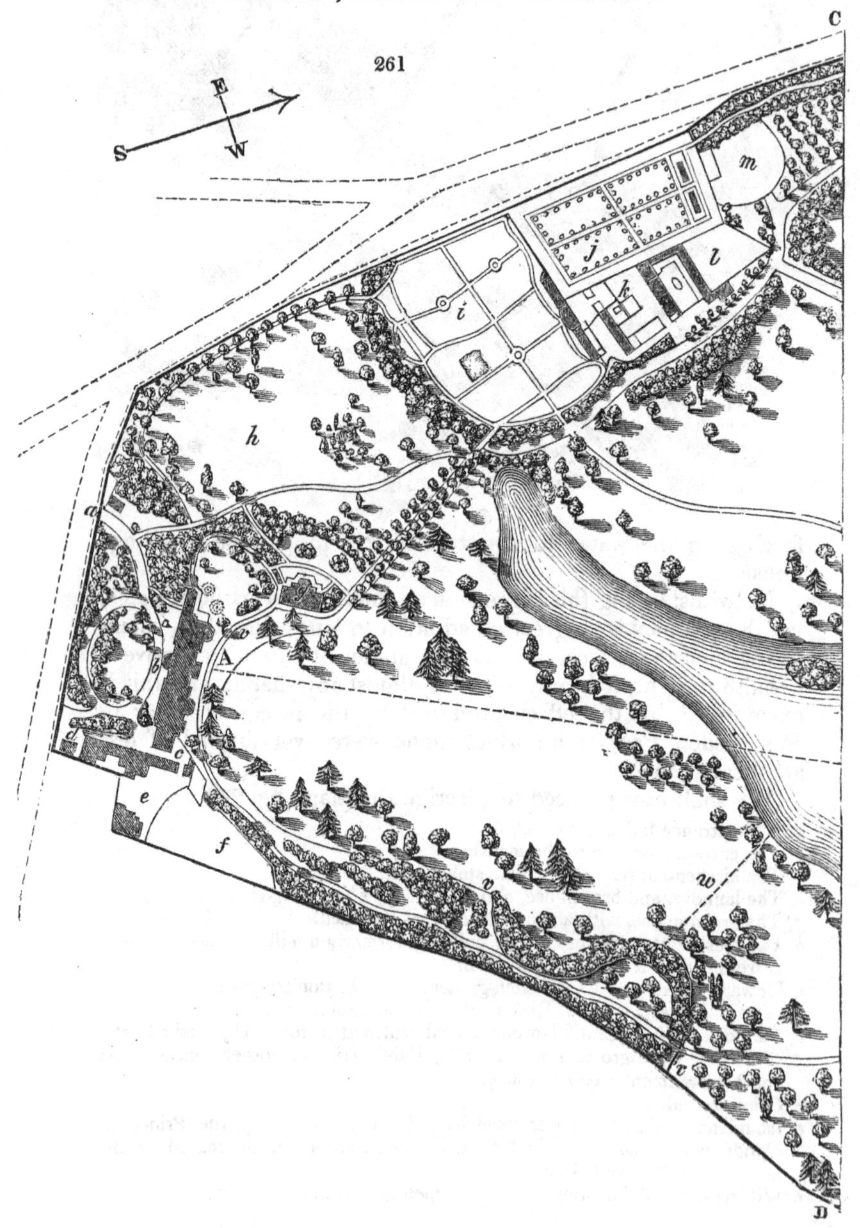

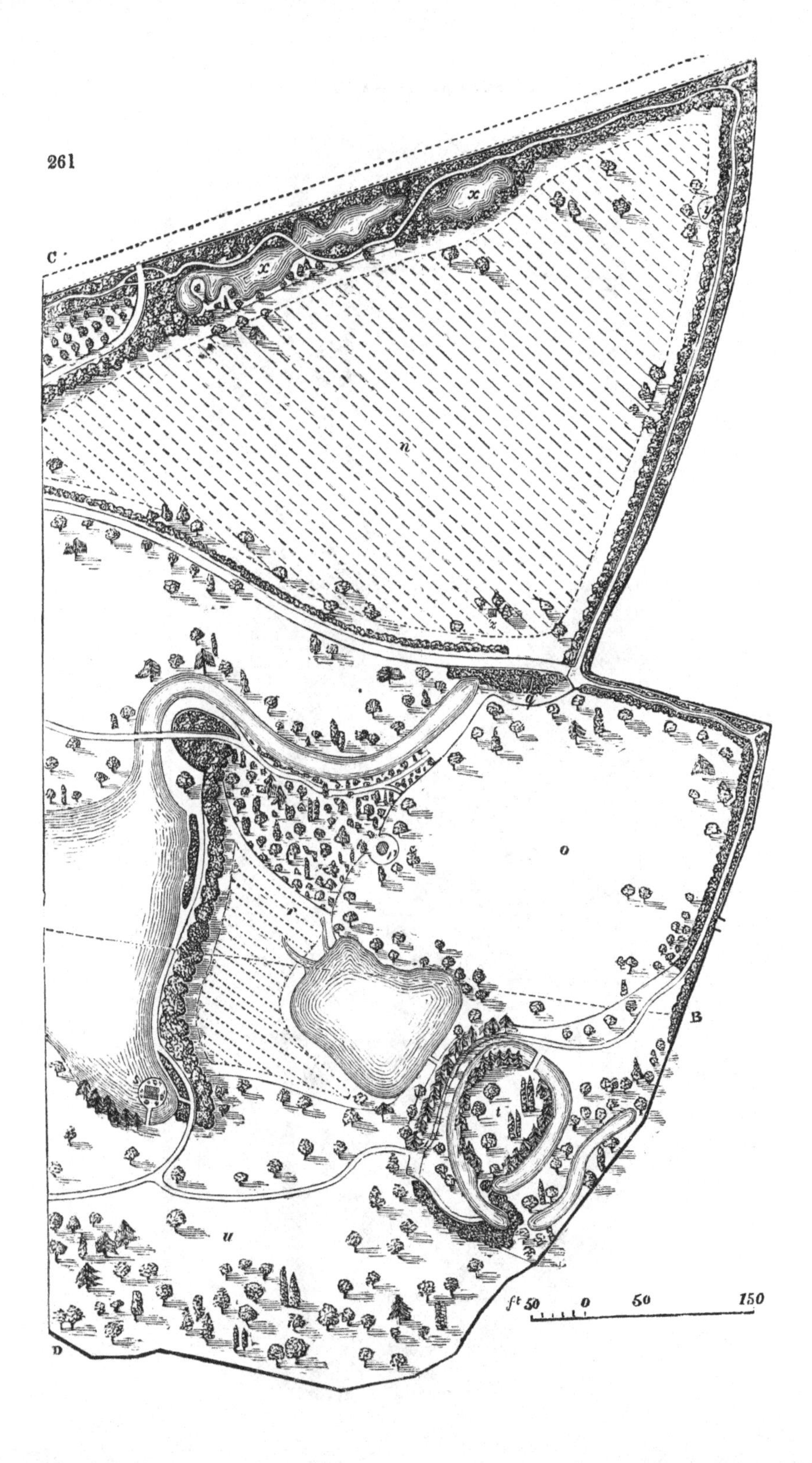
261
C
x
x
x
y
n
z
q
o
r
B
s
t
u
D
ft 50 0 50 150

262

Geometrical Section of the Ground, and Bird's-eye View of the Park, Flower-garden, and Kitchen-garden, at Wimbledon, on the Line A B, looking to the West.

263

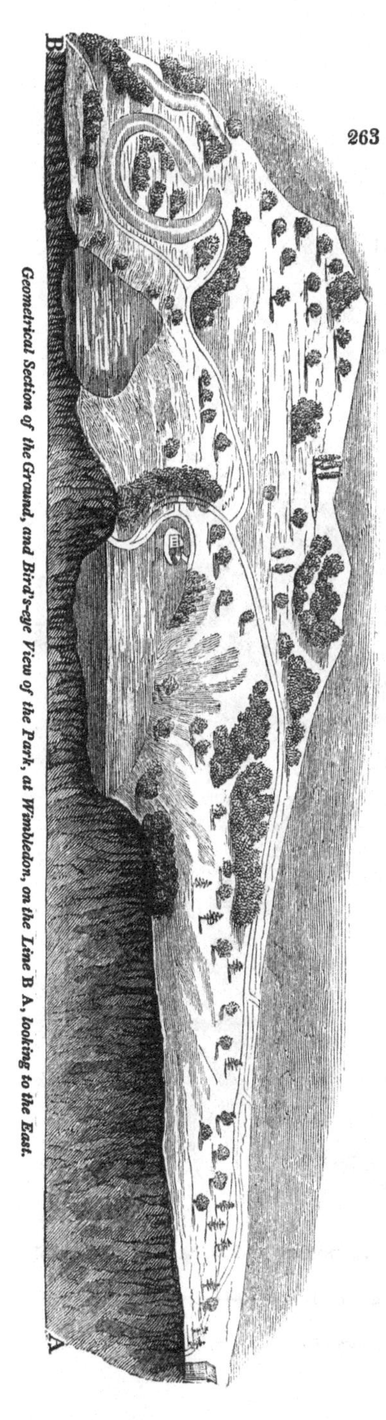

Geometrical Section of the Ground, and Bird's-eye View of the Park, at Wimbledon, on the Line B A, looking to the East.

Fig. 262. is a geometrical section of the ground, and bird's-eye view of the park, looking to the west.

Fig. 263. is a section and bird's-eye view, taken on the same line, looking to the east.

Fig. 264. is a section and bird's-eye view, looking to the north-east.

Fig. 265. is a section and view, looking to the south-west, and showing the front of the house. It will be observed from the lines indicating walks, that there is one which makes the entire circuit of the residence, besides several cross walks. It will also be observed that the carriage communication between the farm-yard and the stable offices at the house is by the public road, though there is a private carriage road, in a more direct line, through the mill paddock.

Fig. 268. p. 642, 643. is a ground plan of the flower-garden and kitchen-garden on a larger scale.

a, The entrance to the flower-garden from the house, through the avenue *b*; the doorway of the garden exhibiting a rustic arch, as shown in the sketch *fig.* 267. From the point *b*, a walk branches off through the mill-paddock to the entrance front of the house, so that persons may come and go to the gardens without going through the house.

b, Elm tree avenue, decorated, during summer, with a row of China vases, placed at regular distances along each side of the gravel walk, and each containing a choice flower.

c, A superstructure of rustic-work, with an elevated rustic vase in the centre, and surrounded by beds, with borders of rustic-work. The circumference consists of an arcade, connected by a column clothed with creepers in the centre, by garlands of creepers twined round chains; and having at a distance the appearance shown in *fig.* 270.

d, Pond, with a fountain in the centre, containing numerous aquatics of different species, and surrounded by rockwork as shown in *fig.* 273. *e*, Marble basin and fountain.

f, Bower of trelliswork.

g, Elevated platform, ornamented on the side next the flower-garden with a screen of rustic-work, and with rustic vases filled with flowers.

h, Arcade, covered with creepers.

i i, Conservative wall, on which are many very interesting exotic shrubs.

k k, Plant-houses, with rockwork in front planted with select ornamental herbaceous plants and undershrubs. The central house (*l*) is a green-house, and those to the right are also green-houses, while those to the left are hot-houses. At the back of one end of the central green-house is a door, which opens to the private or reserve court (*m*), in which there is a pelargonium-house, with a range of pits along the front, for Cape bulbs; and in the interior of the court are various other pits and frames.

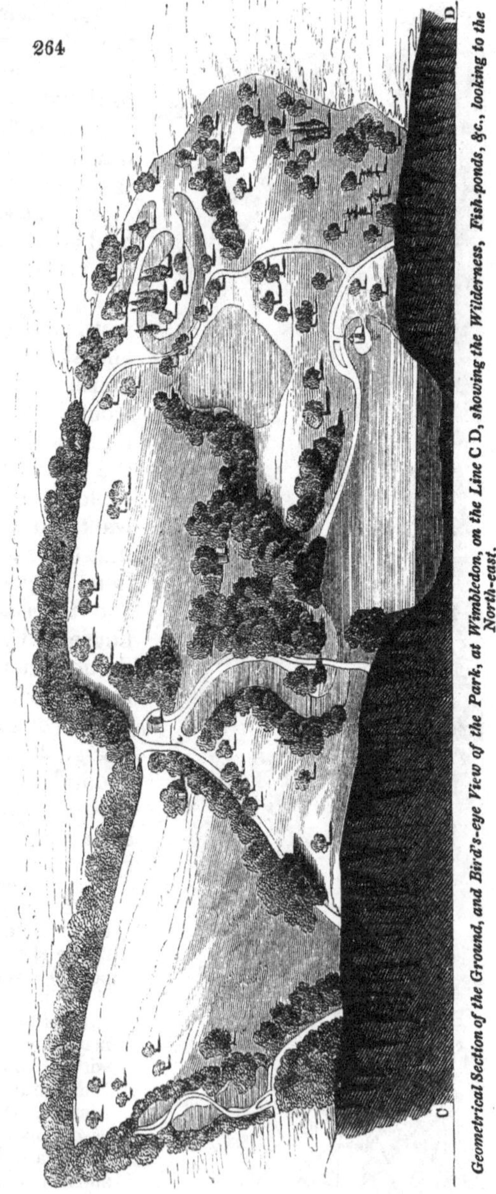

264

Geometrical Section of the Ground, and Bird's-eye View of the Park, at Wimbledon, on the Line C D, showing the Wilderness, Fish-ponds, &c., looking to the North-east.

n, Large reserve-ground, surrounded by the potting and working sheds, tool-houses, seed-room, men's room, &c.

o, Farm-yard, the details of which will be found in *fig.* 271.

p, Poultry-yard.

q, House of the gardener and general manager. *r*, Barn.

s, Yard for rubbish, pea-sticks, &c.

265

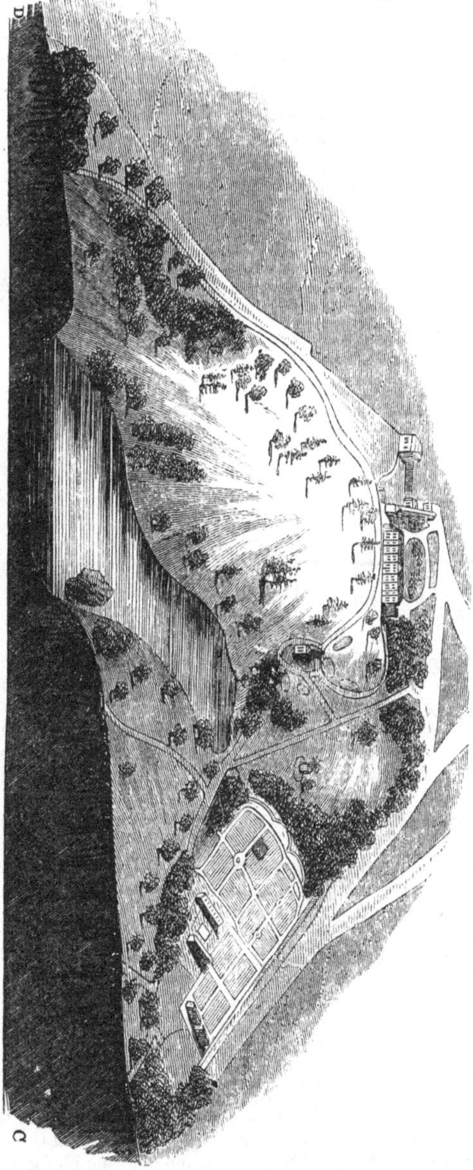

D C

Geometrical Section of the Ground, and Bird's-eye View of the Park, at Wimbledon, on the Line D C, including the House, Flower-garden, Kitchen-garden, Farm, &c., looking to the South-west.

t, Pine-pits in the kitchen-garden.

u u, Part of the private road from the mill-paddock to the farm-yard, rick-yard, and arable field; and which crosses the elm avenue, near the entrance to the flower-garden: but this road being little used, and that only early in the morning, no perceptible marks of it are seen in the elm avenue. This avenue having a wire fence on both sides, a gate is placed in each fence, one opposite the other; and these gates are opened whenever carts are to pass.

v, East entrance to the flower-garden.

w, Road from Wimbledon to London.

x, Wimbledon Common. *y*, Upper park, or lawn.

z, North park, adjoining the arable ground.

Fig. 266. shows the ground plan of the farm-yard and poultry-yard, and also of the range of plant-houses, on a larger scale than in the preceding plan.

a, Orchideous house.

b, Plant stove.

c, *d*, and *e*, Green-houses, with vines trained under the rafters.

f, Aviary. *g*, Working-shed. *h h*, Furnaces.

i, Situation of the cistern on the top of the brick wall of the green-house, to which water is raised by the hydraulic ram, mentioned in p. 634.; whence, it is conducted to the different plant-houses, and to the fountains in the open garden, by pipes.

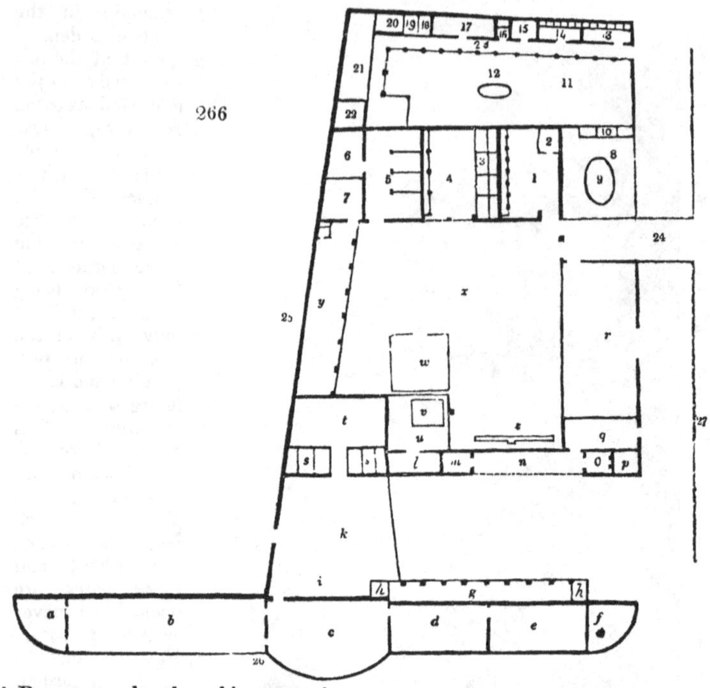

266

j, Reserve-yard and working-ground.
k, Reserve-ground for plants in pots. *l*, Coal-shed. *m*, Tool-shed.
n, Pot-shed. *o*, Man's room. *p*, Seed-room.
q, Carpenter's shop. *r*, Barn. *s s*, Pits for Cape bulbs.
t, Pelargonium house. *u*, Enclosed yard. *v*, Liquid manure tank.
w, Granary, supported on stone pillars, with caps to prevent the ascent of rats and mice. *x*, Open yard.
y, Cattle-shed, with fodder-rack. *z*, Pump and drinking-trough.
1, Cow-house. 2, Calf-pen. 3, Pigsties.
4, Open cattle-shed, with fodder-racks. In the back wall there is an opening, with a shutter, to allow fowls to pass through from the poultry-yard, so as to pick up what food they can, in the cattle-yard.
5, Stable for cart-horses.
6, Hay-room, and place for cutting straw into chaff.
7, Pigsty, near which there is a privy for the farm servants.
8, Yard for ducks, enclosed by trelliswork.
9, Oval pond, paved with brick, into which is thrown the food for the ducks.
10, Houses for the ducks. 11, Poultry yard.
12, Stone basin, to contain water for the poultry.
13, Laying-house for common fowls. 14, Laying-house for hen-turkeys.
15, General fattening-house. In this house, common fowls, turkeys, and ducks, are all kept loose, and fattened together. It is found that this mode contributes generally to their fattening; because, whenever one bird goes to eat, the others are stimulated to imitate it.
16, Place for fattening quarrelsome fowls, one fowl being fattened in it at a time.
17, Rabbit-house, in which the rabbits are kept in hutches.
18, Turtle-doves, formerly kept for ornament in a building at one end of the range of hot-houses, but the space is now occupied with a house for *Orchídeæ*.

Entrance to the Flower-garden at Wimbledon House.

19, House for pheasants; which are bred here, and afterwards turned into the park.

20, Pigeons. These have their principal entrance in the roof, and have always a salt-cake placed on the floor.

21, Fowls' roosting-house. Here the fowls roost on horizontal strips of board, about 2 in. on the side, and raised, so as to form a slope, from the floor in the front of the house, to the top of the back wall, in the manner of a green-house stage. The object of this is to facilitate the ascent of the fowls, and to prevent them from dirtying one another when at roost.

22, Pigsties.

23, Covered way to the poultry-houses, paved with Dutch clinkers, as are all the houses in this yard. 24, Entrance to the farm-yard.

25, Kitchen-garden.

26, Flower-garden, of which a general view, showing the main walk and the plant-houses, is given in *fig.* 269.; and views of the entrance, rustic structure, and fountain, in *figs*. 267. 270. 273. 27, Park.

Fig. 271. in p. 646. shows the plan of the principal floor of the house, and of the domestic offices.

There are several other minor conveniences in this and the three preceding plans, such as water-closets, &c., which we have not thought it necessary to particularise; and in the plan *fig*. 261. there are several rustic seats, benches, and resting-places, which have not been noticed. There is also a reserve ice-house near the head of the upper lake.

In order to convey an idea of the general effect of the park and garden scenery of Wimbledon House, and to afford some relief to the reader after the fatigue of perusing so many details,

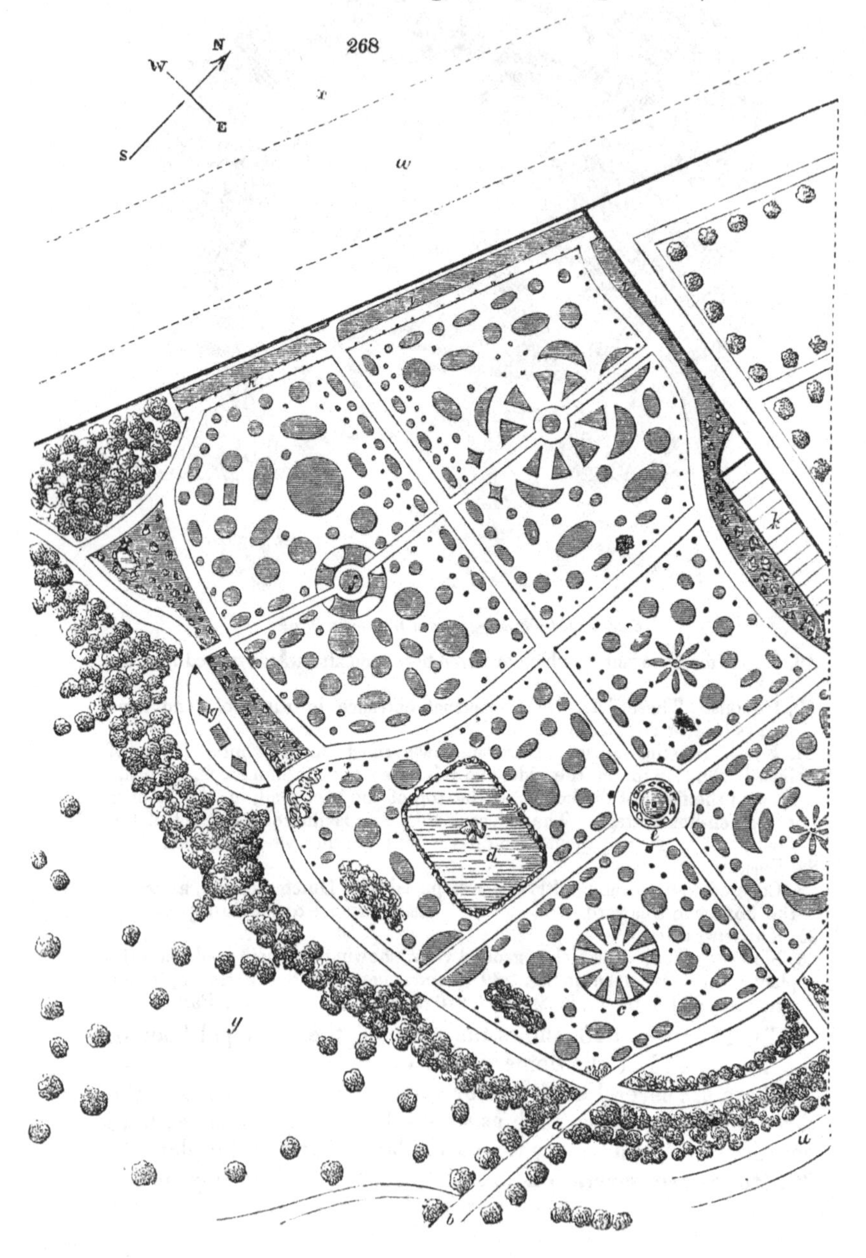

268

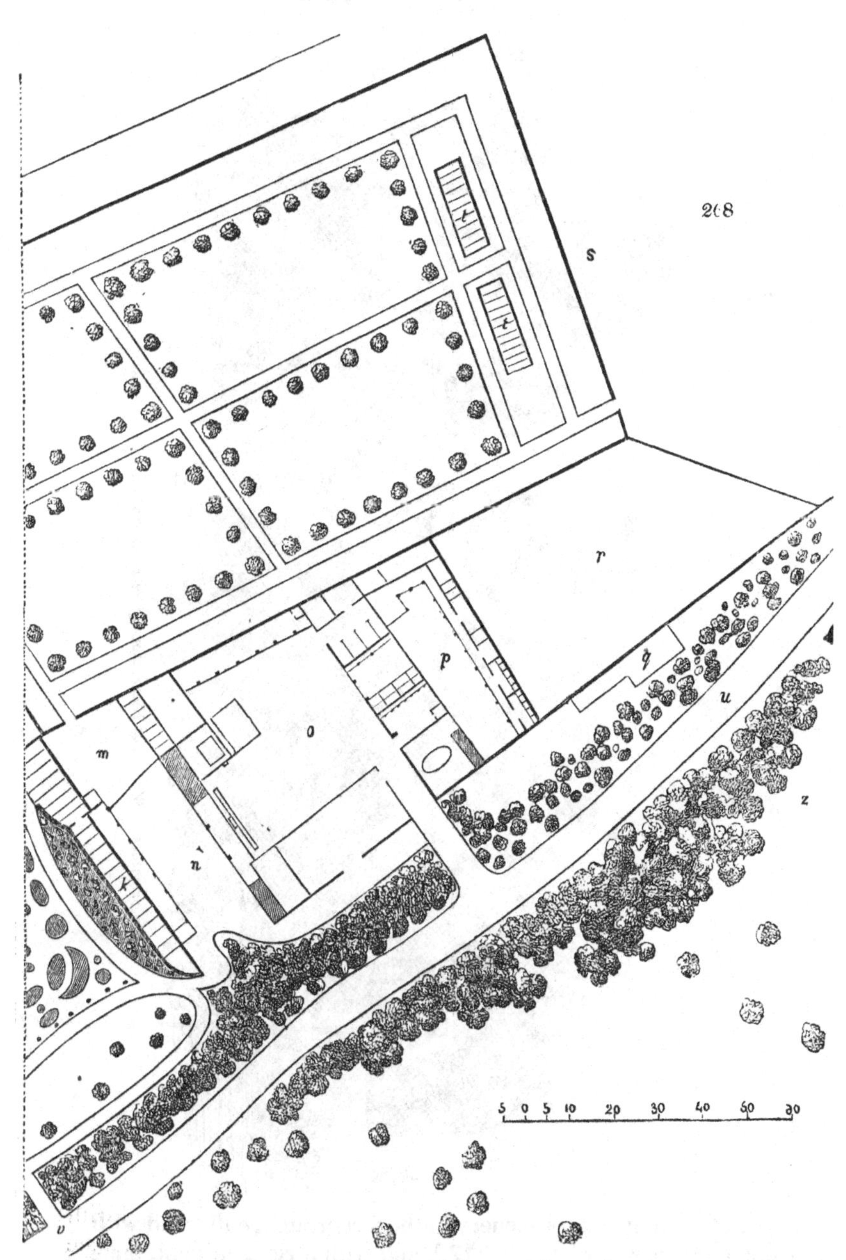

we shall briefly notice the general impressions the scenery made on us when we first saw the place some years ago.

Passing through the entrance hall and the saloon to the gravel walk on the lawn front, a magnificent panoramic view presents

269

Flower-garden at Wimbledon House, showing the main Walk and Plant-houses.

itself; consisting of park scenery in the foreground, enlivened with a fine piece of water (see *fig.* 272.), and with a rich, verdant, tame country in the extreme distance. Among the trees in the park are some fine old oaks and beeches, some very large evergreen oaks, and several large pines, firs, and cedars. The stranger, if he proceeds along the gravel walk to the left, will find it lead him to

270

Flower-garden at Wimbledon House, showing a Rustic Structure and Vase.

an ancient green-house, or orangery, and thence to the flower-garden, the kitchen-garden, the farmery, and the farm. If, on the contrary, he walks to the right, he will pass through a variety of scenery, at first highly polished, with a smoothly mown lawn, and afterwards of a more rustic description; till, having walked upwards of a mile, he arrives at the flower-garden, kitchen-

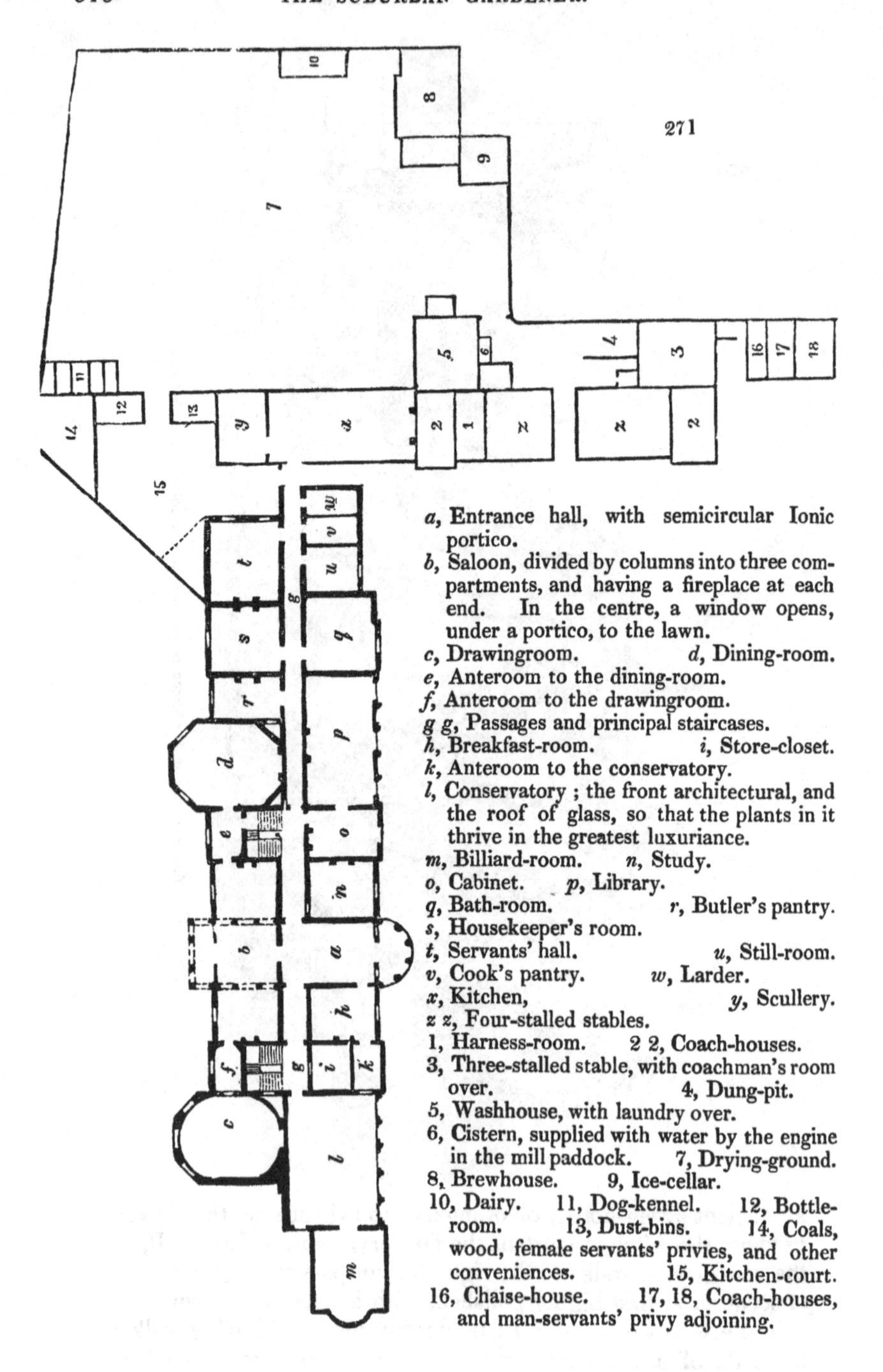

a, Entrance hall, with semicircular Ionic portico.
b, Saloon, divided by columns into three compartments, and having a fireplace at each end. In the centre, a window opens, under a portico, to the lawn.
c, Drawingroom. *d*, Dining-room.
e, Anteroom to the dining-room.
f, Anteroom to the drawingroom.
g g, Passages and principal staircases.
h, Breakfast-room. *i*, Store-closet.
k, Anteroom to the conservatory.
l, Conservatory ; the front architectural, and the roof of glass, so that the plants in it thrive in the greatest luxuriance.
m, Billiard-room. *n*, Study.
o, Cabinet. *p*, Library.
q, Bath-room. *r*, Butler's pantry.
s, Housekeeper's room.
t, Servants' hall. *u*, Still-room.
v, Cook's pantry. *w*, Larder.
x, Kitchen, *y*, Scullery.
z z, Four-stalled stables.
1, Harness-room. 2 2, Coach-houses.
3, Three-stalled stable, with coachman's room over. 4, Dung-pit.
5, Washhouse, with laundry over.
6, Cistern, supplied with water by the engine in the mill paddock. 7, Drying-ground.
8, Brewhouse. 9, Ice-cellar.
10, Dairy. 11, Dog-kennel. 12, Bottle-room. 13, Dust-bins. 14, Coals, wood, female servants' privies, and other conveniences. 15, Kitchen-court.
16, Chaise-house. 17, 18, Coach-houses, and man-servants' privy adjoining.

272

View from the Lawn Front of Wimbledon House.

garden, &c., last, instead of first. We took the right-hand walk from the house, and, passing along it, the first remarkable objects which struck us in the foreground were some very large evergreen oaks, with trunks 2½ ft. in diameter. On the right was a rural fruit garden, planted with the common summer fruits, such as the cherry, gooseberry, raspberry, strawberry, &c., for the indulgence of the young people, who were allowed to gather them here for themselves. This may be called a children's fruit-garden; and it appears to us one of the most amiable features, next to children's gardens (that is, gardens which they are to lay out and cultivate themselves), that can be formed in a suburban residence. Near this rural fruit-garden was a garden for British plants, managed solely by Miss Marryatt; and beside it was a dell filled with rhododendrons, which, from the moistness and shadiness of the spot, had attained an extraordinary size, and were growing with great luxuriance. Adjoining the evergreen oaks was a large cork tree, a very fine *L*igústrum lùcidum, gigantic common and Portugal laurels, a large red cedar, a *R*hododéndron pónticum upwards of a hundred feet in circumference, and immense masses of ivy supported on pollard oaks, which looked like some strange kind of evergreen tree; and, farther on, were several lofty silver firs. Passing through a small wicket in the wire fence which separates the pleasure-ground from the park, but still following the walk, we found ourselves in comparatively open scenery; and, the atmosphere being tolerably clear, we observed the towers of Westminster Abbey and the dome of St. Paul's in the extreme distance. From some points of view, Wimbledon Park, the seat of Earl Spencer, which borders Wimbledon House on two sides, contributes to the effect of the scenery in the park belonging to the latter, more especially by the aid of a beautiful sheet of water; and, in one part, the windings of the Thames itself are seen from a seat under an immense beech tree, one branch of which measured, in 1829, 75 ft. in length. Passing through a rustic gate, we arrived at a spacious ivy-covered summer-house, situated on a slight eminence, and furnished with a large table and matted seats, capable of accommodating a party of twenty. This summer-house overlooks that part of the grounds called the Wilderness (see *t* in *fig.* 261.), in which there are several pieces of water, overhung with weeping willows, and inhabited by numerous wild aquatic fowl. Descending from the summer-house, we passed a rocky cascade, studded with alpine plants, and ascended, through a shrubbery, to the grotto (*p* in *fig.* 261.); from the entrance of which there is the retired home view shown in *fig.* 274. This grotto was formed by Bushell, the most celebrated grotto and cascade artist that ever appeared in England. The grotto at Pain's Hill is considered his *chef-d'œuvre*; and next to it ranks the one at Oatlands, which the grotto at Wimbledon resembles in general character. Entering an open grove of forest trees, we next come to a light iron bridge, passing over which, we noticed an immense pollard oak some centuries old, covered with ivy, and a very large Magnòl*ia* acuminàta, which

Flower-garden at Wimbledon House, showing the Pond and Fountain.

was an imported plant, and brought from America by the botanist Fraser. Here are, also, *Pinus serótina*, and some other American trees, which were planted here when they were first introduced into England. From this open grove, which is situated near the largest piece of water, the stranger may either proceed to the gardens and farm-buildings by an open path directly through the park, or by a circuitous course in the boundary plantation, which forms two sides to the arable land.

Arrived at the farmery and the poultry-yard, their details will be found remarkably complete, as the plans and descriptions which we have already given will prove. The kitchen-garden is chiefly remarkable for producing excellent crops; but the grand feature of the place, in a gardening point of view, is the flower-garden, which occupies upwards of three acres, and contains above 200 beds of flowers. These beds are of different shapes and sizes, and they are scattered over the surface with very little regard to regularity or symmetry; the object, apparently, being to get as many beds as possible into the given space, allowing a small strip of grass between them, in order to keep them distinct, and to admit of walking round them. In point of general design, therefore, this flower-garden has nothing to recommend it; but, from the great number of beds, and almost endless variety of the kinds and colours of the flowers they contain, it presents a dazzling surface, of the most brilliant colours, mingled together in confusion.

This garden is admirably managed by Mr. Redding, who keeps an ample stock of plants in pots, in order that, as soon as one crop of flowers begins to fade, the plants may instantly be removed, and replaced by others just coming into bloom. The number of species cultivated in this garden exceeds 750, exclusive of varieties of florist's flowers, such as dahlias, heartseases, tulips, &c. The first appearance of gaiety in this garden usually commences in February, with the appearance of crocuses, snowdrops, primroses, and Russian violets. In March, to these are added narcissi of various kinds, numerous species of *S*cílla, and other early bulbs, together with *S*axífraga oppositifòlia; and, in April, the different varieties of hyacinth make a splendid show, and perfume the garden with their fragrance. Primroses, double and single, *A*'rabis, Aubriè*tia*, *A*lýssum, *O*'robus, the common wallflower, and a number of other plants, produce a brilliant effect during this month; and, from this period, the brilliancy and beauty of the garden never fades during the whole of the summer. In the autumn, the last flowers are those of the dahlia, some of the Mexican and Californian annuals, and the chrysanthemum; and, after these are removed, in consequence of being blackened by frost, the garden depends for its beauty during the winter on the evergreen flowering shrubs, such as *A*'rbutus, laurustinus, *E*rìca, *D*áphne, &c.; on the beds of evergreen herbaceous plants, such as pinks, carnations, saxifrages, sweetwilliams, &c.; on the green turf, on the rich yellow gravel, on the clear water, and on the state of neatness in which the whole is kept.

As the park abounds with hares and rabbits, it is necessary to

274

View from the Grotto at Wimbledon House.

enclose the flower-garden on the sides next it, and the mill paddock, with a wire fence, hare-proof; and this fence is so artfully concealed among the shrubs, as scarcely anywhere to be seen, except at the two entrances to the garden, where there are, of necessity, hare-proof wickets.

The whole management of the park, gardens, and farm, is committed to Mr. Redding, who is an example of a most judicious, careful, and successful farmer, as well as gardener; two pursuits which, considering the great number of plants in the flower-garden and green-houses, one would think almost incompatible.

Remarks. There is a simplicity, combined with a certain degree of grandeur, in the view from the lawn front, which makes a considerable impression on the stranger. The most complete parts of this residence are the dwelling-house and domestic offices, and the farm offices. The latter are neither regular nor symmetrical; but they are designed on the best principles, they are quite suitable for a small farm where the object is the consumption of the produce, and they are substantially executed; circumstances which show that neither regularity nor symmetry is an essential ingredient where the main object is use. The relative situation of the farm-yard, poultry-yard, kitchen-garden, and back courts to the flower-garden, is also good; and, though there is no convenient connexion, within the grounds, between these and the stable offices, this is rendered comparatively unnecessary by the situation of the public road. The ground plan of the house is well calculated both for splendour and habitableness. The dining-room and drawingroom are spacious rooms, admirably connected, and yet sufficiently calculated for display in passing from the one to the other. When the family are alone, there is a sufficient number of smaller living-rooms to serve the purpose of a dining-room, drawingroom, &c., for more than a dozen persons; without using the principal rooms. The grand central passage through the house being broad, and well lighted both from the ends and the roof, is also a great source of comfort and convenience; as are the domestic offices, from their extent and number.

A Villa of from 50 *to* 100 *Acres in the Geometrical Style* (*figs.* 275. to 277.)—We have given this design to show how the ancient style of laying out grounds can be adapted to modern uses. We have repeatedly observed that we are not among those who consider the ancient, or architectural, style of laying out grounds as fit only for the ages in which it first appeared; and that, on the contrary, we consider it as much a style among other styles of landscape-gardening, as the Gothic is among other styles of architecture. The suitableness of this style for a country in a wild state must, we think, be obvious to every unprejudiced mind, from the contrast which its clearly defined lines and forms afford to the irregularity of the surrounding scenery, and from the obvious expression of art and refinement which they produce. In travelling over a flat country, like great part of the interior of Poland, which may be described as one interminable forest (the cultivated lands and pastures being merely extensive glades), it is delightful to arrive at a post-house with a walled garden, and perhaps a field or two enclosed by clipped hedges. In the neighbourhood of Moscow, the greater part of the country residences are, or at least were in 1814, laid out in the geometrical style, and the principal things that were admired in them by the inhabitants were the long straight avenues and walks, bordered by clipped hornbeam hedges of great height; and the immense spruce fir hedges, which were found in other parts of the grounds and near the mansions. The geometrical style, besides being suitable for all countries not regularly lined out by walls and hedges, and cultivated in regular fields or ridges, is also in an eminent degree adapted for public gardens and parks, in which masses of people, or of horses and carriages, are to congregate together. If the reason be asked why straight walks, avenues, and formal lines of plantation, are more suitable for places of public resort than the circuitous walks, scattered groups, and single trees of the modern style, the answer is, that such straight walks and broad avenues are better adapted for displaying crowds of people, and long cavalcades of horses and carriages, to advantage; and that the great object of those public walks is display. Such scenes are not for solitude, says Byron, speaking of the gardens of Versailles; and the truth of this remark must have been felt by every one who has seen such gardens. We do not say that it is necessary for all the walks of a public garden to be straight, and all the trees in rows, or in formal masses; we merely take these features as characteristic of the style, freely allowing that, when the curves of roads and walks in the modern style are so large as to present large portions of them to the eye at the same time, the effect comes to be nearly the same as in the geometrical style in its utmost rigour. Hence, the somewhat curved roads in the Regent's Park, and in Hyde Park, in London, are nearly as effective in displaying the company assembled in them as the straight roads in the Mall in St. James's Park, in the avenues in the Champs Elysées in Paris, and in the walks of the gardens at Versailles and Schönbrunn. It will be recollected, however, that the walks and roads in all the places mentioned, from their comparative straightness, belong more to the geometrical than to the modern style.

To pass from carriage-drives to walks adapted solely for persons on foot, if we take a review of all the public gardens in Europe, we shall find that the most effective display, on holidays, is always made in broad straight walks: for example, in the broad north and south walk in Kensington Gardens; in the broad avenue through the Regent's Park, opposite Portland Place; the broad walk opposite the Luxemburg, in Paris; that opposite the palace of Schönbrunn, near Vienna; and those in the gardens of the summer palace, in Petersburg. The truth is, that the modern style is essentially calculated for solitude and retirement, while

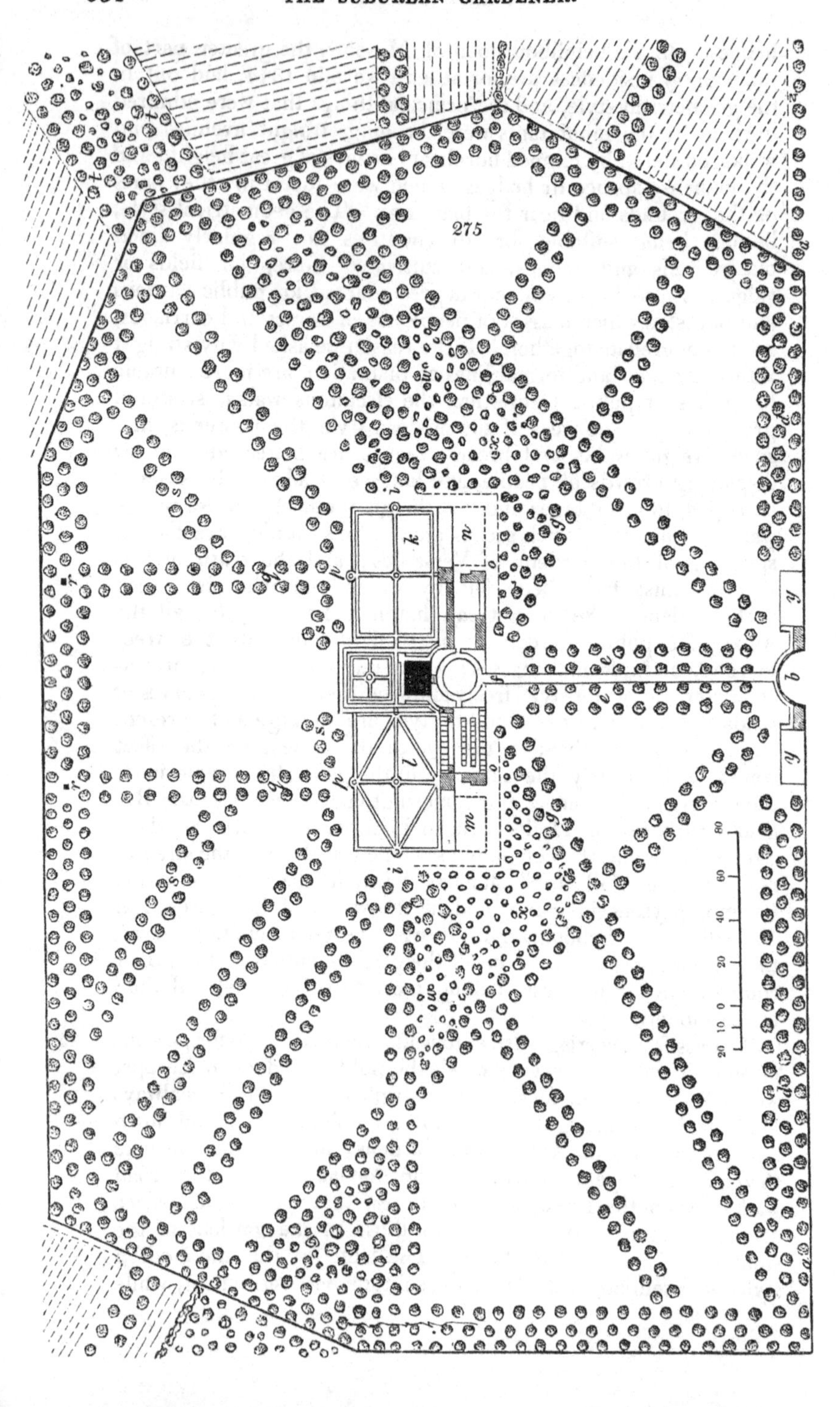
275

the geometric style is especially calculated for publicity and display. The numerous windings of the walks in the modern style, and the various groups into which its woods are thrown, occasion, to a spectator walking along them, a perpetual change in the scenery, and never allow masses or long columns of people to be seen at once. Every thing connected with the ancient style has exactly a contrary effect: there can be no privacy in straight avenues.

Since then the geometrical style is chiefly adapted for an unenclosed or wild country, and for public parks or gardens, on what grounds, it may be asked, is it recommended for suburban residences? Our answer is, to accommodate the taste of particular individuals; to introduce as a contrast to suburban residences in the modern style; and to suit newly peopled, and thinly inhabited, countries, such as the back settlements of America or Australia. Besides these reasons, which refer to what may be called the relative beauty of the style, we contend that it has some positive beauties, which are peculiarly its own, and on which account alone it deserves occasionally to be introduced. Among these are, the grandeur of its avenues, and the consistency with which this expression is maintained throughout a whole place; the masses of light and shade which are produced by these avenues; the succession and uniformity of the trees which compose them; the idea of distance given by their lengthened vistas; and the feeling of shelter and protection which intersecting avenues and rows of trees always produce. To these beauties ought to be added those of the scenery in the immediate vicinity of the mansion, the terraces, the embroidered parterres, the fountains, the statues, and the comfort produced by the near vicinity of the kitchen-garden.

Taking both the ancient and the modern styles, however, and examining them impartially, we readily allow that much more variety and interest can be created by the latter style than by the former; and, further, that the modern style is better adapted to modern circumstances, not only on account of the contrast which its lines and forms produce to the lines and forms of a country every where cultivated, but on account of its affording more ready means of displaying the numerous foreign trees and shrubs, which at the present day are become comparatively common in the gardens of temperate climates. We shall now describe the plan before us.

Passing along the turnpike road (*a a* in *fig.* 275.), indication is given of a gentleman's seat by the triple row of trees within the boundary wall. Arrived at the point *b*, we there find an open iron gateway, in the centre of a semicircular wall; and, looking through the gateway along the avenue, we observe at its further end a lofty square house (the mansion) on a raised basement. Between the centre gateway and the lodges on each hand, there are two openings of the same width as the gateway; and, like it, they are filled in with iron-work, so as also to resemble gates. Through these the eye looks along the avenues *c c*, which are well calculated to give an idea of extent. After passing through the gates, if we look to the right and left, we see the avenues at right angles to the approach avenue (*d d*), and, on each side of the avenue leading to the house, the side avenues (*e e*). The centre avenue has a broad gravel road, but all the other avenues are in grass. Arrived at *f*, which is the entrance to the court of honour, if we look to the right and left, we have the avenues *g g*; though these are chiefly calculated for making an impression on a stranger when he comes out of the court of honour on returning from the mansion. Having passed through the court of honour (*h*) (in the centre of which is a statue, a sundial, or an obelisk), and also directly through the hall of the house, we descend the steps from the platform on the garden front, and arrive at a terrace walk, which extends on each side as far as *i i*. Directly in front, there is an ancient parterre, with a raised terrace walk round it. On the right is the kitchen-garden (*k*); and on the left, the bosquet (*l*), which is analogous to the modern shrubbery. To the north of the bosquet, there are a conservatory, summer-room, frame-ground, and reserve-garden (*m*), all of which will hereafter be explained more in detail; and, to the north of the kitchen-garden, there are the stable offices, cow-yard, poultry-yard, laundry, and drying-ground (*n*), as will also be hereafter more fully explained. There are the back roads (*o o*) to the offices, gardens, &c., proceeding from the main approach at *f*; so that no servant need ever have occasion to cross the court of honour, except when in attendance on his master, or to open the court gates. From the points *i i* and *p p*, and from the walks in the kitchen-garden, and in the bosquet, leading to these points, vistas are obtained along the avenues in the park, which are shown by the plan so distinctly, that it is unnecessary here to enter into further details. Two of these avenues (*q q*) are shorter than the others; but, to direct attention from this circumstance, which would show the boundary fence rather nearer than is desirable, obelisks are placed near their termination, at *r r*. Two other avenues (*s s*) are interrupted, by omitting the tines of trees in that portion of them which would pass the front of the house, land would have obstructed its view; but this interruption will not materially injure the effect from the points *p p* in the kitchen-garden and the bosquet. Certain avenues are carried through the arable lands exterior to the park or paddock, in order to show that the proprietor's property extends on every side. At *t t*, these avenues border the natural wood; but, in the other places, they are carried through the middle of corn fields, the ground under the trees being kept in grass. At *v v* are woods, that is, standard trees, with coppice-wood beneath them for the protection of game; at *w w* are hop-grounds; at *x x*, orchards; at *y y*, gardens to the entrance lodges, (not shaded by the trees of the avenues, as such gardens too frequently are, but fully exposed to the south); and at *z* is a single row of trees, which is carried along the public road as far as the property extends. The farmery in this design is supposed to be situated on the opposite side of the road, and is not shown in the plan; but several of the arable fields belonging to it are seen on the east and west boundaries of the park.

Fig. 276. shows the house, gardens, and offices of the residence just described, on a larger scale than in the plan *fig.* 275.

a, The court of honour, with three gateways, and an obelisk in the centre.
b, The stable-court, on the north side of which are a six-stalled stable, harness-room; with groom's sleeping-room over, and two coach-houses; and in the north-east angle there is a turret with a clock, a corresponding turret being formed on the opposite side of the court of honour. *c*, Cow-yard.
d, Open shed, for roots and other food for the cows in winter, and where calves or a pig may be fattened, and one or two donkeys kept.
e, Cow-house for five cows.
f, Wash-house, with laundry over. *g*, Fruit and root room.

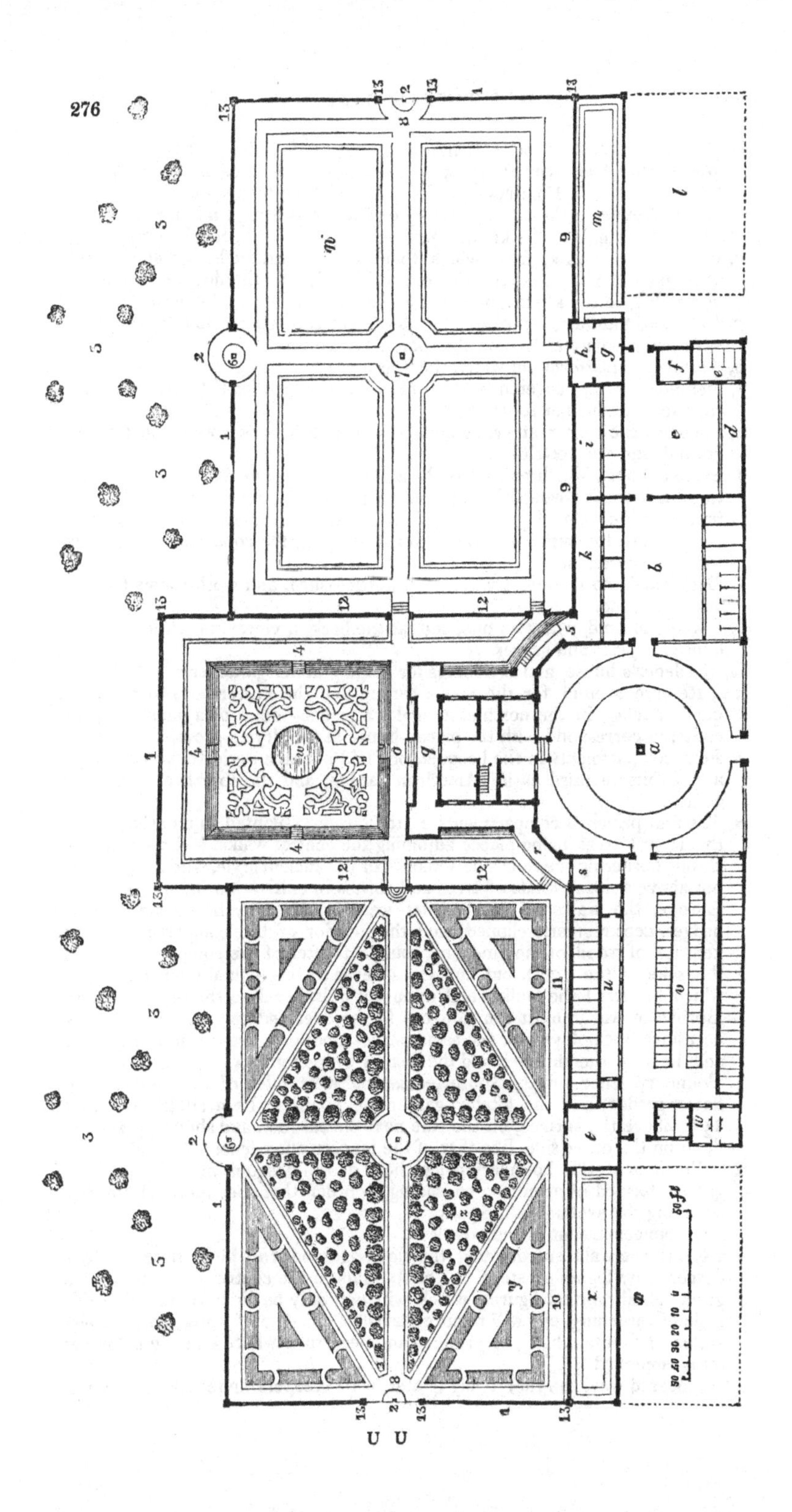

h, Banqueting-room, or summer-house, for eating fruit in, with rooms on the first floor. A cistern for supplying the fountains is immediately under the roof. *i*, Poultry-yard. *k*, Kitchen-court.
l, Drying-ground, in the north-west angle of which is a pigeon-house.
m, Reserve ground for the kitchen-garden. *n*, Kitchen-garden.
o, Grand terrace walk, from which there are several flights of steps; one ascending to the house terrace, one descending to the flower-garden, one descending to the kitchen-garden, and one descending to the bosquet.
p, Basin and fountain, in the centre of the flower-garden; the fountain supplied with water from a cistern on the roof of the house.
q, Terrace, or platform, on which the house stands.
r, Veranda, or way covered with a glass roof, which leads to the conservatory and the summer room (*t*).
s, Lobby to the conservatory, communicating with a back room, the forcing-ground, and a water-closet.
t, Summer room, which may either be a museum, a room for pictures or statues, a room for reading in, a school-room, or a banqueting-room. On the first floor are rooms for books or pictures; and immediately under the roof is a cistern for supplying water to the bosquet, the conservatory, and the forcing department.
u, Back sheds to the conservatory and potting-shed, and work-rooms for the forcing-ground.
v, Forcing-ground, with two pits, and, at the back, a vinery, peach-house, and house for pines and grapes.
w, Gardener's house, and two rooms for lodging under gardeners.
x x, Reserve ground for the flower-garden and the flower-beds in the bosquet; having, in the north-east angle, a pigeon-house, or a house for peacocks, to correspond with the pigeon-house in the drying-ground.
y, Four compartments in the bosquet, one laid out as an ericetum, another as a rosarium, a third as an American garden, and the fourth as a flower-garden.
z, The four principal compartments of the bosquet, planted as an arboretum; the largest trees being placed adjoining the central walks, and the smallest shrubs bordering the four side walks. In the ancient style, when there were not above a dozen kinds of trees and shrubs in cultivation in any garden in Europe, the walks of the bosquet were bordered by hornbeam or yew hedges, kept regularly clipped; and the interior of the compartments was kept full of wood, of the kind most abundant (often of the common hazel, for the sake of the nuts), but cut so as never to rise higher than the hedges which bordered the walks, and often cut quite level on the upper surface. Hornbeam was in most common use for bosquet hedges; but the yew was occasionally employed for the same purpose, both in France and England; and, in Italy, the phillyrea and the ilex, or evergreen oak.
1, Boundary fence on the south, east, and west sides of the bosquet, the flower-garden, and the kitchen-garden, supposed to be a sunk wall, rising 4½ ft. above the surface, on the side next the bosquet, and showing a sloping ditch on the other side, like that of the modern sunk fence.
2 2, Portions of the boundary fence, both of the bosquet and the kitchen-garden, formed of open iron palisading, in order that spectators within may see along the avenues in the park.
3 3 3, Commencement of the avenues.
4 4 4, Terrace walk surrounding the flower-garden, to which there are four descents by flights of steps. The fountain in the centre may consist of a group of allegorical figures, which will not only be in harmony with this style of gardening, but will require a smaller quantity of water, and less elevation of fountain-head, to produce an effect, than where a single columnar jet is employed.
5, A covered way, the roof being glazed, with creepers underneath, to corre-

spond with the veranda (*r*). Underneath this covered way there are two passages; one to the kitchen and other underground offices (which are all lighted from windows, only half the height of which is above the level of the platform), and the other up a flight of steps to the terrace; the object of this last passage being to admit of the master or mistress descending to the kitchen-court (*k*), and thence to the poultry-court (*i*), and to the fruit-room, banqueting-room, &c., by the most direct mode.

6 6, Situations for sun-dials.
7 7, Situations for statues.
8 8, Situations for obelisks.
9, Peach and fig wall, 12 ft. high, with broad border.
10, Conservative wall, 12 ft. high, with a narrow border.
11, Conservative border, for Cape bulbs, in front of the conservatory.
12, 12, Holly hedges, 10 ft. high, with stone piers, surmounted by vases, at the doorways, angles of intersection, &c. For the holly hedges, brick walls may be wholly or partially substituted, and covered with fruit trees or ornamental shrubs.
13 13, Stone piers, surmounted by vases.

277

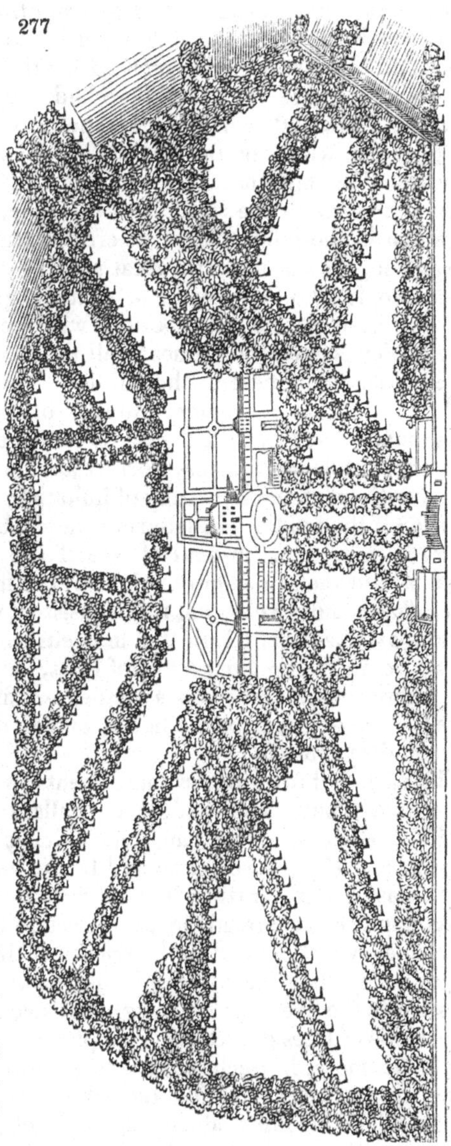

Fig. 277. is a slight sketch, showing a bird's-eye view of the whole place.

Remarks. In this design we have purposely omitted a feature common in the geometrical style; viz. a moat of water round the house, offices, and gardens; and the reason we have done so is, that we suppose the situation to be flat in a flat country, and, consequently, that so much stagnant water would be unwholesome. In many cases, also, a broad straight canal would, in ancient times, have been carried on from that part of the moat which was in front of the flower-garden to the boundary of the park; and this also we have omitted for the same reason. The bosquet, in some cases, would have had one of its compartments laid out as a labyrinth, which we have not shown, because we prefer devoting the whole of the bosquet to a collection of trees and shrubs. We would not, however, omit the labyrinth, not only because it is one of the features of the style, but because we observe that in those few places where labyrinths at present exist in England, they are always great sources of amusement to the young people who are allowed to visit the grounds. Every one who has been at Hampton Court will recollect the labyrinth there, which is open at all times to all the world, and is the source of perpetual amusement to the public. There is also a very well kept labyrinth at Chevening, the seat of Earl Stanhope, where the grounds have been laid out by the present earl, chiefly in the ancient style, and in very correct and appropriate taste; and where, with an example worthy of imitation, they are at all times open to all the public. A labyrinth might be introduced at *w* in *fig.* 275. in p. 655., instead of one of the hop-grounds. The woods *v v*, in the same figure, might be cut into alleys, stars, and *pattes d'oie*; and it would not be inconsistent with this style, if the two side avenues (*c c*) terminated in obelisks. Directly in front of the house, in the boundary row of trees, or immediately within it, a column surmounted by a statue, or an obelisk, might be placed, as an object from the house, and to divert attention from the boundary fence.

The park is thrown into compartments by the intersections of the avenues; but, as no hedges or walls are supposed to exist within it, there is a free range for animals and sportsmen over the whole. If, however, it should be desired to make the most of the pasture, by grazing it with cows or sheep, the compartments might be separated, in the direction of the lines of trees, by hurdles, which, joined to the trunks of the trees, and partially concealed by their heads, would not be in the slightest degree offensive to the eye. In planting these avenues, there are only a few kinds of trees that could be employed; such as the oak, elm, beech, sycamore, horsechestnut, sweet chestnut, and a few others. Some might prefer planting the whole with the black Italian poplar, which would produce a splendid effect in ten years after planting; others might prefer the Lombardy poplar, which would also soon produce a striking effect, and which would do very little injury to the pasture below. The young trees should, if possible, be procured from the nurseries with clear stems, not less than 6 or 8 feet high, and not less than 2 in. in diameter at the collar. To insure the possession of plants of this size, it may be advisable to order them a year or more beforehand, or to procure them, and rear them in a private nursery in the ground, till they are sufficiently large for planting out. Whichever mode is adopted, it must be borne in mind that all the trees of the same kind intended for avenues must be of the same size, and of the same vigour of growth, when planted where they are finally to remain; and, also, that no more than one kind of tree can be planted in each avenue. The reasons for these dicta are sufficiently obvious; viz., to insure uniformity of progress and appearance in every part of each avenue. For the same reason, the soil, in every part of an intended avenue, must be trenched, or otherwise prepared, to the same depth; and, by the addition of manure or compost, rendered every where of the same quality. The trees, after being planted on little hillocks, may be protected in Mr. Lawrence's manner, already described (p. 555.); or, if it be considered less expensive, each avenue may be enclosed by lines of fencing along its exterior sides. The kitchen-garden will be in strict accordance with the ancient style if surrounded by a hornbeam, beech, holly, or yew hedge; but stone and brick walls were also used in this style, and, in modern times will doubtless be preferred, on account of the facilities which they afford of growing the more delicate fruits. No slip is shown round the kitchen-garden, because, in the plans, the boundary fences are hedges with stone piers; but the introduction of a regular slip round the walls of the kitchen-garden has nothing in it inconsistent with the geometrical style, and may therefore be adopted at pleasure. The compactness of all the garden scenery in the design before us, and in the geometrical style generally, is attended with some economy in management, because the whole is more immediately, and at all times, under the eye of the master, every part being overlooked from the window of the house; and it may also be more conveniently examined by the female part of the family, and by invalids.

Kenwood, the Seat of the Earl of Mansfield, at Hampstead. (*figs.* 278. to 288.)—This is, beyond all question, the finest country residence in the suburbs of London, in point of natural beauty of the ground and wood, and in point also of the main features of art. The park may be said to consist of an amphitheatre of hills; the house being situated on one side, backed by natural oak woods rising behind it, and looking across a valley, in which there is a piece of water, to other natural woods, also chiefly of oak, which clothe the opposite hills: and which, combined, give the name to the place; *ken*, being derived from *kern*, the ancient British name for an acorn. In consequence of this natural disposition of the grounds, and of the woods, all exterior objects are excluded; and a stranger walking round the park would never discover that he was between Hampstead and Highgate, or even suppose that he was so near London. It is, indeed,

278

View on entering the Gates at Kenwood.

difficult to imagine a more retired or more romantic spot, and yet of such extent, so near a great metropolis. This impression is felt the moment we enter the gates, and proceed along the the approach, in a hollow between deep banks, rendered dark by overshadowing trees, as shown in *fig.* 278. The impression is not lessened when we come within sight of the house, a part of the entrance front of which is represented in *fig.* 279. ; or when, passing through a walk covered with trelliswork, in the flower-garden, to the lawn front, we look down the declivity to the water, at the foot of the rising woods on the opposite bank, as shown in *fig.* 280.

Kenwood is one of those places of which but a very imperfect

279

View at Kenwood, looking towards the House.

idea can be given by a ground plan, and little more than some notion of the inequalities of the surface, by geometrical sections of the ground, and sectional views. The beauty of the oak trees, and the simple and sylvan grandeur of the scene, can only be represented by landscapes on a tolerably large scale; or best of all by a panoramic view, taken from a central point in the lowest part of the grounds.

The ground plan (*fig.* 281.) is taken from a published map of the parish of St. Pancras, in which the property lies, with some additions, trees, &c., from memory. The sections (*figs.* 282. to 285.) are also from memory; but the views (*figs.* 278. to 280. and *figs.* 286. to 288.) are faithful copies from nature. The sectional views are far from doing justice to the beauty of the grounds, but these, together with the views, may serve to convey to the reader some general ideas of the style of beauty which prevails in the grounds at Kenwood.

Proceeding from Hampstead on the road towards Highgate, at the end of Hampstead Heath we arrive at the tavern known as the Spaniards (*fig.* 281. *a*), where the Kenwood property com-

280

mences. Here there is a considerable descent along the road, while the ground on the right hand rises gently, and that on the left falls considerably. At *b* is a lodge, which leads to the farm offices, which are in the form of an octagon, and were built from a design given by the celebrated agriculturist Marshall. The arable farm lands lie chiefly on the opposite side of the road. At *c* is the principal approach to the mansion; and at *d* there is another lodge, which may be called that of the Highgate approach, and also that of the road to the domestic and stable offices. The following are further details of the plan:—

e, The house. *f*, The domestic offices. *g*, The gardener's house.
h, The stable offices. *i*, The kitchen-garden. *k*, The flower-garden.
l l, Terrace walk.
m m m, Wire fence, separating the mown ground from the sheep pasture.
n, Lime tree avenue, through which the terrace walk passes.
o, Main walk, from no one point of which is any object seen that does not belong to the woody scenery of the park.

281

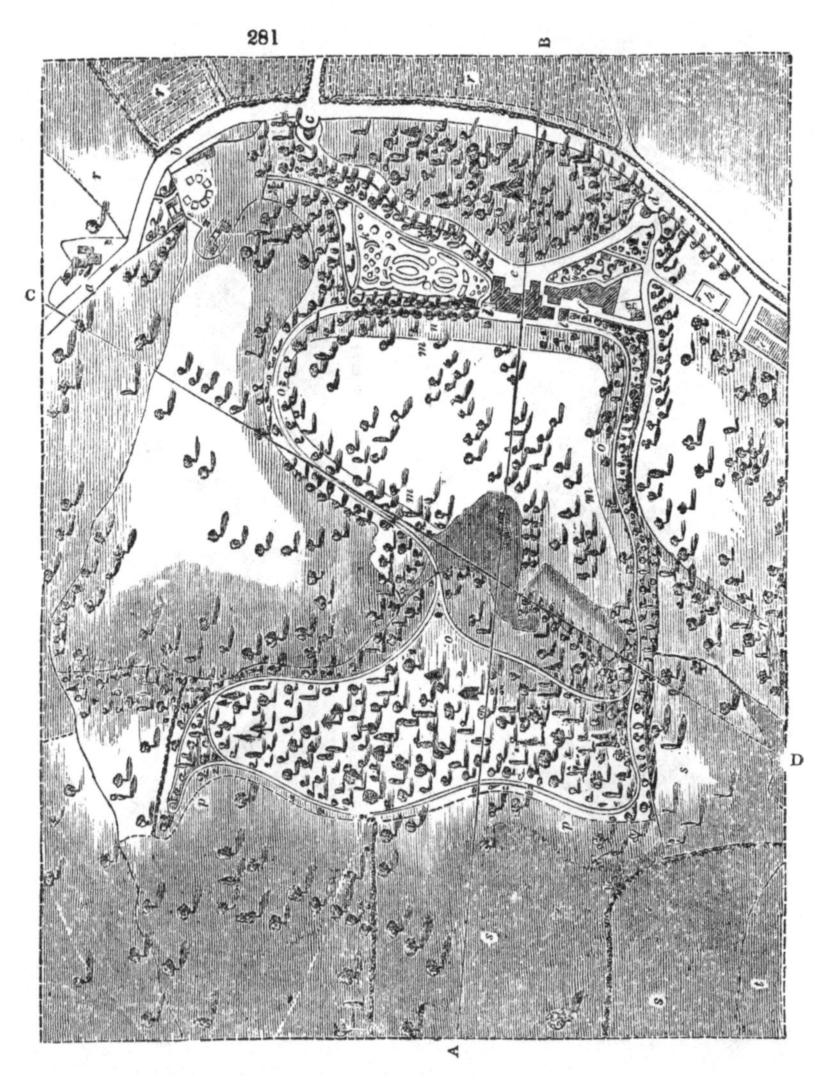

p p, A broad terrace walk of turf, or rather moss, overhung by immense trees, on the outskirts of the park, and from some points in which magnificent views of London are obtained.

q q, Private approach road, commencing in Kentish Town, and continued upwards of 2 miles entirely through the Kenwood property. This approach, if widened, and properly planted, would form the noblest avenue to a gentleman's seat in the neighbourhood of London, and come nearer to the magnificent idea of the Duke of Chandos, of having a straight avenue from Chandos House, in Cavendish Square, to Canons House, near Edgeware, than any other that we have heard of.

r, Arable lands occupied as the home farm. *s*, Grass lands let on lease.

t t, Ponds, supplied from the ponds in the park.

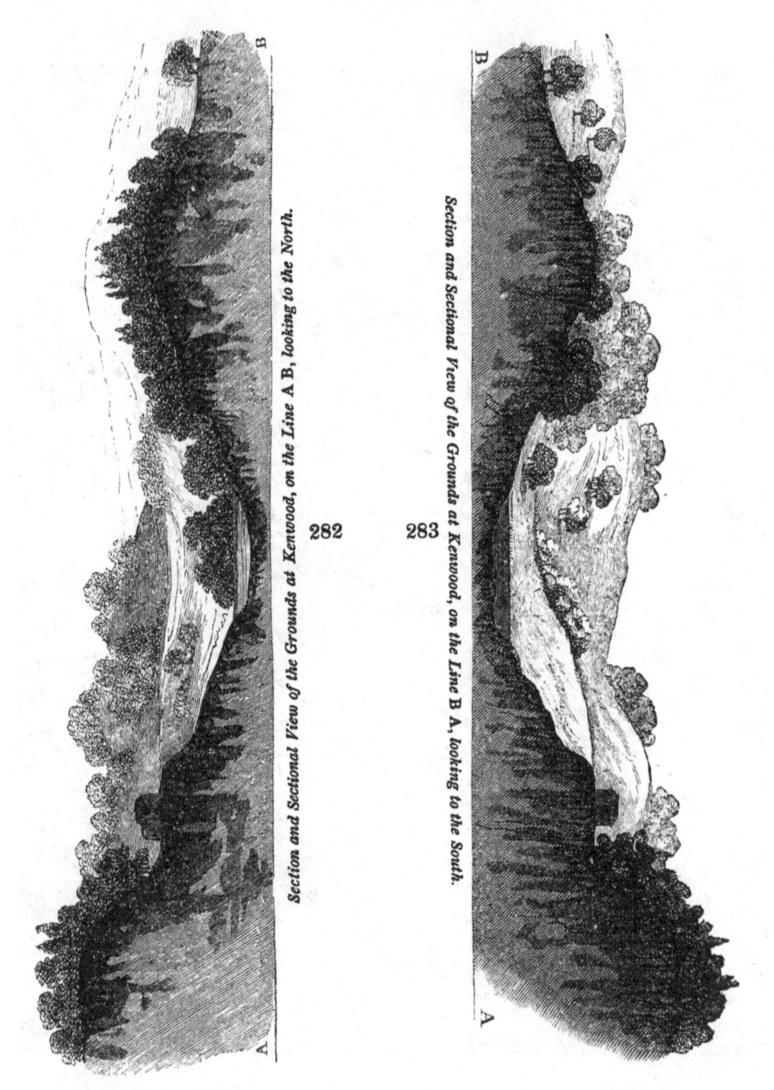

282 Section and Sectional View of the Grounds at Kenwood, on the Line A B, looking to the North.

283 Section and Sectional View of the Grounds at Kenwood, on the Line B A, looking to the South.

Fig. 282. Section and sectional view on the line A B in the plan, looking to the north.

Fig. 283. Section and sectional view on the line B A, looking to the south.

Fig. 284. Section and sectional view on the line C D, looking from the house.

Fig. 285. Section and sectional view on the line D C, looking towards the house.

The house, the kitchen offices, and the flower-garden, stand on a nearly level platform, which is gradually united to the rising

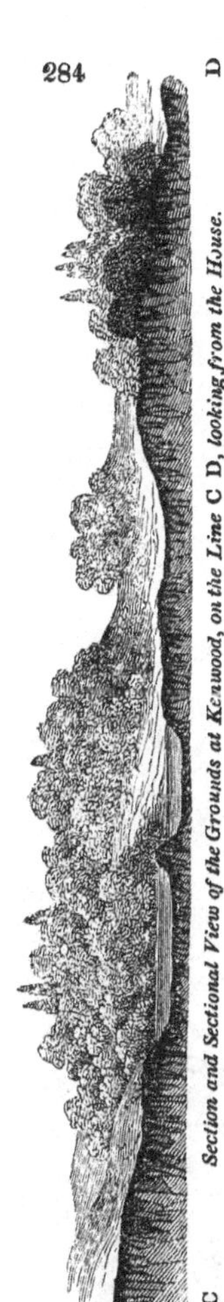

284 Section and Sectional View of the Grounds at Kenwood, on the Line C D, looking from the House.

285 Section and Sectional View of the Grounds at Kenwood, on the Line D C, looking towards the House.

ground on the approach side, and terminates on the lawn front, in a broad terrace walk, which is carried along the brow of an artificial slope, slightly indicated in the plan. This terrace walk is one of the finest artificial features of Kenwood. It is 20 ft. wide in front of the house; and this width is continued both to the right and left, as far as the walk remains in a straight line; it then becomes imperceptibly narrower, till, in the lowest parts of the grounds, where it passes the ponds of water, it is no more than 8 ft. wide; and it is continued at this breadth through the woods. In consequence of this terrace walk being so much above the wire fence which separates the mown ground from the pasture ground, the fence is never seen till we begin to descend the declivity, when, a good impression having been made by its absence in front of the house, it is not in the slightest degree offensive. Indeed, we scarcely ever knew a place where, from the existence of so high a terrace, there was so slight an appearance of fencing or confinement. But the great value of Kenwood to a landscape-gardener, is the perfect unity of expression which prevails in the views obtained in every part of the grounds. These views, though all are equally decided in expression, are of three distinct kinds: first, the views from the entrance front of the house along the approaches, and those along the approaches to the entrance front; secondly, the views in the flower-garden, which is surrounded by trees on every side except that next the house, and the views in which are, consequently, confined to flowers and flower-beds; and, thirdly, the views from the lawn front of the house to the grounds, and from the grounds to the lawn front, in which there are no flowers to be seen. If it were desired completely to spoil the characteristic beauty of Kenwood, as far as could be done without removing any of the trees, the way would be to place beds of flowers along the terrace

286

View at Kenwood, from the Terrace Walk near the House.

walk, and on each side of that walk, from the commencement of the terrace to the end of the wood. It argues great good taste in the proprietor that this has not been done; for nothing can be more natural than for a gardener, who wishes to make the most of the pleasure-ground, to think of doing so by adding more flowers. A person must have acquired some feeling for general effect, before he can fully understand where flowers will be useful, and where they will be injurious, in a scene. *Figs.* 286. and 287. are views of the grounds from the terrace walk near the house; and *fig.* 288. is a view of the house from the main walk in the bottom, near the water.

When Flower-beds may be introduced on the Lawn Front, and when they are better omitted. There are some places where the scenery, as viewed from the lawn front of the house, has little or no natural expression or character, and cannot be made to have any by the proprietor, either in consequence of a flat surface, or of the intervention of other property. There are other residences, in which the view from the lawn front is so confined, that it scarcely can be made to form a whole; and some, where, from offensive objects, the view from the lawn front may be positively disagreeable. In these and similar cases, that is, wherever there is no marked expression, or an indifferent or bad expression, flowers and flower-beds may be introduced in the foreground of the lawn front. On the other hand, wherever the view from the lawn front has a decidedly marked expression that is agreeable, whether by its grandeur, its picturesque beauty, or its peaceful rusticity, flowers in the foreground ought to be avoided.

As examples of what may be called grand views from the lawn front, we may mention Syon House, in the neighbourhood of London; and, if we recollect the situation correctly, Dreghorn Castle, in the neighbourhood of Edinburgh. As an example of picturesque beauty, none, that we know of, in Britain, can be compared to Redleaf, the seat of William Wells, Esq.; and, as an example of peaceful sylvan beauty, nothing can surpass Kenwood. There are many places in the neighbourhood of London in which flower-beds are placed immediately under the drawingroom windows, where they would be much better omitted; and others, where, though they may not be required, and are not introduced, they might be so without destroying any expression better than that which they would give; but, on the contrary, creating an interest, which could not be given by any other means. Bedford Lodge, Camden Hill, may be referred to as an example of the use of beds of flowers on a small scale; and Chevening, in Kent, as a similar example of their judicious employment on a large one.

It may be laid down as a general principle, that it is always more or less dangerous to introduce flowers in the foreground,

287

View at Kenwood, from the Terrace Walk near the House

when there is an extensive distance as a termination to the view; and, on the contrary, that, where there is little or no distance, and a view limited in extent, flowers may generally be safely introduced. To avoid error in these and other similar matters, a person must either have studied the subject so far as to be able to exercise his reason on it, or he must have a natural feeling or taste for the beautiful in landscape. These remarks will not be without their use, if they induce persons to think before they introduce flowers into particular parts of pleasure-grounds where no flowers have been before; and to examine whether flower-beds already existing might not be better removed. It has often struck us with surprise, that the proprietors of the finest residences in England, noblemen and gentlemen of high education and refined taste in other things, possessing collections of the finest pictures, and whose eyes must consequently be familiar with all that is noble and beautiful in landscape, should yet commit the laying out of their grounds to their gardeners; or, at all events, permit them to make alterations and additions in whatever relates to flower-beds, flowering shrubs, and rockwork; forgetting that the life of the gardener has been devoted to the study of the culture of plants, and not to that of the composition of forms, and their effect in landscape scenery. Hence it is that many of the most beautiful places in England are at this moment disfigured by flower-beds, either placed where there ought to be none, or put down of such shapes, and in such a manner, as neither to form a whole among themselves, nor with the other objects near them. How rarely do we find pieces of rockwork, or rocky cascades, in England, which a man who had profited by the study of pictures could take pleasure in looking at? It is clear to us, that the possessors of pictures in general derive very little benefit from them, as regards the improvement of their taste in landscape. How few landed proprietors can, like the late Sir Uvedale Price, and the present William Wells, Esq., of Redleaf, transfuse the spirit of the finest landscape into the artificial scenery which they create in their grounds? Many country gentlemen are in the habit of having artists at their houses, to take portraits, views, &c.; and these being, in many instances, the guests of the family for weeks together, we often wonder how it happens that they do not point out the grosser errors of want of connexion and unity of expression, with which they must so frequently be shocked in passing through flower-beds and pleasure-grounds; but we suppose that gentlemen do not think of asking the opinion of a landscape-painter on any point connected with gardening; forgetting that the composition of forms is the business of the landscape-painter, and that his eye has been educated by a long course of study and observation, so that he can detect what is right or wrong at a single glance. There are some

288

Kenwood House, from the main Walk in the Bottom, near the Water.

proprietors who have studied the subject themselves, or who, fortunately knowing their own ignorance of it, have had the wisdom to consult such artists as Gilpin, Nesfield, &c.; and we only wish that those who do without such aid could see their places as they are seen by men of real taste.

To return to the subject of flowers and flower-beds, we may remark that the flower-garden at Kenwood is the only defective part of the place. It is naturally shaded and confined by a lofty lime tree avenue on the one hand, and by a rising hill of oak wood on the other; and the area of the garden contains by far too many small trees and shrubs among the flowers: in consequence of this, the turf is almost always damp on the surface; and the flowers come up with slender and etiolated stems, and pale colours. Most of the flower-beds, also, are too large; and they do not combine so as to form a whole. Were it ours, we should clear the whole area, and lay out a new combination of figures, chiefly along the centre, planting them solely with flowers, and keeping between them and the boundary a broad margin of turf, so as to insure that airiness, dryness, and sunshine, which are at present so much wanted.

The Variety of Trees and Shrubs in the Grounds at Kenwood is not very great; nor is it desirable that it should be so, except in the more secluded parts of the place, where they would not interfere with the general effect. Adjoining the flower-garden, and bordering a walk which leads from it to the dairy and farm, a number of new species of ligneous plants have recently been introduced, and a small pinetum planted. This walk, from the botanical variety which it exhibits, forms an agreeable contrast to the main walk on the lawn front of the mansion, though along that walk, also, there are a few very fine specimens of foreign trees and shrubs. Of some of these; such as a cedar of Lebanon, which, in 1836, was 90 ft. high, and which was planted by the celebrated Judge Mansfield, with his own hands, about ninety years before; a larch of the same age and size; and a Robín*ia* Pseùd-*A*càcia; we have, with the permission of the present earl, given portraits in our *Arboretum Britannicum.* The oak woods, which are probably the oldest about London, are remarkable for being composed almost entirely of *Q*uércus sessiliflòra.

The Farm and Farmery are conducted in the Scotch manner, under a Scotch bailiff, who raises admirable crops of turnips, potatoes, and clover, the soil being a deep sandy loam. *L*upìnus polyphýllus has been tried here, as an herbage plant, with success.

Remarks. Kenwood, being at no season of the year shown to strangers, we regret to think that so few of our readers will have an opportunity of studying there the effect of unity of expression in landscape, and of feeling the powerful impression made by scenery so decidedly simple, rural, and sylvan, in the immediate neighbourhood of London. The contrast is powerfully felt, not only between this place and a crowded city, but between it and the extreme artificialness of most other suburban residences. Gardeners, however, can always visit gardeners, and they may profit from perusing these remarks, and comparing them with the impression made on them by a visit to Mr. Cockburn, the gardener at Kenwood. One grand cause of the beauty of Kenwood, though it is one that scarcely admits of imitation, consists in the prevalence in it of natural oak woods, and the manner in which they are displayed by the hilly and undulating surface of the ground. The same extent of wood on a flat surface could never have presented more than a side view to the eye of a spectator walking through the grounds; and the beauty of the individual trees in the interior of the wood must, consequently, have been entirely lost. Whether a wood on a flat surface were a mere strip, or a mile in depth, the effect to a stranger would be the same; but in the amphitheatre at Kenwood the trees are raised one above another, they are in no part crowded together, and not only display great extent of wood as a whole, but a degree of grandeur and beauty in the individual trees, which they could not exhibit on any other character of surface. Hence the impossibility of conveying an equal expression of naturalness and sylvan grandeur in any place of smaller extent, or in any place (whether large or small) having a flat surface.

CHAP. IV.

SUPPLEMENTARY DETAILS.

In the preceding pages we have, agreeably to the plan on which we set out, introduced the discussion of principles and the details of execution, occasionally, and apparently incidentally, as we thought they were most likely to make an impression on the uninitiated reader. On looking over the work, we find we have omitted no topic of discussion of much importance, and few details except such as we have stated should be given in a future chapter. The present is that chapter, and in it we propose to supply everything that is wanted to render the work what it professes to be, viz., scientific instructions, popularly explained, for choosing, laying out, planting, and managing a suburban residence. We shall commence these supplementary remarks with the dwelling-house, including the bath-room and plant cabinet; and take in succession the domestic offices, including the stable offices; the farm buildings including the cow-house, dairy, poultry-yard, rabbit-house, and piggery; the aviary, apiary, fish-ponds, and icehouse; and, lastly, the residence considered as a whole.

SECT. I. *The Dwelling-house.*

IN looking over the contents of Chap. II. Sect. IV. p. 82., we find we have said nothing of the bath-room or of the nursery, on each of which we now propose to submit a few hints; and we shall add a few observations on the use of having a gentleman's dressing-room on the ground floor.

The Bath-room, when for the use of the family generally, may be on the ground floor, or on the bed-room floor of the house, as may be most convenient; but a bath should never be placed either in a bed-room or in a sitting-room, because, when a hot bath is used, the quantity of steam produced fills the atmosphere of the room, and, condensing on whatever furniture it may contain, necessarily renders it damp. The advantage of having the bath-room on the ground floor is, that hot water may be more conveniently supplied to the bath from the cistern at the back of the kitchen fireplace; and the advantage of having it on the bed-room floor is, the convenience and saving of time in using the bath in the morning before dressing, or at night before going to bed. Where there are dressing-rooms to the bed-rooms, a bath may be placed in each of these, or in such of them as may be thought necessary; and the water may be heated in the room by a fireplace and small boiler, properly constructed for that purpose; or it may be pumped up from the boiler at the back of the kitchen fireplace, or brought down in pipes from a boiler placed in a room in the floor above, or in a garret. (See *Encyc. of Cott. and Villa Arch.*, § 1858.) *Fig.* 289. shows a bath-room placed behind

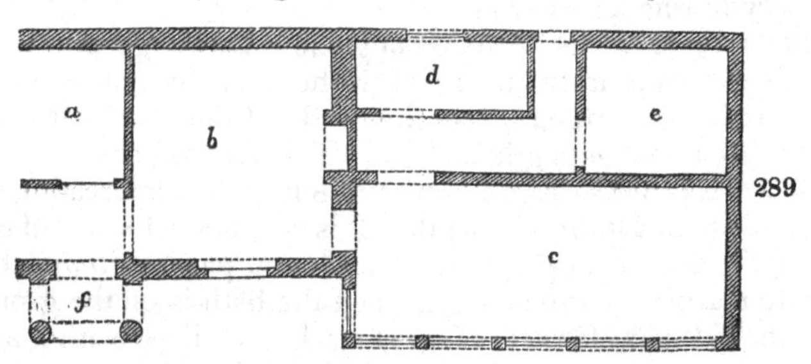

a conservatory, the bath being heated by the same fire, and from the same boiler. In this plan, *a* is the dining-room; *b* the drawingroom, from which there is a glass door into the conservatory (*c*); and *d* is the bath-room, between which and the conservatory there is a passage connecting it with the boiler-room (*e*). The entrance porch of the dwelling-house is shown at *f*. The boiler is placed in the room *e*, on a level with the bath as it stands in the bath-room; and the bath is heated by a coil of pipes at one end. The water of the bath might have been heated in a simpler and less expensive manner by causing it to circulate directly through the boiler; but, as the water in a boiler which is in constant use for heating a conservatory, or other building or apartment, generally contains a good deal of sediment, it is not fit for using as a bath.

The size of the bath-room need not be large, because it is not understood to contain a swimming-bath, but only one for immersion. For this purpose, as the bath need not be larger than 8 ft. in length and 3 ft. in breadth, a room 8 ft. by 10 ft. in the clear might suffice, the bath being along the narrowest end, the fireplace on one side, a window on that opposite, and a door on the side opposite the bath. A room 10 ft. or 12 ft. square, however, will be more convenient; and, if it is to contain a vapour-bath and a shower-bath, as well as a common water bath, it should be somewhat larger. The height of the room, if it is to contain a shower-bath, should not be less than 10 ft.; but, for a common bath or vapour-bath, 7 ft. or 8 ft. will be sufficient; and in this case the bath-room may be placed as a mezzanine, or half-floor room, like the *entresol* of the French. Adjoining the bath-room, and as an anteroom to it, there should be a dressing-room with a fireplace, and this fireplace may be so arranged as to heat the water in the bath-room. There should be a bell in the dressing-room, communicating with a bell-pull, suspended from the wall at one side of the bath, near the head; it may also be convenient to have a bell appropriate to the bath-room hung along with the other house bells, because it sometimes happens that invalids are seized with cramp while in the bath, and find it necessary to ring for assistance.

The bath, or vessel for containing the water, is generally half or a quarter sunk in the floor; or, if the floor does not admit of this, there is a step raised to half the height of the bath in front, in order to facilitate getting in and out. The dimensions of the bath are from 6 ft. to 8 ft. in length, from 1 ft. 8 in. to 3 ft. in breadth, and from 1 ft. 6 in. to 2 ft. 6 in. in depth. It is commonly formed of inch boards, lined with copper or sheet-lead, and painted in imitation of white marble; but sometimes, when the bath is on the ground floor, the sides are formed of brickwork, and lined with glazed tiles. Sometimes, also, they are formed of polished stone or marble, or slabs of slate. The sides of the bath are generally made perpendicular, or nearly so or, somewhat wider at top than bottom; the end for the head generally slopes at an angle of from 45° to 50°, but the foot is commonly perpendicular, or slightly sloped, like the sides. One pipe from the boiler or other source of hot water, and another from the cistern or other source of cold water, are introduced at the back of the bath; but so as not to project into it, the cocks being placed against the wall, with their nozzles turned outwards, so as barely to come over one of the sides of the bath. A plate should be fixed in the wall beside each cock, the one inscribed with the word "hot," and the other with the word "cold;" and the distance of the cocks from the head of the bath should be such, as to allow a person while seated in the water to reach them without inconvenience. In the bottom of the bath, at the foot end, there must be a waste-pipe for emptying the bath, which may either be stopped with a plug, having a grating beneath to prevent the escape of any matters which would choke up the pipe; or the waste-pipe may terminate under a grating or plate pierced with holes in the floor of the bath, near which there may be a stopcock, concealed by a small trapdoor, by turning which the water may be let off at pleasure; or, what is preferable, if the situation admits of it, the opening of the stopcock may be in the floor of the room at the foot of the bath.

The simplest Form of a Vapour-Bath is to have an open boiler with a fire beneath, and over the boiler a grated platform, on which a chair is placed for the bather to sit upon, while a large cloth or sheet is thrown round him in such a manner as to enclose the whole of the grated platform, and every part of his body except his head. The water being kept in a boiling state, the steam ascends under the cloth, raising the temperature of the bather to a high degree. If thought advisable, herbs may be thrown into the boiler, which will communicate their odour to the vapour.

A Shower-Bath is always of cold water; and, as it produces no steam, when of the common portable kind, it may be used in any room that is most convenient. When fixed, there may be a supply-pipe, with a stopcock above it, from some adequate source. When this is the case, the shower-cistern may be placed over the middle of the common bath, and the discharge-pull may hang down close by the wall.

Heating. A bath in a bath-room may be heated in various ways. 1. There may be a fireplace in the room, or in the ante-room, enclosed in a small copper boiler, which need not be much larger than a teakettle; and from this boiler there may be a going and returning pipe to the bath, in a casing at one end of which the pipe may form a coil. The quantity of cold water required, being let into the bath, water put into the boiler, and the fire lighted, the water in the bath will be heated by the coil to any degree which may be required, short of the boiling point. 2. There may be a fireplace at one end of the bath, to communicate heat directly to the water, without the intervention of a boiler or a coil, on the principle adopted in the common portable baths. 3. If there be a boiler at the back of the kitchen fire, and the bath is below the level of that boiler, the hot water may be introduced into the bath by a pipe and cock; but, if the bath should be above the level, then, if the height does not exceed

30 ft., the hot water may be pumped up by a small pump in the bath-room. 4. There may be a cold-water cistern either in the bath-room, or in any room over or adjoining it, provided it is a few feet higher than the bath, from which a pipe may be conducted to the boiler behind the kitchen fire, where it must form a coil in the water, and from which it must afterwards be conducted back again to the bath. Whenever a warm bath is wanted, in this case it is only necessary to raise the plug in the cold-water cistern, so as to permit the cold water to rush down the pipe to the coil in the hot-water boiler, where, being heated by the time it takes to pass through the coil, it will rise up to the bath moderately warm. In this way, a person, without troubling servants, may take a warm bath at a minute's notice, at any hour of the day or night: all that is essential is, that the boiler at the back of the kitchen fire should be kept constantly filled with water, as the fire requisite for cooking will be quite sufficient for keeping that water warm. (See *Encyc. of Cott. Arch.*, § 1749.) 5. There may be a fireplace and boiler in any room on a higher level than the bath, and from the boiler there may be a pipe direct to the bath. In general, however, such a fireplace and boiler should be placed in the floor immediately over that in which the hot water is to be used; because, if placed much higher, the pressure upon the delivery-cock will be so great as to cause it to leak. This we consider to be by far the best mode of supplying hot water to bed-rooms and dressing-rooms, a pipe being conducted from the boiler to each place where the hot water is wanted. Indeed, hot water, in this way, may be distributed over the whole of any house, provided the delivery cocks are of the conical kind invented by Siebe*, and of

* *Siebe's Self-pressure Cock* (*fig.* 290.) deserves to have a preference not only wherever there is any great pressure of water, but even in the case of common boilers and cisterns. It combines security against leakage, simplicity

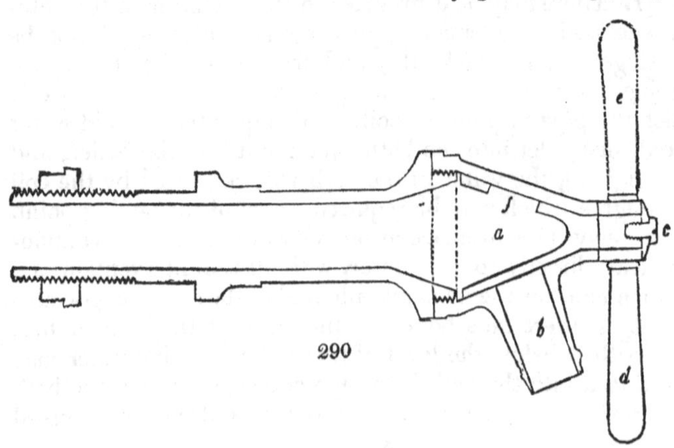

290

sufficiently good workmanship to prevent leakage; or that there are small intervening cisterns with ball-cocks in the middle floor of the house, to receive the hot water for the lower rooms, from which it would be drawn off as wanted. It is needless to add that cold water might be distributed all over a house exactly in the same manner. (See *Encyc. of Cott. Arch.*, § 1857. 1861.)

Where a cold bath only is wanted, it may be convenient, in some cases, to introduce it in a bed-room or dressing-room, half sunk in the floor, and concealed by a sofa, or by the bed being drawn over it; the bath, in either case, having a closely fitting wooden cover; or it may be raised above the floor, and disguised as a commode, chest of drawers, or dressing-table. Where a hot bath is only wanted occasionally, one of the portable kind, such as are sold by ironmongers, with a small boiler and fireplace placed at one end, and pipes to conduct the smoke up the chimney, is the most economical mode; but still such baths should not be used very often in a bed-room, on account of the condensation of the steam from the heated water upon the bedclothes.

A Gentleman's Dressing-room on the ground or living-room floor is a great convenience in the case of small houses. Where all the members of a family dress regularly every day before dinner, and where, when a stranger guest arrives, he can, if necessary, be shown to a spare bed-room, a common dressing-room may be dispensed with: but, in the case of families who live in a humbler style, when a guest arrives unexpectedly, or when he is asked to dinner, after having been on horseback, or walking in the dust perhaps, the greater part of the morning; or when he has been engaged for some hours before dinner in examining books, plants, or minerals, or other objects in a library or museum, or, perhaps, gardening with you; a common dressing-room is a very great convenience. If possible, there ought to be a

---

of construction, great durability, and facility of repair should it at any time go wrong. "The plug (*a*) being a hollow cone, the sediment from the boiler or cistern rests inside the cone, instead of against the wearing part of the plug (as with the common cock); and, immediately on opening the water way (*b*), it is washed through, without injuring the barrel or plug. As the water is constantly pressing on the inside of the cone or plug, consequently the greater the pressure from the boiler or cistern, the tighter the cock holds, which is the reverse with all other cocks. This conical plug is not liable to stick fast, as it gives way to the expansion of the metal occasioned by heat. As the plug wears, so it tightens; and, having but one opening, it has a greater lock than those in common use. The screw in front (*c*) is for the purpose of adjustment when the plug turns too easily. The water flows, on lifting up the handle (*d*) to the position shown by the dotted lines (*e*), which brings the two openings (*b* and *f*) in contact. These improved cocks are made (by the inventor) of the best gun metal, and not of the common pot metal (which contains a large proportion of lead), and are sold at prices varying from 9*s*. to 1*l*. 1*s*. each. (*Gard. Mag.*, vii. p. 85.) This cock can be taken to pieces to clean or repair without removing the shank from the boiler or cistern, which is requisite with a common cock, and is at all times inconvenient.

water-closet attached; and, as in this case there would be a cistern, a pipe from it would supply cold water to the wash-hand stand. In some families, this dressing-room might serve also as a bath-room; and in others, perhaps, as a smoking-room. Whatever may be its uses, it ought to have a fireplace; and, if it be intended for a bath-room, a small boiler should be fixed on one side of the fire.

The Nursery is a room set apart for the children, till they are three or four years of age. In first-rate houses, a room may be attached, to serve as a place for cooking the babies' food, and washing and drying their clothes, and washing and bathing all the children. There should be a constant supply of cold water from a cistern not liable to be frozen, and of hot water from a boiler at one side, or at the back of the fireplace in the room. Adjoining the nursery, there should be a water-closet conveniently fitted up for infants; and, if there were a gallery, a loggia, or veranda, open in front, but covered above, in which the children might be carried backwards and forwards, or allowed to run about in the air when they could walk by themselves, so much the better. Children are always particularly fond of playing in a balcony or veranda, and often, when they are unwilling to try to walk in a room, will be delighted to run up and down there.

The School-room for the older children may join the nursery, and should be sufficiently large for the children to take their dancing lessons in it, and to serve for them to play at battledore and shuttlecock in, or take other exercise during inclement weather.

An additional Room, either as a sitting-room, a place of business or as one of the domestic offices, is often wanted to houses in the country: and sometimes there is great difficulty in knowing where to place it. By reflecting on the various plans which have been had recourse to in cases of this kind, the difficulty may frequently be got over without doing violence to the elevation of the house in point of effect; but very commonly such additions are more deformities than beauties. As hints, we may observe that we have seen a billiard room disguised as a conservatory, as a lodge to the entrance gate, as a summer-house, and, in one instance, as a stable. Cellars, and underground offices, are generally easily obtained; because, at the most, the surface of the ground over them only requires to be raised. Bed-rooms may be obtained in a detached tower, connected with the main body of the house by a glazed veranda, which may serve as a conservatory or green house; and so on.

The Plant Cabinet has been already slightly mentioned in several places. It may be described as a case framed in wood, or of wood and iron, and glazed on the sides (or at least on the top and one side); and placed exterior to the house, and communicating with it by some door or window. As it is chiefly intended for preserving plants which have been brought forward elsewhere, it is of no great consequence whether it is placed against the north, west, or east side of a house; though the south and south-east sides are doubtless the most favourable, and the north-east the least so. In street houses, it is often very conveniently projected from a staircase window, either on the parlour floor, or the floor above. It is also sometimes joined to the back parlour, or placed over the entrance porch; and, occasionally, it forms a projection, supported on pillars, from the back drawingroom. Very commonly, it is formed on the flat roofs of some attached out-buildings, such as a back kitchen, wash-house, or rubbish place. In short, there is no situation where there is a door or a window in the house, and where perpendicular light is obtainable outside, in which a plant cabinet may not be formed. However irregular the plan may be in point of outline, and however uneven the roof or roofs which are to form the floor, the situation is still eligible for a plant cabinet, or a small green-house; on the same principle as an irregular piece of ground is for laying out a flower-garden in the picturesque manner. In the plant cabinet, as in the flower-garden, the whole depends on the contrivances for displaying the flowers. The great art in arranging an irregular plant cabinet consists in the disposition of wires and rods, in the form of trelliswork, arches, and arcades, for climbers; and of imitations of rockwork, banks, or benches of stones, for receiving bushy or creeping plants in pots, such as pelargoniums, mesembryanthemums, &c. The rockwork, banks, benches, &c., may be made of bricks and cement, stained or dashed with paint in such a manner as to represent different kinds of stone or spars; or natural crystallisations of different kinds may be procured. The smallest, the most irregular, and apparently the most unfitting situation for a plant cabinet may be rendered interesting by means of climbers on perpendicular props, no matter how irregularly placed, plants rising from groups of rockwork on the floor, and trailing plants suspended in pots or baskets from the ceiling. In some cases, the effect of a picturesque grove of climbers of this kind may be heightened by the introduction of a little stained glass in the roof; but this ought to be used most sparingly, and not in a larger portion in one place than a star a couple of inches in diameter, half a dozen of which will suffice for the roof of a plant cabinet containing upwards of 100 square feet of glass. In the evenings, on particular occasions, two or three coloured lamps may be introduced; but these also should be used very sparingly. Whatever attracts more attention than the plants should be avoided, as interfering with the main object of the structure.

Wherever the plant cabinet is placed, and in whatever manner it may communicate with the house, one point only in its construction is absolute; which is, that it should be at least as lofty, from the floor to the glass of the roof, as the living-rooms of the house. When this is not the case, it has an appearance of meanness, which, instead of an elegant ornament, renders it rather a disagreeable excrescence. The form of the ground plan must, of course, be in a great measure determined by the situation; but, in general, a parallelogram, placed with its narrow end to the house, will have the best effect. One side and the roof should, at all events, be glazed; but, if both sides be glazed, the effect is much better than if one is opaque; provided, however, that the roof is glazed, and the width of the house is as great as its height, or nearly so. Plants will thrive in a house with all its sides opaque; it being understood that the house is as wide as the side walls are high, and that the plants are placed on a stage, or on the floor, so that the light may fall in direct lines on the upper surface of their leaves. The sides, when of glass, may be framed and glazed in any mode considered as in character with the windows of the house; and the roof may be glazed like a common span-roofed hot-house: but, if the panes of glass are above 8 in. wide, they ought to be of extra-thick crown glass, such as is now sold in London for conservatories; or of plate glass. One or more sashes in the side or sides, or one sash in the farther end, ought to be made to open at top and bottom, for the sake of ventilation; but this may be accomplished without having the sashes hung with cords and pulleys, by having two narrow sashes made to slide past each other, or even by having a pane in the upper and lower part of each window to open. In general, all hinged sashes or panes should open outwards; because then they are not in the way of the plants within.

Where a plant cabinet faces the north, and the situation is much exposed to north winds, it would be very desirable if the glazed sides and roof were made double. This construction would retain the heat much better in the winter time; and during summer the inner sashes might be taken away altogether, and used for growing cucumbers or melons in the garden or yard behind the house, or, if there were no room there, on the roof, if that were suitable. In the case of plant cabinets facing the north, where the expense of the double-glazed roof and sides is considered too great, arrangements should be made for forming a temporary inner roof of matting or canvass, for the purpose of retaining the heat during the night, and even sometimes during the day in severe weather. An inner covering of matting or canvass, at 6 or 8 inches' distance from the glass, is always much more powerful in retaining heat, than if the same covering had been placed outside the glass; because in the inside it is kept dry, whereas on the outside it will be liable to become saturated with wet; and in that state it would carry off much more heat by evaporation, than can possibly take place from water running down the smooth surface of the glass. There is another reason against all outside coverings except those of boards, which is, that they are apt to be deranged, and to break the glass during high winds. The inside covering may be made to roll up like a window-blind, and it may rest on iron rods, placed parallel to the roof and to the sides, and about 8 in. distant from them. In many cases, shutters may be contrived for the roof, and put on from an upper window; and this covering, when the weather is extremely severe, may be left on for two or three days at a time; it being understood that light is freely admitted from the sides. Provided that the roof of a green-house is securely protected from perpendicular cold, as it is called, a covering for the sides is comparatively of little importance. We may add that, among nurserymen and commercial gardeners, a substitute for these different coverings is found in tying or nailing bast mats to the trellis or rafters of the roof and sides, inside the house; and, as such coverings are seldom wanted above a month or two in the year, during the depth of winter, it is hardly worth while going to the expense of having them made of canvass for the inside, or of boards for the outside.

In the eagerness of many persons to render a plant cabinet as ornamental as possible, they introduce stained glass and ground glass, along with the clear glass, in the roof and sides. In our opinion, this is in bad taste, because any other light than the clear light of day is injurious to vegetation: the strong tint of stained glass overpowers and destroys the delicate tinges of colour in the flowers, and the associations connected with it belong rather to churches and chapels than to green-houses; and ground glass weakens the light (certainly never too strong in London) which passes through it, and besides gives the idea of something that requires to be concealed. In London, ground glass in a plant cabinet is particularly objectionable; because in many houses this glass is placed in the doors or windows of water-closets.

The placing of the plants in the interior of a plant cabinet should vary according to the size of the cabinet and its situation. Where the cabinet is 10 ft. or 12 ft. wide, and of an equal or greater length, a narrow stage in the middle, with a shelf a foot broad round the sides, will display the plants to the greatest advantage; but this supposes that there is nothing disagreeable in the exterior scenery, which will be seen through the side windows. Where it is not desirable that the external objects should be seen, the stage, instead of being in the middle, should be ranged along the sides and the farther end, and in that case no more of the sides require to be glazed than what is above the highest shelf of the stage. As the main object is to display the plants to the spectator within, and as, when placed in close ranks on a stage, such plants can only be seen on one side, the admission of light to their other side, as it would chiefly strike the under sides of the leaves, is comparatively of little use; and, hence, that portion of the side and end walls which is under the level of the top shelf of the side and end stages may always be constructed of opaque materials, such as brick, lath and plaster, &c., which will be a considerable saving in first cost. In general, wherever there are objects exterior to the plant cabinet which it is not desirable to see, the stage should be placed against the side walls and the further end; but, where there is abundance of light on every side, and nothing without to conceal, the best effect to the eye will be produced by bringing the glass down to the ground on every side and the farther end, and by having the stage in the centre.

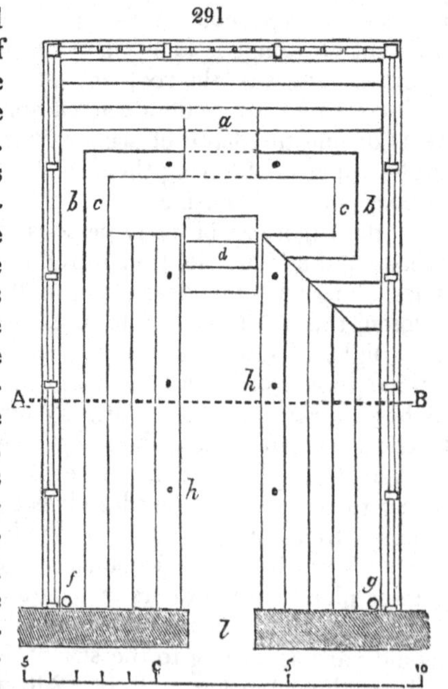

Fig. 291. is a plan, and *fig.* 292. a cross section, of a plant cabinet, in which the stage is placed along the sides and against one end. The stove for heating it is placed at *a*, from which hot-water pipes proceed to the right and left under the side stages; the two shelves *b b* and *c c* being for the purpose of concealing these pipes. The stove is concealed by the portion of the stage *d*, which is on casters, and draws out with the pots on it, to admit the operator to the stove. The hot-water pipes are seen in the section (*fig.* 292.) at *e e*. The smoke from the stove may be conducted away in a tube under one of the stages, and carried up the side of the house, as indicated at *f*. The water of the roof may be collected and conducted to a cistern under the stage by the pipe *g*, placed in the opposite angle to the smoke-tube. Props for climbers may be placed as indicated at *h h*, &c.; and, during the severest weather in winter, matting may be made fast to the iron tie-rods (*i i* in *fig.* 292.) of the span roof. The exterior guttering is shown at *k*.

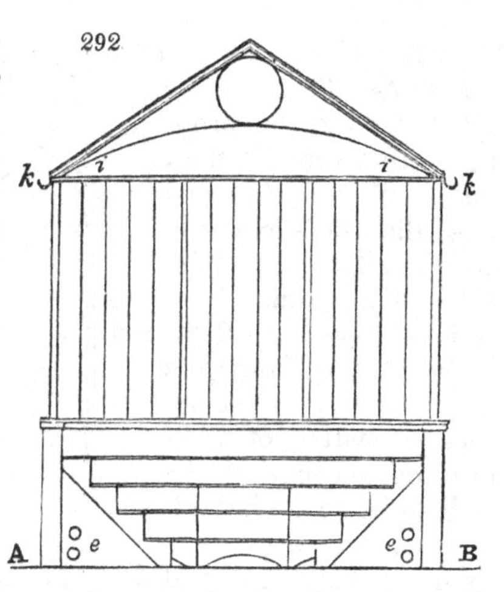

Fig. 293. is the plan, and *fig.* 294. a cross section of a span-roofed plant cabinet, with the plant stage in the middle. The stage may either be finished at the end next the entrance with right angles, as at *a*, or rounded off, as at *b*; and the further extremity both of the sides of the house, and the end of the stage, may either be semicircular or square, at pleasure. If it were square, a portion of the stage might be made to draw out right and left, to admit the attendant to the fireplace; and if semicircular, as in the plan, the semicircle might be formed into two quadrants, each turning on a pivot at the angle, and opening outwards to a sufficient extent to admit the operator to the stove, as shown in *fig.* 293. at *c*. One advantage of this arrangement is, that the stove can never be seen by a visiter; because the attendant, when he is at work at *c*, cannot escape from that place without shutting at least one of the quadrants. Another advantage is, that, while both quadrants are open to the fullest extent, as at *d d*, the pots being on the shelves, any person entering the cabinet will rather consider the appearance as the peculiar termination of the stage, than as a convenient arrangement for getting at the stove. There are props for plants at *e*.

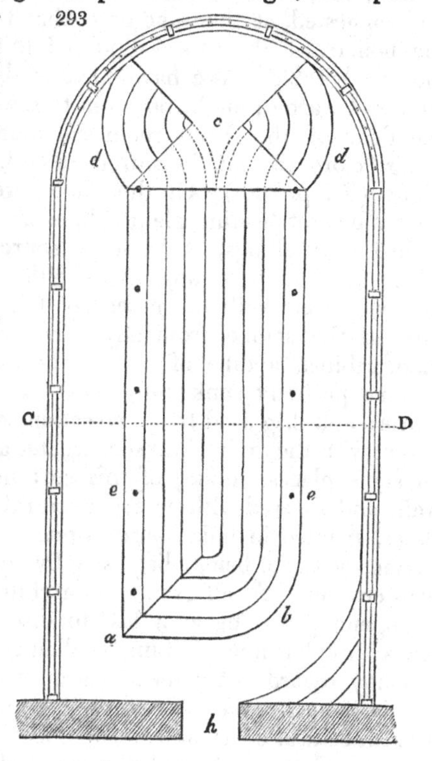

The water-pipes are shown at *f*, in *fig.* 294., and the arched rods to the props at *g*. In severe weather, mats can be stretched along the whole extent of the roof, attached to these arches. The entrance from the house is at *h*, *fig.* 293.

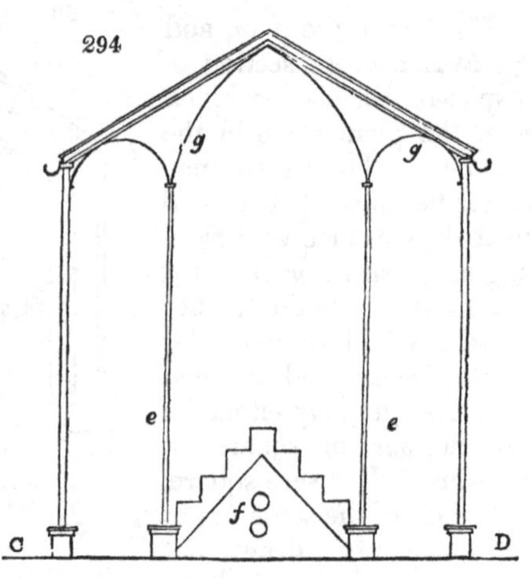

The Mode of heating a Plant Cabinet is sometimes a matter of difficulty, on account of the small space to be heated, and the large surface exposed to the external air. We have already mentioned the practicability of heating such places from the kitchen fire, or from a fire or boiler placed in another story; and suggested, at the same time, that the necessary pressure on the boilers, or the tubes, rendered this mode of heating by no means advisable. We have mentioned, too, that in some places sufficient warmth might be given to the plant cabinet by opening the door of the sitting-room communicating with it the last thing before the family retired to bed, and leaving it open all night. (See p. 360.) On the whole, we are of opinion that the best mode of heating plant cabinets, or small green-houses attached to dwellings, is by some source of heat on their own level; and not from any source either above or below. If no flue has been built in the wall of the house suitable for carrying off the smoke from any stove or fireplace made in the plant cabinet, a tube of cast or sheet iron, or of earthenware, may be partially sunk into the outer face of the wall of the house; and disguised by a projection so designed as to be consistent with the architectural character and effect of the elevation. In some places, tubing of this sort may be placed against the wall, and covered with an architectural case of boards, metal, or slates, painted in imitation of stone; and sometimes these projections admit of being disguised by common ivy, or the Virginian creeper. At all events, no architect of the slightest degree of ingenuity will be at a loss to discover the proper situation for a circular flue of 3 in. in diameter; and no builder who has any regard either for appearances, or the free ascent of the smoke, in such a flue, will ever put it up, without a casing to give it architectural effect, and to serve as a nonconductor, and thus to preserve what may be called a lining of heat round it, to favour the ascent of the smoke. The situation for the flue having been fixed on, the next thing is to determine the mode of heating; and this, we are of opinion, ought, in most cases of small plant cabinets, to be by a hot-water stove placed within the cabinet, and heated by a register fire-pot within, like that of Dr. Arnott. The fuel used may be coke on ordinary occasions; anthracite when a greater heat was wanted; and, perhaps, charcoal in the most severe winter nights, when the heat required was very considerable. A stove of this kind, properly constructed, may be kept burning night and day, regulating the admission of air to the fire according to the heat required. For this purpose, the stove may either have a hand regulator, as in the imitations of Dr. Arnott's stoves, a thermometer one being unnecessary; or, in order to insure a draught, the air may be brought to the stove by a leaden pipe of 1 in. in diameter within, from a lower level, either immediately under the house, or from the open air; or from any place from which it is desirable to extract the air for the purpose of ventilation. In all these cases, the air admitted to the fire may be regulated by a common stopcock, like that in use for common water-pipes. We think it not unlikely that the smoke, or products of combustion, where the kind of fuel we have recommended is used, might be conveyed away in a horizontal direction, or perhaps even downwards to a drain, in a tube of not more than double the diameter of that used for supplying air to the fire; but, never having seen this mode put in practice, we cannot venture to recommend it. We have seen the common smoke-flues of hot-houses discharge their smoke horizontally (see *Gard. Mag.*, vol. i. p. 152.); but it is always attended with a waste of heat.

The mode in which water is heated by a small stove is now so well known to ironmongers, that it seems scarcely necessary to describe it. Supposing the fire-pot, or fireplace, surrounded by fire-brick, to occupy a cubic foot in the centre; then enclose this on three sides and over the top with a square or circular double cylinder, water-tight. The side not cased with water must contain the furnace door for supplying fuel, the ash-pit door for withdrawing the ashes, and an opening immediately under the top, or cover, for the insertion of the tube to convey away the smoke. On any of the three sides cased with water two tubes must be joined, one at the bottom, and the other at the top; and these may be conducted on a level to any distance from the stove that may be desirable, being joined at the farther extremity, either by a vertical tube, or by the ends of the horizontal tubes being inserted into an open cistern. This being done, and the fire lighted, the circulation will go on in consequence of the difference in specific gravity between cold water and hot water.

Fig. 295. is a section which will give a general idea of this kind of stove: *a* is the fire-pot; *b* is the furnace door; *c*, the ash-pit door; *d*, the situation of the pipe which supplies air; *e*, the

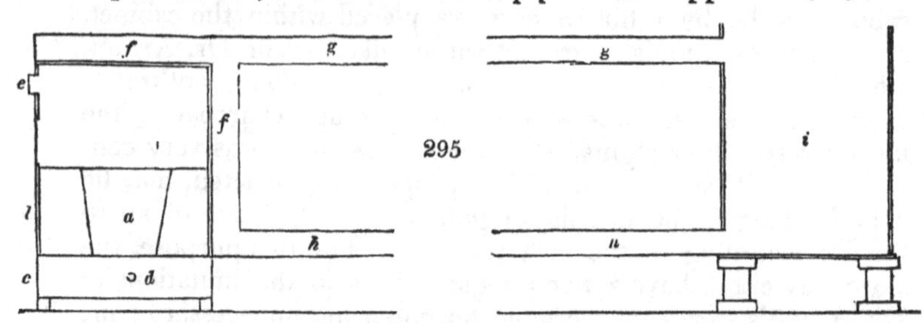

nozzle to which the smoke-pipe is attached, and which may be turned in any direction, except downwards, that may be suitable to the situation; *f f*, the casing of water which surrounds the boiler; *g*, the upper pipe; *h*, the under pipe; and *i*, the cistern which forms the junction between both pipes at the farther extremity. Stoves and pipes of this description may be placed under the stage of plants, so as to be completely concealed from the eye; a portion of the stage being made to separate from the rest, and to draw out, so as to admit the attendant to supply fuel, &c. In order to save time, and to look well by never requiring the plants to be taken off the shelves, the movable portion of the stage may be on casters and small wheels, the latter running in grooves as far as the area of the stage extends, and the casters serving to make it run easily on the paths where the grooves would be unsightly. In plant cabinets where there is no stage the pipes may be concealed by rockwork, or by some other suitable contrivance; or the casing of water may be enlarged, so as to form a reservoir of heat sufficient for the demands of the house, without any pipes. In some cases, the stove and reservoir of water might be covered with a small stage of plants, with rockwork, with sculpture, or with statuary; or the stove might be rendered ornamental in its form, so as to be regarded as a handsome piece of furniture: but this last plan would, we think, render the plant cabinet too much like a living-room.

Fig. 296. is a portable hot-water apparatus for plant cabinets and small green-houses, invented by Mr. Joshua Major of Knowstrop, near Leeds. This apparatus may be made of tin or copper; the latter, though more expensive at first, being from its durability much the cheaper in the end. Charcoal is employed as fuel for this apparatus: oil lamps and gas have been tried instead of charcoal, but with not nearly so powerful an effect. When charcoal is used, it is necessary to employ pipes to conduct the effluvium arising from it out of the place to be warmed; and it will be advisable, in order to abstract all the heat possible

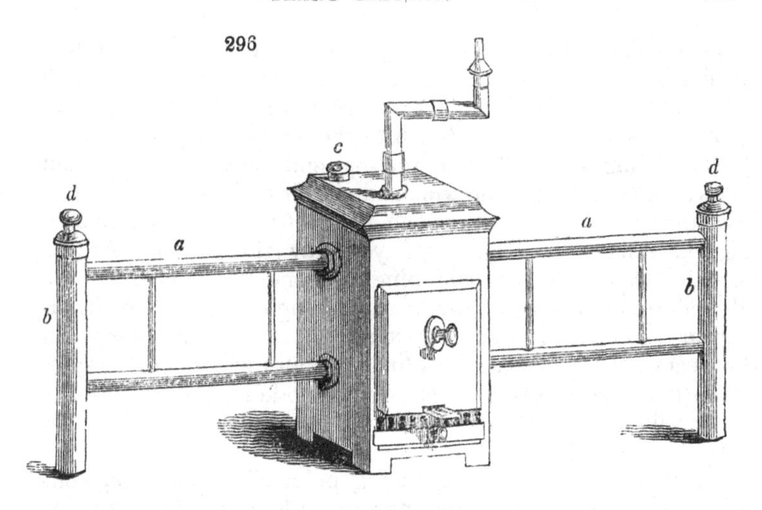

from this smoke-pipe before it reaches the outside of the house, to have it of a considerable length. In order to render the smoke-tubes suitable for any situation, it is necessary to have elbow pipes, like those used for turning corners, and several lengths of straight pipes, by means of which the piping may be lengthened, and turned in any direction that may be required. The largest-sized apparatus should not be more than 8 ft. long; as, if longer, it would be inconvenient to move about. The size of the one which Mr. Major found the most useful is as follows: — The whole height of the centre portion of the apparatus, comprising the boiler, &c., is 15 in., and the width 5¼ in. by 7½ in.; the fire-pan is 5¾ in. by 4½ in., and 3½ in. deep; surrounded on three sides by a boiler, in the form of a casing, half an inch in diameter, which becomes more spacious upwards, as the fireplace diminishes. The opening necessary for the reception of the fire-pan, and for supplying it with fuel, is 6 in. wide by 5½ in. deep. At the top of this opening the fireplace begins to taper, and, consequently, the water in the boiler expands more immediately over the fire; the smoke-pipe takes its regular width (1½ in.) in the boiler, about an inch below where the lid unites; the horizontal water-pipes (*fig.* 296. *a*) are each 28 in. long, by 2 in. in diameter; the end pipes (*b*) are 14¼ in. high, by 3 in. in diameter; a feeder (*c*) is added, in case it should be thought better to have the lid fixed tight on the boiler. In order to promote the circulation of the water, small holes are to be perforated in the top of the lids (*d d*), which are also intended to be fixed tight. The apparatus may either be placed on the floor of the house to be warmed, or raised by bearers, or suspended by wire or cord; the two latter methods making the fire to burn more freely. (*Gard. Mag.*, vol. x. p. 23.)

Joyce's Apparatus for heating by Steam. A mode of heating a small green-house by steam, from a portable apparatus placed within the house, has lately been invented by Mr. Joyce. The apparatus is a copper cylinder, with the fire placed in the centre, the fuel being supplied from the top, and the ashes coming out below, through the grating which admits air to the fire. The fuel is charcoal, and the little smoke which it produces is delivered into the same tube which conveys away the steam. At the farther extremity of the steam-pipe the fumes of the charcoal are allowed to escape outside the house, through a tube; which, for ordinary apparatus, need not exceed an inch in diameter. As the steam-pipe is placed so as to return all the condensed water to the boiler, the loss of heat by this mode is extremely small; but it will not anwer well for any other fuel except charcoal, which is expensive.

Ventilation is the only remaining point of importance, connected with plant cabinets or small green-houses, which remains to be touched on, and this is highly essential in every situation. In houses 20 or 30 feet long, and from 10 to 15 feet broad, the sashes of the roof should be made to slide, so that the upper ones may be let down at pleasure, and the lower ones drawn up; or they may be hinged at the upper end, and made to lift up. In addition to this, the upright sashes should either be made to open outwards, by being hinged at one side (which, in general, is the best mode) or at the top; or they may slide in two grooves, so that the one can be pushed past the other. These modes are applicable to green-houses 20 or 30 feet in length; but, for those under 20 feet in length, it will generally be sufficient to have one or more small openings, for the admission and exit of air, in the roof or sides; and through these, when the door is open, there will be as much ventilation as plants usually require. One of the simplest and most elegant modes of having an opening in the roof is by having a cap to be raised by means of a vertical rod, with a line and pulley, as shown in *fig.* 297. In this figure, which is a cross section of a span roof, *a* represents the cap; *b*, the rod by which it is raised; *c*, a cross piece of iron, in which the rod *b* works; *d*, the cord passing over the pulley *e*, for raising and lowering the rod; and *f*, the rafter (forming part of the span roof) which supports the whole. The rod and bars of the cap are generally of iron; and the panes of glass small, to lessen the risk of breakage. Sometimes the cap is balanced by a weight attached to a cord which passes over a pulley fixed on the other side of the rod opposite to *e*; and which, by counterbalancing the greater

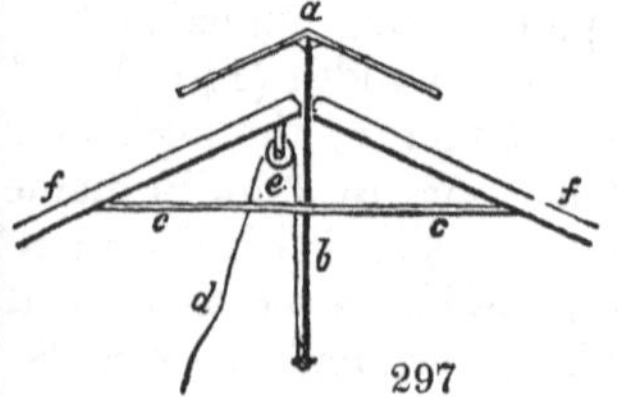

297

part of the weight, and leaving little more than the friction by the rod passing through the bar *c* and the rafter *f* to be overcome, renders it easy to raise a cap of the largest size, either by hand, or by a self-acting apparatus to be hereafter mentioned. Though we have shown in *fig.* 297. this cap on the ridge of a span roof, yet it may be constructed with equal ease on the upper part of the slope of any pent roof, or even in any part of that slope, by introducing on the upper side of it a cross bar, or a flashing of lead, to throw off the rain to the two sides. Air may also be admitted through the side sashes, by hinging a flap in the upper part of any of the sashes, with a lever and pulley to open it, as shown in *fig.* 298. In this figure, *g* is a lever, which, when pulled by the cord *h*, raises the flap at pleasure to any point not beyond *i*. Whenever ventilators of this kind are used, they ought to be placed in the highest part of the house; because as the hottest air always ascends, it consequently collects there, and will pass off by any opening with greater rapidity than it would through an opening on a lower level, admitting a counter current of fresh air to supply its place.

298

When the sloping sashes of a span roof are made to slide, the operation of opening them may be greatly facilitated by balancing them with a weight in the manner shown in *fig.* 299. In this figure, *a* represents cords attached to two opposite sashes, and passing over pulleys fixed on the ridge-piece; after which, under the stage (*c*), they are joined together at *b*, which represents a weight attached to a pulley, and which weight is sufficiently heavy to balance the two sashes. By this arrangement, either sash or both sashes may be let down at pleasure, to any length desired. When they are drawn closely up, the weight (*b*) is within 2 in. of the floor (*d*); but, if it were necessary, an opening might be made in the floor for the descent of the weight to the extent of 2 or 3 feet. On inspecting the figure, it will

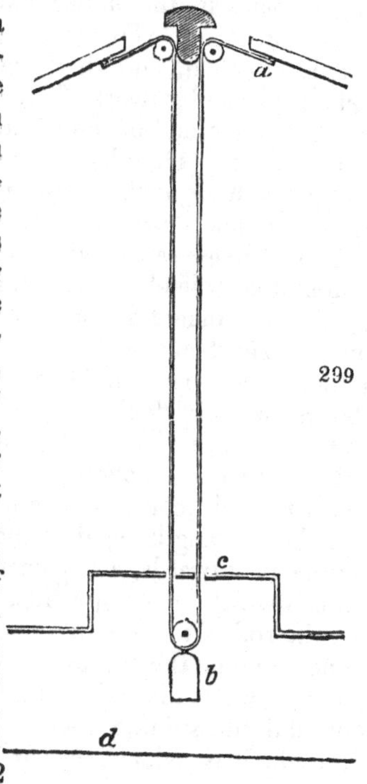

299

appear evident that a cap, such as that described in *fig.* 297., may be balanced and raised in a similar manner, as shown in *fig.* 300., and as practised in the case of chandeliers in churches and ball-rooms. In this figure, the weight (*e*) may have a basket or saucer attached to it, so as to admit of disguising it, by surrounding it with pots of hanging plants.

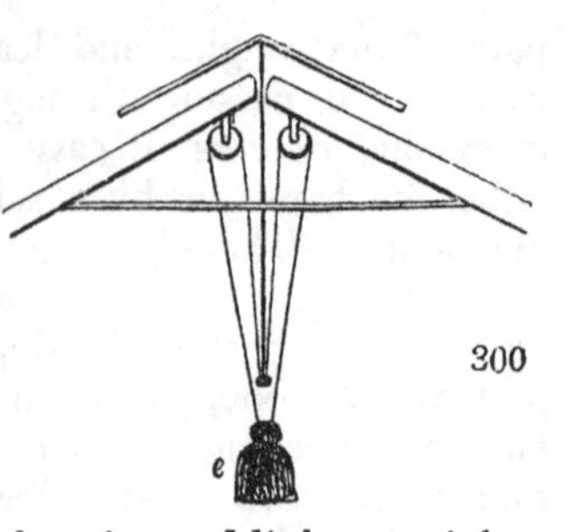

300

In this case, however, there should be heavier and lighter weights to hook on at pleasure, so that, with the addition of the pots of plants, the total weight should be no more than just sufficient to balance the cap.

Regulation of Temperature. In the case of small green-houses, where a regular gardener is not kept, a provision for preventing the house from becoming over-heated, which will act independently of human assistance, is very desirable. There are many contrivances of this kind; some, such as Kewley's automaton gardener, calculated for regulating the temperature to the greatest nicety; and others for merely preventing great extremes of heat. The simplest and most economical mode is to have a hinged pane in the upper part of one of the upright sashes, opening inwards. To the lower part of this pane one end of a cord is attached, and the other is joined to a netting which encloses a bladder nearly filled with air, and air-tight. A string from the other end of the bladder is attached to any fixed point near at hand. The bladder should be moderately distended, by filling it with air in the house, when the air of the house is at the highest temperature required; and, consequently, when any increase of temperature takes place, the air in the bladder will expand and completely distend it, so as to change its shape from that of an oval to that of a globe; shortening the long axis and lengthening the short one. The effect of shortening the long axis is to pull in the hinged pane of glass, and thus to admit the external air. This is, without doubt, a very rude mode, and will not regulate the temperature with any degree of nicety; but for ordinary purposes, and where economy is a great object, it is quite sufficient. Another mode consists in applying a hollow brass tube, of 5 or 6 feet in length, against the lower part or valve of a window that opens outwards, or against the lower end of the vertical rod of a cap which opens upwards. Thus, in *fig.* 301., if we imagine the rod *f* to be a hollow brass tube, touching the floor (*g*) at one end, and the spindle (*h*) of the ventilating cap at the other; and that, when the rod is in this state of contact with both the floor and the spindle, the house is at the maximum temperature required; it is evident that any increase of heat, by expanding,

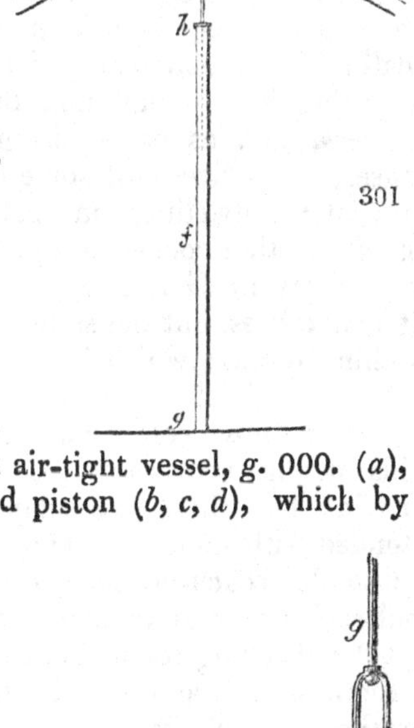

and consequently lengthening, the brass rod, would raise the ventilating cap; and the extent to which this might be done, would be greatly increased by placing the brass rod on the end of a lever. Brass rods, when required to open the sashes or ventilators of a green-house, may also be applied in a great variety of other ways, which any ingenious mechanic will readily discover. The last mode which we shall mention is one which has been adopted with success by an eminent horticulturist, John Williams, Esq., of Pitmaston. This ventilator operates by the expansion and contraction of air in an air-tight vessel, g. 000. (*a*), communicating with a cylinder and piston (*b*, *c*, *d*), which by means of a rod (*g*), operates on the hinged pane or sash to be opened. The use of the water or other fluid is to confine the air; and, by that means, when the air expands or contracts, it operates upon the piston. By means of an adjusting screw, the register may be made to open at any required degree of heat. The air-vessel should contain several gallons, according to the size of the valve or register to be opened. When first used, the vessel must be heated sufficiently to expand the internal air; water is then to be poured in at the top of the cylinder, so as to give the required motion to the float; and about half an inch of fine oil must be laid on the top of the water, to prevent evaporation. (*Gard. Mag.*, vol. i. p. 419.)

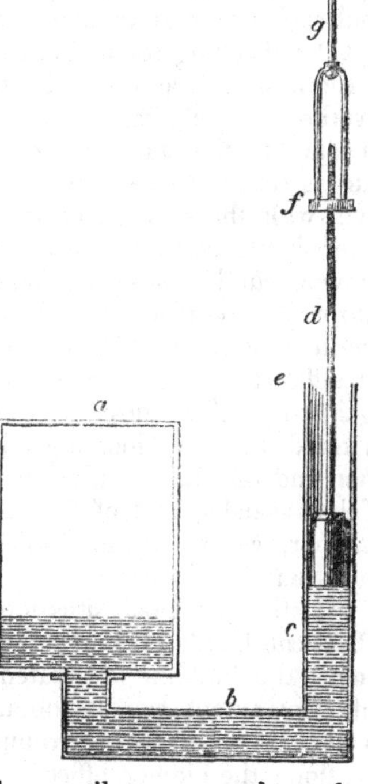

In a plant cabinet, such an instrument may be conveniently placed under the stage, so as to have the rod (*g*) directly under the ventilator or sash to be opened.

SECT. II. *The Domestic Offices.*

THE Domestic Offices may be considered as including the kitchen offices, and the stable and other out-door offices. The former, in small houses, are generally placed under the living-rooms, not only as costing less in building, but as saving labour in household business, and as economising heat during winter. In large houses, the offices are sometimes placed partly under the main body of the dwelling, and partly adjoining it, in what are called wings, or other connected appendages. On the Continent, and particularly in Germany, it is common to have not only the kitchen offices, but the stable offices, in the same building as the dwelling-rooms; which is not only a considerable saving of first cost, but of labour to the servants afterwards; and, during winter, in those cold countries, it is a great economy of heat. This arrangement cannot be proposed for general adoption in Britain; but there are many situations where it would be attended with solid advantages. In open, elevated, airy places, the stable, coach-house, and sometimes even the cow-house and poultry-house, may be placed on the same level with the kitchen, and the dwelling-rooms built over them, for the sake of economy in the first cost, and in order to produce a large mass of building for the sake of effect. In arranging a plan of this kind, the entrance to the stable and cow-house should not be on the same side as the entrance to the house; and, while the former is on a level with the surface, or not more than one step above it, the latter should be two or three steps, or the height of a whole story above ground; the stable and cow-house doors, and also the pits into which the manure is thrown, should be on the side opposite to that from which the prevailing winds blow, in order that all effluvia may be carried from the house rather than driven against it. If the designer of such a house will constantly bear in mind, that the chief objection against the placing of the kitchen and the stable under the dwelling-house is, the production of bad smells, and of flies and other noxious insects during summer, various details which it is unnecessary here to mention will occur to him.

But the principal situation in which the stable and kitchen offices can be placed under the dwelling-house with advantage is, where the building is situated on the side of a steep slope. If sufficient extent can be obtained to form a carriage entrance to the lowest part of the ground, then the stable offices may form one floor, the kitchen offices a floor over it, and above these may be the living-rooms, &c., in the usual manner. In general, however steep the surface may be on which a house is to be built, by judicious contrivance no part of the walling need be lost: for it must always be recollected that, however steep the surface may be, the building will always be wholly open on one side, and partially so at the two ends; so that, if the house be not made broader than the depth of one room, several stories may be obtained under one another, all well lighted from one side. Hence, in particular situations, where a house on a steep slope is to be entered from above, the living-rooms may form the upper floor, and the stairs may descend to the bed-rooms, instead of ascending to them. Much might be said on this subject; but we only advert to it for the sake of impressing on the minds of such of our readers as may be about to build, that they ought never to consent to bury much extent of walling in order to produce the commonplace forms and arrangements of dwellings; and to assure them that anomalous situations will always call forth appropriate designs from a man of taste and genius, and be favourable, rather than otherwise, to the production of beauty and character.

Situation of the Offices. In suburban houses built twenty or thirty years ago, and subjected to various repairs and additions, nothing is more common than to find the offices placed at the wrong end of the house, or perhaps partly at one end, and partly at the other. There is a positive loss of the labour of the domestics in this arrangement; nor can they, when this is the case, be considered as under the full command of the master and mistress, because they cannot be expected to hear the bells, when the offices are divided, unless there were two sets, one on each side; while there is a constant passing and repassing of the entrance front, or probably of the lawn front, by persons having business in the kitchen or stable-court. The seclusion of the lawn front, in cases of this kind, is totally destroyed; and neither the living-rooms of the house, nor the walks in the pleasure-grounds, can be said to possess that complete privacy which is generally considered to be one of the greatest luxuries of a residence in the country. An example of this kind of house was noted by us, some years ago, in the neighbourhood of Chertsey, at Lyme Grove; and, as we were on a visit for upwards of a fortnight to the family who at that time occupied it, we had ample opportunity of studying its inconveniences. The house is beautifully situated in the midst of grounds much diversified by nature, not badly planted, and extending so far on every side, that the boundary is nowhere seen; but the house, though it contains some spacious rooms, has great faults: the store-room and bath-room (*a*, *b*, *fig.* 303.) can only be entered through the dining-room (*c*) or drawingroom (*d*); while the offices (*e*), by being at the wrong end, occasion the road to the kitchen-court (*f*) to pass across the lawn before the garden front (*g*). Had the general arrangement been reversed, as shown in *fig.* 304., and the bath-room

303

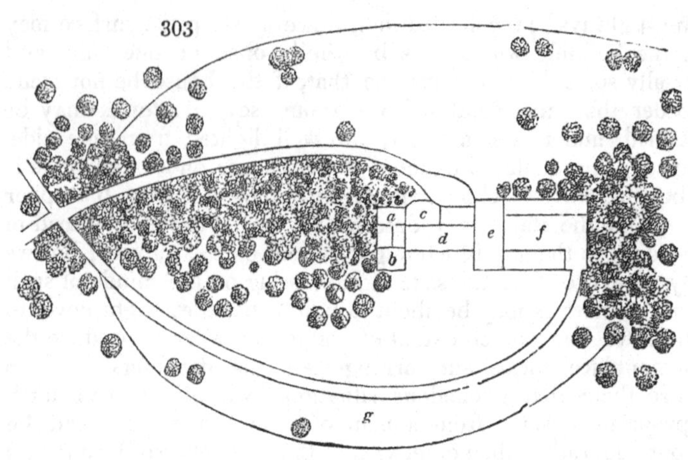

and store-room been put on the side next the offices, the latter might have been conveniently entered without coming within sight of the house; and the privacy of the lawn front would thus have been complete. Among the petty evils of a mal-arrangement of this kind are, the number of gates, and the extent of fencing which it requires; and, though these make little appearance in the description of a place, or in its plan on paper, yet, in its actual working (so to speak) by the occupier, they are important items. The difference between a bad arrangement of

304

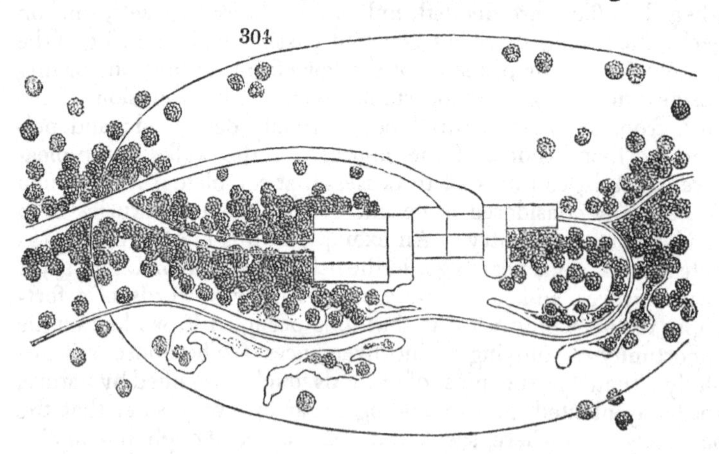

fences and walks, and a good one, may occasion the opening of a gate or gates fifty times a day, instead of five times; increasing the risk of admitting cattle, pigs, or poultry, where they ought not to be admitted, in a corresponding ratio.

The Kitchen Offices. — In designing these, it is of more importance to bear in mind general principles, than to adhere to particular forms and dimensions. Security from vermin is an essential requisite in the construction of the floor and walls of every part of a house, and more particularly of those parts where provisions are kept. The power of thorough ventilation of the atmosphere of every apartment, even of the cellars, when requisite, is another desideratum. Light is more or less essential to every office, except the cellars for wines and liquors; and it is in an eminent degree required for the kitchen, and all those places where food is prepared for the table. A larder, if thoroughly ventilated, may preserve meat without much light; but a pantry requires abundance of light, to insure cleanliness. Light, in the case of meat kept in safes, or in other situations, is unfavourable for the preservation of the meat, as it collects together flies, which are never found in an active state in the dark; and in such cases it ought to be under command; but, in general, it is much more desirable to have light in excess, than to have a deficiency of it. A command of the temperature is a desideratum in the offices of every dwelling, no less than in the living-rooms. Among the first requisites to this are, that of having all the walls, floors, and ceilings made of non-conducting materials; and that of having the ground on which the building stands rendered thoroughly dry by underground drainage, and by a vacuity all round the outside of the foundation walls.

We recall these principles to the minds of our readers, in order to lead them to think on the causes why some parts of dwellings are arranged in one way, in preference to another, rather than to lead them to dwell on particular forms and dimensions; and to induce them to require, from the architects or builders whom they employ, reasons for every thing which they propose, and such reasons as can be made intelligible to all persons of common sense.

The Stable Offices. — In designing these, a recurrence to first principles is of equal importance as in arranging the kitchen offices. The ground on which a stable or coach-house is built ought to be dry, either naturally or by drainage; and the walls ought to be of sufficient thickness to exclude the extremes of heat in summer, and of cold in winter. There ought to be abundance of light admitted, as well for cheerfulness as for cleanliness; and, independently of the doors and windows, there ought to be suitable openings for ventilation. In the stable there ought to be a wooden tube, at least 6 in. square, placed over every stall, and reaching from the under surface of the ceiling to the ridge of the roof, with a cap on the outside to exclude rain; and a slide in the ceiling, at the lower end of the tube, to regulate the quantity of air admitted, so as to keep the atmosphere of the stable at about 50° in winter, and from 60° to 65° in summer. This flue will do for the stable

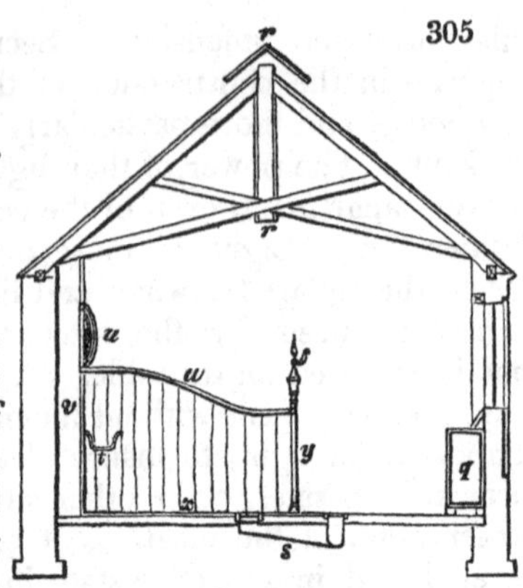

of the horse, what the chimney-flue does for the dwelling-room of the human being. To give a palpable idea of this mode of ventilating, we may refer to *fig.* 305.; in which the ventilating tube, with its protecting cover, is shown at *r r*: *q* is a corn-bin, placed under one of thewindows; *s* is the drain under the floor of each stall, which leads to a main drain under the gutter behind the horses; these drains having grated openings, each with a bell-trap, to prevent the ascent of bad smells: *t* is the cast-iron manger; *u*, the bull's eye cast-iron rack, the hay to which is let down from the loft above, through the open space behind. The mode of keeping hay in lofts over the horses, and letting it down through a space over the racks, which is continually open, is very properly objected to, as contaminating the hay by the breath of the horse; but, when there is a raised ceiling with a ventilator, as in the section before us, the breath of the horse rises to the ventilator, and does the hay no injury whatever. At the bottom of the rack there is, in the space behind, a grated floor, which retains the hay, while it allows the dust and seeds to drop down into the space (*v*), whence they can be taken out at convenience, by an opening under each stall. There is a cast-iron ramped cap (*w*) to the boarded partition between the stalls; and a cast-iron sill (*x*); both cap and sill having grooves for receiving the ends of the boards which form the partition. The partition-post (*y*) is also of cast iron. That part of the floor of the stable on which a horse stands should always be made perfectly level, with a grating and bell-trap under it, in the centre, for drainage. In general, the floor of the stalls should never be covered with litter in the daytime, or when the horse is not expected to lie down; because the litter retains moisture, harbours insects, and produces an unequal surface for the horse to stand on. Stables, as they are commonly kept, contain an atmosphere charged with ammoniacal gas from the urine, and carbonic acid gas from the lungs, of the horse, which, with moisture from the floor, and other aeriform matters, are extremely disagreeable to man; but, if the stables were properly constructed, ventilated, and drained, kept free from litter during the day, and amply lighted, in the manner we have recommended, they would be as wholesome for a human being to enter, and to remain in, as the living-rooms of a dwelling-house. For various modes of constructing stables, and all their details on the best principles, we must refer the reader to our *Encyclopædia of Cottage, Farm, and Villa Architecture*, p. 373. to p. 383.

The Coach-house should not only be dry, and well ventilated, from its situation and construction, but there should be a fireplace in it, or, rather, a flue for a stove; because it must be recollected, that a coach is composed of wood, iron, cloth, stuffing, &c., and is as liable to be injured by changes in the tempe ature, or by moisture, as chairs, tables, sofas, beds, or any other articles of furniture in the dwelling-house. Nobody would think for a moment of sitting or lying on a sofa which had been standing for some time in a damp out-house without fire; and yet, if there be any difference between the quantity of damp likely to be imbibed by a sofa and a coach, it would undoubtedly be greater in the latter. The lining, stuffing of the back, and cushions, all act as sponges, and become charged with moisture; which, when the coach is used, is drawn forth by the animal warmth of the persons it contains. Hence, the unknown cause of colds, inflammations of the lungs, and many other diseases, with which those who keep close carriages are often attacked.

The Harness-Room, as we have already stated (p. 82.), should have the walls lined with boards in those parts where the harness is to be hung, and should have a fireplace or stove, with ample means of ventilation and lighting.

The Groom's or Coachman's Room is very properly placed over, or adjoining to, the stable for the horses; because the latter are liable to various kinds of accidents in the night-time. There should be a stove (one of Arnott's, for example) in the stable sleeping-room, for use during the most severe weather of winter; because such rooms are, in general, close under the roof, and liable to be very cold; but, by a little arrangement, it might be so contrived that the flue from the fireplace of the harness-room might give all the heat to the coachman's bed-room that was requisite; and, indeed, when the harness-room is placed, as it ought to be, between the stable and coach-house, with doors communicating with each, all that is necessary is to place the coachman's room over it. By opening the door of this room, which is necessarily placed at the top of the stoves, the heated air of the harness room may be admitted to ascend to it at pleasure.

The Farm Buildings which are of most interest in a suburban residence are, the cow-house, dairy, piggery, poultry-houses, and rabbit-house; and in the same section with these we shall include the aviary, the apiary, fishponds, &c. On farm buildings generally, as required for a suburban residence, we have made some remarks in p. 626., which we consider sufficient; but to what we have already said on the cow-house and dairy, in p. 548., we shall add some remarks.

The Cow-house, in Britain, is a rude neglected building, compared with what it is on the Continent, among proprietors who take a pleasure in having fine cows. In Holland, the cow-houses, even of the common farmers, are kept in a very superior degree of cleanliness, and are lighted and ventilated in a much better manner than the cow-houses of villa residences are in England, though the greatest attention be paid to the dairies. One reason of this may be, that, in many parts of England, the winter is so mild, that the cows are kept during that season in open yards, having sheds attached, in which they are left to take shelter at pleasure; thus no skill is required for constructing the cow-house, and no art for its proper management. In Germany, however, from the severity of the climate, the milch cows are obliged to be kept in closed houses half the year; and hence the greater attention that has been paid in that country to the construction of cow-houses. It may also be observed, that, as the pastures in most parts of Germany are of a very thin and inferior description, the cows kept by proprietors who pay much attention to this subject are generally stall-fed throughout the year. We cannot, however, recommend this practice for England; since cows so treated can never be in so healthy a state as those which are allowed to take free exercise in the open air, at least a portion of every fine day. When only two or three cows are kept, instead of tying them up, we would allow them to run at large in a yard which had a proper cow-house attached to it, for them to shelter in; and we would only tie them up when they were being milked, or in extremely severe weather, when it might be necessary to shut them up in the cow-house.

As a few points which deserve particular attention in the construction of cow-houses, we may mention the following. The floor on which the cow stands should be perfectly level; and behind the stall there may be either an open gutter, or, what is preferable, one covered with an oak board pierced with holes, or with an iron grating, the holes of which are not larger than an inch across. Through this grating, or prerced plank, all the urine and recent dung will easily drop into the gutter, more especially with the occasional assistance of a broom; and the gutter may be either cleaned out once or twice a day, by lifting up the grating or plank, and applying the broom, or by turning on water at one end; and both the dung and the water should be conveyed by the gutter to the liquid manure tank. If the cows stand along the house, it should not be less than 18 ft. wide within, which will allow room for a passage at the cows' heads, as well as behind them. There ought to be windows with luffer-board shutters, that is, with shutters like Venetian blinds, on both sides of the house; and ventilating tubes from the sides of the walls, or the centre of the ceiling if there be one, to the ridge of the roof. The breadth allowed for each cow should not be less than 5 ft., and the length of the stall 8 ft.; and there ought to be a boarded or slate partition between every two cows, or so as to include two cows in each stall. The manger should be a trough formed of board, stone, or slate, raised so that the upper edge may be from a foot to 18 in. above the surface of the ground, or about the height of a cow's knees. As this manger is to contain green food, or grains when they can be obtained, as well as roots, hay, &c., it should not be less than 18 in. or 20 in. wide, and a foot deep. Where cows are intended to stand in the cow-house during great part of the year, the manger for each cow should be in three divisions; a large one in the centre for hay or herbage, a small one on one side for water, and one on the other side for roots, grains, or prepared mashes. A hay-rack for a cow-house is not only unnecessary, but even dangerous, unless placed as low as the manger; because it is found that when cows are in calf, raising their heads much above the natural level is apt to make them slip their calves. So strongly are the farmers of Normandy impressed with this idea, that they not only take care to have the floors of their cow-houses perfectly level, and the manger or hay-rack not higher than the cows' knees, but, when they turn the cows out to grass, they always harness them with a bridle and brechin, to prevent them from tossing up their heads. We have already (p. 550.) referred to the Harleian system of dairy management, and we strongly recommend whoever intends building a cow-house, to consult the notices of that system, and of various others, which will be found given at length in our *Encyclopædia of Cottage, Farm, and Villa Architecture*, § 757. to § 764.

The Dairy, in warm climates, or in such as are extremely cold, requires much more care in its construction, than in the temperate climate of England. A building with thick stone, mud, or turf walls, or with hollow brick walls, with a thatched far-projecting roof, standing east and west, and with its door and windows in the north side and in the ends, will suffice for a dairy, in the warmest part of our island. The windows should have glass sashes for use in winter, they should open inside, and they should be covered on the outside with wirecloth, to exclude flies and other insects during summer, when they are kept constantly open. There ought also to be outside shutters, to exclude the sun, in the mornings and afternoons, from the east and west windows, and to keep in the heat during the severe weather in winter. A thermometer should be kept in the dairy, and the temperature should never be allowed to rise above 55° or fall below 48° or 50. In winter, the temperature is kept up by the warm milk being brought in, when newly taken from the cow; but, when necessary, this heat may be increased by placing vessels of hot water on the floor. Every dairy requires to have a scullery attached, with a fireplace and boiler, as the vessels which hold the milk require to be frequently washed in scalding hot water. Cleanliness is essential in every place where human food is to be kept, but it is particularly so in a dairy; as, if the milk be put into unwashed or carelessly washed vessels, it will soon turn sour; and, if the churn and other utensils used in preparing butter be not frequently scalded, they will give the butter an unpleasant taste, and prevent it from keeping. The dairy scullery may very conveniently be placed at one end, or on one side, of the dairy, taking care to have them separated by double doors, with windows for ventilation in the passage between, or with a window in the ceiling. This prevents the transmission of heat from the scullery to the dairy in summer; while, by shutting the windows and opening the doors in winter, heat may be admitted from the scullery to the dairy, so as to keep it at the proper temperature. These hints will be sufficient to show the importance of the subject, and to lead those who can deduce rules from general principles into the right mode of thinking respecting dairies; while such as require more minute details may have recourse to our *Encyclopædia* already quoted.

The Piggery.—The great requisites here are, cleanliness, dryness, warmth in winter and shade in summer, a convenient mode of keeping and supplying food, and a place for the pigs to take exercise. These requisites are very well attained in the ordinary pigsties, which consist of a small open court for the pigs to feed and take exercise in, and a covered building for them to sleep in, The principal defect in pigsties of this kind is, a want of sufficient drainage in their courtyards, in consequence of which much more moisture is retained there, than is desirable for the health and cleanliness of the pigs. There ought to be to every pigsty an underground drain with a grating and trap, (the latter as much to keep out dirt, as the bad smell,) communicating with the liquid manure tank, and to be washed out occasionally by the water used in cleansing the floor of the pigsty. Though pigs are generally considered to be dirty animals, yet they have some habits of cleanliness which are remarkable: for example, that of never doing anything dirty in their sleeping-place, but always choosing a particular corner of their sty for this purpose. Hence, if this natural tendency were properly seconded, a pigsty might be kept as clean as a cow-house or stable. The food for pigs is either of a dry or of a liquid kind. The latter ought always to be kept in a large tub, or in a tank under cover; where it is kept till it becomes in a state of fermentation; it being found to be far more nourishing for the pigs when sour, than when it is sweet. For this reason, there ought to be two tubs or tanks, that the one may be filling with the liquid refuse from the kitchen, while the other is using; and they may be put under cover, or sunk in the soil, in order that fermentation may not be checked by the cold of winter. Where a proprietor feeds his own pigs, the greatest improvement that we can suggest is, that of having the pigsties connected with a grass field of one or two acres in extent, or an orchard; and of having various aromatic herbs sown among the grass; this would be a near approach towards making the flesh of the common pig as highly flavoured as that of the wild swine. Indeed, we cannot too strongly impress it on the mind of the reader, that the great improvements wanting to almost all the modern systems of fattening animals are, arrangements for allowing them to take abundance of exercise while they are fattening, and supplying them with a greater variety of food, and especially of aromatic kinds.

Poultry.—Next to the kitchen-garden, there is no accompaniment of a country residence that contributes more to the comforts of its inhabitants than the poultry-yard; and, considerable as these comforts are, if the management of poultry were properly understood, they might be greatly increased. It is not our intention, in this article, to go at length into the subject, because it would occupy more space than we can afford, and, indeed, would be less useful as a guide, than some of those small popular works devoted entirely to the subject, which may be readily procured. One of the latest of these is entitled *Poultry, their Breeding, &c.*, by R. B. Dickson, 1838; and there is another, of equal or perhaps greater merit, entitled *On the Breeding, Rearing, &c., of Poultry*, published by Ridgway, also in 1838.

In order to determine the principles on which poultry-houses should be constructed, it is necessary to bear in mind that domestic fowls are natives of different countries and climates; that some are aquatic, and, consequently, have very different habits from those that naturally perch on trees or other lofty objects; that, in a state of subjection to man, they are, like other domesticated animals, liable to more diseases than when they are in a wild state, and that these diseases are generally incurable by art; and, finally, that they may be stimulated by rich food and heat, so as to produce eggs earlier, and in greater abundance, than they ever do in a wild state, or in cold seasons, in a state of domestication; while, on the contrary, the season of laying may be retarded by scanty food, and cold. Experience has shown that the taste of the flesh of fowls depends very much on the quality of the food which they eat. Musty or damaged grain, of any kind, will communicate a peculiar taste to the flesh of the common fowl, the goose, the turkey, or the duck, that has been fed on it, however carefully it may be dressed by the cook. All the different species of poultry are exceedingly pugnacious, not only with other species, but even with their own kind; and all fowls, except the aquatic, have the greatest dislike to damp, and suffer sooner from it than from any other cause. All fowls are more or less omnivorous, and they consequently require, not only corn and green vegetables, but insects, worms, and similar vermin, or the flesh of animals minutely divided, so as to serve as a substitute for these objects. The most remarkable circumstance connected with the common fowl is, that, though it is domesticated in every part of Europe, it is a native of the warmest parts of Asia; and from this circumstance, independently of other considerations, it requires greater warmth in its lodging-place than either the duck or the goose, which are natives of Europe; or the turkey, which is a native of North America. Hence, the necessity of having a separate house for every kind of fowl which is to occupy the poultry-yard, unless we except one or two kinds of similar natures, such as the duck and the goose, which, in some cases, may lodge in the same building.

The Poultry-yard. The situation of the poultry-yard should be open and airy, but at the same time thoroughly sheltered from the prevailing winds of the locality, and, in England, more especially from the north, north-east, and north-west. In general, the best situation is one which slopes to the south-east, with the poultry-houses in the upper part, with water for the aquatic fowls in the lower part, and a lofty open shed, in a central position, for feeding the poultry under during rain, and for the peacocks, peahens, and Guinea fowls to roost under in severe weather. This shed should also be so far extended as to include in it a few square yards of surface, to be dug out to the depth of a foot or more, and filled up with dry sand, for the fowls to take what may be called a sand bath, which is the principal means they have of ridding themselves of the body vermin with which they are infested. In addition to this shed, the poultry-yard, if it be large, may contain two or three trees, as well to afford shade and shelter to the common fowls, as for a roosting-place to the pea fowls and Guinea fowls during summer, and to harmonise the poultry-yard with the woody scenery of the place. There ought also to be, in a proper situation, a covered tank for receiving the washings of the poultry-houses, and retaining the liquid till it ferments, and thus becomes fit for manure. The soil is, perhaps, of more consequence than even the situation; for, unless that be naturally quite dry, or rendered thoroughly so by art, however suitable the situation and treatment of the fowls may be in every other respect, nothing can be more certain than that they will not thrive. The poultry-yard should, of course, vary in size according to the number of fowls which it is intended to keep; but more still according to the extent of ground which the fowls are allowed to run over, beyond the limits of the enclosure. If the fowls are never to be allowed to go beyond the poultry-yard, then we should say that the very smallest space in which the common hen can be kept in perfect health, and so as to lay wholesome eggs, even though the poultry consisted of not more than a dozen fowls, is the sixth part of an acre. The reason why so much space is required for so few birds is, that, to be maintained in health, they require a good deal of exercise, and they also require to eat grass, and different other kinds of green food, and to pick up insects and sand, and other objects. Now, though all these articles might be supplied to them in the poultry-house, or in an enclosure in front of it of a few yards square, yet it is certain that the fowls, when so confined, would not keep their health. The criterion of the size of the poultry-yard, therefore, where the poultry are never to go beyond it, is the growth of the herbage on its surface; for, unless this be produced in a certain degree of vigour, it will neither support insects nor snails, and, consequently, the fowls will be deprived of their animal food; and every cultivator knows that grass and other herbage will not grow with vigour where it is much trodden on by fowls. Where different kinds of poultry are kept, for the supply of even a moderate-sized family, the poultry-yard, where there was no other range allowed for the fowls, would require to be two or three acres in extent; but, as this would be expensive, and in many cases inconvenient, the poultry-yard ought always, if possible, to be so contrived as to have a free communication with the open fields, or even with the nearest public road. There is never any danger of losing the fowls, unless by thieves; for not only will they return home every night, at roosting-time, but at any time during the day, on the appearance of rain or thunder, or at any hour at which they are accustomed to be fed. When the poultry-yard is at some distance from a field or the public road, and a place intervenes into which the poultry cannot be admitted to run at large, a pathway through it may be fenced off, with open latticework on each side, and with a projecting roof to throw off the rain; or, if the distance does not exceed a few yards, a dry tunnel may be made under ground, or an enclosed way over a wall, or through any out-building. We have frequently seen tunnels of this sort under barn floors, to admit fowls from the farm-yard to the rick-yard, without entering the barn; and we conceive an enclosed gangway fixed to a garden wall, or to the fence of a frame-ground (such as is shown in several of Mr. Rutger's designs, in preceding pages of this work), would answer the purpose of conducting the fowls from the poultry-yard to the public road.

All Poultry-houses should have the following particulars attended to in their construction. They should all face the south or south-east; and all should have walls hollow, or of sufficient thickness to exclude extreme cold in winter, and excessive heat in summer. A 9-inch brick wall, like that employed in England, is not sufficiently warm, even for the climate of London, unless the bricks be kept so far apart as to form a vacuity of at least 2 in. in the interior of the wall. The height of the side walls should not be less than 7 or 8 feet; but 9 or 10 feet would be preferable, as admitting more perfect ventilation by openings, covered with latticework or wire, immediately under the roof. The side walls should be plastered with cement, or with sound lime and sand, so as not to harbour insects, and to admit of being washed to destroy their eggs, if any should have been deposited on them. The ceilings should, in general, be treated in the same manner, and the floors should be laid with smooth pavement, so as to admit of being washed out as clean as any kitchen floor. In every poultry-house there ought to be a window, with the sill about 3 or 4 feet from the ground; the sashes being glazed, and constructed so as to slide past each other in grooves. There should be inside shutters to these windows, for use during winter; and outside shutters, luffer-boarded, to close during summer, when the sashes are kept open all night. Close under the eaves, there ought to be an opening for ventilation, exactly over the window, of the same breadth, and a foot in depth, with a wire grating, and a shutter on the inside, to slide in grooves, and which should have a cord attached to each end, hanging down within reach, so as to draw it on or off at pleasure. The roofs of all poultry-houses should project sufficiently to protect from rain the side walls, and the outside stair or ladder, which is made for the fowls to get up to their roost; and there ought always to be gutters, to collect the water from the roof, so as to keep the ground immediately in front of the poultry-house doors as dry as possible. Each kind of poultry should have two apartments: one for roosting in, and the other for laying and sitting in; and, in large establishments, there may be two or three other houses, for fattening in, for confining individuals rendered unfit to be at large from their pugnacity or from disease, and also for rearing young chickens, unless there should be plenty of room for this purpose in the laying-houses.

Fattening Poultry. With respect to fattening poultry of every kind, it may be proper to observe, that the preceding recommendations as to fattening-houses have been given on the supposition that every kind of cramming and confinement in coops for fattening is to be given up; and that the fowls are to be kept constantly in good condition, and only confined for a week or two in a feeding-house, with a small yard attached, immediately before being killed. We may also observe that fowls, so far from fattening better alone and in solitary coops, actually eat more, and consequently fatten faster, when several are kept together, and free access allowed them to food at all hours of the day. A corresponding system has for some years past been adopted for feeding sheep and cattle, and with similar success. Having thus noticed the requisites common to all poultry-houses, a very few words will suffice to state what is peculiar to each.

The Hen-house, as generally lodging the most numerous and useful class of poultry in a yard, requires to be the largest. The roosting-house should contain, at least, a square foot of area for every fowl that is to roost in it; this average allowing rather more than a square foot for full-grown fowls, and less for chickens. The perches for roosting on may either form a slope from within 3 ft. of the ground, to within the same distance of the ceiling; or they may form a floor 4 ft. below the ceiling; the perches being placed across the house, and about 18 in. apart. The form of these perches ought to be square or angular in the section; for gallinaceous fowls cannot bend their toes so as to grasp a round perch. The entrance for the fowls should be on a level with the roosting-floor, from which the ascent and descent on the outside may be by a ladder formed by nailing fillets of wood at regular distances across a board, to serve as steps; or, which is more architectural, by neat brick or stone steps, projected from the wall. In the inside, there may be a portable wooden ladder, to enable any chickens which may have fallen from the roost during the night to get up again in the morning, so as to go out with the other fowls, at the opening at the top of the outside stair. There ought to be a shutter to this opening, which should be carefully closed every night, after the fowls have gone to roost, in order to exclude vermin. The great advantages of having the perches all on one level, on what is called a roosting-floor, instead of having them sloping like the stage of a green-house, are, that the whole of the fowls roost in the upper, and consequently warmest, part of the house; that there is no scrambling among the stronger fowls to get to the highest perch, in consequence of which the weaker ones are often thrown down and hurt; and that there is no temptation to the stronger fowls to fly up to their perches at night, or fly down in the morning, which always deranges the weaker ones that are obliged to go up and down to their perches by the ladder. The shutter to the opening at the top of the outside ladder should be opened every morning at sunrise; or, when the sun rises earlier than five or six o'clock, at whatever hour the workmen are accustomed to go to work, in order that the fowls may get out to pick up snails, worms, and insects, while the dew is on the grass, and before these vermin have returned to their holes in the soil. Afterwards, the large door and all the windows of the roosting-house should be opened, and left in that state till towards the time when the fowls usually go to roost. The floor, in the meantime, should have been carefully washed out, so that, on the return of the fowls, they may find their roosting-house perfectly clean and sweet in every part.

The Laying and Sitting House for the common fowls may be of the same dimensions as the roosting-house; but, instead of being furnished with a framework of perches near the ceiling, there should be a number of boxes about 18 inches square, each with a little doorway cut in it in front, for the hen to enter and come out. A row of these boxes should be ranged along the side and back walls on the floor, for sitting-boxes; above these, if many fowls are kept, may be another row of boxes for laying in, with similar doorways, and a ledge along them in front, broad enough to allow the hens to walk along it, and which may be ascended to by a ladder (like that of the roosting-house) at each end. When the hens seem inclined to sit, the requisite number of eggs is put into one of the lower tier of boxes for the hen to sit on; and she should be supplied with food and water in the house, to prevent her from leaving her eggs for more than a few minutes at a time. Some authors direct the laying-boxes to be always raised 3 ft. from the ground; but this is unnecessary if only a few fowls are kept, and the floor is quite dry, and where the hens have a different house for laying in, from that in which they roost. As the floor of the laying-house will not be so dirty as that of the roosting-house, and as a moist floor would occasion cramp, and consequently death, to the young chickens, it should never be washed, but only strewed with clean sand every day, the dirty sand of the previous day being first swept out.

Warming the Houses for common Fowls. Wherever it is intended to have an abundant supply of eggs in the winter season, provision should be made for supplying artificial heat to all the houses occupied by the common fowls. The temperature required should not be less than 60°, because in no case do fowls lay more abundantly than in common cottages, where they may be said to live with the family, and where the temperature varies in the winter season from 50° to 70°. The means of supplying heat may be a smoke flue conducted along the back and sides of each of the houses (but not along the front, as it would interfere with the entrance); or hot-water pipes may be used in a similar manner; or there may be a small stove in each house. By a little contrivance one fire may be made to heat three or four houses. It may also sometimes happen that the common fowl-houses may be built over a cow-house, or over a dung-pit; or what might frequently be the case, a lining of dung might be heaped up, when the artificial heat was wanted along the outside of the back wall of the fowl-houses. In general, the difficulty is not so much in contriving means of heating, as in convincing persons who keep fowls, that artificial heat is necessary to induce them to lay. Every body knows that eggs are cheaper at midsummer than at

Christmas; though but few people are aware that the only reason for this is, the increased warmth of the former season.

A Fattening-house for common Fowls may be of the same dimensions as the roosting-house and the laying-house. Where the fowls are fattened in coops, these should be placed round the walls; but, as we recommend the free method of fattening, that is, allowing them to be at liberty, with plenty of food to eat when they please, we consider no furniture necessary, farther than some roosting-poles, which may be placed round the walls at the distance of 3 or 4 feet from the ground, and vessels for containing the food and water, which may be set in the middle of the floor. In front of the fattening-house there should be a space enclosed as a yard, for the fowls to take exercise in; and in this yard they may always be fed in mild weather.

The Turkey-house may very properly adjoin the common poultry-house, and it need not differ from it in any particular, except that all the details of fitting up may be one fourth part larger. The opening at the top of the outside stair should be one fourth higher, and one fourth wider; and the roosting-poles should be 2 ft. apart. There should be a laying-house, in which the boxes may be 2 ft. square, and also a feeding-house with a feeding-yard.

Houses for Guinea Fowls and Pea Fowls are not wanted, as these birds can seldom be brought either to roost or lay in any regular house. An open shed or a tree will generally be found their roosting-place, and if the poultry enclosure be surrounded by a thick hedge, or if it contain clumps or thickets of shrubs, their nests will commonly be found there.

Houses for Aquatic Fowls. Where only a few geese and ducks are kept, one house will be sufficient for them, provided it be divided into three parts; one for the common lodging-place, one for laying and sitting, and the other for fattening. No other furniture or fittings up are requisite than boxes for the laying and sitting house; those for the ducks may be 18 in. wide, and 2 ft. long; and those of the geese proportionately larger. It may be here observed, that, in order to keep ducks and geese, it is not necessary to have either large deep ponds, or running water. A basin of a few yards in diameter, and deep enough to admit of their swimming in it, will be found quite sufficient, provision being made for a frequent supply of fresh water. Where geese are kept in any quantity they require a yard by themselves, and an extensive range of pasturage, as they are fond of grass, and it appears essential to them; and as, when confined with other fowls, they become very pugnacious and very much harass hens and turkeys.

Fowls may be kept on a small Scale, so as to supply the family with eggs, by purchasing hens in a laying state, and furnishing them with a portable wooden house, containing a sufficient number of nests; placing this in a warm situation, as in a stable or cow-house, or adjoining a kitchen or other room having constantly a fire in it; and contriving free ingress and egress from the public road, or from some extensive space, in which the fowls can run about and find vegetables and insects. Besides this house for laying in, there would require to be another adjoining it, and of the same size, with perches for the fowls to roost on: a wooden house 6 or 8 feet high, and about the same length and breadth, would be sufficient for roosting ten or twelve full-grown fowls, and one of half that size would be sufficient for containing nests for them.

A very common mode in which small fowl-houses of this description might be heated is by a pipe of hot water, communicating with the cistern at the back of the kitchen fire; and many other methods might be suggested. In short, if the reader will bear in mind, that the common fowl, in order to lay abundance of wholesome eggs, requires abundance of farinaceous food; an extensive range of surface for exercise, and for picking up green meat, insects, worms, and other animal food, and the small stones and gravel necessary for digestion; and that when the fowl is not in active exercise, it should be in a temperature of between 50° and 60°, he will be at no loss for contrivances not only to keep fowls, but to insure an abundant supply of eggs during the winter season, whether his residence be large or small.

The Pigeon-house, or Dovecot. — The common pigeon, of which there are many varieties, may be kept in a small house, in a manner similar to common fowls; but it succeeds better in buildings somewhat elevated, or in low buildings in which the place of entrance is made in the roof; because pigeons fly higher than any other domesticated birds. A very convenient situation is a loft over some other building, or when there are various out-buildings, a turret may be added where it will have a good effect in an architectural point of view, and the interior turned into a place for pigeons. All the fitting up requisite is to place nests against the wall; these nests consist of open boxes, about a foot square, with the lower side projecting 3 or 4 inches for the birds to light upon. The best food for pigeons is peas, but they will eat oats and barley like the common fowl. Where pigeons have not an extensive range, it is common to place in the pigeon-house a lump composed of lime rubbish, loam, salt, and other matters, for them to peck at occasionally, and which it is found prevents them from eating the lime from chimney tops, and other parts of buildings which they frequent. In general, a pigeon-house may always be made an ornamental object, whether it forms a detached building, or an appendage to other offices. As young pigeons are highly valued in the kitchen for pies, and full-grown ones for substantial, savoury, and stimulating food, a pigeon-house is a very desirable appendage to every suburban residence that has land attached; but, where there is no land, it is scarcely fair to keep pigeons, as they must necessarily live on our neighbours.

The Aviary is an enclosure which should always, if possible, be of considerable extent, containing houses for different kinds of birds. These houses may be small rustic structures, each with an enclosed court, and covered with netting or wirework, to prevent the birds from flying away, as well as to prevent the intrusion of other birds in a wild state. In these small enclosures the young are reared; and, when full-grown, they are allowed during the day to go at large in the aviary ground. An aviary, where there are five or six different kinds of birds to be kept, should never be attempted where there is less than an acre of ground; because few things disfigure either grass or plants so much as a superabundance of fowls among them, while the injury to the birds themselves, for want of space for exercise, is proportionately great. The birds commonly kept in aviaries in the neighbourhood of London are, gold and silver pheasants, partridges, quails; red, black, and wood grouse; pigeons, turtle-doves, Muscovy ducks, the Canadian goose, the bustard, gulls, and curious varieties of the common fowl, such as bantams, &c. There is however no limit to the number of species and varieties of birds that may be kept in aviaries, provided there are abundance of space, and ample pecuniary means. A number of the exotic birds require artificial heat in their roosting-places during the winter season; but this is not the case with the kinds mentioned above as commonly kept in aviaries. The handsomest places of this kind that we know, near London, are at Woburn Abbey in Bedfordshire, and at Cobham Hall in Kent. There was formerly a tolerably complete aviary at the Duke of Devonshire's villa, at Chiswick; but it has been given up for the last ten years. The most complete aviary in England for singing birds, is at Knowlesly Hall, near Liverpool.

The swan is one of the greatest living ornaments to a piece of water of two or three acres in extent; and it is useful in keeping the surface of the water free from weeds. Swans require very little care from man, except that of feeding them during severe weather in winter, and having a small house for the protection of their young. Whenever there is a considerable piece of water in the grounds of a residence, there may not only be swans, but ducks of different kinds, curious geese, gulls, and other aquatic fowls. The swans, as we have above observed, are useful for keeping the water free from weeds, and the others for keeping the adjoining grounds free from worms, slugs, snails, &c.

Whoever intends to pay much attention to birds, whether

those adapted for the aviary or common poultry, will resort to some book or books on the subject, as guides. One of the most recent on common poultry is Dickson's *Poultry*, already mentioned. Bechstein on *Singing Birds* is a useful work, as is Thompson's *Bird-Fancier*. In general, it may be observed, that the keeping of birds of any kind is a great source of recreation and amusement to females living in the country, where there is abundance of room; but where this is wanting, and the birds have not sufficient range, however abundant the supply of food may be, the result will only be disappointment, filth, and disease.

The Rabbitry. — Rabbits may frequently be kept in the aviary, though, unless the extent of ground be considerable, they are better in an area by themselves. There are two kinds of rabbits reared for the market; one is the common wild rabbit, which is kept in warrens, and the stock taken by nets or traps, as it is wanted to be used or sent to market; and the other, the animal in a highly artificial state, kept in hutches or small boxes, and fattened like fowls for killing: the flesh of rabbits of the latter kind cannot be considered as wholesome; and, leaving the treatment of them to those who advocate cramming fowls and stall-feeding oxen, we shall close what we have to say on this subject, by recommending the keeping of rabbits, of what are called the wild or common kind, in a house with a small yard attached, to serve as a warren or place of exercise. The house may form one of the same range as the poultry-houses; and the warren may either be close to the rabbit-house, or at any convenient distance from it, communicating with it by a covered way. The artificial warren may be 50 or 60 feet square: if the subsoil be dry, the surface or floor of the warren may be sunk 2 or 3 ft., and surrounded by a wall rising 3 or 4 feet above the natural surface of the ground; but, if the subsoil be moist, then the warren may be made in the form of a raised ridge, and surrounded by a dwarf wall, surmounted by an open fence. The use of the warren is for the rabbits to take exercise in by burrowing; and hence, when the soil is not sufficiently free for this purpose, it ought to be made so by deep trenching, and by the addition of sand. In order to admit a free circulation of air to the warren, and also to admit of seeing it at pleasure, it should never be surrounded by a high close fence; but a sunk wall will always be necessary, to prevent the rabbits from burrowing into the adjacent grounds. The rabbits should never be fed in the warren, but always in the rabbit-house; and this should be done at stated times, in order that the rabbits may acquire the habit of appearing there regularly. See, on this subject, the *Encyclopædia of Agriculture*, 2d ed., p. 1073., and the *Quart. Jour. of Agr.*, vol. ix. p. 30. The article referred to in the latter work is entitled " A Domestic

Rabbit Warren," and will be found well worthy of perusal by every one who intends keeping rabbits.

Guineapigs are amusing little animals, and may be kept in a small house with a little yard to run about in.

The Apiary is another source of interest to all who live in the country, and fortunately it may be indulged in by the humblest labourer, no less than by the wealthiest citizen, provided there are fields and gardens in the neighbourhood containing flowers. A beehive, when there is no room for it any where else, may, like a pigeon-house, or even a garden of pots, be placed on the roof of the house. Much has been, and continues to be, written on the subject of bees; and the kinds of hives are proportionately numerous. Instead of pointing out what we consider to be the merits and defects of the principal of these, we shall limit ourselves to observing that, where little or no attention can be paid to the bees, except perhaps at the swarming season, the common hive of the country, whatever that may be, for example the straw hive in Britain and on the Continent generally, the trunk or pipe hive in Poland, and the cork hive in Spain and the Canaries, will in our opinion be found the best, because every body understands it; but that, where there is leisure, and a disposition to attend to bee culture, Nutt's hives are by far the best that have been yet invented. It has been a great object with the inventors of hives to devise means for taking the honey without killing the bees; and Mr. Nutt not only effects this, but what is of incomparably more importance, he prevents young bees from being generated, except when they are wanted, and consequently prevents swarming with all its attendant troubles. The principle upon which all Mr. Nutt's improvements are founded, is that of regulating the temperature of the hives, so that the bees may breed in one temperature, and make their honey in another. Under a certain degree of heat, the queen bee will not lay eggs, nor will these eggs be hatched; while the process of collecting and storing up honey goes on without much reference to temperature, provided the sun shines. Nutt's hive requires to be placed under some description of cover or bee-house. This should,

306

307

in general, be so contrived as to leave free access to the hive behind, and hence it can never be placed against a wall or against a house. It may be in a detached building, consisting of a rustic structure covered with bark; or it may be placed under a roof open on every side, the props being rustic pillars, and the roof being covered with thatch, reeds, woodman's chips, spray, bark, heath, or similar materials. *Fig.* 306. shows a handsome bee-canopy of this kind, covering one of Nutt's hives, which stands in a recess in the pleasure-ground at Chipstead Place, in Kent. At Bayswater, our Nutt's hive is placed in the front of a veranda (see *fig.* 307.), in a line with its pillars, and is consequently protected from perpendicular rain; but as the excessive heat of summer is equally injurious with rain, it is protected from that, and from the sudden influence of either heat or cold in winter, by a casing of broom and heath. The back of the hive, where the doors are, on opening which the bees may be seen at work, is most conveniently examined from the veranda.

The Fishponds. — The custom of keeping fish in the grounds of country residences is much less common now, than it was in the days when, from the whole country being Catholic, fish was essential as an article of food two days in every week; and when the communication between the interior and the sea was so slow, as to be unavailable for the transport of fish. Nevertheless, fishponds, wherever they can be made, are not only sources of beauty in the landscape, but of interest and use with reference to the fish that may be reared and fed in them. In every garden, however small, and even in every green-house or conservatory, there may be a vase or small basin for gold and silver fish. These require very little care, whether in the house or in the open air, except, in the latter case, breaking the ice in winter, to admit air when the water is frozen over. Should the fish, however, be intended to breed, the pond must be in a warm situation, fully exposed to the sun, so as to raise the temperature of the water early in the season; and the margin must be shallow and sandy, as it is only in shallow water on a sandy bottom, or on roots or bundles of sticks, that fish will deposit their spawn. The carp, the tench, and the perch are the most convenient fishes for managing in artificial ponds; and, throughout Europe, they are more used for this purpose, than any other kinds. Of these, the carp is incomparably the best, on account of its astonishing fecundity, its large size, and the rapidity with which it grows when well fed, notwithstanding the great age which it has been known to attain. To manage carp properly, three ponds are requisite; one for breeding, another for rearing, and a third for feeding. In the spawning or breeding pond, full-grown fish should be put early in spring; the season for spawning being from the latter end of May till the beginning of July, the time varying according to the warmth of the season. After spawning, the old fish are put back into the feeding-pond, and the young fry left to themselves till the spawning season approaches in the following year. They are then removed to the nursing-pond, where they remain about two years, the time varying according to their growth. Every season the nursing-pond is drawn, and all the fish which are above 5 in. in length, put into the feeding-pond; whence the largest are taken out as wanted for use. When wanted of extraordinary size, they may be kept in stews, and fed with garbage, boiled potatoes, bread, boiled rice, or any soft substance which does not require mastication. Their natural food consists of the larvæ of insects, worms, and soft aquatic plants. The pond in which the feeding fish are kept, should be rather deep, and have a soft marly or muddy bottom, and a warm exposure; the water should be soft, and this it generally is on marly soils: when carp are fed in stews, they should be kept in rain water. A carp will usually attain the weight of 3 lbs. in six years, and 6 lbs. in ten years. The largest ever caught was not quite 20 lbs. weight. They are in season from October to April. The whole business of stocking ponds, and raising and fattening carp, is reduced to a regular system, which is practised extensively in the interior of France and Germany, and more particularly of Prussia. But in suburban gardens, it may be tried with a single pond; taking care to reduce the number of fish by using some of the largest every year in the proper season. From 300 to 400 carp to an acre, is the number allowed in the feeding-ponds in Prussia; but in these ponds very little food is given, and hence the range required is the greater. The tench is generally kept in the same pond as the carp, and requires the same treatment. It will, however, thrive, and even attain an extraordinary size, in situations, and in stagnant fetid water, where no other fish would live. The tench is very tenacious of life, and requires less oxygen than any other fish. Mr. Yarrell, in his excellent work on the *British Fishes*, says that the tench can breathe when the quantity of oxygen is reduced to the 5000th part of the bulk of the water; ordinary river water usually containing the 100th part of oxygen. Hence, tench will thrive in deep muddy holes, where no other fish could exist; though, in this case, they should be kept a week or ten days in clear water, before using for the table, in order that the muddy taste may go off. The tench is covered with a thick slimy matter, which is said to have a healing property for other fish, if they rub themselves against it when they are wounded; and, hence, the tench is vulgarly called the fishes' physician.

The perch requires clearer water than either the carp or the tench, and will thrive in rivers, where the current is not too rapid. When kept in stews, there should be a stream of water constantly running through these. The perch eats worms, flies, and young fish of its own kind, and also minnows, and small roach, dace, &c. The eel thrives in deep, muddy, shady ponds, where, however, the water must not be stagnant. Eels attain the greatest size in millponds, or in muddy rivers, in the soft banks of which they can bury themselves 12 or 16 inches deep, while the stream continues constantly running its course over the mud, and where they are frequently found, and dug out or speared, in the winter. In stews they may be fed at discretion, with snails, frogs, worms, &c.; and will attain a large size, but they are seldom kept in ponds, as they destroy other fish. It is not known to many persons that the eel, in the summer season, frequently quits the water during the night, and wanders among grass in search of slugs, frogs, and worms; yet this is the case, and where there is a mill-pond, or a dam between two ponds, large eels may often be seen in a warm summer's evening, when the meadows are wet with dew, making their way, with an undulating, or wriggling, though not very rapid motion, through the long grass, in search of food, or from one pond to the other. The pike grows to a large size in stews or ponds, where it requires clear and hard water; but it is so voracious that its keep is very expensive, as it devours all the other fish, and even all the fry of its own species that come within its reach. "Eight pike," says Mr. Jesse, "about 5lbs. weight each, consumed nearly 800 gudgeons in three weeks." "In default of other fish," says Mr. Yarrell, "pikes will seize moor-hens, ducks, or indeed any animals of small size, whether alive or dead." They swim rapidly, and dart at their prey with great fierceness; and from this, their extraordinary strength and boldness, and their voracity, they afford more sport than most other kinds of fish to the angler. The roach and dace are sometimes kept in ponds, and have a beautiful appearance in the water, but their flesh is insipid and woolly. Gudgeons, and other common small fish, are usually found in rivers; as are trout, which, however, may be kept, and will attain a large size, in ponds which have a hard rocky bottom, clear water, and a constant stream running through them.

Other Animals that may be kept in Suburban Residences. — Those who wish to add to their sources of recreation and amusement may keep land tortoises, much in the same manner as we have proposed to keep rabbits, but with a smaller exercising ground, and they may feed them with corn, meal, and succulent vegetables, and use them as food; or they may keep them in the open garden, being fed with lettuces, cabbages, &c., and allowed to burrow in the soil during winter. Tortoises, however, cannot be kept, except in mild situations, not much above the level of the sea, for they are very impatient of cold. The tree frog, which feeds on flies, might be kept in cages, or at liberty, in greenhouses and vineries; either merely as an ornament, or to destroy the flies and other insects: and the large green frog might be kept in ponds, and fed for culinary purposes, as it is in France, Germany, and Italy. The edible snail (Hèlix pomàtia *L.*) might be kept, as it is near Vienna, in large pits covered with boards, and fed with cabbage leaves and other vegetables.

The river crawfish (*A'*stacus fluviátilis) is not very common in Britain, but it abounds in some parts of the Continent, particularly in the marshy meadows of the Vistula, near Warsaw. These fish are reckoned a great delicacy on the Continent, either boiled and eaten cold like shrimps, or put into soup. In England they are found in meadows on the margins of the Trent; and it might amuse a curious suburban resident desirous of making experiments, to try to subject them to cultivation. The medicinal leech, which lives chiefly on the spawn of fish or of frogs, might easily be kept in small ponds, provided care were taken to prevent it from getting to the fishponds; during summer, the leeches come out on the grass in search of snails and other food during night, and during winter they go into deep water. The silkworm, wherever there is a white mulberry tree, or abundance of lettuces, may be reared, and silk produced as an amusement. The eggs, which may be procured in Covent Garden market in May and June, require to be hatched in a temperature of from 60° to 80°, on dry shelves, kept clean and well ventilated, and the worms require to be carefully tended and well fed. The silk, however, produced in England, is of no value but as a curiosity, from its want of tenacity; the thread only acquiring the necessary toughness to enable it to be reeled and spun with a view to profit, in a hot and dry climate.

The Icehouse. — Ice may be kept in a dry cellar with as much ease as coals, wine, or beer. All that is necessary, is to have the walls and roof of extra thickness, so as to exclude heat; or to have them built double, or battened, and lathed and plastered. By the last process, a vacuity is formed completely round the sides and roof of the ice-chamber; and a similar vacuity should be formed under the floor, communicating with a drain having a trap, so as to convey away any water that may collect from the thawing of the ice, without admitting fresh air by the drain. This cellar should either have double doors placed 2 or 3 feet distant from each other; or, when the ice is put in, an ample space should be left between it and the door, in order to allow room for a large quantity of straw, to serve as a nonconducting medium to the heat that would otherwise pass through the chinks of the door. In filling an ice-cellar, the ice, having been first collected and laid down near it, is broken into small pieces, and then pounded till it becomes a powder composed of particles not larger than those of sand or coarse salt. It is then carried into the cellar, and laid up in a heap, beating each layer as deposited, so as to form the whole into a compact mass, and occasionally sprinkling a little water over it, in order to consolidate it. An improved method consists in using water impregnated with salt, by dissolving 10 lbs. of common salt in 10 gallons of cold water, and pouring it on the ice through a common garden watering-pot, every 2 or 3 feet in thickness, as the cellar is filling. The ice, in cellars filled in this manner, will be found when opened in summer, to consist of one solid mass of ice, which cannot be broken without the pickaxe. It will keep much longer without thawing in the cellar, and also much longer when exposed to the open air; because salt water, and consequently salted ice, has a much less capacity for heat than fresh water or fresh ice. This mode of keeping ice was first adopted by Mr. James Young, gardener at Willford House, Nottinghamshire. He was induced to try it, because the icehouse there was placed in a very exposed situation, the sun shining on it from rising to setting. Before the use of salt, it was found impossible to keep the ice above a year; but, afterwards, it was kept three years, and the last of it was found to be as good as the first. (See *Gardener's Magazine*, vol. iii. p. 138.)

The cool temperature by which the ice must be surrounded, to make it keep, has suggested the idea of forming niches or shelves within the icehouse, or in the passage leading to it, for keeping articles cool in the summer season. This has been tried with great success in the case of common icehouses; and it would in all probability succeed in the case of ice-cellars, though the difficulty in this last case would perhaps be increased from the want of ventilation. We have already, in this work, expressed our opinion, that it would be more convenient to have the place for keeping the ice connected with the dwelling-house, either as a cellar under the living-rooms, or as a cellar under some of the outdoor offices, and to place it under the care of the butler, than to form it in the park or pleasure-grounds, and place it under the care of the gardener. However, as the common prejudice is in favour of icehouses built in the usual manner in the open grounds, we shall here describe one of these formed according to the best principles.

Out-of-door Icehouses. — Where the soil is wet, and does not admit of underground drainage, the icehouse may be constructed above the natural surface, and afterwards surrounded and covered with a mound of earth; it being farther protected from the heat by trees or shrubs, or what is perhaps preferable to either, by a covering of giant ivy. The common form of outdoor icehouses is an inverted cone, and the supposed advantage of this form is, that as part of the ice thaws, the remainder slides down the sides of the cone, and still keeps in one compact body. This is no doubt true; but the advantage by no means compensates for the difficulty of constructing an icehouse in the conical form. A plain square room, with double side walls, say a foot apart, a double arch over, and a double floor under, which can be built with the same ease as any common cellar, will, all other circumstances being alike favourable, keep the ice as long as any conical form whatever. Where there is a doubt of being able to exclude the heat, treble walls, roof, and floor may be resorted to; and the entrance, which should always be three or four yards in length, instead of being straight, may be made crooked, with a door at each turn. The space between at least two of the doors should always be filled with straw; and, to render the removal of this straw easy, when there was occasion to pass from the outer door to the ice, it might be put into two or more canvass bags, like immense cushions, which might be hooked to the ceiling and the sides, so as to close up every interstice. The space between the second and third doors may be widened, and fitted up with shelves for holding articles which require to be kept cool, but not to the same degree as if they were placed in the icehouse. *Fig.* 308. is a section, and *fig.* 309. a ground plan, of an icehouse on the inverted cone principle, but of an improved construction. This form, as usually employed in English country seats, frequently fails in keeping the ice, from not having double walls and double or treble doors, or from imperfect drainage; but the plan now submitted is free from these defects, and will keep ice throughout the year in any climate, if covered with a sufficient thickness of earth or straw. In this design, *a* is the well or cellar for the ice; *b*, a drain from its bottom, for carrying off such water as may be produced by the gradual thawing of the ice; *c*, a trap in this drain to prevent the external air from communicating with that of the icehouse; and *d*, a leaden pipe from this trap, connected with a small pump at *e*. The object of this pump is to allow of making use, during the heat of summer, of the very cold water produced by the thawing of the ice, either for cooling wines or water; and in order that there may always be a considerable quantity of the water produced by the melted ice in the trap, it may be made of considerable size, so as to serve as a reservoir or well. This water would be very nearly as cold as the ice, and might be used for all the same purposes; while it

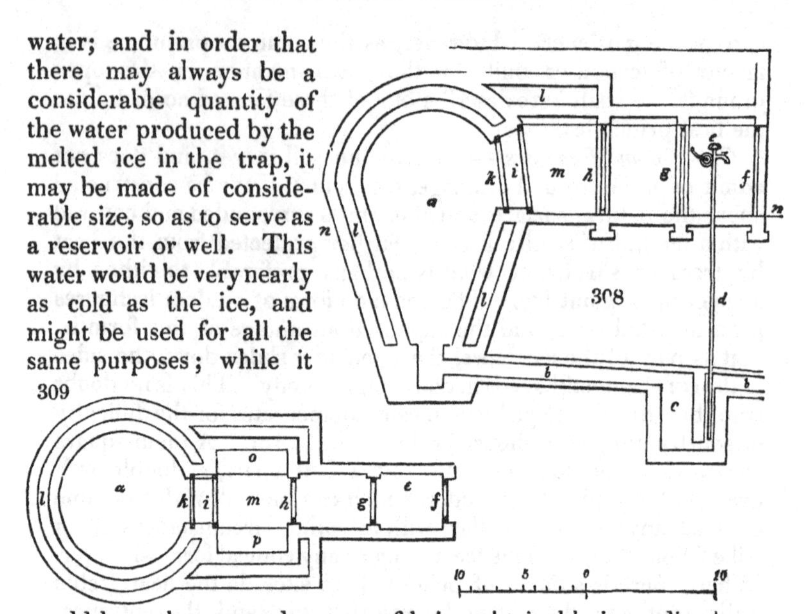

would have the great advantage of being obtainable immediately, whenever it was wanted, and with very little trouble. There are five doors to this icehouse, at *f*, *g*, *h*, *i*, and *k*; and a vacuity (*l*), one foot wide, between the two walls surrounding the cellar, and covering the inner division of the passage (*m*). This passage may be fitted up with shelves, as a place for keeping vegetables and various other articles of food (raw or cooked) fresh during summer. The cone containing the ice being supposed to be under ground, the natural level of the surface is shown at *n n*; and the whole superstructure may be covered, in Britain, to the depth of 2 or 3 feet with earth, planted with ivy and surrounded with trees. In warmer climates the depth of earth ought to be increased to 8 or 10 feet. The size of the well ought also to be enlarged, and there might be even a third vacuity round it. The space between the doors *i* and *k* should be filled up by a barley-straw cushion, and it would be well to have similar cushions against the doors *g* and *h*, at least during summer. The two recesses, *o* and *p*, are here shown only 2 ft. in depth; but, by making the walls behind them of brick, that depth may be increased at pleasure. It may also be observed, that, in situations where brick is cheaper than stone, all the walls in this design may be built hollow; and that the side walls of the ice-well need not be more than the width of a brick in thickness, the one wall being tied into the other. It may be further observed, that, if it should be inconvenient to cover the building with earth, a covering of straw or reeds, or even planting ivy against the outside walls, and surrounding the whole with a few trees, will be equally efficient in keeping out the heat. If trees cannot be planted, on account of the soil, or of shutting out any view, a slight roof elevated on props of any sort will have the same effect. One of the most effective icehouses that we have ever seen, was sheltered in this manner by a wire trellis, covered with Ayrshire roses, honeysuckles, clematis, and Virginian creeper. Icehouses, we are persuaded, would become much more general, were country gentlemen fully aware of the fact, that they might be built square just as well as round; and would be, at less expense, as effective above the surface as under it. A square icehouse above ground, or sunk 3 ft. into it, may have treble hollow floors formed of bricks on edge, covered with foot-tiles or flagstones; and its side walls may be treble also, and built of bricks on edge, To form the roof, a 9-inch semicircular arch may be first thrown on these walls, and over this arch three vacuities built of bricks on edge and tiles: there may be five doors, as in *fig.* 309.; and the whole may be covered with a cone of earth or thatch 4 ft. thick, and clothed with giant ivy. In fact, by the use of framework, or a stratum of loose stones to keep the ice from the ground and to allow the water produced by its thawing to drain away from it, and abundance of thatch, ice may be kept above ground just as easily as potatoes.

SECT. III. *The Scenery of a Suburban Residence.*

EVERY place may be considered as producing three leading impressions on the mind of a stranger: that called forth by the entrance lodge, gates, and the portion of the approach which is seen through the latter; that which is produced by the first good view of the house, obtained from the approach road; and that which is obtained from the drawingroom windows, after entering the house. There are many houses in romantic or other extraordinary situations, to which these remarks may not be strictly applicable; but that they are so to all country houses in comparatively flat or tame countries, we think few persons of observation will doubt. In this section, therefore, we shall submit a few supplementary observations on each of these points, commencing with the entrance lodge and gates.

The Entrance Lodge and Gates. — Good sense dictates that the architecture of the entrance lodge and gates should be in harmony with that of the house; and hence, if the dwelling-house and offices have any marked character of the Roman, Italian, Gothic, or other styles, the same ought also to be obvious in the lodge. On the other hand, if the architecture of the house and offices be mixed, or anomalous, so may be that of the lodge; or the lodge may be wholly in any one of the styles which is mixed up with the others in the mansion. That in every case the lodge ought to be in an humbler style of design than the more important buildings of the place, is sufficiently obvious from its more humble use. Independently altogether of style, care should be taken that the magnitude and seeming importance of the lodge be not such as to raise false expectations of the magnitude and importance of the house; at the same time, no lodge whatever ought to be so small as to raise doubts in the spectator of its being a comfortable habitation for a human being. This idea ought to be borne constantly in mind, as well as those which respect style; for few things connected with an elegant country residence are more offensive to a humane mind, than to see boxes set down as lodges, not more, perhaps, than 10 ft. or 12 ft. square, with rooms not above 7 ft. high in the clear. When this is the case, and the occupant has children, he is sometimes obliged to live with half his family in a box on one side of the gate, and to send the other half to sleep in a box on the other side.

Gates. Whether a gate ought to be of openwork that may be seen through, or of close boards so as to obstruct the view, is a question that has reference chiefly to places in the immediate vicinity of towns. In all country residences, there can be no objection to the gates being open; and there is an advantage to the public in their being so, as it enables the passing stranger to form some idea of the place, and also tends to ornament the country. On the other hand, it is equally clear to us, that the gates of most small places near towns ought to be close; because one great object of every country residence is privacy, and, in a small place, without close gates, privacy is impossible. At the same time, in residences near towns, where the house forms part of a street or row, and where the entrance door is only a few yards from the road, an open gate allows the servants to see any person who may ring, either from the entrance door or the kitchen window; and thus enables them to avoid losing their time by going to open the gâte for beggars and other intruders. In similar situations, it is often convenient, when there are close gates, to have a small grating, with a slide within, in order to permit the servant to speak to any person calling at the gate before opening it. A door is also sometimes made in the carriage gate, to facilitate the ingress and egress of persons without opening the entire gate. Close gates may, also, be sometimes desirable in point of effect, even where they are not required to insure privacy. Wherever there is any very striking object within the grounds, which can be seen from the entrance gate to better advantage than it can be from any other situation, then we should say that a close gate was desirable; in order that the effect of the object alluded to might not be marred to a stranger, by his seeing it imperfectly before he entered the grounds; and thus getting his first impression of it disfigured by the bars of the open gate. At Blenheim, a more striking view of the lake and bridge than is to be obtained in any other part of the grounds, bursts upon the eye of the stranger, when the close gates at Woodstock are thrown open; but if open bars were substituted for these heavy doors, the spectator would see the lake and bridge first through them, and the striking effect, which now produces a strong emotion of admiration and astonishment, would be frittered away.

The style of the entrance gates ought to be as much attended to as that of the lodge. They ought neither to be so plain as to resemble common field gates, nor so covered with ornament as to be disproportionate to the style of the house. Whatever style is employed in the house and lodge, the lines, forms, and

Entrance Lodge, at Chequers, Buckinghamshire.

ornaments of that style ought alone to be employed in the gates. In the case of houses in the Elizabethan, or in the Gothic style, this would give rise to many rich and beautiful designs; very different, indeed, from those cast and wrought iron patterns manufactured by wholesale, and set up without the slightest regard to the style of the lodge or house to which they belong; but which, from their cheapness, are now so very generally adopted by country gentlemen. The sight of these gates, at the entrance to a place, forbids all idea of taste pervading the interior of the grounds. The great point for an artist or an amateur to bear in mind is, the difference between a gate which is merely to serve as the movable part of a fence, for the purpose of allowing persons and things to pass from one side of the fence to the other, and a gate which is to serve as a portal to a human residence.

The humblest cottage, if this idea were borne in mind, would be approached through a gate superior in style of design to the gate or door into the adjoining field; while, at the same time, it would be inferior to the gate of a farm-house. We shall now give two or three designs for entrance lodges and gates, referring those of our readers who may wish to know more on the subject to our *Encyclopædia of Cottage, Farm, and Villa Architecture*, p. 997. to p. 1006.

Fig. 310. is a lodge in the old English manner, designed by E. B. Lamb, Esq., and executed at Chequers, in Buckinghamshire, the seat of Sir Robert Frankland Russell, Bart. The walls are composed of stone and flints, both found on the spot, the materials being placed alternately, so as form a chequered surface, in allusion to the name of the place. The roof is thatched with Kyanised straw.

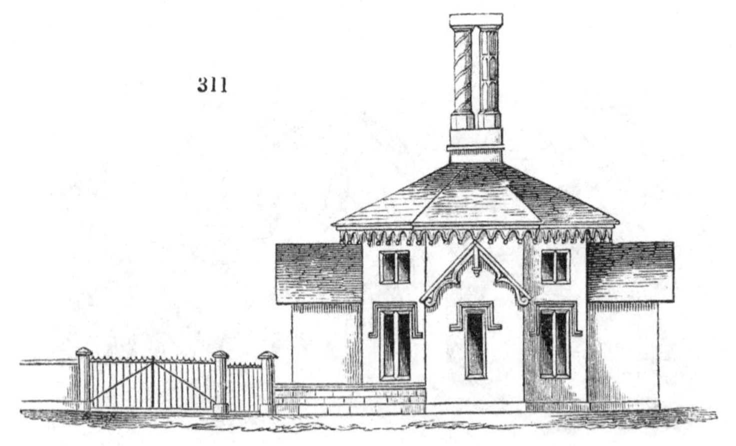

Fig. 311. is the elevation, and *fig.* 314. the plan, (both to a scale of 1-14th of an inch to a foot,) of a Gothic entrance lodge of the smallest size, by Edward Brigden, Esq., Architect, Bristol. The ground plan consists of a living-room (*fig.* 314. *a*), with a large recess (*b*) next the road. The entrance is through the porch *c*, and *d* is intended for a scullery; *e* shows a staircase leading to an octagonal bedroom over the whole. The most appropriate material, says Mr. Brigden, "for the walls of this cottage, would be flint or ragstone; the latter laid in random courses, and neatly tuck-pointed, has a very good effect. The dressings might be of Yorkshire stone, or any other which would harmonise with

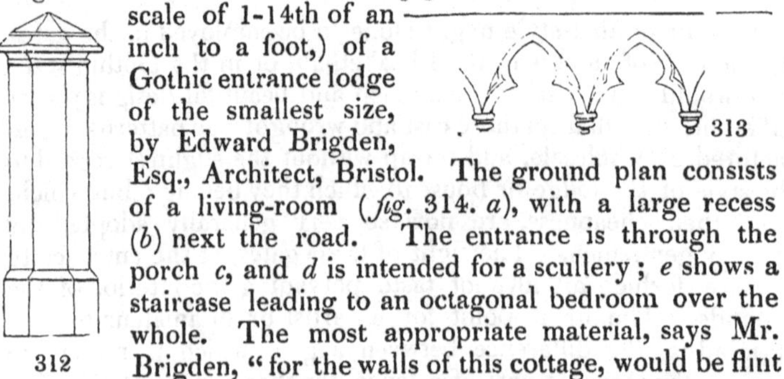

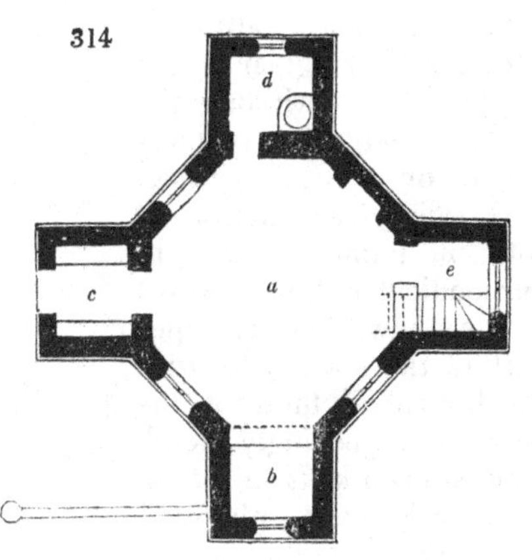

the colour of the wall. The same stone might be used for the gate piers; or, if magnesian limestone could be obtained in sufficiently large blocks, it would answer the purpose well (it being properly polished). The form of these piers may be as shown in *fig.* 312. The roof of the lodge may be covered with slate, or with stone tiles. The flues should be conveyed to the centre of the building; and the chimney stacks, which are ornamental, may be of Austin's artificial stone. *Fig.* 313. on a larger scale, shows the eaves-boards which may be made of deal."

Fig. 315. is the elevation, and *fig.* 316. the ground plan, of an old English cottage, formed of wooden framing, raised on a

cyclopian substructure. This cottage was designed by William Wells, Esq., and is erected on his estate of Redleaf, in Kent.

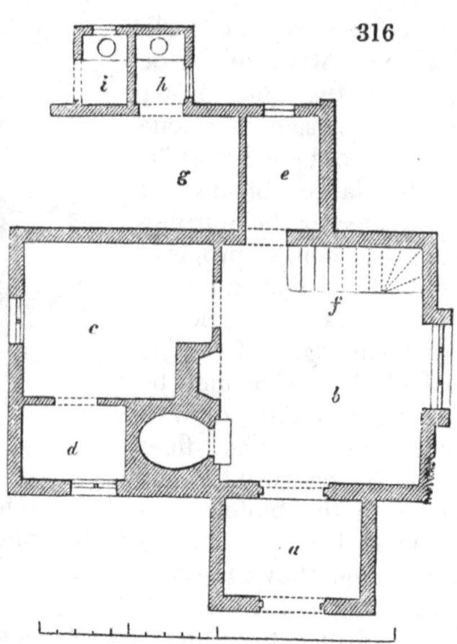

316

The cyclopian walls are "formed of irregular blocks of sandstone, without the slightest indication of horizontal or vertical courses. The effect, as contrasted with the numerous straight perpendicular lines formed by the studwork in the upper part of the walls, and with the horizontal lines of the roof, is exceedingly good. The studwork is filled in with brickwork plastered over; the smoothness and finished appearance of which, as contrasted with the rudeness of the cyclopian part, is forcible, and at the same time pleasing. A great beauty in this cottage results from the horizontal division of the upper part of the roof, which projects slightly over the lower part. The chimney top is massive and original. The whole was executed by local carpenters and masons, from the sketches of Mr. Wells, out of the timber and stone produced by the estate. The plan (*fig.* 316.) shows a porch (*a*), kitchen (*b*), parlour (*c*), light closet (*d*), pantry (*e*), and a staircase (*f*) to two good bedrooms above, and to a cellar under the parlour below; also an open shed (*g*) for fuel: *i* is a water-closet for men, and *h* one for women and children." The oven in the kitchen is sufficiently large to admit of its being heated with faggots. The roof of the lodge is covered with pantiles; but in America, or any country where wood is cheap, it is, from its high pitch, particularly well adapted for shingles. Where stone is not plentiful, the lower part of the walls may be of brick or mud, on a foundation of masonry.

Fig. 317. is an elevation of what might form a very suitable lodge for a thatched villa, perhaps of such a style and magnitude as that shown in the design, *fig.* 318., by Mr. Lamb. We do not give the ground plan of either, the object being simply to show that where the villa is not in a very high style of art, or on a large scale, the lodge is not to bear the same relation to it, that it would do to a mansion, but, on the contrary, that there ought to be

317

318

much less difference between them. In such a case, while the proprietor lived on the most beautiful part of the estate, in such a cottage as is shown in *fig.* 318., the superintendent might live at the farm-yard, or at the entrance to the private road, in such a cottage as is shown in *fig.* 317.

319

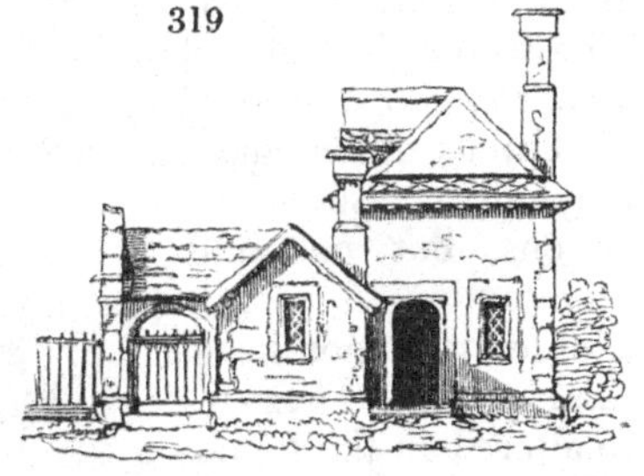

Fig. 319. is a design for an entrance lodge to a villa, where the house is in a somewhat higher style of architecture. This design, the preceding one, and the three following ones, are reduced from Robinson's *Rural Architecture*, published in 1829, a work which has materially contributed to improve the taste of the country.

320

Fig. 320. is a design for a gate lodge in a simpler style, but very well adapted for its purpose.

321

Fig. 321. is a gate lodge in the Swiss style, which, in a hilly situation, backed by a hanging grove of larch trees, would have a powerful and characteristic effect.

322

Fig. 322. is a design for a gate lodge in the Italian style. By contrasting this design by Mr. Hunt, with *fig.* 325. by Mr. Lamb, both architects of taste, it will be seen that massiveness is a leading feature in Italian lodges.

The first two of the following designs in the Italian style are from Hunt's *Picturesque Domestic Architecture;* a work which displays great taste in the elevations, but not sufficient attention to comfort and accommodation in the ground plans. These we have altered in the following figures, to supply what we thought was wanting in the originals.

323

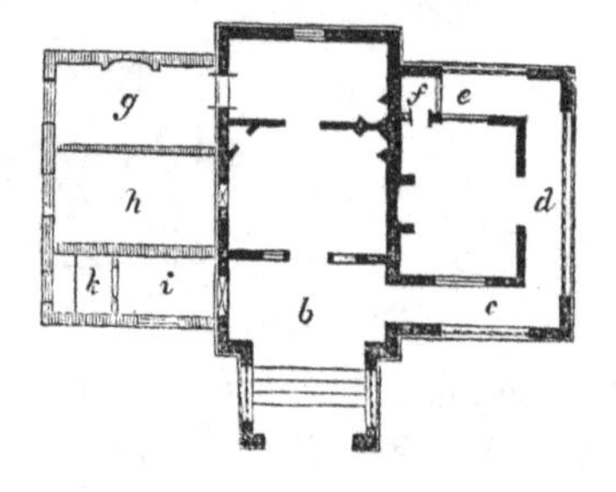

Fig. 323. shows the plan and elevation of what would form a very handsome gate lodge, and serve at the same time as a gardener's house. The porch is seen at *b*; *c* and *e* show an open gallery or veranda, embracing three sides of a schoolroom, which is entered at *d*, and adjoining which there is the closet *f*. The kitchen and back kitchen are in the centre of the house, and require no explanation; *g* and *h* are the two principal bedrooms; *i* is an open gallery; and *k* a water-closet.

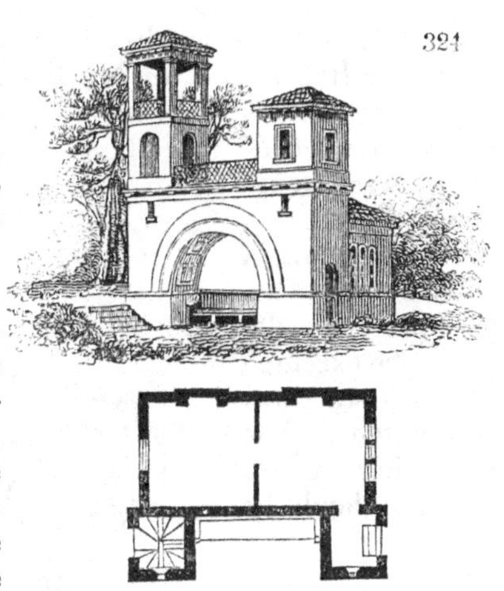

324

Fig. 324. is a gate lodge, the upper part of which forms a prospect tower. The gate and approach road are supposed to be on the right hand, and the sleeping-rooms to be over the arch (under which there is a seat), and in the lower tower. Such a lodge is perhaps suitable for an entrance gate on an eminence, where it may be seen from a great part of the surrounding country.

325

Figs. 325. and 326. form a design for an entrance lodge in the Italian style, by Mr. Lamb. The ground plan (*fig.* 326.) contains a porch (*a*), a living-room (*b*,) a sleeping-room (*c*), back kitchen (*d*), cellar for fuel, lumber, &c. (*e*), and store closet and pantry (*f*.) Behind, or at one side, there is supposed to be a sunk area, enclosed by a dwarf wall, as in the case of the lodges at Hyde Park; and in this area are contained several necessary appendages, and a door to the cellars under the house.

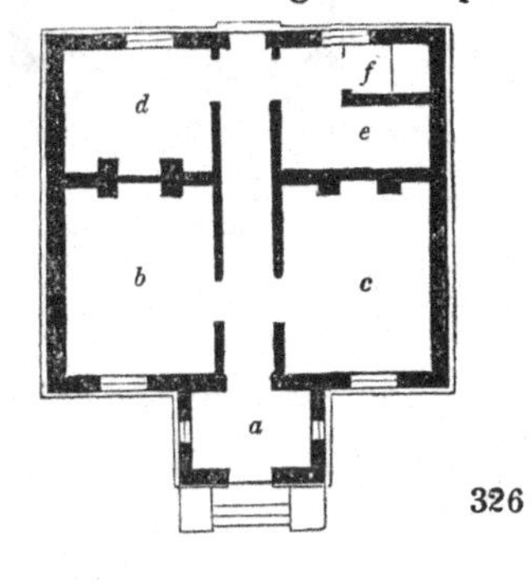

326

Figs. 327. and 328. are the ground plan and elevation of a gate lodge in the Italian style, by Mr. Brigden. The plan (*fig.* 328.) shows a porch (*a*), living-room (*b*), bed-room (*c*), wash-house (*d*), water-closet (*e*), and staircase, which leads to a room in the tower (*f*). This design may be executed in brick and stuccoed, or in freestone. The roof, being of a low pitch, should be slated; and the floors should be boarded, the sleepers being of oak, or of fir Kyanised.

327

Fig. 330. is a design for a lodge in the Grecian style, by Mr. Lamb. It is intended for the very handsome Grecian villa of

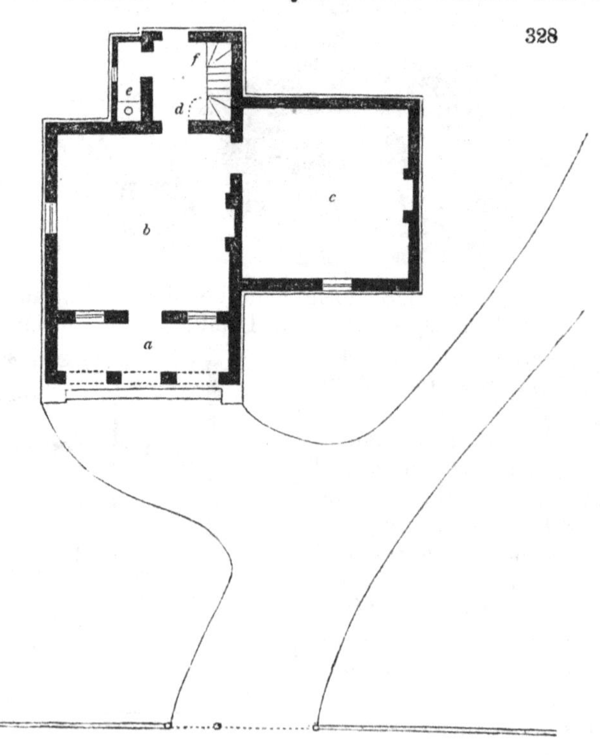

328

329

which *fig.* 329. is an elevation; (the ground plan, and other particulars relating to which, will be found in the *Arch. Mag.*, vol. v., for 1838;) and which is given here to show in a palpable way how the style and manner of the lodge ought to correspond with those of the mansion.

330

The Approach from the lodge to the entrance front should display the features of the grounds through which it passes, to the greatest advantage; or, if there be no striking feature worth looking at between the entrance and the house, the road ought to pass through a dense wood, or a grove. When it passes through open park-like scenery, trees should be sprinkled along it, in such a manner as to form a kind of running foreground, both to what is before, and to what is on each side.

The first view of the entrance front should, in general, be at an oblique angle, in order that two sides of the building may be seen at once; or, at all events, that so much of two sides may be visible, as to make it evident that the building is a solid mass, and not a mere screen wall. Where the approach is a straight avenue, advancing to the entrance front at right angles with it, a view showing two sides of the house cannot be obtained; and the building must in that case depend, for the effect it produces, on the height or breadth of its elevation, and on the circumstance of its being with or without open porticoes, projecting towers, or wings. There is no fixed distance, at which it is more desirable than at any other, to see the entrance front; but there is one fixed circumstance which especially requires the attention of the artist in laying out and planting the approach. This is, that the first view of the entrance front of the house which is obtained within the grounds ought to show it as the leading feature of the landscape, and to be, in fact, the best view of the entrance front that can be any where obtained. Before entering the lodge gates, if the house be seen at all, it has no right to appear otherwise than as a feature among other features of the general scenery of the country; but, within the lodge, the house is the main object of attraction, to which every thing else ought to be kept subordinate.

The View from the Drawingroom Front.—Having entered the house, and been shown into the drawingroom, the next grand impression made on the stranger ought to be by the view which he sees from its windows. This should be the most striking and the most beautiful view or prospect which the place affords. In order to be striking, it ought, if possible, to be considerably different from any views obtained between the lodge and the entrance front. The beauty which it exhibits may be of different kinds: it may be grand from the extent of prospect; bold and abrupt from the strong contrast and irregularity of its outlines; or it may be simply beautiful as a home view, from the undulations of its surface, the smoothness of its green turf, and the luxuriance of its trees, shrubs, and flowers. One of the most common modes of producing a striking view from the drawingroom windows of a house in a flat country is, by having the living-room floor of the house on a level of 10 or 12 feet above the natural surface of the ground; and by ascending to this level on the entrance front, not rapidly by a ramp or by steps, but gradually and insensibly by artificially raising the general surface, so as to give the house, on the entrance front, the appearance of standing on a natural knoll; and to prevent the stranger from suspecting that the ground on the lawn front, is not on the same level as that on the entrance front. When he reaches the drawingroom, therefore, and looks down on the lawn, the effect of the scenery is greatly heightened, by the commanding situation in which he finds himself so unexpectedly placed.

As circumstances which seldom fail of producing beauty in the view, we may mention extreme smoothness and high polish in the lawn, the branches of the trees and shrubs being allowed to recline on the ground, which shows that no cattle are introduced there; extent in every direction, but more particularly in the front, which prevents the idea occurring of confinement and limited property; irregularity in the boundary to the lawn, produced by scattered trees and bushes, creating variety and intricacy, which detain the eye and excite the imagination; and an architectural foreground, immediately below the windows of the house, which serves to connect and harmonise it with the grounds. We do not here mention the flower-garden, which is often placed on the lawn front of the house, because it may be sometimes desirable to have one in that situation and sometimes not; but the above requisites can seldom be dispensed with.

One of the most common faults in the view from the drawing-room front of the house is, want of breadth in the foreground. This commonly arises from too many objects being placed there; from these being too uniformly distributed over the whole; or from a clump, a walk, a pond, a tree, or some other object being placed exactly in the middle. A second fault, very commonly met with is, want of proportion between the foreground and the distance. Perhaps the foreground may be

331

View from the Lawn Front of Wimbledon House.

covered immediately in front of the windows with beds of flowers or of shrubs, which may occupy too large a space, or which may have grown so high as to shut out great part of the middle distance; or, in contradistinction to this, there may be no effective object in the foreground at all, when of course it will be overwhelmed by the large proportion of the view occupied by the distant scenery.

By effective object we mean the architectural appendages of the house, flower-beds, shrubs, or any other objects which rise up from the surface, and produce shade; which, as every one who has ever drawn a landscape on paper knows, is essentially necessary to a foreground; and hence, when artists have nothing in nature which they can copy into the foreground, they introduce the shadows of supposed clouds, or other objects, or human figures or animals, as in the view shown in *fig.* 331.: which view, without the horsemen and their dogs, would be nothing; while with them it is an absurdity, as the walk in the foreground, on which the horsemen are, is that on the lawn front of Wimbledon House, and is never used but for foot passengers. In *fig.* 332., which is a view from the entrance portico of the mansion at Kenwood, by E. B. Lamb, Esq., the figures are introduced with more propriety; though even they attract rather too much attention from the noble oak trees in the back ground.

A third fault, and one almost as common as the two that have been mentioned, is, want of harmony between the foreground and the distant scenery; not in point of extent, but in point of style of scenery, or of ornament. Thus we sometimes find an extensive lawn in the front of the house, which is continued in the same style of smoothness and high keeping till it terminates abruptly on a common, or in the hedgerow of a corn field, or an extensive wood, or some other uniform surface or mass of similar or greater extent than itself.

Perhaps the most common of all faults in the views from a country residence, next to want of breadth in the foreground, is, the monotony or deformity of the lines and shapes produced by hedgerows and plantations in the middle or third distance. When these faults are in the grounds of an adjoining proprietor, they of course may be considered as beyond the reach of correction; but in this case they are frequently at such a distance from the eye as to be inconspicuous, or to admit of being somewhat disguised by a few trees in the foreground. There are certain faults of this kind peculiar to every style of country. In some parts of Middlesex we have a monotony of hedges and pollard trees, with a total absence of ploughed fields. In some parts of Kent we have ploughed fields, with a total absence of pasture; and, in many parts of the lowlands of Scotland, we have the beautiful slopes of the hills cut across by stone walls, hedges, or belts of plantation, thus abruptly separating the arable plains from the hill pasture. To overcome or to mitigate difficulties of this kind, is one of the most common purposes for which a landscape-gardener is employed; and, as our friend Mr. Nesfield particularly excels in this department of his profession, we shall, with his permission, give an example which recently occurred in his practice.

332

View from the Entrance Portico at Kenwood.

Fig. 333. shows the outline of a range of distant scenery, from the drawingroom front of a suburban residence near Stafford. On the hill which forms the distance, there is a long belt or plantation, running most offensively parallel to its ridge; and swelled out in the middle, so as to form a clump-like protuberance there. This clump was made to conceal a high broken bank of

333

Distant Scenery near Stafford.

p, Slope of hill, descending from the back of the belt. *w*, Portion of a more distant ridge. *y y*, Thorn hedges. *z z*, Single thorn trees.

334

Same View, as proposed to be altered.

a, Steep rocky bank. *b*, Patches of furze. *c*, Conical hill, rising in front of the more distant ridge *d*.

rock, which was considered by the planter, or his employer, as an object that ought to be shut out; probably, because it was neither productive of grass nor trees: for persons who have no idea of the enjoyments of taste, can see no beauty in anything that is not applicable to common purposes of utility. This plantation is shown surrounded by a thorn hedge, which, having been long left uncut, has attained a great height, and thereby renders the outline of the plantations as conspicuous and hard as possible. The outline at *p* indicates the slope of the hill descending from the back of the belt; and the line at *w* shows a portion of a more distant ridge, which, it is proposed, partially to plant; in order, by the contrast of the trees with the line of the unplanted ground at *p*, to render the ridge more conspicuous to the eye. By planting this distant ridge, a mass will be produced, which will appear to retire behind the belt, and thus increase the perspective effect, and raise in the imagination, the idea of this plantation extending beyond the other, and even down the other side of the ridge. At *y* are high and very conspicuous thorn hedges. At *z z* are single thorn trees, which were part of some hedgerow lines that marked the outlines of former fields; which, by taking out some, and adding others, may be thrown into ornamental groups.

Fig. 334. shows the same view as proposed to be altered. In this view, *a* shows the steep rocky bank as it will appear when the trees are cut away; *b*, patches of furze, left for the encouragement of game, at the request of the proprietor; *c*, a conical hill rising in front of the more distant ridge *d*. As in this scene, as it actually exists, there are numerous trees of transplantable size, of thorns, horsechestnuts, limes, elms, and sycamores, all very favourably circumstanced for removal, most of the additional outstanding trees in groups, shown in *fig.* 334., may be transplanted from one part of the grounds to another, so as to produce immediate effect at comparatively little expense.

The Connexion of the House with the Grounds.— It is allowed by all authors who have written on landscape-gardening, since the days of Uvedale Price, that the want of artistical connexion between the house and the grounds is the most glaring defect in English country residences. Houses are very generally seen rising abruptly from a surface of naked gravel or turf, or partially disguised by a few trees or shrubs, accompanied perhaps on the lawn front by some flower beds. The offices at one end of the house, and the green-house or some other projection at the other, are commonly half or entirely hidden by masses of plantation, so that nothing is seen of the main body of the edifice but the two fronts; and even these have their architectural effect often greatly injured by single trees immediately before them.

The principal circumstance which has led to the want of connexion between the house and grounds in English residences is, the sudden and inconsiderate revolution introduced by the modern system of landscape-gardening; by which all the ancient terraces and mural barriers were indiscriminately swept away, in order to give place to the characteristic features of the new style, viz. the undulating surface, and apparently unlimited extent of smooth lawn, the winding approach road, and the ah ! ah ! * ditch, or the invisible wire railing, which separates the lawn from the park. The powerful influence of novelty rendered this arrangement satisfactory at the time; but, after a certain period had elapsed, the incongruity of so much art being found in juxta-position with what was avowedly simple nature, was felt to be absurd. Attempts were now made to connect the house with the grounds by means of trees, partly scattered round it, close up to the doors and windows, and partly in masses of close plantation. In the course of twenty years, the trees and shrubs having grown up, this also was found an unsatisfactory mode; the trees obstructing the views of the surrounding landscape from the house, and of the house from the surrounding landscape; and, besides, rendering the house gloomy and damp within. It was now felt that it was not only necessary to connect the house with the grounds, but to connect it in a harmonious manner; and, instead of bringing the two extremes of highly refined architecture and simple nature in immediate contact, by planting trees close to the house, it was thought advisable to introduce, as a medium between the trees and the architecture, certain architectural appendages, either useful or ornamental; and gradually to unite these with the woody scenery of the place. The most general appendage of this sort, and one which is applicable to the smallest houses as well as the largest, and which, in our opinion, scarcely any country house ought to be without, is, a plinth round the outside wall, from which the walls of the house should appear to rise. This plinth, on the smallest and simplest scale, may be from 1 ft. to 2 ft. high, projecting from 2 in. to 4 in., and be continued round the house: beyond this there may be a platform 6 or 8 feet wide of gravel, from 18 in. to 2 ft. higher than the surrounding surface, and bounded by a slope of turf at an angle of 45°. Opposite the entrance door, this platform may be ascended to by three stone or brick steps, in the slope; and there may be similar steps descending to the lawn on the pleasure-ground front. In the case of houses on a larger scale, the platform may be much wider; and, instead of being laid with gravel, it may be paved with flagstones; for the slope covered with turf, a low wall may be substituted surrounded by a balustrade or other

* This foss, or ditch, is commonly termed a ha ha; but, on turning to the writers at the end of the 17th century, it will be found spelled ah, ah ; and Switzer gives it this name, even as late as 1729, when he published his *Introduction to a General System of Hydrostatics*, see vol. i. pl. 3. p. 130. of that work.

ornamental openwork. The flight of steps may also have side or spandril, walls, appropriately finished and decorated.

In the case of some houses, in particular situations, there may be a second terrace or platform exterior to the first, on a lower level, and considerably broader. The walk on this second platform may be of gravel, and there may be a strip of turf between it and the first platform, which may be ornamented with shrubs or flowers, or it may be without these, according to circumstances. On the outside of the gravel walk, the lawn may extend indefinitely, and may either terminate in an architectural boundary at a few yards' distance, or it may extend to an ah! ah! or to a wire fence, the situation of which is not observable from the house. In short, while the first platform is in width and character chiefly influenced by the magnitude and style of the house, the outer or lower one ought to be jointly influenced by the style of the house, and the natural character of the surface of the ground. Where the character of the surface is at all marked by bold undulations, steep slopes, or abrupt transitions, these features ought to influence the lower platform more than the character of the mansion; but, on the contrary, where the grounds are comparatively flat, then the character of the outer or lower platform should partake of that feature in the scenery which is the most conspicuous, and which of course will be the house. Platforms or terraces, therefore, may be said to form the groundwork of the appendages to a house in the country. The appendages themselves are various, as, independently of the offices, which, we contend, ought to be only partially and not wholly concealed, there are the green-houses or conservatories, architectural seats or loggias, covered or open; connecting verandas, sundials, vases for flowers, basins for fountains, architectural baskets, and other mural compartments for plants or flowers; and, to mansions of a certain size, there may even be architectural watch-boxes. In the management of these appendages, an artist, without some invention and good taste, will be greatly at a loss; and therefore the proprietor of a house already built, who wishes to harmonise it with the grounds, should consider well on whose advice he acts. A mere architect is no more competent to advise in such a case, than a mere gardener or a mere landscape-painter. It must be a person who has directed particular attention to the subject, and who unites the knowledge of the architect, as far at least as the exterior of the buildings is concerned, with that knowledge of composition of general scenery which is necessarily possessed by the landscape-painter, and a considerable share of that knowledge of cultivation which is essential to the gardener.

Water is so desirable a feature in landscape, and so useful for various purposes in every country residence, that we think it advisable to add a few supplementary observations on the subject. Every dwelling-house requires a supply of water, and, when this cannot be obtained naturally by an overflowing spring, a river, or a lake, recourse must be had to some artificial process. The most common is that of digging a well, the mode of doing which we need not detail, as it is now an operation as well understood as building a house. A well is for the purpose of retaining water, which it collects either in consequence of penetrating to a perpetual spring, or of cutting through what are called surface or land springs. There is scarcely any situation in which a well of the latter description may not be formed. Both kinds of wells may sometimes have their supply of water increased by boring in the bottom of the excavation till a spring is reached; and boring may in some cases be employed alone, so as to bring the water either to the surface, or to within the depth of 10 or 12 feet, when it may be retained in a well. Boring, unlike well-digging, is only successful when it penetrates into a stratum containing water; which water communicates with, or is continued to, another part of the stratum, which acts as a reservoir, and which is sufficiently high to enable the water to rise, at the spot where the boring takes place, to the surface of the ground. The operation of boring, therefore, is but of partial application; as in many cases the subsoil is not stratified, and in others there is no reservoir sufficiently high, for the water, rising to its level, to reach the surface of the ground. Nevertheless, in favourable situations, boring will afford an abundant supply of water throughout the year; but such situations can only be discovered by persons having some knowledge of the local geology. Those who wish for farther information on this subject may consult Jameson's *Mechanics of Fluids for Practical Men*, chap. xiv.; or the *Architectural Magazine*, vol. v. p. 174. to p. 178., which contains a succinct but clear view of the entire process.

When water cannot be procured by any of the means mentioned, it may be collected from the roof of the house, or from the surface of the ground, and preserved in tanks or cisterns. Where the water is not to be used for culinary purposes, but merely for cleaning, watering the garden, or as a resource against fire, the cistern or tank in which it is contained may be kept above ground; but where it is desirable to keep the water cool, the tank should be sunk so deep, as to be beyond the influence of the sun. When the water that falls on the roof is to be collected, but not for culinary purposes, the cistern to contain it may be placed immediately under the roof, which will be attended with various advantages: the water will not be liable to freeze during severe frost, because all the waste heat ascends to the roof; and it will be in the most desirable situation for extinguishing fire, or for supplying baths, water-closets, or wash-hand stands, throughout the house. As the quantity of rain which falls in Britain varies in different places from 2 ft. to 3½ ft. in depth in a year, the water that may be collected on the roof of even the smallest house, is much greater than we are apt to imagine. Even if this water were used for no other purpose than supplying a fountain in a flower-garden, it would be worth collecting; and we shall hereafter give data, by which the curious reader may calculate the number of hours in a year which a fountain would play, if only fed with the water collected from a roof of any given surface. Rain water, however, is so exceedingly useful for washing, for water-closets, for cold baths, and for shower baths, that, where the expense is not an object, no house ought to be without the means of procuring it. The safest mode of forming cisterns under the roof, for collecting the rain water, is, to have them shallow, broad, and divided into a good many compartments; so that the pressure may be distributed regularly over the walls and partitions of the house, and that, if any compartment should leak, no more water would descend than what was contained in it.

In many situations it becomes necessary, or is found desirable, to raise water from a well or from a stream, for the purpose of supplying the house; or to bring water into the garden or pleasure-grounds, either for the purposes of culture or for fountains. When water is to be raised from a well or pond, the common mode is by a forcing pump; which may either be worked by hand, or by a machine driven by a horse or donkey: but, where it is to be raised from a stream, or from a pond on the side of a hill supplied by springs, provided there be a fall of 3 or 4 feet, the hydraulic ram may be had recourse to. This is considered to be one of the most economical, as well as the most ingenious, modes of raising water; the quantity raised being equal to a half or more of the quantity expended. The machine (which will be found described under the head "Hydraulics" in the seventh edition of the *Encyclopædia Britannica*, and in the article on Hydraulics published by the Society for the Diffusion of Useful Knowledge) is found not liable to go out of order, and of itself does not cost more than 10*l.* or 12*l.*; but the total expense, including pipes of conveyance, must of course depend on the distance of the hydraulic ram from the reservoir into which the water is to be delivered.

Fountains are universally acknowledged to be great ornaments to gardens; chiefly perhaps from the appearance of life and motion which they communicate to the surrounding scenery. Many persons are deterred from indulging in *jets-d'eau*, or in some other description of fountain, from an idea that they cannot be obtained without great expense, and without an immense supply of water; but the height to which a jet will rise does not depend on the quantity of water in the reservoir, but on its height above the surface from which the jet rises; and, when it is considered that no fountain need be allowed to act in the nighttime, or even in the day at certain seasons of the year, the quantity of water that will require to be raised to the reservoir is much less than at first sight will appear necessary. The greater number of fountains in suburban gardens are indeed only played off when visiters are present; and this can be effected with a very moderate-sized reservoir, and with no more use of the forcing pump than may be necessary to fill up the leisure time of a house servant, or a gardener's labourer. Though no person will undertake to erect a fountain without employing an engineer, or a plumber who possesses skill and experience in this branch of hydraulics; yet it may be useful to state a few particulars, with a view of enabling the proprietor of a suburban residence to judge how far a fountain may be practicable in his case; and to form some notion as to whether the person whom he intends employing to erect it has a competent knowledge of the subject.

Wherever a fountain is forced, the jet may be made to rise nearly as high as the reservoir. If the reservoir be 5 ft. 1 in. above the level of the surface from which the *jet-d'eau* is to ascend, the jet will rise 5 ft. If the reservoir should be 10 ft. 4 in. high, then the jet will rise 10 ft.; and if it be 51 ft. 9 in. high, it will rise 45 ft. In order to procure a jet of 100 ft. in height, it is necessary that the reservoir be 133 ft. 4 in. high. Now the jets will rise at this rate relatively to the reservoir, whether the latter contains only one cubic yard of water, or 10,000 cubic yards; all the difference being, that in the former case the jet would not last longer than a minute or two, because the cubic yard of water would be immediately exhausted, while in the latter it would last several days. The next point which a person about to erect a fountain would desire to know is, the diameter of the pipe which is to convey the water from the reservoir to the jet. Now this diameter may be as large or as small as the party chooses: it may be a foot in diameter, or it may be only an inch; but, whatever its diameter may be, this condition is essentially necessary, viz. that the orifice from which the jet issues shall be small as compared with the diameter of the supply pipe, in proportion as the jet is to rise high. If the jet is to rise to a maximum of height, that is, as nearly as the nature of things will admit, to the height of the reservoir, then the orifice must not exceed a fourth part in diameter of the orifice of the conducting pipe. If, instead of a fourth part, the orifice were made of the entire width of the conducting pipe, the water would hardly rise above the surface; and this, which is very remarkable, would be the case, whether the reservoir were 10 ft. or 100 ft. high. Thus, when the reservoir is 10 ft. 4 in. high, the height to which it is wished to have the jet to rise is 10 ft., and the diameter of the pipe which conveys the water to it from the reservoir is 2¼ in., then the diameter of the orifice must not be larger than ½ in. The following tables contain useful data on this subject. The first is taken from Switzer's *Introduction to a General System of Hydrostatics*, &c., vol. ii. p. 126.; and the second from the excellent article on Hydraulics in the *Encyclopædia Britannica*, 7th edition, already referred to.

Height of the Reservoir.		Diameter of the Pipes.		Thickness of the Metal.		Diameters of the Orifices.		Height the Water will rise to.	
Feet.	Inches.	Inches.	Lines.	Lines.		Lines.	Parts.	Feet.	Inches.
100	0	7	0	15½	16	12	15	80	0
86	4	6	0	14	0	12	14	70	0
72	0	5½	0	12	13	10	12	60	0
58	4	5	0	9	9	8	10	50	0
45	4	4½	0	7½	8	7	8	40	0
33	0	3	0	6½	7	7	0	30	0
21	4	2½	0	5½	6	6½	0	20	0
15	9	2¼	0	4	5	6	0	15	0
10	4	0	25	3½	4	5	0	10	0
5	1	0	22	3	3½	4	0	5	0

Altitude of the Jet.	Altitude of the Reservoir.		Quantity of Water discharged in a minute from an Adjutage or orifice six Lines in diam.	Diameters of the horizontal Tubes suited to the two preceding columns.	Thickness required for the Metal of the Pipes in Lead.
Feet.	Feet.	Inches.	Quarts.	Lines.	Lines.
5	5	1	32	21	1½
10	10	4	45	26	1¾
15	15	9	56	28	2
20	21	4	65	31	2¼
30	33	0	81	34	3
50	58	4	108	39	4
70	86	4	131	43	5
100	133	4	163	49	6

The pipes that convey the water from the reservoir to the jet should be laid, as nearly as possible, in a straight direction; and that part which is under ground should be sufficiently deep to be out of the reach of frost, and of one uniform slope from the reservoir to the jet. In proportion as the direction of the supply pipe is circuitous, and the level irregular, will the friction of the water be increased when in motion, and consequently the power of the jet diminished. There are two other evils which result from not laying the pipes on a uniform slope: viz., that air will collect in all the higher parts of the pipe, and mud in all the lower parts; so that, for each deviation upwards from the uniform slope, it is essentially necessary to have a small pipe with a cock to permit the escape of the air from time to time; and in the lowest part of every deviation from the uniform slope downwards, a pipe is also necessary with a cock to admit of occasionally, say once a year, washing out the mud. Few persons are aware how very soon air collects in water pipes; or how much sediment is deposited by even the purest water; which sediment is generally difficult to remove, in proportion to its fineness. We have known fountains that in a very few months after they were put up have ceased to act, owing solely to the accumulation of air in their pipes; and in other places, where old fountains, after having played for many years, have ceased to act, the cause is to be found in the accumulation of air and sediment. Much more might be said on this subject, as a matter of science, but we trust enough has been advanced to show the necessity of employing a competent person, whenever it is desired to erect a fountain.

Fountains, as matters of Taste, like every other object by which the fancy is called into exercise, admit of a great variety of designs. Our guide in the choice of these must be principally the character of the scenery in which the fountain is to be placed. In a Dutch, French, or other architectural garden, we see no objection to figures of animals spouting water (see *figs.* 335, 336.); to that of a nymph wringing her robe (*fig.* 337.), or

335 336

her hair (*fig* 338.); or to Caus's very elegant design, *fig.* 339. In an English garden, however, where the scenery around is not so decidedly artificial, a simple jet from a rock, or a group of shells, as in the wood at White Knights (see *fig.* 341.); or from the centre of a basin, as at St. Cloud (*fig.* 342.), or in the design by W. Mason, Esq. (*fig.* 340.) In these last two designs, the tube from which the jet issues is to be considered as proceeding from a mass of rock, just seen above the water. *Fig.* 343. is a view of a piece of rock and shell work, with fountains, lately erected in the garden of Mr. Thomas, at Peckham, in Surrey, by Mr. Benjamin Andrews, gardener there. "The basin, which is about 12 ft. long, is of an irregular shape, somewhat

337

338

approaching to that of an oval; and a jet of water rises in its centre by means of a leaden pipe, carried up through a convolvulus-shaped figure. Above this, and springing from its centre, is a second smaller convolvulus, to conceal the extremity of the pipe. Various adjutages in brass, forming different devices for throwing out the water, and which may be screwed on the leaden pipes, may be obtained in great variety from the plumbers. The rockwork round the basin is composed of vitrified bricks, flints, spars, &c., partly projecting into the water; and it is contrived so as to present as rough and inartificial an appearance as possible. Creeping plants are introduced in some places; and a large *Yúcca* gloriòsa occupies the space between the two principal masses of rockwork.—*B. Andrews. 6. South Grove, Peckham.*" This kind of fountain and its

339

340

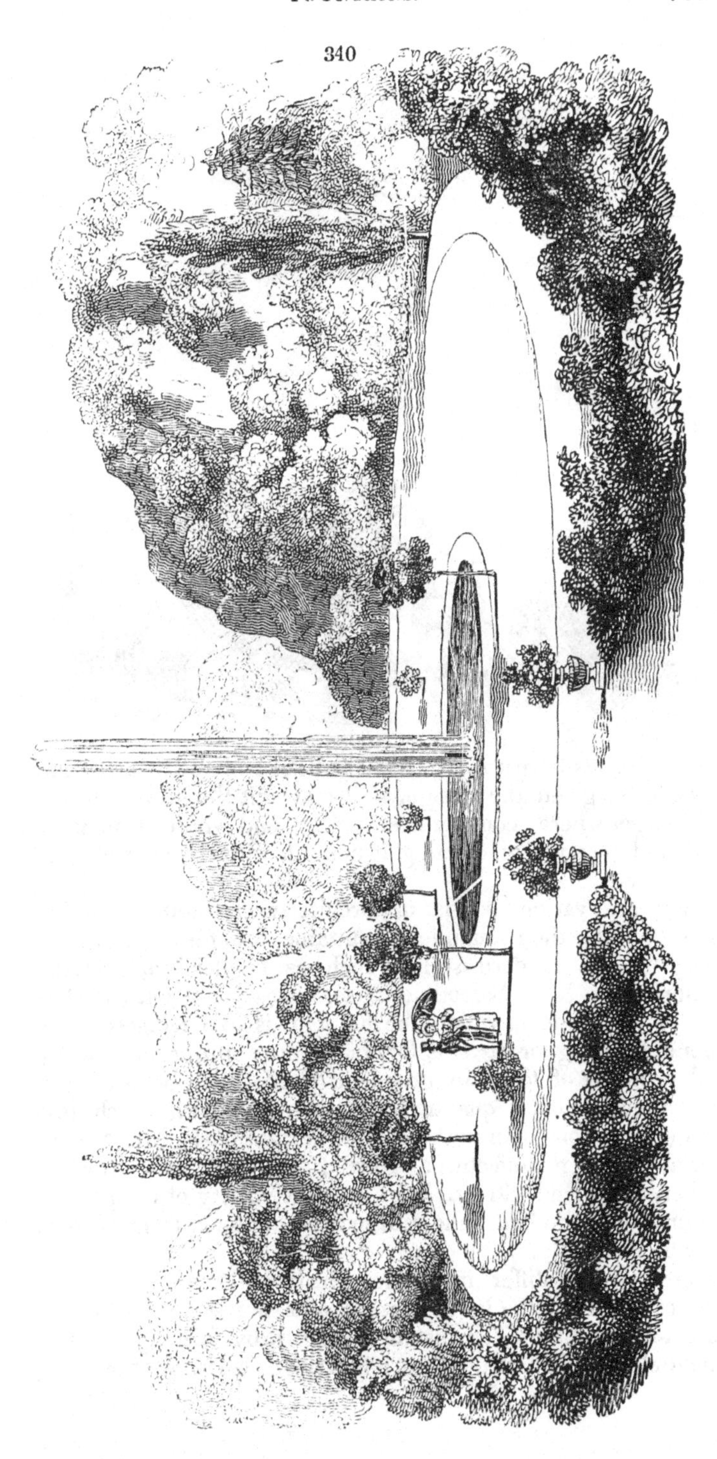

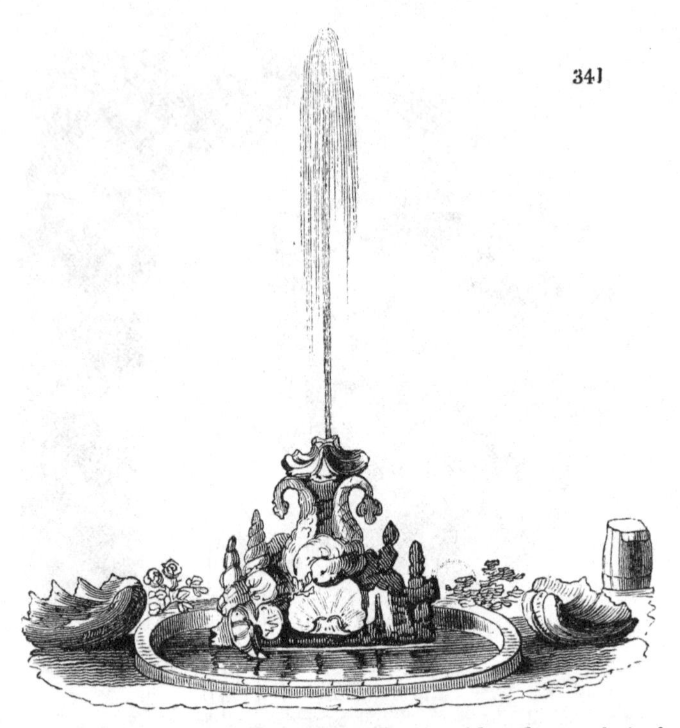
341

accompaniments have no pretensions to be considered as artistical; but we have given the design, as it may suit the flower-garden of a cottage, where economy is an object, and where an amateur would find pleasure in erecting it himself, with the assistance of his gardener.

There are various kinds of architectural fountains which are, also, from their simplicity, suitable for pleasure-ground scenery in the modern style; such as candelabra, tripods, columns, vases, &c. Sometimes, also, instead of a jet, the character of the scene in which the fountain is to be placed may render a drooping, or cascade, fountain more appropriate; for, by this description of fountain, grandeur of effect can be produced with a much smaller quantity of water, and with much less height of reservoir, than in the case of upright jets. Mr. Austin, artificial stone manufacturer, New Road, and Mr. Rowley, plumber, Howland Street, have a great variety of designs for fountains; and the hydraulic ram may be obtained from Messrs. Bramah, Piccadilly.

Flower-gardens differ from those beds of flowers which are often scattered on the lawn, in front of a house (so as to constitute a foreground to the distant scenery), by forming a whole of themselves. For this purpose, a flower-garden is, or should be,

342

always surrounded by some sort of boundary, or enclosure, to separate it from the general scenery of the place to which it belongs. This boundary may be variously formed; it may be a border of shrubs; a fence of trelliswork, either as an espalier or as an arcade; a conservative wall; a walk covered with trelliswork; a clipped hedge of some kind of evergreen, or an unclipped hedge of various shrubs. Whatever kind of boundary is adopted, it is essentially necessary that it should not be so high as to exclude the sun from the area of the garden, or to prevent the free circulation of air round the plants. What the suitable height of the boundary fence may be, to insure the admission of sun and air, depends on the extent of the garden, the elevation or depression of its surface, and its aspect; and, if it be of any other form than a circle or a square, whether its longest diameter is in the direction of east and west, or of north and south.

A case can hardly occur in which a flower-garden should be surrounded by trees: though, in very exposed places, these may be introduced on the north, north-east, and north-west sides; but this should be at the distance of at least three times their height from the boundary fence of the garden. Nothing con-

343

tributes more to the beauty of flowers, than abundance of solar light, and the free circulation round them of air. Even if trees are planted only on the north side, but so near as to exclude the free action of reflected light from that side, the flowers will be weak and drawn towards the sun. All near fences have a tendency to produce this kind of etiolation, as any one may observe in the case of plants growing on the south side of a common hedge; and this etiolation is directly the reverse of that bushy, tufted, or spreading appearance, which always accompanies vigorous growth in open airy situations.

The outline of the flower-garden should always be such, as that on the first view of the interior some general form may be recognised in the area; for example, it may approach to a circle, to a broad or narrow oval, to a square, to a parallelogram, or to a triangle. A long narrow strip, either crooked or straight, would not be satisfactory to the eye; nor would a space, broad in some places and narrow in others, here straight and there crooked, afford the spectator half so great a degree of satisfaction, as he would experience if he recognised some known and definite figure. Whether the boundary line be regular or irregular is a matter of no consequence, with reference to this end; provided there be ample length and breadth included by the outline, to throw the interior into some definite shape, when seen in perspective.

A flower-garden is formed of beds in which flowers are cultivated, placed on turf or gravel, or surrounded by both of these materials. The surface of the ground may be level, of one uniform slope, or several slopes; undulated, or of abrupt transitions from one form of surface to another. The flower beds may be solely planted with flowers, or they may contain also shrubs; and these shrubs, when of the flowering kinds, such as dwarf roses, azaleas, &c., may be planted in masses, and occupy entire beds; while other shrubs, and sometimes low trees, may be treated as single objects, and either planted in the beds, or on some of the open spaces on the lawn. The beds, when formed on turf, may be enriched with edgings of box or of other plants, of wickerwork or wire; or of earthenware, slate, tiles, brick, stone, &c.; or, instead of edgings, there may be a mere line of decoration between the dug ground and the turf. The beds, when surrounded by gravel, must of necessity have an edging of some sort, which may be of any of the kinds mentioned.

The features above enumerated may be considered as the essential materials, or component parts, of a flower-garden; and to them may be superadded various ornaments, such as vases, statues, spars, and other objects of curiosity, natural or artificial; rustic baskets, vases, or other contrivances for containing plants; trelliswork, arcades for climbers, open and covered seats, summer-houses, fountains, aquariums, rockwork, rootwork, grottoes, and grotesque objects; and, lastly, a green-house, conservatory, or some other building for plants. Now, these materials and ornaments being given, the next point is to lay down the principles for using them in the composition of a flower-garden. In this, as in every other species of design, the leading principle is the formation of a whole. For this purpose, there must be a consistency in every part, with reference to culture, and also with reference to lines and forms. One style of form and line must pervade the entire garden. In the simplest gardens curved lines or straight lines, and acute angles or right angles, may prevail both in the ground and ornaments; but, in the more artistical kinds, the beds and walks may display the scrollwork or the embroidery of the French gardens. Whatever form or style is adopted as the groundwork of the garden, that is, as the shape of the beds and walks, that form or style must pervade all the ornaments with which the whole is decorated. Where the beds are in scrollwork, the vases and the pedestals on which they stand, the trelliswork of the arcades, the statues, and the seats and summer-houses, should all be of the style of Louis XIV. A simple classical vase or statue, in such a garden, would be quite out of place; and equally so would be an Italian summer-house, or an English thatched rustic seat. The artist is not limited to introduce only those component parts, or those ornaments, which were known in the time of Louis XIV.; but, such as he does introduce, he is bound to compose in that style.

As remarks applicable to the composition of flower-gardens generally, we may observe that ornaments should not be distributed equally over the garden; and that, as far as practicable, the ornaments should appear to arise out of something that is of use. For example, at the intersection of two principal walks, as conspicuous points in the design which attract the notice of all spectators, there may be a basin and fountain; and the parapet which forms the margin of this basin may be ornamented with vases. In the intersection of other walks there may be a sun-dial on an architectural basement, or a statue on a pedestal. Where there is a terrace-walk connected with the house, or with some other building, it may be bordered by a parapet; and this parapet may be ornamented with statues, vases, or other architectural or sculptural ornaments, according to circumstances. In these, and similar cases which might be mentioned, there seems to be a reason for the placing of the ornaments; that is, they appear as a finishing to the works of art to which they are attached, and they harmonise with all the associations connected with them. On the other hand, when a vase or a statue is set down on naked turf, in the midst of a bed of flowers, or in a group of shrubs, it seems in a false position, and loses its effect; not only from the want of proper accompaniments, such as an architectural basement, &c., but from the presence of such accompaniments as are improper, such as a basement of dug ground and flowers.

A covered way of trelliswork and creepers often forms a very suitable boundary for a flower-garden; and it is almost always an excellent approach to one from open scenery, or as a passage from one flower-garden to another. The reason is, that it confines the eye, and slightly darkens the vision; and this, being succeeded by a view of the open garden, its ornaments and the brilliancy of its flowers produce by contrast a more lively effect on the spectator. Trelliswork in the centre of a flower-garden, as a bower to sit in, is also useful; more especially if a portion of it should be continued for a few feet along each of the radiating walks, so that each may serve as a sort of telescope to the views of the garden, as seen by a spectator seated in the bower.

The laying out of flower-gardens is too commonly intrusted to persons who have no knowledge of the principles of design; which is a proof that the possessors of them do not fully comprehend the beauties of which flower-gardens are susceptible. Were this the case, an artist would be considered as essential in this department, as he now is in any department of architecture, or of landscape gardening.

THE END.

London:
Printed by A. Spottiswoode,
New-Street-Square.